Early Life on Earth

Early Life on Earth

Nobel Symposium No. 84

Stefan Bengtson, editor

Columbia University Press   New York

Columbia University Press
New York   Chichester, West Sussex

Copyright © 1994 Columbia University Press

Library of Congress Cataloging-in-Publication Data

Nobel Symposium (84th : 1992 : Karlskoga, Sweden)
    Early Life on Earth / Nobel Symposium No. 84 ; Stefan Bengtson, editor.
      p. cm.
    Includes bibliographical references and index.
    ISBN 0–231–08088–3
    1. Evolutionary paleobiology — Congresses.   I. Bengtson, Stefan.   II. Title.
QE721.2.E85N63   1994
560—dc20
                                        94–3822
                                          CIP

♾

Printed in the United States of America

c 10 9 8 7 6 5 4 3 2 1
p 10 9 8 7 6 5 4 3 2 1

Contents

Theme 3: Multicellularity and the Phanerozoic Revolution

As language, genes, and culture provide insights into human history, so do all living organisms contain phenotypic and genotypic clues to their own evolutionary history. This has long been recognized in the study of comparative morphology and physiology, and more recently it has been demonstrated in molecular studies. The molecules of heredity, RNA and DNA, and their protein products hide a treasure of information now tapped by molecular biologists. The new molecular data, in concert with morphological and physiological data from living and fossil organisms, are providing revolutionary insights into how the multifarious lineages of life have evolved.

History is not only narration – in addition to the "what," we also want to know the "how" and "why," even the "if . . . then what." The study of early life on Earth requires that we investigate contemporary chemical, geological, and biological processes. This is sometimes the only approach, for the direct documentary record in the rocks may be miniscule and ambiguous. The origin of life is the most obvious and significant such field where direct evidence is hard to come by. The question here is "What could have occurred?" rather than "What occurred?" If the ensuing hypotheses are well formulated, they may be used to help us find supporting or disproving evidence from other sources, such as the rock record.

Thus questions of the origin and early history of life on Earth have become the concern of a variety of scientists with disparate training, attitude, and goals. Notwithstanding valuable efforts to bring these disciplines together (e.g., Schopf 1983b; Schopf & Klein 1992), the task to develop and maintain productive interdisciplinary approaches remains a constant challenge. This was the rationale for the 84th Nobel Symposium, "Early Life on Earth," held at Alfred Nobel's manor, Björkborn, in Karlskoga, Sweden, on May 16–21, 1992. Forty-four chemists, geologists, paleontologists, and biologists from 12 countries gathered for a week of debate on life's origin and early evolution. It was an exciting and rewarding week. Draft manuscripts had been prepared by participants and in most cases were distributed before the meeting as a basis for discussions. Lectures stimulated questions, questions led to answers that generated more questions, and the sessions typically continued into discussions in small and large groups in the halls of the Nobel manor, on the green lawns and lush garden (where the Swedish early summer was generously cooperative), and at the hotel Svartå Herrgård, a beautiful 18th-century manor by one of Sweden's thousand lakes.

The symposium was organized around three great themes: the origin and early diversification of life during the Archean Eon (before 2½ billion years ago), the maturation of life and the Earth during the almost 2 billion years long Proterozoic Eon, and the explosive diversification of multicellular life that marks the dawn of the Phanerozoic Eon around 550 million years ago. It was evident that although remarkable progress is being made in all these fields, some of the key questions are yet unanswered, or, rather, answers have not been found that satisfy all data or all scientists. For example: Did the origin of life require a concentrate of organic molecules synthesized through abiotic processes – a "primordial soup" – or could a self-reproducing system get started directly from inorganic compounds? Why are the living representatives of the most ancient branches of the evolutionary tree thermophilic organisms: Does it reflect life's hot cradle or only the selective survival of heat-resistant or vent-dwelling organisms during the last major bolide impact to cook the planet's surface? Do the earliest known fossil organisms include oxygenic phototrophs, or were

Preface: Approaches to understanding early life on Earth

Alone in our solar system, Earth harbors life. What chain of events led to this remarkable feature? Did life emerge early in Earth's history or only after a long period of planetary development? How has life evolved through time, and to what extent can the history of life be reconstructed? How should we go about trying to understand evolutionary processes that acted billions of years ago?

The questions concern a chain of events that occurred only once, a long time ago, and cannot be repeated. They therefore deal with history. In tracing the course of human history, anthropologists, archaeologists, and historians gather information from past documents and artifacts and from cultural and, increasingly, genetic comparisons among peoples. So we must do to understand life's history – only with different documents and a much broader comparison among organisms.

Our documents are written in rock. They include the remains of life forms that died and sank into the Earth's skin of sedimentary rocks. Through time, as biotas changed, successive layers of sediment incorporated different remains. There is no more palpable evidence of evolution than Earth's layered skin of fossil-bearing rocks, in which the same succession of forms can be followed in remote corners of the Earth, layer upon layer upon layer. Equally, there is no more compelling evidence for the immense antiquity of life than the presence of simple fossils in some of the oldest sedimentary rocks yet discovered. Paleontologists seek out these layers, reading them as archaeologists read the pieces of bones, pottery, and kitchen debris that accumulate wherever humans have lived.

Fossils can be bones and shells, but they can also be soft tissues, cuticles and spores, cell walls and other parts of cells, and organic molecules made by organisms and preserved in sediments for upwards of a billion years. The record also includes burrows and trails left in the mud and metabolic calling cards printed in the chemical composition of rocks. Even the composition of the atmosphere and oceans reflects and influences biological activity. Earth's corrosive, oxygen-laden atmosphere shows that life abounds here, even to an observer too far away to see life's individual manifestations. The oxygen, released by photosynthesis, strongly influences the ways that organisms live and how they are distributed on the planet. It also affects the weathering of rocks and formation of sediments, thus linking biology and environment to Earth's sedimentary skin. Another link comes from biomineralization. When organisms started to make minerals for their own use, they began to litter the sea floor with mineral grains, changing even the gross composition of many sedimentary rocks. The chemistry of the Earth's surface is thus linked with that of the biosphere, and a lot of what we know, or think we know, about life's early history on Earth comes from the study of geochemistry and biochemistry.

Bengtson, S. (ed.) 1994: *Early Life on Earth. Nobel Symposium No. 84.* Columbia U.P., New York

they all anoxygenic? Regardless of when oxygenic photosynthesis originated, when did the atmosphere accumulate free oxygen in concentrations sufficient to support aerobic metabolism? Is the chemical evolution of the Earth's surface mainly driven by geological processes, or does life exert a decisive influence on the global environment? What is the nature of the famous Ediacaran biota – an aborted branch of the evolutionary tree, the first glimpse of extant animal phyla such as cnidarians, or a mixture of the two? Why did the first evolutionary radiation of animals at the Proterozoic–Phanerozoic transition take place so late and so rapidly? What are the exact relationships among the major groups of animals, and how can this phylogenetic information help us to disentangle the jumbled threads of the Ediacaran and Cambrian fossil records?

These yet-unanswered questions may seem serious flaws in the fabric of science, but science is by its very nature more questions than answers. The fact that these particular questions have not yet been answered to general satisfaction suggests that maybe at other levels of inquiry we have not yet found the right answers or even asked the right questions. What we do have, as the result of ongoing research in a variety of fields, is a stirring picture of life's early history that, although crude and with many blank patches, represents an incomparably better knowledge than we had one or two decades ago. We can now say with confidence that Earth accreted by collisions of smaller bodies about 4½ billion years ago; that phototrophic bacteria had evolved 3½ billion years ago; that the last common ancestor of all currently living things was a hyperthermophilic prokaryote similar to those that live in hot springs today; that the eukaryote branch appeared more than 2 billion years ago, and that endosymbiosis with different lineages of prokaryotes was a major factor in subsequent eukaryote evolution; that at least 2 billion years ago oxygen levels in the atmosphere had built up to a significant proportion (more than one-tenth) of the present atmospheric level; that the microbial mats that were such a dominant feature of early life environments (see the cover of this book) started to decline in abundance and diversity about 1000–900 million years ago concomitant with the first appearance of identifiable green, red, and chromophyte seaweeds; that the earth became episodically glaciated from about 850 to 590 million years ago, sometimes with glaciers covering even tropical lowlands; and that, coincident with a breakup of a giant continental mass into smaller continents, unicellular and multicellular eukaryotes diversified strongly around 550 million years ago, creating a biosphere essentially like today's, except that it lacked a diversified terrestrial biota.

The time framework is depicted in Fig. 1. The datings are mainly based upon measurements of the radioactive decay of various isotopes, in combination with relative-age correlation of rock units based on physical, chemical, and biological signatures as well as their spatial interrelationships. The chapters in this book make frequent references to the major time divisions – the Archean, Proterozoic, and Phanerozoic eons and their subdivisions; absolute ages are given in the units Ma (mega-annum, million years, 10^6 years) and Ga (giga-annum, billion years, 10^9 years). (The beginning of the Phanerozoic is given with some spread owing to the fact that this boundary, unlike the earlier ones, is defined in rock rather than in time and has been poorly dated. Recent datings [e.g., Compston *et al.* 1992; Bowring *et al.* 1993] show that the age of 570 Ma found in most textbooks is probably more than 20 Ma too old.)

The framework is like a sketch map of a new territory; the details remain to be filled in, and some major features are still lacking. This is what the book is about: a collection

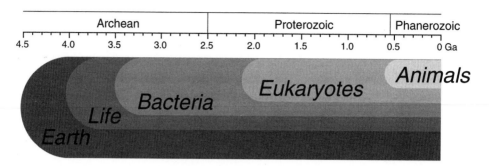

Figure 1 Time scale for Earth and life history. The book deals with the time up to about 0.5 Ga (billion years), when animals had appeared. (Dinosaurs became extinct between the 0.1 and 0 lines, and hominids are known only from time representing one-third of the thickness of the zero line.)

of fresh contributions from scientists in the various associated fields describing how they currently view the picture. If you look to this book for final answers about life on Earth, you will be disappointed – it is a time document. The chapters reflect the views, knowledge, and inevitable biases of the individual authors or teams of authors, and they are written to convey the experience and excitement of active research projects. The chapters have not been squeezed into a common shape, except in the trivial typographic sense. You will thus find contradictions in the book, which is as it should be in any vital field of research. Such a multitudinous approach should also make the book more enjoyable to read – each chapter has its own flavor.

Imagine a similar book written ten years from now. Some questions on life are more or less eternal and will still be asked in that book. Some questions in the present book will have been answered more or less to everyone's satisfaction, and some will have been dropped as unanswerable or, for other reasons, uninteresting. The only thing that can be safely predicted about that imaginary book is that it will deal with new questions that few of us have even thought of today. So read this book now – the fizz may be gone in ten years! But by all means read it in ten years' time as well – a bouquet may develop.

Acknowledgments. – We express our sincere gratitude to the Nobel Foundation, Stockholm, for its full and constructive support of this symposium through its Nobel Symposium Fund, and to the Alfred Nobel's Björkborn Foundation, Karlskoga, for the congenial planning and splendid execution of the practical arrangements under Gertie Ågren's leadership. Sif Samuelson managed the financial affairs of the symposium with bravado. The various chapters in the book have profited from reviews by different persons: in addition to the various types of feedback given during and after the meeting by the participants and the personal reviews specified by the authors in the respective chapters, the combined book manuscript was reviewed wholly or partly by the following: David J. Bottjer (University of Southern California, Los Angeles, CA), David Des Marais (NASA, Moffett Field, CA), Douglas H. Erwin (Smithsonian Institution, Washington, DC), Robert Horodyski (Tulane University, New Orleans, LA), Gerald F. Joyce (Scripps Research Institute, La Jolla, CA), Charles R. Marshall (UCLA, Los Angeles, CA), and Norman R. Pace (Indiana University, Bloomington, IN). Joanna Fancy (Los Angeles, CA) and Christina Franzén (Stockholm) provided good help with bibliographic matters. Finally, the editor wishes to acknowledge the enthusiastic and professional support from Columbia University Press throughout the planning, editing, and production of this book.

Herrick Baltscheffsky, Stefan Bengtson, Jan Bergström, Andrew H. Knoll, Gonzalo Vidal

Introduction: The coherence of history

Stephen Jay Gould

Museum of Comparative Zoology, Harvard University, Cambridge, Massachusetts 02138, USA

I n May 1992, the 84th Nobel Symposium met at Björkborn, Karlskoga, Sweden, the home and laboratory of Alfred Nobel. "Rightness" of place is a vital component for any prospect of success or sense of well-being. Our ancient literary and moral sources surely recognize this principle, as in the opening of Psalm 127 ("Except the Lord build the house . . .") or the tale of the Shunammite woman (2 Kings 4:13), who refuses Elisha's offer of a reward for her kindness because she already has full satisfaction: "I dwell among mine own people." Experienced conferees also know the pleasure of historically appropriate venues, but I was particularly impressed with a symbolic aspect of rightness for this linkage of Nobel's private house with a conference named "Early Life on Earth."

Several conferees noted the most overt appropriateness of holding a meeting, focused (for a day at least) on the Cambrian Explosion, in the home of a man who invented dynamite and funded both his house and his famous prizes with the proceeds. But I walked into Alfred Nobel's private library one morning and found a source of more fundamental and pervasive connection. Nobel read at least five languages fluently, and he clearly loved history above all subjects. For his shelves are filled with all the great multivolumed classics of 18th and 19th century historiography, many extensively annotated in Nobel's hand – Weber's *Allgemeine Weltgeschichte*, Guizot's *Histoire de la civilization en France*, and, of course, Gibbons's *Decline and Fall of the Roman Empire*.

I had, until perusing these shelves and reading their lesson for us, been bothered by what I perceived as a lack of cohesion among the varied topics of this conference – an especially pressing issue for me, since I had been given this assignment of writing a summary. Two aspects troubled me. First, although we all believe in the virtues of interdisciplinary contact, the practice thereof often sows difficulty based on our own lamentable limitations. That is, many of us simply don't understand enough about other represented fields to make much sense of the proceedings. I, for example (and shamefully), am a chemical ignoramus and therefore could not grasp many of the early papers on the origin of life. Second (and now invoking the more important issue of intrinsic rather than personal limitations), issues of scaling and different focus often convert a subject that seems unitary in statement into a nonmeshing set of disparate concerns breeding confusion when forced together.

Bengtson, S. (ed.) 1994: *Early Life on Earth. Nobel Symposium No. 84.* Columbia U.P., New York

For example, I once had to review a volume of papers from a conference on the apparently unitary topic of "extinction." The topics ranged in scale from death of local populations in restricted ecosystems to extirpation of faunas in geological mass extinctions. We use the same word – extinction – for such events of death across all scales; but does such an amalgamation help or obfuscate our search for causes? I finally concluded, with some regret, that the unity was false and the compendium both forced and harmful. We have an explanatory habit in science, born of reductionism, that leads us to render events at grand scale in terms of causes operating at smaller, observable levels. Thus, nearly all participants assumed that the smallest-scale events dubbed "extinction" in the vernacular must be models or prototypes for the great geological episodes of mass dying (also called "extinctions"). But the causes may be quite different (ordinary natural selection in the biotic and competitive mode at small scale, with fortuity of survival through environmental catastrophe in at least some events at grandest scale) – even though death be a common result. By using the same word across all scales, and by assuming that small must extrapolate to large, our quest for explanation was derailed by a false unity.

Initially, I had the same fears for this disparate set of papers about early life on Earth. Explanatory structures were so varied, ranging from chemical experimentation to computer reconstruction of phylogenies to taxonomic ordering of fossils. Even more troubling, the basic questions were often so incommensurate: Most strikingly, questions surrounding the origin of life (largely an issue of chemistry and the physics of self-organizing systems) are so different from problems in subsequent evolution (issues of natural selection and mechanisms of change). The two subjects grade into each other temporally (for origin does lead to evolution!), and the public certainly views them as united (leading to all manner of political trouble in America, as creationists accuse Darwinism of bankruptcy because natural selection doesn't explain the origin of life). But origin and evolution pose different problems; are studied with different methods, disciplines, and even languages; and, in some deep sense, simply don't belong together.

I was therefore discouraged until I walked into Alfred Nobel's library and found the key that should have been obvious from the start. There is true unity among us, a common focus and concern that couldn't be more central or important. Our point of union is the greatest and, in many ways, the most obvious subject of all: Nobel's personal favorite of *history itself*. We are using methods as diverse as any in science; we speak languages and follow traditions that span the entire range of disciplines usually called scientific. But we are trying to apply all this varied expertise to one common struggle: we are trying to unravel the early history of life as it unfolded on an Earth so different from ours, and so complexly changing. We are historians of most of our planet's time – the long and difficult period before the last ten percent or so, when skeletonized multicellular life left us a conventional sort of historical archive in the form of complexly changing uniqueness through time.

Yet this powerful unity is sometimes hard for scientists to grasp, for two reasons. First, our conventional parsing of disciplines places history into a domain far from science (even into a different faculty of humanities in most universities). Second, scientists tend to denigrate history for a variety of reasons that, although both false and narrowly parochial, lead us to place a stamp of inferiority upon narration, even for topics (like life and Earth) usually placed in the domain of science. Both these reasons are false and harmful, and I dedicate this summary to fostering a respect for history and an accommodation of its different kind of rigor into the central activity of science. Without this integration, the papers of this conference make no sense as a unified inquiry. With this perspective, we convert a disparate set of studies (whatever their individual excellences) into a worthy, integral fabric.

Let me respond briefly to each of these reasons for avoiding and denigrating history. On the first point of *avoidance* – that history, however worthy, is fundamentally different from science – I simply remind everyone (for the point has often been made) that many sciences are intrinsically historical, for their data are sequences of complex temporally ordered events, each occurring but once in all its detailed glory. Earlier taxonomies from the dawn of modern science understood this principle well, for the historically based sciences (from cosmology to archaeology, including, along the way, geology, paleontology, and anthropology) were then linked with the conventional kind of history now taught in faculties of the humanities (rulers, wars, and conquerors from Rome to now) into a comprehensive study of meaning and directionality in time – see, for example, Thomas Burnet's *Sacred Theory of the Earth* (1690s) or Giambattista Vico's *Scienza Nuova* (1725). Our current taxonomy of disciplines is historically contingent, not logically fixed.

On the second point of *devaluation*, scientists often consider history as a "lesser" subject in a continuum of rigor ranging from "hard" disciplines like physics and chemistry (quantitative and experimental) to "soft" studies like sociology and psychology (particularistic and descriptive). History is said to be a playground for our biases (true enough in many cases, but is science any more exempt from the same failing?). History is saddled with the charge of lacking rigor or producing no generalities (in being, as the old saying goes, "just one damned thing after another").

My rebuttal to this bypassing and devaluation of history is twofold: first, to outline the two distinctive features of history that should make scientists proud to participate in the enterprise; second, to argue that the distinctive methods of science can add enormous value to historical inquiry, and that the interdisciplinary ideal should therefore work its finest practice in this arena.

First distinctive feature: the defense of narrative. Why should we be apologetic about the documentation of temporal sequences ("one damned thing after another")? We value the story of an individual life, and biography is surely a noble enterprise. How much more should we treasure the intricate story of our planet and its life – the grandest history of all. How can any thinking person be indifferent to such a tale of extensive, complex, intricate, and often episodic change – to continents forming, coalescing, and breaking; to atmospheres building and changing; to cosmic interruptions; to a tale of life somehow forming from simpler chemistry and then, while retaining a Moneran mode, expanding its anatomical and environmental range to petunias, hippopotamuses, albatrosses, and sargasso weed.

Our individual items of information about temporal sequence have been won with great intellectual struggle (and often physical effort); they are precious beyond pricing. What is it worth to know that a macroscopic creature like *Grypania* lived so far back in the Proterozoic; that genetic sequencing and cladistic methods can formulate a tree of branching representing the main events of life's temporal order; that the mysterious bits and pieces of the lowest Cambrian "small shelly fauna" have now, in some cases, been linked to complete and articulated animals (*Halkieria* and *Microdictyon*); that ten thousand specimens of *Marrella* from the Burgess Shale allow us to reconstruct an odd Cambrian arthropod in intricate detail. Historians revere, and the world knows, comparable items like the Rosetta Stone, the Magna Carta, and the Constitution of the United States. Our items are as full of messages, just as thrilling, and worthy of equal renown.

At the Björkborn conference, Malcolm Walter began his talk on the history of Precambrian stromatolites by recalling a question (actually a stupid, wise-ass crack) that I had made following a similar presentation he had given at Harvard some twenty years ago. He had told a well-documented story of the grand-scale directional history of stromatolites from initial construction by bacteria alone, through the addition of cyanobacteria, then microalgae, and finally to the dénouement as evolving metazoans became their predators and undoers – and I had asked the equivalent of "so what?" I made this dumb comment from my own blinkered perspective of twenty years earlier, when I thought that all science demanded timeless statements based on universal laws, and that meticulously documented narrative could only be a step to that end. In other words, I was, early in my career, a victim of the very arrogance and narrowness that I now deplore (for I had been taught the restrictive stereotype of science as quantified experiment in the search for universals). I began my Björkborn presentation with a formal, if belated, apology to Malcolm for ever making such a stupid statement. His finely documented, interestingly directional story of the major pattern in what is, after all, five-sixths of life's history constitutes its own rationale and illustration of scientific excellence. Pure narrative, well done, ranks among the highest forms of science.

Just compare what we know now with the situation in 1950, at midcentury (and just a metaphorical second ago, even on the short time scale of human history). Plate tectonics didn't exist, and continental drift was haughtily dismissed by most earth scientists. We weren't even sure that any Precambrian life had been documented at all (and the term *Azoic*, meaning lifeless, was still respectable as a name for the earliest geological era). The fine structure of the Precambrian–Cambrian transition (Ediacara to Tommotian to the Burgess Shale faunas, with all associated geographic variation and climatic change) was a future dream. The Burgess Shale fossils lay fallow in drawers in Washington, D.C., still underpinned only by Walcott's false reading of their taxonomic and anatomical variety.

At the Passover service, we Jews offer a prayer (and sing a song) called "Dayenu," meaning "that would have been enough." It describes a sequence during the Exodus, stating for each happy event: "If God had done this for us alone, Dayenu. (But he did even more.)" I feel the same way about our increasing, purely narrative knowledge of this first five-sixths of Earth's history. If all we had gained since 1950 were this magnificent, orders-of-magnitude expansion in factual knowledge about the basic sequence of events, then Dayenu. This simple growth of an empirical record would be enough all by itself, and precious beyond measure. But we have gained even more.

Second distinctive feature: the different principles of history. Historians (and historical scientists) must stop apologizing for working in a domain of uniquely complex events, where the discovery of timeless and universal laws, and the prediction of all occurrences under their guidance, cannot be an expectation or even a desideratum. The historian's different mode of explanation does not express a limit or a failing relative to this supposed ideal of science, but rather records the different nature of complex and unrepeatable historical events versus the relatively simple phenomena of traditional science (Cambrian quartz is like Pleistocene quartz, and Proterozoic balls presumably would have rolled down inclined planes just as in Galileo's experiments, but *Opabinia* existed only once and for a brief time, and North America in the late Cretaceous cannot ever be reconstituted).

The organizing principles of history are *directionality* and *contingency*. Directionality is the quest to explain (not merely to document) the primary character of any true history as a complex but causally connected series of unique events, giving an arrow to time by their unrepeatability and sensible sequence. Contingency is the recognition that such sequences do not unfold as predictable arrays under timeless laws of nature but that each step is dependent (contingent upon) those that came before, and that explanation therefore requires a detailed knowledge of antecedent particulars. Each complex state has a multitude of possible outcomes, and any alteration early in the sequence sends history cascading into a different, but equally sensible, channel. History is therefore unpredictable before the unraveling of a series of events, but just as explainable as any kind of science thereafter.

This asymmetry of prediction and explanation reflects the character of complex and unique historical events, not a limit or the sign of a "lesser" mode of inquiry. As a primary difference from conventional science, historical events must be explained, in large measure, by the unique sequence of actual antecedent states, not by timeless laws of nature. If Peter the Great had never been born, or had died in childhood, or had ever been outmaneuvered in his numerous intrigues, or had accidentally drowned while learning to be a shipwright in England, there would never have been a battle of Poltava, and Russian and Swedish history would have been irrevocably different (but equally sensible). But these events did not befall Peter, the battle did occur – and we can explain it and its consequences! This is the meaning and essence of contingency – not random-ness and despair, but a different kind of sensibility. Similarly, if *Pikaia* and its relatives had died, if the minor lineage of crossopterygians had succumbed, if the late Cretaceous impactor hadn't hit (and dinosaurs hadn't died), if greater dryness had not enveloped eastern Africa some 2.5 million years ago, if a string of many thousand more contingent "ifs" hadn't occurred as they did, *Homo sapiens* would not have arisen, and we would not now be engaged in this quest for understanding.

This different, but equally rigorous, character of historical inquiry should be clear, but it often gets buried in a series of false and hurtful dichotomies that pose a denigrated history to a triumphant science: mere story vs. explanatory rigor; biased tale-telling vs. objective description; art vs. science. Academic historians themselves, beguiled by the success of science, have been guilty of fostering these false and ultimately demeaning divisions. A striking example can be found on Alfred Nobel's bookshelves in the volumes of four leading figures in England and France, all of whom wrote their major works in the Darwinian heyday of the mid-19th century (1850 to 1870). Lord Macaulay in England and François Guizot in France were the leading apostles of grandly biased, old-fashioned history as a heroic tale of steady progress for their own kind of people. (Both, by the way, are specifically mentioned and satirized in W.S. Gilbert's *Patience*, as Colonel Calverley rattles off all the "popular elements in history" that make up a heavy dragoon guard.) In his preface, for example, Guizot (who had held high political office) speaks of "the pleasure of assisting with the laborious but powerful development of my country, to watch it grow and shine across all obstacles, efforts and pains . . . History destroys impatient pretensions and sustains long hopes" (my translation from Nobel's copy of the 1872 edition of *Histoire de la civilisation en France*). "Il en coûte cher pour devenir la France," Guizot concludes.

Macaulay's perspective is even more grand (for white Europeans), and we can easily grasp why he is considered the father of the notorious "Whig interpretation" of history (biased telling as a tale of steady and inexorable progress toward higher moral ends). Again quoting from Nobel's copy of *The History of England* (1873 edition):

> *I shall relate . . . how our country, from a state of ignominious vassalage, rapidly rose to the place of umpire among European powers; how her opulence and martial glory grew together; how by wise and resolute good faith, was gradually established a public credit fruitful of marvels which to the statesmen of any former age would have seemed incredible . . . The general effect of this narrative will be to excite thankfulness in all religious minds, and hope in the breasts of all patriots. For the history of our country during the last hundred and sixty years is eminently the history of physical, of moral, and of intellectual improvement.*

Given this pompous (though magnificently expressed) nonsense, it is no wonder that other "modernist" historians of the mid-19th century rebelled against such a tradition of history as pious, self-serving bias. It is also not surprising that these revisionists turned to science as a source of reform, particularly to the highly mechanistic versions of science then in vogue. In this, however, they went too far. By trying to save history from the patent nonobjectivity of a Guizot or a Macaulay, they abandoned history's legitimate distinctness (in questions of directionality and contingency) and tried to ape the norms of science slavishly. They tried to make history the same kind of fully lawlike, predictable, factually descriptive enterprise that the stereotype of science proclaimed. Nobel owned these major works as well and clearly approved of this style, as his annotations indicate. (Nobel rarely wrote in his books, but he highlighted passages that appealed to him with lines in the margin. Most frequently so emphasized are paragraphs that proclaim the predictive and "scientific" character of true history.)

Nobel also owned the volumes of the leading revisionists in France and England – Hippolyte Taine and Henry Thomas Buckle. Taine waxes eloquent about hopes for an objective empiricism – on the value of eyewitness testimony above all, and on the presentation of documents, so that all fair-minded men may judge (*Les origines de la France contemporaine*, 1885 edition). But Buckle, above all, was the apostle of a "scientific" history – meaning (falsely to him) an enterprise that could formulate fully general laws and predict outcomes with objective rigor. Nobel, for example, underlined the following passage in Buckle's *History of Civilization in England* (1891 edition):

> *Every generation demonstrates some events to be regular and predictable, which the preceding generation had declared to be irregular and unpredictable; so that the marked tendency of advancing civilization is to strengthen our belief in the universality of order, of method, and of law.*

Buckle then goes on to wonder whether history has failed in this effort because its practitioners were inferior or its phenomena more complex. Concluding that a bit of both lay behind the lamentable tradition of Macaulay and his ilk, Buckle laid out his program for a fully scientific history:

> We shall thus be led to one vast question, which indeed lies at the root of the whole subject, and is simply this: Are the actions of men, and therefore of societies, governed by fixed laws, or are they the result either of chance or of supernatural interference?

Buckle opts for a fully lawlike "scientific" approach, relying mainly on the "social physics" of statistical analyses in Quetelet's mode. He argues, for example, that murder and suicide rates are as predictable as the tides, and that frequencies of marriage correlate perfectly with the price of corn!

Thus, Buckle, in trying to imitate an inappropriate form of science, established the harmful and false notion that any explanation must invoke predictability under constant law – and that all else is "blind chance" and therefore formally inexplicable. Contingency provides the exit from this intolerable dilemma (for if Buckle were correct then we would be doomed, for we cannot have his predictability, and therefore under his scheme we could not attain any explanation worthy of the name). History grants us the possibility of rigorous explanation for its unpredictable events – and the methods of contingency, with emphasis on directional sequence and resolution by understanding antecedent events and immediate uniquenesses, must be our guide. (Incidentally, when Buckle was all the rage, Darwin sought and obtained a meeting with him. Darwin was vastly disappointed and concluded that Buckle was basically a humbug. I am cheered that the greatest 19th-century biologist, one of the few men who truly understood the nature of contingent explanation, could thus see through Buckle's pretensions for a scientific history.)

Science can add so much to the study of history – and it seems such a shame that we do not generally seek the union that this conference and volume so clearly demonstrate, and with such excitement. Consider the enormous contribution that conventional scientific approaches can make to the study of contingency. We can, first of all, apply our own methods to establish the fundamental framework of all history – the proper temporal sequence of complex events. A major theme of many papers in this book – the use of genetic data to build cladograms for establishing a phyletic order of temporal branching – amounts, fundamentally, to providing the primary datum of history.

We can also, and secondly, use the powerful tools of multiple criteria from our many disciplines. We are students of physics, chemistry, atmospheric science, geology, genetics, systematics, and several other fields. All of our data converge on the common problem of elucidating history. And since history is, by its very definition, the study of the most complex events of all, here, especially, we need the integrated expertise of numerous disciplines. To cite just one example as a symbol among hundreds that made this conference so exciting: Max Taylor pointed out that phylogeny of unicellular life has been well-nigh intractable because so much classical anatomical data is based upon loss of structures – and loss is both notoriously homoplastic and intrinsically non-revealing (as an absence of data!). But with the possibility of DNA sequencing, we now have positive information to make the same problem tractable.

We can supply multiple criteria from our diverse storehouses of expertise – and history requires this effort. (We can – and should – also pursue our more conventional scientific task of seeking the undertones of timeless generalization amid the uniquenesses of history. If extraterrestrial impact, for example, turns out to be a general cause of the largest mass extinctions, then the general physics of impact lies illustrated within the particulars of each event. But we must not say that the generality is more important and represents the "real" science, with individual extinctions merely ranking as differing examples that permit us to extract the residue of timelessness. We must also, I would say above all, treasure and document the differences among events – and recognize that their contingent results have set the major pattern of life's history. In this sense, the conventional science of general events establishes a proper partnership with contingent history.)

As we unite our methods and expertises to converge upon an understanding for the complex history of early life on Earth, we recognize that our story becomes ever more interestingly intricate – "not only queerer than we suppose, but queerer than we can suppose," in Haldane's famous quip. Consider the two major increments in intricacy. Lyell, in the 1830s, envisioned an earth in steady state, with no phenomena of directionality among its ceaselessly cycling changes. We came to understand that the earth's history contains important components of profoundly altering directionality – making the Precambrian, in particular, such a different place from our current world, and therefore more challenging to encompass. But, following the progressivist legacy of Guizot and Macaulay (and of most 19th-century culture and science), we then thought that these changes unfolded in a gradual, steady, and accumulative way (Lyell's later position after he abandoned his original uniformity of state). We have now come to understand the unpredictably episodic (and sometimes truly catastrophic) character of much shaping change in the earth's history – a theme of so many papers in this volume.

Shall we then be depressed at our necessary acknowledgment of all this contingent complexity? Not at all, for we have the tools of narration and explanation amid our interdisciplinary expertise. Malcolm Walter talked of the earth's "salad days." He meant this quite literally, as he described the feast (of algal mats) available to the first grazing organisms. But consider Shakespeare's metaphorical coining of this famous phrase. In *Antony and Cleopatra*, he speaks of a person's youth as "my salad days, when I was green in judgment." I suggest that we apply this perspective to our planet as a symbol of the tractability of our subject – early life on Earth. Do not think of the Precambrian as impossibly distant and ineffably ancient – with scientists as poor youth trying to understand such antique profundity. As Bacon first noted in his famous paradox of 1605, we must invert this perspective. "Antiquitas saeculi, juventus mundi," Bacon wrote – the old days were the world's youth. The earth was but a babe in the early Precambrian. We are the greybeards, the inhabitants of an ancient planet, now more than 4½ billion years old. We have the accumulated wisdom and experience of the ages. We are the results of all the contingent history that came before. With enough struggle and enough application of all our diverse expertises, we can surely understand the earth's salad days of its extreme and bumptious youth.

Theme 1:
Life's Gestation and Infancy

The Earth accreted some more than 4½ billion years ago, and life originated fairly soon thereafter. Paleontological, sedimentological, and geochemical studies leave no doubt that complex ecosystems characterized our planet's surface at least 3½ billion years ago and possibly earlier. The recognition that organisms emerged rapidly once the planetary surface became habitable means that life was born of the same processes that shaped the Earth.

What were those processes? Despite major advances during the past decades, the origin and early evolution of life remain poorly understood. The chemistry that gave rise to the first cells occurred on the surface of the young Earth, either broadly or in discrete environments such as rift vents. For this reason, hypotheses about chemical evolution are constrained by knowledge of Earth's accretion, internal differentiation, degassing, and postaccretion impact history. This section of the book consequently starts with three chapters on the early geological and chemical environment of the Earth (Chang; Lowe; Towe). The following nine chapters (Oró; Lazcano, 2 chapters; Baltscheffsky & Baltscheffsky; Gedulin & Arrhenius; Deamer, Harang Mahon & Bosco; Wächtershäuser; Buss; Ohta) deal with the very heart of the matter: What chemical and selective processes may have been at work to produce entities capable of energy conversion and reproduction?

Once organisms had become established on the surface of the Earth, they began to deposit a historical record in their own molecules and as fossils in rock. The chapters by Stetter, Kandler, Pierson, and Sogin describe how the study of living microorganisms helps us understand how the early diversification of life took place and how the early ecosystems evolved. Finally, Schopf presents the oldest records of life preserved in rocks.

The Archean world remains sketchily known. But the difficult task of extracting reliable information on life's origin and earliest evolution from experiments, geological observations, and comparative biology can easily be justified, because it is the events of this eon that imparted to Earth its biological countenance.

The planetary setting of prebiotic evolution

Sherwood Chang

Planetary Biology Branch, NASA Ames Research Center, Moffett Field, California 94035-1000, USA

Except for major short-term perturbations in surface environments caused by a declining flux of impactors, equable conditions for prebiotic evolution could have existed as early as 4.4 Ga. Giant impacts undoubtedly constrained the timing of life's origin, but quantitative statements about when the clock was set await stronger consensus on impactor fluxes and more refined theoretical models. Organic matter surviving impacts or synthesized in impacts would have augmented the inventory of compounds produced in surface environments. The oxidation state of the prebiotic atmosphere remains controversial, but little question exists about the reduced state of the early ocean, which may have provided a more productive medium for chemical evolution than the atmosphere. Submarine hydrothermal systems and the wind-mixed layer of the ocean are specific settings that may have favored prebiotic evolution. Especially interesting is the ocean–atmosphere interface, where a complex set of physical and chemical processes operated continuously: collection of gas, aerosols, and dust from the atmosphere; recycling of organic and inorganic solutes between the ocean and atmosphere through bubble formation and bursting; organic synthesis by UV radiation, cavitation, and other energy sources; and formation, dissipation, and reformation of surface-active monolayers and bilayer vesicles. The intersections of the wind-mixed ocean layer with shorelines of volcanic platforms and shallow-marine hydrothermal systems may have been key sites for prebiotic evolution.

Although life originated in the first billion years of Earth's history, only a fragmentary record of Earth's earliest surface environments is preserved in the 3.8 Ga metasediments of Isua, Greenland. These rocks indicate surface temperatures below 100°C; an extensive body of liquid water; carbon dioxide, water vapor, and presumably nitrogen in the atmosphere; higher heat flow and more intense volcanism than now; the beginnings of continental growth; weathering and, by inference, hydrologic and carbon geochemical cycles. Within this global context, the 3.5 Ga sediments of Western Australia containing the earliest compelling evidence of life (Schopf, this volume) record a shallow, marine environment dominated by episodic island volcanism and hydrothermal activity (Barley *et al.* 1979; Groves *et al.* 1981). Such settings probably existed earlier than 3.8 Ga, and their occurrence as early as 4.4 Ga is not contradicted by any theories or observations. Except for probably higher frequency and larger size of objects impacting Earth, conditions at 4.4 Ga may not have been vastly different from

Bengtson, S. (ed.) 1994: *Early Life on Earth. Nobel Symposium No. 84.* Columbia U.P., New York

those at 3.5 Ga (Lowe, this volume). Impacts, however, could have exerted beneficial as well as detrimental effects, both of which only recently have been studied. Whether surroundings of the sort recorded in Western Australia spawned the first biota is not evident, nor are the factors that made them particularly favorable. A useful gauge of the potential of an environment for the origin of life is whether it is capable of producing and maintaining the organic compounds necessary for prebiotic evolution.

Accretion, core formation, and evolution of the atmosphere–ocean system

According to several recent models of Earth origin, a differentiated planet with metallic core, highly convective mantle, molten surface, and massive steam atmosphere formed as the direct result of accretion (Stevenson 1983; Matsui & Abe 1986; Zahnle et al. 1988; Abe & Matsui 1988). As the accretionary energy input declined, surface temperatures dropped, and water rained out to form an ocean. Surface temperatures at or below 100°C could have prevailed as early as 4.4 Ga. Early catastrophic outgassing is supported by isotopic systematics of radiogenic and primordial noble gases in terrestrial materials (Thomsen 1980; Allegre et al. 1987).

Because N_2 behaves much like a noble gas, it is expected to have outgassed at the same time as the noble gases. Hydrogen (Boettcher et al. 1975) and carbon (Berg 1986; Tingle et al. 1988; Katsura & Ito 1990) are partially lithophilic elements and would have partitioned between the mantle and the atmosphere–ocean system early on, as they have since then (Javoy et al. 1982; Walker 1983b). The time at which an ocean of the present size (1.4×10^{24} g) formed remains uncertain, although the very early existence of a substantial body of liquid water seems highly probable. The noble-gas data are also consistent with a scenario in which the atmosphere formed by impact from a late-accreting veneer of volatile-rich material (Anders & Owen 1977). As a result of a declining impactor flux, the bulk of the ocean would have accumulated early in the period 4.4–3.9 Ga.

To keep the early ocean from freezing, a global greenhouse was necessary to offset the effects of a faint young Sun dimmer than today by 25–30%. Climate models confirm that 100–1,000 times the present atmospheric level of CO_2 would have sufficed (Kasting et al. 1984). If an ocean and equable temperatures were the only limiting requirements, life could have arisen at any time between 4.4 Ga and 3.5 Ga. The availability of organic compounds may have been a more stringent requirement. If the gas phase was a source of organic matter, synthesis would have been favored by a highly reducing atmosphere (see Chang et al. 1983).

Composition of the prebiotic atmosphere

Volcanism and impacts fueled formation of the prebiotic atmosphere. The composition of gases released by volcanism is governed by the redox state of the final mineral assemblages with which the gases chemically equilibrate before expulsion to the atmosphere. Thus measurements of the oxygen fugacities of ancient igneous rocks may provide clues to past atmospheric compositions. The recent ascendancy of a neutral redox model for the atmosphere (N_2, $CO_2 > CO >> CH_4$, $H_2O >> H_2$, $SO_2 > H_2S$) at the time of life's origin is based on theoretical arguments and redox measurements on rocks younger than 3.8 Ga (Chang *et al.* 1983; Holland 1984). For lack of a geological record to the contrary, a highly reducing atmosphere (N_2, $CO > CH_4 > CO_2$, $H_2O \sim H_2$, $H_2S > SO_2$) cannot be excluded, but its continued existence beyond the accretionary epoch required ongoing chemical equilibration with metallic iron, which early core formation would have removed from the mantle and crust. Measurements of mantle xenolith oxygen fugacities indicating a highly reduced environment (Arculus & Delano 1980, 1981) now appear to have been in error (Mathez 1984; Virgo *et al.* 1988). Measurements by a more reliable method (Wood & Virgo 1989) point to a redox state buffered close to the fayalite–quartz–magnetite system consistent with a neutral redox atmosphere and characteristic of basalts throughout the geological record.

In analogous fashion, the gases resulting from an impact reflect the oxygen fugacity and composition of the impact plume produced by shock-driven partial melting and vaporization of the projectile and excavated target (e.g., Zahnle 1990). Atmospheric CO/CO_2 ratios could have been enhanced over that in a neutral-redox atmosphere ($CO/CO_2 \sim 5\times10^{-4}$) by high-temperature reactions of carbon with other elements in the impact plume. Impactors containing metallic iron or organic carbon would have generated pulses of reduced atmospheric gases (Kasting 1990). Kasting (1990) incorporated impact generation of CO in a photochemical model and showed that $CO/CO_2 > 1$ could have occurred during the first several 0.1 Ga of Earth's history, provided that CO was not rapidly dissolved in oceans and converted to bicarbonate. Photochemistry in atmospheres containing CO_2 or mixtures of CO and CO_2 yields formaldehyde as a major product (Pinto *et al.* 1980; Bar-Nun & Chang 1983; Kasting 1990).

Sulfur-containing gases, H_2S and SO_2, would have been released into the early atmosphere by impact outgassing as well as by volcanism. In oceanic hydrothermal systems, precipitation of highly insoluble iron sulfides by abundant aqueous ferrous iron would have served as a highly effective sink for H_2S (Walker & Brimblecombe 1985). In prebiotic atmospheric photochemical models, SO_2 persists as the major S-containing gas. It is accompanied by minor amounts of sulfuric acid and elemental sulfur, S_8, which could have provided a protective screen against harmful ultraviolet light (Kasting *et al.* 1989).

Despite the critical roles attributed to ammonia and HCN in chemical evolution (see Oro, this volume; Bada & Miller 1968), no atmospheric process has been shown capable of producing them efficiently under neutral redox conditions. Nor can NH_3 in the atmosphere survive rapid photochemical destruction (Ferris & Nicodem 1972). Rutile, TiO_2, in desert sand catalyzes the photochemical reduction of N_2 by sunlight and water (Schrauzer *et al.* 1979, 1983). Today, this source of NH_3 amounts to one third of the N_2 fixed by lightning discharges; it would have been much smaller with less land surface early on. In a neutral redox atmosphere in which the $C/O < 1$, shock heating by lightning or impacts produces predominantly nitric oxide, NO, rather than HCN (Chameides & Walker 1981; Fegley *et al.* 1986). Hydrolysis of HCN rained into an ocean would yield NH_3, but a strongly reducing atmosphere is required for HCN production (Zahnle 1986; Stribling & Miller 1987). Thus syntheses of NH_3 and HCN pose major challenges for scenarios of chemical evolution that rely on atmospheric origins of these key compounds. Ammonia may have been more efficiently produced in the ocean (see below), but sources for HCN remain problematic.

Early seawater

If an ocean formed so rapidly that weathering could not keep up, a highly acidic solution could have resulted from dissolution of outgassed HCl (Holland 1984). Since the rate of weathering on the primordial Earth is unknown, the lifetime of a hypothetical acidic ocean is also unknown. Bicarbonate concentrations would have depended on temperature, weathering rates, and atmospheric CO_2 levels, which could have been as high as tens of bars immediately after accretion and up to several bars thereafter until 3.8 Ga (Walker 1985). The possible persistence of initially high and slowly declining CO_2 partial pressures was attributed to slow weathering rates in the absence of much continental surface. Greenhouse calculations (e.g., Kasting *et al.* 1984) and the sedimentological record (Lowe, this volume; Grotzinger, this volume) point to a warm (80–100°C), bicarbonate-rich ocean with pH perhaps as low as 6 before 3.8 Ga.

The redox capacity of an early ocean far outweighed that of the atmosphere, and its reducing power would have been replenished on time-scales shorter than 10^7 years (Wolery & Sleep 1976) by seawater circulation through hydrothermal systems (Veizer 1983; Derry & Jacobsen 1990). Aqueous ferrous iron, Fe^{2+}, extracted from hot igneous rocks during such circulation would have been the major reductant (Cloud 1973). Estimates of dissolved iron exceed 10^{-4} moles/liter for the seawater from which banded iron formations precipitated (Holland 1973). Precipitation of FeS would have kept H_2S and sulfide ion concentrations at negligibly low levels. Rain out of sulfate produced by SO_2 oxidation in the atmosphere could have yielded concentrations as high as 10^{-3} moles/liter (Walker & Brimblecombe 1985).

Despite the surety of a reduced primordial ocean, the prebiotic chemistry of Fe^{2+} in seawater has not been extensively explored. Irradiation of aqueous Fe^{2+} with ultraviolet light results in release of H_2 and precipitation of iron hydroxides (Braterman *et al.* 1983; Borowska & Mauzerall 1988). This reaction could have provided an important source of reducing equivalents in the upper meters of the ocean. Photostimulated reduction of bicarbonate to formaldehyde in aqueous Fe^{2+} at neutral pH has been reported (Borowska & Mauzerall 1988), but the results have been found irreproducible (Borowska & Mauzerall 1991). Other workers have reported ferrous-iron-mediated photoconversion of carbon dioxide to formaldehyde under acidic conditions (Getoff *et al.* 1960; Åkermark *et al.* 1980). Formaldehyde produced photochemically in the atmosphere would have rained out into the ocean and, in the absence of sinks, could have accumulated to concentrations about 10^{-4} moles/liter after a million years (Pinto *et al.* 1980). Nitrogen oxides fixed in the atmosphere and rained into the oceans as nitrite would have been reduced to NH_3 by aqueous ferrous iron. Model calculations based on estimated production rates of atmospheric nitric oxide (Kasting 1990) and measured rate constants for nitrite reduction at pH 7.4 and 25°C suggest maximum, steady-state, ammonia concentrations between 3.6 and 70 micromoles/liter depending on whether photochemistry or hydrothermal systems were active sinks for ammonia (Summers &

Chang 1993). Lacking additional sources, pathways for organic synthesis at low concentrations of formadehyde and ammonia (e.g., Miller & Van Trump 1981), possibly in the absence of HCN, remain to be evaluated. Additional investigations of the chemistry of Fe^{2+}-rich seawater under photochemical and dark conditions should be undertaken to further explore its potential for chemical evolution.

Influence of impacts on the origin of life

Little doubt now exists about the importance of impact processes in the history of Earth. They provided a mechanism for accretion of the terrestrial planets (Wetherill 1980), formation of the Moon (Hartman & Davis 1975), the delivery of Earth's atmophilic and biogenic elements (Arrhenius *et al.* 1974; Benlow & Meadows 1977; Lange & Ahrens 1982), and natural selection (Alvarez *et al.* 1990). Just as environmental perturbations caused by impacts are thought to have influenced biological evolution, analogous changes must have affected chemical evolution as well. Not surprisingly, the role of impacts on the origin of life is a topic of considerable recent interest (Fig. 1).

Calculations based on impactor size and frequency distributions gleaned from the lunar cratering record have dealt with time constraints on the origin of life imposed by giant impacts capable of destroying Earth's ecosystem. Assuming the origin of life required 10^5–10^7 years, Maher & Stevenson (1988) showed that a time interval of this length could have occurred between ocean-vaporizing impacts as early as 4.2–4.0 Ga. The results of other calculations suggest that the last such catastrophic impact could have occurred as early as 4.4 Ga (Sleep *et al.* 1989) or later than 3.7 Ga (Oberbeck & Fogelman 1989, 1990). Since the frequency and size of impacts increase backward in time, ecosystems at or below the deep ocean–sediment interface had the best prospect for early survival, while life in the photic zone would have been chancy until late in this interval. If life arose at 3.8 or 3.5 Ga, the maximum time available for chemical evolution since the previous giant impact is calculated to be 6 or 165 Ma, respectively (Oberbeck & Fogelman 1990). Undoubtedly, impacts constrained the timing of the origin of life; quantitative statements about when the clock was set must await stronger consensus on impactor fluxes and more refined theoretical models.

Since organic compounds appear to be difficult to synthesize in the neutral-redox model of the prebiotic atmosphere favored by geochemists (Chang *et al.* 1983; Stribling & Miller 1987), investigators have evaluated extraterrestrial sources of prebiotic organic matter as an alternative. Today, interplanetary dust particles (IDP; Mackinnon & Rietmeijer 1987) and meteorites enter Earth's atmosphere, and many are collected essentially intact. Stony meteorites a few meters in diameter can be decelerated by the atmosphere and survive, but larger asteroidal and cometary objects are expected to suffer catastrophic destruction (Chyba *et al.* 1990).

Early in Earth's history, the flux of IDP and material injected by asteroidal and cometary impactors would have been much higher than now, and intact extraterrestrial organic compounds would have augmented those produced by terrestrial processes (Oró 1961; Chang 1979; Chyba *et al.* 1990). Using estimates of organic carbon in the infalling objects and scaling to the lunar impact flux, Chyba & Sagan (1992) estimated the flux of extraterrestrial organic carbon delivered to Earth over the period 4.4–3.0 Ga.

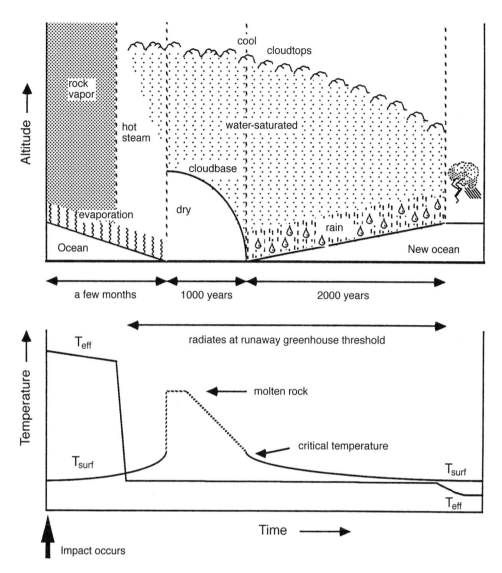

FIGURE 1   Thermal evolution of Earth's surface following an ocean-vaporizing impact. A collision with a 500 km object results in production of 100 atmospheres of rock vapor at several thousands of degrees, radiation from which eventually evaporates the entire ocean, forming a steam atmosphere. Cooling by radiation to space occurs, and a new ocean is rained out after several thousand years. (Courtesy of K. Zahnle.)

The IDP contribution dominated that from comets and meteorites by orders of magnitude with fluxes decreasing from 10^9 to 4×10^5 kg organic carbon/yr. On a vastly different size scale, Clark (1988) suggested atmospheric deceleration during a grazing cometary impact as a possible mechanism for delivery of an intact comet to Earth's surface, the slow melting of which was speculated to provide an organic-rich pond for chemical evolution.

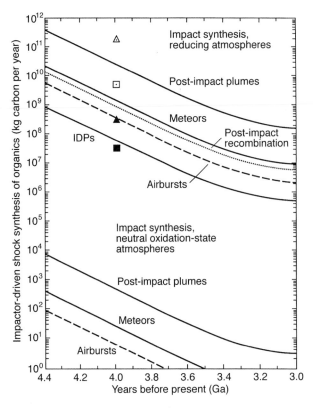

FIGURE 2   Rates of impact-shock synthesis, exogenous delivery, and endogenous production of organic carbon as a function of time. Upper set of curves for a reducing atmosphere: $CH_4 + (N_2$ or $NH_3) + H_2O$; lower set for neutral redox atmosphere: $CO_2 + N_2 + H_2O$. Dashed lines indicate upper bounds, and the dotted line represents a poorly understood estimate perhaps independent of atmospheric redox state. Open triangle and square, respectively, indicate production rates from lightning and UV light (absorption by H_2S) in highly reduced atmosphere; filled symbols apply to neutral atmosphere (absorption by CO_2). UV production in a neutral atmosphere is assumed to be 100 times less efficient than in a reduced atmosphere; no organic nitrogen is photochemically produced, however. (After Chyba & Sagan 1992.)

Impacts also provided an energy source and mechanisms for production of prebiotic organic matter: synthesis in atmospheric shocks (Bar-Nun & Shaviv 1975; Fegley *et al.* 1986) and synthesis by molecular recombination in impact plumes (Mukhin *et al.* 1989). Chyba & Sagan (1992) incorporated both mechanisms and included shocks from meteors, airbursts, and post-impact vapor plumes in a time-dependent calculation. Their results are summarized in Fig. 2, where impact synthesis and delivery of exogenous organics by IDP are compared with endogenous syntheses at 4.0 Ga. In a highly reduced atmosphere, abiotic syntheses yield more organic carbon than intact delivery. In a neutral redox atmosphere with $H_2/CO_2 \sim 0.1$, photochemical synthesis of organic carbon (formaldehyde only) appears to be more productive than delivery by IDP and synthesis by lightning. For atmospheric H_2/CO_2 ratios less than 0.1, as supposed in some other studies of synthesis in a neutral atmosphere (e.g., Pinto *et al.* 1980; Kasting 1990), correspondingly lower organic production rates from lightning and photochemistry would have prevailed. Today, this ratio is about 0.3 in gases emitted at hydrothermal systems of midocean ridges (Welhan 1988) and 0.05 in Hawaiian volcanic gases (Walker 1977). Fig. 2 provides a status report rather than the final word on the relative contributions of various sources of organic carbon. Mechanisms for organic synthesis remain to be discovered, and large uncertainties exist in the calculations, as do unaccounted factors in the modeling of the complexities of the natural world in laboratory and computer experiments. Among such factors are processes acting as sinks for organic matter, for example, destruction of organic compounds by ultraviolet irradiation of surface waters (Dose 1974) or sequestration in insoluble polymers (Nissenbaum *et al.* 1975).

Oberbeck & Aggarwal (1992) incorporated a photochemical sink (Dose 1974) in modeling the time dependence of amino acid concentrations following the collision of a 10 km comet (Fig. 3) into a shallow sea. Two amino acid sources were assumed: ordinary electric-discharge synthesis and shock synthesis in the impact plume. Their calculations showed that the concentration of comet-derived amino acids decreased over time, eventually falling below the level maintained by synthesis from discharges. This crossover point occurred after 90 or 140 years, depending on the discharge-energy flux used in the calculations. Thus for ~100 years after each impact, shock-synthesized amino acids would have made the predominant contribution. Since 10% of initial amino acids were photochemically destroyed in 38 years, while cycling of an ocean through hydrothermal systems would have taken at least 10^6 years, the authors concluded that pyrolysis in hydrothermal systems was not a significant sink. They suggested that adsorption on clay minerals (e.g., Lahav & Chang 1976; Hedges & Hare 1987) might have occurred faster even than photolysis. Calculations like those of Oberbeck & Aggarwal (1992), although model-dependent, take into account both sources and sinks and represent a valuable approach for assessing the contribution of any process to chemical evolution.

FIGURE 3   Time-dependent concentrations of amino acids produced by comet impact and by corona discharge in a neutral redox atmosphere. Oberbeck & Aggarwal (1992) estimated an initial concentration of 10^{-7} moles/liter for amino acids produced by shock synthesis. Their calculations were based on the assumptions that the abiogenic a-amino-isobutyric acid in K/T boundary sediments (Zhao & Bada 1989) was synthesized in the impact, and the inland Cretaceous sea at Stevns Klint, Denmark, had a 100 m maximum depth (Alvarez *et al.* 1990). Production from discharges was caculated using energy fluxes of 12.5 Joule/cm^2 (Stribling & Miller 1987) and 0.4 Joule/cm^2 (Chyba & Sagan 1992). (Oberbeck & Aggarwal 1992; courtesy of Kluwer Academic Press.)

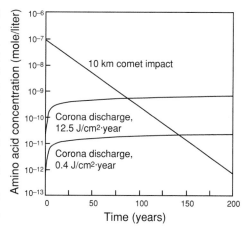

Interfacial environments

The importance of interfacial environments on the prebiotic Earth has been emphasized (Chang 1988). Most relevant for the origin of life may have been the ubiquitous, geophysically active regions at the ocean–atmosphere interface and at the ocean–crust interface in marine hydrothermal systems. These realms contain phase boundaries between gas, liquid, and solid states where disequilibrium resulting from gradients in physical and chemical properties are maintained by physical and chemical energy fluxes. Within these environments, small-scale interfaces are provided by aerosols, volcanic and cometary dust, hydrothermal minerals, chemical precipitates, and vesicle-like structures of organic or mineral chemical composition. Inasmuch as life itself must have emerged as a phase-bounded system, the formation, dissipation, and reformation of small-scale interfaces must have been a prerequisite for the origin of life.

Hydrothermal systems

Hydrothermal systems have been proposed as sites for organic synthesis and the origin of life by a number of authors (e.g., Ingmanson & Dowler 1977; Corliss *et al.* 1981; Wächtershäuser 1988a, b; Shock 1990a, b, 1992). The main objection to these sites is the perceived difficulty of synthesizing and preserving the organic compounds necessary for the evolution of cellular life (Miller & Bada 1988). Energy sources, metal ions, and iron-bearing minerals as catalysts and mechanisms for organic synthesis have been proposed (Holm 1985; Arrhenius 1986; Wächtershäuser 1988a, b, 1990a, this volume),

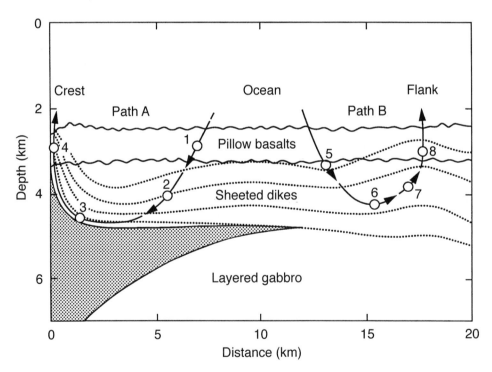

FIGURE 4 Water-circulation paths in hydrothermal systems. Path A shows solutions taken deep into the vicinity of the magma chamber and emerging through black smokers at the ridge crest. Fluids taking Path B penetrate and emerge from the flanks of the system following a lower temperature trajectory. Dotted lines indicate isotherms. The numbered locations correspond to points for which activities of organic compounds were calculated at defined temperatures and oxidation states: **1:** 100°C, hematite/magnetite buffer (HM); **2:** 200°C, pyrrhotite–pyrite–magnetite buffer (PPM); **3:** 400°C, fayalite–magnetite–quartz buffer (FMQ); **4:** 350°C, PPM; **5:** 100°C, HM; **6:** 250°C, FMQ; **7:** 200°C, PPM; **8:** 150°C, PPM. (Shock 1992; courtesy of Kluwer Academic Press.)

but little research into these materials and mechanisms has been conducted under hydrothermal conditions (see, however, Yanagawa *et al.* 1988; Shock 1990a; Bloechl *et al.* 1992; Hennet *et al.* 1992; Yanagawa & Kobayashi 1992).

A thermodynamic basis for organic synthesis in hydrothermal systems has been advanced by Shock (1990b), who proposed that abiotic synthesis could have occurred in metastable states as seawater circulated through hydrothermal systems. At stable C–H–N–O equilibrium and redox states buffered by mineral assemblages in these systems, activities for relevant organic compounds are negligible. Below about 600°C, however, kinetic barriers to thermodynamic equilibrium are observed to occur, allowing attainment of metastable states and much higher activities of organic compounds.

The scenario proposed by Shock (1990b, 1992) for organic synthesis in hydrothermal systems starts with C and N as predominantly CO_2 and N_2 at high-temperature equilibrium with mineral assemblages deep in hydrothermal systems. As fluids containing these species circulate to lower-temperature regimes, they cool under the influence of mineral-buffered hydrogen fugacities. As they cool, they move from conditions under which CO_2 and N_2 predominate to those in which CH_4 and NH_3 dominate. If kinetic barriers to CO_2 and N_2 reduction are surmounted, organic compounds are produced and maintained in metastable states. Further conversion to CH_4 and NH_3 is supposed to be inhibited by other kinetic barriers.

In Fig. 4 two paths are shown, illustrating the circulation of water through hydrothermal systems. The calculated activities of aqueous organic compounds and ammonia were found to be significant and higher at the final point of Path B than of Path A (Fig. 5). Notably, the activity of ammonia suggests a significant source of reduced nitrogen. Since flow through potentially destructive Path A is estimated to be less than 5% of that along Path B, the flanks of these systems represent sites of high potential for synthesis of organic carbon. Shock (1992) estimated this source (kinetic barriers permit-

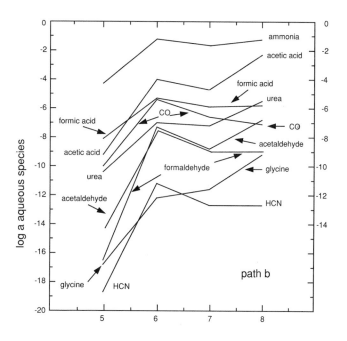

FIGURE 5 Plots of the log activities of selected aqueous species at points indicated along Path B in Fig. 4. (Shock 1992; courtesy of Kluwer Academic Press.)

ting) at 2×10^8 kg per year, which is comparable to other estimates of production rates in a neutral redox atmosphere (Fig. 2). If the hypothesis of hydrothermal synthesis is verified by experiments, hydrothermal systems could emerge as sources for organic compounds and ammonia rather than sinks.

Synthesis of organic compounds is only the first step in chemical evolution, however. Along the way to more complex chemical systems, condensation reactions are necessary to form phosphate esters, lipids, peptides, and oligonucleotides from monomers (see chapters in this volume by Oró; Gedulin & Arrhenius; Deamer *et al.*). These reactions require the intermolecular elimination of water, and plausible mechanisms remain to be demonstrated under hydrothermal conditions (cf. Flegmann & Tattersall 1979; Siskin & Katritzky 1991; Yanagawa & Kobayashi 1992).

For any synthesis scheme to gain credibility, however, it must go beyond mere plausibility; its efficacy must be evaluated within the constraints imposed by the particular environment. Among the general constraints applicable to hydrothermal systems is the time available for chemical evolution of organic compounds before diffusive loss to the bulk ocean or burial with minerals in sediments below levels where useful chemical transformations could take place. Extensive metalliferous sediments occur over the crests and flanks of ocean-ridge hydrothermal systems today. Based on estimated fluxes of hydrothermal particulates in the northeast Pacific (Baker *et al.* 1985; Dymond & Roth 1988), accumulation of a 1–7 cm thick sediment layer would occur in 10^4 years. In addition, the particulate phases (colloidal iron oxides and hydroxides, and iron sulfides) act as scavengers of phosphorus (and other elements) from seawater (Arrhenius 1952; Feely *et al.* 1990). Coprecipitation with iron hydroxides has also been used as a means of stripping dissolved marine organic compounds from seawater (Garrett 1967). Diffusion, burial or irreversible adsorption, and destruction by secondary reactions were probably limiting factors for chemical evolution of organic compounds in all planetary environments.

The ocean–atmosphere interface

For this discussion the ocean–atmosphere interface is taken to include a zone from the ocean-surface film to about a hundred meters below. This zone is continuously mixed by winds and waves. Residence times in the upper few hundred meters of the ocean are about 100 years. A packet of water transported to depth will have a residence time in the deep ocean water of about 10^3 years. Assuming similar residence times early on, the deep ocean would have acted as a sink for nonbuoyant and non-surface-active materials. If life arose in the wind-mixed layer, and if 100 years is an inconceivably short time for the origin of life, then surface-active substances must have played a critical role in sustaining chemical evolution.

Various processes acting on and within the uppermost layer of the mixed zone would have made it a complex, physically and chemically active environment perhaps well suited for chemical evolution (Lerman 1986). Material in the atmosphere from all sources (oceanic, detrital, volcanic, as well as extraterrestrial) would have fallen onto the ocean surface and mixed with organic compounds synthesized in situ. In turn, substances in the mixed layer would have been exposed to wave action, diurnal solar and cosmic radiation, electrical discharges, and other energy sources. Concentrations of reactants and products delivered from the atmosphere would have been highest at the ocean surface and would have decreased with depth of mixing. Airborne polyphosphates produced by volcanism (Yamagata *et al.* 1991) and deposited at the ocean surface could have helped fuel phosphorylation reactions. Surface-active compounds capable of forming bilayer membranes could have been supplied by extraterrestrial inputs (Deamer 1985). Disruption of surface monolayers could have enhanced the formation, dissipation, and re-formation of the bilayer vesicles that were prerequisites for the origin and evolution of cellular life (Stillwell 1980; Koch 1985; Morowitz *et al.* 1988; Deamer *et al.*, this volume). Preferred molecular orientations and ordering at the atmosphere–monolayer, bilayer–water and atmosphere–water interfaces (MacIntyre 1974a; Pohorille & Benjamin 1991; Wilson & Pohorille 1991) could have promoted molecular self-assembly (Whitesides *et al.* 1991) and selectively enhanced some reaction pathways over others. Today the pH of rain lies in the 3–5 range, resulting from dissolution of CO_2, sulfuric- and nitric-acid aerosols, and organic acidic air pollutants (Chameides & Davis 1983). Acid rain on the primitive Earth would have originated from atmospheric photochemistry (Kasting *et al.* 1989) and aerosols of volcanic (Snetsinger *et al.* 1987) and impact origin (Prinn & Fegley 1987). Possibly, a slightly acid ocean-surface film could have benefited chemical evolution, for instance, allowing photoreduction of bicarbonate or carbon dioxide by ferrous iron (Åkermark *et al.* 1980).

At any given moment, bubbles cover 3–4% of the ocean surface (MacIntyre 1974a). Bubble formation occurs in the upper portion of the mixed zone as the result of wind and wave action, and bubble bursting at the surface ejects aerosols and particulates into the overlying atmosphere. The importance of bubble formation and bursting in accounting for differences in chemical properties between the thin surface skin of the ocean and underlying waters has been known for years (MacIntyre 1974a, b; Liss 1975; Wu 1981; Tseng *et al.* 1992). Among these differences are strong enrichments in organic compounds (particularly surface-active material), dissolved phosphate, Cr, Cu, Fe, Pb, and Zn. Apparently, surface-active and hydrophobic organic compounds complexed with other species diffuse to and concentrate at the hydrophobic inner surface of bubbles as they rise to the surface. Bursting of bubbles ejects a proportion of the inner wall into the atmosphere as aerosol droplets (Fig. 6), which can undergo other processes and be transported downwind or higher in the atmosphere.

Return of aerosols to the ocean causes chemical enrichments in the surface film relative to bulk water. Phosphate ion enrichment in drops from breaking bubbles can attain factors of 600 over the bulk solution (MacIntyre & Winchester 1969); organic matter can be enriched by more than a factor of 10^3, with organic/salt mass ratios of 0.3–0.9 (Berg & Winchester 1978). Airborne material today contains mineral grains, organic compounds, and sea salts, all of which undoubtedly had ancient analogs. Modern estimates of the global ocean emission of sea salt alone approach 10^{12} kg per year (Chester 1986). If 30% of this amount is organic carbon, 3×10^{10} kg per year, or 100 times as much, is processed through marine aerosols today as is estimated to have fallen to the ocean surface 4.0 Ga ago from all sources in a neutral redox atmosphere and all but the photochemical source in a highly reduced atmosphere (Fig. 2). Since wind and wave action then was unlikely to have been much less intense than now, prebiotic organic matter in the ocean-surface layer would have been similarly reprocessed in marine aerosols. Interestingly, total oceanic biomass amounts to ~3×10^{12} kg today.

Only recently has the possible relevance of ocean bubble phenomena to the origin of life been suggested. According to Lerman (1986) material scavenged from the mixed zone by bubble formation could have taken part in multiple cycles of processing during repeated ejection into the atmosphere and return to the surface as aerosols and particulates. Supposedly, aerosols could have undergone evaporative dehydration, rehydration as condensation nuclei for rain and snow, and exposure to energy sources (sunlight, lightning, and coronal discharges) before redeposition in the ocean. Could condensation reactions and synthesis of organic phosphates have occurred first in such cycles? Oberbeck *et al.* (1991) also suggested that condensation reactions of organic compounds supplied by or synthesized during injection of extraterrestrial material in the atmosphere could have been enhanced by wetting and drying cycles in cloud drops.

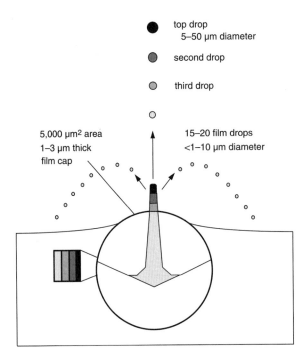

FIGURE 6   Composite schematic cross section of a 500 µm bubble bursting and aerosols generated by the burst at the ocean–atmosphere interface. Ejected aerosols come from the thin layer peeled off the inner wall of the bubble. Each successive drop originates from increasingly deeper onion-like shells in the bubble wall. The enlargement of the bubble wall is intended to show the origin of the jetting drops, which can be accelerated at >10³ G, and the less energetically accelerated film droplets. (After MacIntyre 1974a and Berg & Winchester 1978.)

The enormous amount of energy in wave action would have had other effects in addition to bubble formation. Evidence that cavitation and sonochemistry accompany wave action has been reported by Anbar (1968), who also suggested a role for these processes in prebiotic synthesis. Heating and cooling rates during cavity collapse are estimated to be billions of degrees C per second with peak temperatures above 5,000°C (Suslick 1989; Flint & Suslick 1991). Scavenging of sonochemically produced hydroxyl radicals and other oxidants by aqueous Fe^{2+} may eliminate side reactions and favor organic synthesis. The possibilities for prebiotic synthesis in the exotic cavity microenvironment are essentially unexplored.

There are also detrimental aspects to the processes acting at the ocean–atmosphere interface. Ultraviolet photochemistry can be destructive to organic compounds at the surface, and photodecomposition cannot outpace synthesis. In addition to producing H_2, the photochemistry of aqueous ferrous iron also yields colloidal ferric oxide–hydroxide, which could have scavenged organic compounds as it settled to the seafloor. Bubble phenomena also enhance particle aggregation, thereby accelerating the removal of particulate-bound organic matter, phosphate, and trace metals from the mixed zone (Wallace & Duce 1978; Sackett 1978). Clearer insight into the chemical-evolution potential of the wind-mixed layer will be gained from a more comprehensive characterization and evaluation of sources and sinks for organic compounds.

Consider the implications of the fluxes of organic carbon shown in Fig. 2. If transport to the deep ocean on a 100-year time scale was the only sink, upper limits for the total concentration of organic carbon in the top 100 m of the ocean would have been 5×10^{-8} moles per liter under a neutral redox atmosphere and 5×10^{-5} moles per liter under a highly reduced atmosphere. Since typical prebiotic energy sources (e.g., lightning) produce a variety of compound types, concentrations would have been much lower for amino acids or any particular class of compounds deemed necessary for prebiotic evolution. The dilution of any critical compound would have been far worse in the absence of a compound-specific synthetic pathway. Because prebiotic evolution in the ocean at such high dilution is problematic, evaporation of water and cycles of wetting and drying in shoreline environments or in bodies of water with restricted access to the ocean have been proposed to overcome this dilemma (e.g., Lahav & Chang 1976).

Mechanisms combating dilution would have been necessary at the ocean–atmosphere interface as well. Rather than dispersing by dissolution, surface-active compounds would have accumulated at the surface film as a result of buoyancy, surface tension, and hydrophobic molecular interactions. Higher steady-state organic carbon concentrations could have been achieved in the upper several meters by bubble phenomena. Higher concentrations of key organic compounds in and near the surface film could have been maintained by adsorption on and transfer among surface-active monolayers or within bilayer vesicles. Evidence that some of these phenomena are at work today can be found in the chemical oceanographic literature cited above. Finally, more productive and selective mechanisms of organic synthesis may have existed on surface films and within vesicles than in seawater solution. Although these processes offer potential solutions to the problem of dilution, their effectiveness in a prebiotic setting remains to be evaluated.

Clearly the ocean–atmosphere interface is a dynamic environment worthy of much future study as a site for prebiotic evolution. Its intersection with the shorelines of volcanic platforms and shallow-marine hydrothermal systems – where possibly other sources of ammonia, organic compounds, mineral catalysts, and chemical energy were available – could have been extremely important for the origin of life. Perhaps the geological setting associated with the earliest record of ecosystems was indeed a spawning ground for life.

Acknowledgments. – I thank the organizers of this Nobel Symposium for inviting me to contribute this manuscript. G. Arrhenius provided a constructive review of an earlier draft. Louis Lerman recognized and brought to my attention the roles that sea-surface phenomena could play in chemical evolution. Support for this work was provided by the Exobiology Program of the National Aeronautics and Space Administration.

Early environments: Constraints and opportunities for early evolution

Donald R. Lowe

Department of Geology, Stanford University, Stanford, California 94305, USA

A speculative picture is developed of Archean environmental conditions and evolution based on the integration of observations about the characteristics of the early lithosphere, atmosphere, hydrosphere, and biosphere. The Archean earth was dominated by oceanic lithosphere on which the principal land areas were unstable volcanic islands and small microcontinental blocks. A CO_2 greenhouse maintained temperatures of 30–50°C at 3.5–3.2 Ga. Because there was no polar shelf ice, the oceans were permanently stratified into deep, stagnant, iron-rich and shallow, wind-mixed, iron-poor layers. Ocean stratification and small land area kept the surface-layer nutrient depleted. Bacterial plankton productivity, O_2 production, and sedimentation of banded iron formations (BIFs) were largely restricted to areas of dynamic upwelling; benthic bacterial mats covered adjacent shallow platforms. Biogenic O_2 was rapidly lost to local (organic C), atmospheric (H, S), and deep-ocean (Fe) sinks. A steady-state carbon cycle balanced subduction loss and mantle outgassing. High atmospheric CO_2 resulted from high outgassing rates, inefficient weathering, and low biological productivity. Early continent formation at 3.3–3.1 Ga caused a drop in atmospheric CO_2 but no major climatic reorganization. The formation of enormous continental blocks at 2.7–2.5 Ga triggered large-scale CO_2 depletion, greenhouse collapse, and ocean mixing. The pre-3.8 Ga Earth never included significant continental crust and was warmer than the post-3.8 Ga world, perhaps buffered at temperatures of 90–100°C. Wherever they evolved, the earliest organisms were moderate to extreme thermophiles.

Twenty years ago, Sagan & Mullen (1972) observed that Archean sedimentary rocks testify to the presence of liquid water on the earth 3.2 billion years ago. Although considerably more was known about the Archean environment even then, the past 20 years have seen enormous advances in our understanding of surface conditions, processes, and life on the early Earth. The following discussion will examine the terrestrial surface environment between 3.8 and 2.5 Ga in terms of the lithosphere, atmosphere, hydrosphere, and biosphere and their interactions, with an emphasis on information from the geologic record, and will then consider the implications of this environmental picture for understanding conditions before 3.8 Ga under which life may have evolved.

Bengtson, S. (ed.) 1994: *Early Life on Earth. Nobel Symposium No. 84.* Columbia U.P., New York

Lithosphere

Although a wide range of scenarios has been proposed for the growth of the global mass of continental crust over time, considerable evidence suggests that the pre-3.0 Ga Archean earth was dominated by oceanic lithosphere (see Veizer & Jansen 1979, 1985; Abbott & Hoffman 1984; Taylor & McLennan 1985; and Armstrong 1981 for discussions of crustal growth models) and that the Archean continental inventory consisted mainly of microcontinental blocks totaling less than 5% of the area of the present continental crust (Lowe 1992a). The composition of Archean seawater was controlled by its interaction with the oceanic crust and mantle (Veizer *et al.* 1982; Veizer *et al.* 1989b), whereas that of post-Archean seawater reflects weathering and erosion of continental crust (Holland 1984). A similar change is seen in the average composition and depositional setting of sedimentary rocks across the Archean–Proterozoic boundary (Veizer 1988b). Archean sedimentary rocks occur mainly within greenstone belts and are mainly volcaniclastic rocks, cherts, and immature graywackes and mudstones, whereas Proterozoic sedimentary rocks are largely continent-derived quartzose sandstones, shales, and shelfal carbonates. Archean and Proterozoic greenstone belts appear to represent a precratonic stage of continental growth characterized by the assembly of subduction-related accretionary complexes (Abbott & Hoffman 1984; Hoffman 1989b), and Proterozoic and Archean cratonic sediments represent a postcratonization stage of sedimentation on and around continental blocks. The Archean-to-Proterozoic change in sediment composition represents the transition from an ocean-dominated Archean Earth to a continent-dominated Proterozoic Earth, probably without any fundamental change in lithospheric dynamics (Burke *et al.* 1976; Abbott & Hoffman 1984; Hoffman 1989b; and many others).

An inventory of the age of formation of present-day Precambrian continental crust provides another means of evaluating Precambrian crustal growth patterns (Lowe 1992a). It indicates that 50–60% of the Precambrian crust formed in the Late Archean, mainly between 2.7 and 2.5 Ga, and that less than 5% existed before 3.0 Ga (Fig. 1). The enormous areal extent of Late Archean continental crust and decreasing abundance of younger Precambrian crust is the opposite of the trend expected if crustal age distri-

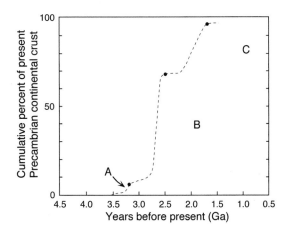

FIGURE 1 Growth curve of the Precambrian continental crust based on the present age distribution of Precambrian crust. Phanerozoic crust was not included in the inventory because of uncertainties in the age of continental basement beneath many younger orogenic zones and in the long-term survivability of many Phanerozoic domains. Present crustal age distribution suggests three main episodes of continental growth: **A:** 3.3–3.1 Ga, when about 5% of the present Precambrian crust formed, **B:** 2.7–2.5 Ga, 58%, and **C:** 2.1–1.6 Ga, 33%. (From Lowe 1992a.)

bution has been controlled by recycling (Veizer & Jansen 1979, 1985). In a recycling-controlled system, younger crustal blocks would be more abundant because of the gradual destruction of older blocks with time. If continental crust of all ages is equally resistant to recycling, an assumption that may be only qualitatively correct (Hoffman 1989b), 50–60% represents a minimum estimate of the proportion of continental crust formed in the Late Archean.

The changing pattern of lithospheric tectonics over Precambrian time also reflects the increasing importance of continental blocks (Lowe 1992a). The Archean Earth was characterized by a tectonic cycle in which continental blocks were bit players. Rifted and passive margin sequences are rare. Late Archean continents were assembled as accretionary complexes made up largely of subduction-related volcanic arcs (green-stones), tonalite–trondhjemite–granodiorite (TTG) arc intrusive suites, and associated largely volcaniclastic sedimentary sequences (Langford & Morin 1976; Hoffman 1989b; Card 1990; Lowe & Ernst 1992). Older microcontinental blocks are found along the margins of these cratonized accretionary complexes. Some apparently acted as rigid blocks against which accretion commenced. Others terminated the accretionary cycle when they collided with the active margin of the accretionary complex, closing the intervening oceans. Continental terranes were rarely incorporated into the accretionary complexes, and late-stage microcontinent collision seldom led to suturing of the converging continental blocks and obliteration of the intervening accretionary belts. The Archean tectonic cycle led to the construction of enormous new blocks of continental crust, largely because older continental blocks were small and few in number and played a mainly passive role.

The large continental blocks formed in the Late Archean became active participants in the Proterozoic tectonic regime. As during the Archean, a number of large accretionary terranes were assembled as new pieces of continental crust between 2.1 and 1.6 Ga, such as the 2.1–2.0 Ga Birrimian basement block of west Africa (Abouchami et al. 1990), the 1.8–1.6 Ga basement block of southwestern and central USA (Bickford 1988), and the 1.8–1.7 Ga Svecofennian Province of Scandanavia (Gorbatschev & Gaal 1987). In contrast to the Archean, however, Early Proterozoic orogenesis widely involved the formation of narrow belts of new continental crust between converging Late Archean continents, such as the 1.9–1.8 Ga Trans-Hudson Orogen (Hoffman 1989b), and in many areas converging Late Archean continental blocks collided and sutured with little or no new crustal production. Rifting of the Late Archean continents and deposition of thick rift and passive margin successions were widespread between 2.5 and 1.7 Ga.

With more than 90% of the present Precambrian crust in place by 1.6 Ga, Late Proterozoic tectonics were dominated by continental blocks. The Late Proterozoic saw widespread continental rifting, deposition of enormous rift and passive margin successions, and formation of linear orogenic belts. This represents the classic Wilson cycle of continental rifting, ocean formation, ocean closure, and continental suturing that characterizes plate-tectonic regimes dominated by macrocontinental blocks. Except in the Arabian–Nubian shield area of northeast Africa and Arabia (Kroner et al. 1987; Dixon & Golombek 1988), few large accretionary complexes and little new continental crust formed in the Late Proterozoic (Lowe 1992a).

The principal land areas on the Archean Earth were mafic volcanic islands developed along spreading centers and subduction zones and over intraplate thermal

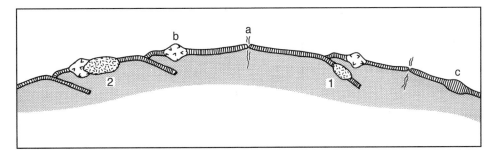

FIGURE 2 Schematic sketch of the principal elements of the Archean lithospheric system. The principal land areas included volcanic islands formed over spreading centers (a), subduction zones (b), and intraplate hot spots (c) and microcontinents (1–2). Before about 3.8 Ga, lithospheric recycling was complete, and even the most buoyant blocks, probably represented by microcontinents formed from subduction-related volcanic-intrusive complexes, were recycled (1). After 3.8 Ga, however, accretionary complexes began to grow (2) and stabilize to form larger blocks of nonrecyclable continental crust.

plumes or hot spots (Fig. 2). The higher rate of heat loss on the early Earth probably resulted in such islands being more numerous in the Archean than they are today, and the low viscosity of komatiitic magmas implies that these islands formed as enormous, low-relief simatic shields. If the eastern Pilbara Block, Western Australia, represents a single, large, deformed 3.5–3.3 Ga old volcanic platform (Barley *et al.* 1979; Lowe 1983), it was at least 300 km across. The tops of these island platforms commonly lay at or near sealevel and hosted a range of subaqueous, shallow-water, and, locally, terrestrial depositional environments (Lowe & Knauth 1977; Barley *et al.* 1979; Lowe 1980a, 1982, 1983). These were unstable, magmatically and tectonically active platforms charac–terized by rapid fluctuations in water depth and subject to inundation by lavas and pyroclastic debris. They hosted bacterial communities during breaks in eruptive activity and served as sites for the deposition of a variety of sediments, mainly volcaniclastic detritus but including carbonaceous biogenic deposits and a variety of orthochemical sediments (Lowe 1980a, 1982). Few of these sediments were carbonates. Virtually all carbonate in both the Barberton Greenstone Belt, South Africa, and eastern Pilbara greenstone belts (3.5–3.2 Ga) formed by seafloor weathering and hydrothermal alteration of submarine volcanic rocks and sediments (de Wit *et al.* 1982; Lowe & Byerly 1986; Veizer *et al.* 1989a, b). The abundance of iron-rich dolomite and ankerite as alteration products in these older greenstone sequences indicates that seafloor alter-ation was an important sink for iron and carbon in the Archean.

Atmosphere

Models of atmospheric evolution suggest that the Archean atmosphere was composed mainly of CO_2, H_2O, and N_2 (Rubey 1951, 1955) with perhaps 100–1,000 times the present atmospheric level (PAL) of carbon dioxide (Owen *et al.* 1979; Kasting 1987; Walker 1990; see also chapters by Holland, Schopf, and Towe, this volume). This composition is consistent with the inventory and probable evolution of terrestrial volatiles (Holland 1984; and many others) and with the inference that, because of the Archean sun's lower luminosity, above-freezing temperatures were maintained on the early Earth by an enhanced greenhouse effect (Owen *et al.* 1979; Walker 1982, 1990; Kasting 1987). Surface temperatures of 30–50°C have been suggested (Ohmoto & Felder 1987; Kasting 1987) and are consistent with geological data. There is wide evidence, for instance, for effective chemical weathering during the Archean, in spite of the dominance of tectonically and magmatically unstable crust and of compositionally immature sediments in Archean greenstone belts. The 3.26–3.22 Ga Fig Tree Group in the southern part of the Barberton Greenstone Belt contains thick layers of first-cycle sandstone and conglomerate deposited in alluvial, fan-delta, and surrounding subaqueous settings (Nocita & Lowe 1990). The detritus formed through weathering and erosion of nearby uplifted blocks of the underlying greenstone sequence, composed largely of komatiitic and basaltic volcanic rocks. The conglomerates, however, are composed almost exclusively of the hardest, silica-rich components of the greenstone sequence, including black chert, white chert, banded chert, jasper, and silicified dacitic and komatiitic tuff. These clasts represent silicified sedimentary horizons constituting less than 20% of the greenstone sequence. The products of weathering of unsilicified basalt and komatiite, which make up more than 80% of the greenstone sequence, occur as clays in Fig Tree mudstones (Danchin 1967). The overlying 3,000 m thick Moodies Group (3.2–3.0 Ga), derived largely by weathering of TTG plutonic rocks and infolded greenstone remnants (M.P.A. Jackson *et al.* 1987) and felsic volcanic rocks, shows a great enrichment in quartz and chert over the source rocks. Similarly, younger first-cycle cratonic sediments, such as the 3.0 Ga Pongola Supergroup and the 2.8–2.7 Ga Witwatersrand Supergroup, South Africa, consisted largely of coarse-grained, quartz-rich, feldspathic sandstones. Compared to the probable granitic source rocks, these sediments had most feldspar removed during weathering. The extensive removal of labile components from first-cycle orogenic clastic debris suggests an effective weathering system, consistent with high levels of atmospheric CO_2 and warm climate.

Altered evaporites are also abundant in 3.5–3.2 Ga greenstone sequences (Lowe & Knauth 1977; Barley *et al.* 1979; Groves *et al.* 1981; Lowe 1983; Worrell 1985; Buick & Dunlop 1990). While evaporites themselves are not necessarily indicative of warm climates, their abundance in relatively unstable Archean greenstone settings suggests high evaporation rates that would have been favored by warm climatic conditions. The wide development of gypsum (Barley *et al.* 1979; Lowe 1983; Worrell 1985) suggests temperatures generally below 58° C (Walker 1982).

The existence of an Archean global greenhouse is also consistent with the absence of unambiguous glacial deposits until late Transvaal and Huronian time (2.4–2.2 Ga). Diamictites in the 2.8–2.7 Ga Witwatersrand Supergroup interpreted to be of possible glacial origin (Wiebols 1955) are probably debris flows related to tectonism within and around the Wits basin rather than to glaciation (Martin *et al.* 1989).

Although some investigators have suggested that free oxygen was a major component of the Archean atmosphere (e.g., Dimroth & Lichtblau 1978), most evidence argues for very low levels. The presence of easily oxidized detrital uraninites and pyrites in the 2.8–2.7 Ga Witwatersrand Supergroup, the lack of redbeds before about 2.0 Ga, and the leaching of iron during early soil formation suggest atmospheric oxygen levels well below those of today (see reviews in Grandstaff 1980; Walker *et al.* 1983; Holland 1984; Walker 1990; and Holland & Beukes 1990). However, the abundance of Archean oxide-facies iron formation indicates that O_2 was available at least locally in seawater (see also Towe, this volume). Although some O_2 could have originated by photodissociation in the upper atmosphere (Berkner & Marshall 1965; Schopf 1975; Towe 1978), biological photosynthesis is generally regarded as the principal source of oxygen on the early Earth (Cloud 1968b, 1976a; Schidlowski 1976; and many others).

The availability of O_2 in surface waters by 3.5 Ga is suggested by the abundance of Archean sulfates. Silicified or baritized gypsum is widespread in both the eastern Pilbara and Barberton greenstone belts (Lowe & Knauth 1977; Barley *et al.* 1979; Groves *et al.* 1981; Lowe 1980a, 1982, 1983; Worrell 1985; Buick & Dunlop 1990). Barite occurs in shallow-water deposits in the eastern Pilbara, where it replaces gypsum (Buick & Dunlop 1990). In the Fig Tree Group in Barberton, it occurs as diagenetic bladed crystal aggregates, possible primary precipitates, and cross-bedded barite sands (Heinrichs & Reimer 1977; Reimer 1990). Hanor & Baria (1977) have argued that barite in the Mississippian Stanley Group of the Ouachita Mountains, Arkansas, was deposited when faulting accompanying orogenesis released barium-bearing fluids into the sulfate-rich ocean. Similarly, Fig Tree barites formed during the earliest stages of orogenesis in the Barberton Belt and probably reflect release of barium into a sulfate-bearing ocean rather than vice versa. The ubiquity of sulfate minerals in shallow-water Archean deposits suggests that sulfate was widely present in ocean surface water and that barium, not sulfate, was the limiting component for barite deposition. The abundance of sulfate evaporites in sediments deposited on the tops of oceanic volcanic islands makes it unlikely that all represent local, isolated, sulfate-rich ponds or lagoons in a globally sulfate-poor ocean as suggested by Lambert *et al.* (1978). Rather, it seems likely that volcanogenic sulfur was oxidized by O_2 in the atmosphere and surface layer of the ocean to form sulfate (Ohmoto & Felder 1987).

Hydrosphere

There is substantial evidence that the Archean oceans were strongly and permanently stratified, including a deep anoxic bottom layer and a thin, wind-mixed, upper layer (Klein & Beukes 1989 and many others). This interpretation derives from the implications of climatic modeling, the distribution and sedimentation of Archean banded iron formation (BIF), and the oxidation state of shallow-water Archean sediments. Climate plays a major role in ocean mixing. While winds stir the upper 300–500 m of modern oceans, deep mixing results from the sinking of cold, saline water beneath shelf ice in polar regions (Berry & Wilde 1978). The dense water flows along the seafloor to lower latitudes, forming a worldwide bottom layer of cold, oxygenated water. Local mixing may also occur where midlatitude surface waters sink as their salinity and density are increased by evaporation (Arthur & Natland 1979; Wilde & Berry 1982). If surface temperatures at 3.5–3.2 Ga were 30–50°C, polar temperatures would have been high enough to prevent the formation of marine shelf ice and, hence, rapid deep-ocean mixing. Similar conditions during Cretaceous global warming also led to deep-ocean anoxia (Wilde & Berry 1982). The mixing time of Archean oceans may have been measured in hundreds of thousands or millions of years instead of about 1,000 years today.

Lowe (1980a, 1982) has summarized the distribution of Archean sedimentary rocks. Iron formation is ubiquitous in deep-water deposits not overwhelmed by coarse clastic debris. Oxide-facies prodominates in major BIF units, mainly in distal settings relative to volcanic centers. Large sulfide and siderite deposits occur locally close to volcanic centers and hydrothermal vents. In deep-water portions of both 3.5–3.2 and 2.9–2.6 Ga old greenstone belts, many thin interflow cherty units are mixtures of fine-grained siderite, clay, fine ash, silica, and carbonaceous matter (Lowe, submitted). They appear to have formed in areas of reduced sedimentation by the slow accumulation of precipitates, clay, organic matter, and distal airborne pyroclastic material. The wide distribution of siderite in deep-water Archean deposits suggests that it was a common hemipelagic sediment.

The ubiquity of iron formation as a deep-water sediment contrasts with its virtual absence in shallow-water deposits (Lowe 1980a, 1982). Local oxide-facies BIF in shallow-water sequences, such as in the 3.26–3.22 Ga Fig Tree Group, the 3.0-Ga Pongola Supergroup, and the 2.8–2.7 Ga Witwatersrand Supergroup, South Africa (Beukes 1973), reflect temporary flooding of the mainly shallow-water platform surfaces by deeper water.

The formation of oxide-facies BIF requires separate sources and reservoirs of ferrous iron and oxygen. Oxygen was produced either in the atmosphere or upper 100–200 m of the water column. Iron was probably released through deep-water hydrothermal vents and by seafloor alteration, although Holland (1984) has suggested surface weathering as the main source. Although slow iron sedimentation may have occurred globally due to diffusion and low-intensity mixing across the interface between deep- and shallow-water masses, major BIF units would have formed mainly below sites of large-scale dynamic mixing, such as areas of unusual geostrophic or wind-driven upwelling and within and above hydrothermal plumes over oceanic volcanic vents. Suggestions by Walker (1987), Cloud (1968b), and others that iron formation marks areas of high bacterial productivity may be true if there were plankton and if, as in modern oceans, productivity were highest in areas of large-scale mixing between deep, nutrient-rich and shallow, nutrient-depleted waters. In a strongly stratified Archean ocean with minimal input of continent-derived dissolved solids, the surface layer might have been a nutrient-depleted biological desert in which organism and oxygen "blooms" occurred only in areas of strong ocean mixing.

Mixed-ocean models of BIF sedimentation (Walker 1987) or those that derive iron through subaerial weathering (Holland 1984) fail to account for the absence of shallow-water Archean iron formation. If O_2 originated through shallow-water biological photosynthesis, the locus of iron precipitation would have been the mixing zone between Fe- and O_2-bearing fluids over or adjacent to areas of highest organic productivity. While highly productive areas may have been associated with dynamic upwelling, shallow-water platforms were also sites of biological activity throughout the Archean. The absence of shallow-water BIF on these platforms indicates that the upper, wind-mixed layer of the ocean contained little or no iron. Early Proterozoic Superior-type shallow-water BIF provides a better example of iron sedimentation in a mixed or mixing ocean.

Not only is there little or no shallow-water Archean iron formation, but Archean cratonic sediments are strikingly similar to their Phanerozoic analogs in terms of other oxidized and reduced components, including organic carbon (Schidlowski 1988; Holland 1984). In the 2.6–2.3 GaCampbellrand Subgroup of the Transvaal Supergroup (Beukes 1987; Klein & Beukes 1989), shelfal sediments deposited in quiet water below wave base include highly carbonaceous shales and carbonates. Carbonate sands and stromatolitic carbonates deposited in wave- and current-agitated settings generally contain little organic carbon. Supratidal and intertidal carbonates, even where deposited in low-energy settings, are generally light gray in color and carbon poor. A similar distribution of carbon is present in the 3.0 Ga Pongola Supergroup (Matthews 1967; Beukes & Lowe 1989), where shallow-water, intertidal, and supratidal stromatolitic carbonates representing wave- and current-active settings are essentially carbon-free (Beukes & Lowe 1989).

Sediments in 3.5–3.2 Ga greenstone belts show a very different distribution of organic matter. Carbonaceous matter is abundant in thin, shallow-water sedimentary units between volcanic flows in the Barberton and eastern Pilbara greenstone belts. Silicified, fluffy, sand-sized, carbonaceous granules as well as ripped-up chunks of carbonaceous mats are abundant in current-deposited volcaniclastic sands (Lowe & Knauth 1977, Fig. 11a; Lanier & Lowe 1982, Figs. 5 and 9a; Walsh 1989; Walsh & Lowe, submitted). In areas of low volcaniclastic input, many shallow-water black cherts contain layers composed almost exclusively of current-deposited carbonaceous particles (Lowe & Knauth 1977, Fig. 12a; Lowe 1983, Fig. 13c). Associated shallow-water to intertidal evaporites also contain carbonaceous particles and carbonaceous bacterial mats that drape evaporite layers (Lowe 1983, Fig. 12).

The lack of shallow-water Archean iron formation suggests that iron was stripped from upwelling deep water before it spread onto adjacent shallow shelves, probably as a result of oxidation by O_2 generated by bacterial plankton blooms within the mixing water masses. On relatively stable continental shelves, which became abundant only in the latest Archean and early Proterozoic, organic productivity was also high, and organic carbon was removed from sediments flushed by large volumes of near-surface water, probably through oxidation by O_2 released by benthic bacterial mats. However, on unstable volcanic islands, long-term productivity was low, silicification of deposited sediments was rapid, and there was little long-term flushing. The sediments were often preserved with much of their carbonaceous matter intact. O_2 released to the atmosphere above productive zones in the surface layer was probably rapidly consumed by the oxidation of volcanogenic hydrogen to form water and sulfur to form sulfate. With so many large oxygen sinks, there was probably no net accumulation in the atmosphere.

Biosphere

Carbonaceous cherts in the 3.5–3.3 Ga Onverwacht Group in the Barberton Greenstone Belt, South Africa, and Warrawoona Group in the eastern Pilbara greenstone belts, Western Australia, have yielded a sparse though critical record of carbonaceous sediments, microfossils, and stromatolites older than 3.0 Ga (Schopf 1983b; Awramik 1984; Walsh & Lowe 1985; Nisbet 1985; Byerly et al. 1986; Schopf & Packer 1987; chapters by Schopf and Walter, this volume). During breaks in volcanic activity, the flat surfaces of the simatic island platforms were flooded by shallow seas and transformed temporarily into enormous bacterial meadows. In the Barberton Belt, preserved bacterial mats are largely restricted to shallow-water deposits (Walsh 1989; Walsh & Lowe, submitted), suggesting that mat-forming bacteria grew mainly within the photic zone. The organisms were probably phototactic, photoautotrophic prokaryotes (Walter 1983). In view of the probable presence of O_2 in shallow Archean seas, some were probably capable of oxygenic photosynthesis. Deep-water Archean sediments also contain abundant carbonaceous material, mainly structureless, nonlaminated kerogen mixed with clay, ash, silica, and/or fine carbonate (Walsh 1989; Walsh & Lowe, submitted). The wide deposition of deep-water carbonaceous shale and apparent absence of non-carbonaceous shale suggests the presence of planktic bacteria in the water column.

Discussion

Archean environmental scenario (Fig. 3)

Archean surface temperature, weathering, and other climate-related factors were, as today, closely related to the carbon cycle and especially to the level of atmospheric carbon dioxide. In the absence of continental blocks and with full lithospheric recycling, the largest Archean carbon reservoirs would have been the mantle and atmosphere (Javoy et al. 1982). Even today large amounts of mantle carbon are being released into the exogenic system (Des Marais 1985), and during the Archean the mantle would also have been an important interactive element of the carbon cycle (Javoy et al. 1982).

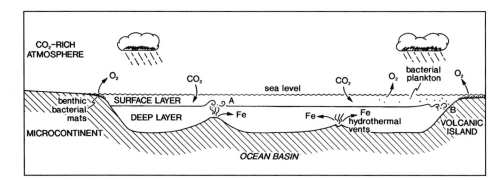

FIGURE 3 Schematic diagram of the Archean environmental system. A strongly and permanently stratified ocean is maintained by the CO_2-mediated global greenhouse. Because of the relatively small land area and stratified ocean, the surface layer is a nutrient-starved desert. Biological activity is focused in areas of dynamic mixing of deep, nutrient-rich and shallow, nutrient-poor water masses, such as above volcanic and hydrothermal vents (**A**) and along the margins of microcontinents and large volcanic islands (**B**). Biological productivity in areas of upwelling releases oxygen that is quickly consumed by combining with (1) iron in the upwelling waters, resulting in the deposition of iron formation below the upwelling sites, (2) the carbonaceous remains of organisms in the water column and on adjacent shelves, and (3) reduced gases and sulfur in the atmosphere.

Smaller reservoirs were represented by the biosphere, sediments, and diagenetic and hydrothermal carbonate in the seafloor. Today's large carbon reservoirs in metamorphic and sedimentary rocks (Hunt 1979) are strictly continental accumulations and were absent in the Archean (Walker 1990).

On a steady-state Archean Earth with complete lithospheric recycling, carbon loss through subduction would have balanced carbon outgassing from the mantle. Carbon dioxide levels in the atmosphere would have adjusted to equalize these fluxes (Walker 1990). The rate of carbon loss through subduction was ultimately controlled by the rate of carbon removal from the atmosphere through weathering and biological activity leading to sedimentation, and seafloor alteration. If the rate of mantle outgassing were higher, land area smaller, and biological productivity lower on Earth before 2.6 Ga, high levels of atmospheric CO_2 would have been required to drive the transfer of carbon to the seafloor through the relatively inefficient processes of weathering/sedimentation and seafloor alteration. If, however, outgassing rates were low, atmospheric CO_2 levels would have been correspondingly low. Geological evidence for the period 3.5–3.2 Ga favors the former scenario.

Early stages of global environmental change commenced about 3.3–3.1 Ga ago with formation of several large blocks of continental crust, including the Kaapvaal Craton of southern Africa and the Pilbara Block, Western Australia. These were initially young, high-standing orogens subject to deep weathering. Between 3.1 and 2.6 Ga, these new continents hosted deposition of the oldest platform carbonates on Earth, including those in the 3.0 Ga Pongola Supergroup, South Africa (Matthews 1967; Beukes & Lowe 1989) and in the 2.75 Ga Fortescue Group, Western Australia (Hickman 1983). In contrast to greenstones older than 3.0 Ga, 3.0–2.5 Ga greenstone belts in Canada, Zimbabwe, and Siberia contain stromatolitic carbonates. All, however, are thin and locally developed units. Thick, platformwide carbonate units did not appear until after 2.6 Ga. If these Late Archean carbonates represent CO_2 removed from the atmosphere by weathering of the 3.3–3.1 Ga continental crust, either a great deal of carbonate was deposited in the deep sea and lost through recycling or the net loss of atmospheric CO_2 was small. The latter alternative is consistent with the inference that only a small amount of continental crust formed 3.3–3.1 Ga ago (Lowe 1992a). Although drawdown of atmospheric CO_2 began during the Late Archean, the basic dynamics of the Archean surface system, overall high atmospheric levels of CO_2, global greenhouse, and stratified ocean appear to have persisted until the Early Proterozoic.

The environment before 3.5 Ga

The two principal controls on the Archean environment 3.5–2.5 Ga ago were the small area of continental crust and high atmospheric CO_2. The small amount of continental crust reflects the more-or-less complete recycling of the lithosphere. There is little reason to suppose that before 3.5 Ga there was ever any greater amount of continental crust or that lithosphere recycling was slower or less efficient.

Although there is geological evidence for liquid water on the earth after 3.8 Ga, similar evidence does not exist for the period from accretion to 3.8 Ga, the time of deposition of the oldest preserved sedimentary rocks. There is as yet no geologic evidence, therefore, regarding atmospheric CO_2 and surface temperatures before

3.8 Ga. From accretion to about 4.0 Ga, rapid crust–mantle interchange through tectonic recycling and impact gardening would probably have released large amounts of mantle carbon to the atmosphere. At the same time, the small land area and low biological productivity would have restricted the rate of carbon flow to the lithosphere. Seafloor alteration may have been the principal process of carbon removal from the atmosphere. If so, atmospheric CO_2 levels were probably high, and elevated surface temperatures could have been maintained by a global greenhouse. If greenhouse-driven surface temperatures ever approached 100°C, the oceans would have begun to boil, exposing the seafloor to rapid weathering and accelerating carbonate precipitation in the open ocean. Before 4.0 Ga, the exogenic system may have been buffered to maintain a surface temperature somewhat below 100°C. Short-lived excursions to higher global temperatures may have occurred early in Earth's history as a result of large impacts (Sleep *et al.* 1989).

After 4.1–3.9 Ga, impacts declined, and a more stable surficial tectonic system was probably established. As the earth cooled, recycling slowed and microcontinental blocks formed. Carbon loss from the mantle would have also slowed, and the rate of atmospheric carbon removal by biological and weathering processes would have increased because of increasing land area. The net result was probably a decline in atmospheric CO_2. This scenario would suggest that surface temperatures may have declined from 90–100°C at 4.1–4.0 Ga to 30–50°C at 3.5–3.2 Ga.

The post-Archean environment

Although global environmental change commenced 3.1–2.6 Ga ago, the results did not substantially alter the dynamics of the surface system. However, enormous areas of continental crust formed between 2.7 and 2.5 Ga. Weathering of this new continental crust after 2.6 Ga was accompanied by the deposition of thick, widespread sequences of Early Proterozoic shelfal carbonate, such as the 2.6–2.3 Ga carbonates of the Transvaal Supergroup, South Africa (Beukes 1987). These shallow seas also hosted enormous bacterial mats, which both consumed CO_2 through photosynthesis and released O_2 to the atmosphere (Knoll 1979). The depletion of atmospheric CO_2 by weathering, biological activities, and sedimentation on these new continental blocks and the attendant cooling is reflected in Huronian glaciation 2.4–2.1 Ga ago. Cooler climatic conditions were accompanied by the formation of polar shelf ice and cold-water sinking that initiated ocean mixing, iron precipitation, and the destruction of the stratified ocean.

Conclusions

The Archean earth 3.5–3.2 Ga ago was dominated by oceanic lithosphere and microcontinents having an area of less than 5% of that of the present continental crust. Lithospheric recycling was rapid, and the average age of the oceanic lithosphere was young (Burke & Dewey 1973; Burke *et al.* 1976; Abbott & Hoffman 1984). Mafic and komatiitic volcanic islands constituted the principal land areas throughout all but the last 100 million years of the Archean. Although the volcanic portions of greenstone belts may represent only some of the volcanic settings on the early Earth, land areas probably included volcanoes developed over spreading centers, subduction zones, and intra-plate hot spots. They were of low relief and often submerged and provided platforms for growth of early bacterial communities and deposition of the oldest preserved sedimentary rocks.

Higher atmospheric CO_2 levels maintained a global greenhouse, with surface temperatures of perhaps 30–50°C at 3.5–3.2 Ga. Because there was little or no polar shelf ice, cold-water sinking, and deep-ocean stirring, the Archean ocean was strongly stratified. The deep ocean was a reservoir of iron, barium, and other mantle- and seafloor-derived materials. The shallow, wind-mixed surface layer contained dissolved CO_2, small amounts of O_2 in areas of high organic productivity, and abundant SO_4^{2-}. The Archean biosphere included prokaryotic benthic and probably planktic bacteria. Because of ocean stratification, biological productivity was severely nutrient-limited. The principal productive areas were probably open-ocean sites of mixing above hydrothermal plumes and coastal areas of dynamic upwelling. Oxygen produced by biological photosynthesis was rapidly cycled into local (organic C), overlying atmospheric (H, S), and underlying deep-ocean (Fe) sinks.

The Archean climate was controlled by a steady-state carbon cycle in which subduction loss balanced mantle outgassing. High atmospheric CO_2 was maintained by a combination of rapid outgassing and inefficient carbon transfer to the lithosphere through weathering/sedimentation and seafloor alteration. Before 4.0 Ga, surface temperatures may have been buffered at 90–100°C. During the waning stages of large impacting, as the lithosphere cooled and plate-tectonic systems were stabilized, atmospheric CO_2 was gradually reduced through a decline in outgassing and the accelerated removal of CO_2 from the atmosphere by increased weathering and biological activity. By 3.5 Ga, the surface temperature may have declined to 30–50°C. Although estimates are poorly constrained, surface temperatures before 3.5 Ga were almost certainly high, and it seems inescapable that the first organisms evolved under warm and possibly hot surface conditions. Whether life evolved around deep-sea hydrothermal vents (Corliss *et al.* 1981) or not (Miller & Bada 1988), within the water column, or on shallow-water volcanic platforms, the earliest organisms would have been moderate to extreme thermophiles.

The formation of several larger blocks of continental crust at 3.3–3.1 Ga increased weathering and atmospheric CO_2 drawdown. This is reflected in the deposition of thin Late Archean carbonate units on the 3.3–3.1 Ga cratons and in Late Archean greenstone belts. This depletion, however, was not enough to alter substantially the dynamics of the surface system. With the formation of immense areas of continental crust in the Late Archean, between 2.7 and 2.5 Ga, accelerated CO_2 depletion due to weathering of and biological activity on the continental crust resulted in large-scale carbonate sedimentation, collapse of the global greenhouse, cooling, and mixing of the stratified oceans.

Acknowledgments. – This research was supported by Grant Nos. NCA2-332 and NCA2-721 from the NASA Exobiology Program, Grant No. EAR89-04830 from the National Science Foundation, and Grant No. NAG9-344 from the NASA Planetary Materials and Geochemistry Program. I am grateful to Ms. Ronadh Cox for reviewing the manuscript and offering many helpful suggestions.

Earth's early atmosphere: Constraints and opportunities for early evolution

Kenneth M. Towe

Department of Paleobiology, Smithsonian Institution, Washington, DC 20560, USA

It is a widely held speculation that early life evolved on an Earth beneath an atmosphere virtually devoid of free oxygen. Such an anoxic environment suggests constraints and opportunities for early life that are distinctly different from those on a primitive Earth where atmospheric oxygen and ozone were present (at lower levels than today) but where local anoxic environments also obtained. Evidence from the rock record and the biology and biochemistry of primitive prokaryotes appears to be little consistent with evolution at vanishingly low global levels of free oxygen in the Archean. It is probable that free oxygen has been a part of the Earth's atmosphere and an important factor in the early evolution of life since the time that the oldest known sedimentary rocks were laid down about 3.8 Ga ago.

There have been no direct measurements of the Earth's early atmosphere. Few hard facts exist concerning either its qualitative or quantitative composition. Most of what is "known," even about the qualitative composition of the early atmosphere, is based on speculation, and the opinions and conclusions derived from this often model-dependent speculation vary widely. Many would agree, however, that from among the various possible gases that may have existed, none is likely to have had as profound and varied an impact on both the origin and evolutionary progression of early life, and the rock record that preserves it, as free molecular oxygen (O_2) would have had. Clearly, an atmosphere containing free oxygen would have provided quite different constraints and opportunities for early life than would one devoid of O_2.

In the introductory portion of his now-classic book *Earth's Earliest Biosphere*, J. William Schopf (1983a) remarked that "... models or working hypotheses that have become widely accepted as organizing principles for the field ... may lead us to ask the wrong questions ... or to disregard significant lines of evidence simply because they seem inconsistent with our model-dependent predilections."

In 1974, Preston Cloud wrote: "The purpose of evolutionary models is to help integrate present knowledge and identify critical areas for future research – not to arrest our kaleidoscopic view of the evolving system or to embalm our current prejudices, however insightful they may seem at the time."

What follows here is an assessment of what I feel are significant lines of evidence that seem inconsistent with current prejudices and widely accepted, model-dependent

Bengtson, S. (ed.) 1994: *Early Life on Earth. Nobel Symposium No. 84.* Columbia U.P., New York

predilections, not only about the anoxygenic composition of the Earth's early atmosphere but also its effect on early life. The assessment is, like others before it, little more than speculation. The conclusions and opinions derived from my speculations are based on what few facts are available from the early rock record, combined and integrated with a simple uniformitarian approach toward the biochemical needs and physiological responses displayed by presumably primitive prokaryotic organisms.

Constraints for early life with atmospheric oxygen

Prebiotic materials

For many years the virtual absence, or "vanishingly small" concentration, of free molecular oxygen ($<<0.001\%$ O_2) in the Earth's early atmosphere has been one of the accepted environmental constraints and conditions for the early evolution of life. This cardinal tenet of origin-of-life studies has been popular, even "obvious," primarily because oxygen will inhibit the synthesis of prebiotic organic matter. In the presence of even traces of free oxygen, Miller–Urey-type experiments fail to produce meaningful yields of the primitive building-block amino acids and nucleotide bases necessary for conventional life. This seemed a powerful argument necessitating the absence of free oxygen from the early Earth's atmosphere (Cloud 1983; Miller 1992).

Without a supply of basic raw materials to the early Earth, the opportunities for life to originate would have been limited. It is becoming increasingly clear, however, that the oxygen-sensitive Miller–Urey-type syntheses were not the only possible source of these materials for the early Earth. A variety of primitive organic compounds are known to occur in carbonaceous chondrites (Anders *et al.* 1974) and have been presumed to occur in comets (Oró 1961; Oró *et al.* 1980; Delsemme 1984). It has been repeatedly suggested that this supply could have been important for the origin of life (Bernal 1954; Sylvester-Bradley 1971; Rasool *et al.* 1977; Degens 1978; Towe 1981a; Anders 1989). These materials arrive safely on Earth today in small quantity, and recent evidence suggests that they were likely made available to the early Earth in greater abundance (Chyba & Sagan 1992; Chang, this volume). For me this means that the inhibitory effect of oxygen on Miller–Urey-type experiments is no longer the fundamental constraint it once was. The existence of organic matter in a variety of extraterrestrial objects (asteroids, comets, chondrites, interstellar dust particles) presents a viable alternative to the Miller–Urey hypothesis as it is applied to Earth. Obviously, it still applies to wherever the extraterrestrial organics were originally formed.

The degradation of primeval organic matter (from whatever source) in the presence of oxygen or ozone has also been viewed as another compelling constraint in favor of an anoxic atmosphere (Miller 1992). Such oxidation reactions are, of course, thermodynamically favorable. However, organic compounds are rapidly oxidized on the surface of the Earth, not by inorganic oxidation but primarily as a result of aerobic microbial activity. On a prebiotic Earth, without this biogenic activity the kinetics of *abiotic* oxidation of organic compounds would have been slower. In addition, the primeval oxidation of primitive organic matter could have been buffered by ferrous iron (Garrison *et al.* 1951) or other inorganic materials such as the surfaces of sands and silts (Suzuki *et al.* 1979).

Essentially unattenuated UV fluxes to the Earth would have been the case for atmospheres with oxygen levels below about 0.002%, since no ozone screen exists at these levels (Levine *et al.* 1980; Kasting 1987). There are serious dangers for early life in the complete absence of an ozone screen, the presence of other UV-absorbing entities notwithstanding (Sagan 1973). The evolutionary development *and growth* of meaning-ful biological macromolecules is very difficult to visualize deep underwater, where the danger of spontaneous hydrolysis (Dickerson 1978) would offset the protection that might have been available against the higher-than-modern levels of UV radiation that are now known to have impinged on the early Earth from the young Sun (Canuto *et al.* 1982). After all, polymerization, bond formation, and chain extension, all leading ultimately to nucleic acids and proteins, are inhibited by liquid water. It is heating and drying that favors *pre*biotic nucleotide- and peptide-bond formation as well as oligo-nucleotide and polypeptide extension and growth. The dangers that surround both the formation and growth of prebiotic nucleic acids and proteins are therefore not trivial.

The same holds for the establishment of photosynthetic life itself. The inhibitory effect of UV radiation on prokaryotes can be severe. A variety of prokaryotes, even those strains that have evolved significant UV resistance, succumb rapidly to today's levels of unattenuated UV radiation (Bhattacharjee & David 1977; Rambler & Margulis 1980; Horneck *et al.* 1984). The diminution of ozone protection may be detrimental to Antarctic phytoplankton beneath today's ozone hole (El-Sayed 1988). Without oxygen there would have been an Archean "ozone hole" of global proportions, and UV inhibition would have been even more detrimental to early evolving life, especially photoautotrophic life.

Interestingly, the inhibitory effect of oxygen on prokaryotic life is much less signifi-cant than is the inhibitory effect of UV radiation. A number of so-called strict anaerobes not only tolerate modern 21% oxygen levels (Talley *et al.* 1975; Wall *et al.* 1990), but many can sustain growth at 0.2% levels (Loesche 1969). A value of 0.2% oxygen is approxi-mately the same level of oxygen necessary to provide a moderate ozone screen (Levine *et al.* 1980; Kasting 1987). The oxygen tolerance of modern anaerobes is, of course, in large part the result of Fe- and Mn-containing superoxide dismutase enzymes (Rolfe *et al.* 1978), but as is discussed below, these enzymes are homologous and of great antiquity (Runnegar 1991).

I prefer to believe that the dangers from ultraviolet radiation to the origin and early evolution of life should have been much greater than the dangers from low levels of free oxygen (Towe 1988). The presence of an early ozone screen would therefore represent an important opportunity for early life to originate, proliferate, and evolve rapidly toward the light-gathering photosynthetic lifestyle that the isotopic and stromatolitic fossil records seem to mandate (Schidlowski 1988; Schopf & Packer 1987; Schopf, this volume).

Sulfur and photosynthesis

The very first life forms are widely believed to have been strict anaerobes. It follows, then, that the presence of oxygen in the environment around them could have been detrimental. Sulfides represent a case in point. Sulfides support anoxygenic photosyn-thesis. Without widespread sulfide availability, the anoxic photosynthetic bacteria that are thought by many to represent the earliest primary producers should have had difficulty dispersing away from local sources of sulfide to populate global environments. Therefore, oxygen should not have been present, for it would have tended to eliminate, or at least severely limit, the supply of these biochemically important reduced substances, whether in the atmosphere or dissolved in the oceans.

On the face of it, this seems a compelling argument in favor of global anoxia. On the other hand, there are several observations that serve to compromise it or at least alter its global significance. The earliest photoautotrophs may indeed have been sulfide-dependent, but the currently available evidence from biochemical sequence studies on a number of prokaryotes (Woese 1987b; Lake 1988; Stetter, this volume) emphasizes the importance of elemental-sulfur respiration as an equally primitive method of biological energy conversion on the early Earth. Many of these so-called sulfur-loving bacteria are both anaerobes and acidic thermofiles (Brock 1986; Stetter *et al.* 1987; Stetter, this volume), and some are even aerobes (Jones *et al.* 1987). Oxygen may indeed oxidize and eliminate sulfides, but without oxygen the supplies of elemental sulfur (and sulfate as well) would have been neither easily formed in quantity nor readily sustainable over time to support early life (Towe 1988). Elemental sulfur is most stable under mildly oxidizing conditions at low pH; it is less so under reducing conditions at oceanic pH (Krauskopf 1979, Fig. 10-2). On the present-day Earth these sulfur-dependent organisms minimize the dangers of oxygen by living at fumaroles (acid hot springs), where the low solubility of molecular oxygen at the elevated temperatures helps mediate the dangers of its atmospheric presence, even at today's 21% levels (Brock 1986).

The early Archean opportunities for the biological utilization of sulfur (and sulfate) would have been difficult under anoxic atmospheric conditions. The presence of atmospheric oxygen, however, would have made elemental sulfur a sustainable quantity yet may not have seriously impeded the development and growth of organisms in localized hot-spring environments. The presence of atmospheric oxygen does not prevent their growth in these locales today, and there is no reason to suppose that it would have done so in the early Archean, especially if the levels were lower than modern ones.

The metabolic requirements of the recently discovered, deeply branching, eubacterial hyperthermophile *Aquifex pyrophilus* are quite consistent with these arguments (Burggraf *et al.* 1992). These marine organisms grow optimally at temperatures near 85°C utilizing small amounts of oxygen (<0.5%) to oxidize hydrogen (R. Huber *et al.* 1992; Stetter, this volume).

Banded iron formations and sulfide availability

The presence of oxide-facies banded iron formations (BIFs) in the Archean, especially in the 3.75 Ga Isua rocks, implies that dissolved ferrous iron was readily available for oxidation. Where *excess* dissolved ferrous iron is available for oxidation, dissolved sulfide is not (except at very low equilibrium levels); sulfide precipitates quantitatively to form "pyrite." Thus, for ferrous iron to have been both mobile over large distances and available in large quantities, either dissolved in rivers and streams (Garrels 1987) or from hydrothermal sources in the deep ocean (Holland 1973), dissolved sulfides could not also have been present. Atmospheric hydrogen sulfide levels could not have been maintained for any length of time (Levine & Augustsson 1985), but even if they could have been, the legendary insolubility of metal sulfides would have made the availability of dissolved sulfides for photosynthesis virtually nonexistent where iron oxides were forming. The loss of reducing power to the Photosystem I activity of the early photosynthetic bacteria by the loss of sulfides *due to atmospheric oxygen* is, therefore, less of a concern because substantial losses of sulfide would have taken place anyhow, because of the rapid photochemical destruction of H_2S in the atmosphere and the necessity of having excess ferrous iron available for iron-formation deposition.

Furthermore, the loss of sulfide would represent an evolutionary opportunity. Life would have been quickly stimulated by the absence of sulfide to experiment with Photosystem II activity very early in Earth's history. The early Archean evolution of cyanobacterial oxygenic photosynthesis is therefore likely, regardless of how sulfides were eliminated. An early origin of oxygenic photosynthesis makes the origin of oxides in iron formations and the origin of organic carbon in sediments around the world easier to comprehend. Cloud (1974, 1983) even went so far as to impute the early existence of oxygenic photosynthesis from the very presence of BIFs in the 3.75 Ga Isua rocks. The existence of Archean sedimentary sulfates (Walter, Buick & Dunlop 1980; Lowe & Byerly 1986), awkward under anoxic conditions, also becomes more easily understandable.

Iron formations and rare-earth elements

Today, oxidation in the oceans of the rare-earth element cerium appears to be a biologically mediated process taking place in surface waters (Moffett 1990). The net result of this oxidation is the development of a negative cerium anomaly in seawater due to the partitioning and removal of insoluble Ce(IV) colloids by adsorption onto hydrous Fe–Mn oxides (Elderfield 1988; German & Elderfield 1990). Regardless of whether or not a biological oxidation of cerium is necessary for the early Archean, data from carefully restudied banded iron formations (Beukes & Klein 1990; Derry & Jacobsen 1990), even the earliest ones (Dymek & Klein 1988), seem to imply that a negative cerium anomaly for Archean seawater is necessary. This is implied because the ferric oxide minerals in the BIFs carry the negative Ce-anomalies consistent with their having been precipitated in waters of this anomalous composition. Thomson *et al.* (1984) state: "The REE pattern of the authigenic component is, in general, the mirror image of the most common pattern for sea water." It should be emphasized, therefore, that the BIF authigenic iron-oxide precipitates could not have formed in shallow surface waters. Had they done so, they would have scavenged the insoluble Ce(IV) colloids, and the BIF oxides would show *positive*, not negative, cerium anomalies. When the negative cerium anomalies are placed together with the positive europium anomalies indicative of a hydrothermal influence (Klein & Beukes 1989), it is almost certain that the BIF oxides were formed at some depth beneath a surface-oxidized ocean (Towe 1991; Lowe, this volume). If oxygen did not exist in the surface waters of the global Archean oceans, it is difficult to visualize the oxidation of cerium, the subsequent separation of the insoluble tetravalent cerium Ce(IV) onto colloids (other than iron), and the development of a basinwide negative cerium anomaly.

Opportunities with ammonia

Ammonia, as the most reduced form of nitrogen, is life's ultimate nitrogen source. Unique to prokaryotes, biological nitrogen fixation is an energy-expensive process whereby ammonia may be generated from atmospheric nitrogen when exogenous supplies of ammonia are absent. Nitrogen-fixing enzymes (nitrogenases) are homologous and primitive and are found among Archaebacteria (Souillard *et al.* 1988) as well as Cyanobacteria (Postgate 1982). They are very sensitive to the presence of ammonia. They are less sensitive to the presence of oxygen, and most prokaryotes have evolved a variety of adaptations to protect themselves from oxygen (Postgate 1982). In the presence of ammonia at very low levels ($\sim10^{-4}$ M), both the nitrogenase gene expression (Helber *et al.* 1988) and the process of dinitrogen fixation (Drozd *et al.* 1972) are inhibited. The early evolution by prokaryotes of the energy-expensive process of obtaining ammonia via dinitrogen fixation would be redundant, had the ammonium ion been present. Regardless of its ultimate source (Chang, this volume), rainout of ammonia is the dominant atmospheric loss mechanism (Levine & Augustsson 1985). Therefore, this should have made the very soluble ammonium ion widely available in the early ocean, thus mitigating the necessity to evolve N_2-fixation early on under anoxic conditions. Oxidative loss of ammonia, however, would have been an environmental pressure favoring the evolution of dinitrogen fixation. The opportunities for the evolution of nitrogenases would thus seem plausible under *local* anoxic conditions where global oxygenic conditions had helped to eliminate the reduced nitrogen supplies. It would seem awkward, however, to evolve such an energy-expensive process in a *globally* anoxic world where ammonia should have been available and where it would seem wasteful to evolve adaptations to a molecule that supposedly did not exist at levels high enough to affect the principal enzyme involved (e.g., $O_2 > 10^{-3.5}$ P.A.L.; Towe 1985).

Protection from oxygen

Superoxide dismutases are important oxygen-protective enzymes that occur very widely in aerobes (Cammack *et al.* 1981). These enzymes have now also been found in a variety of evolutionarily primitive anaerobes, including the anoxygenic photosynthetic bacteria and the methanogens (Asada *et al.* 1980; Kirby *et al.* 1981). As they function to offset the detrimental aspects of biological oxidations, the presence of these homologous antioxidants in primitive anaerobes would seem to imply the presence of oxygen early in Earth's history (Runnegar 1991). Furthermore, their presence in prokaryotes more primitive than the Cyanobacteria implies that some *pre*-photosynthetic source of oxygen existed to provide the opportunity for this early evolution (Towe 1988). It is difficult to understand the evolution of an enzyme designed to protect an organism from something that supposedly did not exist. Unless they originally evolved with some other as-yet-unknown function, superoxide dismutases would have made little sense in an anoxic world.

Organic carbon sources and sinks

Reduced carbon occurs in varying amounts widely distributed in rocks of Archean age, including the oldest rocks at Isua, Greenland. There are three possible origins for this carbon: (1) abiotic synthesis (terrestrial or extraterrestrial), (2) anoxygenic photosynthesis, and (3) oxygenic photosynthesis. The isotopic composition of Archean carbon (except possibly for the carbon found in the oldest rocks at Isua, Greenland), its association with stromatolites (evidence for life), and the fact that its distribution is facies-controlled speak in favor of a biotic, photoautotrophic origin. Independent of the fossil evidence supporting it (Schopf & Packer 1987) and the geochemical evidence consistent with biological carbon isotope fractionation (Schidlowski 1988), the very early evolution of oxygenic photosynthesis (Cyanobacteria) is likely. This is simply because, as stated earlier, it is difficult to imagine a globally distributed source of sulfide for Photosystem I that would be sustainable for at least 500 Ma across all the various facies in which reduced carbon occurs. Once again, this is especially so if the oceans and/or rivers and streams had been replete with dissolved ferrous iron awaiting oxidation. The global use of molecular hydrogen, rather than sulfide, could be an alternative for Photosystem I activity. This process would, however, require the use of enormous quantities of volcanic hydrogen (1000 times the present amount) for hundreds of millions of years. And, of course, all the other evidence of oxygen cited above would have to be entirely overlooked. Only if the process of biogenic sulfate reduction had evolved almost simultaneously with photoautotrophy could sulfides have been widely distributed over a variety of sedimentary environments. This, of course, would require a large, sustainable source of sulfate, which under anoxic conditions is equally problematical (Walker & Brimblecombe 1985).

With abiotic or anoxygenic mechanisms unlikely as the source for the bulk of the reduced carbon found in Archean sediments, the early evolution of oxygenic photosynthesis remains. This process is likely to have produced the oxygen that was responsible for the oxides in BIFs (Cloud 1974), the negative cerium anomalies in BIFs (Dymek & Klein 1988), the oxide coatings on aluminosilicate clays (Veizer 1978), the oxygen to support methylotrophy leading to strongly ^{13}C-depleted carbon isotopes (Hayes 1983), the ferric iron substituted for calcium and magnesium in calcites (Veizer 1978), and the iron oxides found preserved in terrestrial weathering profiles (Shegelski 1980; Schau & Henderson 1983; Holland 1984; Grandstaff et al. 1986) and seen in the submarine weathering of basalt pillows (Dimroth & Lichtblau 1978).

In place of oxygenic photosynthesis, a photochemical origin for some of the oxidized iron has been suggested (Braterman et al. 1983; François 1986). It is a difficult proposal to accept. First of all, an abiotic photolytic explanation certainly requires that ferrous iron be present in surface waters in a variety of facies. This would once again eliminate the surface sulfide necessary for Photosystem I to operate, thereby leaving no explanation for the origin of organic carbon without Photosystem II. Surface-water oxidation of iron, as described above, is also at odds with the observed cerium anomalies in BIFs. The high UV fluxes of the early Archean would favor photochemical oxidation of iron but would also work against photoautotrophy, especially in the shallow, intertidal regions where stromatolitic growth was vigorous (chapters by Golubic and Walter, this volume). And finally, the amounts of iron oxidized globally would have been enormous, e.g., 1.8×10^{14} mol, or 10^{16} g Fe/year (François 1986). A photosynthetic origin of oxygen and carbon is much easier to fit in with all the evidence.

Carbon and iron

Given oxygenic photosynthesis, iron has been the commonly accepted "sink" for early Precambrian oxygen (the so-called Cloud model [Cloud 1973]). Together with volcanic gases, vast quantities of "virtually limitless" iron are supposed to have kept the oxygen levels in the hydrosphere and atmosphere at "vanishingly small" levels (below about 0.000002% O_2) until the oceans had been swept free of iron (Kasting & Walker 1981; Kasting 1991). This model is simple, consistent with a classical prebiotic chemist's view, and very appealing (Lovelock 1988, Chapter 4). It is, from my perspective, awkward to support.

The impact of the evolution of oxygenic photosynthesis would have been dramatic and its effects widespread and daunting (Tappan 1968; Schidlowski 1988). Photosynthesis operates on a daily cycle; it works every time the Sun comes up. The biological production of oxygen is rapid and is in marked contrast to the slow, plodding geological production and availability of dissolved ferrous iron. The approximate turnover rate of the deep ocean is a sluggish 1,000 years. To compare "biology" and "geology," two basic equations are relevant:

$$H_2O + CO_2 \rightarrow \text{"CH}_2\text{O"} + O_2 \text{ (oxygenic photosynthesis)} \tag{1}$$

$$4Fe^{2+} + O_2 + 4H_2O \rightarrow 2Fe_2O_3 + 8H^+ \text{ (ferrous iron oxidation)} \tag{2}$$

Note that for each mole of organic carbon synthesized, a mole of oxygen is produced. To "sink" one mole of oxygen requires the oxidation of four moles of ferrous iron to "hematite." The initial pressure on geology (the oxygen sink) to keep up with biology (the oxygen source) should have been imposing. An ocean saturated with ferrous iron at 3–6 ppm would contain about 10^{17} mol of Fe. Dividing by 4 and then by the 1,000-year turnover rate gives the *maximum* number of moles of oxygen that the oceanic iron reservoir could absorb per year: 2.5×10^{13}. This same number of moles of photosynthetic carbon would have been produced each year. Multiplying by 12 yields 3×10^{14} g C/yr. Dividing by the area of the oceans (3.6×10^{14} m^2) yields an average global marine primary productivity of ~0.8 g C/m^2 per year. This average global productivity is a very low figure, virtually barren by modern standards. It is 175 times lower than today's global marine average (Martin *et al.* 1987), 30 times lower than the suggestion of Kasting *et al.* (1983) for the Archean, and completely at odds with assumptions about the luxuriance of early life (Tappan 1968; Schidlowski 1988; Rothschild & Mancinelli 1990).

A carbon burial rate of 10^{13} mol C/yr has been accepted for the Archean (Kasting 1987; 1991). Applying this global burial rate to the global productivity would require that fully 40% of it be buried and only 60% of it be recycled, a very inefficent recycling process by any standards. Regardless of carbon recycling efficiency, the iron that should have been oxidized and deposited each year to balance even this very low average productivity would have been a staggering 5.6×10^{15} g Fe [(10^{17}/1,000)×56]. If the entire global sedimentation rate had been 10^{16} g/yr (Veizer & Jansen 1985), the *average* Archean rock should then contain a plausible, if rather high 1.2% C, but an absurd 56% total Fe. Even with allowance for an annual replenishment from midocean hydrothermal sources (about 2×10^{14} g Fe/yr), the oceans would have been swept free of iron in thousands of years rather than hundreds of millions of years, leaving enormous oxide deposits behind. The clear result is that there is simply not enough iron buried in Archean sediments to account for the oxygen that is represented by the *net* carbon produced by oxygenic photosynthesis and found buried in these same rocks (Towe 1990, 1991).

As another example, the usual figures quoted for *average* organic carbon and *average* total iron are ~0.5% C and ~5% Fe, respectively. These figures are, of course, poorly constrained and uncertain. Nevertheless, for reasons involving the isotopic composition of carbon, those for organic carbon cannot be too far off (Holland 1984, p. 353) and may be used to approximate the role of iron as the dominant sink for the oxygen produced by photosynthesis. And because the percentage figures are values for an average rock, this calculation is independent of sedimentation rate. If 0.5% C is an *average* figure for Archean sedimentary rocks, the photosynthesis that this buried carbon alone represents would require that more than 9% total iron be present in the average Archean rock to account for the oxygen released (almost double the Holland value). The specific calculation is:

$$[0.5\% \ C/12] = 0.042 \ \text{mol} \ C = 0.042 \ \text{mol} \ O_2 \times [4Fe \times 56] = 9.3\% \ Fe$$

It is important to emphasize that this simple calculation assumes that *none* of the primary productivity was recycled (used by other organisms), i.e., 100% of the *net* photosynthetic organic carbon was buried. This is, of course, completely unreasonable. Clearly, some recycling must have taken place, if for no other reason than to recycle phosphorus, erase biological imprints (morphology), and produce the finely disseminated carbon we see in rocks.

Aerobic recycling

How much carbon was recycled and how much was buried? Nobody knows, but any recycling only aggravates the problem by requiring an increase in the productivity and an increase in the oxygen produced, thus amplifying the iron demand further. If 90% of the carbon had been recycled, then the average 0.5% buried represents a 10% preservation rate for the productivity. In this example the *average* Archean rock would now have to contain ~93% Fe, if iron had been the principal oxygen sink:

$$[0.5\% \ C/12] = 0.042 \ \text{mol} \ C \ \text{buried} = [0.042/10\%] \ \text{mol} \ C \ \text{produced} = 0.42 \ \text{mol}$$
$$O_2 \times [4Fe \times 56] = 93\% \ Fe$$

No plausible geological processes can operate quickly enough to sink the incessant daily production of oxygen via *net* photosynthesis. Only another biological process, *net* oxygen respiration (the reverse of equation 1, above), can keep up with and brake this solar-driven oxygen machine. Aerobic heterotrophic recycling of organic carbon should therefore have been in place to sequester most of the oxygen to avoid an oxygen runaway (Towe 1990, 1991). Iron and volcanic gases would have been secondary sinks, as they are today. The minimum level of atmospheric oxygen necessary to support aerobic respiration is not well known, but a value around 0.2–0.4% oxygen has been proposed (Chapman & Schopf 1983). Early oxygen availability (from whatever source) was therefore an opportunity for life to evolve rapidly toward the clearly advantageous, energy-efficent process of aerobic respiration and heterotrophy. The Archean rock record seems to require, if not demand, its presence.

Constraints and opportunities with methane

In the absence of free oxygen and aerobiosis, the predominant method of recycling organic carbon would have to have been anaerobic fermentation and methanogenesis. In this process, at completion, a mole of organic matter ("CH_2O") yields a half-mole of methane and a half-mole of carbon dioxide. A number of authors have considered this subject, and global methane atmospheres ranging from 30 to 100 ppm CH_4 have been visualized for the early Archean (Kasting *et al.* 1983; Lovelock 1988; Hayes, this volume). However, the geologic record provides a necessary constraint on the reality of such methane atmospheres. As an example, a global methane atmosphere of 100 ppm methane (Lovelock 1988, Table 4.1) resulting from organic recycling under anaerobic conditions would require a sustained global methane flux at the surface of about 4.7×10^{11} molecules/cm^2 per second, or 1.25×10^{14} mol CH_4 per year. Assuming no losses of methane due to photooxidation, such a flux would have been supported by a global methanogenic biota recycling (to completion!) twice this amount (2.5×10^{14} mol) of photosynthetically produced organic carbon per year. We may calculate the primary productivity responsible for this organic matter by adding to the recycled carbon the carbon that escaped to become buried in sediments. The generally accepted burial value is 10^{13} mol/yr (Kasting 1987), thereby yielding a minimum net global productivity of 2.6×10^{14} mol C_{org}/yr. Equation 1 above requires that 2.6×10^{14} mol of oxygen also be produced. In the absence of aerobiosis, the iron required to mop up this oxygen would have been ×4 and ×56, or 5.8×10^{16} g Fe/yr (Equation 2). This is clearly an absurd value because it is, by itself, 3–5 times the total annual sedimentation rate (Veizer & Jansen 1979). A global methane atmosphere thus appears difficult to reconcile with the rock record.

On the other hand, the Archean rock record of organic carbon displays unusually isotopically light (^{13}C-depleted) kerogens that require explanation (Schidlowski 1988; Hayes 1983, this volume). Without oxygen in the Earth's atmosphere and in the surface waters of the hydrosphere the two most relevant biological processes involved would have been limited: (1) the oxidation of biogenic methane to isotopically light CO_2, followed by its photosynthetic refixation, and (2) the one-step direct utilization of isotopically light methane for carbon fixation by methylotrophs, bacteria that are themselves aerobes (Hayes 1983). Limiting these processes makes the widespread evidence for the strongly ^{13}C-depleted organic carbon in early Precambrian rocks rather awkward to explain. The widespread presence of free oxygen, however, represents a facile opportunity to accommodate either of these mechanisms.

Sources of pre-photosynthetic oxygen

At present, the only plausible alternative to cyanobacterial photosynthesis as a source of free oxygen is photolytic dissociation of water vapor in the upper atmosphere accompanied by the loss of hydrogen to space. Although there are some hydrogen-isotopic data consistent with this mechanism (Anders & Owen 1977; Ferronskiy & Polyakov 1982), there are also atmospheric models and other calculations (Walker 1978; Kasting *et al.* 1979) purporting to demonstrate that photodissociation of water vapor on the early Earth would produce only trivial and inconsequential amounts of free oxygen. Because there is a natural tendency to believe that the output of a computer model is somehow more reliable than an opinion (Ellsaesser 1991), many of these models have been given considerable weight and priority by geologists and biologists alike. The fact that the models are extensive, complex, and computerized may make their output precise, but, of course, the conclusions are no more or less accurate or reliable than the assumptions upon which they are based. In spite of their mathematical precision and chemical rigor, they are only speculation too. In fact, other model calculations for the early atmosphere using different assumptions have "demonstrated" the opposite (Brinkmann 1969; Walker 1976; Carver 1981). Indeed, if late-stage bombardments had "blasted" the early oceans, significant water vapor may have entered the upper atmosphere and may have provided a parallel to the photolytic oxygen-producing steam atmospheres modeled by Walker (1976).

Conclusions

Geologists, geochemists, and paleobiologists have been faced with an imposing citadel of no-oxygen constraints provided by prebiotic chemists, biochemists, and atmospheric scientists. They have searched the rock record diligently, if not obligingly, for supporting evidence. Early Archean stromatolites were found, some with actual microfossils preserved within them (Schopf, this volume). Their morphology is consistent with present-day photosynthetic prokaryotes capable of living under anoxic conditions. But the morphology is also consistent with an aerobic lifestyle (Schopf 1983a). The famed, if enigmatic, banded iron formations seemed to make sense only if they had formed under a global atmosphere virtually devoid of oxygen (Cloud 1973; Garrels 1987). After all, even traces of oxygen would have severely limited the movement of dissolved ferrous iron in rivers and streams or over hundreds of kilometers in oceanic surface waters. But acceptance of a deep-ocean hydrothermal source of ferrous iron (Drever 1974; Holland 1973), now supported by the positive europium anomalies in BIFs (Klein & Beukes 1989), has changed this perception. Reduced uranium minerals (uraninites) were found preserved in ancient stream beds and seemed to belie an oxygenic atmosphere (Holland, this volume). Yet their occurrence in modern stream beds (Palmer *et al.* 1987) and studies of their dissolution revealed that they can and do survive variable, if uncertain, levels of oxygen (Grandstaff 1980). Terrestrial redbeds (oxidized sediments formed under continental conditions) appeared to be absent from early sedimentary locales, but the requirements for redbed formation are more than just a function of oxygen partial pressure (Dunbar & Rodgers 1957, pp. 209–218), and Archean examples do exist (Shegelski 1980). Sediments interpreted to be Archean soil profiles (paleosols) have chemical signatures that, with important assumptions about diagenesis and life on land, imply low but significant oxygen levels (Holland 1984, this volume; Grandstaff *et al.* 1986).

In short, the accumulated evidence that once seemed to be compelling added support for the prebiotic and early Precambrian no-oxygen constraints on early life has been gradually but persistently reduced in importance. The realization that extraterrestrial sources of prebiotic organic matter are plausible (Chang, this volume) further reduces the necessity for theorizing that global anoxic conditions existed in the beginning. I hope that the arguments presented here have given the reader additional pause to at least reconsider current views. Paraphrasing the introductory quotes of Cloud and Schopf: a too-ready acceptance of dogma may entice us to disregard significant lines of evidence simply because they seem inconsistent with a model-dependent predisposition or they threaten current prejudices. I have made my prejudices clear. I believe there are numerous, legitimate difficulties and disadvantages with an oxygen-free, reducing-atmosphere model for the origin and early evolution of life. I believe there are numerous advantages to be gained by rejecting it. If many of the difficulties are not outright problems, they are at least impediments to our clear understanding of what happened more than 3.5 Ga ago. To this extent, the advantages of an oxygenic atmosphere are at least as deserving of serious consideration and discussion as its disadvantages seem so readily to have been.

Early chemical stages in the origin of life

Juan Oró

Department of Biochemical and Biophysical Sciences, University of Houston, Houston, Texas 77204-5500, USA

The triple alpha process occurring in the interior of stars at the temperature of 100 million degrees is responsible for the nuclear formation of carbon. The synthesis of such a remarkable nuclide is a unique event without which we would not be able to talk about early life on Earth. Once carbon is synthesized, all other biogenic elements are formed, and their sophisticated chemical reactivities are unleashed to generate an organic Universe. Indeed, 75% of the molecules detected in interstellar space are organic. I summarize here the six early chemical stages that preceded the appearance of life on Earth. Each one of these stages is a sort of paradox, an ordered or low-probability event, working against the thermodynamicist's arrow of time. They start with (1) the formation of circumstellar and interstellar organic molecules, and (2) the accumulation of terrestrial volatiles, primarily from comets, by late-accretion processes. They are followed by (3) the nonbiological synthesis of biochemical monomers on the primitive Earth, (4) the prebiotic condensation of monomers and synthesis of biopolymers, and (5) the self-assembly of prebiotic membranes. The final, most enigmatic, chemical stage of self-organization is (6) the encapsulation of coding and catalytic biomolecules within the structure of a protocell, which is triggered into action, or into life, by the high-energy bond of pyrophosphate or ATP, presumably generating Darwin's ancestral cell. Whether this is an optimist's or a romantic's point of view, it is nonetheless certain that the bottom line is: We are made of stardust!

Certain things appear to be unavoidable: death, taxes, and the often devastating effects of the second law of thermodynamics. It is therefore paradoxical that we live in an universe plethoric with highly ordered, exquisitely complex structures such as spiral galaxies, stars, living organisms, ecosystems, and even human societies. How did they come into being? The possibility of a sudden emergence of highly ordered structures and systems from random, completely unorganized precursors is now being discussed by mathematicians, theoretical physicists, and other scientists. However, although such sudden changes may have taken place during the evolutionary history of terrestrial life, it was the application of Darwin's ideas and their development by Oparin and Haldane that led to the contrasting suggestion that the appearance of life was the result of a process of gradual change that began with chemical and protocellular evolution.

The hypothesis of chemical and protocellular evolution has allowed scientists to break this evolutionary process into several major steps (Oró 1965). It begins with the nuclear synthesis of carbon, nitrogen, oxygen, phosphorous, and other biogenic elements (Aller 1961). It continues with the origin of extraterrestrial organic compounds,

Bengtson, S. (ed.) 1994: *Early Life on Earth. Nobel Symposium No. 84.* Columbia U.P., New York

the formation of our planet, the abiotic formation of monomers and polymers of biochemical significance, and their self-assembly into precellular systems from which the first living systems emerged. The purpose of this chapter is to describe our current knowledge of this series of stages. They form part of a rather remarkable genealogy that begins inside stars long ago disappeared and has had a spectacular outcome: the appearance of the life on our planet.

First stage: The formation of extraterrestrial organic compounds

We live in a highly reducing universe in which hydrogen, the simplest possible atom, is overwhelmingly abundant. Compilations of elementary cosmic abundances, however, show that with the exception of the noble gases He and Ne, the four most abundant elements are hydrogen, carbon, oxygen, and nitrogen, precisely the four major constituent elements of organic matter. It was therefore logical to conclude in the early 1960s that the molecules formed from these elements should be the most abundant compounds in the universe, and that the prevailing chemistry in the cosmos should be based on carbon; in other words, that it is essentially an organic cosmochemistry (Oró 1963, 1972).

Given the extreme physical conditions of space, it is easy to understand the initial reluctance of some to accept the possibility that more or less complex organic molecules could exist in the interstellar environment. However, this skepticism was quickly superseded when formaldehyde, the first simple organic molecule, was detected in interstellar clouds. This was followed very soon by the discovery of a large array of almost one hundred different molecules (Irvine & Knacke 1989). No major biochemical compounds like amino acids or purines have been detected in the interstellar medium, although a significant percentage of the interstellar molecules are organic and include many of the precursors and functional groups of compounds of biological significance. Moreover, one of the most remarkable surprises that emerged from the radioastronomical observations was that the interstellar compounds include many that had been used in laboratory simulations to obtain most of the biochemical monomers found in contemporary living systems (Table 1).

How are these interstellar molecules formed? As shown by the presence of CN, C_2, and other simple chemical species in the solar upper atmosphere, simple diatomic or triatomic combinations may persist at temperatures of the order of 6,000 K. There they exist as the result of the combination of the chemical elements formed in stellar interiors by nucleosynthesis, which rapidly diffuse out or convectionally reach the surface of the stars. The discovery of the linear $HC_{11}N$ (Bell et al. 1982) and C_3 (Hinkle et al. 1988) molecules in the circumstellar envelope of the cool carbon star IRC+10216 suggests that these stars eject material into the interstellar medium, where it can undergo further chemical changes on the surface of silicate, graphite-rich interstellar grains.

The abiotic synthesis of biochemical compounds during the time of formation and early evolution of the solar system, or even during presolar epochs, is also supported by the detection of a number of organic molecules and radicals (C_2, C, CH, CO, CS, HCN, CH_3CN) and other simple molecules (NH, NH_3, H_2O, etc.) in comets. Comets are

TABLE 1   Biomonomers, biopolymers, and chemical properties that can be derived from interstellar and cometary molecules.

Molecule	Formulae	Biomolecules and chemical properties
Hydrogen	H_2	reducing agent
Water	H_2O	universal solvent
Ammonia	NH_3	catalysis and amination
Carbon monoxide	$CO (+H_2)$	fatty acids
(Linear nitriles)	$H(C)_nCN$	(fatty acids)
Formaldehyde	CH_2O	ribose and glycerol
Acetaldehyde	$CH_3CHO (+CH_2O)$	deoxyribose
Aldehydes	$RCHO (+HCN+NH_3)$	amino acids
Hydrogen sulfide	H_2S (+ as above)	cysteine and methionine
Hydrogen cyanide	HCN	purines (e.g., adenine)
Cyanacetylene	HC_3N (+cyanate)	pyrimidines
Phosphate*	PO_4^{3-} (+nucleosides)	mononucleotides
(PN)		(e.g., ATP)
Cyanamide	H_2NCN (condensation)	biopolymers: peptides and oligonucleotides

*detected in interplanetary dust particles of possible cometary origin and in meteorites

thought to be the most pristine bodies in our planetary system, with nuclei that may include adenine and other purines, as well as large, dark HCN polymers (Kissel & Krueger 1987; Matthews & Ludicky 1987). Furthermore, a large array of proteinic and nonproteinic amino acids, carboxylic acids, purines, pyrimidines, hydrocarbons, and other molecules has been found in the relatively primitive carbonaceous chondritic meteorites. Recent analyses of samples of the Murchison meteorite (Cronin 1989) have shown the presence of racemic mixtures of 74 different amino acids: 8 that are present in proteins, 11 with other biological roles (including, quite surprisingly, some neurotransmitters!), and 55 that have been found almost exclusively in extraterrestrial samples.

Second stage: The origin of terrestrial volatiles

Since most of the interstellar organic compounds have been detected in dense, cool, interstellar clouds where star formation is taking place, it is reasonable to assume that the primordial solar nebula had a similar molecular composition. However, the Earth was formed in a part of the solar system almost completely devoid of volatiles. There is convincing evidence suggesting that the organic molecules from which the first organisms arose were not necessarily derived directly from the earliest terrestrial atmosphere. This conclusion is supported both by detailed theoretical models that predict a molten early Earth (Wetherill 1990), which imply an extensive destruction of any primordial organic compounds, as well as by the well-known terrestrial depletion of noble gases relative to solar abundances (Chang, this volume).

Moreover, according to the single-impact theory of the origin of the Earth–Moon system, a body near the size of Mars collided with the Protoearth, whereupon all of the iron from the impactor was injected into the core of the Protoearth. The impact of the collision was such that both the impactor and most of the Earth's mantle were melted, and a substantial mass was ejected into Earth's orbit, subsequently coalescing into the Earth's only natural satellite (Cameron & Benz 1991). As a consequence of such an impact, our planet must have lost into space practically all of the water and most of the biogenic elements (H, C, N, O, S and P) that the protoplanet may have previously retained (Oró et al. 1992a, b).

How then could life appear in a barren, volatile-depleted primitive Earth? When I asked Professor A.G.W. Cameron for an explanation of the terrestrial volatiles according to his single-impact theory, his answer was simple: "They were brought in by comets." In fact, I suggested this hypothesis more than thirty years ago (Oró 1961). After accomplishing the prebiotic synthesis of adenine, amino acids, and other biochemical compounds from HCN (Oró 1960), and also being aware that the same chemical precursor is present in cometary nuclei and that the 1908 Tunguska event in Siberia was probably caused by a comet fragment (10^{10} g), I came to the conclusion that cometary collisions may have provided the primitive Earth with an important source of volatiles. Thus, I reasoned that these volatiles, including the water and the cometary organic compounds, were probably the starting point for the nonbiological synthesis of biochemical molecules that preceded the first organisms.

By assuming that cometary nuclei had densities in the range of 0.01–0.5 gm^{-3} and diameters in the range of 1–10 km, I estimated, based on Urey's (1957) collisional probability, that the Earth had accreted during its first 2×10^9 years up to 2×10^{18} grams of cometary material, which could have led to localized primitive environments with high concentrations of chemical precursors of prebiotic significance (Oró 1961). The original estimates were revised with the valuable collaboration of Professor A. Lazcano when he was at the Instituto de Astronomía (UNAM), and we obtained a significantly higher value ($\sim10^{23}$ g; Oró et al. 1980). In recent years the latter estimates have been essentially confirmed and upgraded by more complex and detailed models and calculations involving a wide range of dynamical parameters, lunar cratering rates, and different masses for the Oort cloud. This has led to suggestions that during its first 10^8 years our planet may have acquired as much as 10^{24-26} g of cometary material (Chyba et al. 1990). Accordingly, even if only 10% of the bodies colliding with the primitive Earth were comets, they would still account for all the oceanic waters (Chyba 1987; Delsemme 1992a, b) as well as all the buried carbon in the Earth's sedimentary shell (Arrhenius et al. 1974; Berner & Lasaga 1989).

There are several independent lines of evidence supporting this hypothesis, but perhaps one of the most recent and surprising comes from observations of the β Pictoris system, around which the Infrared Astronomy Satellite (IRAS) detected a circumstellar disk. These infrared observations were confirmed when visible light images were obtained showing a highly flattened disk, formed by solid particles in nearly coplanar, low-inclination orbits extending more than 400 astronomical units. This disk is considered to correspond to a planetary system in the early stages of formation (Smith & Terrile 1984).

The favorable edge on orientation of the β Pictoris disk allows the direct spectral observation of its gaseous components, and it has been possible to detect strong variations in several spectral lines (Vidal-Majdar et al. 1986; Lagrange-Henri et al. 1988). As argued by Lagrange-Henri et al. (1988), the most likely explanation of this phenomenon is that 10^{13-17} g cometlike bodies with radii ranging from 0.3 to 6 km are falling at high velocities towards β Pictoris, and are being vaporized by the stellar radiation. This model requires 10–100 impacts per year, and its consistency with current descriptions of the collisional history of the early solar system (Lagrange-Henri *et al.* 1988), supports the original contention that comets and other similar volatile-rich minor bodies contributed significantly to the formation of the terrestrial ocean and atmosphere.

Third stage: The nonbiological synthesis of biochemical monomers

Scientists should always be careful when making definite statements. "Art cannot combine the elements of inorganic nature in the manner of living nature," wrote the Swedish chemist Jöns Jacob Berzelius in 1827 (cited by Leicester 1974, p. 154). Just one year later his friend and colleague Friedrich Wöhler demonstrated that urea could be formed by the rather simple reaction between ammonia and cyanic acid (Leicester 1974). Thus began an era in the development of chemical studies. Eventually, this led, by the turn of the 20th century, to many laboratory syntheses of biochemical compounds, including sugars (Butlerow 1861; Baly 1924; Baly *et al.* 1927) and even amino acids (Loeb 1913), as part of experimental attempts to study the mechanisms of photosynthesis and other metabolic pathways (Lazcano *et al.* 1992).

Inspired in part by the presence of meteoritic hydrocarbons, the detection of the CN lines in cometary spectra, and Mendeleyev's theory on the origin of petroleum (which has since been proven to be essentially wrong), Oparin (1924) suggested that the first cells had been preceded by a long period of chemical evolution that had given rise to complex organic compounds on the primitive Earth. Oparin's ideas were to remain relatively unknown until a 1938 English translation of one of his books became available in the West. This led Urey (1952) to propose a quantitative model of the Earth's primitive atmosphere and eventually led to the successful experimental study of prebiotic synthesis of amino acids and other biochemical compounds by Miller (1953; Miller & Orgel 1974), using experimental devices such as the one shown in Fig. 1.

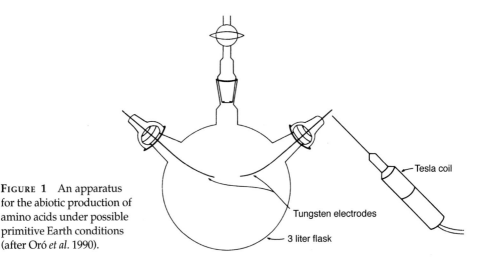

FIGURE 1   An apparatus for the abiotic production of amino acids under possible primitive Earth conditions (after Oró *et al.* 1990).

Tesla coil

Tungsten electrodes

3 liter flask

Having been acquainted with the work of Charles Darwin and Ernst Haeckel since my school days, I had independently arrived at the same basic ideas, which were not fully developed until after my discussion with Severo Ochoa in New York in 1952. Later, in 1955, these ideas were tested when I went to the University of Houston (Oró 1976). After reacting different solutions of formaldehyde and nitrogen derivatives, we were able to demonstrate the nonbiological synthesis of amino acids (Oró et al. 1959). Also, thanks to my previous expertise in the commercial uses of HCN, I began studying its polymerization products. The experiments followed a simple design: hydrogen cyanide was bubbled through an aqueous solution of NH_4OH, and the mixture was refluxed overnight before being analyzed. Since those were the romantic days before the full development of gas chromatography, mass spectrometry, HPLC, and other modern analytical methods, a fraction of the supernatant obtained was spotted on a filter paper for chromatography. Concentrating the supernatant severalfold eventually allowed the detection of adenine and a host of other compounds, including the biological intermediates for purine synthesis (Oró 1960).

We know now that HCN is a prebiotic Pandora's box. Its direct involvement in interstellar chemistry (Table 1) and as precursor not only of amino acids and purines but also of pyrimidines, prebiotic condensing agents, and many other biochemical compounds makes it a molecule essential to our present-day understanding of the process that eventually led to the emergence of life (Oró & Lazcano-Araujo 1981). In retrospect, the examination of its unusual condensation into adenine is almost inevitable (Fig. 2); after all, the empirical formula of adenine is $H_5C_5N_5$, i.e. corresponds to

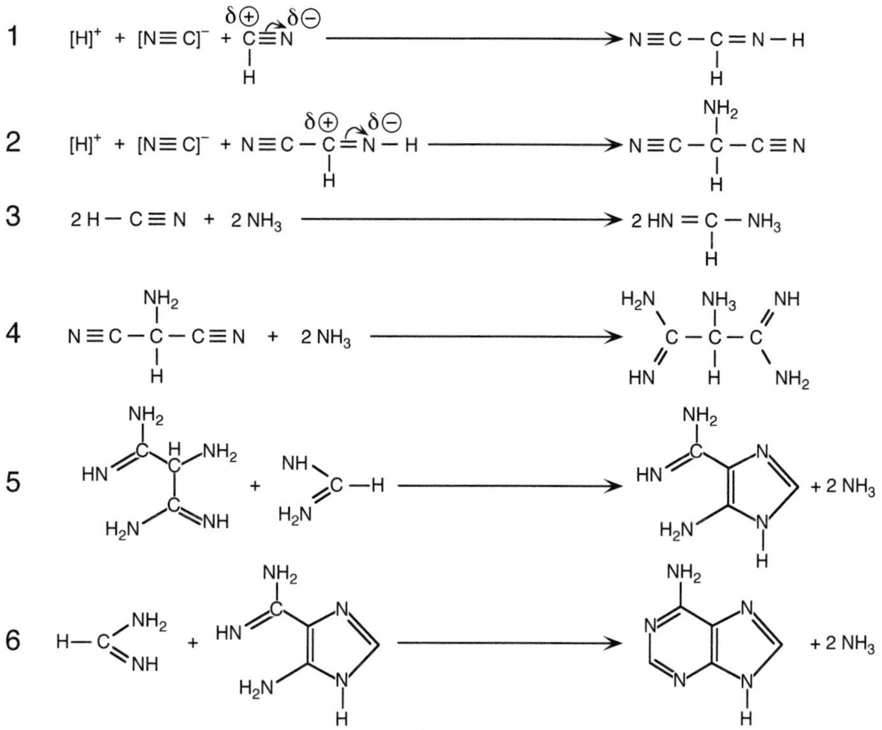

FIGURE 2   The prebiotic synthesis of adenine. Overall reaction: 5HCN = adenine. (After Oró 1965.)

that of pentameric hydrogen cyanide. Nonetheless, it is a paradox that one of the most toxic, poisonous substances to the majority of living organisms today was probably the prebiotic precursor of one the most important molecules to life, since it is present in RNA and DNA as well as ATP and many other coenzymes.

Several routes for the abiotic formation of amino acids and purines have been described (Oró *et al.* 1990), including the hydrolysis of HCN oligomers, which also yields orotic acid, the biochemical precursor of uracil in contemporary organisms (Ferris *et al.* 1978). Pyrimidines can also be synthesized by the condensation of cyano-acetylene with urea. Recently we have achieved the abiotic synthesis of histidine, an important component of the catalytic site of many enzymes (Shen *et al.* 1990a), and of several major coenzymes (Mar & Oró 1990, 1991). As shown by Butlerow (1861), sugars are readily formed when formaldehyde polymerizes under basic conditions, while the polymerization of formaldehyde, or, better yet, glyceraldehyde with acetaldehyde, yields significant amounts of deoxyribose. Although the formose reaction yielding ribose has been criticized because of the simultaneous formation of more than 50 different sugars (Joyce 1989), the recently described synthesis of a ribose derivative from a mixture of phosphorylated glycoladehyde and formaldehyde (Müller *et al.* 1990; Arrhenius, this volume) suggests the existence of as yet undescribed mechanisms to account for the formation and accumulation of specific nucleic acid components in the prebiotic environment. Additional information on the synthesis and accumulation of other organic compounds of biochemical significance has been discussed elsewhere (Miller & Orgel 1974; Oró *et al.* 1990). A summary of the key role of HCN in the formation of biomonomers and other compounds is shown in Fig. 3.

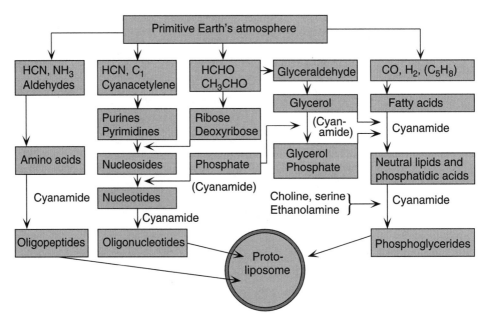

FIGURE 3   The processes of chemical and precellular evolution and the formation of a protocellular system.

Fourth stage: Prebiotic condensation reactions

Liquid water is an essential component both of living beings and of their environment. As noted above, many of the prebiotic chemical pathways are completed by the hydrolysis of different precursors, as in the case of the Strecker synthesis of amino acids and in the hydrolysis of the HCN polymer studied by Ferris *et al.* (1978). The opposite is true in the case of nucleotide synthesis: the formation of nucleosides and their further phosphorylation requires several dehydration reactions. Therefore, we are confronted with a rather vexing question: How could the prebiotic oligomerization reactions take place, if their monomers were in aqueous solution?

The simplest of all dehydration mechanisms is thermal condensation, which has been extensively studied in amino acids by Sidney W. Fox and his coworkers (Fox & Dose 1977). However, the products of high molecular weight that are obtained by this method are primarily not linear polypeptides with alpha–alpha bonds but have other abnormal ones. It is not easy, either, to imagine conditions under which the prebiotic formation of such polymers could actually take place. Several other mechanisms for the condensation of amino acids and nucleotides have been studied under abiotic conditions, including coupling reactions using activated derivatives, condensation reactions using polyphosphates, and condensation reactions using organic agents derived from HCN (Oró & Lazcano-Araujo 1981).

Cyanamide is the simplest and probably one of the most effective prebiotic condensing agents. Inspired by the effectiveness of dicyclohexylcarbodiimide in the oligonucleotide anhydrous synthesis achieved by H.G. Khorana, I suggested the potential prebiotic role of cyanamide for the formation of polymeric compounds when I wrote my first review on organic cosmochemistry (Oró 1963, 1976). In fact, carbodiimide in the form of its tautomer cyanamide is now known to be present in interstellar space and was presumably formed in the primitive Earth environment by the action of ultraviolet light on HCN. Furthermore, it is relatively stable in aqueous solutions, and its dimer derivative dicyandiamide is also an effective condensing agent.

Did prebiotic oligomers form within an aqueous environment? On February 1st, 1871, Charles Darwin wrote to his friend Hooker a letter stating that "if (and oh, what a big if) we could conceive in some warm little pond with all sort of ammonia and phosphoric salts, -light, heat, electricity, &c present, that a protein compound was chemically formed, ready to undergo still more complex changes, at the present day such matter would be instantly devoured, or absorbed, which would not been the case before living creatures were formed." In fact, we have been rather successful with the laboratory simulations of Darwin's warm little pond. The primitive environment was constantly undergoing periodic changes because of the succession of night and day, tidal cycles, seasonal changes, and so on. The evaporating-pond model provides a geologically plausible and realistic model by means of which cyclic changes in humidity and temperature in the presence of condensing agents (cyanamide, pyrophosphate, clays, etc.) may have yielded significant amounts of oligopeptides, oligonucleotides, and phospholipids (Oró *et al.* 1990). In typical experiments, aqueous solutions of the monomers and the condensing agents are evaporated and heated just to be dehydrated once more. Repeating this simple process as in diurnal cycles has yielded peptides of glycine, alanine, and other amino acids (Hawker & Oró 1981), and significant amounts of the catalytic dipeptide histidyl–histidine (Shen *et al.* 1990b, c).

Cyanamide is also an excellent condensing agent in the synthesis of deoxynucleotides. The evaporating-pond model has allowed us to achieve the nonenzymatic synthesis of oligothymidilic acids containing 2–8 condensed monomers linked primarily by the biological 3',5' phosphodiester bonds (Sherwood *et al.* 1977). Nonenzymatic template-dependent reactions have been studied extensively by Leslie Orgel and his collaborators (see Orgel 1987). In typical experiments, Zn^{++} has been used as a catalyst for the template-dependent polymerization of activated nucleotides, in which a high fidelity in the incorporation of complementary bases was observed (Bridson *et al.* 1981). Additional comparable experiments involving nucleotides activated with 2-methyl imidazole derivatives, and even nucleoside analogues that lack ribose or deoxyribose, have been described by Orgel (1987) and by Schwartz & Orgel (1985), respectively.

Fifth stage: The self-assembly of prebiotic membranes

How did the organic compounds of biochemical significance organize themselves into the first living entities? An essential element for the origin of cells must have been the emergence of double-layered membranes formed by lipidic molecules. Starting from simple precursors and following a sequence of step-by-step reactions, high yields of several lipidic molecules were obtained, including amphiphilic phosphatidic acids, phosphatidylethanolamine, and phosphatidylcholine (cf. Oró *et al.* 1978). It should be pointed out that, with the exception of the work of David Deamer and his associates (this volume) and the work in our laboratory, few other experiments of significance have been carried out on the synthesis of these compounds. In fact, the presence of lipids in the prebiotic environment is further supported by the existence of membrane-forming nonpolar molecules extracted from samples of the Murchison meteorite (Deamer & Pashley 1989).

We need not invoke any major force to account for the formation of precellular membranes. Lipidic molecules can easily self-assemble into bilayered or multilayered liposomes under physiological or prebiotic conditions. Aside from the fact that these structures are readily formed by self-organization of amphiphilic molecules, liposome structures are very stable and have properties typical of plasma cell membranes, such as the semipermeability to certain ions and weak bases (Deamer & Oró 1980). Encapsulation of nucleic acids and other molecules has been achieved by hydration–dehydration cycles (Deamer, this volume) and is enhanced by metallic cations (Baeza *et al.* 1987). Therefore, it is conceivable that equivalent processes may have led to the enclosure of prebiotic replicating polynucleotides and catalytic peptides like histidyl–histidine within them.

It is at this point that the processes of prebiotic synthesis begin to connect with the ones that generate boundary structures and cooperative molecular interactions. These are the organization steps of precellular evolution that eventually gave rise to cells (Oró & Lazcano 1990; Lazcano, this volume). As argued persuasively by Tanford (1978), biological organization may be viewed as consisting of both the biosynthetic and the assemblage stages. Accordingly, the appearance of precellular systems can be divided into (1) the nonenzymatic synthesis of biochemical monomers and oligomers and (2) their assembly into polymolecular systems bounded by lipidic membranes formed spontaneously by the self-assembly of their amphiphilic components.

Sixth stage: The triggering of life

What has been described in the previous pages is primarily based on observation and experimentation. However, it would be pretentious to describe with any degree of certainty how the triggering of life took place. The following is a speculative account of five steps of synthesis and interaction that may have taken place within the boundaries of prebiotic liposomes during the transition from chemical to biological evolution:

- Synthesis of a protoenzyme, i.e. an abiotically formed, linear oligopeptide such as histidyl–histidine or the active center of an ATP synthetase, able to exhibit substantial stereospecific catalytic activity.

- Synthesis of protogenes, i.e. linear, homochiral, informational, replicating molecules (proto-RNA), which would have stored in their sequence cryptic information that could not have been decoded until specific interactions had been established between their base sequences and other compounds, either amino acids or adenosyl derivatives of amino acids.

- Synthesis of proto-tRNA, i.e. small oligonucleotides with code-translating properties, with amino acids covalently bound to their terminal adenosyl group with ester bonds. These molecules would have been able to translate the information coded in the protogenes into the amino acid sequences of protoenzymes.

- Synthesis of protoribosomes, i.e. primitive peptide-synthesizing complexes where the formation of the peptidic bonds between the amino acids linked to the charged proto-tRNA probably took place. The evidence of the catalytic properties of RNA (Cech & Bass 1986) and the inability to ascribe catalysis of peptide bond formation to a particular ribosomic protein (Noller 1991) suggest that the protoribosome may have been made solely of RNA molecules.

- The operational self-starting of a protocell. This could have been initiated by a transmembrane proton gradient within a liposome containing some of the above molecules. The protonization of a peptide, similar to the conserved region of the active center of a reversible ATPase, may generate the high-energy bond of a pyrophosphate or ATP (Baltscheffsky & Baltscheffsky, this volume). In this way, the functional self-starting of a relatively simple protocellular structure would have taken place, provided the primordial soup was rich in precursor molecules. This is the key functional process most characteristic of life, i.e., autobiopoiesis (Fleischaker 1990). This process can only be perpetuated if a continuous source of chemical energy is available. The origin and evolution of early biological energy conversion has been studied in detail by Baltscheffsky & Baltscheffsky (this volume). They suggest that a "pyrophosphate world" may have been a necessary prerequisite for an "RNA world" (Joyce 1989; Lazcano *et al.* 1992; Lazcano, two chapters in this volume).

In summary, the encapsulation of the above four types of molecules within the lipidic boundaries of liposomes would produce an entity with practically all the essential attributes of life (Fig. 3). The cooperative, synergistic interaction between these components triggered by the synthesis of pyrophosphate or ATP caused by a proton gradient, and the subsequent increased reaction rates, would create the preferential accumulation of their products and precursors, thus differentiating the living system from the inanimate world. Once several systems like this were formed, the most efficient in the cooperative processes of catalysis and self-duplication would have prevailed in accordance with Darwinian evolution. These entities may be considered as the protobionts that gave rise to the emergence of Darwin's ancestral cell. Needless to say, the processes described in this section are those of which we know the least.

Summary and conclusions

As a result of the general processes of organic cosmochemistry, life has appeared and developed into a complex system, the history of which begins sometime during the early Archean and continues until today. According to the ideas discussed in this paper, the appearance of life in our planet may be understood as the result of evolutionary processes that involve the following major steps:

1 The stellar thermonuclear synthesis of the biogenic elements other than hydrogen (C, N, O, P and S), their dispersal into space, and their subsequent incorporation into circumstellar and interstellar molecules.

2 The formation of the Earth and its subsequent acquisition of volatiles by collisions with comets and other similar minor bodies, which provided the precursors of biochemical molecules.

3 The prebiotic synthesis and accumulation of amino acids, purines, pyrimidines, lipids, and other monomers in the primitive terrestrial environment.

4 The prebiotic condensation reactions involving the synthesis of oligomers such as oligonucleotides and oligopeptides, with replicative and catalytic activities, respectively.

5 The synthesis of lipids, their self-assembly into double-layered membranes and liposomes, and the sequestering of prebiotic replicative and catalytic molecules within their boundaries.

6 The synergistic, cooperative interaction between small catalytic peptides, replicative molecules, proto-tRNAs, and protoribosomes inside liposomes to produce a protobiont, an immediate precursor to the first living systems.

All the above chemical stages are truly negentropic or ordered processes that move against the unstoppable arrow of time, but they do not work against the second law of thermodynamics. That means that whereas order is created in each one of them, they are temporal creations, and someday they will cease to exist, as the universe continues towards its thermal death. In that sense we could call them fantasy, but what a beautiful fantasy, even if temporal, the fantasy of life.

Someday, DNA, that selfish molecule, will also become dust again. After all, we are made of stardust, and into stardust we will return, or as my compatriot the poet Joan Maragall would say, challenging the heavenly fantasies of eternal life in another world, "Be my death, a greater rebirth, into this beautiful world."

Acknowledgments. – I would like to thank Professor Severo Ochoa for encouraging me to undertake the experimental chemical studies on the origin of life, which I had dreamed of in my school days. I also thank Professor Antonio Lazcano and Thomas Mills for their continued collaboration. This work has been supported in part by NASA Grant NAGW 2788.

The transition from nonliving to living

Antonio Lazcano

Departamento de Biología, Facultad de Ciencias, UNAM, Apdo. Postal 70-407, Cd. Universitaria, México 04510, DF, Mexico

Where, when, and how did life appear? Although we have no detailed answers for these three equally alluring questions, if current interpretations of the significance of the properties of RNA molecules for the origin and early evolution of life are correct, then a major step towards the appearance of cells was the emergence of a liposome-bounded system in which energy conversion became associated with template-directed ribonucleotide polymerization, producing catalytic and replicative RNA molecules within its lipidic membranes. The attributes of the first forms of life are unknown, but preliminary insights gained from molecular phylogenetic analysis are providing unequivocal evidence that the organisms that preceded eubacteria, archaebacteria and the eukaryotic nucleocytoplasm component were ancestral prokaryotes with simpler ATP-synthetases and protein-synthesis machinery. This suggests that a large number of (not necessarily slow) uncharacterized evolutionary changes took place between the origin of life itself and the last common ancestor of all extant life.

All the organic beings which have ever lived on this Earth," wrote Charles Darwin in the *Origin of Species*, "may be descended from some one primordial form." But how did this common ancestor come into being? What was its nature? Although Darwin never overcame his reluctance to discuss in public the appearance of life, it was within the framework of his ideas that seventy years later A.I. Oparin and J.B.S. Haldane suggested a possible explanation for the emergence of the first living systems, based on the hypothesis that the earliest organisms were fermentative, obligately anaerobic bacteria that had been preceded by a long period of chemical abiotic synthesis of organic compounds.

Alternative routes to biopoiesis have been suggested (Wächtershäuser, this volume), including the possibility that the origin of life was concomitant with the fortuitous formation of a single replicating ribozyme, i.e., a catalytic RNA molecule self-assembled from unorganized prebiotic raw material in which lifelike properties were completely absent. However, the possibilities of scientific inquiry are much broader if the study of life's emergence is approached by assuming a procession of changes, through stages of gradually increasing complexity, until a system that can be recognized as living is attained (Oró, this volume). If this scheme is valid, then there must have been a turning point in this evolutionary process during which the transition from nonliving to living took place. The study of this crucial but largely undefined stage is the subject of this short essay.

Bengtson, S. (ed.) 1994: *Early Life on Earth. Nobel Symposium No. 84.* Columbia U.P., New York

What is life?

"Music," once said Isaac Stern, "can be described, but not defined." Perhaps the same is true of life itself. An all-embracing, generally agreed-upon definition of life has proven to be an elusive intellectual endeavor, but any explanation of the origin of living systems should attempt the definition of a set of minimal criteria for what constitutes a living organism, including the extremely elementary basic characteristics with which the first living beings were endowed. What are these essential attributes? As argued forcefully by Gail R. Fleischaker (1990), there is a categorical distinction between nonliving and living, and the latter can be characterized by operational criteria that account not only for the internal structure, organization, and operation of organisms but also for their interactions with their environment. Despite the spectacular molecular acrobatics performed by viruses, viroids, catalytic RNAs, and many other subcellular systems, extant life is generally identified at the very minimum with cells, i.e., with dynamic membrane-bounded systems incessantly exchanging matter and energy with their environment, with the common imperative operations involved in basic metabolism, self-maintenance, heredity, and reproduction with variation. The history of change and continuity between the earliest forms of life and extant organisms implies, says the Cambridge University philosopher Harmke Kamminga (1992), that "the first living organisms took part in the evolutionary process – in other words, that they had descendants unlike themselves."

Unfortunately, the inability to discriminate between traits that may have resulted from truly abiotic processes and those that are outcomes of biological evolution has led to the frequent misconception that modern cells are perfect models for the first forms of life. Evolutionary criteria have frequently been absent in the chemical approach to the origins of life. For instance, the nonenzymatic synthesis of deoxyribose, thymine, and many different oligodeoxyribonucleotides has been achieved in several laboratories. Do these results imply that wriggling DNA molecules were floating in the waters of the primitive ocean, ready to be used as primordial genes? Of course, this is unlikely. From a biological perspective the presence of DNA in contemporary cells can be explained not in terms of prebiotic chemistry but rather as the end-product of an ancient metabolic pathway that evolved in early Archean cells possessing RNA genomes in which translation had already appeared (Lazcano *et al.* 1992).

Given adequate expertise and experimental conditions, it is possible to synthesize almost any organic molecule. The fact that a number of molecular components of contemporary cells can be formed nonenzymatically in the laboratory does not necessarily mean that they were also essential for the origin of life or that they were available in the prebiotic environment. The primitive broth must have been a bewildering organic chemical wonderland, but it could not include all the compounds or the molecular structures found today in even the most primitive prokaryotes – nor did the first bacteria spring completely assembled, like Frankenstein's monster, from simple precursors present in the prebiotic soup.

The RNA world revisited

"RNA and DNA are the dumb blondes of the biomolecular world," wrote Francis Crick in *Life Itself* (1981), "fit mainly for reproduction (with a little help from proteins) but of little use for much of the really demanding work." This may still be true for DNA, but it is true neither for blondes nor for RNA. The independent discovery of ribozymes by Thomas Cech of the University of Colorado and Sidney Altman of Yale University barely one year after Crick's book was published quickly raised RNA from a humble biochemical position as a molecular handyman and mere go-between to a central character in the early evolutionary drama (Fig. 1).

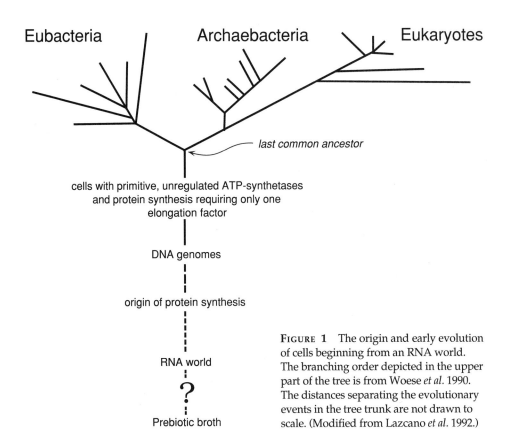

FIGURE 1   The origin and early evolution of cells beginning from an RNA world. The branching order depicted in the upper part of the tree is from Woese *et al.* 1990. The distances separating the evolutionary events in the tree trunk are not drawn to scale. (Modified from Lazcano *et al.* 1992.)

It is unlikely that the intron self-splicing reaction discovered by Cech and the catalytic abilities of the RNA moiety of ribonuclease P described by Altman are truly vestigial activities (Cech & Bass 1986). However, the disclosure of RNA-mediated catalysis led Harvard's molecular biologist Walter Gilbert (1986) to suggest that the starting point for life's history on Earth had been the so-called RNA world, an early stage during which alternative life forms based on ribozymes existed before the development of protein biosynthesis and DNA genomes. Somewhat similar proposals have been made independently by a number of authors, but together with Wally Gilbert, Bruce Alberts (1986) and I (Lazcano 1986) specifically argued that the development of the translation machinery had begun in the snug, organic-rich microenvironment defined by the lipidic boundaries of cells whose metabolism and reproduction had been mediated until then by ribozymes. Thus, although the abiotic formation of peptide bonds under possible primitive conditions is well documented (Oró *et al.* 1990), it may have no direct relevance to the origin of protein biosynthesis. According to this rather liberal definition of primordial life, the basic selection pressure for the origin and stabilization of the primitive translational apparatus was the enhancement of the catalytic activities of these RNA-based cells in order to increase their dynamic stability and reproductive fitness. More is said about the RNA world in my other chapter (Lazcano, this volume, pp. 70–80).

The RNA world is not a radical, totally unheard-of hypothesis without historical precedent. The existence of a primitive replicating and catalytic apparatus devoid of both DNA and proteins and based solely on RNA molecules was suggested in the late 1960s by Carl Woese (1967), Francis Crick (1968), and Leslie Orgel (1968). As pointed out by the British biologist Norman W. Pirie in 1953, "If we found a system doing things that satisfied our requirements for life but lacking proteins, would we deny it the title?" However, in part because of deeply rooted biochemical prejudices, the idea of a RNA world has met with some healthy resistance, especially from those who argue that protein synthesis is such an essential characteristic of cells that its origin should be considered synonymous with the emergence of life itself. This pervasive view has been strongly challenged by the increasing evidence that RNA molecules are efficient and versatile catalysts. Ribozyme-mediated peptide-bond formation has not been demonstrated, but there are strong indications that the RNA substrate repertoire may include amino acids, as hinted by several observations, including (a) the discovery of an arginine binding site in a ribozyme from the ciliate *Tetrahymena* (Yarus 1988a), (b) the ability of a slightly modified form of this same ribozyme to catalyze the hydrolysis of the aminoacyl bond between formylmethionine and an oligonucleotide (Piccirilli *et al.* 1992), and (c) the nearly conclusive evidence that the formation of peptide bonds can be catalyzed solely by ribosomal RNA in the absence of proteins (Noller *et al.* 1992).

God may not play dice – but Nature can be tricky. Are the proponents and adherents of the RNA world an evolutionary sect seduced by false coincidences? The answer is probably negative. In addition to its dual ability as a catalyst and as an informational macromolecule, RNA is a resilient, chemically reactive, abundant structural component of all cells, as of the ribosome where it appears to be more than a simple molecular scaffold (Noller 1991; Noller *et al.* 1992). Ribonucleotides are essential metabolic precursors in the biosynthesis of deoxyribonucleotides, and they also play a central role in contemporary cellular metabolism (as ATP, for instance, or as ribonucleotide derivatives like NADH, acetyl CoA, and FADH, among others), which may be a reflection of their ancient origin. Moreover, approximately 50% of all catalogued enzymes cannot function without coenzymes, many of which are small ribonucleotide-like molecules whose biosynthesis is intimately linked to the metabolism of RNA and its monomers (White 1982). As argued long ago by Orgel & Sulston (1971), these coenzymes may be vestiges of pre-genetic-code catalysts present in early RNA cells before the appearance of true proteins and the development of enzymes.

Although as of October 1993 a template-dependent ribozymic RNA polymerase had not been discovered, the application of powerful molecular-biology techniques has led to the synthesis of an artificial ribozyme able to catalyze repeatedly the ligation of short oligonucleotides complementary to a RNA template (Doudna & Szostak 1989). The possibility of simple prebiotic ribozymes is supported at least in part by the discovery of the rapid, highly specific cleavage reaction catalyzed by a small, synthetic 19-nucleotide RNA molecule under physiological conditions described by Uhlenbeck (1987). Nonetheless, the RNA-world model confronts several serious challenges, including the lack of plausible primitive abiotic mechanisms to account for the formation and accumulation of ribose, which is a minor, relatively unstable component of a complex array of products formed simultaneously from the putative prebiotic self-condensation of formaldehyde (Shapiro 1986). Indeed, the problems involved with prebiotic synthesis of ribose and other nucleic-acid components (Ferris 1987) have led to the suggestion that RNA itself may have been preceded by genetic polymers with a simpler backbone, in which ribose presumably was replaced by acyclic, flexible compounds like glycerol, a stable, three-carbon chain compound (Joyce *et al.* 1987).

Attempts to bury the RNA world together with the spoils of other ephemeral scientific speculations may be premature and should be met with caution. In spite of the appeal of a pre-RNA era based on nucleic-acid-like molecules, (a) ribose analogs may not be suitable at all, as suggested by recent experiments showing that nucleic-acid double helixes are easily destabilized by the presence of even a few flexible glycerol-nucleosides (Schneider & Benner 1990); (b) ribozymic activity is strongly dependent on the ribose 2'-OH group, which plays a direct role in different hydrolytic, phosphorylation, and condensation reactions, as well as in intron self-splicing. The absence of an equivalent hydroxyl group in glycerol and other acyclic ribose analogues seriously hinders their possible catalytic activity (Lazcano *et al.* 1992); and (c) recent experiments by A. Eschenmoser and his associates at the ETH in Zurich have shown that high yields of a ribose derivative are easily obtained by the formaldehyde-mediated phosphorylation of glycoaldehyde (Müller *et al.* 1990), suggesting that a ready prebiotic synthesis of RNA components may exist. This possibility is supported by the work of Gustaf Arrhenius and his associates (Gedulin & Arrhenius, this volume), who have shown that hydrotalcite, an abundant hydroxide mineral, concentrates the glycoaldehyde phosphate and catalyzes the formation of sugar phosphates at mild conditions. What may be actually required is an experimental redefinition of the chemical conditions for a truly prebiotic synthesis of polyribonucleotides.

Natura non facit saltus?

In spite of its inherent limitations and reductionist overtones, the RNA-world hypothesis has an intrinsic heuristic value that cannot be overstressed. At the very least, it is a causal narrative, i.e., a logical, evolutionary-oriented, plausible sequence of events that attempts to explain the transition from a simple replicating system based on RNA to one involving DNA and proteins. The idea that life began with the appearance of RNA-based cells may help to overcome the historical dispute between those who argue that the first living system was a self-replicating nucleic-acid molecule and those who identify the origin of life with the emergence of a membrane-bounded, polymolecular, heterogeneous system endowed with basic metabolic properties. But how did the hypothetical RNA cells suggested by Alberts, Gilbert, and me come into being?

Although we are still far from a complete understanding of the processes that may have led to the formation of the so-called prebiotic broth, the presence of a large array of organic compounds in carbonaceous meteorites and the astonishing ease with which amino acids, adenine, lipids, and many other molecules can be synthesized under primitive conditions imply that a large set of biochemicals was readily formed on the early Earth (Oró et al. 1990). In particular, the chemical synthesis of amphiphilic molecules (Deamer 1986a), the nonenzymatic template-directed polymerization of nucleotides and nucleoside analogs (Orgel 1987), and the abiotic synthesis of histidyl–histidine and other small catalytic peptides (Shen et al. 1990c) suggest that replication-like processes, chemically active peptides, and the self-assembly of membranes from lipidic components were possible before the emergence of life.

Primitive liposomes were probably relatively simple structures, formed by small, single-chain, ionic linear fatty acids, which could easily sequester catalytic and replicative molecules (Deamer, this volume). Under laboratory conditions this process can take place in the presence of histidine, cyanamide, and several other prebiotic condensing agents and is enhanced by basic polypeptides and metallic cations (Oró & Lazcano 1990). Mixtures of different lipids are known to produce liposomes with nonselective pores, and complexes between nucleotides and metal ions could have facilitated diffusion, leading to rudimentary transport mechanisms across primitive membranes.

Precellular evolution was not a continuous, unbroken chain of progressive transformations steadily proceeding to the first living beings. Many prebiotic culs-de-sac and false starts probably took place. Emergence of the first living beings must have required the simultaneous coordination of many different components in a confluence of processes. Thus the mere encapsulation of ribozymes, amino acids, oligopeptides, and many other potential cofactors and substrates involved in RNA catalysis within liposomes was a necessary but not sufficient condition for the origin of RNA cells. Perhaps the first forms of life did not require membranes (Lamond & Gibson 1990), but if our interpretation of the evolutionary significance of the properties of RNA molecules is correct, then a major decisive step towards the appearance of life was the emergence of a system in which energy coupling associated with membrane and proton gradients was used in template-directed ribonucleotide polymerization to produce catalytic and replicative RNA molecules within the lipidic boundaries of primitive liposomes.

The work of Peter Gogarten and his colleagues (1989) has shown that proton pumps producing H^+ gradients appeared very early in cellular evolution, before the evolutionary divergence of eubacteria and archaebacteria. These complex oligomeric enzymes may have been preceded by a simpler proton-pumping pyrophosphatase with H^+-PPase and PP_i synthetase activities (Baltscheffsky & Baltscheffsky, this volume), but how proton gradients actually originated and became coupled with ions and directionality is still an unsolved problem. Very little is known of the origin of biological energy-conversion mechanisms, although it is likely that ion channels and proton-selective pores appeared before the origin of life. Small, simple synthetic amphiphilic oligopeptides long enough to span the hydrocarbon phase of lipid bilayers have permeabilities and lifetimes resembling those of proton-selective channels and the acetylcholine receptor (Lear *et al.* 1988). Although these experiments have not been performed within an evolutionary context or under primitive conditions, they illustrate how small oligopeptides of prebiotic origin, or those synthesized by primitive cells with limited coding capabilities, could have been involved in ion transport across membranes (Lazcano *et al.* 1992).

According to the scheme discussed in this chapter, survival and reproduction of primordial RNA cells depended on ribonucleotides, lipids, and other compounds of prebiotic origin and must have been hindered by the exhaustion of this supply. It is this direct uptake of organic molecules from the primitive environment, and not the universal distribution of glycolysis, that should be interpreted as the defining feature of the heterotrophic nature of the first cells. In spite of its simplicity, central metabolic position, and ability to function under anaerobic conditions, glycolysis as such requires a set of enzymes too complex to be expected in the first organisms. Extant bacteria have various relatively inefficient mechanisms that allow them to use organic molecules from external sources, including nitrogen bases and nucleosides (Kornberg & Baker 1992). Some of these salvage pathways may be analogous (or perhaps even homologous) to the uptake by primitive heterotrophs of nucleic-acid components from the external milieu.

Early biological evolution: The molecular chronicles

As shown by the Warrawoona fossil assemblage, an abundant, complex, and highly diversified microbiota which may have included cyanobacteria existed only 10^9 years after the Earth had formed (chapters by Schopf and Walter, this volume). Life is probably much older than these early fossils, but how long did it take for it to appear and become established? Almost nothing is known about the time scales required for the origin and evolution of bacterial metabolic pathways. The primitive environment was no microbial Eden, but the Archean paleontological record shows that once life emerged, it was rapidly able to endure, diversify, and adapt itself to the stinky, harsh environmental conditions of the early Earth.

It is unlikely that the paleontological record will ever provide direct evidence of the transition from prebiotic organic molecules to the earliest cells, nor will it tell us much about the nature of the first biological systems. However, as shown more than twenty-five years ago by Emile Zuckerkandl and Linus Pauling (1965), nucleic acids and protein sequences are an extraordinarily rich source of evolutionary information. Although a cladistic approach to the origin of life is not feasible, the comparison of ribosomal RNA sequences has become an important tool in understanding the early stages of cellular evolution and has had a significant impact on our interpretation of bacterial relationships. A major achievement of this approach was the construction of a trifurcated, unrooted, universal evolutionary tree in which all known organisms can be grouped in one of three major lineages: the eubacteria, the archaebacteria, and the eukaryotic nucleocytoplasm (Woese 1987b; Sogin, this volume).

The immediate predecessor of these three cellular lines was already a rather complex organism, much like extant bacteria in many ways. Few genes found in the three major cellular lineages have been sequenced and compared, but the sketchy picture that is already emerging of the last common ancestor of eubacteria, archaebacteria, and eukaryotes shows that it was a rather sophisticated cell with complex ribosome-mediated translation, membrane-associated H^+-ATPases engaged in active transport, ability to synthesize histidine and purine, and a set of enzymes involved in glycolysis, pyruvate oxidation, and other mainstream heterotrophic anaerobic metabolic pathways.

It is likely that genetic recombination appeared early in evolution, suggesting that Archean microbes led a life that was not totally chaste. However, gene duplications followed by further sequence divergence were probably the most important mechanism by which early cells increased their hereditary endowment. These evolutionary innovations arose in individual organisms and then spread rapidly through ancestral bacterial populations, becoming fixed before their divergence into the three cellular lineages. In contrast with orthologous genes, which are duplicate sets that diverge through speciation, paralogous genes are those that diverge after a duplication event. Paralogous genes are extremely useful in rooting evolutionary trees, since one set of sequences can be used as an outgroup for the other one. As discussed by Gogarten *et al.* (1989) and by Iwabe *et al.* (1989), the sets of paralogous genes unequivocally identified in all three cellular lineages are those coding for (a) the two elongation factors that assist in protein biosynthesis and (b) the two components of the hydrophilic ATP-synthesizing unit of ATP synthetase, a ubiquitous membrane-associated protein complex that harvests the energy associated with proton gradients, forming ATP from ADP and phosphate.

In spite of the intense dispute over the taxonomic significance of rooted universal trees derived from paralogous genes, the upper part of Fig. 1 clearly shows that the earliest detected branching event led to the eubacterial line on the one hand and to the archaebacterial–eukaryotic line on the other (Woese *et al.* 1990). Millions of years later, during the Proterozoic, some of the descendants of these two lines would meet once more, becoming forever associated in intimate symbioses – but that, of course, is another chapter in the saga of cell evolution (Margulis & Cohen, this volume).

What is important for the present discussion is to recognize that if the last common ancestor of eubacteria and archaebacteria had two sets of duplicate homologous genes coding for elongation factors and for the ATP-synthetase units, then it must have been preceded by a simpler cell with a smaller genome, in which only one copy of each of these genes existed (Fig. 1). In other words, the ancestor of the eubacterial and archaebacterial lines was a prokaryote in which ATP synthesis and protein biosynthesis were both less complex than those of the even simpler extant life forms (Lazcano *et al.* 1992). Evolutionary biologists have long argued that organisms simpler than extant bacteria existed, but such claims were based on highly evolved entities such as viruses, mitochondria, mycoplasma, and others, none of which are free-living. These analogies are useful, but now an even more important task is the identification and characterization of additional (if any) sets of such genes from ancestors of the last common ancestor that can lead to a more complete understanding of the biological attributes of these bygone Archean prokaryotes.

Conclusions

"What we do not know today we shall know tomorrow" concluded A.I. Oparin in his 1924 book *The Origin of Life*. "A whole army of biologists is studying the structure and organization of living matter, while a no less number of physicists and chemists are daily revealing to us new properties of inanimate things. Like two parties of workers boring from the two opposite ends of a tunnel, they are working towards the same goal. The work has already gone a long way and very, very soon the last barriers separating the living from the nonliving will crumble under the attack of patient work and powerful scientific thought."

In spite of the spectacular results achieved by this two-way approach, there is still a huge, insurmountable gulf between the results achieved by laboratory simulations and our present-day understanding of the essential features of a truly minimal living being. Elegant experiments that combine selection and mutation of catalytic RNA molecules have been performed (Beaudry & Joyce 1992), but given the present state of both prebiotic chemistry and molecular biology, it is probably preposterous to attempt the laboratory synthesis of RNA-based life. However, the RNA-world hypothesis is amenable to experimental analysis, including the development in vitro of ribozymes with new substrates, the study of simple membrane-associated energy-harvesting molecules, and the characterization of RNA-replicating systems within liposomes. Using the techniques of molecular phylogenetic analysis, we are peeking into the molecular intimacies of the primitive organisms that preceded the bifurcation of archaebacteria and eubacteria. The preliminary results discussed here already suggest that a long series of evolutionary changes took place after the origin of life itself but before the first speciation event separating the ancestors of eubacteria from those of archaebacteria.

We may be able to see even further back in time. Our molecular remembrance of things past may soon allow us to gaze into the evolution of proteins older than DNA itself. As pointed out twenty years ago by the late Margaret Dayhoff (1972), most amino-acid sequences can be classified into relatively few families. Recent development of sequence databases of genes and gene products has confirmed her early insight (Gilbert 1986; Doolittle 1990). It is possible that these few families resulted from amplification processes of ancestral genes coding for proteins whose basic functional properties and structural constraints were established before the emergence of DNA genomes, a hypothesis that could be tested by a detailed statistical analysis of the available databases (Lazcano *et al.* 1992). This approach can be complemented by the cloning and sequencing of ancient genes, such as those coding for thymidilate synthetase and ribonucleotide reductases, the enzymes directly involved in deoxyribonucleotide biosynthesis. The world of cells with RNA genomes in which protein biosynthesis had already appeared is not totally lost.

We have gained some insights into the biological processes that took place in the early Archean world. Nevertheless, we are still very far from understanding the origin and nature of the first living beings. These are still unsolved problems – but they are not completely shrouded in mystery, and this is no minor scientific achievement. Why should we feel disappointed by our inability to even foresee the possible answers to these luring questions? As the Greek poet Konstantinos Kavafis once wrote, Odysseus should be grateful not because he was able to return home, but because of what he learned on his way back to Ithaca. It is the journey that matters.

Acknowledgments. – I am deeply indebted to the organizers of the 1992 Nobel Symposium "Early Life on Earth" for their kind invitation to participate in this meeting. This chapter was written during a short leave of absence in which I first enjoyed the hospitality of the Ovando Foundation (México) and then of Dr. Juan Oró and his associates at the University of Houston (NASA Grant 44-005-002). I thank Drs. Gail R. Fleischaker and Juan Oró, as well as Mary Alpaugh, for their critical reading of this manuscript and many suggestions, and Drs. Harmke Kamminga, Thomas R. Cech, Gustaf Arrhenius, and Joseph Piccirilli for kindly providing me with results of their work before publication. Work reported here has been supported in part by UNAM.IN 105289.

The RNA world, its predecessors, and its descendants

Antonio Lazcano

Departamento de Biología, Facultad de Ciencias, UNAM, Apdo. Postal 70-407, Cd. Universitaria, México 04510, DF, Mexico

According to the RNA-world hypothesis, primordial cells based on catalytic and replicative polyribonucleotides of abiotic origin were the phylogenetic forerunners of the complex functional relationship between DNA, RNA, and proteins that drives extant life. This is a Russian-doll scheme that assumes that nucleic-acid-directed protein synthesis was selected for in primordial cells because of the enhanced adaptability that protein catalysts brought with them. It is in such an enzyme-rich intracellular environment that DNA genomes eventually evolved, confiscating the genetic role that until then had been performed by RNA molecules. The process of increasing complexification assumed by the RNA-world model is consistent with many examples of biological evolution in which new traits are added without the complete loss of previous characteristics. Problems involved with the prebiotic synthesis and accumulation of RNA have led to the suggestion that the RNA-based biosphere was preceded by a pre-RNA world in which genetic macromolecules were nucleic-acid-like polymers of nucleoside analogues. No explanation has been offered for the transition from such a hypothetical archaic system to the RNA world, which remains the best working model for explaining the origin and early evolution of life.

The RNA-world hypothesis states that during the long-forgotten primordial times when protein biosynthesis and DNA genomes had not yet evolved, the reproduction and metabolism of the earliest cells depended on the catalytic and replicative properties of RNA molecules (Fig. 1). This idea, which was sparked by the startling discovery of the catalytic activities of RNA molecules, has opened the possibility of understanding the origin of extant nucleic-acid-directed protein biosynthesis by suggesting that both DNA and proteins are the evolutionary outcomes of RNA-based cells (Alberts 1986; Gilbert 1986; Lazcano 1986).

RNA may be unique among biomolecules because of its dual ability to serve as repository of genetic information (as in the case of infamous viruses associated with diseases like polio, AIDS, and others) and to perform catalytic activities – a property that until a few years ago was believed to be found exclusively in proteins. RNA molecules are also conspicuous structural components of all cells, where they play a key role in protein synthesis, DNA replication, and RNA processing. In one known case the polar moiety of a eubacterial membrane component is a ribonucleotide, and, at least in some eukaryotes, RNA is also involved in transcription and in protein translocation

Bengtson, S. (ed.) 1994: *Early Life on Earth. Nobel Symposium No. 84.* Columbia U.P., New York

across membranes. It is unlikely that all these activities are relicts of a vanished RNA world. Nonetheless, they demostrate the multiple roles that polyribonucleotides can play in biological processes and thus provide indirect support to the idea that RNA-based life once existed.

As argued in the accompanying chapter, it is conceivable that early Archean RNA cells were the ancestors of contemporary cells (Lazcano, this volume, p. 60). Did their appearance also mark the beginnings of life? The strong points of this idea have their corresponding weaknesses, including the nontrivial issue of the prebiotic availability of RNA molecules. The purpose of this chapter is to discuss the advantages and limitations of the RNA-world hypothesis as well as to review the evidence indicating that rudimentary protein synthesis once existed and evolved through biological mechanisms and not because of chemical events in the prebiotic environment.

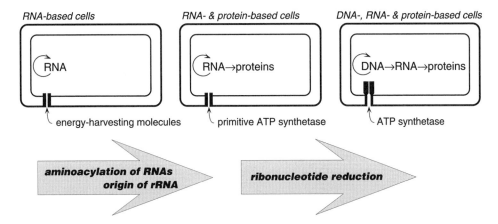

FIGURE 1   The three major steps in cellular evolution, beginning from an RNA world. (Modified from Alberts 1986.)

Imagining life in an RNA world

Despite some initial skepticism, it is now generally accepted that the cut-and-trim activities involved in the hydrolysis and transfer of phosphodiester bonds first described in self-splicing introns were the first evidence of the extraordinary versatility of ribozymes, i.e. catalytic RNA molecules (Cech & Bass 1986). Previous claims about the catalytic properties of the RNA moiety of a polysaccharide-branching enzyme (Shvedova *et al.* 1987) have not been sustained, but there is a growing tide of experimental results showing that ribozymes are truly astonishing catalysts that are not confined to nucleic-acid substrates. Is it possible to construct a reasonable working model of primordial RNA-based cells?

An RNA world requires ribozymes capable of replicating RNA templates. The existence of such molecules is supported by the work of Jennifer A. Doudna and Jack W. Szostak (1989), who have engineered a catalytic RNA with RNA-joining activity that uses external templates. As noted by the Scripps Research Institute evolutionary biologist Gerald F. Joyce (1991), this ribozyme is not completely equivalent to proteinic RNA-dependent RNA polymerases: it does not move processively along the template, nor can it copy self-structured regions. However, it is reasonable to assume that the discovery of a catalytic RNA overcoming such limitations is only a matter of time.

It is unlikely that a template-dependent ribozymic polymerase can achieve absolute copying fidelity. In such a case, populations of molecules with considerable genetic variation would exist, because of both point mutation and RNA-mediated rearrangements of polynucleotide sequences. Selection of some strands over others, i.e. Darwinian evolution at the molecular level, can thus be expected (Joyce 1989). That such a system can be designed is supported by experiments in which RNA templates are copied by Qβ replicase, a viral RNA-dependent proteinic RNA polymerase (Spiegelman 1971; Biebricher *et al.* 1982). After approximately 100 replication cycles, several smaller variants of the original templates accumulated, as they were selected by the elimination of segments unnecessary to the Qβ replication process (Mills *et al.* 1973). A similar phenomenon can be predicted in a system with catalytic RNA-mediated RNA replication, but template shortening would not have been advantageous in an RNA world. In other words, ribozyme-mediated replication was a necessary but not sufficient condition for the emergence of RNA-based life, whose existence and maintenance would depend on the dynamic equilibrium between replication, the intake of raw material from the external environment, and the utilization of high-energy bonds from energy-harvesting molecules (Lazcano, this volume, p. 60).

If RNA were present in the primitive environment, it could not have been engaged in chemical soliloquies. Company would have been kept by a wide range of potential cofactors and substrates including metal ions, amino acids, polypeptides, sugars, lipids, and many other molecules of prebiotic origin (Oró *et al.* 1990). The coexistence of ribozymes with other chemical species could have led RNA to the acquisition of additional functional groups. Examples of such molecular interactions are provided by contemporary biochemistry. For instance, the Pb^{++}-dependent self-cleavage of a transfer RNA (Sampson *et al.* 1987) strongly suggests the existence of primordial metalloribozymes (Gilbert 1987). Another particularly interesting case is that of a membrane component of the purple non-sulfur bacterium *Rhodopseudomonas acidophila*, which is formed by a hydrophobic terpenoid covalently linked to a ribonucleotide (Neunlist & Rohmer 1985; Ourisson, this volume). This unusual compound suggests that a direct association between lipids and RNA may have existed, one that could have facilitated both encapsulation and chemical reactions in the lining of primordial liposomes.

The abiotic synthesis of several coenzymes has been achieved in the laboratory (Mar & Oró 1991), but it is also possible that contemporary nucleotide-like coenzymes are molecular remnants of an RNA world (Orgel & Sulston 1971; White 1982). This alternative possibility is strengthened by (a) the highly ubiquitous presence of pyridine nucleotide coenzymes and other ribonucleotide prosthetic groups and (b) the fact that in the absence of their corresponding protein, many coenzymes can catalyze chemical reactions similar to those in which they take part as mere cofactors. According to the Delaware University biochemist Harold B. White (1982), even histidine, an imidazole-bearing amino acid which forms part of the active center of many enzymes, may be a ghost of its former self. Although the prebiotic synthesis of histidine has been reported

(Oró, this volume), it is the only amino acid whose biosynthesis begins from a phosphorylated sugar and a ribonucleotide. This unusual pathway has led to the idea that histidine may be the molecular descendant of a catalytic ribonucleotide derivative (White 1982). This is not a far-fetched suggestion – as a matter of fact, the disappearance of biological traits that fade away, leaving behind only the shadow of their grin, is a well-documented evolutionary phenomenon (Margulis & Cohen, this volume).

In the RNA world, some biochemical reactions may have been spontaneous, and others could have depended on the catalytic effect of transition metals, but RNA was the major catalyst. It was probably never a very efficient one – otherwise, ribozymes would have been discovered long ago by biochemists. The RNA world was in constant risk of sailing over the edge, since self-maintenance and reproduction of RNA cells were probably hindered by the insidious hydrolytic cleavage of the RNA phospho-diester backbone. By assuming that Darwinian mechanisms began operating earlier than previously thought, it has been argued that the enhancement of the catalytic activities of RNA cells was the basic selective pressure underlying the origin and stabilization of the translation apparatus (Alberts 1986; Gilbert 1986; Lazcano 1986). The sequence of events leading to RNA-directed protein synthesis probably began with simple chemical interactions between amino acids and ribozymes, but it would eventually seal the fate of RNA-based cells. How this may have happened is discussed below.

A world without mirrors: The molecular predecessors of RNA

A major assumption underlying the hypothesis that life began with RNA-based cells is that catalytic and replicative RNA molecules were brewed in the prebiotic environment from random abiotic condensation reactions of ribonucleotides. Although this view is mildly supported by the discovery of rather small ribozymes (Uhlenbeck 1987), the initial optimism surrounding the possibility of an RNA world has been challenged by an increasing awareness that current evidence does not support the abiotic formation of RNA molecules. One case in point, as cogently argued by Joyce (1989), is the prebiotic synthesis and accumulation of pyrimidines, nucleosides, and nucleotides, which face several major obstacles. Moreover, the formose reaction, traditionally invoked to account for the presence of ribose on the primitive Earth, requires unrealistically alkaline conditions and yields a complex array of many different sugars, of which ribose is only a minor and relatively unstable component (Shapiro 1986; Ferris 1987).

Although the existence of a nucleotide-rich primitive broth is debatable, recent experiments have shown that high yields of a ribose derivative are obtained by the formaldehyde-mediated condensation of phosphorylated glycolaldehyde (Müller et al. 1990). It is not known if this chemical pathway can lead to the accumulation of biological nucleotides (Fig. 2). However, it does suggest the existence of as-yet-unaccounted prebiotic reactions and geochemical settings that could have favored the formation of D- and L-nucleosides (Gedulin & Arrhenius, this volume). Even if we assume that

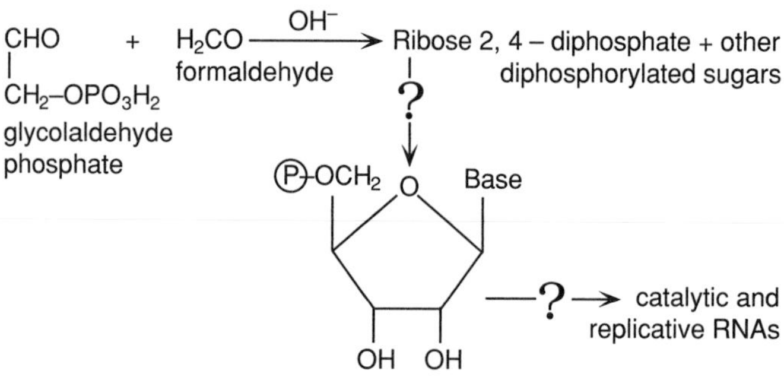

FIGURE 2   A possible
prebiotic synthesis of
ribonucleosides. (Based
on Müller *et al.* 1990.)

$$\underset{\substack{\text{glycolaldehyde}\\\text{phosphate}}}{\overset{\displaystyle\text{CHO}}{\underset{\displaystyle\text{CH}_2\text{-OPO}_3\text{H}_2}{|}}} + \underset{\text{formaldehyde}}{\text{H}_2\text{CO}} \xrightarrow{\text{OH}^-} \text{Ribose 2, 4 - diphosphate + other}$$

diphosphorylated sugars

ribose was present in the primitive soup, important objections can still be raised against the RNA-world hypothesis. On the one hand, the abiotic condensation of ribose and nitrogen bases leads to a mixture of nucleosides with different configurations, including not only those with the β-glycosidic linkage of contemporary nucleic acids but also the slightly twisted geometry of the nonbiological alpha form. On the other hand, while biological systems use only D-ribose, both the formose reaction and the condensation of glycolaldehyde–phosphate produce racemic mixtures, i.e., equal amounts of D- and L-ribose and their derivatives. This is a major obstacle, since under such racemic conditions, nonenzymatic template-dependent polymerization reactions are halted by the so-called enantiomeric cross-inhibition, which rapidly stops chain elongation (Joyce 1991).

As argued forcefully by Joyce *et al.* (1987), some of these problems can be avoided by assuming that the most archaic genetic system was based not on RNA but on simpler polymers of prochiral open-ring nucleoside analogues such as glycerol, a simple three-carbon chain alcohol (Fig. 3). Since such molecules are prochiral, i.e., lack mirror images, enantiomeric cross-inhibition would not have prevented their replication. Support for this possibility has been provided by the pioneering work of Alan W. Schwartz and Leslie L. Orgel (1985), who have shown that nucleoside analogues can undergo template-dependent polymerization reactions. These experiments have been extended to a large variety of compounds, including phosphoramidates, acetic-acid derivatives, deoxynucleoside diphosphates, and many other molecules that yield oligomers with rather unusual backbone structures (Rodriguez & Orgel 1991). We are thus faced with the dazzling possibility of a pre-RNA world based on informational polymers akin neither to RNA or DNA, molecular structures that lacked ribose, phosphates, and perhaps even the conventional nitrogen bases found in contemporary nucleic acids!

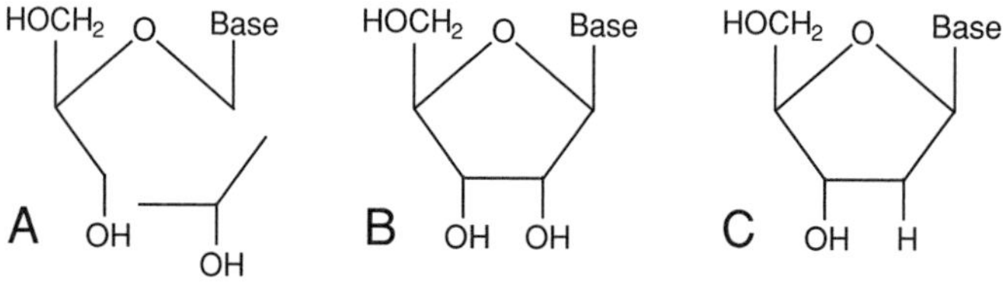

FIGURE 3   A: Glycerol-
based nucleoside analog.
B: Ribonucleoside.
C: Deoxyribonucleoside.

At the very least, the study of protonucleic acids suggests that many different molecules may have been involved in replication-like reactions in the primitive environment. Did such reactions form part of a biological prelude to the RNA world? Such a possibility pushes further back the origin of life and touches other issues like the antiquity of protein synthesis and the genetic code, which may be older than RNA itself (Orgel 1987). As argued in the accompanying chapter, the pre-RNA-world hypothesis is not without problems of its own (Lazcano, this volume, p. 60). Acceptance of a pre-RNA scenario requires: (a) evidence of catalytic activities in nucleic-acid-like molecules; (b) a convincing explanation of its evolutionary transition into an RNA world, which may have involved a stepwise process through several intermediates (Orgel 1987); and (c) a primordial ribozyme-based metabolism with the ability to synthesize ribonucleotides, if these cannot form abiotically. Is such an RNA-based system feasible? Answers to this and other equally significant questions lie in the largely untreaded ground of the chemistry of nucleoside analogues and their polymers. Until adequate solutions are offered to overcome these problems, the RNA world may be considered a reasonable model for understanding the transition from nonliving to living (Lazcano, this volume, p. 60).

The origin of protein biosynthesis

The possibility that DNA genomes evolved before protein synthesis cannot be completely discarded. Nonetheless, according to the RNA-world hypothesis, ribosome-mediated protein synthesis evolved in membrane-bounded systems in which RNA molecules had functioned until then, both as the source of inheritable genetic information and as principal catalysts. It has been difficult, however, to exorcize the looming legacy of academic emphasis on the proteinic nature of biological catalysts. There is still some resistance to the idea that RNA could have played a primordial role in biological catalysis in the absence of proteins. It is frequently argued that nucleic-acid replication and genetic coding of proteins coevolved, i.e. that protein synthesis and replicative nucleic acids emerged as a result of the interactions between catalytic peptides and replicative RNA templates (Oró, this volume). Are proteins and nucleic acids molecular Siamese twins, as implied by the different coevolution theories on the origin of translation?

I find it difficult to support this possibility. Of course, there is considerable evidence suggesting that amino acids and oligopeptides were present on the primitive Earth (Oró et al. 1990). Histidyl–histidine and other small catalytic oligopeptides that can use nucleotides or oligonucleotides as substrates have been synthesized under plausible prebiotic conditions (Brack & Barbier 1989; Shen et al. 1990c). It is doubtful, however, that the promiscuous coexistence of replicative oligonucleotides and catalytic peptides would lead to an inheritable liaison between them. Even if ("and oh, what a big if") we assume that huge amounts of oligopeptides with catalytic and structural characteristics necessary for polynucleotide replication were available in the prebiotic environment, sooner or later this supply would have been exhausted or hydrolyzed in the absence of translation. Regardless of how many different peptides were formed on the primitive Earth, nucleic-acid-instructed protein synthesis could never have evolved without a replicating mechanism insuring the maintenance, stability, and diversification of its basic components.

Protein synthesis is an elaborate, exquisitely tuned process requiring more than a hundred distinct components, which include ribosomal RNA (rRNA) and proteins, transfer RNAs (tRNAs), aminoacyl-tRNA synthetases, and proteins such as the initiation and elongation factors. How the actual formation of peptidic bonds takes place inside the ribosome is still a matter of conjecture. Tampering with rRNA has dramatic effects on translation, but there is no evidence that ribosomal proteins have any kind of catalytic activity (Nomura 1987). No ribosomal protein is known to catalyze the formation of a covalent bond between adjacent amino acids, and recent experiments (Noller *et al.* 1992) have confirmed previous speculations that this ability lies in rRNA itself (Nomura 1987; Noller *et al.* 1990). The observations by Noller *et al.* (1992) add considerable credibility to the idea that protoribosomes were devoid of proteins (Woese 1967, 1980; Crick 1968). How did the ancestral rRNA originate? Harry F. Noller (1991), molecular biologist of the University of California at Santa Cruz, has suggested, from observations of von Ahsen *et al.* (1991) on the inhibition of self-splicing ribozymes by antibiotics that are also known to affect translation, that ribosomes may have evolved from an RNA related to the so-called group I catalytic introns. As noted by Peter B. Moore (1988) of Yale University, there is a certain chemical similarity between the transesterification reactions in ribozymes (Cech & Bass 1986) and the transpeptidation event that takes place during protein synthesis.

Interactions between amino acids and RNA in the primitive environment were almost unavoidable, and it is generally agreed that the attachment of amino acids to polyribonucleotides was one of the earliest steps of protein synthesis to evolve. According to Yale University molecular biologists Alan Weiner and Nancy Maizels (1987), transfer RNAs may be derived from terminal tRNA-like structures that tagged primordial genomes at their 3' end, marking an initiation site for ribozymic-mediated RNA replication. As argued by Orgel (1989), the linkage of amino acids or dipeptides to RNAs could have provided a transcription initiation site, while according to Wächtershäuser (1988b), aminoacylated RNA molecules could have associated to cationic mineral surfaces and undergone further changes. An additional possibility has been raised by the Hong Kong University biochemist J. Tze-Fei Wong (1991), who has noted that the bonding of amino acids to ribozymes may have increased their catalytic activities by adding more chemically active functional groups. These ideas are not mutually exclusive. Aminoacylation may have originated as a tagging process and then been maintained and further refined and exploited because of the additional capabilities gained by RNA cells.

There is no experimental evidence showing that ribozymes can react with activated amino acids and catalyze the formation of peptide bonds, but there are several indications that such reactions are feasible. A statistically significant correlation between the polarity and the hydrophobicity of amino acids and their anticodon nucleotides has been described, hinting at a primordial interaction between RNAs and amino acids that may be related to the origin of the genetic code (Lacey & Mullins 1983). The specific binding of arginine to a catalytic RNA (Yarus 1988b) and of aromatic amino acids to fragments of phenylalanine tRNA (Bujalowski & Porschke 1988) are also well documented. Even more encouraging are the recent results that have expanded the known repertoire of ribozyme chemistry by showing that peptide-bond formation depends on the catalytic properties of ribosomal RNA (Noller *et al.* 1992) and that RNA also has the ability to hydrolyze the bonds that join amino acids to RNA – suggesting that the reaction can also proceed in the opposite direction if the equilibrium conditions are changed (Piccirilli *et al.* 1992).

Although it is not known how protein synthesis began, there are several independent but complementary lines of evidence showing that a rudimentary version of this process once existed. Data supporting this possibility include:

1 Experiments in vitro in which peptide-bond formation is catalyzed solely by a ribosomal RNA moiety (Noller *et al.* 1992) or can take place in systems with complete ribosomes but in the absence of a number of proteinic components of the translation apparatus, such as initiation and elongation factors (Spirin 1986). Protein synthesis probably took place in a distant past without these molecules, which may be related to the appearance of regulatory or optimization mechanisms (Woese 1980; Spirin 1986; Moore 1988).

2 An extensive comparison of primary and secondary structures of cellular, mitochondrial, and plastid rRNAs has led to the recognition of conserved, highly defined cores significantly smaller than the typical eubacterial ribosomal RNAs. Since these minimal rRNAs contain most of the functional sites involved in translation, they may be the oldest recognizable functional portion of extant ribosomes (Gray & Schnare 1990).

3 The genes coding for the two elongation factors in eubacteria and their homologues in the other two major cellular lineages are the result of an ancient gene duplication thought to predate the separation of prokaryotes into eubacteria and archaebacteria (Iwabe *et al.* 1989). This suggests that a simpler process of protein biosynthesis, proceeding with only one elongation factor, may have taken place (Lazcano, this volume, p. 60).

As argued persuasively by Carl R. Woese (1967), primitive translation must have been an ambiguous, error-ridden process, with triplets coding probably not for individual amino acids but for classes of amino acids with similar physicochemical properties. It is unlikely that the first RNA-directed proteins emerged fully endowed with catalytic properties. Tertiary structure plays a major role in extant enzymatic activity, but the earliest coded proteins must have been rather small oligopeptides. Their properties probably depended more on their primary structure than on the limited secondary and tertiary conformations available to them. The first proteins may have been nucleic-acid-binding oligopeptides involved in RNA unwinding or in ribozyme stabilization. A good contemporary example of such interactions is the functional interplay between the components of the RNA-protein complex of ribonuclease P (Lazcano 1986; Westheimer 1986), in which the stabilizing effect of the basic protein over the RNA moiety enhances its catalytic effects (Altman 1984). Such hybrid intermediate stages in which RNA and proteins acted together may have existed, but it is unlikely that they lasted for long: with the appearance of protein synthesis the RNA world ran full stride towards evolutionary oblivion.

Into the DNA world

Protein synthesis can take place without DNA but not in the absence of RNA. It is thus reasonable to assume that DNA cellular genomes are the result of an early molecular takeover that took place well after the origin of life, when protein synthesis was already established and different enzymes were available. In the words of Bruce Alberts (1986), "All DNA functions must have evolved in an intracellular environment rich in protein catalysts, where RNA catalysis had become largely obsolete." This conclusion is supported by the fact of deoxyribonucleotide biosynthesis, which removes the highly reactive 2'-OH group from a pool of preexisting ribonucleotides (Lammers & Follmann 1983). It is possible that this enzyme-mediated process may have been acquired in cells with RNA genomes as a final step in nucleotide biosynthesis (Lazcano *et al.* 1988), a mechanism that is consistent with the pathway in which the DNA-specific deoxy-thymidilate is formed by adding a methyl group to a deoxy-derivative of the RNA-specific base uracil (Kornberg & Baker 1992).

That DNA had displaced RNA genomes long before 3.5 Ga ago is suggested by the morphological complexity of the Warrawoona microfossils (Schopf, this volume). No archaebacterial ribonucleotide reductase gene has been cloned and sequenced, but it is likely that DNA cellular genomes are a monophyletic trait that evolved before the divergence of the three main cellular lineages. Sequence similarities shared by ancient proteins found in all three lines of descent suggest that considerable fidelity already existed in the then-operative genetic system (Lazcano *et al.* 1992). Given the chronic high mutation rates of RNA genomes, it is unlikely that such fidelity could have been achieved if the last common ancestor of eubacteria, archaebacteria, and eukaryotes lacked DNA and repair mechanisms insuring its genetic integrity.

As shown by its amazing recovery from a magnolia leaf between 17 and 20 Ma old (Golenberg *et al.* 1990), double-stranded DNA is an extremely resistant macromolecule. It is generally agreed that DNA genomes were selected over RNA for a very simple reason: the latter are fragile, reactive polymers that undergo many chemical changes, including their almost complete hydrolysis. Genetic information stored in RNA de-grades because of the cytosine's strong tendency to deaminate to uracil and the lack of a correcting enzyme. Furthermore, the lack of substantial amounts of free atmospheric oxygen and the consequent lack of an ozone shield would have led to a high ultraviolet flux in the early Archean, leading to deleterious rates of UV-induced mutations in cells with RNA genomes. Since DNA-repair systems depend on the duplication of the genetic information contained in the complementary strands of the duplex DNA molecules (Friedberg 1985), the emergence of double-stranded DNA would have stabilized the earlier irreparable system, leading to the selection of mechanisms to correct damage caused by UV light (Lazcano *et al.* 1988).

The appearance of DNA led to more stable ways of storing genetic information in hitherto highly mutating RNAs with limited coding abilities. Since DNA genomes are error-correcting because of their double-stranded structure, their presence opened the possibility of increasing genome size by gene duplication. Of course, multiple copies of the same gene may have existed in cells with RNA genomes. However, no RNA polymerase is endowed with the proofreading activity that DNA polymerases possess. RNA replication is an intrinsically noisy process, one that limits template size, since the number of accumulated point mutations is proportional to their template length. As shown by some contemporary viruses, this limitation may be overcome in part by segmented genomes (Reanney 1982). It is reasonable to assume that in some intermediate stage of early cellular evolution, genomes were disaggregated and rapidly mutating RNAs, in which several copies of the same gene existed. Because of the difficulties in insuring genetic identity of their offspring, such cells would be rapidly selected against.

The evolutionary emergence of double-stranded DNA genomes and of DNA polymerases with editing properties allowed the drastic development of large cellular genomes with increased coding potential. The appearance of DNA unleashed the enormous catalytic potential of proteins. Tinkering of exons and of the products of gene duplications made proteins truly malleable commodities, developing their catalytic prowess and enhancing the fitness of primitive cells. The use of new substrates and the regulation of metabolic pathways became possible, leading to intricate webs of reactions involved in basic metabolic networks and well-attuned biochemical processes. The world of modern cells with DNA, RNA, and proteins was well on its way.

Conclusions

Catalytic RNA may be a molecular remnant of a bygone early stage of biological evolution, but were the first forms of life actually based on ribozymes? As summarized by Joyce (1991), there is an unbridgeable gulf between our current descriptions of the primitive environment and the biochemical properties of RNA molecules. In spite of this limitation and of its inherent panselectionist explanations, the RNA-world hypothesis has the advantage of readdressing the problem of the origin of proteins and DNA from an articulate novel perspective amenable to empirical analysis.

From this standpoint, future major insights on the emergence of nucleic acid-directed protein synthesis can be expected to result not from chemical simulation experiments but from detailed characterizations of ribozymes and the development of RNA systems evolving in vitro. Further understanding of the origin of extant DNA genomes will be provided by phylogenetic comparisons of genes from the major cellular lineages coding for ribonucleotide reductases, thymidilate synthases, DNA polymerases, primases, and other proteins involved in DNA replication. Results from such research can be expected to influence other fields of biological inquiry. For instance, are RNA viruses, retroviruses, and DNA viruses related to the evolutionary transition from RNA to DNA cellular genomes, as argued by Weiner (1987b), or should we seek an explanation of their origin in more recent biological processes?

The study of the origins of life has focused mainly on the appearance of proteins and nucleic-acid replication. A wealth of data has accumulated, but a broader approach is needed. More emphasis should be given to the long-neglected question of the emergence of basic metabolic pathways. The notion of a retrograde evolution, suggested many years ago by the California Institute of Technology biologist Norman H. Horowitz (1945) to explain the appearance of metabolic pathways, is not supported by current evidence. In fact, the alternative idea that metabolic pathways evolved through the "patchwork" assembly of primitive proteins with broad substrate specificity

(Jensen 1976) is consistent with the properties of early nucleic-acid-coded peptides discussed in this chapter, and it is also supported by the homologous character of different enzymes involved in widely separated biosynthetic processes (Parsot 1987; Lazcano *et al.* 1992).

In the past decade considerable progress has been achieved in our understanding of the emergence and early evolution of living systems, but we are still haunted by major uncertainties, the magnitude of which is matched only by our ignorance. Even though the sequence of evolutionary events discussed in this chapter may be correct, continuing inquiries into prebiotic chemistry, paleobiological analysis, and molecular phylogenetic comparisons are required to fully validate it. Empirical evidence is the ultimate bonfire of our theoretical vanities.

Acknowledgments. – I wish to express my gratitude to the organizers of the 1992 Nobel Symposium "Early Life on Earth" for their kind invitation to contribute with this additional chapter. I am indebted to Dr. Stefan Bengtson for his continuous encouragement and patience, and to Drs. Gail R. Fleischaker and Juan Oró, who painstakingly read the manuscript and made many helpful and constructive comments. I thank Drs. Thomas R. Cech, Gustaf Arrhenius, and Joseph A. Piccirilli for kindly sharing with me the results of their work before publication. This manuscript was finished during a short leave of absence at the University of Houston, under the auspices of NASA Grant 44-005-002 to Juan Oró. Work reported here has been supported in part by UNAM.IN. 105289.

Molecular origin and evolution of early biological energy conversion

Herrick Baltscheffsky and Margareta Baltscheffsky

Department of Biochemistry, Arrhenius Laboratories, Stockholm University, S-106 91 Stockholm, Sweden

Processes involving energy-transfer reactions were prerequisites for the emergence of pre-RNA, RNA, ribozymes, and the polynucleotide–polypeptide link and its kind of information storage and processing, which allowed genetic continuity from early biological evolution up to the present time. The earliest energy-coupling reactions may have involved the energy-rich phosphate compound inorganic pyrophosphate (PPi), which possibly preceded adenosine triphosphate (ATP) as a biomolecular "energy currency." The suggested central role of PPi in early biological conversion, coupling, and conservation of energy is based on its formation in bacterial photophosphorylation, its capabilities as biological energy and phosphate donor, its comparatively uncomplicated structure, its occurrence as mineral, and its formation from hot volcanic magma. Furthermore, the PPi synthetase is less complex structurally than the ATP synthetase. In addition, the energy-linked reactions catalyzed by PPi synthetase involve less free energy and have a less complex mechanism of energy conservation than those catalyzed by the ATP synthetase. Finally, the electron-transport coupled formation of PPi is faster than that of ATP under conditions where the energy supply is limited. Both a "pyrophosphate world" and a "thioester world" may well have been prerequisites for an "ATP world" and an "RNA world."

Biological energy conversion may be described in general terms as the processes whereby radiant (often visible light) or chemical energy is converted in such a manner that the products can be utilized to drive the cellular energy-requiring reactions. Both the energy-converting and the energy-requiring reactions are well known to be an integral part of the life of the cell.

The molecular machinery involved in biological energy transfer and energy coupling and the "energy currency" that they produce for subsequent consumption have long been central objects of investigation in bioenergetics. Another pivotal point, of special significance in connection with this symposium, is the question: Which was the original biological energy currency? Of great relevance for our attempts to learn more about early life on Earth are also the detailed properties of enzymes catalyzing those biological energy-coupling reactions in living cells that may be regarded as "primitive," and the possible molecular similarity between such enzymes and their corresponding prebiological and early biological counterparts.

This presentation will give a broad but brief general background and thereafter will more specifically consider the origin and early evolution of biological energy conver-

Bengtson, S. (ed.) 1994: *Early Life on Earth. Nobel Symposium No. 84.* Columbia U.P., New York

sion and energy coupling. It has long been our view (Baltscheffsky 1971) that prebiological energy-conversion reactions as a necessary prerequisite to the origin of life included energy coupling that was at least somewhat related to that found in living cells. In other words, coupled energy flows involving inorganic and organic molecules are assumed to have been in operation before, and to have made possible, the emergence of, say, an early "RNA world" with ribozymes, evolving to the nucleic acid-to-protein-type system, with its genetic continuity found in all known forms of life.

It should be emphasized that with respect to both the early evolution and the molecular mechanism of biological energy coupling, there exists very little solid knowledge. Both energy transfer and energy coupling in living cells are extremely complex processes. This contributes to the difficulty in establishing their molecular origin and early evolution. Fortunately, however, several recent results in this area appear to increase the confidence with which one attempts to extrapolate forward from chemistry and geology and backward from biology.

We shall discuss here the apparent evolutionary consequences of some of these new results. In particular, our earlier picture about the possible role of inorganic pyrophosphate (PPi) as a very early molecular energy currency involved in prebiological and biological energy coupling (Baltscheffsky 1971) will be developed further, not least on the basis of a remarkable variety among the new pieces of information.

PPi is not the only candidate for a central role in prenucleotide, prebiological, and early biological energy coupling. De Duve (1987) has proposed that prebiotic oxidations may have been coupled to formation of energy-rich thioesters and has suggested that a "thioester world" may have provided the earliest energy coupling. The recent demonstration that formation of pyrite (FeS_2) from FeS and H_2S is linked with hydrogen evolution under anaerobic conditions (Drobner et al. 1990) is also of interest in connection with early oxidation–reduction-linked energy coupling, pyrite formation having earlier been suggested to be the first energy source for life (Wächtershäuser 1988a) and very recently to be feeding the first cellular bioenergetic process by providing a proton motive force across a cell membrane (Koch & Schmidt 1991). A common denominator is that all the energy coupling suggested above could well have occurred before the emergence of nucleotides, especially ATP, whereas the only energy-rich phosphate compound, suggested above to be involved in early coupling, is PPi.

General background

It may be useful, at a symposium as multifaceted as this, to give a brief overview of biological (1) energy sources, (2) energy-liberating (exergonic) reactions, and (3) energy coupling involved in providing the energy-rich conditions and compounds that are capable of driving the energy-consuming (endergonic) reactions and (4) the actual energy-rich conditions and compounds.

1. Energy sources

a *Radiant:* Early: cosmic radiation, UV, visible light and IR. Today: visible light and IR.

b *Chemical:* Inorganic and organic compounds: often typically

    I oxidation–reduction compounds ("redox" energy)

or

    II C-, H-, O-, N-, P- and/or S-containing compounds with "substrate" energy, however, with no strict demarcation line existing between types I and II.

c Heat, for example in combination with dehydration, shock waves.

2. Energy liberation

a Light-driven ("photosynthetic"):

    I Series of oxidation–reduction reactions following chlorophyll energization (cyclic or noncyclic pattern).

    II Carotenoid energization (trans–cis isomerization).

b Chemical, "dark" ("oxidative"):

    I Series of oxidation–reduction reactions between "substrate" and "acceptor."

    II "Substrate level" (exo- or endo-) oxidation–reduction.

3. Energy coupling

Coupling events may involve transient protein conformation changes, possibly also between low- and high-energy forms.

a Electron-transport-coupled ion-gradient formation.

b Cis–trans-isomerization-coupled ion-gradient formation, involving Schiff base deprotonation–protonation.

c Ion-gradient-coupled phosphorylation.

4. Energy-rich conditions and compounds

a Conditions.

Ion gradients: Protons (H^+) and other ions can provide an electrochemical potential difference between two bulk phases separated by a membrane ($\Delta\mu H^+$).

$\Delta\mu H^+ = F \times \Delta\Psi - 2.3RT \times \Delta pH$, where

ΔpH = the pH difference between two bulk phases separated by a membrane, and $\Delta\Psi$ = the membrane potential, i.e. the electrical potential between two such bulk phases.

Protein conformations: Transient energy-rich protein conformations may exist, as mentioned above.

b Compounds.

Inorganic or organic, often containing atoms and bond sequences such as

$$O = \overset{|}{\underset{|}{P}} - O - X = \text{ or } O = \overset{|}{C} - S - \text{ where X is P or C:}$$

Examples:

$$O = \overset{\overset{\displaystyle OH}{|}}{\underset{\underset{\displaystyle OH}{|}}{P}} - O - \overset{\overset{\displaystyle OH}{|}}{\underset{\underset{\displaystyle OH}{|}}{P}} = O \qquad \text{inorganic pyrophosphate (PPi)}$$

$$\text{adenyl-ribosyl} - O - \overset{\overset{\displaystyle O}{\|}}{\underset{\underset{\displaystyle OH}{|}}{P}} - O - \overset{\overset{\displaystyle O}{\|}}{\underset{\underset{\displaystyle OH}{|}}{P}} - O - \overset{\overset{\displaystyle OH}{|}}{\underset{\underset{\displaystyle OH}{|}}{P}} = O \qquad \text{adenosine triphosphate (ATP)}$$

$$O = \overset{\overset{\displaystyle CH_3}{|}}{C} - O - \overset{\overset{\displaystyle OH}{|}}{\underset{\underset{\displaystyle OH}{|}}{P}} = O \qquad \text{acetylphosphate}$$

$$O = \overset{\overset{\displaystyle CH_3}{|}}{C} - S - \text{CoA} \qquad \text{acetyl-coenzyme A (CoA-SH = coenzyme A)}$$

PPi and early energy coupling

Three sets of results obtained in our studies of photophosphorylation in chromato-phores from the purple "nonsulfur" photosynthetic bacterium *Rhodospirillum rubrum* drew our attention to the possibility that PPi rather than ATP may have been the original energy-rich compound formed at the expense of radiant energy in an early version of bacterial photosynthesis: our discoveries of light-induced formation of PPi (H. Baltscheffsky *et al.* 1966) in a reaction that was not inhibited by oligomycin and thus did not involve any ADP–ATP reaction (Baltscheffsky & von Stedingk 1966), of a membrane-bound PPase (M. Baltscheffsky *et al.* 1966), and of energy-requiring reactions, which were driven even more rapidly with PPi than with ATP (M. Baltscheffsky *et al.* 1966; Baltscheffsky 1967). Here, a first alternative energy-rich phosphate compound, PPi instead of ATP, could be formed in a biological photosynthetic system and could be used in it as an energy donor. These results appeared to add experimental support to the contention of Lipmann (1965) that "On the phosphate side, it may be reasonable to assume that generation of the phosphate group potential might have originated with inorganic pyrophosphate as the primitive group carrier." Evidence for the involvement of PPi as a donor of energy and phosphate at the substrate level of intermediary metabolism had earlier been provided, first with the discovery of the phosphoenolpyruvic carboxytransphosphorylase reaction in *Propionibacterium shermanii* (Siu & Wood 1962) and more recently in several other reactions (Wood 1985).

Before discussing PPi in more detail we would like to refer, for additional general information, to several publications that have recently appeared on origin, evolution, and mechanisms of biological conversion and coupling of energy (de Duve 1991); oxidation–reduction involving pyrite (FeS_2) formation, suggested to be of early bioenergetic significance (Wächtershäuser 1988a, b; Drobner *et al.* 1990); a model system for early light-induced energy conversion (Goncharova & Goldfeld 1990); energy coupling in archaebacteria (Deppenmeier *et al.* 1990) and light-induced proton transport by bacteriorhodopsin (Oesterhelt *et al.* 1991; Mathies *et al.* 1991; Lanyi *et al.* 1992); proton circuits in biological energy conversion (Williams 1988); PPi and PPases (Baltscheffsky & Baltscheffsky 1992); and general bioenergetics, emphasizing chemiosmosis (Nicholls & Ferguson 1992).

PPi, as will be further discussed below, can be formed both in prebiotic conditions from hot volcanic magma and, in present-day biology, at the expense of light energy in bacterial photosynthesis. This energy-rich inorganic phosphate compound would appear to remain an excellent choice as the first molecular energy currency for the origin and early evolution of life, as it can be formed independently of which answer will finally be given to the question about the original immediate energy source: radiant or chemical? The sun, with its radiation of light quanta of appropriate energy contents continuously reaching the Earth, is an attractive candidate. But so are inorganic and organic molecules with suitable oxidation–reduction properties or with a satisfactory content of group-linked free energy.

A major problem with current photosynthetic systems involving chlorophyll is that the transformation of light energy to biologically useful chemical energy is by far too complicated to allow detailed, meaningful extrapolation backwards to the earliest such systems. On the other hand, the best-known biological mechanism for the transformation of radiant energy to proton-motive force is that of light-induced proton transport in bacteriorhodopsin (a 26 kD, seven-transmembrane helix protein with its Lys-216 bound to a retinal chromophore moiety in the form of a protonated Schiff base) of the halobacterial purple membrane. It includes a sequence of at least five discrete steps, in addition to the occurrence of structural changes in the protein during the transport of protons (Lanyi *et al.* 1992):

1 Light-induced isomerization of the retinal around the 13–14 carbon bond from trans- to cis-form. This isomerization displaces the Schiff base away from Asp-212 towards ionized Asp-85 (Mathies *et al.* 1991). About 16 Kcal/mol of the initially absorbed light energy is stored and subsequently used for proton transport (Birge *et al.* 1989).

2 Transfer of the Schiff base proton to Asp-85 and release of this or another proton from a nearby residue to the external side.

3 Proton transfer from Asp-96 to the Schiff base.

4 Reprotonation of Asp-96 from the cytoplasmic side and reisomerization of the retinal.

5 Return of the bacteriorhodopsin to the original state.

This sequence illustrates that discernible acidic or basic groups in the membrane-bound protein, by undergoing protonation–deprotonation, can perform the proton transport.

It also highlights some of the spatial complexity necessary in this unique physiological light-induced proton transport. One may observe that the ensuing energy coupling between the obtained proton gradient and the ATP formation catalyzed by the membrane-bound ATP synthetase (H$^+$-ATP synthetase, ATPase, F$_o$F$_1$ATPase) has not been taken into account. Considering the apparent complexity of the still-unknown mechanism of energy coupling in the ATP synthetase reaction, it is tempting to speculate about the possible existence of a PPi synthetase also in some type of bacteriorhodopsin-containing or similar "primitive" membranes. In this connection, one should also recognize the attempts to contribute with model experiments to an understanding of the origin of light-induced formation of energy-rich phosphate (Goncharova & Goldfeld 1990).

The more we learn about the membrane-bound PPase from bacterial chromatophores, about its structure and functions, the more it emerges as a comparatively uncomplicated but competent converter and coupler of chemical energy. The relative simplicity of this PPase and its reactions, as compared with the ATPase and its reactions, can easily be elucidated at both the enzyme and the reaction level. The enzyme consists of a single polypeptide (56 kD) and is probably a dimer, which carries the catalytic function and the proton channel and the possibility for coupling proton transport with phosphorylation (Nyrén $et\ al.$ 1984; Nyrén $et\ al.$ 1991). This structure may be compared with the ATPase (F$_o$F$_1$ATPase), which contains a minimum of eight different subunits in two distinct supercomplexes, the catalytic part, F$_1$ (3α, 3β, γ, δ, ϵ), and the proton-translocating part, F$_o$ (a, 2b, 10–12c).

There may well be a causal relationship between the simplicity of the PPase and its substrate PPi, as compared to the much more sophisticated ATPase and ATP and the less complex kinetic properties of the PPase. No regulatory properties are known for the PPase, unless one includes the thermodynamic limitation of the ability of the enzyme to pump protons against an existing proton gradient. The synthesis of PPi, but not of ATP, occurs even at very low levels of electron transport, such as at low light intensities or by inhibition with antimycin (Nyrén $et\ al.$ 1986). This is in accordance with the fact that ATP synthesis and hydrolysis require an activating threshold level of $\Delta\Psi$, whereas PPi synthesis and hydrolysis do not. In experiments where artificial pH gradients and/or diffusion potentials with K$^+$-ions in the presence of valinomycin were the driving force for, respectively, ATP and PPi synthesis, the two enzymes behaved very differently. As was shown by Leiser & Gromet-Elhanan (1974), synthesis of ATP does not occur even at a high ΔpH in the absence of a diffusion potential, which presumably is needed to activate the ATP synthetase. In addition, the hydrolysis of ATP by this enzyme requires a proton motive force, as exemplified by the inhibition of ATPase activity by high uncoupler concentration (Edwards & Jackson 1976) or by the rapid hydrolysis of ATP after a short light flash (Baltscheffsky & Lundin 1979). The membrane-bound PPase, on the other hand, is able to synthesize PPi with a ΔpH as the sole driving force, and the hydrolysis of PPi is not inhibited by high uncoupler concentration (Strid $et\ al.$ 1986). The proton-pumping capacity of the two enzymes also differs. While the ATPase for synthesis has been found to require 3–4 H$^+$/ATP and the hydrolysis yields the translocation of at least 2 H$^+$, the PPase does not appear to involve more than 1 H$^+$ in either direction (Moyle $et\ al.$ 1972).

The low energy requirement of the PPase is also evident from results with various inhibitors. Uncoupling agents are needed in higher concentrations for PPi synthesis than for ATP synthesis, a result of the requirement for a lower proton gradient. In other words, total uncoupling of ATP synthesis may occur at a $\Delta\mu H^+$ that is below the level necessary to activate the ATPase but still large enough to support a low level of PPi synthesis (Nyrén *et al.* 1986). In all known instances when there is only a low energy supply, the rate of PPi synthesis is as great as, or greater than, the rate of ATP synthesis.

Scarcity of water-soluble phosphates on the primitive Earth has long been considered a basic problem in connection with the origin of life, and so has the question of what mechanisms may have been available for the continuous synthesis of energy-rich phosphates or other compounds suitable for driving energy-requiring reactions. As PPi, the simplest existing energy-rich phosphate compound, has been shown to function in both substrate level and electron-transport-coupled energy transfer and phosphorylation, new geochemical results concerning PPi appear to be of special significance in this connection. Attention had earlier been drawn to the lack of PPi-containing minerals, and only recently has such a mineral been found, namely canaphite, $CaNa_2P_2O_7 \cdot 4H_2O$ (Rouse *et al.* 1988). This demonstration that PPi can be stable under natural conditions elucidates the fact that a kinetic barrier can prevent a thermodynamically plausible reaction ($PPi + H_2O \rightarrow 2Pi$) and contributes to diminish the problem concerning the possible presence of PPi on the early Earth. Early experiments by Ponnamperuma & Chang (1971) have shown that prolonged heating to 65°C of Ca and Na salts of orthophosphate resulted in partial conversion to pyrophosphate.

A potential source of more or less continuous prebiotic production of PPi and higher inorganic polyphosphates has recently been found both in experiments that simulate magmatic conditions and from analysis of the volatile condensates from hot volcanic gas (Yamagata *et al.* 1991). As was pointed out, magmatic P_4O_{10} obtained in volcanic activity can through its partial hydrolysis produce water-soluble PPi (and polyphosphates). In fact, higher polyphosphates have long been considered as prebiological and prenucleotide as well as physiological alternatives to PPi and ATP, as has been well reviewed by Kulaev (1979). Yamagata *et al.* (1991) emphasize that so far this seems to be the only identified route for large-scale production of PPi on the primitive Earth. They may well have provided a viable solution to the problem concerning the availability of an energy-rich phosphate compound for primordial evolution and the evolution of early life on Earth.

A more detailed treatment of the geochemistry of phosphates is given by Gedulin & Arrhenius (this volume).

Acetylphosphate – a link between a "pyrophosphate world" and a "thioester world"?

As has been suggested recently (de Duve 1991; Baltscheffsky & Baltscheffsky 1992), acetylphosphate, which is an energy-rich metabolite of bioenergetic significance in many bacteria, could well have been an early molecular link between PPi and thioesters in prebiotic and early biotic metabolism, as shown by the scheme:

PPi ⇌ Acetylphosphate ⇌ Acetyl-S-R

The reactions involved occur in living cells, which strengthens the argument for their possible prebiotic and/or early biotic existence. Gilbert (1986) has, on the basis of the fact that RNA has emerged as the molecule capable of both information and function, contemplated an "RNA world" with no proteins involved and only RNA molecules catalyzing the synthesis of themselves. In the light of present knowledge it does not appear to us to be too far-fetched to imagine that a pyrophosphate world and possibly also a thioester world were necessary prerequisites for the emergence of such an RNA world. Fig. 1 presents this possibility within the more general picture that a pre-nucleotide world preceded and facilitated the emergence of a nucleotide world.

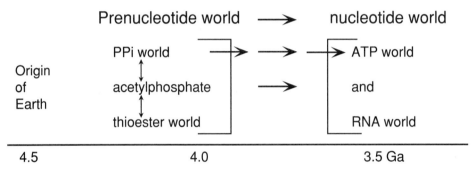

FIGURE 1 Scheme illustrating the suggested existence of an early prenucleotide world as a prerequisite for a subsequent nucleotide world. Energy-rich compounds such as PPi and thioesters, which still play central roles in the metabolism of living cells, would appear to have been plausible carriers for the conversion and coupling of energy in a prenucleotide world on the early Earth. Acetylphosphate is the logical molecular link between a PPi world and a thioester world. It is an open question whether a stepwise evolution (indicated with three arrows), perhaps still discernible, from PPi metabolism to ATP metabolism has taken place. The time indicated for the prenucleotide world and its transition to a nucleotide world is very tentative.

PPase, H⁺-PPase, H⁺-PPi synthetase

In the last few years there has been a great increase in our knowledge about the structures of different inorganic pyrophosphatases (PPases), which catalyze the reaction:

PPi + H₂O ⇌ 2 Pi

and which may be either cytoplasmic (PPase) or membrane-bound (PPase, H⁺-PPase, H⁺-PPi synthetase). In contrast to the photobacterial PPase of the chromatophore membrane, which has been shown to function both as a proton-transducing H⁺-PPase and a proton-transducing and energy-coupling H⁺-PPi synthetase, the plant-vacuolar membrane-bound H⁺-PPases have only the former function but no energy-coupled PPi formation.

Yeast cytoplasmic PPase is the only PPase that has been crystallized. Its three-dimensional structure has been determined to 3Å resolution, which has allowed a rather detailed view of its active site (Kuranova 1988). The primary structures of several cytoplasmic, one mitochondrial, and one vacuolar PPase have recently been elucidated. Their structural and other properties have been reviewed in considerable detail (Baltscheffsky & Baltscheffsky 1992).

The first (and so far only) known amino acid sequence of a mitochondrial PPase, from the yeast *Saccharomyces cerevisiae* (Lundin *et al.* 1991), is about 50% homologous to the cytoplasmic PPase sequence from the same organism, both lacking extended hydrophobic stretches. However, the first (and also so far only) known sequence of a plant-vacuolar H^+-PPase, from *Arabidopsis thaliana* (Sarafian *et al.* 1992), is not homologous to other known PPases. It contains 13 extended hydrophobic stretches and is, as the bacterial H^+-PPi synthetase, an integrally membrane-bound protein.

Our recent demonstration (Nore *et al.* 1991) that the proton-pumping H^+-PPi synthetase from *R. rubrum* shows immunological cross-reactivity with mung bean vacuolar H^+-PPase indicates significant structural similarities between the energy-coupling PPase from eubacterial chromatophores and the energy-transducing PPase from vacuoles of a eukaryotic plant. Closer structural characterization of both these enzymes may be expected to reveal new molecular characteristics of structural requirements for energy transduction and, of particular importance in the present context, energy coupling.

The "potential 15 amino acid Cys–Cys loop" of the mitochondrial PPase (Lundin *et al.* 1991), which functionally appears to occupy a position somewhere between the cytoplasmic and the two membrane-bound PPases discussed above, may be a structure of particular interest in this connection. It appears to contain several amino acids of importance in the active, PPi-binding, site and is remarkably similar to a corresponding loop in many ligand-gated ion-channel polypeptides. There could, conceivably, be an evolutionary connection between such a loop, maybe even at the level of oxidation–reduction (2 -SH $\rightleftharpoons$ -S-S- + 2H), and "primitive" energy-linked oxidation–reduction reactions of inorganic and/or organic compounds.

Since active sites of PPases seem to contain several, especially negatively, charged amino acid side groups, and since at least three Mg^{2+} ions appear to be present in the functional enzyme, a picture has emerged with the negatively charged PPi substrate "surrounded" by positively charged Mg^{2+} ions held in place in a specific molecular pattern involving several negatively charged amino acid side groups (mostly Asp^-) of the protein. One may seriously consider exploring this situation in various model experiments with clays, minerals, or other active surfaces, in attempts to simulate what may have been very early prebiotic energy conversion and coupling.

Summary and outlook

Oró *et al.* (1990) recently emphasized that "Neither a single energy source nor a single process can account for all of the organic molecules that were found in the prebiotic Earth. The importance of a given energy source is determined by the product of the energy available and its efficiency for synthesis of organic compounds or their intermediates . . ." and that "Unfortunately, very little is known about the origin and early evolution of transport mechanisms and membrane-bound bioenergetic systems," which is "an area that requires intensive study in order that we may develop an understanding of the origin of ion pumps and bioenergetic processes and the development of metabolism."

Indeed, there seems to be general agreement that current knowledge about the origin and evolution of early biological energy conversion is shrouded in uncertainty, to say the least. On the other hand, PPi has long been suggested to be an attractive candidate for the role as an early "energy currency," based on both biological and chemical evidence. The recent geochemical and additional biological results obtained would appear to have added considerable support for the assumption that this uniquely simple energy-rich inorganic phosphate compound may have played a central role in prebiological and / or early biological conversion and coupling of energy.

Acknowledgments. – Supported by the Nordic Yeast Research Project (H.B.) and the Swedish Natural Science Research Council (M.B.). Written during our "sabbatical" at the Department of Biochemistry, University of Oxford, U.K. We thank Professor George K. Radda and his staff for their active interest and support.

Sources and geochemical evolution of RNA precursor molecules: The role of phosphate

Benjamin Gedulin and Gustaf Arrhenius

Scripps Institution of Oceanography, University of California San Diego, La Jolla, CA 92093-0220, USA

Life depends entirely on energy storage in bonds of phosphate anydrides. Organic phosphate ester formation can be traced to the most primitive organisms known, and the exceptional features of the phosphate system suggest that its control reaches further back into the prebiotic era, including the emergence of life. The high ionic charge is responsible for the efficient concentration of phosphate species from dilute solution into anion-exchanging minerals, which induce selective formation of aldohexose sugar phosphates. An extensive search has been made for geochemical processes that induce phosphate condensation. Phosphate anhydride formation in the solid state, mediated by structural protons and transition metal ions in minerals, are found to be efficient mechanisms in this respect. Precipitation of phosphate from sterile aqueous solutions, in the compositional range of natural seawater, yields protonated Ca and Mg–Ca phosphate minerals at pH <8.5. Upon heating to a few hundred degrees, these minerals produce condensed oligophosphates, providing a potential energy and material source for phosphorylation of simple organic compounds. Such phosphorylation is an essential metabolic process today and is generally thought to have been an integral step in the emergence of life in a prebiotic ocean. In contrast, the basic calcium phosphate, apatite, is a magnesium-free mineral, which at the high magnesium content and near-neutral pH of seawater is deposited only by cellular organisms. The emergence of life must in all likelihood have depended critically on the availability of the abiotically produced protonated and condensable phosphates that by the arrival of life and biogenic apatite were relegated to minor importance as sedimentary minerals. The identification of authigenic sedimentary apatite in the oldest deposits known on Earth, along with other features indicating a microbial origin, provides substantial support for the notion that cellular life existed on Earth 3.7–3.8 Ga ago.

Models for emergence of the genetic material must take into account that such a complex molecule as RNA could hardly have arisen in a one-step process. For each of the functional parts of the molecule questions then present themselves: How was it synthesized; how was it linked to a larger complex; at what point in evolution did it arrive at its present configuration and function; and how could an evolutionary force be maintained in the interim, driving the system toward its ultimate functional form? Tracking the predecessors of the present design is essentially a problem in reverse

Bengtson, S. (ed.) 1994: *Early Life on Earth. Nobel Symposium No. 84.* Columbia U.P., New York

engineering. In trying to understand the origin of the recognition mechanism, now relying on purine–pyrimidine base pairing, researchers in this field take heart from the fact that a compound like adenine may arise spontaneously from aqueous cyanide solution (Oró 1960; Sanchez *et al.* 1967; Ferris & Hagan 1984; Schwartz *et al.* 1992). However, an often ignored fact is that effective pairing of bases requires their incorporation into a finely tuned tertiary macromolecular structure with a hydrophobic interior. On their own in aqueous solution, the bases, overwhelmed by hydration, are incapable of forming mutual hydrogen bonds. If the precursors of RNA were based on similar recognition principles, one might therefore have to search for more primitive structures that could have provided internal hydrophobicity. In modeling the possible evolution of nitrogen-based recognition molecules, the fact has also to be taken into account (Schlesinger & Miller 1983) that hydrogen cyanide and cyanide ion could not have coexisted, because they would have formed glycolonitrile. Further consequences of this reaction, and of cyanide complex formation with Fe^{2+}, are discussed by Arrhenius (1990) and Arrhenius *et al.* (1993).

Similarly, problems surround the formation of sugars or sugar derivatives as components of precursor forms of RNA. Identification of a plausible process for selective formation of one specific backbone molecule has been an elusive goal. Progress toward solution of this problem is, however, indicated by the Eschenmoser glycolaldehyde phosphate reaction (Müller *et al.* 1990), which, in strongly alkaline solution and at high reactant concentration, selectively produces specific aldohexose phosphate diastereoisomers. Geochemical plausibility is added to this scheme by the discovery of a corresponding heterogeneous reaction, where aldolization of monomeric glycolaldehyde phosphate in dilute solution and under mildly alkaline conditions is induced in the interlayer of common anion-exchanging minerals (S. Pitsch,

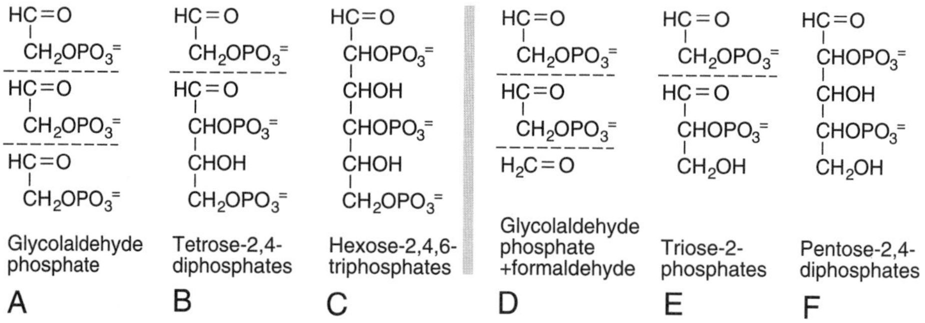

FIGURE 1   Mineral-induced selective sugar phosphate formation from glycolaldehyde phosphate in the interlayer double-layer hydroxide minerals. Glycolaldehyde phosphate (**A**) forms hexose-(primarily altrose) 2,4,6-triphosphate (**C**) via tetrose-2,4-diphosphate (**B**).

   In absence of the sorbent double-layer hydroxide mineral, the aldolization reaction proceeds only under more alkaline conditions and at higher reactant concentration. In the presence of formaldehyde (**D**) it yields pentose-(mainly ribose-) 2,4-diphosphate (**F**) via triose-2-phosphate (**E**) in addition to hexose phosphate (**C**). In the mineral-free homogeneous reaction (Müller *et al.* 1990), the dominant hexose product is allose-2,4,6-phosphate, in contrast to altrose phosphate, which is the main diastereoisomer produced in the mineral reaction discussed here.

A.E. Eschenmoser, B. Gedulin, S.Y. Hui, and G. Arrhenius, unpublished). In this mineral-induced aldomerization, racemic aldohexose 2,4,6-phosphates form as the major reaction product with preference for altrose (Fig. 1). The facile selective synthesis of aldohexose phosphate has led to the implication of hexose-based, nonhelical RNAs (cf. Fig. 2) as "potentially prebiological natural products" (Eschenmoser 1991; Eschenmoser & Loewenthal 1992; Eschenmoser & Dobler 1992).

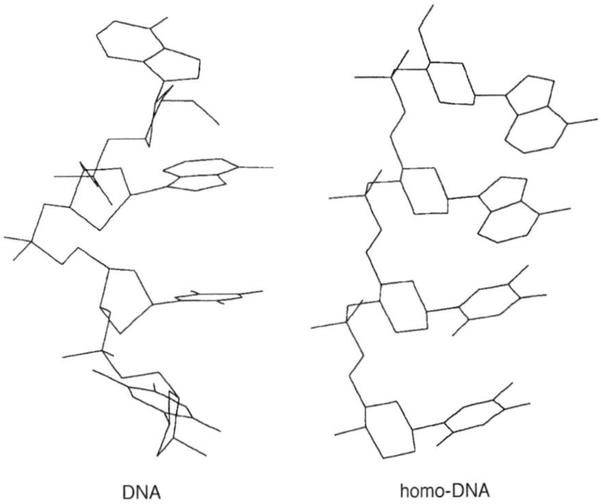

FIGURE 2 Schematic depiction of a DNA single strand, compared to homo-DNA in their pairing conformation, the latter with a hexapyranosyl moiety replacing the naturally occurring pentofuranose sugar ring. In its stable conformation, homo-DNA double strands assume a near-linear ribbon structure. (From Eschenmoser & Loewenthal 1992.)

DNA   homo-DNA

The entry of phosphate in biopoesis

Stringent organic chemical and geochemical requirements are tied to the incorporation and function of phosphate as a structural moiety in RNA. It was proposed by Westheimer (1987) that nature selected phosphate as a linking unit in nucleic acids because of its unique chemical properties. With its trivalent charge, orthophosphate ion is able to link two adjacent nucleosides while remaining ionized; this property is even more pronounced in the diphosphate-linked nucleotide oligomer synthesized by Schwartz *et al.* (1987). The anionic nature of the charge deters nucleophilic attack in the vicinity of the phosphate such that the rate of hydrolysis of phosphodiester bonds is slow enough to maintain relatively error-free nucleotide sequence. The protection by the anionic charge against nucleophilic attack and hydrolysis is also important in thermodynamically unstable phosphoric anhydrides such as ATP and inorganic diphosphate. The rate of hydrolysis of these molecules in near-neutral aqueous solution, unless catalyzed, is thus very slow. Furthermore, at the cellular stage of evolution, the charged surfaces of nucleotide polymers prevent their loss by diffusion through cellular membranes (Westheimer 1987; Deamer 1985, 1986b).

The question left open is when and how inclusion of phosphate took place in the evolving system. Was it an early development, or could it have waited until ribozyme or enzyme evolution, steered by some type of phosphate-free proto-RNA, had arrived at efficient kinases of the type relied on today? To bring the latter possibility in from the realm of speculation, one would require the experimental demonstration of a phosphate-free molecular system capable of replication and mutation and of operation in an aqueous medium.

To Westheimer's list of desirable properties of phosphate esters may be added others that contribute to the geochemical plausibility of phosphate as an early component in biopoesis. One is the capability of phosphates and phosphate esters for concentration from dilute solution by surface-active anion exchange minerals (Arrhenius 1987; Kuma *et al.* 1989; Holm *et al.* 1993). Without such selective concentration, most prebiotic scenarios come to a halt, since the early concept of a highly concentrated organic "probiotic soup" has become implausible for several reasons (e.g., Sillén 1965), and the dilution expected in the absence of efficient concentration mechanisms would appear prohibitive for most reactions.

Phosphate-aided sorption on external or internal mineral surfaces may also lower the activation energy for intermolecular reactions; effective concentration followed by condensation is exemplified by the mineral-induced aldomerization reaction discussed above, which leads selectively to the formation of hexose- (mainly altrose-) triphosphate. The notion of a primitive, phosphate-linked recognition molecule (proto-RNA) receives further support by this role of phosphate in controlling a selective sugar synthesis and by the geochemical availability of achiral phosphate anhydrides reported in this paper. Also in modeling reactions of the modern form of RNA, sorption on anion-exchanging minerals has been found effective (Holm *et al.* 1993).

Probably the most ancient preserved trace of early dependence on condensed phosphate by organisms, preceding the modern ADP–ATP system, is the production and utilization of pyrophosphate (diphosphate) by primitive bacteria, discovered by H. and M. Baltscheffsky and coworkers (H. Baltscheffsky *et al.* 1966, 1986; M. Baltscheffsky *et al.* 1985). In this case, the energy for the enzyme-assisted phosphate condensation is provided by light. The antiquity of the process, combined with an argument for conservatism in evolution, provides yet another probability argument for the role of phosphate, particularly condensed species, in biopoesis.

Mineral sorption of phosphate

The anionic phosphates and phosphate esters are, because of their high charge density, efficiently sorbed by minerals with positive surface charge. These include iron(3) oxyhydroxide polymorphs, which play an important role in the phosphate cycle in nature (Holm 1987), particularly under present-day oxidizing conditions, and potentially at photo-oxidation of iron(2) at the surface of the anoxic Archean ocean (Braterman *et al.* 1983; Holland 1984; Holm *et al.* 1993; Sloper *et al.* 1983). Sorption of orthophosphate ion on iron(3) hydroxide (ferrihydrite) is practically irreversible, because of the development of a strong chemical bond; reversal can be achieved by reduction to Fe(2), the process responsible for the seasonal cycle of release of phosphate from lake and basin sediments (Holm 1987).

Minerals that are capable of concentrating and internally binding organic source compounds are of particular interest as possible scaffolding for evolving biomolecules (Arrhenius 1984). Zeolites, clay minerals, manganates, oxyhydroxides, and double-layer hydroxides fall in this category. Zeolites suffer from the rigid size limitation for host and product molecules, imposed by the size invariance of the molecular tunnels formed by their three dimensional silicate networks, and from the fact that the negatively charged framework attracts only cationic species. This property is shared also with smectite clay minerals and manganates, whose freely expanding sheet structures permit intercalation complexes of practically unlimited size. Manganate(4) minerals (Giovanoli & Arrhenius 1988) may be of more restricted prebiotic interest than ferric oxide hydroxides, because of the high oxidation state.

Clays have been at the forefront of interest since Bernal (1949) and Goldschmidt (1952) proposed the mediating effect of such minerals and Cairns-Smith (1965) suggested the possible role of mineral–organic complexes as the first living organisms. Many studies have been made of the interaction of clay minerals with amino acids (see, e.g., Katchalski 1951; Lahav & Chang 1976; Lahav & White 1980; Theng 1974). Catalytic phosphate diester bond formation was demonstrated by Ferris *et al.* (1990) and Ferris & Ertem (1993), who achieved oligomerization of imidazole-activated nucleotides by interaction with smectite clays, presumably by surface sorption of the protonated nitrogen bases.

Since the source molecules for RNA – carbonate (reducible to formaldehyde and sugar), cyanide (monomeric component of bases), and phosphate (nucleoside linkage ion) – all are anions, expanding sheet-structure minerals with excess positive charge have attracted interest in studies of the possible evolution of RNA precursor molecules (Arrhenius 1987; Arrhenius *et al.* 1989; Kuma *et al.* 1989; S. Pitsch *et al.*, unpublished). Prominent among expanding sheet-structure minerals with anion-exchange properties are the double-layer hydroxide minerals. These owe their name to the alternation in the structure of main sheets of metal hydroxide with interlayers of exchangeable hydrated

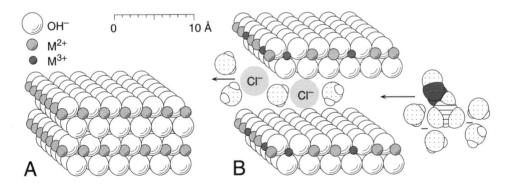

FIGURE 3   A: Simple bivalent cation hydroxide sheet structure, exemplified by minerals such as $Mg(OH)_2$, $Fe(OH)_2$, and $Mn(OH)_2$. B: Replacement of a fraction of the M^{2+} ions in A with trivalent cations, such as Fe^{3+} or Al^{3+}, introduces excess positive charge on the sheets, which separate and admit an interlayer of negative ions. Each anion is associated with three to four water molecules; together, they form a concentrated, two-dimensional solution weakly bonded to the catalytic metal hydroxide sheets. Ions such as Cl^- are readily displaced by higher-charge density anions such as phosphate or metal-cyanide complexes. The figure shows schematically the replacement of two monovalent chloride ions by one much larger trivalent glycolaldehyde phosphate ion entering the interlayer with associated water, and expanding the interlayer width from 3.6Å for the chloride complex to 6.4Å for the polar glycolaldehyde phosphate.

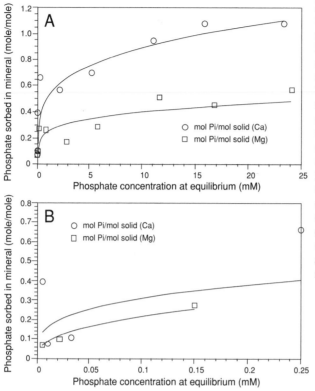

FIGURE 4 **A:** Sorption isotherms for orthophosphate ion in calcium aluminum hydroxide (the mineral hydrocalumite) and magnesium aluminum hydroxide (hydrotalcite). The isotherm plot shows on the y-axis the number of moles of phosphate sorbed per mole of solid and on the x-axis the equilibrium phosphate concentration in millimoles per liter. **B:** Isotherm plot from **A** expanded 100-fold to show the details of sorption in the range of micromolar equilibrium concentration of phosphate – at these low concentrations the analytical data scatter becomes marked.

anions or anion complexes (Fig. 3). The anion migration into the interlayer is due to excess positive charge on the main hydroxide sheets caused by population of about one-third of their cation sites with trivalent species such as Al^{3+}, Fe^{3+}, or Cr^{3+}, while the remaining sites are occupied by divalent ions such as Mg^{2+}, Ca^{2+}, Zn^{2+} or transition-element cations such as Mn^{2+}, Fe^{2+}, Co^{2+}, or Ni^{2+}, each cation combination and intercalation species giving the opportunity for a special mineral name. Among the most common minerals of this type are hydrotalcite with Mg^{2+}–Al^{3+} hydroxide main sheets, pyroaurite (Mg^{2+}–Fe^{3+}), and green rust (Fe^{2+}–Fe^{3+}), the last probably abundant in the primordial anoxic ocean as the source of the huge Archean banded iron formations. Common interlayer ions in these minerals are carbonate, sulfate, sulfide, and chloride.

Orthophosphate, condensed phosphate ions, and phosphate esters, because of their high charge, are among those species that are particularly efficiently sorbed in the interlayer of the double-layer hydroxide minerals, as demonstrated by the steep sorption isotherms (Fig. 4). Simple organic compounds that contribute the source for more complex biomolecular assemblages can, after phosphorylation, be brought to high concentration and into the reactive environment of the mineral interlayer. Use of glycolaldehyde phosphate as a source compound rather than the underivatized aldehyde thus makes it possible to efficiently concentrate the solute from micromolar solution to a 7-molar, two-dimensional quasisolution in the mineral interlayer, where aldomerization to tetrose and hexose phosphate is rapidly induced (S. Pitsch *et al.*, unpublished). The exclusion of all ketoses is due to blocking of the β-keto group by the phosphate, while the selection of altrose and, in the homogeneous reaction, ribose phosphates over other aldohexose and pentose phosphates is due to kinetic effects (Müller *et al.* 1990).

Geochemical phosphorylation mechanisms and condensed phosphates in nature

Much effort has, over the years, been devoted to exploration of geochemically plausible phosphorylation mechanisms (see review by Gabel 1990). Some of the most successful experiments in phosphorylating nucleosides at temperatures in the range of 100°C, with yields as high as 30%, are those of Österberg *et al.* (1973) and Handschuh *et al.* (1973). Miller & Parris (1964) and Neuman & Neuman (1964) were also able to demonstrate condensation of orthophosphate with the aid of cyanate on the surface of hydroxyapatite.

In view of the potential efficiency of condensed phosphate species in ester formation, investigations have been devoted to such reactions (e.g., Gabel 1968; Gabel & Thomas 1971). In this context, the question arises if efficient mechanisms exist in nature for condensation of orthophosphate, stabilizing the condensed species against hydrolysis and bringing them into reactive situations, and if, as a consequence, metastable condensed phosphate minerals occur in nature. On theoretical grounds, it has been proposed that this would not be possible (Liebau & Koritnig 1969); on the other hand, at least one finding of calcium sodium pyrophosphate, canaphite, has been reported in the literature (Rouse *et al.* 1988). The crystals occur as overgrowth on a zeolite in cavities in a Triassic basalt. The circumstances and extent of the formation of the mineral have, however, not been further investigated; until the paragenesis and the generality of the occurrence have been verified, it seems prudent not to invoke it as analogue to a prebiotic source of condensed phosphate.

Our search for pyrophosphate mineral sources has followed three lines: investigation of suspect minerals from phosphate-bearing hydrothermal deposits; analysis of minerals claimed in the literature as "metaphosphates," on structural or compositional grounds; and evaluation of phosphate minerals known or postulated to contain structural protons as a source of condensed phosphate at thermal metamorphism.

White Mountain in Mono County, California, offers an unusual section through schists that have been highly metamorphosed by acidic magmatic solutions. Leaching of the parent rock has extensively mobilized and removed soluble cations and silica, leaving behind a residue of andalusite ($Al_2[O \mid SiO_4]$), locally reaching ore grade (Champion Mine). The mineral associations have been described in detail and discussed in the context of origin by Wise (1977). The phosphate-rich hydrothermal solutions have deposited a variety of phosphate minerals associated with different temperature stages (Moore 1973; Wise 1977). Aside from various orthophosphates, the assemblage includes a sulfate-phosphate (woodhouseite) and phosphate-silicate (the zeolite viséite).

Considering that the pyrophosphate mineral canaphite is found in a hydrothermal association, we undertook to analyze a suite of hydrothermal minerals kindly provided by W.S. Wise from his collection of minerals from the Champion Mine. The samples were assayed for condensed phosphates by ion chromatography after extraction by grinding in 1 M NaOH. None of these minerals yielded clear evidence of the presence of pyrophosphate. An analytical difficulty is due to the lack of complete solution of the minerals under leaching conditions that would not destroy any condensed phosphate by hydrolysis.

A method has recently been devised by Sales *et al.* (1986, 1992) for extraction of phosphate from minerals with low solubility, such as calcium and magnesium phosphates, and for quantitative detection of condensed phosphate oligomers. The extraction technique utilizes complexing agents in aqueous solution made alkaline to prevent the hydrolysis of released pyrophosphate during the extraction. Separation is achieved by gradient elution from an anion exchange column, followed by postcolumn hydrolysis and detection of the fractionated and subsequently hydrolyzed oligomers by the phosphomolybdate blue technique. In collaboration with B. Sales, we repeated our analyses of the mineral assemblages from the Champion Mine, the Lovozero Massif in the Kola Peninsula, and the Varuträsk pegmatite in Sweden and demonstrated the lack of measurable amounts of pyrophosphate in all but two of the suspect minerals. The two exceptions are alluaudite ($(Na,Ca)(Fe^{2+},Mn^{2+})Fe_2^{3+}(PO_4)^3$) and hureaulite ($H_2(Mn,Fe)_5(PO_4)_4 \cdot 4H_2O$). Hureaulite, suspected of containing structural protons and thus yielding pyrophosphate on heating, gave about 1% of diphosphate upon heating to 800°C, while alluaudite, heated to 550°C in N_2, yielded about 30% diphosphate. The Oak Ridge group has also demonstrated (Chakoumakos *et al.* 1990) that α-particle irradiation in the U–Th-bearing pegmatite phosphate mineral griphite has resulted in extensive oligomerization (11%) of the original orthophosphate groups.

These results indicate that mechanisms indeed exist for generating pyrophosphate in structures originally crystallized as orthophosphates. In the case of alluaudite, this is probably achieved by coupling to a redox reaction involving iron; in the case of hureaulite, by reaction of structural protons with orthophosphate, removing oxygen. At high temperatures, such as in igneous melts, the equilibrium between orthophosphate and diphosphate is sufficiently displaced toward the latter so that a fraction can be quenched in at cooling and appears as diphosphate ion in low concentration (about millimolar) in hydrothermal steam (Yamagata *et al.* 1991).

All of these processes, although of interest in principle, are with the currently found parageneses and extent of condensation unlikely to have been important from a geochemical point of view, unless mechanisms are demonstrated that preferentially concentrate pyrophosphate over orthophosphate from dilute weathering – or hydrothermal solutions. An alternative process of greater potential importance is discussed below.

Phosphate mineral formation in a live ocean

The Mg/Ca ratio is on the average 0.4 in river water and >5 in seawater (Broecker & Peng 1982). Within this range the magnesium concentration is sufficiently high, relative to calcium, to influence the structure and composition of the minerals forming from solutions in the pH range 6.5–9.0. None the less, apatite (calcium hydroxy fluorocarbonate-phosphate) is, besides iron phosphates, the most common sedimentary phosphate mineral. Sedimentary occurrences of apatite are, by all indications discussed below, of biogenic origin, some more obvious in the form of skeletal debris of various organisms, others requiring high-resolution techniques to demonstrate characteristic biogenic features (Prévôt & Lucas 1986). Such deposits include excretionary products and microbially induced formations, in the marine environment mostly as cryptocrystalline masses (collophane). Experimental data, discussed below, suggest that biogenic apatite must form by cellular processes, selecting calcium over magnesium, independently of the composition of the water in which the source organisms live while depositing the mineral. One such mechanism, utilized by a variety of bacteria, employs ammonium ion to bind magnesium, leaving calcium ion to form apatite. Magnesium ion is sequestered as a complex, which at sufficiently high concentration precipitates as the mineral struvite ($MgNH_4PO_4 \cdot 6H_2O$).

The role of living organisms in aqueous formation of apatite has been investigated by numerous authors; results and overviews are given, e.g., by Le Geros (1981), Reimers *et al.* (1991), Jahnke *et al.* (1983), and most extensively by Lucas and collaborators (references below). Particular emphasis has been placed on the biological induction of apatite both in fresh water and seawater, where the high Mg/Ca ratio precludes inorganic formation of this mineral except under extreme conditions (pH >8.5). Lucas and coworkers (Lucas & Prèvôt 1981, 1984, 1985; Prévôt *et al.* 1989; Hirschler *et al.* 1990a) employed for such studies a heterogeneous model system consisting of calcium carbonate in the form of aragonite and magnesian calcite, in contact with solutions of RNA as a source of phosphate. In the presence of microorganisms, the solid carbonates were completely converted to apatite, isomorphously replacing the original mineral structures – a phenomenon commonly observed in nature. In these experiments, apatite formation was accompanied by the precipitation of struvite. In contrast, no apatite or struvite formation took place under sterile conditions in otherwise identical, long-term experiments.

On the basis of these observations and the fact that Mg^{2+} ion in solution prevents inorganic apatite formation in the natural pH range, Lucas and Prévôt proposed magnesium immobilization by ammonium ion as one of the mechanisms for biogenic apatite precipitation; ammonium ion, released by microorganisms, precipitates magnesium as struvite. The formation of this mineral effectively reduces the concentration of magnesium ion in solution and, as a result, brings the system into the stability field for apatite. This mechanism can operate in nature under circumstances in which high ammonia concentration can be generated and in which the magnesium reservoir is strictly limited, such as in pore solutions in sediments. Isomorphous carbonate replacement by apatite as well as direct phosphorite precipitation is commonly observed in sediments, occasionally in association with ostentatious bacterial activity (Reimers *et al.* 1991) and the formation of struvite (Böggild 1907, 1909). Detailed microbiological investigations of the struvite reaction have been carried out by Hirschler *et al.* (1990a, b) and Rivadeneyra *et al.* (1993).

Gulbrandsen *et al.* (1984) carried out nonsterile crystallization experiments that are unique with regard to the length of time of maturation of the initial precipitates. After ten years, these showed the presence of microcrystalline apatite, together with another phase, tentatively identified as bobierrite, $Mg_3(PO_4)_2 \cdot 8H_2O$. These experiments demonstrate that under conditions where microbial activity is not excluded by sterilization, apatite may form, not only as replacement of solid calcium carbonate but also at microbial interaction with the magnesium–calcium phosphate gel that characteristically forms from natural waters. The magnesium phosphate (bobierrite) observed as a segregated phase in these long-term experiments may result from eventual denitrification of the magnesium ammonium phosphate, struvite.

Phosphate mineral formation in a lifeless ocean

Attempts to establish the phosphate phases forming from sterile solution with seawater composition have encountered the difficulty that precipitates forming rapidly in non-equilibrium solutions with the appropriate magnesium-to-calcium ratio generally have low crystallinity and are inaccessible by X-ray diffraction; they are referred to as "X-ray amorphous." Magnesium-free solutions at pH 7 and higher, on the other hand, yield apatite or metastable crystalline calcium phosphate precursor phases. As the magnesium concentration is increased in such precipitation experiments, apatite formation is slowed down until it vanishes at the expense of an X-ray-amorphous solid. This situation has frequently been described as magnesium ion "inhibiting" the crystallization of apatite (Martens & Harriss 1970; Van Cappellen & Berner 1991), thus assuming that the magnesium ion does not enter and control the structure and that the gel is an amorphous solid with apatite composition (Gulbrandsen *et al.* 1984). In contrast, we find that magnesium is a structural component of the phosphate minerals formed from the precipitate, to the exclusion of apatite.

In the absence of microorganisms, apatite forms as a single phase at pH 9 or higher in a seawater medium, and in the two-phase whitlockite–apatite region from pH ~8.5 to 9.0. Preformed crystals of apatite placed in neutral or slightly alkaline solution with the Mg/Ca ratio of seawater convert to whitlockite. The high activation energy for this phase transformation requires high temperature or long time for measurable amounts of the conversion product to appear – typically a month at 300°C for the order of 10% conversion, and a week for microcrystals of Mg–Ca phosphate to appear on the surfaces of large single crystals of apatite at 80°C.

To further clarify the stability relationships, we have undertaken experiments aimed at identifying the phases forming from solutions with magnesium-to-calcium ratios characteristic of natural waters in the open hydrosphere, as a function of temperature, Mg/Ca ratio, pH, pCO_3, and pF. In order to increase the crystallite size in the precipitates into a range where their structures are accessible by electron- and X-ray

TABLE 1   Crystalline phases forming in the range of compositions and conditions specified.

Mineral	Composition	pH
brushite	$CaHPO_4 \cdot 2H_2O$	6–7.5
whitlockite	$Ca_{18}Mg_2H_2(PO_4)_{14}$	7–9
magnesium phosphate pentahydrate[a]	$Mg_3(PO_4)_3 \cdot 5H_2O$	7–9
apatite	$Ca_5[OH, F \mid PO_4, CO_3)_3]$	>8.5

[a] Mazghouni *et al.* 1981.

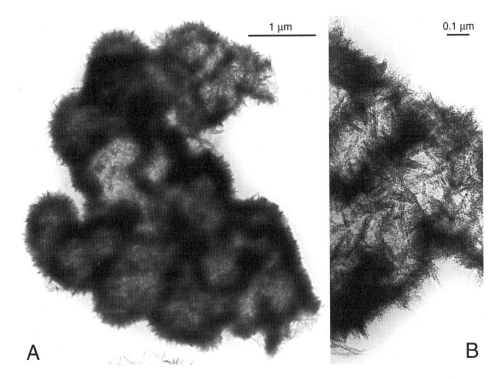

FIGURE 5   Calcium–magnesium–hydrogen phosphate formed by slow oxidation of phosphite ion to orthophosphate in artificial seawater at room temperature and pH 7.5. The acicular aggregates, forming globular masses, are too small for identification by standard X-ray diffraction methods but, after annealing in water at 130°C, show the interlayer spacings characteristic of $Ca_{18}Mg_2H_2(PO_4)_{14}$ (whitlockite) (Fig. 6), together with those of magnesium phosphate pentahydrate (Fig. 6). **A:** Aggregate of globular masses. **B:** Detail of **A**, showing domain of larger, needle-shaped aggregations.

diffraction techniques, a method was developed for slow and controlled homogeneous precipitation. This was achieved by dissolving sodium phosphite in source solutions of seawater composition and slowly oxidizing phosphite ion to phosphate by bubbling a stream of argon with a small fraction of ozone through the solution. With the precipitation rate and crystal growth controlled in this fashion, mineral formation results in solids that in favorable cases are amenable to direct structural study by electron or X-ray diffraction. It is also possible to grow the amorphous or microcrystalline phases to crystallite sizes of the order of 10~50 μm by annealing the solids either in water vapor or in their source solution in the range of 50–130°C. The resulting crystals are characterized by electron-beam microanalysis, electron and X-ray diffraction, and optical microscopy. The precipitation experiments were generally undertaken in the absence of sulfate ion, in order to model the composition of Archean preoxic seawater, and both in the absence and presence of carbonate and fluoride ion.

The resulting phase distribution is given in Table 1; representative electron micrographs and X-ray diffraction records of coexisting phases are shown in Figs. 5–7 and Tables 2–3.

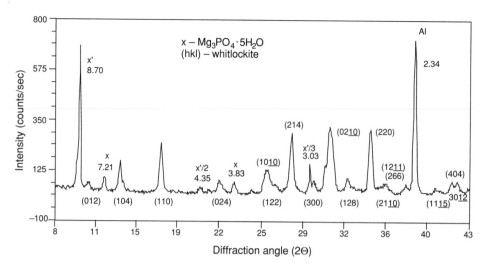

FIGURE 6 X-ray diffractogram of Ca–Mg–H phosphates (Fig. 5), precipitated from aqueous solution of synthetic, sulfate-free seawater and annealed in saturated water vapor at 130°C. Crystal growth from the initially X-ray-amorphous gel confirms the presence of whitlockite (diffraction maxima marked with Miller indices [hkl]) together with magnesium phosphate pentahydrate (marked X) as the two phases present. Second- and third-order reflections of the fundamental reflection X' are marked X'/2 and X'/3; interplanar spacings indicated in Ångström units. X-ray penetration through the sample gives rise to diffraction marked (Al) from (111) reflection of aluminum in the sample holder. The observed interplanar spacings and diffraction intensities for whitlockite in this preparation are given in Table 2, together with corresponding data from the literature.

TABLE 2 X-ray diffraction data for whitlockite, $Ca_{18}Mg_2H_2(PO_4)_{14}$.

| Line number | Interplanar spacing | | Intensity | | Miller index |
	observed	literature[a]	observed	literature[a]	
1	8.13	8.01	7	20	012
2	6.55	6.35	25	30	104
3	5.18	5.22	58	45	110
4	4.04	4.02	16	30	024
5	3.433	3.407	31	55	1010
6	3.34	3.304	1	10	122
7	3.187	3.160	73	65	214
8	2.986	2.974	2	10	300
9	2.867	2.837	100	100	0210
10	2.734	2.719	15	25	128
11	2.587	2.572	83	80	220
12	2.499	2.491	15	10	2110
13	2.387	2.379	11	20	1211, 226
14	2.243	2.230	9	20	1115, 042
15	2.171	2.171	15	20	404
16	2.150	2.143	15	30	3012

[a]Fisher & Volborth 1960; Gopal *et al.* 1974; Joint Committee on Powder Diffraction Standards, *Powder Diffraction File*, no. 13-404, Swarthmore, Pennsylvania.

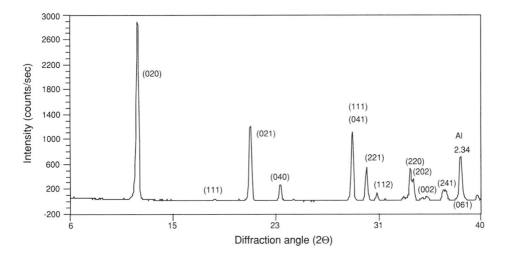

Figure 7 X-ray diffractogram of brushite ($CaHPO_4 \cdot 2H_2O$), formed at room temperature from sulfate-free, artificial seawater under sterile conditions and in the pH range 6–7.5. Miller indices are shown in parentheses and in Table 3 together with present measurements and data from the literature. The diffraction maximum marked Al is the (111) reflection from the aluminum sample holder.

Table 3 X-ray diffraction data for brushite, $CaHPO_4 \cdot 2H_2O$.

Line number	Interplanar spacing		Intensity		Miller index
	observed	literature[a]	observed	literature[a]	
1	7.61	7.57	100	100	020
2	4.93	4.93	1	1	111
3	4.24	4.24	42	100	121
4	3.80	3.80	9	7	040
5	3.63	3.63	1	1	131
6	3.047	3.05	38	75	111, 041
7	2.928	2.928	18	50	221
8	2.854	2.855	4	9	112
9	2.799	2.797	>1	3	200
10	2.669	2.670	2	5	150
11	2.622	2.623	18	50	220, 151
12	2.605	2.603	12	30	202
13	2.553	2.554	1	5	002
14	2.527	2.520	2	5	132
15	2.433	2.434	6	15	241
16	2.267	2.268	3	5	061

[a]Beevers 1958; *Powder Diffraction File*, no. 9-77.

Protonated sedimentary phosphates as source of condensed phosphates

One of the most striking features of phosphates found to precipitate from solution of seawater composition without interference of live organisms is the formation from them of protonated minerals including whitlockite and, at lower pH, brushite (calcium hydrogen phosphate). It has long been known that hydrogen-containing phosphates at heating react to form diphosphate (pyrophosphate) and water and that condensed phosphates are efficient phosphorylation agents. This dimerization of orthophosphate is the first step in a condensation sequence that includes several higher oligomers (Chang *et al.* 1970). A widespread formation of protonated phosphates in the primordial lifeless ocean may therefore have been a factor in prebiotic molecular evolution.

Sales *et al.* (1992, 1993) provided new insight in the solid-state chemistry of condensation of protonated phosphates by demonstrating the formation not only of lower oligomers, including the dimer pyrophosphate, at dehydration of crystalline protonated phosphates, but also of a sequence of higher phosphates, up to 13 P–O–P units (Fig. 8). The protonated phosphates synthesized from seawater media in our work yielded oligomeric sequences (up to 7 for brushite) after treatment at temperatures of a few hundred degrees (Fig. 9) (B. Sales, B. Gedulin & G. Arrhenius, unpublished).

We consequently visualize as perhaps the most prolific source of pyrophosphate in the prebiotic era thermal condensation of hydrogen phosphate in sediments. The enhanced solubility of condensed phosphates over orthophosphate leads to their selective separation from the residual, less soluble Ca–Mg orthophosphate component of the transformed mineral (Sales *et al.* 1992). This selective dissolution may provide a

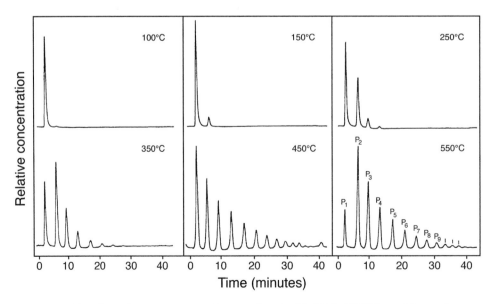

FIGURE 8   Phosphate oligomer spectrum from newberyite ($MgHPO_4 \cdot 3H_2O$) heated for one hour at temperatures between 100 and 500°C. (From Sales *et al.* 1993.)

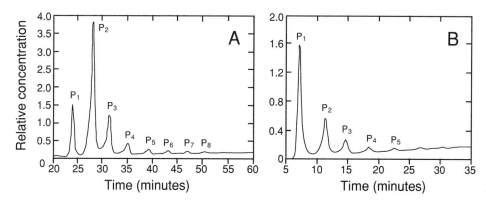

FIGURE 9   Chromatograms of oligophosphate species formed at 500°C in protonated phosphate, precipitated at room temperature from synthetic, sulfate-free artificial seawater. **A:** Brushite, crystallized at room temperature and pH 6.5. **B:** Whitlockite and magnesium phosphate pentahydrate, grown at 130°C from gel precipitated at room temperature and pH 7.5.

relatively concentrated source of dissolved, reactive pyrophosphate, with the possibility for further concentration into the reactive interlayer of anion-concentrating double-layer hydroxide minerals, discussed above.

Apatite as a biomarker

The results discussed above suggest the potential use of sedimentary apatite as an indicator of the presence of life in an ocean with Mg/Ca ratio of the order of magnitude of the present value, with an ammonium ion concentration lower than isomolar with Mg^{2+}, and with pH in the range below 8.5.

Although there is considerable uncertainty about the properties of the Archean ocean at the time when the oldest known sedimentary deposits were formed, these particular conditions were probably satisfied. No mechanism is known that is likely to constrain the Mg/Ca ratio to substantially lower values; on the contrary, the prevalence of eruptive rocks with high magnesium content during the early Archean and the presence of a substantial magnesium carbonate component in Archean sedimentary carbonates suggest, if anything, a higher relative magnesium concentration than in the present-day ocean (O.A. Christophersen, personal communication, 1991).

In regard to the ammonium ion concentration in the early Archean ocean, model calculations (Miller & Van Trump 1981) suggest an upper limit of 10 mM. Mineral synthesis experiments employing this ammonium ion concentration failed to generate struvite or apatite at pH 8.0. A content of ammonia in the primordial ocean sufficient to remove magnesium as struvite would appear necessary to permit inorganic apatite formation at pH lower than 8.5.

Similarly, it is difficult to justify an assumption of pH in the range above 8.5 or even 8 in the Archean ocean. Current models including the controlling effects of atmospheric carbon dioxide (Kasting & Ackerman 1986) and the unbuffered hydrogen ion excess from degassing of HCl and HF suggest (Holland 1984) an originally moderately acidic (pH 3–5) and gradually more weakly acidic ocean (pH 6–7) with a relaxation time depending on the exposure to weathering of differentiated continental rocks above the sea surface (Högbom 1894). Current considerations of Archean tectonic evolution (Veizer, this volume; de Wit *et al.* 1992; Kröner & Layer 1992) suggest that continental nuclei were suboceanic plateaus with little subaerial exposure. Progressive titration of excess hydrogen ion from mantle degassing would, under such conditions, be limited to relatively slow reaction with submarine eruptives (Holland, at this symposium).

In conclusion, it appears unlikely that conditions in the early Archean ocean would have permitted abiogenic formation of authigenic sedimentary apatite; instead, whitlockite and brushite would be expected to have been the major sedimentary phosphate minerals in the absence of life. In addition, phosphates would tend to form by reaction with iron and manganese (Arrhenius 1952; Berner 1973). Whitlockite, like apatite, is a refractory mineral, resistant to metamorphism. At elevated temperature, it undergoes slight structural changes [to the β-$Ca_3(PO_4)_2$ structure] due to the proton-induced partial condensation of the phosphate but remains otherwise intact until melting above 1,200°C (Gopal *et al.* 1974). Whitlockite, even after leaching of the more soluble condensed phosphate component, should thus remain recognizable in metasedimentary rocks on the basis of structure, magnesium content, and calcium/phosphorus ratio. In contrast, the dehydration product of brushite, monetite [$CaH(PO_4)$], is not observed in thermally metamorphosed rocks, presumably because of the complete conversion to more soluble condensed phosphate minerals above 150°C (Sales *et al.* 1992); the low-temperature pyrophosphate mineral canaphite may possibly derive from dissolution and secondary precipitation of such calcium oligophosphates.

The appearance of cellular life would, as demonstrated by the microbial experiments discussed above, have dramatically changed this situation by bringing about local conditions permitting the precipitation of apatite. This would have been achieved primarily within the sedimentary surface layer, where today microbial discrimination mechanisms effectively prevent magnesium ion from interfering with calcium hydroxyphosphate formation. The phosphate mineral distribution in aquatic sediments on Earth and Mars consequently offers a potential for determining the appearance of cellular life, provided that assertions or determinations can be made of Mg/Ca ratios at least of the same order of magnitude as in present-day seawater, ammonia concentration less than isomolar with Mg, and pH below 9 in the body of water where the sediments were formed. The finding of ubiquitous sedimentary apatite, also with other features indicating a microbial origin, in the metasedimentary Isua formation (Arrhenius *et al.* 1993) consequently would seem to support indications of more uncertain nature that life existed on Earth 3.7–3.8 Ga ago.

Acknowledgments. – The authors wish to acknowledge help in obtaining the critical mineral samples from Professors W.S. Wise, Yu. Menshikov, I. Tolstikhin, Yu. Shukolyukov, G. Gladyshev, K.M. Lipkina, and K. Boström; from the Swedish Museum of Natural History by U. Hålenius; and from the National Museum of Natural History, Smithsonian Institution, by Dr. P.J. Dunn. Professors A.E. Eschenmoser and P.B. Moore provided valuable discussion. The experimental work was greatly facilitated by equipment grants from NSF (EAR 89-16501) and the W.M. Keck Foundation; the condensation studies draw extensively on collaboration with Dr. B. Sales. The generous research support from the National Science Foundation through grants EAR89-16467 and from the National Aeronautics and Space Administration through grants NAGW1031 and 2881 is gratefully acknowledged, as well as the opportunity offered by the Nobel Foundation and the symposium organizers for discussion of the results at Symposium 84.

Self-assembly and function of primitive membrane structures

David W. Deamer, Elizabeth Harang Mahon, and
Giovanni Bosco

Department of Zoology, University of California, Davis, California 95616, USA

A minimal role of membranes in the origin of primitive cells is to encapsulate replicating/catalytic macromolecules. It is reasonable to expect that membranes might also mediate bioenergetic functions such as selective solute permeation, light transduction, and the development of chemiosmotic potentials. The chapter discusses possible sources of prebiotic amphiphiles capable of forming membranes. In particular, certain components of carbonaceous meteorites can self-assemble into membranous structures, suggesting the presence of simple membranes on the prebiotic Earth. Mechanisms for encapsulation and permeation of solutes must therefore have been available. Investigations of amino acid and phosphate permeability show that lipid bilayers present a substantial barrier to diffusion of these important nutrients. This barrier must be taken into account when considering growth processes in early cellular systems. In regard to membrane-related bioenergetic functions, two primitive pigment systems – ferrocyanide and polycyclic aromatic hydrocarbons – can accept light energy and produce proton gradients across lipid-bilayer membranes. Both systems are plausible components of the prebiotic environment and represent useful models for investigating possible chemiosmotic energy sources during the origin and evolution of protocells on the early Earth.

The cell consists of numerous half-living chemical molecules suspended in water and enclosed in an oily film. When the whole sea was a vast chemical laboratory the conditions for the formation of such films must have been relatively favourable.

Haldane, The Rationalist Annual, 1929

By definition, even the simplest cells are bounded by membranes. If the first forms of life were cellular systems of encapsulated replicating/catalytic macromolecules, it follows that we must understand not only the macromolecular systems that have dominated most research on the origin of life but also the origin of membrane structure. Relatively few laboratories have approached the latter question, and therefore a brief discussion of requirements for boundary membranes in primitive cells is appropriate.

The most obvious function fulfilled by a boundary membrane is to provide an enclosed volume that maintains macromolecular components within the same micro-

Bengtson, S. (ed.) 1994: *Early Life on Earth. Nobel Symposium No. 84.* Columbia U.P., New York

environment. In the absence of a boundary membrane, the components would diffuse away into the surrounding bulk phase, and the potential for interactive systems would be lost.

An important corollary is that boundary membranes provide a means by which the components of a macromolecular system can vary. In the absence of an encapsulated microenvironment, macromolecules would largely exist as random mixtures, and selective processes could not take place.

A third aspect of primitive membrane function concerns bioenergetics. Contemporary membranes play a central role in this regard: membrane-associated pigment systems capture and transduce light energy, and coupled electron-transport systems produce chemiosmotic ion gradients that serve as primary energy sources for the cell. In the absence of membrane-boundary structures, light energy would not readily be captured, and ion gradients obviously could not develop.

Finally, membranes provide the potential for selective transport of specific nutrients. It seems likely that early cells would have made use of such transport processes and may even have required them in order to sort out and concentrate useful components of the highly complex organic inventory that presumably existed on the prebiotic Earth's surface.

The questions we will address here are related to the points outlined above:

1 What were the sources of prebiotic membrane-forming molecules?

2 What self-assembly processes might have been available to allow organized membranes to appear from the complex mix of organic compounds in the prebiotic environment?

3 How could primitive membranes capture macromolecules to produce encapsulated systems with catalytic replicative abilities?

4 What pigment systems were available, and how were they coupled to the transduction of light energy into energy resources for primitive cell function?

5 What is the role of a permeability barrier in early bioenergetic functions?

Sources of prebiotic amphiphiles

A significant current question concerns primary sources of organics in prebiotic evolution. It is generally agreed that smaller biologically relevant molecules like amino acids, purines, and simple carbohydrates were probably available on the early Earth, at least in highly dilute solutions. (See Miller *et al.* 1976 and Ferris *et al.* 1978 for reviews.) Their synthesis through Miller–Urey reactions has been extensively studied, and the discovery that organic compounds are present in carbonaceous meteorites confirmed the conjecture that relatively complex organic substances can be produced by abiotic reactions. A broader possibility is that the infall of extraterrestrial material delivered significant amounts of organic carbon to the Earth's surface. Chyba *et al.* (1990), Chyba & Sagan (1992), and Anders (1989) have explored this question in some detail. Conservative estimates are 10^6–10^7 kg per year for cometary delivery and 10^8–10^{10} kg per year for delivery in the form of interplanetary dust particles. These rates are in the same range as estimates of atmospheric formaldehyde production by photochemical reactions (Pinto *et al.* 1980). To give a perspective, over the period of 10^8 years of late accretion following Earth's primary accretion process, the total organic carbon added by meteoritic and cometary delivery would have been 10^{16}–10^{18} kg, several orders of magnitude greater than the total organic carbon content of existing organisms, estimated to be 6×10^{14} kg. It therefore seems likely that cometary and meteoritic infall was a significant source of organic carbon on the early Earth.

Given such a source, what kinds of molecules might have been added to the prebiotic environment? The organic content of carbonaceous meteorites represents the only pristine sample we have of compounds synthesized by nonbiological processes in the early Solar System. In typical carbonaceous meteorites several percent of the mass is organic carbon. Table 1 summarizes the mineral and carbon composition of the Murchison meteorite, which has undergone intensive analysis in the years following its fall in 1969. The most abundant organic material is a complex aromatic hydrocarbon polymer (~90% of total organic content), followed by a variety of organic acids. Aliphatic and aromatic hydrocarbons, amino acids, ureas, ketones, alcohols, aldehydes, and perhaps purines are present in smaller quantities, measured in parts per million. Similar patterns of organic content have been observed in other carbonaceous meteorites (see Cronin *et al.* 1988 for review).

Assuming that a significant amount of cometary and meteoritic infall survives atmospheric entry, the major fraction would presumably enter the oceans, with a smaller amount accumulating on land masses. It is not at all clear how organic components would be released from the various forms of extraterrestrial infall, and much would depend on the rate of infall, the mix of cometary and meteoritic material, rates of degradative turnover, and the site of delivery. Assuming that there was some mechanism by which organics were released into the environment, water-soluble compounds such as amino acids would ultimately form dilute solutions in lakes and oceans, while relatively low-density hydrocarbons and their derivatives would

TABLE 1   Organic and mineral composition of carbonaceous meteorites. (Modified from Cronin *et al.* 1988.)

Class	PPM	Carbon atom content
Monocarboxylic acids	>300	2–12
Dicarboxylic acids	>30	2–9
Amino acids	60	2–7
Amides	55–70	—
Aliphatic hydrocarbons	>35	1–23
Aromatic hydrocarbons	15–28	6–20
Aldehydes/ketones	27	1–5
Hydroxycarboxylic acids	15	2–6
Alcohols	11	1–4
Amines	8	1–4
N-heterocycles	7	–
Purines/pyrimidines	1.3	–
Anhydrous mineral content >97%		
Kerogen-like substances 1.5% (polymers of complex PAH)		

accumulate at air–water interfaces, forming a kind of prebiotic oil slick. The effects of wind and tide would concentrate the films at intertidal zones, just as today. It follows that likely sites for the origin of cellular life are cycling environments such as tide pools associated with intertidal zones: Darwin's "warm little pond."

Because all organic compounds are altered by hydrolysis and photochemical reactions, it is uncertain what concentrations might have built up in the oceans. On a global scale, organic concentrations were probably very dilute. It has been calculated that even under the most favorable conditions of Miller–Urey synthesis, amino acid concentrations in the ocean would reach little more than millimolar ranges (Stribling & Miller 1987) . Therefore local conditions that would concentrate organic compounds were probably essential to the origin of cells. Some obvious possibilities include adsorption to mineral surfaces, hydration–dehydration cycles in tide pools, and concentration of hydrocarbons and surfactant molecules at air–water interfaces.

Constraints on prebiotic self-assembly processes

All membranes of contemporary cells incorporate a lipid bilayer as the primary permeability barrier. Bilayer barriers are as essential to life as replicating–catalytic macromolecules, and it is reasonable to ask whether it is plauseible that such structures were available on the prebiotic Earth. There are several requirements if an organic molecule is to self-assemble into stable bilayer membranes. First, it must be an amphiphile, with a hydrophilic "head" and a hydrophobic "tail" on the same molecule. A variety of hydrophilic heads are present in membranogenic amphiphiles, including phosphate, sulfate, carboxylate, and amine groups, the only requirement being that they are highly polar or ionized. The hydrophobic tail must be a substantial part of the molecule but again can be quite variable as long as certain basic requirements are fulfilled. Earlier work from our laboratory (Hargreaves & Deamer 1978) showed that single-chain amphiphiles such as alkyl phosphates, alkyl sulfates, and even fatty acids assemble into bilayer membranes if they contain 10 or more carbon atoms in their hydrocarbon chains (Fig. 1). In more recent work (unpublished) we have made similar observations with phospholipids. For instance, phosphatidylcholine (PC) is a common membrane lipid and is often used to produce lipid-bilayer vesicles called liposomes. We found that PC with 12-carbon chains produces stable lipid-bilayer membranes that provide a partial permeability barrier for ions, albeit relatively permeable compared with the 16- and 18-carbon phospholipids of biological membranes. PC with 8-carbon chains cannot assemble into bilayers at all, while PC with 10-carbon chains forms lipid vesicles that can be visualized by phase microscopy but are unable to provide a significant permeability barrier to flux of ionic solutes.

A final constraint on membranogenic amphiphiles concerns the physical state of the membrane they form. Cell membranes cannot function if their component lipid chains are in the gel state (Singer & Nicolson 1972), and all contemporary organisms maintain their membranes in a fluid state by controlling chain length, unsaturation, and, in certain prokaryotic cells, chain branching. There is no reason to think that a primitive cell membrane could escape this requirement.

Given that hydrocarbon chains at least 10–12 carbon atoms long are required for the formation of bilayer membranes with minimal stability, is there a plausible source for

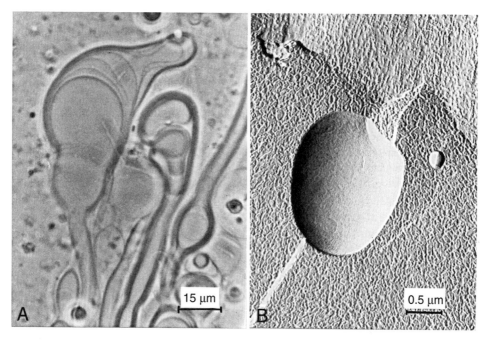

FIGURE 1   The simplest organic molecules capable of self-assembling into bilayer membranes are single-chain amphiphiles. **A:** A 10-carbon fatty acid (n-decanoic acid) at relatively high concentrations (100 mM) with the pH adjusted to the pK of decanoate under these conditions. Membranes can clearly be seen but are unstable and do not provide a diffusion barrier to solute flux. **B:** Freeze-fracture microscopy confirms the presence of lipid bilayers.

such chains in the prebiotic environment? In fact, long-chain hydrocarbons are relatively difficult to synthesize under simulated prebiotic conditions. Until recently, it was believed that long-chain hydrocarbons were present in meteorites, and it followed that synthetic pathways were presumably available in the prebiotic environment. However, Cronin *et al.* (1990) have shown that the long-chain hydrocarbons originally believed to be present in meteorites are in fact terrestrial contaminants, and that only cyclic aliphatic hydrocarbons appear to be present in significant quantities. Mono- and dicarboxylic acids have been observed, but the longest chain reported so far is octanoic acid (Lawless *et al.* 1979).

It has been suggested that the Fischer–Tropsch synthesis could perhaps provide long-chain hydrocarbons and their derivatives (Studier *et al.* 1972), but the reaction conditions require a gaseous form of carbon (such as carbon monoxide) and water vapor to be passed over a hot iron catalyst. Although long-chain hydrocarbons and various derivatives are produced, it is not easy to imagine how such specialized conditions might have been established on the prebiotic Earth. Furthermore, even if amphiphilic hydrocarbon derivatives could be synthesized, the chains must undergo condensation reactions with linker species (glycerol, phosphate) to form more complex species such as phospholipids required for self-assembly of stable membranes (see Hargreaves *et al.* 1977; Oró *et al.* 1978).

The last point is that phosphate is an essential component of most contemporary membrane lipids (phospholipids), but a primary source of inorganic phosphate on the prebiotic Earth is not immediately apparent. Phosphate is an essential but rare component of the contemporary biosphere and is often a limiting growth factor in lakes and seas. If all known phosphate were entirely dissolved in the oceans, including that trapped in mineral deposits and biological organisms, the phosphate concentration of seawater would increase only slightly above its present 3 μM. There is no reason to think that phosphate was more abundant globally on the prebiotic Earth, although one might imagine that certain limited environments could have accumulated significant amounts through various geological processes. (See Arrhenius, this volume, for further discussion of possible prebiotic reactions involving phosphate.)

Self-assembly of primitive membranes

We can now return to the original question: How did membrane-enclosed systems arise from the prebiotic milieu? In the absence of the catalyzed biosynthetic pathways of modern organisms, the earliest systems must have self-assembled from the mixture of organic species available. As noted earlier, amphiphilic molecules can self-assemble into molecular aggregates called bilayers, but long-chain hydrocarbons appear to be difficult to synthesize under plausible prebiotic conditions and are not present in at least one carbonaceous meteorite that has been examined. Furthermore, there are no clear pathways from hydrocarbons to oxidized derivatives like fatty acids, and phosphate is a rare species. It might therefore seem unlikely that amphiphilic compounds would have been available for the assembly of prebiotic membranes.

Nonetheless, there is some evidence suggesting that the presence of membrane structures is more plausible than might be expected. In approaching this question, we have been guided by the organic mixture of organic compounds present in carbonaceous meteorites, under the assumption that such compounds represent a plausible example of components available in the prebiotic environment, either delivered by meteoritic and cometary infall, as discussed earlier, or synthesized by unknown pathways. In past work the question of whether amphiphilic compounds were present in the Murchison carbonaceous meteorite has been explored (Deamer 1985; Deamer & Pashley 1989). Findings showed that surface-active substances were relatively abundant and that at least one class of minor components can assemble into vesicular membranes (Fig. 2). The membranes represent true barriers to ionic diffusion, since charged fluorescent molecules like pyranine can be trapped for several minutes and visualized in intravesicular compartments by fluorescence microscopy.

This fraction appears to be a complex mixture of oxidized hydrocarbons that can be partially resolved by standard GC–MS methods. Fig. 3A shows several components, but other peaks remain to be identified. Of particular interest are the octanoic (C8) and nonanoic (C9) monocarboxylic acids and the aromatic carboxylic acids. Although these compounds are not sufficiently amphiphilic to produce bilayers that are stable at a variety of pH and concentration ranges, we are now able to report that nonanoic acid can form membranous structures at concentrations above 100 mM and at pH values near its pK (6.8 at this concentration), where half the carboxylic-acid head groups are charged and the other half protonated. A light micrograph of the nonanoic acid membranes is shown in Fig. 3B. It is likely that nonanoic acid is among the components of the membrane structures previously observed in meteoritic extracts (Deamer 1985).

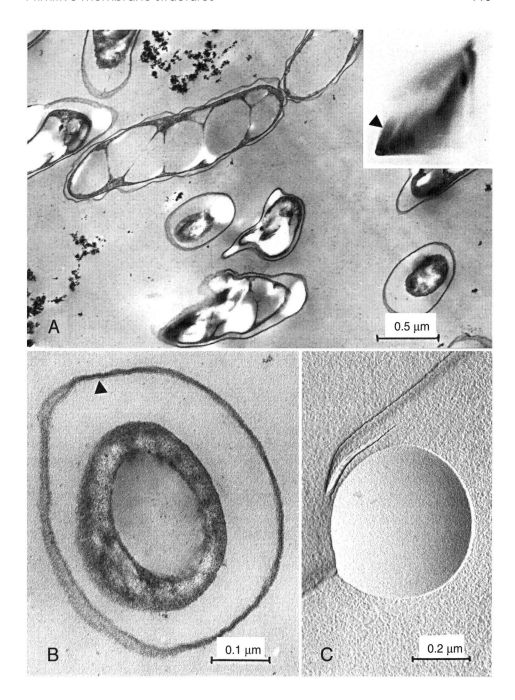

FIGURE 2   Amphiphilic compounds isolated from the Murchison meteorite have the ability to self-assemble into membranous vesicles. **A:** Low magnification; the inset shows a negative image of fluorescent material on a 2-D TLC plate. The membrane-forming compounds were isolated from the fluorescent region just above the origin at lower left. **B:** At higher magnification the trilaminar image characteristic of lipid bilayers can be seen (arrow). **C:** Freeze-fracture images confirm the presence of bilayer strutures.

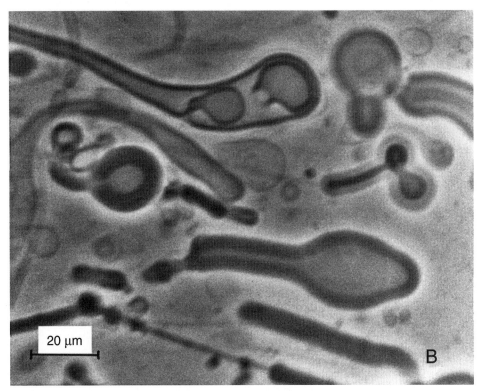

FIGURE 3   The membrane-forming fraction of the Murchison extract has been analyzed by gas-chromatography mass spectrometry (unpublished results by Mautner, White, Ellerbe & Deamer). **A:** Identified aliphatic and aromatic acids. **B:** One of the organic acids – nonanoic acid – at a concentration of 100 mM, pH 6.8. Under these conditions nonanoic acid readily forms membranous structures.

Lipid-bilayer membranes readily encapsulate large molecules

Assuming that primitive membranes self-assembled at some point to produce proto-cellular structures, we can ask how such structures could have encapsulated larger molecules involved in catalysis and information processing. When a small volume is enclosed by the permeability barrier of a membrane, the same membrane that provides an encapsulated microenvironment would also exclude macromolecules. It follows that there must be some reversible process by which the barrier first can be broken, allowing entry of large molecules, then resealed.

An important property of fluid lipid bilayers in this regard is that such barrier membranes are able to break and reseal, thereby encapsulating whatever solutes leak in during the breakage. Liposomes, defined as small vesicles composed of lipid in the form of bilayers, are useful laboratory models of membrane-bounded environments. From our understanding of liposomes, there are several processes that would lead to encapsulation. The first is simply mechanical disturbance: when liposomes are agitated in aqueous dispersion, they break and reseal, thus capturing any solutes in the surrounding solution. A related process involves osmotic gradients. It has been known for many years that various membranous structures swell and burst when exposed to hypotonic solutions. For example, if osmotic salt-concentration gradients are produced across liposome membranes, swelling and rupture of the bilayer occurs. If this is carried out in the presence of solute molecules, they leak inward through the transient rupture. When osmotic equilibrium is reached, the bilayer seals and thereby captures the molecules. It is not difficult to imagine capture of macromolecules on the prebiotic Earth by a similar osmotically driven rupture and resealing of primitive membranes.

Probably the most robust encapsulation mechanism involves drying–wetting cycles. When liposomes are dried, they tend to fuse into multilayered structures that sandwich any solutes present. Upon rehydration, the lipid layers form vesicles containing encapsulated macromolecules, as much as half of the original solute (Fig. 4). Similar drying–wetting cycles must have occurred in the prebiotic environment, particularly at intertidal zones, so that encapsulated systems of large molecules might have been reasonably common if membranogenic amphiphiles had been available.

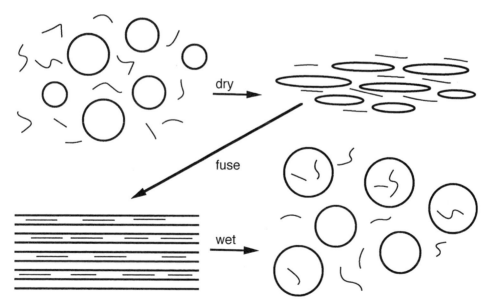

FIGURE 4   Large molecules such as nucleic acids and proteins are efficiently encapsulated by lipid vesicles undergoing cycles of drying and wetting. During drying, vesicles become highly concentrated and finally fuse. The result is a lamellar sandwich of lipid and solute molecules, indicated here as short lines representing nucleic-acid molecules. Upon rehydration, vesicles are reconstituted from the lamellar phase, and approximately half the solute molecules are encapsulated. (Deamer & Barchfeld 1982.)

Permeability constraints on primitive cell functions

As noted in the previous discussion, a membrane-enclosed volume provides the encapsulated microenvironment necessary for cell function, but at the same time limits access to nutrients required for cell growth. How would an early cell have solved the problem of selective permeability?

First, it might be helpful to have some perspective on typical membrane permeability ranges. For example, if we compare the rates at which water and ions diffuse across a lipid bilayer, we observe that small, uncharged water molecules undergo passive transport at rates approximately a billion times faster than ions. This means that water can equilibrate across liposome membranes in microseconds, while a sodium-ion gradient would require hours to reach equilibrium. The immense permeability barrier to ionized solutes is essential, since life processes depend on the lipid-bilayer barrier to maintain ion gradients required for cell function.

Most common nutrients in solution are ionized, primary examples being amino acids and phosphate. The permeability of liposome membranes to amino acids and phosphate has recently been investigated (Chakrabarti & Deamer 1992). Table 2 summarizes some of the permeability coefficients, which were in the range of 10^{-11}–10^{-12} cm·s^{-1}, similar to values for sodium, potassium, and chloride ions. Such extremely low permeability coefficients could present a significant problem to early cells. For example, bacteria such as E. coli can reproduce every twenty minutes, which means that they must transport enough phosphate across their membranes to double the amount of nucleic acid present in that time interval. To perform this remarkable growth, bacteria have carrier enzymes that capture phosphate and transport it across the bilayer barrier. If primitive prokaryotic cells on the early Earth depended on passive phosphate transport across bilayer membranes that lacked a carrier enzyme, the low permeability coefficient would inhibit growth to the extent that they could only reproduce once every year or so. Given geological time scales and essentially no competition for resources, this might not seem to be a significant difficulty. However, it should be kept in mind that self-assembled molecular structures are subject to a variety of dispersive and degradative effects, so that they are metastable at best. It seems improbable that a primitive cell could maintain a given structure if growth processes were too time-consuming.

Fortunately, it is not difficult to imagine solutions to this problem. Our measurements of amino acid and phosphate permeability coefficients used purified phospho-

TABLE 2   Permeability coefficients of ionic solutes. The amino acid measurements (Chakrabarti & Deamer 1992) were carried out at pH 6.0 in liposomes composed of egg phosphatidylcholine (EPC) or synthetic dimyristoylphosphatidylcholine (DMPC). The phosphate measurements were performed with EPC liposomes at pH 4.0, where phosphate exists as a monoanion. Permeability values for sodium and chloride ions are from Mimms et al. 1981 and are given for comparison. Units: P = cm s^{-1} × 10^{-12}.

	Gly	Lys	Ser	Try	Phe	Pi$^-$	Na$^+$	Cl$^-$
EPC	5	3	10	410	250	5	0.95	76
DMPC	16	18	11	–	–	–	–	–

lipids with chain lengths around 18 carbon atoms. Simply shortening the hydrocarbon chains to 12 atoms would increase permeability by several orders of magnitude. Furthermore, early amphiphiles were likely to be present as complex mixtures, and it is well known that modest amounts of impurities also increase permeability manyfold. We suggest that the earliest membranes were self-assembled mixtures of amphiphiles containing short chains, about 12 atoms. Such membranes would be relatively permeable to ionic solutes and relatively impermeable to macromolecules. Nutrient solutes would then have access to the cellular microenvironment to take part in polymerization reactions involved in growth, while larger polymeric molecules would be maintained in the vesicle interior.

Pigment systems in the prebiotic environment

Sunlight is the most abundant energy source available on the present Earth and was presumably equally abundant at the time of life's origin. Light energy today is captured by the pigment systems of plants and results in the loss of an electron from the electronic structure of chlorophyll. The electron carries with it the energy of the original photon of light that was absorbed. When it is donated to an acceptor molecule in the chloroplast membrane, the acceptor acquires chemical energy. This seemingly simple reaction is the energy source for all life on Earth.

Was light the original energy source driving prebiotic evolution toward the first living systems? If so, pigment molecules must have been available in the prebiotic environment. However, plausible pigments are not at all obvious. Chlorophyll itself is a highly evolved porphyrin–magnesium complex. Simpler porphyrins may have been precursors during cell evolution (Mercer-Smith & Mauzerall 1984), but there is no prebiotic reaction pathway to porphyrins for the ancestral protocell.

Contemporary photosynthetic systems have the ability to capture light energy, then use the energy to strip electrons from water and donate them to carbon dioxide. Although simple in concept, this has proven to be very difficult to reproduce in the laboratory, let alone in plausible prebiotic scenarios. It is easier to imagine that a prebiotic pigment system might be able to carry out individual steps of modern photosynthesis, with the expectation that the reaction would make some form of energy available that could be used by a developing protocellular system. For example, if a pigment is present in the lipid environment of an early membrane, upon illumination a proton gradient could be produced by a relatively simple photochemical reaction. The gradient could then be used to drive a number of useful transport processes.

Two laboratory models of such a system have been established. The first involves illumination of an encapsulated complex ion, ferrocyanide (Deamer & Harang 1990). Complex ions of iron are plausible components of early oceans, and their chemistry has been the subject of numerous investigations. Although ferrocyanide is not usually considered to be a pigment, ferrocyanide solutions in fact absorb near-UV light in a photochemical reaction that causes cyanide radicals to be lost from the iron complex. Because cyanide is a weak acid (pKa=9.2), it associates with protons at pH ranges between 6 and 9, thereby causing the pH of the solution to increase. The pH change is proportional to the photon flux and ferrocyanide concentration. For instance, the pH of an unbuffered 1.0 mM potassium ferrocyanide solution increases from 6 to 9 within seconds upon illumination.

This preparation offers a convenient system for generating pH gradients across membranes and for studying their possible role in primitive energy-transduction processes. In particular, it would be interesting if an illuminated ferrocyanide solution could provide sufficient energy to drive covalent bond formation. To this end, we initiated a research collaboration with Herrick and Margareta Baltsheffsky in Stockholm and were able to demonstrate (unpublished results) that illumination of ferrocyanide in the external medium of isolated *Rhodospirillum rubrum* chromatophores can result in pH gradients up to 3.5 pH units across the chromatophore membrane. The pH increase occurs in the external medium surrounding the chromatophores, so that hydrogen ions move outward through the proton-ATPase and proton-PPase, thereby providing an energy source for the synthesis of adenosine triphosphate (ATP) or pyrophosphate (PPi). (See Baltscheffsky & Baltscheffsky, this volume, for a review of primitive photosynthetic pathways related to pyrophosphate bond formation.) ATP and PPi synthesis was measured using the novel stabilized luciferin/luciferase method of Nyrén & Lundin (1985). After two minutes of illumination in 0.1 M ferrocyanide, during which a 3.5 pH unit gradient was generated, ATP production was twice that of the controls. No ATP synthesis occurred in 10 mM ferrocyanide, which produced a gradient of two pH units. However, PPi production does not require as much chemiosmotic energy for its synthesis, and we found that PPi synthesis did occur, increasing by 33% within two minutes. These results are summarized in Table 3.

TABLE 3 ATP and PPi (pyrophosphate) synthesis was measured in *R. rubrum* chromatophores using pH gradients established by ferrocyanide illumination. Ferrocyanide was illuminated with white light under conditions where pH shifts of 2 pH units (10 mM ferrocyanide) and 3 pH units (100 mM ferrocyanide) produced pH gradients of the same magnitude across the chromatophore membranes. The gradients provided chemiosmotic energy for ATP and PPi synthesis.

$Fe(CN)_6^{4-}$	ΔpH	ATP (moles $\times 10^{-11}$)	PPi (moles $\times 10^{-10}$)
10 mM	2	0	7.1
100 mM	3	2.7	11

We conclude that chemiosmotic energy produced by illuminated ferrocyanide solutions can provide an alternative to the more highly evolved photosynthetic machinery present in *R. rubrum* chromatophores. Although it is not clear how chemiosmotic energy would drive high-energy bond formation in a hypothetical protocell, the energy made available by chemiosmotic gradients could be used in several other ways. For instance, sufficiently permeable weak acids or weak bases respond to pH gradients across lipid-bilayer membranes. As shown in Fig. 5, a buffered proton gradient of two pH units can produce a hundredfold concentration gradient of a solute across a membrane (Deamer *et al.* 1972). This means that weak base or weak acid concentrations in the micromolar range can approach millimolar ranges inside lipid vesicles maintaining a pH gradient. Because pH gradients represent one of the most readily available forms of chemiosmotic energy, it follows that such gradients could serve to concentrate

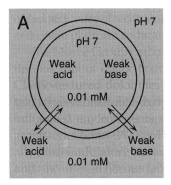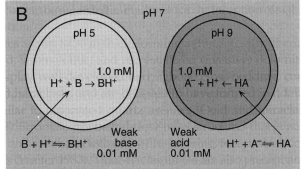

FIGURE 5   Electrochemical proton gradients in the form of pH gradients have free energy that can be used to drive transmembrane transport. In the absence of a pH gradient (**A**) low concentrations of weak acids and bases come to equilibrium across a bilayer membrane. If a pH gradient is present, weak acids and bases accumulate (**B**). For instance, a gradient of 2 pH units can drive a hundredfold concentration of a weak acid or weak base (Deamer *et al.* 1972). The mechanism depends on the fact that the neutral form of the acid or base readily permeates a membrane barrier, while the same barrier is relatively impermeable to the ionized form. From Henderson–Hasselbach theory it can be calculated that the total accumulation of weak acid or base in a vesicle interior will approximately match the magnitude of the pH gradient. Since many organic components of the prebiotic environment would be weak acids or bases, a vesicle system could potentially use pH gradients and selective permeation as a concentrating mechanism.

"nutrients" for primitive cells (Deamer & Oró 1980). One such nutrient is phosphate, presumably an essential component of early metabolic pathways (Westheimer 1987). It would be highly significant if a plausible mechanism for phosphate accumulation could be demonstrated in a vesicle system that uses the chemiosmotic energy of light-dependent pH gradients as an energy source.

Other potential pigment molecules in the prebiotic environment are the polycyclic aromatic hydrocarbons (PAH). As noted earlier, PAH derivatives, largely in the form of kerogenlike polymer, represent more than 90% of the organic material of carbonaceous chondrites. Several examples are shown in Fig. 6. If there was substantial

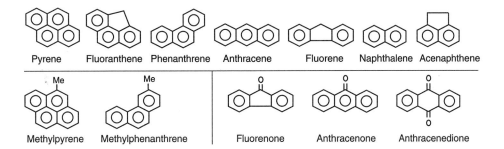

FIGURE 6   Polycyclic aromatic hydrocarbons and their derivatives are present in the Murchison meteorite. (Data from Basile *et al.* 1978 and Krishnamurthy *et al.* 1992.)

survival of the organic content of meteoritic and cometary infall during late accretion, PAH would presumably be major components of the organic inventory on the prebiotic Earth. It is probable that PAH and other hydrocarbons would float as thin films at air–water interfaces and be acted upon by sunlight, with reaction products accumulating in intertidal zones. If some of the products were amphiphiles, they would presumably self-assemble into membrane structures with the potential for further light-driven reactions.

Polycyclic aromatic hydrocarbons and their derivatives all absorb light in the near UV. PAH molecules partitioned into membranes have the potential to capture light energy, either by donating electrons to produce molecules with higher chemical potential or by generating proton gradients. Two such reactions for 1-naphthol and pyrene are illustrated in Fig. 7. Upon illumination, 1-naphthol enters an excited state having a lower pK value, so that protons are transiently released. Such photochemical protons could accumulate in vesicle systems to form chemiosmotic proton gradients.

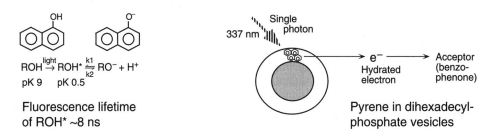

$$ROH \overset{light}{\rightarrow} ROH^* \underset{k2}{\overset{k1}{\rightleftharpoons}} RO^- + H^+$$

pK 9   pK 0.5

Fluorescence lifetime
of ROH* ~8 ns

337 nm — Single photon

e⁻ → Acceptor (benzophenone)
Hydrated electron

Pyrene in dihexadecyl-phosphate vesicles

FIGURE 7   Polycyclic aromatic compounds undergo photochemical reactions that may be relevant to early bioenergetic processes. For example, the pK of 1-naphthol of excited state is much lower than that of the ground state, so that protons are transiently released (Forster 1950; Harris & Selinger 1983). Pyrene, a common component of meteoritic organics, can be placed in lipid bilayers and is then able to photoreduce an acceptor molecule such as benzophenone (Escabi-Perez *et al.* 1979).

The second reaction is equally interesting. Escabi-Perez *et al.* (1979) found that pyrene in dihexadecyl phosphate bilayers absorbs light energy and releases a hydrated electron, which in turn is taken up by an acceptor such as benzophenone. Pyrene is normally unable to participate in light-initiated redox reactions. Apparently the hydrophobic environment of the bilayer lowers the ionization potential by 2.49 electron volts, thereby making possible the loss of an electron to an acceptor. This reaction, in which a hydrophobic membrane environment plays an essential role, represents a photochemical pathway with clear relevance to light-energy capture by a primitive pigment system.

With this in mind, we have investigated two PAH model systems. The first consists of mixtures of PAH with alkanes, which are dispersed as microdroplets in dilute salt solutions (Fig. 8). Upon illumination of the alkane system, marked acid pH shifts were readily observed. For instance, 2-ethyl anthracene, dissolved in dodecane and dispersed by sonication in 0.1 M NaCl, produced pH shifts up to two units upon illumination with a Zeiss 75 W filtered mercury arc lamp (Fig. 9). Other PAH derivatives also released protons but were somewhat less reactive.

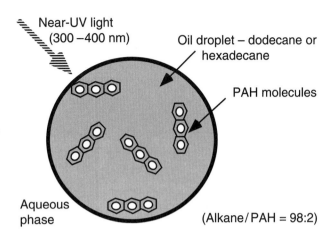

FIGURE 8 As a model system for studying the photochemistry of polycyclic aromatic hydrocarbons, compounds such as pyrene, fluoranthene, and alkyl anthracenes are dissolved in alkanes, then dispersed by sonication. Potential reactants in such a system include PAH, alkanes, water, and molecular oxygen, and potential products include oxidized alkanes, oxidized PAH, and hydrogen ions.

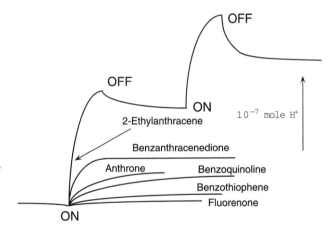

FIGURE 9 The pH of the system illustrated in Figure 8 was followed with a glass electrode, using a variety of PAH derivatives as potential light-energy acceptors. Of the PAH tested so far, 2-ethyl anthracene was the most active in hydrogen ion production.

The second system was designed to test whether protons derived from PAH could be used to transduce light energy into a useful form. To this end, PAH derivatives were included in lipid-bilayer membranes (liposomes) prepared from phosphatidylcholine (Deamer 1992). Upon illumination of the system, protons were produced and accumulated inside the vesicles, with the result that substantial pH gradients were established across the membranes, acid inside.

The production of protons in an anaerobic aqueous system of dispersed hydrocarbons was unexpected, and we do not yet understand the underlying mechanism. Nonetheless, it is clear that PAH dissolved in hydrocarbon environments can absorb light energy, then undergo a photochemical reaction that ultimately releases protons. If PAH derivatives are included in the bilayer membrane of lipid vesicles, the protons accumulate within the membrane-bounded volumes to form proton gradients. This system provides a useful model of a primitive photochemical reaction in which light energy is transduced into potentially useful chemical structures and ion gradients.

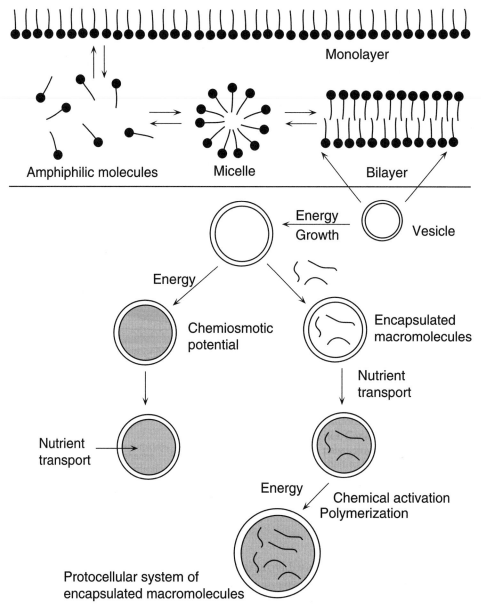

FIGURE 10   The role of amphiphiles and encapsulating vesicles in prebiotic evolution is summarized here. Amphiphiles were presumably present in the prebiotic environment as self-assembled structures such as the monolayer, micelle, and bilayer shown in the upper figure. The structure most pertinent to the origin of cellular life would be vesicles formed from bilayers. To participate in the earliest life forms, such vesicles must have had the capacity to grow. An energy-dependent process in which amphiphilic compounds are added to the membrane is shown here. A second process is shown in the left branch of the lower figure, in which a chemiosmotic potential is generated that can be used to drive some form of nutrient transport. In the right branch, vesicles are shown first encapsulating macromolecules, as illustrated in Fig. 4, then generating a chemiosmotic potential that is coupled to nutrient transport and chemical activation. If the macromolecules could use the activated monomers to drive further polymerization reactions, such a bounded system would be an important step toward the first forms of cellular life.

Summary

We can now attempt to incorporate the observations described above into a series of steps that would lead to a protocell with bioenergetic functions (Fig. 10).

The first step is the formation of amphiphiles in aqueous environments. Primary sources of amphiphiles are not yet established, but small amounts of amphiphilic compounds have been demonstrated in carbonaceous meteorites, as discussed earlier, so it is reasonable to expect that amphiphiles might have been available on the prebiotic Earth. The amphiphiles would have self-assembled into any of several supramolecular structures, including monolayers, micelles, and bilayer vesicles.

Given a vesicle with a boundary membrane composed of amphiphilic bilayers, some form of growth process is necessary if that vesicle is to participate in the growth of a more complex bounded system of macromolecules. The simplest possible growth process is an energized addition of amphiphiles to the vesicle. For example, long-chain hydrocarbons might act as a kind of "nutrient" for the vesicle system, first being oxidized to amphiphilic derivatives such as alcohols and carboxylic acids and then incorporated into the vesicle by partitioning mechanisms. A vesicle system capable of growth has been explored in concept by Morowitz et al. (1988, 1991).

The first protocell was by definition a membrane-bounded system of catalytic and replicating macromolecules, and it follows that a mechanism for capturing macromolecules would have been required. In the earlier discussion, we described how lipid-bilayer membranes readily encapsulate macromolecules by hydration–dehydration cycles that would have been common in the prebiotic environment.

Vesicles must also have some way to capture energy. The most direct pathway would be absorbtion of light energy by a pigment system associated with the vesicle, and two such model pigment systems – ferrocyanide and PAH – were described here. Both systems capture light energy in the form of a chemiosmotic proton gradient.

Finally, for growth of the encapsulated system to occur, some energized process for transport of potential nutrients is necessary. A plausible source of the required energy is a transmembrane proton gradient, and such gradients can drive the accumulation of potential nutrients in the form of weak acids and weak bases from the surrounding environment (Deamer & Oró 1980).

Acknowledgments. – The original research discussed here was supported by NASA Grant NAWG-1119. G. Bosco was supported by a NASA Planetary Biology Internship during preliminary investigations of amino acid permeation of bilayer membranes.

Vitalysts and virulysts: A theory of self-expanding reproduction

Günter Wächtershäuser

Tal 29, D-80331 Munich, Germany

A theory of the earliest mode of evolution is presented. As the point of departure, it assumes an autocatalytic reproduction cycle that multiplies the food acceptor with every turn. This primordial organism persists as long as nonreproductive branch reactions do not exceed reproduction. An inheritable variation occurs by a branch product with a dual catalytic feedback into the production cycle and into its own branch pathway. Each such inheritable variation constitutes an expansion of the reproductive network. Evolution is seen as consisting of a concatenation of such expansions. The coenzymes, the nucleic acids, and the genetic machinery are all seen as examples of catalysts with such a dual feedback.

The scientific marriage between chemistry and biology is a most protracted affair. Chemists of the 17th century classified their field into mineral chemistry, vegetable chemistry and animal chemistry. Berzelius (1806) was impressed by the chemical similarities between plants and animals. He combined plant and animal chemistry into organic chemistry, which he distinguished from inorganic chemistry. Based on this distinction, he formulated the central dogma of chemistry of his time:

The generation of organic compounds from inorganic compounds, outside a living organism, is impossible.

It was believed that the synthesis of organic compounds requires a special life force: *vis vitalis*. With Wöhler's urea synthesis (1828) and the subsequent triumphs of organic chemistry, this dogma was abolished. "Organic chemistry" became simply a synonym for "the chemistry of carbon compounds."

Pasteur (1861) established the central dogma of biology:

The generation of a whole living organism from chemical compounds, outside a living organism, is impossible.

It is today commonly believed that Pasteur is right, right for all known forms of life. But it is also commonly believed that Pasteur is wrong, wrong for a hypothetical first organism, which must have sprung from nonliving matter. The first scientific approach to demonstrate this point is mainly due to the work of two men, the theoretician Oparin (1924) and the experimentalist Miller (1953): the concept of an origin of life in a prebiotic

Bengtson, S. (ed.) 1994: *Early Life on Earth. Nobel Symposium No. 84.* Columbia U.P., New York

broth. This whole theory is based on another dogma, which I shall call the central dogma of the prebiotic-broth theory:

The generation of the first living organism from inorganic compounds is impossible.

It is believed that the first living organism could have arisen only from sufficiently complex organic building blocks. This means that the first organism must have been a heterotroph. Gerald Joyce has stressed this dogma as recently as 1988: "It is almost inconceivable that the first organism was anything but a heterotroph." But where did the building blocks for such a hetero-origin come from? The answer: from the cauldron of the prebiotic broth, where a special kind of chemistry was going on: prebiotic chemistry.

It is the purpose of my paper to combat this view. I will argue that an autotrophic origin is indeed conceivable, an origin of life from inorganic scratch, and that the bogey of a prebiotic chemistry can be abolished.

The primordial driving force of evolution

If we ever create the first form of life in a test tube, by what will we know it? We will know it by the one property that ties together all the known forms of life: *variable reproduction.* This is what the biologist would say. The chemist would call it by a different name: synthetic chain reaction with branchings. Such a chain reaction is frequently called *autocatalytic cycle.* A synthetic autocatalytic cycle means reproduction, and branching reactions are related to the variations for evolution.

Let us first consider primordial reproduction. Here we come across a fundamental alternative. It is the alternative between two ways of life: autotrophy and heterotrophy. This gives biology its primary problem. Was the first organism an autotrophic producer (auto-origin) or a heterotrophic consumer (hetero-origin)? Any testable theory within one of these two alternatives has to specify a synthetic autocatalytic cycle, a reproduction process, not in formal terms but in specific chemical terms. The hetero-origin scenario has dominated biology now for more than 60 years. But all the specific suggestions regarding autocatalysis are limited to nucleic acid replication: The first organism is supposed to be something like a "living RNA molecule," an extreme food specialist – more extreme than any heterotroph known today. It requires a broth of activated nucleotides. This suggestion has come to be known under the trademark *RNA world* (Gilbert 1986).

The auto-origin alternative postulates an autocatalytic carbon-dioxide fixation cycle. It therefore requires a strong reducing agent that must at once be specific enough to be nondestructive and potent enough to support an evolution of a reductive metabolism. This brings me to my main thesis:

The first organism is a chemoautotroph. It uses the formation of pyrite from hydrogen sulfide as a source of electrons and as its energy source.

$FeS + H_2S \rightarrow FeS_2 + 2H^+ + 2e^-$

This reaction has a standard potential of $E^{o'} = -620$ mV, more than enough for all biochemical reductions. This energy source has an important characteristic. It is a redox energy source. It can therefore easily exhibit a kinetic inhibition. Such an inhibition is required for maintaining a high chemical potential. The first energy source for life was a continuously available, endless battery (Wächtershäuser 1988a, 1990b).

Now I come to an important consequence. The proposed energy source produces a pyrite crystal. Such a crystal has positively charged surfaces. The immediate products of carbon-dioxide fixation are anionic. They are negatively charged. This means that they will accumulate on the pyrite surface in *statu nascendi*. Here they can undergo subsequent (recursive) reductive reactions. This establishes a two-dimensional surface-reaction system. By the laws of chemistry the bonding between positively charged ferrous ions and negatively charged groups ($-S^-$, $-COO^-$, $-OPO_3^{2-}$) is particularly strong, akin to chemisorption. This causes a degree of thermodynamic isolation. It establishes a flow-through system with an input of inorganic nutrients, an output of detached organic products of decay, and an "internal" surface-reaction system. Such a system tends to favor surface-bonded polymers over surface-bonded monomers. It shows an increased thermal stability (compared to a solution). It even requires a higher temperature for kinetic reasons. It is inherently orderly due to the vectorial nature of surface bonding. And it may give rise to biochirality by a chiral pyrite crystal structure (Wächtershäuser 1988b, 1991).

The primordial reproduction process

Let us now turn to my specific proposal for an autocatalytic carbon-dioxide fixation. The field of known organic chemistry does not give any clue in this direction. Therefore, we have to resort to biochemistry.

If we follow the anabolic pathways backwards, we arrive at several carbon-fixation pathways: the Calvin-Benson cycle, the reductive citric acid cycle, the reductive acetyl-CoA pathway, the formation of glycine, and the purine pathway. Of all these pathways the reductive citric acid cycle is most promising. It is found in archaeal and bacterial organisms. It is most central, providing the starting materials for almost all anabolic pathways. And it is autocatalytic. It doubles the number of CO_2 acceptors with every turn.

Succinate → citrate	(growth and reduction)
citrate → oxaloacetate + acetyl-CoA	(cleavage)
acetyl-CoA → oxaloacetate	(growth and reduction)
2 oxaloacetate → 2 succinate	(reduction)
succinate → 2 succinate	

This cycle can be best understood as a molecular growth process that leads to structures of increasing instability. At the level of citrate the inherent instability causes cleavage. The RNA replication cycle has the opposite character. The growth of the double-stranded RNA leads to a decreasing tendency to cleave by strand separation.

I propose that an archaic carbon-fixation cycle can be reconstructed from the extant reductive citrate cycle by applying the following principles of retrodiction.

1 Replace all reducing agents with FeS/H_2S.

2 Replace thioester activation with thioacid activation.

3 Replace carbonyl groups with thioenol groups.

With these rules we arrive at an archaic autocatalytic carbon-fixation cycle. It is inherently synthetic. The proposal is of course speculative, just like the proposal of an enzyme-free RNA-replication cycle. But it does not require any geochemically implausible assumptions such as that of a prebiotic broth. It merely requires geochemically unproblematic starting materials: FeS, H_2S, and CO_2 (Wächtershäuser 1990a).

Inheritable variation and selection

It has frequently been assumed that the replication of a polymer sequence with replication errors is the only possibility of an inheritable variation. It will now be shown that this principle of variable replication is a special case of a more general principle of inheritable variation. Let us begin with a well-known fact of chemistry. Most chemical reactions of organic compounds have side reactions. From the point of view of an autocatalytic cycle, such side reactions are reactions of decay. They terminate the reaction chains. If they predominate, the autocatalytic cycle will be killed.

Many products of side reactions are not (or are only poorly) capable of surface bonding. They disappear into the water phase. This means that the pyrite surface has a self-sorting effect. Differential surface bonding is the earliest mode of selection. This explains the polyanionic character of so many constituents of the central metabolism: polycarboxylates, anionic peptides, phosphorylated sugars, RNA, DNA, polyanionic coenzymes. Polyanionic constituents are excellent surface bonders for the positively charged pyrite surfaces. They are seen as the result of chemical selection (Wächtershäuser 1988b).

Surface-bonded side-reaction products are detrimental in two ways. They are products of decay for the autocatalytic production cycle, and they are surface repressors. Occasionally, however, one of the side-reaction products may be converted into a derivative, A_i, which exhibits a catalytic effect i for a slow step in the autocatalytic production cycle (Fig. 1A). Such a conversion will be induced by the chemical environment. It turns a negative feedback into a positive feedback. Now the autocatalytic production cycle turns with an accelerated pace. This has important consequences. Temporally speaking, the primitive primordial cycle is replaced by an expanded network comprising the primitive cycle and a grafted positive-feedback loop. Spatially speaking, the production network can now persist in areas in which the primitive primordial cycle could not persist on its own.

The expanded production network can only persist as long as the catalytic product A_i is being produced – as long as the chemical environment is favorable for A_i. As soon as this favorable chemical environment disappears, the expanded network shrinks back to the primitive cycle. To overcome this limitation, we introduce an additional

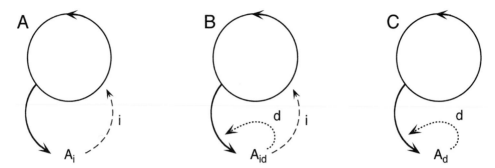

FIGURE 1   **A:** Simple catalyst A_i with indirect positive feedback. **B:** Vitalyst A_{id} with indirect positive and direct positive feedback. **C:** Virulyst A_d with direct positive feedback.

assumption. We assume that a side product, A_{id}, is catalytic for a whole class of reactions, and we assume that this class of reactions includes not only a slow step in the primordial production cycle but also a slow step in the branch pathway to A_{id} itself (Fig. 1B). While A_i is only indirectly autocatalytic by promoting the production cycle, A_{id} is also directly autocatalytic by promoting its own branch pathway. This means that the expansion by A_{id} is truly persistent (Wächtershäuser 1992).

Returning for a moment to the generation of A_{id} de novo, we note that this event may be described as rare in two senses. In a constant chemical environment A_{id} may be generated de novo by a low-propensity reaction. In a fluctuating chemical environment A_{id} may be generated de novo by a high-propensity reaction, caused by a low-propensity fluctuation in the chemical environment. In either event, A_{id} (and the expanded network), once ignited de novo, will persist by virtue of the direct feedback d of A_{id} into its own branch pathway. This has far-reaching consequences. The rare generation of A_{id} de novo can be explained by the actual chemical condition alone. Subsequently, however, the existence of A_{id} (and of the expanded network) cannot be explained by the actual chemical conditions alone. The explanation has to refer additionally to past events under past chemical conditions. The chemist calls such an effect a memory effect. It is the physical basis for inheritance. In this way the catalysts A_{id}, with dual feedback, form the material basis for the evolution of life. Therefore, they will be called "vitalysts." Each new vitalyst produces an expansion of the autocatalytic network.

So far we have considered only simple catalysts A_i with an indirect positive feedback i and vitalysts A_{id} with an indirect positive feedback i and a direct positive feedback d. Now we complete the picture by a third form of catalysts, A_d, with a direct positive feedback d into their own branch pathway, but without an indirect positive feedback into the production cycle (Fig. 1C). They promote the process of decay, and, therefore, they must be considered a self-aggravating burden for the autocatalytic production network – a destructive parasitic loop. Catalysts of this kind will be called "virulysts." Chemical spaces giving rise to a virulyst bring the organism into a diseased state, ultimately killing it.

As long as the organisms of early evolution are open autotrophic surface organisms, new vitalysts, once ignited, will rapidly spread through the biosphere. This establishes the unity of biochemistry. With the invention of cellular organization and biochemical isolation, the biosphere becomes increasingly diversified and segregated into individuals. This establishes the basis for ecological interaction. Opportunistic heterotrophs develop a practice of taking in organic food produced by others. Obligate heterotrophs are incapable of producing certain necessary vitalysts. They are dependent on taking them in with their food. Such an external vitalyst will be called "vitamin." But virulysts may also be taken in from external sources. Such an external virulyst will be called a "virus."

All the catalytic constituents – simple ones (A_i), vitalysts (A_{id}), and virulysts (A_d) – are products of the production network. They have a minimum complexity owing to two functional requirements: they must have a surface-bonding polyanionic moiety for anchoring on the mineral surface and a catalytic moiety.

Now we should see that the definition of a vitalyst, as a specific catalyst promoting a specific reaction *within* the autocatalytic production cycle(s), is too restrictive. It excludes a whole class of feedback reactions that are positive for maintaining the production network. Let us consider a special charging reaction, that between carbon dioxide and a CO_2 carrier, A'H. This brings us to the following reaction sequence:

$$CO_2 + A'H \rightarrow A'-CO_2H; A'-CO_2H + B \rightarrow B-CO_2H + A'H$$

The carrier A'H is a carboxylation catalyst. Strictly speaking, it operates outside the production cycle(s). It catalyzes a feeder pathway that will be called the "anaplerotic pathway." We may readily see that longer and longer anaplerotic pathways may evolve. A'–CO_2H may be converted in a variety of ways before the C_1 unit is transferred into the production network. It may be reduced as in the (methano)pterin pathways: first to the level of formate; then to the level of formaldehyde; then to the level of methylmercaptane; and finally, by reductive carboxylation to an acetylthioester, which is fed into the reductive citrate cycle.

The coenzymes: Vitalysts from earliest evolution

The coenzymes of extant metabolism are most exquisite catalysts. Each catalyzes a huge class of reactions. And nearly each class of coenzyme reactions comprises steps in the biosynthesis of the coenzyme as well as steps elsewhere in the metabolism. This means that coenzymes are vitalysts. A few examples will illustrate this point:

a Coenzyme A is a modified cysteine isopeptide. It shares with the Cys-units of enzymes a catalytically active sulfhydryl group. One of the many catalytic functions of this group consists of intermediary thioester formation. Thioesters play an essential role not only in the reductive citric acid cycle but also as group activation for the reduction of carboxylate groups to the mercaptan level.

b Biotin is an important CO_2 carrier. Its function is anaplerotic for an important step in the reductive citric acid cycle (pyruvate + CO_2 → oxaloacetate). Its biosynthesis involves several carboxylations, notably the formation of pimeloyl-CoA from malonyl-CoA, which in turn requires biotin for its biosynthesis.

c Thiamine is an Umpolung-catalyst. It promotes the two reductive carboxylations in the reductive citric acid cycle (acetyl-CoA → pyruvate; succinyl-CoA → 2-keto-glutarate). It also promotes aldol-condensations, including the formation of 1-deoxy-2-pentulose in its own pathway.

d Pyridoxalphosphate is the most important coenzyme for biosynthesis or modification of amino acids. It seems to be involved in the biosynthesis of 4-hydroxy-threonine in its own pathway and in the synthesis of glycine, which is a biosynthetic precursor for 4-hydroxythreonine.

e Adenosylmethionine is the universal methylating agent. It carries out eight methylation reactions in the biosynthesis of cobalamine, which in turn is a methyl-transfer agent in the biosynthesis of methionine. Thus, adenosylmethionine and cobalamine are both vitalysts.

f NAD(P)H is a universal hydride-transfer agent. It acts as a reducing agent in the formation of aspartate by the reductive amination of oxaloacetate. Aspartate in turn is the biosynthetic precursor in all three biosyntheses of NAD(P)H.

g The flavins are universal redox mediators between the one-electron redox reactions of the ferredoxins and the two-electron redox reactions (e.g., hydride transfers). Such hydride transfers are involved in the biosynthesis of guanine, the biosynthetic precursor of the flavins.

h The pterins have among their functions the transfer of formyl units. Three of these reactions are involved in the biosynthesis of guanine, one in the glycine synthase reaction and two in the establishment of the purine ring. Guanine is the biosynthetic precursor of the pterins.

The complex coenzyme structures, and their no less complex biosynthetic pathways, reflect their long evolutionary history. This evolution is seen as having had two aspects: (1) an evolution of the catalytic function by functional improvement or functional takeover; and (2) an evolution of the anchoring function. The theory of a pyrite-pulled chemoautotrophic origin of life traces the evolution of coenzyme anchoring back to a polyanionic anchoring on a pyrite surface. The eight anionic groups of siroheme may serve as an example. However, the rate of a surface metabolism is determined by the slow rate of a two-dimensional diffusion on the surface. This is seen as having been overcome by various strategies of mobility increase. One such strategy is the insertion of long pivotal arms between the distal catalytic end and the proximal anchoring end of the coenzyme.

With the enzymatization of the metabolism we see a changeover of the pyrite surface anchoring of the coenzymes to a covalent enzyme anchoring. This occurs either by a co-translational incorporation of a catalytic group into the sequence of an enzyme (e.g., Cys, His) or by post-translational covalent anchoring of a coenzyme. This covalent enzyme anchoring is brought about by anhydride activation (between carboxyl and phosphate groups or between two phosphate groups). The nucleotide groups (e.g., adenosyl) of coenzymes (e.g., CoA) are primarily groups for such activations. In the case of CoA, biotin, and lipoate this is obvious. In the past these activating groups have received an alternative interpretation: as having an origin in RNA structures of a make-believe RNA world (White 1976). There is no biochemical basis for such an assumption.

Quite late in evolution, with the appearance of well adapted and precisely folded enzymes, some covalent anchoring of coenzymes in enzymes comes to be replaced by noncovalent anchoring. This greatly increases coenzyme mobility and metabolic versatility. In coenzymes like flavins, molybdopterin, and NAD(P)H an activating function of the nucleotide group does not seem to be any longer in service of covalent anchoring. Here, noncovalent anchoring has completely taken over. Thus, the overall evolution of coenzymes seems to go in the direction of increasing freedom by modularization and mobilization.

Nucleic acids as glorified coenzymes

By the principles of my theory, nucleic acids are seen as vitalysts, as glorified coenzymes. Their evolution should be understood as following the dual track of coenzyme evolution: an evolution of catalytic function and an evolution of anchoring function. The origin of the anchoring function is easily understood. The polyanionic character of RNA and DNA suggests an origin in a stretched-out pyrite-bonded state. A still earlier nucleic acid has been suggested. It does not consist of a hydrolytically labile phosphodiester backbone. Rather, it consists of a hemiacetal backbone of surface-anchored phosphotrioses. The pendant phosphate legs serve as surface bonders. This hypothetical precursor structure has been termed "tribonucleic acid (TNA)" (Wächtershäuser 1988b).

The catalytic functions of the archaic nucleic acids are attributed to the pendant bases. These are seen as being originally erected not onto nucleotides but rather postcondensationally onto a pre-established polymer backbone. Nucleotide modules, modular replication, transcription, etc. are all seen as latecomers – as late windfall profits of an archaic catalytic function of bases. The nature of the earliest catalytic function is revealed by the purine pathway: an aminoimidazole base, catalytic for acid–base reactions. Similar imidazole structures are exhibited by 3-bonded purines. Much later this function is taken over by the histidine units of enzymes.

With the advent of purines (by postcondensational base modification), base pairing and base stacking appear. Now the nucleic acids can take off from the surface by helix formation and folding. Such folding processes withdraw the catalytic nucleic acids from the surface metabolism. This means that the first nucleic-acid folding process must have been virulytic, a burden for the metabolism. But it produces the material for the ignition of the ultimate vitalyst: the genetic apparatus. This apparatus comprises most complex components, the translation machinery, the transcription machinery, and the replication machinery. The temporal order of the appearance of these components forms one of the great clusters of unsolved problems of biology. Here we can only attempt to sort out and conceptualize some of these problems.

It is frequently assumed that the earliest form of reproduction is an RNA replication of sorts. This view has come to be known by the trademark "RNA world." It assumes that life began with RNA replication and that the process of translation appeared later.

This means that RNA replication must at first have been nonenzymatic. Attempts, however, to demonstrate such enzyme-free replication have convincingly failed (Joyce 1989). Then came the recognition that RNA can be catalytic (ribozymes [Cech 1986]). It can catalyze transesterifications. This gave rise to the speculation that life began with something like a ribozyme for RNA replication (Weiner 1987b). It should not be forgotten that this proposal was made in the context of a prebiotic soup of activated nucleotides (Gilbert 1986). The chance assembly of a replication-ribozyme molecule is assumed to mark the ignition of the first autocatalytic cycle – the RNA-replication cycle. In this RNA-world scenario, however, the assumption of a soup of activated nucleotides fails to make chemical sense. Therefore, we are now experiencing a redefinition of the term "RNA world" as *"life invents* RNA replication before *life invents* translation."* Let us see what this could mean in the context of a chemoautotrophic origin of life.

We have seen that for an autotrophic surface metabolist, the folding of functionless RNA is a cause for decay. Now, a folded ribozyme, catalytic for the replication of its own kind but for nothing else: this epitomizes the concept of a virulyst – a selfish, parasitic RNA that would tend to kill the autocatalytic production cycle of its host metabolism. Before translation and coded enzymes, there would have been only one way of turning such a virulyst into a vitalyst: the replication ribozyme catalyzes not only its own production but also the production of ribozymes that are catalytic for steps in the production cycle(s). However, such an assumption presupposes a high template-copying fidelity. Without such a high fidelity the RNA vitalyst would rapidly relapse into a virulyst by replicating RNA mutants devoid of a positive feedback into the production cycle(s). But the first replication ribozyme would of necessity have a high error rate. Thus, it is hard to see how such a ribozyme-catalyzed RNA replication could have ever taken hold.

There may be a radical solution to this problem. A primitive process of nucleic-acid-catalyzed peptide formation, a placeholder for translation, could have started before a process of replication. For such a process a surface-bonded nucleic acid (later mRNA) would merely be a lateral positioning gadget, devoid of any sequence significance. Such a machinery could be a robust vitalyst. It could produce peptide catalysts for the production cycle(s) and for the branch pathways to the nucleic acid. It would later turn into the ribosomal machinery, which is clearly a vitalyst, producing all proteins, including its own ribosomal proteins.

The machinery for replication (RNA replication, DNA replication; DNA–RNA transcription; reverse RNA–DNA transcription) would become installed after the primitive process of "translation" is already in place. The machinery of replication could then be a vitalyst from the start. It could also give rise to functionless nucleic acids, even to selfish nucleic acids, and in the worst case to a runaway process of virulytic replication. For such a primitive genetic machinery, sequence fidelity would not be an early requirement but, rather, a late achievement.

Protocell life cycles

Leo W. Buss

Departments of Biology and Geology and Geophysics, Yale University, New Haven, Connecticut 06511, USA

Models of the origin of cellular life have traditionally required the simultaneous evolution of metabolism and replication. This quandary arises only if one assumes a life cycle in which the proliferative stage is cellularized. This chapter outlines an alternative life cycle, wherein the origin of a self-maintaining metabolism is decoupled from the evolution of a self-reproducing cell.

Hypotheses regarding the origin of life fall into two broad categories: those emphasizing the evolution of cellular metabolism (e.g., Oparin 1924; Fox & Dose 1977; Dyson 1985) and those emphasizing the evolution of self-replication (e.g., Haldane 1954; Miller & Orgel 1974; Eigen & Schuster 1979). The two perspectives have traditionally been difficult to reconcile, as the "metabolic school" views self-replication as necessarily dependent upon cellular metabolism, and the "genetic school" views the evolution of metabolism as dependent upon the pre-existence of a heritable material (Dyson 1985). This difficulty has been partially resolved by the discovery of the catalytic capacities of RNA, in that RNA catalysis makes it conceivable that a single molecule could both carry genetic information and perform metabolic function (Kruger *et al.* 1982; Guerrier-Takada *et al.* 1983).

A major challenge, nevertheless, remains in understanding how *cellular* life might have arisen. How might a self-replicating molecule, even one capable of catalysis, have come to be encased within a membrane and yet continue to proliferate? As a minimum, any such molecule would require metabolic machinery for importing and exporting materials across the cell membrane, for orchestrating cell division or budding, and for insuring that buds were populated by daughter molecules. While a variety of materials display properties that might partially satisfy these requirements (Fox & Dose 1977), it is difficult to conceive of a protocell acquiring all these properties instantaneously. However, the full suite of functions is surely required for the protocell to behave as an evolutionary unit.

I argue that these difficulties may be simply circumvented by hypothesizing a particular form of life cycle for the protocell. The requirement of the instantaneous origin of metabolism arises only if the cellular phase of the life cycle was, as it is in modern cells, the proliferative phase. If, however, the proliferative phase of the life cycle was a "free-living" molecule and the cellular stage a nonproliferative dispersal stage, the requirement of a pre-existing metabolic apparatus is removed. The cellular phase of the protocell life cycle may have originally served as an adaptation for dispersal.

Bengtson, S. (ed.) 1994: *Early Life on Earth. Nobel Symposium No. 84.* Columbia U.P., New York

Geochemical considerations are relevant here. Modern geochemical interpretations suggest that habitats favorable to the origin of life were not ubiquitous (e.g., Corliss *et al.* 1981; Nisbet 1985). A self-replicatory molecule arising in a rare hospitable environment would not be expected to have survived in the mileu surrounding the uniquely favorable site of origin. To traverse the vast spaces surrounding favorable sites, a self-replicatory molecule must have been buffered from external environmental influences. The suggestion that the protocell served as an adaptation for dispersal through unfavorable environments is not only an appealing resolution of the metabolism problem but it may also be an interpretation effectively mandated by geochemistry.

The proposed life cycle is not merely hypothetical. It occurs today in RNA viruses. The encapsulated form, little more than a protein-coated RNA molecule, is the non-proliferative dispersal phase, and the self-replicatory molecule proliferates only in the presence of environmental replicases. RNA viruses live out this life cycle in the only environment on earth currently rich in environmental replicases – that of living cells. While RNA viruses are surely not the literal phyletic precursors of modern cellular organisms, their life cycle testifies to the inherent plausibility of such a life form.

While the life cycle proposed here provides no indication of how a transition from a nonproliferative cellular stage to a proliferative one might have occurred, it removes the requirement of an effectively instantaneous evolution of metabolism. Understanding the origin of life may be profitably explored by decoupling of the origins of different features of life. The life cycle posited here suggests that the origin of self-maintenance and the origin of self-reproduction need neither have been coincident nor have involved similar preconditions.

Acknowledgments. – I thank N. Blackstone, D. Bridge, M. Dick, and an anonymous reviewer for comments and the U.S. National Science Foundation (BSR-88-05691, OCE-90-18396, 9215183) for support.

Early evolution of genes and genomes

Tomoko Ohta

National Institute of Genetics, Mishima 411, Japan

In the early evolution of genes and genomes, ribozymes and reverse transcriptase probably played important roles. In such a system, reverse flow of genetic information may have been more frequent than in the present DNA world. The interaction between natural selection and reverse flow of genetic information might have provided an efficient environmental feedback to genes and genomes.

It is now almost certain that life was based on RNA at some stage in evolutionary history, which is feasible, since RNA can replicate and catalyze (Lazcano, this volume, pp. 70–80). This theory has led to the startling discovery of various surviving molecular fossils of the ancient world (Joyce 1989). According to the theory, a self-replicating RNA appeared in the prebiotic soup, and this RNA genome acquired various functions that facilitated its own survival. One such function might have included RNA enzymes (ribozymes) facilitating intermediary metabolism and synthetic pathways. However, since proteins are more efficient than ribozymes, the utilization and stable supply of proteins would have been advantageous. Thus, the RNA code system came about in the RNA world. The present code system is highly complicated, and most likely it evolved step by step in the ribonucleoprotein (RNP) world. The first tRNA, primitive tRNA synthetase, and primitive ribosome would have dealt with only one particular amino acid. The RNA genes for these three would have diverged to use other amino acids, since more efficient proteins could thus be made. As the number of RNA genes increased, the RNA genome was converted into the more stable DNA genome.

A very attractive theory states that the protoribozyme and the primitive tRNA synthetase would have evolved by gene duplication of the first replicase and that tRNA would have been a genomic tag that provided an initiation site for replication and that also functioned as a telomere (Weiner & Maizels 1991). From this first replicating RNA, numerous genes have arisen by gene duplication. This scenario is reminiscent of the origin of various gene families observed in present chromosomes of eukaryotes. The members of the present gene families, however, show sequence homologies and have related functions, and they are the products evolved after the genetic machinery was established. In contrast, the origin of the translation system took place in a primitive world, and an enormous period of time would have been needed.

I have been studying evolution by gene duplication from the standpoint of population genetics. The process of acquiring beneficial gene families is highly stochastic, and

Bengtson, S. (ed.) 1994: *Early Life on Earth. Nobel Symposium No. 84.* Columbia U.P., New York

the interaction of natural selection and random genetic drift is important. This chapter examines the theory on primitive genomes just outlined, in view of various facts of contemporary genome organization and within the framework of population genetics, with special reference to the duplication of genetic information.

Evolution by gene duplication

The genomes of higher organisms are quite complex, having long spacers, introns, and multigene families. This complicated organization is the product of evolution, indicating that duplication and other forms of illegitimate recombination are important. In particular, the functional diversity of genes is almost always related to the duplication of genes or gene segments. Numerous examples include the immunoglobulin superfamily, serine proteases, protein kinases, rhodopsin superfamily, homeobox-containing family, steroid hormone receptor family, cytochrome P450, etc. In almost all these examples, functional diversity is enhanced by the modification of gene products through such mechanisms as proteolytic cleavage and differential splicing. Thus, an enormous amount of diversity, like that required for the immune reaction, owes much to multigenic organization.

I performed extensive simulations to show how a useful gene family may be attained. Two basic assumptions of the model are that mutations are random and that both deleterious and beneficial mutations occur. The deleterious mutations act to preserve the structure of gene products and occur more frequently than the beneficial ones, which are for acquiring the new function. A crucial quantity is the ratio, R, of the rate of spread of useful mutations to that of detrimental mutations, as follows:

$$R = \frac{u_+ v_+}{u_- v_-}$$ (1)

where u_+ and v_+ are the fixation probability and mutation rate of beneficial mutations, and u_- and v_- are the fixation probability and mutation rate of detrimental mutations. This ratio is an indicator of how much junk DNA accumulates as the price of acquiring useful genes. In actual gene families, R should not be very small, otherwise only junk would accumulate. Fig. 1 shows the results of my simulations on how new genes accumulate, starting from a single gene. The simulation studies have led to the following conclusions (Ohta 1988a, b):

1 The formation of a useful gene family needs positive Darwinian selection under realistic conditions.

2 The interaction among unequal crossing-over, mutation, random genetic drift, and natural selection is important for the evolution of gene families.

In the next section, let us consider how the above process of acquiring new gene functions would be different in the primitive RNP world.

Properties of RNA genome

One may imagine the properties of the RNA genome as follows:

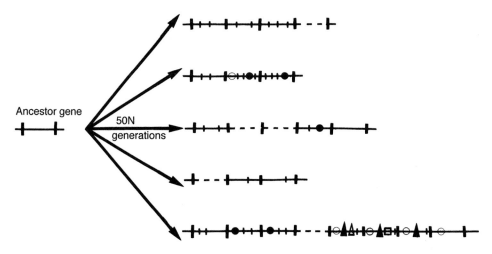

FIGURE 1 Chromosomal organization of genes at the 50Nth generation of simulation experiments, where N is the population size. Each of the five horizontal bars to the right represents one chromosome randomly taken from the population of one run. Broken lines are those having detrimental mutations. Small vertical marks on solid lines are neutral mutations, and circles, triangles, and rectangles represent beneficial mutations. (From Ohta 1988a.)

1 The primitive RNA replicase would have been inefficient because of a high error rate in terms of nucleotide substitutions.

2 The genomic and functional materials are the same, but the two forms of RNA seemed to have become distinguishable at a very early stage. The genomic RNA was likely to have been double-stranded, since the single-stranded RNA is subject to a high rate of spontaneous hydrolysis. Also, the transition from an RNA to a DNA genome would have been possible only if the RNA was a double-stranded structure (Watson *et al.* 1987, pp. 1140–1141). Several different gene functions already existed that had presumably diverged from the first replicase. They might have included a replicase for the RNA genome, a transcriptase that copied genomic RNA into functional RNA, and some metabolic ribozymes. A tRNA-like structure is thought to have served as a genomic tag to punctuate transcription (Weiner & Maizels 1991).

3 The RNA genome size should have increased to allow for more functional diversity, and the RNA replicase itself is suggested to have acted as an RNA recombinase (Watson *et al.* 1987, p. 1124). The duplication of genes or gene segments would have occurred with the help of recombinase that would have improved its function via duplication and differentiation of the replicase gene. An RNA genome would have been more versatile and would have had closer contact with the environment than a DNA one. Of course, less genetic information can be stored in an RNA genome.

In such a primitive world, chance and necessity would have interacted to produce an efficient genome. The most plausible theory states that the first tRNA synthetase and the first ribosome were for a basic amino acid, e.g., lysine, and that these protogenes coevolved to handle other amino acids by duplication and diversification (Wong 1981). In this process, the genetic code was largely determined. It is commonly believed that the genetic code was frozen by accident (Crick 1968). However, an examination of the code table tells us that it is perhaps the "best" code. For example, related codons are used for chemically related amino acids, and the degeneracy in the third position of the codons acts to minimize the deleterious effect of errors at transcription. How could such a remarkable code system have evolved by a frozen accident? One must assume that natural selection had been very efficient even in the primitive RNA world. In the next section, I consider how the efficiency of natural selection might have been increased.

Was reverse flow of genetic information useful?

RNA can be both genomic and functional, and primitive ribozymes were unlikely to be very specific in their functions. Primitive proteins would have helped the catalytic activities of ribozymes. Subject to such a condition, would gene duplication by recombinase, with subsequent accumulation of beneficial base changes, be the sole path for acquiring new genes? Gene duplications and base substitutions are thought to occur randomly, and useful genes are thought to accumulate with the help of natural selection. Here, a very intriguing question arises: Couldn't functional RNA be incorporated into the genome? If a replicase, a recombinase, and a transcriptase are all primitive and not very specific, the incorporation of functional RNA might have happened once in a while. If beneficial base changes happen to occur in a functional RNA, it would help survival of the genome. If the genome lives long, it would have a larger chance of incorporating the functional RNA. Fig. 2 illustrates this situation and compares it with the ordinary process of gene duplication. It is assumed here that, for survival, the

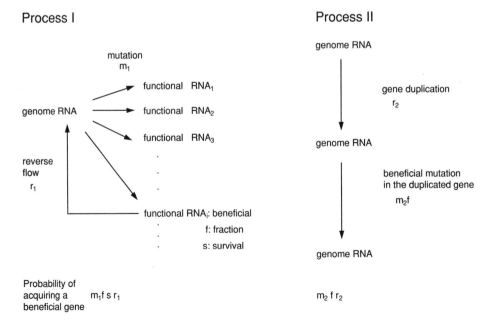

FIGURE 2   The two alternative processes of acquiring a beneficial gene.

beneficial change is required as a new function in addition to the previous one. Let us refer to the acquisition of a new gene by reverse flow as "process I," and to ordinary gene duplication as "process II."

Let m_1 be the rate of mutation at transcription and r_1 be the rate of incorporation or reverse flow of a particular RNA into the genome. For these two parameters and others in the following, I assume that all rates are measured per unit time. Among the mutants, there are various types, and let the i-th type be the beneficial mutant. Let f be the fraction of the beneficial type among the mutants. The genome that has the i-th mutant survives longer, and let s (>1) be the rate of survival of this genome relative to those that do not have the i-th mutant. Then the rate of incorporation of the i-th mutant into the genome per unit time becomes

$$k_1 = m_1 f s r_1 \qquad (2)$$

Ordinary gene duplication with subsequent differentiation (process II) is more commonly accepted as a process for acquiring a new gene. Let k_2 be this rate per unit time. Let m_2 be the genomic mutation rate and r_2 be the rate of gene duplication in a genome. Then k_2 becomes

$$k_2 = m_2 f r_2 \qquad (3)$$

In the primitive world, the error rate of the replicase and transcriptase would not differ much. However, in one cycle of the replication of genomic RNA, more than one transcription would occur, therefore $m_1 > m_2$. The value of s can be much larger than one, if the mutation provides the required function. Then process I would be more efficient than process II, unless r_1 (reverse flow rate) is much less than r_2 (gene duplication rate). Remember that process I is a kind of Lamarckian evolution, since the environmental effect is fed back into the genome.

What about the relative magnitude of r_1 and r_2? I would like to suggest that r_1 is only slightly smaller than r_2, since both genomic and functional RNAs are in a very small primitive cell. Then process I would have been the main path for acquiring new genes. As mentioned above, this is a kind of Lamarckian evolution, but natural selection is also important in that it helps survival of good mutations that have occurred in functional RNA. Hence process I is a combination of Darwinian and Lamarckian evolution. It may also provide a basis for a vague term, *feedback*, that is often used to describe molecular adaptation to the environment.

Does process I still exist in the present DNA world? In higher animals, the germ-cell line differentiates from the somatic cells early in ontogeny, and there would be little opportunity for the reverse flow of genetic information. In lower animals and in plants, however, the germ line may differentiate later from the somatic cells, and such opportunity for reverse flow of information could arise. Of course, for a DNA genome, the reverse flow occurs via reverse transcriptase. There are various examples of reversely transcribed genes, most of which are pseudogenes and not used (Weiner *et al.* 1986). Not many examples are known yet in plants, but reversely transcribed active genes might be found in the future. It should be noted that process I differs from an attractive theory advanced by developmental biologists who emphasize hierarchical selection, such as selection at the somatic cell level, in addition to individual selection (Raff & Kaufman 1983, pp. 336–344; Buss 1987, pp. 171–197; Edelman 1987, pp. 8–22).

Selection at the level of somatic cells would be important in plants in which germ cells differentiate late in ontogeny. In their effects, selection at the cell level and reverse flow might look similar.

A related observation is the evolution of a bacterial plasmid, ColE1. This plasmid has the DNA genome, and the direct RNA products regulate the genome replication. Its evolution is characterized by the nonrandom pattern coming from the RNA folded structure, and the structural change of RNA might possibly have been incorporated into the DNA genome (Tomizawa 1993). The structure of folded RNA is known to have direct effect on survival of the plasmid, and the interaction between DNA and RNA would have been essential for plasmid evolution.

Many basic protein designs appear to have been invented in the primitive world. Efficient metabolisms and anabolisms had been established in conjunction with the introduction of many protein designs. Very efficient and sophisticated biological systems would have been attainable via combined Darwinian and Lamarckian processes in which environmental feedback to the genome would have been easier than in the present DNA world. In the process of acquiring such biological systems, genes had been very versatile, and the step-by-step progress had taken place through feedback. Each step might have required numerous mutations that interacted with each other. Almost all established biological systems are highly complicated interacting sets of reactions, and the simultaneous acquisition of many reactions of any system is impossible. A versatility of the initial stage is the prerequisite for further development by a step-by-step progress.

Versatile genome and function

Versatility is generally observed even in the present highly evolved organisms. Embryologists have long recognized that early ontogeny is rigid and conservative but that late ontogeny is versatile and modifiable in evolution, presumably adding new functions to the slightly changed preexisting systems (but see Raff, this volume). As shown by paleontological studies (e.g., Valentine 1977 and this volume), the basic body plans of metazoans had been invented in the early Cambrian Period, and over the last half-billion years, metazoan development has followed the same basic plans. Thus, it appears that most basic protein designs had been invented early in the unicellular evolutionary stages and that the body plans of animals had been attained by the early Cambrian, during times of great versatility.

It is now being clarified what genetic systems are involved in the body plan. The homeotic genes of *Drosophila* are the most remarkable examples. These genes are activated in a hierarchical fashion and regulate embryogenesis (for reviews, see Dressler & Gruss 1988; Akam 1989). It is now known that the insect homeotic gene complexes and the vertebrate Hox clusters are homologous and that the corresponding genes show the same relative boundaries of expression along the anteroposterior axis of the embryo between the two taxa. Thus, these animals have inherited the same regulatory genetic system from a common ancestor. By comparative studies of these two gene families, a positive correlation between organismal and genetic complexities may be found, i.e. multiple sequences homologous to the homeobox of *Drosophila* have been found in the genomes of echinoderms, annelids, chordates, and arthropods but not in bacteria, slime molds, or yeast (Dressler & Gruss 1988).

Another remarkable example is the immunoglobulin gene superfamily, which emerged along with metazoans. Hood & Hunkapiller (1991) argue that this superfamily may have been one of the driving forces of metazoan evolution, because members of this family perform many important cell-surface interactions that are essential for animal life. These authors provide a model of the branches of the superfamily with relative times of divergence. Fig. 3 presents their model, showing that the basic protein

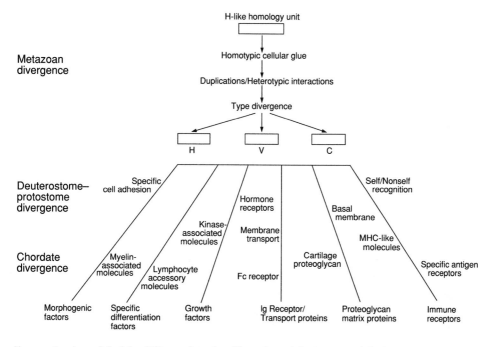

FIGURE 3   A model of the different functional branches of the immunoglobulin gene superfamily with relative times of divergence and emergence of various functions (from Hunkapiller *et al.* 1989).

domains of this family had been established before the deuterostome–protostome divergence, i.e. near the Precambrian–Cambrian transition. They emphasize that the subsequent Cambrian Period contains a bewildering array of different fossils (Gould 1989). In connection with the evolution of complex organization, this superfamily has enormously expanded its function. There are other examples of gene families important for body plan: protein kinases, DNA binding proteins, integrin receptor, and so on. All these examples suggest that the basic protein designs had been invented early in unicellular organisms and that the interaction systems of these proteins started shortly before the Cambrian.

As mentioned before, the reverse flow of useful genetic information seems not to occur in present higher organisms. In the RNP world, process I with a reverse flow of beneficial mutations is thought to be the major path for acquiring a new gene. Was the reverse flow occurring when basic protein designs and body plans were invented? In unicellular organisms without a nucleus, the interaction of genomic DNA and messenger RNA would have been straightforward. The reverse transcriptase, which is thought to have played an essential role for the origin of the DNA world, would still have been active. Thus one may say that a reverse flow of genetic information would have been significant. In contemporary organisms, such a reverse flow results mainly in pseudogenes, simply accumulating junk DNA. However, in early genomes, the proportion of useful genes obtained by reverse flow would have been much higher, because of the interaction between selection and reverse flow. This proportion is thought to have decreased as more complex organisms have evolved.

Selection at various levels

There is interest in examining the origin of mutations in bacteria, and there has been controversy over whether mutations are truly random in a strongly selective environment. Cairns *et al.* (1988) reported that useful mutations occurred more frequently in a strongly selective medium in *E. coli* populations. The special medium contained lactose as a sole energy source, and mutation from Lac⁻ to Lac⁺ occurred more often than expected. These authors speculate that reverse flow of useful messenger RNA might have occurred. Stahl (1988) offered an explanation that, because of the slow repair in nutritionally depleted cells, useful mutations might persist long and be translated. Hall (1990) reports that frequent mutations also occur in the trp operon, if a similar strongly selective environment is provided. He suggests another possibility, namely, that bacteria become hypermutable in the stressed medium and those cells that happen to have good mutations are selected. In any case, such observations indicate that the interaction among genes, metabolism, and environment has some effects on acquiring useful mutations even in contemporary bacteria.

From a different point of view, Holmquist (1989) argued that selection at the molecular level inside a cell or between cells within an individual may be important in higher organisms. For example, copy-number regulation of redundant gene families would be performed by selection among cell lineages. He called such processes "molecular ecology." It may be related to the competition among cell lineages in development (Buss 1987) mentioned earlier. Another related subject is the idea of "neural Darwinism" by Edelman (1987, pp. 8–22). According to this theory, the development of neural networks is accomplished by Darwinian selection among neuron cell lineages. Here the genetic characters of cells in terms of competitive ability interact with the environment, and selection would increase desirable genetic characters in conjunction with the environment.

All these examples force us to reconsider evolutionary theory, i.e. natural selection is not at all as simple as envisioned by neo-Darwinists. It is much more complicated, operating at various levels and in many different ways.

Acknowledgment. – I thank Dr. Stefan Bengtson, Dr. Junichi Tomizawa, and Dr. Christopher Jess Basten for their many valuable suggestions on the manuscript. This is contribution No. 1938 from the National Institute of Genetics.

The lesson of Archaebacteria

Karl O. Stetter

Lehrstuhl für Mikrobiologie, Universität Regensburg,
D-93040 Regensburg, Germany

All the deepest and shortest branches within the phylogenetic tree are occupied by groups of hyperthermophilic Bacteria and Archaea, which appear, therefore, still rather primitive. In addition, these organisms are chemolithoautotrophs. This suggests a thermophilic autotrophic origin of life. Hyperthermophilic autotrophs grow by several types of aerobic and anaerobic respiration, using H_2 or reduced sulfur compounds as electron donors and CO_2, O_2, and oxidized sulfur compounds as electron acceptors. Heterotrophs should be seen as late "opportunistic" consumers of organic matter that had been formed by autotrophic producers.

Today there are several controversial theories about the origin of life and its early mode of expression on Earth (e.g., Oparin 1924; Chang *et al.* 1983; Woese 1979; Wächtershäuser 1988b). Possible strategies of primitive life depended on mainly unknown geophysical and geochemical conditions on the juvenile Earth. Microfossils within early Archean stromatolites demonstrate the existence of life already 3.5 Ga ago (Awramik *et al.* 1983; Schopf & Packer 1987; Schopf 1993, this volume). By their shape, these oldest traces of life show prokaryotic cell organization and resemble some members of recent cyanobacteria. Possibly, life had originated much earlier, probably already after the end of the major period of meteorite impacts about 3.9 Ga ago (Schopf *et al.* 1983). At that time, the Earth had an overall reducing atmosphere and was most likely much hotter than today because of stronger volcanism and radioactive decay (Ernst 1983). Questions arise about possible physiological properties, modes of energy acquisition, and kinds of carbon sources of the earliest organisms. Did a prebiotic broth that contained significant amounts of macromolecules ever exist? In that case, a heterotrophic origin of life would appear most probable, and chemolithoautotrophy should have developed much later. Or, alternatively, were the first organisms already chemolithoautotrophs, able to gain their energy by an inorganic chemical reaction and to build up their organic material from CO_2 as single carbon source? Such an organism was postulated by the "surface metabolism" theory (Wächtershäuser 1988b). Concerning physiological properties of a primitive organism, a heat-loving (thermophilic), possibly halophilic anaerobe appears most probable under the view of the gross global environmental conditions at that time. However, since the scenario of the origin and the first biotopes of life are unknown, locales with moderate temperatures or the availability of traces of (abiotically formed) free oxygen have to be taken into consideration, too (Walker *et al.* 1983; Towe, this volume).

Bengtson, S. (ed.) 1994: *Early Life on Earth. Nobel Symposium No. 84.* Columbia U.P., New York

During the last years extremophilic prokaryotes have been isolated from environments some of which are reminiscent of the primitive Earth and may have remained chemically and physically almost unchanged during billions of years. A great deal of extremophilic organisms are hyperthermophiles that exhibit an optimal growth temperature between 80 and 110°C. Therefore, they represent an as yet unknown dimension of extreme thermophily (Stetter 1983, 1986, 1992; Stetter *et al.* 1990). Powerful molecular techniques have recently been developed to investigate the phylogeny of life (Zuckerkandl & Pauling 1965; Woese *et al.* 1976; Woese *et al.* 1983). As a consequence, for the first time the universal tree of life has been established (Woese 1987b; Iwabe *et al.* 1989; Woese *et al.* 1990). Therefore it is now possible to follow the traits of evolution even for prokaryotes and to identify still rather primitive lineages (Woese *et al.* 1990; Burggraf *et al.* 1991; Burggraf *et al.* 1992). Here, I present evidence about the physiology and metabolic properties of members of the deepest branches within the phylogenetic tree, which are all prokaryotes. Most interestingly, some of them also represent extremely short lineages, indicating a very slow clock of evolution. Therefore, they may still be rather primitive. Most of these organisms belong to the "archaebacteria" (Archaea), several to the "eubacteria" (Bacteria; Woese *et al.* 1990).

Phylogenetic tree and the testimony of hyperthermophiles

The universal phylogenetic tree exhibits a tripartite division of the living world into the bacterial ("eubacterial"), archaeal ("archaebacterial"), and eucaryal ("eukaryotic") domains (Fig. 1; Woese & Fox 1977; Iwabe *et al.* 1989; Woese *et al.* 1990). The root was inferred from phylogenetic trees of duplicated genes of ATPase subunits and elongation factors Tu and G. Deep branches are evidence for very early separation. For example, the separation of the Bacteria from the Archaea – Eucarya lineages represents the deepest and earliest branching point within the phylogenetic tree (Fig. 1). Short lineages indicate a rather slow evolution of the organisms they are harboring. The root of the tree had been determined by sequence comparisons of the duplicated ATPase gene (Iwabe *et al.* 1989; Gogarten *et al.* 1989). In contrast to the Eucarya, the bacterial and archaeal domains within the universal tree exhibit some extremely deep and short branches (Fig. 1). Based on the unique phylogenetic position of *Thermotoga*, a thermophilic ancestry of the eubacteria had been taken into consideration (Achenbach-Richter *et al.* 1987a). Very surprisingly, as a rule, all deep and short lineages are represented by hyperthermophiles (for example: the Thermotogales and Aquificales within the Bacteria; the Pyrodictiales, Desulfurococcales, Sulfolobales, Thermoproteales, Methanopyrales, and Thermococcales within the Archaea; Fig. 1). The shortest branches are represented by the most extreme hyperthermophiles known so far (Pyrodictiales, Methanopyrales, Aquificales). On the other hand, mesophilic and moderately thermophilic Bacteria and Archaea represent long lineages within the phylogenetic tree (for example: gram positives, purple bacteria, Cyanobacteria within the Bacteria, and Halobacteria and Methanosarcina, etc. within the Archaea; Fig. 1). The Methanococcales and Methanobacteriales contain hyperthermophiles in addition to moderately thermophilic (e.g., *Methanococcus thermolithotrophicus*, *Methanobacterium thermoauto-*

trophicum) and mesophilic (e.g., *Methanococcus vannielii, Methanobacterium uliginosum*) species (Fig. 1). However, as a rule they exhibit a faster evolution (as indicated by longer lineages) and were able to adapt already to colder environments. On the basis of these observations, it is clear that hyperthermophiles most likely have existed since the origin of life and may still be rather similar to their primitive ancestors.

In the following, under the view of suggestions and possible conclusions for early life on Earth, I will concentrate on hyperthermophiles, their recent environments, and their modes of life.

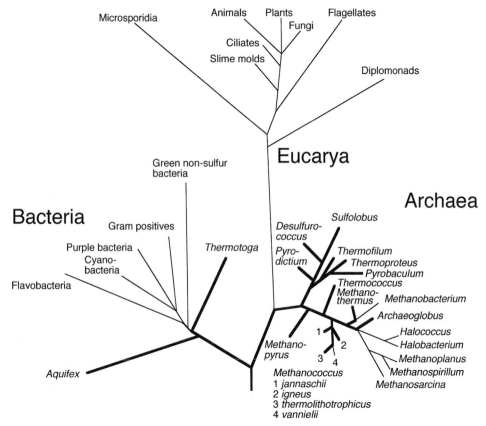

FIGURE 1 Universal phylogenetic tree. Bold lines: Hyperthermophiles. Schematically redrawn and modified from Woese *et al.* 1990; root according to Iwabe *et al.* 1989.

Hyperthermophiles – biotopes, physiology, and metabolism

Biotopes

Hyperthermophilic prokaryotes have been isolated from geo- and hydrothermal areas. Terrestrial biotopes are mainly solfataric fields. They consist of soils, mud holes, and surface waters heated by volcanic exhalations from magma chambers below. Solfataric soils are usually acidic (pH 0.5–6) and rich in sulfate and elemental sulfur at the surface and appear ocher-colored because of the presence of ferric iron (Stetter *et al.* 1986). In the depth, solfataric fields are weakly acidic to neutral (pH 5–7) and have a blackish appearance caused by ferrous sulfide. Marine hydrothermal systems are situated in shallow and abyssal depths. They consist of hot fumaroles, springs, and deep-sea vents ("smokers") with temperatures up to 400°C. Further submarine biotopes for hyperthermophiles are active seamounts (Huber *et al.* 1990). Submarine hydrothermal systems, similar to seawater, usually contain high concentrations of NaCl and sulfate and exhibit a slightly acidic to alkaline pH (5–8.5). Because of the low solubility of oxygen at high temperatures and the presence of reducing gases, most biotopes of hyperthermophiles are anaerobic or contain only traces of oxygen. Major components within volcanic exhalations are steam, CO_2, H_2S, S^0 (formed by oxidation of H_2S at the surface), some CO, hydrogen, methane, nitrogen, and traces of ammonia.

Physiological properties and energy-yielding reactions

Hyperthermophiles are adapted to distinct environmental factors including neutrality, high salinity, and low redox potentials within marine hydrothermal systems or acidity, low salinity, and higher redox potentials at the surface of solfataric fields.

Terrestrial hyperthermophiles. – The hot, oxygen-rich, acidic upper layer of terrestrial solfataric fields almost exclusively harbors extremely acidophilic hyperthermophiles. They consist of coccoid-shaped aerobes, facultative anaerobes, and strict anaerobes (Table 1) and are extreme acidophiles by their requirement of acidic pH (opt ≈ pH 3). Phylogenetically, they belong to the archaeal genera *Sulfolobus, Metallosphaera, Acidianus,* and *Stygiolobus*. Together, these genera form the Sulfolobales order (Fig. 1; Table 1). Representatives of the genus *Sulfolobus* are strictly aerobic chemolithoautotrophs, growing by oxidation of S^{2-} and S^0, forming sulfuric acid (Table 1; Brock *et al.* 1972). Several members of the Sulfolobales, like *Metallosphaera sedula, Sulfolobus metallicus, Acidianus infernus* and *Acidianus brierleyi,* are powerful "metal-leachers" able to oxidize sulfidic ores like pyrite, chalcopyrite, and sphalerite, forming sulfuric acid and solubilizing the heavy-metal ions (Tables 1, 2; Brierley & Brierley 1973; Huber *et al.* 1989; Huber & Stetter 1991). As we found recently, some members of the Sulfolobales are able to grow by hydrogen oxidation (Table 2; G. Huber *et al.* 1992). Members of *Acidianus* are able to grow by anaerobic and aerobic oxidation of H_2, using S^0 and O_2 as electron acceptors, respectively (Segerer *et al.* 1986). Alternatively, *Acidianus* is able to grow by S^0-oxidation, too (Table 2). *Stygiolobus azoricus* represents a new genus of thermoacidophilic, strictly anaerobic, sulfidogens, growing by H_2S formation from H_2 and S^0 (Segerer *et al.* 1991). Many members of the Sulfolobales are facultative, some even strict, heterotrophs, growing by aerobic respiration on a variety of organic compounds (e.g., yeast extract, sugars; Table 3).

Terrestrial hot springs and the depth of solfataric fields harbor slightly acidophilic and neutrophilic hyperthermophiles, which are usually strict anaerobes. They are members of the genera *Pyrobaculum, Thermoproteus, Thermofilum, Desulfurococcus, Methanothermus,* and *Thermotoga* (Table 1). Members of *Pyrobaculum, Thermoproteus,* and *Thermofilum* are stiff, regular rods. Sometimes, spheres are protruding at their ends,

TABLE 1   Biotopes, morphology and growth conditions of hyperthermophiles.

Species	Biotope marine (m) terrestrial (t)	Morphology	Growth conditions Temperature (°C)			pH	Aerobic (ae) anaerobic (an)
			Min	Opt	Max		
Sulfolobus acidocaldarius	t	lobed cocci	60	80	85	1–5	ae
Metallosphaera sedula	t	cocci	50	75	80	1–4.5	ae
Acidianus infernus	t	lobed cocci	60	88	95	1.5–5	ae/an
Stygiolobus azoricus	t	lobed cocci	57	80	89	1–5.5	an
Thermoproteus tenax	t	regular rods	70	88	97	2.5–6	an
Pyrobaculum islandicum	t	regular rods	74	100	103	5–7	an
Pyrobaculum organotrophum	t	regular rods	78	100	103	5–7	an
Thermofilum pendens	t	slender regular rods	70	88	95	4–6.5	an
Desulfurococcus mobilis	t	cocci	70	85	95	4.5–7	an
Staphylothermus marinus	m	cocci in aggregates	65	92	98	4.5–8.5	an
Pyrodictium occultum	m	discs with fibers	82	105	110	5–7	an
Pyrodictium abyssi	m	discs with fibers	80	105	110	4.7–7.5	an
Thermodiscus maritimus	m	discs	75	88	98	5–7	an
Thermococcus celer	m	cocci	75	87	93	4–7	an
Pyrococcus furiosus	m	cocci	70	100	103	5–9	an
Archaeoglobus fulgidus	m	irregular cocci	60	83	95	5.5–7.5	an
Archaeoglobus profundus	m	irregular cocci	65	82	90	4.5–7.5	an
Methanothermus sociabilis	t	rods in clusters	65	88	97	5.5–7.5	an
Methanopyrus kandleri	m	rods in chains	84	98	110	5.5–7	an
Methanococcus jannaschii	m	irregular cocci	50	85	86	3–6.5	an
Methanococcus igneus	m	irregular cocci	45	88	91	5–7.5	an
Thermotoga maritima	m	rods with sheath	55	80	90	5.5–9	an
Thermotoga thermarum	t	rods with sheath	55	70	84	6–9	an
Aquifex pyrophilus	m	rods w. refractile areas	67	85	95	5.4–7.5	ae

possibly indicating a mode of budding (Zillig *et al.* 1981, 1983a; Huber *et al.* 1987). Cells of *Pyrobaculum* and *Thermoproteus* are about 0.50 µm in width and are able to form filaments up to 100 µm long. Cells of *Thermofilum* ("the hot thread") are of similar length but are only about 0.17–0.35 µm in width (Table 1). *Pyrobaculum islandicum, Thermoproteus tenax*, and *Thermoproteus neutrophilus* are able to grow autotrophically by anaerobic formation of H_2S from H_2 and S^0 (Table 2; Fischer *et al.* 1983). Recently, we have isolated a novel rod-shaped hyperthermophile of the genus *Pyrobaculum* (*P. aerophilum*), which surprisingly grows by hydrogen oxidation under microaerophilic conditions (Völkl *et al.* 1993). *Pyrobaculum islandicum* and *Thermoproteus tenax* are strictly anaerobic, facultative heterotrophs, growing alternatively by sulfur respiration on organic substrates like yeast extract and prokaryotic cell extracts (Table 3). Strains of *Thermofilum, Pyrobaculum organotrophum*, and *Thermoproteus uzoniensis* are obligate heterotrophs thriving by sulfur respiration (Table 3; Stetter *et al.* 1990; Bonch-Osmolov-skaya *et al.* 1990). Members of the genus *Desulfurococcus* are coccoid-shaped, strictly

TABLE 2   Energy-yielding reactions in lithoautotrophic hyperthermophiles. Genera marked with asterisks contain opportunistic heterotrophs.

Electron donor	Electron acceptor	Energy-yielding reaction	Genera
H_2	CO_2	$4H_2+CO_2 \rightarrow CH_4+2H_2O$	*Methanopyrus, Methanothermus, Methanococcus*
H_2	S^0	$H_2+S^0 \rightarrow H_2S$	*Pyrodictium, Thermoproteus*, Pyrobaculum*, Acidianus*, Stygiolobus*
H_2	$SO_4^{2-};(S_2O_3^{2-})$	$4H_2+H_2SO_4 \rightarrow H_2S+4H_2O$	*Archaeoglobus**
H_2	O_2	$H_2+\frac{1}{2}O_2 \rightarrow H_2O$	*Aquifex, Sulfolobus*, Acidianus*, Metallosphaera**
S^0	O_2	$2S^0+3O_2+2H_2O \rightarrow 2H_2SO_4$	–"–
(S^{2-})	O_2	$(2FeS_2+7O_2+2H_2O \rightarrow 2FeSO_4+2H_2SO_4)$	–"–

TABLE 3   Chemoorganotrophic modes of life in hyperthermophiles.

Type of metabolism	External electron acceptor	Energy-yielding reaction	Genera
Respiration	S^0	$2[H]+S^0 \rightarrow H_2S$	*Pyrodictium, Thermoproteus, Pyrobaculum, Thermofilum, Desulfurococcus, Thermodiscus*
Respiration	SO_4^{2-} $(S_2O_3^{2-};SO_3^{2-})$	$8[H]+SO_4^{2-}+2H^+ \rightarrow \rightarrow H_2S+4H_2O$	*Archaeoglobus*
Respiration	O_2	$2[H]+\frac{1}{2}O_2 \rightarrow H_2O$	*Sulfolobus, Metallosphaera, Acidianus*
Fermentation	–	Pyruvate→L(+)-lactate +acetate+ +H_2+CO_2	*Thermotoga, Thermosipho, Fervidobacterium*
Fermentation	–	Peptides→isovalerate, isobutyrate, butanol, CO_2, etc. *Pyrococcus:* Pyruvate →acetate+H_2+CO_2	*Pyrodictium, Hyperthermus, Thermoproteus, Desulfurococcus, Staphylothermus, Thermococcus, Pyrococcus*

heterotrophic sulfur respirers (Tables 1, 3; Zillig *et al.* 1982). Exclusively from the depth of solfataric fields in the southwest of Iceland, rod-shaped methanogens of the genus *Methanothermus* have been isolated (Stetter *et al.* 1981). So far, the species *M. fervidus* and *M. sociabilis* are known (Table 1). They are strict chemolithoautotrophs, gaining energy by reduction of CO_2 by H_2 (Table 2). From terrestrial neutral hot springs situated at the base of evaporite mounts at Lac Abbé, Djibouti, Africa, *Thermotoga thermarum* has been isolated (Windberger *et al.* 1989). Members of *Thermotoga* are rod-shaped hyperthermophilic organisms belonging to the bacterial domain (Table 1). Usually, they are thriving in marine hydrothermal systems, where they grow by fermentation of carbohydrates. As an exception, *T. thermarum* grows only at low ionic strength in the presence of up to 0.55% NaCl and is therefore adapted to terrestrial springs with low salinity.

Marine hyperthermophiles. – Many groups of hyperthermophiles are adapted to the high salinity of seawater ($\approx 3\%$ salt). They are represented by the genera *Pyrodictium, Thermodiscus, Staphylothermus, Hyperthermus, Methanopyrus, Pyrococcus, Thermococcus,* and (some members of) *Methanococcus,* all belonging to the Archaea, and by the bacterial genera *Thermotoga* and *Aquifex* (Table 1). The organisms with the highest growth temperatures in the laboratory are members of the genera *Pyrodictium* and *Methanopyrus,* exhibiting an upper temperature border of growth above 110°C (Table 1; Stetter *et al.* 1983). They are so dependent on high temperatures that they are unable to grow below 80 and 84°C, respectively. Cells of *Pyrodictium* are disk-shaped (0.2–3 µm in diameter) and are usually connected by a network of hollow fibers, about 30 nm in diameter. However, there are also some recent isolates that belong to the same genus but do not form fibers (K.O. Stetter, unpublished). As a rule, members of *Pyrodictium* are strict chemolithoautotrophs gaining energy by reduction of S^0 by H_2. Very recently, we have isolated a novel disk-shaped prokaryote (Isolate 1A) growing at temperatures up to 113°C that is closely related to *Pyrodictium* (Blöchl, Burggraf & Stetter, unpublished). Surprisingly, isolate 1A is unable to grow on S^0 but is a microaerophilic H_2 oxidizer. As an exception, *Pyrodictium abyssi* is a heterotroph growing by fermentation of peptides (Pley *et al.* 1991). Its growth is strongly stimulated by H_2 and S^0. *Methanopyrus* is a strictly chemolithoautotrophic methanogen. The cells are rod-shaped (Table 1). It represents the deepest phylogenetic branch within the archaeal domain (Fig. 1). Further marine hyperthermophilic methanogens are *Methanococcus jannaschii* and *M. igneus* within the Methanococcales (Fig. 1). Within this order, these hyperthermophiles represent the shortest phylogenetic lineages (Fig. 1). *Methanococcus thermolithotrophicus* is a moderate thermophile, growing at temperatures up to 70°C (Huber *et al.* 1982). Archaeal sulfate reducers are represented by members of *Archaeoglobus,* which are so far all hyper-thermophiles (Table 1; Stetter *et al.* 1987). *Archaeoglobus fulgidus* and *A. "lithotrophicus"* are chemolithoautotrophs able to grow by reduction of SO_4^{2-} and $S_2O_3^{3-}$ by H_2 (Table 2). *Archaeoglobus fulgidus* is a facultative heterotroph growing on a variety of organic substrates like formate, cell extracts, sugars, proteins, and starch by sulfate respiration (Table 3; Stetter 1988). *Archaeoglobus profundus* is an obligate heterotroph (Burggraf *et al.* 1990). *Archaeoglobus* possesses several coenzymes that had been thought to be unique for methanogens (e.g., F_{420}; methanopterin, tetrahydromethanopterin, methano-furane). In agreement with these similarities, *Archaeoglobus* represents a separate phylogenetic branch within the methanogens (a different order [Woese *et al.* 1991]). Within the Bacteria domain, the deepest phylogenetic branch is represented by the recently discovered *Aquifex pyrophilus* (Fig. 1; Burggraf *et al.* 1992; R. Huber *et al.* 1992). It grows at temperatures of up to 95°C and is therefore the most extremely hyper-thermophilic bacterial member known so far. *Aquifex pyrophilus* is a rod-shaped strict chemolithoautotroph growing by oxidation of H_2 at oxygen concentrations below 0.5% (Tables 1, 2).

Groups of strictly heterotrophic hyperthermophiles are thriving in submarine hot vents, too. *Thermodiscus maritimus* is a disk-shaped heterotroph growing by sulfur respiration on yeast extract and prokaryotic cell homogenates (Tables 1, 3). In the absence of sulfur, it is able to grow by an unknown fermentation (Stetter 1986). Cells of *Staphylothermus marinus* (Table 1) are coccoid and are arranged in grapelike aggregates (Fiala *et al.* 1986). They are highly variable in diameter, from 0.5 to 15 µm. Members of

Staphylothermus are able to ferment peptides, forming fatty acids, alcohols, CO_2, and H_2 (Table 3). Members of the genera *Thermococcus* and *Pyrococcus* are fermenting peptides, too (Fig. 1; Table 3; Zillig *et al.* 1983b; Fiala & Stetter 1986). *Pyrococcus furiosus* is able to ferment pyruvate, forming acetate, H_2, and CO_2 (P. Schönheit, personal communication, 1990). Hydrogen is inhibitory to growth of *Thermococcus* and *Pyrococcus* and can be removed by "gas stripping" or by coculturing H_2-consuming hyperthermophilic methanogens (Bonch-Osmolovskaya & Stetter 1991). In the presence of S^0, H_2S is formed instead of H_2, which does not inhibit growth (Fiala & Stetter 1986). Together with archaeal hyperthermophiles, many submarine hydrothermal fields contain members of the bacterial genus *Thermotoga*, which in view of *Aquifex* represents the second-deepest branch within the Bacteria domain (Huber *et al.* 1986). *Thermotoga maritima* and *T. neapolitana* are rod-shaped hyperthermophiles exhibiting an upper temperature border of growth of 90°C (Table 1). Cells show a characteristic "toga," a sheathlike structure surrounding cells and overballooning at the ends. The toga contains porins and is most likely homologous to the outer membrane of gram-negative bacteria (Rachel *et al.* 1990). Members of *Thermotoga* grow by fermentation of carbohydrates, forming L-lactate, acetate, H_2 and CO_2 as end products (Table 3). Similar to *Pyrococcus*, H_2 is inhibitory to growth. In the presence of S^0, H_2S is formed instead of H_2, which does not inhibit growth (Huber *et al.* 1986).

Conclusions

Natural water-containing high-temperature environments harbor a variety of different hyperthermophilic prokaryotes that consist of primary producers and consumers of organic matter and that form unique, purely prokaryotic high-temperature ecosystems on Earth.

Within the universal phylogenetic tree of life, hyperthermophiles form a "bloc" around the root, occupying all the deepest branches. Therefore, an origin of life at high temperatures appears most probable. By restrictions imposed by the extreme environment, hyperthermophiles may have evolved much slower than mesophiles and are therefore still rather primitive.

By their mode of carbon assimilation, hyperthermophiles consist of heterotrophs and chemolithoautotrophs. Assuming a heterotrophic origin of life, the shortest and deepest branches within the phylogenetic tree would be expected to consist exclusively of heterotrophs. In striking contrast, however, the phylogenetic tree demonstrates that the deepest and shortest branches are represented almost exclusively by strictly chemolithoautotrophic organisms like *Pyrodictium*, *Methanopyrus*, and *Aquifex*. Therefore, autotrophy appears to be a very ancient feature, supporting theories about an autotrophic origin of life (Wächtershäuser 1988b). In addition, an autotrophic origin of life would be favored by elevated temperatures, while a heterotrophic origin would be improbable because of the increased instability of macromolecules. Since all deeply branching autotrophs are hyperthermophiles, the theory of an autotrophic origin of life is further favored.

The energy-yielding reactions in chemolithoautotrophic hyperthermophiles are anaerobic and aerobic types of respiration (Table 3). Molecular hydrogen is widely used as an electron donor. It is present within volcanic exhalations or may originate in the hot environment by the recently discovered anaerobic pyrite formation (Drobner *et al.* 1990), which had been theoretically postulated as a possible source of energy for early life in the theory of Wächtershäuser (1988a). Carbon dioxide, oxidized sulfur compounds, and oxygen serve as electron acceptors. Since molecular hydrogen, CO_2, and sulfur compounds are present within volcanic environments, anaerobic hyperthermophilic autotrophic organisms using them are excellently adapted to this environment. They are completely independent of oxygen and therefore represent life independent of the sun. They could even exist on other planets that possess volcanic activity and liquid water.

The presence of hydrogen oxidizers among hyperthermophiles is a big surprise. Within Archaea, members of the thermoacidophilic Sulfolobales show this property (G. Huber *et al.* 1992); these, however, represent a rather fastly evolving phylogenetic lineage (Fig. 1). Since oxygen appeared globally rather late within the Earth's atmosphere (Schopf *et al.* 1983; see chapters by Holland, Lowe, and Towe, this volume), this metabolic property could still be explained by a rather late adaptation to oxygen of these terrestrial organisms. However, very recently we were able to isolate different hyperthermophilic members of Archaea from marine hydrothermal systems (isolates 1A; *Pyrobaculum aerophilum*) that are hydrogen oxidizers, too. In contrast to the Sulfolobales, they are extremely microaerophilic and grow only in the presence of traces of free oxygen (Blöchl, Huber, Völkl & Stetter, unpublished). Interestingly, they represent members of the genera *Pyrodictium* and *Pyrobaculum* which are deeply branching short lineages (Burggraf, Woese & Stetter, unpublished). A further microaerophilic hyperthermophilic hydrogen oxidizer is *Aquifex pyrophilus*, which represents the deepest branching within the Bacteria domain (Burggraf *et al.* 1992). Alternatively, these deeply branching hyperthermophilic hydrogen oxidizers are able to grow by oxidation of sulfur compounds, again under microaerophilic conditions. The utilization of (traces of) oxygen seems therefore to be a rather ancient feature obvious by examination of hyperthermophiles. Most likely, traces of free oxygen built up photochemically had been present on Earth since the origin of life and could be already used by organisms early in the evolution of life (see also Towe, this volume).

Several autotrophic hyperthermophiles are able to grow on organic matter, too. In view of the evidence for an autotrophic origin of life, they should be seen as opportunistic heterotrophs, while strict heterotrophs appear to be late adaptations as consumers of organic matter within already existing high-temperature ecosystems.

Acknowledgments. – I wish to thank Günter Wächtershäuser and Carl R. Woese for stimulating discussions. The essential help for the preparation of this manuscript of Sigfried Burggraf, Gerhard Frey, Dagmar Uhl, and Marlis Wördemann is highly appreciated. The work presented from my laboratory was supported by grants of the *Deutsche Forschungsgemeinschaft,* the *Bundesministerium für Forschung und Technologie,* and the *Fonds der Chemischen Industrie* to K.O.S.

The early diversification of life

Otto Kandler

*Botanisches Institut der Universität, Menzinger Straße 67,
D-80638 München, Germany*

Life originated with chemolithoautotrophs in a heated Archean ocean, and energy sources diversified early. This statement is based on a retrospective evaluation of the distribution pattern of basic metabolic and structural traits in a universal phylogenetic tree based on sequence comparison of 16S rRNA, ATPases, and protein elongation factors. According to the scenario based on such ideas, the primordial geochemical redox energy (presumably pyrite formation, according to Wächtershäuser's proposal), which incited CO_2 fixation and organic synthesis, was replaced at an early stage of biochemical evolution by redox energy derived from the redox couples H_2/S^0, H_2/CO_2, and H_2/O_2 still employed by extant organisms. The three domains of extant life (Bacteria, Archaea, Eucarya) are thought to have emerged from a multiphenotypical population of primarily chemolithoautotrophic pre-cells by cellularization and further evolution by a process resembling allopatric speciation and parallelophyly.

The nearly exponential progress in molecular phylogeny based on the sequencing of semantophoretic molecules (Zuckerkandl & Pauling 1965) and the discovery of ever more new organisms exhibiting putative archaic traits (Stetter *et al.* 1990; Stetter, this volume) make it possible to construct a universal phylogenetic tree (Woese 1987b; Woese *et al.* 1990; Wheelis *et al.* 1992). The distribution pattern of structural and biochemical traits among extant organisms in such a phylogenetic tree allows us to propose plausible scenarios for the early diversification of life into a multitude of extant phenotypes and the shaping of the three domains: Bacteria, Archaea, Eucarya. (The recently proposed formal names Bacteria, Archaea, and Eucarya [Woese *et al.* 1990] will be used throughout this paper instead of the vernacular names eubacteria, archaebacteria, and eukaryotes, respectively.)

Energy sources and temperature regime of early life

The recently discovered hyperthermophilic (growth optimum >80°C) members of the domains Archaea and Bacteria (Stetter *et al.* 1990; Stetter, this volume) form the deepest branchings of the phylogenetic tree based on 16S rRNA sequence comparison (Fig. 1). Hyperthermophiles are obligately or facultatively chemolithoautotrophic, employing the redox couples H_2/S^0, H_2/CO_2, H_2/O_2, H_2S/O_2, or H_2/SO_4^{2-} as energy sources, or they are organotrophic by fermentation or sulfur respiration. None, however, is

Bengtson, S. (ed.) 1994: *Early Life on Earth. Nobel Symposium No. 84.* Columbia U.P., New York

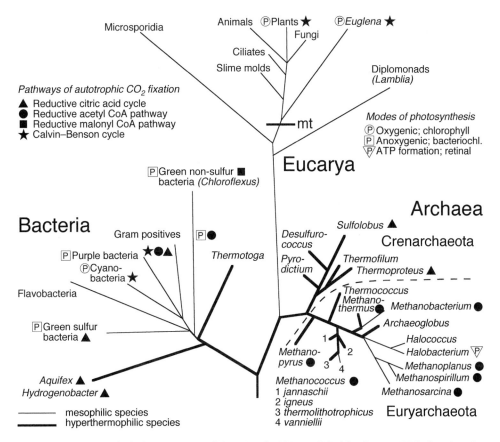

FIGURE 1   Universal phylogenetic tree (Woese *et al.* 1990, modified by Stetter 1993) showing the division of organisms into three domains and the distribution of autotrophic pathways. Branching order and branch lengths are based upon 16S/18S rRNA sequence comparisons. The position of the root was determined by comparing known sequences of paralogous genes (ATPases; protein elongation factor) that diverged from each other before the three primary lineages emerged from their ancestral condition (Gogarten *et al.* 1989; Iwabe *et al.* 1989).

photoautotrophic. These observations call for a chemolithoautotrophic origin of life in a hot (about 100°C) environment and for an early diversification of inorganic energy sources. On the other hand, photoautotrophy based on chlorophyll in some mesophilic branches of the domain Bacteria and photoorganotrophy based on retinal in only one superficial branch of the domain Archaea (*Halobacterium*) seem to be later phylogenetic inventions, with oxygenic photosynthesis being the most recent achievement.[1]

Similarly, the path of carbon in autotrophic CO_2 fixation shows also a distinct relationship with the phylogenetic position of the respective oganisms, since the Calvin–Benson cycle (ribulose bisphosphate pathway) is found in all oxygenic and in some anoxygenic photoautotrophs, as well as in chemolithoautotrophic organisms of

---

[1] Photosynthesis within the domain Eucarya is due to the adoption of cyanobacteria as chloroplasts via endosymbiontic processes.

distant branches of the domain Bacteria, while the anoxygenic photoautotrophs of phylogenetically distinctly older branches and all chemolithoautotrophic hyperthermophiles, bacterial and archaeal alike, fix CO_2 by a variety of other mechanisms (Fig. 1). These organisms use either the reductive citric acid cycle or modifications thereof: the reductive acetyl-CoA, or the reductive malonyl-CoA pathways (Fuchs 1989; Fuchs *et al.* 1992; Strauss *et al.* 1992). This indicates that, phylogenetically, the Calvin–Benson cycle is a later invention than anoxygenic photosynthesis and that it is preceded by the three above-mentioned pathways of CO_2 assimilation. These three pathways are still coupled with H_2-dependent chemolithoautotrophy in extant organisms of both prokaryotic domains or with anoxygenic photosynthesis in early branchings of the domain Bacteria (Fig. 1).

Among the three pathways, the reductive citric acid cycle is probably the phylogenetically oldest one, since both the reductive acetyl-CoA pathway, common in the archaeal methanogens and the bacterial acetogens, and the as yet only partially explored reductive malonyl-CoA pathway, so far only found in the moderately thermophilic green non-sulfur bacteria (Strauss *et al.* 1992), require coupling with the citric acid cycle for the biosynthesis of amino acids and other essential cell constituents. This view is in line with Wächtershäuser's (1990a) proposal that the reductive citric acid cycle was the very first anabolic cycle in the early biochemical evolution of life and only later became the main catabolic and anaplerotic cycle for extant life.

Allopatric speciation and parallelophyly

While the most basic biochemical features (genetic code, set of protein amino acids, etc.) are shared by all three domains, thus testifying to the unity of life on this planet, many important evolutionary features are found in only one or two of the three domains. The distribution of such features found in only two domains is not unequivocally compatible with the branching order at the basis of the phylogenetic tree depicted in Fig. 1. For instance, glycerol fatty-acid esters, catabolic mode of glycolysis, etc., occur only in Eucarya and Bacteria; circular genome structures, Shine Dalgarno sequences, etc., are common characteristics of Bacteria and Archaea; while V-ATPases, glycoproteins, etc., are only found in Eucarya and Archaea. Thus each of the three possible pairs among the three domains appear to be sister groups depending on the characters considered.

Such a quasi-random distribution of characters among the three domains may best be explained by assuming a process resembling allopatric speciation (Mayr & Ashlock 1991) concurrent with a cellularization process in a multiphenotypical population of pre-cells at an early stage of life, as depicted in Fig. 2.

In the scenario underlying the scheme shown in Fig. 2, life was incited in the heated hydrosphere of a primeval inorganic world by CO_2-reduction and organic synthesis driven by inorganic redox energy and electron flow. As pointed out by Wächtershäuser (1988a, b), pyrite formation ($H_2S + FeS \rightarrow FeS_2 + H_2$; $G^0 = -41.9$ KJ/mol), a common geochemical process which occurs spontaneously at elevated temperatures (Drobner *et al.* 1990; Wächtershäuser, this volume), is a reasonable though still hypothetical source of reducing power for primeval CO_2 reduction and organic synthesis.

FIGURE 2   Scheme of early
diversification of life and
allopatric speciation by
parallelophyly. **A, B, C**: Precellular founder groups
undergoing cellularization.
Essential evolutionary
improvements: **1**: CO_2
reduction and organic
synthesis driven by inor-
ganic redox energy and
electron flow. **2**: Primitive
enzymes and templates.
3: Elements of a transcription
and translation apparatus.
4: Formation of pre-cells
(loose envelopes). **5**: Circular
or linear stable genome.
6: Cytoplasmic membrane.
7: Rigid cell wall (murein).
8: Propensity for various
non-murein rigid cell-wall
polymers. **9**: Glycoprotein-
aceous cell envelope or
glycocalyx. **10**: Cytoskeleton.
11: Nuclear membrane and
complex chromosomes.
12: Cell organells via
endosymbiosis.

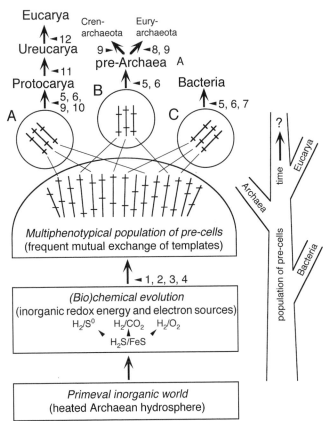

Among the sources of inorganic redox energy still employed by extant chemolithoautotrophic hyperthermophilic members of the domain Archaea, the redox couples
H_2/S^0 and H_2/CO_2 were available in the Archean, at least in the then extensive volcanic
areas, whereas the couple H_2/SO_4, rarely used by Archaea (Stetter *et al.* 1987), may have
been available only in the vicinity of sulfate-rich magmatic-hydrothermal vents in an
otherwise sulfate-poor Archean ocean (Hattori & Cameron 1986). The redox couple
H_2/O_2 utilized by the hyperthermophilic chemolithoautotrophic members of the
domain Bacteria (*Aquifex, Hydrogenobacter*; R. Huber *et al.* 1992; Burggraf *et al.* 1992) may
have become available only when low oxygen concentrations had developed in surface
waters of the otherwise anaerobic Archean ocean or in shallow shelf areas as a result of
the photochemical splitting of water vapor by the then intensive UV radiation (Towe,
this volume).

At a very early stage of biochemical evolution, various redox energy sources based
on hydrogen oxydation may have supplemented and finally replaced the aboriginal
redox energy source, presumably the spontaneous pyrite formation proposed by
Wächtershäuser (1988a, b), when the first primitive hydrogenase activity had evolved.
Thereby a wider range of substrates became available, and the evolving life could

venture into a larger variety of habitats, even into microaerobic habitats in shallow, illuminated surface waters. The continuous changes in the physical environment on the aging and cooling Earth led to further diversification of habitats and favored opportunistic radiation of primitive life into numerous phenotypes on the basis of each of the different chemolithoautotrophies. Concomitantly with the accumulation of organic matter derived from chemolithoautotrophic life, opportunistic and obligate heterotrophic life may also have developed.

The finally reached and often discussed common ancestral condition of extant life, from which the three extant domains emerged, probably consisted of a multiphenotypical population of pre-cells, i.e. metabolizing, self-reproducing entities exhibiting most of the basic properties of a cell but unable to limit the frequent mutual exchange of genetic information to a level that allowed the separation of gene pools and speciation. Among other reasons, the lack of a stabilized coherent genome structure and of a proper cytoplasmic membrane to control sufficiently the export and entry of genetic material may have been responsible for the ongoing genetic communication, which was limited only by structural and functional constraints such as structural, genetic and metabolic incompatibilities. Thus, the population of pre-cells was a quasi-random but not truly stochastic mixture of various strictly chemolithoautotrophic and derived obligately and opportunistically heterotrophic phenotypes.[1]

Only after the transformation of some of the pre-cells into cells (i.e. development of stable circular or linear genome structures and an effective lipid-rich cytoplasmic membrane, to mention only two of the most important evolutionary achievements necessary for cellularization) was it possible for presumably physiologically and/or ecologically separated groups of pre-cells (Fig. 2) to become founder groups of primary taxa. Owing to shared propensities and compatible genetic potentials to evolve in a parallel manner and to develop independent evolutionary changes in various sublineages of the population, these founder populations resulted in the three domains of extant life, each comprising branches exhibiting distinct autapomorphic features. Such a process resembles the formation of new derived taxa in animals which is described as *parallel polyphyly* or *parallelophyly* (Mayr & Ashlock 1991, pp. 256–259). However, the entities forming the founder groups of the domains were not descendants of an individual member of an ancestor, as in the case of parallelophyly in higher organisms, but aboriginal products of biochemical evolution (pre-cells) undergoing cellularization (Fig. 2).

Origin and diversification of the domain Bacteria

A variety of phenotypes of H_2/O_2 chemolithoautotrophic pre-cells resembling the phenotype of members of the extant hyperthermophilic Aquificales (R. Huber *et al.* 1992; Stetter, this volume) and thriving in the "modern" microaerobic and relatively

---

[1] Woese's progenote (Woese 1982, 1987b) may be considered one of the early stages of biochemical evolution preceding the stage of pre-cells, since it is defined as a most primitive entity containing only short pieces of polymers, not yet having evolved a link between phenotype and genotype.

fast-cooling shelf habitats may have been first to reach the stage of cellularization and to function as founder population, giving rise to the domain Bacteria.[1]

In addition to the development of a circular stable genome structure, a cytoplasmic membrane containing fatty acid lipids, and other improvements necessary for cellularization, a murein cell wall – typical of extant Bacteria (Kandler & König 1985, 1993) including the hyperthermophilic Aquificales (R. Huber *et al.* 1992; Burggraf *et al.* 1992) and Thermotogales (Huber *et al.* 1986) forming the two deepest branches within the domain Bacteria – may also have been invented at an early stage of speciation before radiation into a multitude of lineages. The rigid murein sacculus made the cells highly resistant to mechanical and osmotic stress and thus became an important factor for the successful colonization of almost all the different habitats that became available during later stages of the Earth's history.

Another evolutionary breakthrough reached independently by some of the moderately thermophilic and mesophilic branches of the domain Bacteria was the invention of various modes of photoautotrophic growth: Anoxygenic modes of photosynthesis allowed mass production in shallow waters independent of atmospheric hydrogen and oxygen, while the invention of oxygenic photosynthesis paved the way for the development of the extant multitude of aerobic life by enriching the atmosphere with oxygen.

Origin and diversification of the domain Archaea

The contemporary domain Archaea comprises all extant H_2/S^0 and methanogenic H_2/CO_2 chemolithoautotrophs. Since the environmental conditions for these two modes of chemolithoautotrophic life changed only slowly throughout the ages and have remained virtually constant in the volcanic areas up to the present days, the stimulation of evolution exerted by changing environments (an important driving force in the development of the H_2/O_2 chemolithoautotrophs) was minimal, especially in the case of sulfur-dependent life. Thus the evolutionary development of the H_2/S^0 and H_2/CO_2 chemolithoautotrophs was probably rather slow, as indicated by the rather short phylogenetic distances between even the contemporary taxa of the domain Archaea (Fig. 1). As a result, the cellularization of pre-cells of H_2/S^0 and H_2/CO_2 chemolithoautotrophs and the formation of the domain Archaea may have occurred distinctly later than that of H_2/O_2 chemolithoautotrophic pre-cells. Thus the formation of the domain Bacteria may have preceded that of the domain Archaea, although H_2/S^0 and H_2/CO_2 chemolithoautotrophies are most likely older than H_2/O_2 chemolithoautotrophy.

The speciation process leading to the domain Archaea must have been more complex than that leading to the domain Bacteria, since pre-cells of both chemolitho-

---

[1] The domain Bacteria includes also acetogenic H_2/CO_2 chemolithoautotrophs. The mechanisms of energy generation and CO_2 fixation resemble those used by methanogens (Fuchs 1989), but acetic acid instead of methane is formed as "waste" product. Phylogenetically, according to 16S rRNA sequence comparison, the respective bacterial taxa are located close to the basis of the gram-positive radiation. Comparison of amino acid sequences of key enzymes of the acetyl-CoA pathways in both domains might help to clarify if there is any genetic relationship between acetogenic bacteria and methanogenic archaea.

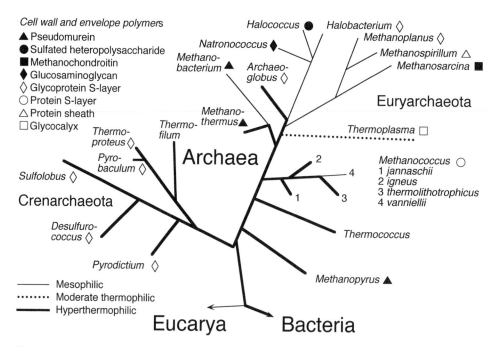

FIGURE 3 Distribution of cell wall and cell envelope polymers among Archaea (Kandler & König 1993).

autotrophic phenotypes had to amalgamate, presumably during the phases of cellularization, when a circular genome and a cytoplasmic membrane containing isoprenyl ether lipids were developed within a mixed founder population. However, before any significant radiation, the thus formed pre-Archaea segregated into two separate founder populations containing either H_2/S^0 or H_2/CO_2 chemolithoautotrophs, which evolved into the kingdoms Crenarchaeota (lineage of sulfur-dependent Archaea, e.g., *Thermoproteus, Sulfolobus*) and Euryarchaeota (lineage of methanogens, halophiles, etc.).

The two kingdoms exhibit distinctly different propensities for developing cell envelopes and cell walls (Fig. 3). While the crenarchaeotes invented elaborated cell envelopes (surface layers) mostly consisting of glycoprotein, various branches of the methanogenic and the extreme halophilic or alcaliphilic euryarchaeotes developed rigid cell walls that functionally and morphologically mimic the murein cell walls of bacteria. However, the chemical composition and biosynthetic pathways of the archael cell-wall polymers differ fundamentally not only from bacterial murein but also among the cell walls of the different branches of the euryarchaeotes (Kandler & König 1985, 1993). Obviously, different cell-wall polymers were independently developed in the various branches. Thus cell-wall formation in euryarchaeotes is a good example of "parallel but independent evolutionary changes in several sublineages" (Mayr & Ashlock 1991), a characteristic of parallelophyly.

Euryarchaeotes show a distinctly wider range of metabolical, structural, and ecological diversification than crenarchaeotes, for instance: adaptation to the use of oxydized compounds as electron acceptors, as in the case of *Archaeoglobus* thriving by H_2/SO_4^{2-} chemolithoautotrophy (Stetter *et al.* 1987; Achenbach-Richter *et al.* 1987b); adaptation to microaerophilic and moderate thermophilic heterotrophic growth as observed in *Thermoplasma* (Searcy & Whatley 1984); and adaptation to a wider range of substrates and mesophilic growth, as found in several genera of methanogens that colonized virtually all terrestrial and aquatic anaerobic habitats of the globe.

A breakthrough to fully developed aerobic and mesophilic life was, however, achieved only by the extreme halophiles and alcaliphiles that colonized the continental salt and soda lakes. They even seem to be on their way to evolving their own type of phototrophy with retinal instead of chlorophyll as light-harvesting pigment. So far, only a few halophilic species are known to exhibit photoorganotrophic growth on ATP formation by light energy (Oesterhelt & Krippahl 1983), but no known halophile is capable of photosynthetic CO_2 assimilation, although ribulose bisphosphate carboxylase (RuBisCo), the key enzyme for CO_2 fixation in oxygenic photosynthesis, could be found in some non-retinal-containing species of halobacteria (Altekar & Rajagopalan 1990). The finding that this enzyme cross-reacts with RuBisCo from spinach chloroplasts suggests its translocation from a member of the domain Bacteria to the archaeal halobacteria by some kind of horizontal gene transfer and makes a genuine evolution of RuBisCo within the domain Archaea most unlikely.

The crenarchaeotes show the least metabolical and ecological diversification. All known members are hyperthermophiles, thrive in volcanic areas, and mostly depend on H_2 and sulfur. However, H_2/O_2 chemolithoautotrophy under microaerophilic conditions has recently been observed in hyperthermophilic members of the Sulfolobales (Stetter, this volume), and mesophilic or even kryophilic relatives of the crenarchaeotes may form a component of marine plankton, as indicated by 16S rRNA sequence data obtained from marine bacterioplankton biomass (Fuhrmann *et al.* 1993). Thus a wider range of phenotypes of crenarchaeotes may be known in the near future.

For the time being, we may conjecture that the extant hyperthermophilic chemolithoautotrophic or sulfur-respiring organotrophic crenarchaeotes and the hyperthermophilic methanogenic euryarchaeotes represent the least modified descendants of the second generation of primitive chemolithoautotrophic phenotypes which replaced the primordial chemolithoautotrophic phenotype employing pyrite formation as energy source, as supposed by Wächtershäuser (1988a, b). Therefore, the domain embracing both kingdoms is named Archaea (Woese *et al.* 1990), although, as mentioned above and as indicated by the branching order at the base of the phylogenetic tree (Fig. 1), the speciation of the domain Bacteria presumably occurred earlier (Fig. 2).

Origin of the domain Eucarya

The domain Eucarya does not contain any genuine chemolithoautotrophs. Hence, in our scenario, the precursors of the domain Eucarya must be delineated from heterotrophic pre-cells that lost their power plant fueled by inorganic redox energy. Instead, they utilized the redox energy stored in biomass produced by the chemolithoautotrophs.

Since the heterotrophic pre-cells were freed from the functional constraints exerted by the chemolithoautotrophic machinery, new, different patterns of phenetic and genetic mosaicism could arise. Thus the pre-cell population may have become enriched by new phenotypes that evolved even after the domains Bacteria and Archaea had been segregated from the mainstream.

The formation of the eukaryotic lineage by a founder population derived from such an advanced pre-cellular stage would explain the intimate mixture of bacterial and archaeal characteristics in the extant domain Eucarya. However, scenarios in this area are especially elusive, as indicated by the various recently proposed alternative scenarios assuming fusion of a protobacterium and a pre-archaeon (Zillig *et al.* 1992) or an engulfment of an archaeon by a protoeukaryotic cell (Sogin 1991, this volume).

Important evolutionary improvements, such as the invention of a linear genome, of a cytoskeleton, or of the formation of cell organells via endosymbiosis, that led to the transformation of the protocarya into the extant Eucarya have been comprehensively treated by Sogin (1991, this volume).

At last, let me pose an intriguing question: What was (is) the fate of the pre-cellular life? Has its evolution stopped after it gave birth to the domain Eucarya? Was it destroyed by changes in the environment; was it devoured by the many hungry "higher"-developed descendants; or can we expect to see an additional domain emerge in the distant future, and, thus, does it make sense to hunt for pre-cellular life in creepy, submarine volcanic areas?

Acknowledgments. – I am indebted to Drs. G. Fuchs, Karl O. Stetter, Günter Wächtershäuser, and W. Zillig for inspiring discussions and communicating unpublished work.

The emergence, diversification, and role of photosynthetic eubacteria

Beverly K. Pierson

Department of Biology, University of Puget Sound, Tacoma, Washington 98416, USA

The oldest fossil record tells us that the phototrophs were well established in shallow benthic environments very early in Earth's history, but it yields little or no information about the metabolism or the ancestry of these ancient bacteria. Were phototrophs the first organisms on earth? Or were phototrophs descendants of chemolithotrophs that formed deep, protected benthic communities long before life could exist in the photic zone? Despite the energetic advantage of photosynthesis, theoretical arguments suggest that the earliest life was nonphotosynthetic. The early surficial environment of Earth was probably a hostile zone bombarded by lethal UV radiation and large objects. Experiments show that the presence of UV radiation was probably not a deterrent to the growth of early phototrophs in the photic zone. Theoretical considerations and experimental data on contemporary microorganisms growing in mat communities suggest that diversification of phototrophs occurred early and was probably driven by intense competition for light and reductants. The key to the elucidation of the origin and path of this diversification lies in molecular phylogeny. Contradictions in the evolutionary relationships among phototrophs obtained from 16S rRNA sequences and from photosynthetic reaction-center sequences emphasize the importance of isolating a larger diversity of contemporary phototrophs to determine their evolutionary history.

I do not intend to argue evidence of the authenticity of the earliest fossil record but merely to accept that sometime between 3 and 4 Ga ago, photosynthesis emerged one or several times. The questions I wish to address are in two major categories.

1 *The emergence of phototrophic prokaryotes.* Who were their ancestors? What was their habitat like? What problems did they face? What was their physiology? Who are their closest living descendants?

2 *Diversification and role of early phototrophs.* What is the nature of the diversity among phototrophs today? What aspects of this diversity are likely to have occurred early in response to environmental constraints? What were the effects of the early phototrophs on the physical and biological environment as they emerged and diversified?

My perspective on the evolution of phototrophy will be restricted to photosynthesis based on a chlorophyll- or bacteriochlorophyll-containing reaction center (RC). I will

Bengtson, S. (ed.) 1994: *Early Life on Earth. Nobel Symposium No. 84.* Columbia U.P., New York

not be considering the bacteriorhodopsin-based photosynthesis characteristic of the halobacterial group within the archaebacteria. The bacteriorhodopsin photosynthesis is very restricted in its distribution, being confined to a small group of extreme halophiles. It is mechanistically quite distinct from chlorophyll-based photosynthesis (Oesterhelt 1989). The latter is restricted, as far as we know, to the eubacteria. Within this group, chlorophyll-based photosynthesis is found in five different and diverse phyla, leading to speculation that the last common ancestor of the eubacterial line of descent may have been photosynthetic (Woese 1987b). The conspicuous absence of chlorophyll-based photosynthesis from the archaebacterial line of descent argues against the presence of such photosynthesis in the common ancestor of both archae-bacteria and eubacteria.

Emergence of phototrophic prokaryotes

The origin of the phototrophs

Two very different series of events could have led to the emergence of the first phototrophic bacterium during the Archean.

(1) Photosynthesis may have originated in primitive cells by a series of abiotic and biotic events leading to the formation of a primitive protoporphyrin-IX based photosys-tem. Such a photosystem may have been based on the association of the protoporphyrin (and in more advanced cases, metalloporphyrins) with simple polypeptides embedded in the membrane. Light absorbed by the pigment molecules would have produced charge separation across the membrane and, hence, energy. This theory ties the origin of photosynthesis to the origin of early cells and to the subsequent origin of electron-transport bioenergetics. Such processes have been proposed by several people and are summarized by Olson & Pierson (1987a) and Pierson & Olson (1989). Critical aspects of this theory are the abiotic and early biotic evolution of the porphyrin–polypeptide structure. Studies referred to in the papers cited above and more recent experiments (Kolesnikov 1991; Masinovsky et al. 1989; Sivash et al. 1991) support the plausibility of such a system. Porphyrin-based photosystems were probably too complex to originate in protocells, and simple molecules might have provided a very early means of using light energy in precellular evolution (Deamer, this volume). The porphyrin theory, however, leads one to the conclusion that the last common ancestor of all contemporary cells was a porphyrin-based phototroph. One would therefore predict evidence of this type of metabolism in both the archaebacteria and eubacteria. While porphyrins and related compounds are present in many archaebacteria, chlorophylls are not present. None of the porphyrins present functions in photochemistry, although they do func-tion in electron-transport reactions. If one wishes to retain this theory, one must construct some explanation for this notable absence of chlorophyll from the archae-bacteria. Two simple explanations can be proffered. Both are dissatisfying. The first is trivial. We simply haven't found the organism yet. We are left with a hypothesis that out there somewhere is an extant archaebacterial phototroph containing chlorophyll (or at least photochemically active porphyrin). It is important to look for such an organism, but it may not exist. The second possibility is that the common ancestor, indeed, was a

protoporphyrin-IX based phototroph that diversified into chlorophyll-based photo-trophs in the eubacterial line but was supplanted with chemotrophs in the archae-bacterial line, the bacteriorhodopsin photosystem evolving quite independently later. This is a scenario that is impossible to verify at this time.

(2) Photosynthesis may have evolved later, after early chemotrophy in the form of lithotrophic anaerobic metabolism was already established in the last common ances-tor (Pace 1991; Wächtershäuser 1988a, b, 1990b; chapters by Kandler and Stetter, this volume). In this case, the emergence of photosynthesis would not have been tied to the emergence of cells and redox reactions but rather would have been a significant alteration in an already established fundamental system of electron-transport bioener-getics. In this scenario, photochemical charge separation by an appropriate protopor-phyrin or protochlorophyll would have conferred an energetic advantage on well-developed cells already competing for electron sources to sustain some primitive electron-transport system. If competition for electron sources or sinks was substantial, such an innovation might have led to the rapid proliferation and diversification of the phototrophic eubacterial cell line, which would have quickly slimed over the light-independent species and fully occupied the awaiting photic zone.

There simply are insufficient data to argue forcefully for either one of these scenarios at present, although I have previously argued for the phototrophic origin of life (Olson & Pierson 1987a, b; Pierson & Olson 1989). At this point, then, I want to assume that by whatever mechanism, the first phototrophs emerged. What were they like? What environment did they face?

The environment of the first phototrophs

The earliest putative microfossils, 3.5 Ga in age (Schopf & Walter 1983; Schopf & Packer 1987; Walsh & Lowe 1985; Awramik 1992; Schopf 1993, this volume), and associated stromatolites (Walter 1983; Awramik 1992) suggest the existence of shallow aquatic microbial mats in a marine marginal environment. The stromatolites appear to have been produced as organosedimentary structures (see chapters by Schopf and Walter, this volume). Evidence has been presented for the presence of chemical precipitation and sedimentation in marine margins in the Archean (Buick & Dunlop 1990), and if phototrophs were present, this is a habitat they were likely to have occupied. While specific microfossils and stromatolites from the Archean are subject to controversy in terms of their biogenicity (Schopf & Packer 1987; Buick 1990), it appears that genuine stromatolites formed sometime in the Archean despite the paucity of the rock record. By the Proterozoic such stromatolites were abundant (Walter et al. 1992).

The temperature of the earliest environments occupied by phototrophs is not well-established (Knoll & Bauld 1989). There does not appear to be overwhelming geologic evidence that these shallow seas were exceptionally warm. The phylogenetic evidence based on 16S rRNA, however, suggests that the ancestral eubacteria were not only phototrophs but also thermophiles (Achenbach-Richter et al. 1987a; Woese 1987b; Stetter et al. 1990; Stetter, this volume). This conclusion is based primarily on the fact that the two deepest divisions in the eubacterial branch of the tree of life contain thermo-philes. One must be cautious, however, since we have so few representatives from these deeper branches to study.

Thus the first phototrophs appear to have inhabited a shallow-marine environment subject to sedimentation. The water may or may not have been very warm. The environment was exposed to sunlight and an atmosphere devoid of oxygen and rich in CO_2 (Knoll & Bauld 1989). Some of the early mats appear to have been in lagoons so shallow that the mats may have been periodically exposed directly to the atmosphere and experienced desiccation (Walter 1983). The amount of organic matter that was present in the water is unknown, as are the local levels of critical nutrients such as sulfide and hydrogen, which serve as reductants for photosynthesis today. Iron (as Fe^{2+} and Fe^{3+}) was also present in some sediments (Borowska & Mauzerall 1988).

A problem facing the emerging phototrophs – UV radiation

The environment faced by the emerging phototrophs was probably high in UV radiation (Kasting 1987). The absence of an ozone shield would have permitted the penetration of high UV-C irradiances to the surface of the Earth. It is possible that other atmospheric screens existed (Kasting *et al.* 1989), but in their absence, phototrophs could have survived and thrived by a variety of mechanisms. Clearly, an atmospheric screen such as ozone was essential before large multicellular plants and animals could colonize the shallow waters or the land. When one focuses on the size range of bacterial cells, however (1 to 10s of μm), it is clear that much more localized "patchy" screens could have been suitable. On the scale of a bacterial cell, the sediments were the most likely source of protection. Bacteria grow and multiply among sediment particles very much larger than themselves. Stromatolites typically formed in evaporitic or sedimentary basins where chemically precipitated and sedimented particles contributed to the organosedimentary structure called a microbial mat (Walter 1983; Pierson 1992). Since sediments and evaporites seemed a plausible shield for phototrophs in microbial mats, we tested the effectiveness of several particulate substrates as UV screens.

The penetration of UV-C radiation from a germicidal lamp (maximum emission at 254 nm) was measured through 1, 2, and 3 mm thicknesses of white quartz sand, black basaltic sand, and $CaCO_3$ (Table 1). The substances were wet to better simulate the conditions prevailing in a moist microbial mat, and the UV irradiance transmitted through each layer was measured and expressed as percent incident irradiance at the surface of the sediment. The incident irradiance was 2.0 Wm^{-2} in all cases (a reasonably high estimate of unattenuated UV-C reaching the surface of the Earth in the Archean over the most damaging range of 240–270 nm). Clear silica gels were also prepared and impregnated with 0.01% or 0.1% $FeCl_3$ to simulate clear sand with known amounts of Fe^{3+} (Table 1). When complexed with various small organic acids, chloride, or hydroxide, ferric iron has a very strong absorbance in the UV-C (Olson & Pierson 1986). Furthermore, the absorbance is broad, covering the UV-A, UV-B, and short-wavelength visible light as well. Although the abundances of Fe^{2+} and Fe^{3+} would have varied in Archean sediments, it seems reasonable to expect minimum levels of Fe^{3+} in the 0.03–0.4 ppt range in some Archean sediments. In some localities the amount of Fe^{3+} complexed as ferric hydroxides in sediments may have been much higher (Borowska & Mauzerall 1988), and the potential UV screening of ferric hydroxides in paleosols has been suggested (Retallack 1990).

Table 1 summarizes the data on UV-C penetration through the various sediments. All sediments were effective to some extent in scattering and absorbing the incident UV

TABLE 1   Transmission of UV-C radiation through moist sediments, expressed in Wm–2 and in percent of incident irradiance. Measurements were made with a UV-sensitive radiometer, Model IL1350, with a calibrated cosine sensor, SED 40/W (International Light, Newburyport, MA). The sensor was placed underneath the quartz-plate support to obtain the incident irradiance without any sediment in place. Various thicknesses of sediments were then layered on the quartz plate, and the irradiance was measured.

	White Sand	Black Sand	$CaCO_3$	Silica Gel (no iron)	Silica Gel (+0.01% $FeCl_3$)	Silica Gel (+0.1% $FeCl_3$)
Incident	2.00 (100%)	2.00 (100%)	2.00 (100%)	2.00 (100%)	2.00 (100%)	2.00 (100%)
1 mm	0.031 (1.5%)	0.035 (1.75%)	0.009 (0.45%)	1.57 (79%)	0.72 (36%)	0.027 (1%)
2 mm	0.011 (0.55%)	0	0	1.13 (57%)	0.18 (9%)	0
3 mm	<0.001	0	0	0.78 (39%)	0.105 (5%)	0

radiation. The clear silica gels were the most transparent to UV-C, but the simple addition of low levels of $FeCl_3$ very much increased their opacity to UV radiation. Only the $CaCO_3$ as a densely packed 1 mm wet layer attenuated the UV radiation to a greater extent than the 1 mm of silica gel containing 0.1% $FeCl_3$. At this concentration most of the $FeCl_3$ was finely precipitated in the neutral gel so that its attenuating properties mimicked those of Fe^{3+} embedded in silt or sand particles. At the scale of a bacterial cell, an iron-bearing sediment or precipitated $CaCO_3$ (seen in mats as "whitings" [Grotzinger 1989]) could provide a very effective UV screen.

An effective UV screen for phototrophs must not only block out substantial UV radiation but must also transmit adequate visible (Vis) and near infrared (NIR) radiation (400–1,100 nm) to sustain photosynthesis. We therefore measured the spectral irradiance through the various sediment layers and integrated the values from 400 to 1,100 nm to determine if photosynthetic activity could occur. The 1 mm thick layer of 0.1% $FeCl_3$ in silica gel transmitted 88% of the incident irradiance in the photosynthetically useful range (400–1,100 nm). Even the thickest layer transmitted 77% of incident radiation in this range. The $CaCO_3$ also transmitted reasonably well in this range (approximately 12% of incident at a thickness of 1 mm and 6.5% at 3 mm). One millimeter of moist white quartz sand transmitted 25% of the visible and NIR radiation, and this value fell to 5% under 3 mm of sand. Although this last value seems low, it is still adequate to sustain photosynthesis, and the visible and NIR irradiance under 1 and 2 mm of sand were certainly adequate. The black sand transmitted very little light. At a depth of 1 mm, barely adequate radiation penetrated (about 2% of the incident) and at a depth of 2 mm the environment was essentially dark (0.04% of incident). The very nature of the stromatolite (an organosedimentary mat) structure intrinsically provided protection against damaging UV radiation while permitting photosynthesis.

The motility of gliding, filamentous, mat-forming phototrophs may have afforded protection from damaging UV radiation by permitting organisms to glide deeper into the sediments as a diel response to escape exposure to intense solar radiation. Others have suggested this mechanism (Margulis *et al.* 1976), and evidence for motile activity and possibly phototactic responses (which can be positive or negative) has been described from the fossil record (Awramik 1992). Recently Castenholz *et al.* (1991) described the downward movement to a depth of 1 mm in the mat of the cyano-bacterium *Oscillatoria boryana* in response to high solar irradiance.

The mat or stromatolitic growth habit itself has been suggested as a protective mechanism from UV damage (Margulis *et al.* 1976). The topmost layer of bacteria absorbs the damaging UV wavelengths, thus protecting the bacteria growing under-neath. The exposed cells are killed and lyse, eventually exposing the previously protected cells. The active growth zone then must be slightly beneath the very surface and must continuously replenish the sacrificial surface screen, enabling the mat as a whole to survive and grow. We have recently demonstrated the potential of this method by using an approximately 0.5 mm thick lawn of *Oscillatoria amphigranulata* grown on agar to protect a culture of the same organism grown in liquid beneath it. Liquid culture grown beneath a layer of plain agar in the absence of UV radiation had a yield measured in optical density (OD) units of about 1.3. Identical cultures grown beneath a layer of plain agar in the presence of 0.01 Wm^{-2} UV-C lysed before reaching an OD of 0.06. Cultures grown under the agar with a lawn of *O. amphigranulata* in the presence of 0.01 Wm^{-2} UV-C reached an OD of nearly 0.3. This extremely thin layer of cyanobacteria conferred substantial protection from exposure to UV-C (Williamson & Pierson, unpublished).

There are other intrinsic biological mechanisms that could have been used by early microorganisms to gain protection from damaging UV radiation.

Biologically produced screens are important today in protecting surface mat-dwelling and other microorganisms from damaging UV-A and UV-B radiation in environments exposed to high solar irradiances. Garcia-Pichel & Castenholz (1991) described the sheath pigment scytonemin found in more than 30 different species of cyanobacteria. Scytonemin was synthesized in response to high-intensity radiation, and UV-A wavelengths (320–400 nm) were particularly effective in stimulating its synthesis. Within cyanobacterial mats, Garcia-Pichel & Castenholz (1991) found scyto-nemin concentration was highest in the top ½ mm, while chlorophyll *a* concentration peaked in the third millimeter down, supporting the role of scytonemin as a protective pigment. Further physiological studies have strongly supported the role of this sheath pigment as a sunscreen (Garcia-Pichel *et al.* 1992). Another UV-A/B absorbing sheath pigment has been described in *Nostoc* (Scherer *et al.* 1988). Cyanobacterial sheaths have long been known to contain pigments, and the prominence of lamellated sheaths in the early fossil record (Schopf & Packer 1987) supports the idea that they may have functioned as early solar UV screens.

While sheath pigments are technically extracellular, intracellular screens may also be effective. Carotenoids are effective protective agents against damage due to expo-sure to intense, short-wavelength, visible and near-UV radiation (Paerl 1984). An early evolution of carotenoids and other internal pigments absorbing more strongly in the far UV may have helped protect cells from damage.

Chloroflexus aurantiacus is intrinsically resistant to UV radiation because of some unknown mechanism. We have studied the intrinsic UV-C resistance of *C. aurantiacus* by growing cultures phototrophically in the presence of Vis light and UV-C radiation and measuring the yield as a function of UV-C irradiance (Pierson *et al.* 1993).

Cultures were grown anoxically in quartz flasks at 55°C in the presence of light supplied by a reflector floodlamp and a germicidal lamp. *Chloroflexus* is a photohetero-troph, and the medium was organic. The UV irradiance was measured at the surface of the culture medium. Cultures were well mixed with continuous sparging with a mixture of N_2 and CO_2. Growth was measured as increase in OD at 650 nm. Fig. 1 summarizes some of the results, showing the growth curves for the control (no UV) and cells grown under continuous irradiances of 0.01 Wm^{-2}, 0.02 Wm^{-2}, 0.05 Wm^{-2}, 0.21 Wm^{-2}, and 0.66 Wm^{-2}. Most surprising was that *C. aurantiacus* grew very well under a continuous irradiance of 0.01 Wm^{-2}, a level of UV exposure that was lethal to *E. coli*. The yields of the *C. aurantiacus* cultures decreased with increasing irradiance. At irradiances above 0.21, cultures still grew initially, but growth leveled off, and the cells lysed after 25–35 hours.

We do not yet know the cause of this intrinsic resistance to UV-C in *C. aurantiacus*, but it is a particularly interesting property given the thermophilic, mat-forming character-istics of this organism as well as its apparent deep divergence in the eubacterial line of descent (Oyaizu *et al.* 1987).

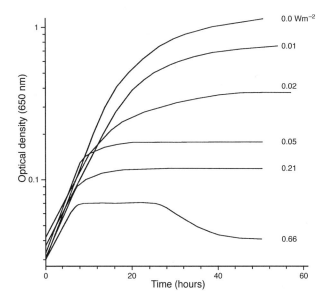

FIGURE 1 Growth curves for *C. aurantiacus* grown in absence of UV-C radiation and in the presence of several different UV-C irradiances. Growth yields (measured as optical density at 650 nm) decreased with increasing UV irradiance. Details of growth conditions are discussed in the text. (Modified from Pierson *et al.* 1993.)

What were the early phototrophs like physiologically, and who are their closest living descendants?

The earliest microfossils and associated stromatolites are frequently attributed to cyanobacteria (Schopf & Packer 1987; Awramik 1992). While morphologically they appear quite similar to extant filamentous and coccoid cyanobacteria, the point has been made repeatedly that on morphological criteria alone the filamentous fossils at least could easily be attributed to a wide variety of prokaryotic groups – including non-phototrophs (Olson & Pierson 1987a; Walter 1983; Knoll & Bauld 1989; Schopf & Packer 1987; Schopf 1993 and this volume). No chemical evidence is available to distinguish among the many possibilities. The presence of microfossils similar to cyanobacteria in structure has been offered as evidence for the presence of oxygenic photosynthesis at a very early stage of the Precambrian (Schopf & Packer 1987; Buick 1992; Awramik 1992). This is such an important point that I wish to pursue it in more depth.

First, I want to summarize recent published work on anoxygenic phototrophic bacterial mats and present recent data on anoxygenic marine phototrophs that support the interpretation that the earliest microfossils and stromatolites were not formed by cyanobacteria but rather by other diverse filamentous phototrophs.

The most abundant microbial mats constructed by phototrophs today in all environments are cyanobacterial (Pierson 1992; Bauld *et al.* 1992; Golubic, this volume). Many of these mats are devoid of anoxygenic phototrophs entirely. Others have well-developed laminated fabrics with lower layers of anoxygenic phototrophs. There are a few limited occurrences of microbial mats composed entirely of anoxygenic phototrophs and devoid of cyanobacteria. These latter mats are totally anoxic (Ward *et al.* 1992).

Ward *et al.* (1989) described some of these mats located exclusively in thermal environments. In Yellowstone National Park two distinct types of anoxygenic hot-spring mats have been studied. One is composed of an autotrophic *Chloroflexus* that forms fairly thick (up to 0.5 cm) laminated mats in sulfide springs. While other non-photosynthetic bacteria may contribute to these mats, cyanobacteria are totally absent. The *Chloroflexus* filaments form a cohesive fabric and are microscopically indistinguishable from many cyanobacteria (Giovannoni *et al.* 1987). In other sulfide thermal springs, a layered mat develops of unicellular *Chromatium tepidum* on the surface and filamentous *Chloroflexus* underneath. Again, cyanobacteria are absent, and no oxygen is ever present. Castenholz *et al.* (1990) have characterized a thermal mat of unicellular *Chlorobium tepidum* that forms in a few slightly acidic sulfide springs in New Zealand. Again, no cyanobacteria were present, and no oxygen was detectable. These mats were not laminated, apparently because of the lack of a filamentous fabric.

The thermal mats described above containing filamentous *Chloroflexus*, if appropriately fossilized, would look like thin stromatolites containing long, sinuous filaments about 1.0 µm in diameter. While these mats are limited in area and arguably rare today, they may well represent a type of anoxygenic microbial mat that was abundant in the Archean. The fact that *Chloroflexus* is the deepest phototrophic branch of the eubacterial tree (Oyaizu *et al.* 1987) makes these thermal mats all the more interesting as representatives of the types of communities formed by the earliest phototrophs.

Anoxygenic, filamentous *Chloroflexus*-like organisms are not confined to a few rare and tiny hot springs. In most alkaline hot springs worldwide, these organisms form a prominent mat layer under a layer of oxygenic cyanobacteria. More recently, *Chloroflexus*-like organisms have been observed in mats in marine intertidal and hypersaline environments (Mack & Pierson 1988; D'Amelio *et al.* 1989; Stolz 1983). In these microbial mats, *Chloroflexus*-like organisms are most often associated with several species of cyanobacteria, purple sulfur bacteria, and *Beggiatoa*. While abundant enough to be recognized, they are not necessarily the dominant organisms in the mat.

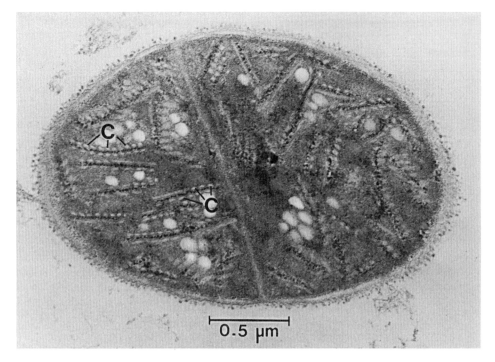

FIGURE 2 Electron micrograph of thin section of undescribed *Chloroflexus*-like organism growing in hypersaline microbial mats. Note internal membranes lined with chlorosomes (C). This organism is currently maintained in crude culture and appears to be photoheterotrophic, like most thermophilic strains.

We have observed marine hypersaline mats composed of *Chloroflexus*, *Beggiatoa*, and the cyanobacterium *Spirulina*, in which the *Chloroflexus*-like organisms were a major, if not *the* major, constituent. From mats such as these we have successfully cultured two very different strains that are maintained as mixed cultures. We have not succeeded in isolating either organism but have done studies on highly enriched populations.

These organisms are relevant to the current discussion because they represent marine hypersaline matformers that have several features in common with the thermophilic *Chloroflexus* species: they are filamentous, contain Bchl *c* or *d*, glide, and have abundant chlorosomes (Fig. 2). The facts that they are mesophiles and hypersaline marine organisms broaden significantly our concept of the *Chloroflexus*-like organisms. Furthermore, in the mat context, these organisms look just like many cyanobacteria ranging in diameter from 1.2 to 2 µm (Fig. 3). The conclusion is that we have today a few hypersaline marine marginal environments that have well-developed microbial mats with abundant anoxygenic phototrophs and few cyanobacteria. We would expect that such mats, devoid of cyanobacteria, would have been abundant in the Archean before the evolution of cyanobacteria and oxygenic photosynthesis.

The absorption spectra (Fig. 4) are of whole-mat homogenates from hypersaline lagoons with salinities of 120–160 ppt. The relatively high absorbance at 753 nm compared with that at 676 nm reflects the high proportion of *Chloroflexus*-like organ-

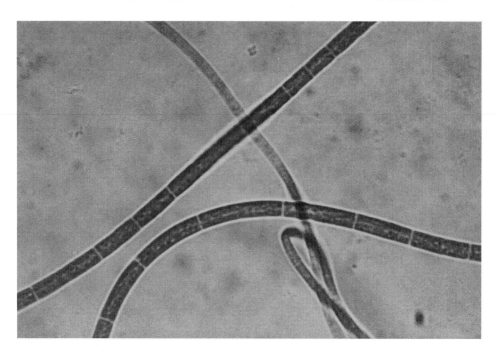

FIGURE 3 Photomicrograph of two marine *Chloroflexus*-like organisms abundant in hypersaline microbial mats. The larger filament (2 μm in diameter) is the one represented in Fig. 2. The thinner filament (1.2 μm in diameter) is an autotrophic strain. Both are maintained in crude cultures.

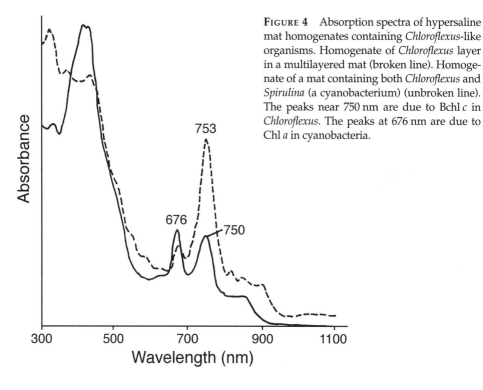

FIGURE 4 Absorption spectra of hypersaline mat homogenates containing *Chloroflexus*-like organisms. Homogenate of *Chloroflexus* layer in a multilayered mat (broken line). Homogenate of a mat containing both *Chloroflexus* and *Spirulina* (a cyanobacterium) (unbroken line). The peaks near 750 nm are due to Bchl *c* in *Chloroflexus*. The peaks at 676 nm are due to Chl *a* in cyanobacteria.

TABLE 2   Anoxygenic filamentous phototrophs usually found in microbial mats. All the organisms in this table were reviewed in more detail in Pierson & Castenholz 1992, where the original references to the data are also cited.

Organism	Morphology	Pigments	Metabolism	Habitat
C. aurantiacus	gliding filaments 0.5–1.5 μm diam.	Bchl a and c	photoheterotrophy, some photoautotrophy (sulfide-dependent or hydrogen-dependent)	neutral to alkaline thermal-springs undermat, surface mat, or unispecific mat
Marine Chloroflexus-like organisms	gliding filaments 1.0–2.0 μm diam.	Bchl a and c	photoheterotrophy photoautotrophy?	marine intertidal salt marsh sediments
Marine Chloroflexus-like organisms	gliding filaments variable diam.	Bchl c	undetermined	hypersaline marine mats, often in association with Microcoleus chthonoplastes
Marine Chloroflexus-like organism	gliding filaments 1.2–2.0 μm diam.	Bchl a + c or d	photoautotrophy photoheterotrophy?	hypersaline marine mats (a) as distinct layer beneath surface layer of cyanobacteria, (b) as surface layer mixed with Spirulina and Beggiatoa
Heliothrix oregonensis	gliding filaments 1.5 μm diam.	Bchl a	photoheterotrophy	alkaline thermal springs lacking sulfide; surface mat above cyanobacteria
Thermophilic red filaments	gliding filaments 1.5 μm diam.	Bchl a	photoheterotrophy	alkaline thermal springs as undermat below cyanobacteria and Chloroflexus
Marine purple filaments	gliding filaments 0.8–0.9 μm diam.	Bchl a	undetermined	hypersaline mats in association with Microcoleus chthonoplastes
Chloronema	gliding filaments 2 μm diam.	Bchl d	undetermined	planktic
"Oscillochloris"	gliding filaments 1.0–5.5 μm diam.	Bchl c	undetermined	surface of freshwater sulfide-containing muds

isms relative to cyanobacteria in these mats. These organisms were clearly autotrophic and functioned as primary producers in this ecosystem.

A survey of the recent literature (Table 2) shows that during the past 10 years many diverse filamentous anoxygenic phototrophs have been discovered in microbial mats. We know very little about these organisms and what role their ancestors may have played in the early evolution of photosynthesis.

In summary, we have demonstrated the existence of dense populations of Bchl-c-containing anoxygenic filamentous autotrophs in marine hypersaline mats as well as in hot-spring mats. In general morphology these bacteria look very much like mat-forming cyanobacteria. It is organisms such as these that are most likely to have accounted for the earliest phototrophic microbial mats in the Archean.

It is also possible that the ancestral non-cyanobacterial mats had already been displaced by largely cyanobacterial or at least mixed mat communities by 3.5 Ga ago and that the microfossils that look so similar to modern cyanobacteria (especially the coccoid forms) are indeed the direct ancestors of these phototrophs. If such is the case, there is not likely to be a fossil record of the microbial mats built by the more ancient anoxygenic phototrophic bacteria. The remaining intriguing question then becomes one of whether or not the organisms fossilized from 3 to 3.5 Ga ago were oxygenic. This is a very important question. Morphology alone does not answer it, and there are several reasons why chemical signals might not, either.

Although the oxidation of iron can be explained by the appearance of biogenic oxygen, it can also be explained by other chemical means (Borowska & Mauzerall 1988) and perhaps by a direct photobiological mechanism (Pierson & Olson 1989; Olson & Pierson 1987a, b). Furthermore, it is likely that the evolution of oxygenic photosynthesis was accompanied very quickly by the evolution of aerobic respiration (Pierson & Olson 1989; Towe 1990, 1991, this volume). Oxygen may have been consumed within the microbial mat community as quickly as it was produced for quite some time. In some contemporary thermal mats, high in sulfide, in which cyanobacteria are sandwiched between layers of *Chloroflexus*, the oxygen produced by cyanobacterial photosynthesis is consumed entirely within the mat, never escaping to the surrounding aquatic phase (Jørgensen & Nelson 1988).

It seems likely that the earliest phototrophs were anoxygenic. The earliest microfossils, however, may represent much more advanced phototrophs capable of oxygenic photosynthesis (Schopf & Packer 1987; Awramik 1992). Evidence is growing for well-established oxygenic photosynthesis before 2.5 Ga ago (Buick 1992). The time of the origin of this process, however, remains elusive.

Diversification and role of early phototrophs

There are two different questions to consider in the discussion of diversification of early phototrophs: (1) How and when did the diversity of extant photosynthetic eubacteria arise? (2) What kind of photosynthetic diversity developed early in the environments of the Archean and Proterozoic? There is a serious problem that plagues the process of answering both of these questions in any meaningful way. We have very limited information about the extent of the diversity that exists among contemporary phototrophs and know even less about the diversity that existed during the genesis of phototrophy. Our limitation in the study of contemporary diversity stems from the fact that few of the extant bacteria of any kind have been isolated and cultured. Any microbiologist who looks at natural habitats is aware of this situation, but few have put it in quantifiable terms. Recently Ward *et al.* (1990a, b), using a 16S rRNA cloning technique on a well-studied thermal mat thought to be of limited species diversity, demonstrated that the isolated and identified microorganisms constituted only a small number of the organisms actually present. On the basis of culture and isolation techniques, it is likely that microbiologists are enormously underestimating species diversity in microhabitats. This generalization, which applies to all prokaryotes, is true of the phototrophs also.

From my own microscopic observations of the more complex (and probably more diverse) habitat of hypersaline microbial mats, it is clear that there are many more phototrophs present than have been isolated. Studies of mats with electron microscopy (Stolz 1983; D'Amelio *et al.* 1987, 1989) have revealed the presence of several phototrophs (recognized by membrane and chlorosome ultrastructural characteristics) that differ significantly from known species. At least five different marine *Chloroflexus*-like organisms have been detected by this means. The uncultured but nevertheless significant organisms *Oscillochloris* and *Chloronema* appear related to *Chloroflexus*, but no 16S rRNA data are available for comparison. In my lab I have worked with four filamentous mat formers: the two hypersaline *Chloroflexus*-like organisms described earlier and two Bchl-*a*-containing organisms from thermal environments (Pierson *et al.* 1984; Castenholz 1984). None of these organisms have been isolated in pure culture. Of the 12–15 filamentous anoxygenic mat-forming phototrophs that have been observed (see Table 2), only one (*Chloroflexus aurantiacus*) has been studied in pure culture and used to help create the eubacterial phylogenetic tree. The other diverse filamentous phototrophs await their placement in the scheme.

Since we must work with what we have, I will consider some of the interesting issues in the origin of the diversity that we see in extant photosynthetic eubacteria. There are three important aspects of contemporary diversity that appear to have some phylogenetic significance and that probably represent major evolutionary events (assuming a monophyletic origin and early evolution of photosynthesis): carbon metabolism, source of reductants for autotrophy, and pigment systems.

Carbon metabolism

Many contemporary photosynthetic bacteria are photoheterotrophs (*Heliobacterium* and its relatives, the purple non-sulfur bacteria, and *Chloroflexus aurantiacus*). Autotrophy is unknown in the *Heliobacterium* group. The pathways of autotrophy in phototrophs and chemolithotrophs have been reviewed by Fuchs (1989). The purple sulfur bacteria are autotrophs and use the Calvin-Benson Cycle for CO_2 fixation, as do the cyanobacteria and chloroplasts of plants and algae. The green sulfur bacteria are obligate autotrophs but apparently lack the enzymes of the Calvin-Benson Cycle, using instead a reductive citric-acid cycle for CO_2 incorporation. Some strains of *C. aurantiacus* can be grown autotrophically and use a simple mechanism of CO_2 fixation quite unlike either of the others used by photosynthetic organisms (Holo 1989; Strauss *et al.* 1992). The CO_2-fixation pathway in *Chloroflexus* involves the carboxylation of acetyl-CoA to form the unusual intermediate 3-hydroxypropionate (Holo 1989), followed by subsequent reduction and another carboxylation step to produce succinate (Strauss *et al.* 1992). Questions still exist regarding the pathway, but it is cyclic and clearly unrelated to other known pathways. The presence of such a novel carbon-fixation pathway in *C. aurantiacus*, coupled with its apparent antiquity among photosynthetic eubacteria, reinforces my conviction that it is necessary to isolate a greater array of phototrophs for comparative biochemical studies. The diversity of autotrophic mechanisms in the phototrophs is interesting, because it is so group-specific. While only three pathways are known in phototrophs, their distribution is very restricted.

It is a common assumption that heterotrophy preceded autotrophy even in photo-trophs (Pierson & Olson 1989), and various mechanisms have been proposed for the abiotic production of organic carbon (Borowska & Mauzerall 1988). Recent arguments have been put forth supporting autotrophy as a primordial mechanism being driven by abiogenic pyrite formation and perhaps by the earliest cells (Wächtershäuser 1990b). It has further been suggested that this primordial autotrophy might have evolved into the reductive citric-acid cycle as an early metabolic pathway for autotrophy (Wächters-häuser 1990a). This particular scheme suggests that autotrophy preceded phototrophy.

The advent of autotrophy, by whatever mechanism, conferred upon phototrophs the all-important role of fixing inorganic carbon into organic forms that could sustain the growing biosphere and establish a carbon cycle. The evolution of autotrophy freed the early phototrophs from dependence on abiogenic organic substrates and probably encouraged a very rapid proliferation of biomass.

Despite various speculations, we do not know which autotrophic pathway arose first in the phototrophs. We also know nothing about the evolutionary relationships, if any, among the various pathways. The evolutionary history of carbon-fixation mecha-nisms is probably polyphyletic.

Although we know that autotrophy must have existed very early to support the burgeoning biomass, without knowing the nature of the earliest pathways, we do not know what contribution each particular pathway might have made to the carbon-isotopic signals in the geologic record (see chapters by Schopf and Hayes, this volume, for extended discussions of the carbon-isotopic record). The $\delta^{13}C$ values for carbon-isotopic fractionation by cultures of autotrophic bacteria using different pathways are significantly different (Fuchs 1989; Preuß et al. 1989). Values range from –36‰ for non-phototrophs using the reductive acetyl CoA pathway to –26‰ for phototrophs and chemolithotrophs using the Calvin-Benson cycle to –3.5 to –12‰ for the reductive citric-acid cycle found in green sulfur bacteria. A value of –13.7‰ has been reported for *C. aurantiacus* and its incompletely elucidated pathway (Holo & Sirevåg 1986).

Although the negative $\delta^{13}C$ values from the Archean record seem likely to be a signal for autotrophy (Schidlowski 1988), assignment to a particular carboxylation mecha-nism is not possible when several different mechanisms exist and others are only just being elucidated. In addition, several environmental factors, in particular local abun-dances of CO_2, can influence the $\delta^{13}C$ values (Preuß et al. 1989).

Once autotrophy was established, overall primary productivity was probably higher than in microbial mats today because of the higher availability of inorganic carbon (Rothschild & Mancinelli 1990; Rothschild 1991). Although carbon may have become limiting in some localized Archean mat environments, it seems likely that once autotrophy evolved, competition for limiting reductants and light were more signifi-cant factors ecologically.

Competition for reductants and the birth of biogeochemical cycles

The origin of autotrophy and the subsequent development of a carbon cycle drove the competition for reductants and enhanced the cycling of other significant elements. Hydrogen, H_2S and other reduced forms of sulfur function as reductants for CO_2 fixation in all groups of autotrophic phototrophs today. Water functions as a reductant only within the cyanobacteria (and associated prochlorophytes and chloroplasts). The presence of photosystem-I-dependent autotrophy using either H_2 or H_2S in some cyanobacteria in the absence of photosystem-II-dependent oxygen evolution (Cohen 1984) is consistent with the idea that these reductants served as the earliest electron donors to photosynthesis.

The photosystem-II reaction center (RC) in cyanobacteria is functionally similar to the quinone-based reaction centers of purple sulfur and non-sulfur bacteria and the *Chloroflexaceae*. It is likely that these RCs share a common ancestry (see discussion below). However, the jump in reduction potential from the redox systems based on H_2S and H_2 in the anoxygenic phototrophs to the water-dependent photosystem II of cyanobacteria is quite large (0.6 V). It has been suggested that electron donors with intermediate reduction levels preceded the selection of the extant high-potential water-oxidizing system (Pierson & Olson 1989). The participation of other reductants in an evolutionary trend of increasing potential has been postulated to account for a gradual succession between hydrogen sulfide and H_2O (Olson 1970). The problem is that no other reductants have as yet been found to sustain photoautotrophy in extant organisms. Among the possible intermediate reductants, perhaps the most intriguing is iron (Fe^{2+}). Low-potential reductants such as H_2 and H_2S were abundant at least locally in the Archean, as was the higher potential Fe^{2+}. In the absence of biogenesis of H_2 and H_2S, the dependence of photosynthetic metabolism on these reductants would have confined substantial growth and accumulation of microbial mats to basins in which these substrates were supplied from some sort of venting system. As venting systems plugged up or were exhausted, or as biomass increased, competition for a limiting supply of low-potential reductants would have provided a selective advantage to phototrophs able to extract electrons from a more ubiquitous subtrate such as Fe^{2+}. Given some degree of variability in the potential of the RCs, those with slightly higher potentials might have been able to oxidize iron or other relatively high-potential reductants such as manganese. Ultimately, of course, such a process would have culminated in the evolution of the highest-potential RC able to extract electrons from the unlimited supply of water. If an organism ever existed that was able to use Fe^{2+} for photoautotrophy and not able to use H_2O, one might expect to find descendants among extant organisms today. While Cohen (1984) demonstrated some possible photo-dependent iron-oxidizing ability in cyanobacteria, it is difficult to prove conclusively that this oxidation does not depend on minute amounts of O_2 production in an oxygen-evolving phototroph. Proof of the existence of photoferrotrophy will be unequivocal if bacteria are found that oxidize Fe^{2+} only and not water. None have been found, but it also appears that no one has looked very hard for such bacteria. Just as H_2S-dependent phototrophs still exist today, photoferrotrophs might also exist and have some advantage for growth in a microbial community devoid of sulfide and hydrogen, devoid of O_2, rich in Fe^{2+}, and exposed to wavelengths of light unusable by cyanobacteria. We are currently searching for such an organism and believe such a search for evolutionary intermediates is particularly important to clarify some of the questions that remain to be answered about the evolution of RCs (see below).

The existence of a photoferrotroph would have provided an anaerobic means of biologically producing Fe^{3+} in the Archean before the advent of oxygenic photosynthesis. Currently some argue for the abiological oxidation of iron and others for an indirect biological oxidation via oxygen produced by ancestral cyanobacteria. The direct biological oxidation by phototrophy is a compelling possibility. It is not clear whether such organisms might have contributed directly to the formation of BIFs (Hartman 1984) and to an iron cycle between photoferrotrophs and early anaerobic respirers able to reduce the biogenically produced Fe^{3+} (Hartman 1984), perhaps later becoming the first aerobic respirers. An early photoferrotroph would have had a distinct advantage in competition for light by depositing its own UV shield in the form of Fe^{3+}-bearing sheaths around its cells, thereby permitting growth in an otherwise hostile habitat at the very surface of microbial mats. As we will see in the next section, light limitation would have been a very important factor in the earliest microbial communities once photo-autotrophy was established.

Pigment systems

Light-harvesting systems. – There is a greater diversity of light-harvesting (accessory pigment) systems than reaction centers in phototrophs. Study of phototrophs growing in natural communities (both aquatic and microbial-mat communities) suggests that the evolution of this diversity may have occurred relatively early.

The vertical stratification of phototrophic communities with different light-harvesting pigments is well known today in the occurrence of planktic "plates" of bacteria in aquatic systems. Similar stratification (compressed to a millimeter scale of distribution) exists in highly layered microbial mats.

Light penetration in microbial mats is particularly relevant to our discussion of the development of Archean photosynthetic communities. Vertical stratification in mats where light of one wavelength range may quickly become limiting is possible only if the different mat layers possess pigments that can absorb different wavelengths of light (Pierson *et al.* 1990). Once sediment communities become densely packed with pigmented organisms, attenuation of light to levels that are limiting to photosynthesis occurs rapidly.

Fig. 5 is a depth profile of spectral irradiance through a microbial mat in a hypersaline pond at Guerrero Negro, Baja California, Mexico. At a depth of 3.5 mm from the mat surface there was no light at wavelengths less than 530 nm. The irradiance at 670 nm (absorbed by Chl a) had fallen to 0.001 μmol m^{-2}s^{-1}nm^{-1}, and cyanobacteria were unable to grow. Yet over the range of 800–900 nm, where bacteriochlorophyll a absorbs, the irradiance was more than two orders of magnitude greater than at 670 nm and sufficient to sustain photosynthesis by the purple sulfur bacteria found at this depth. At depths of 6.0 mm or more, the irradiance at 720–760 nm exceeded that from 800–900 nm, and populations were found of *Chloroflexus*-like organisms, which absorb light in the mid-700-nm range. These anoxygenic phototrophs would not be able to grow in this habitat without their diverse light-harvesting pigments. The photic zone in ancient microbial mats simply could not have developed to any appreciable thickness without similar diversity in pigmentation. In the presence of only one pigment system, mats would have to be thin and productivity relatively low (Pierson *et al.* 1990, unpublished).

Not all layered mats comprise organisms with the same absorption properties. Fig. 6 illustrates some of the diversity of functional light-harvesting systems in modern

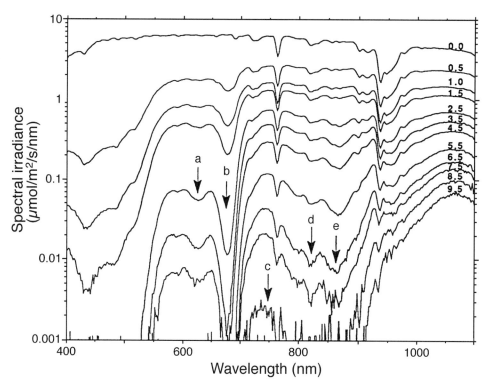

FIGURE 5 Depth profile of spectral irradiance through a hypersaline microbial mat. Depths in millimeters are given at the far right of each spectrum. Troughs in the spectra represent attenuation of radiation by pigments. Arrows indicate positions where particular pigments absorb. a: phycocyanin; b: Chl *a*; c: Bchl *c*; d and e: Bchl *a*.

mats from various environments. In different mats, different pigments allow for layering. Almost all mats have drastic attenuation below 550 nm, and the most significant diversity in pigment systems occurs from 550 to 1000 nm. Most mats transmit considerable radiation in the NIR. Radiation above 1020 nm, however, is too low in energy to sustain photosynthesis. The diversity in light-harvesting systems in microbial mats is due primarily to the presence of Chl *a* and phycobilin pigments in cyanobacteria and Bchls *a*, *b*, *c*, *d*, and *e* in anoxygenic phototrophs. We have not yet detected Bchl *g* in mats.

Reaction centers. – Among all phototrophs the most essential part of the photosynthetic apparatus is the photochemical reaction center (RC), an integral membrane complex of two or more protein subunits and associated redox centers, some of which are chlorophyll molecules. Because of its essential nature and the fact that all reaction centers have certain structural (at least organizational) features in common that result in a similar function of charge separation, it has been suggested that the origin of all contemporary RCs is monophyletic. Arguments for such a monophyletic origin are based primarily on key functional similarities (Mathis 1990; Nitschke & Rutherford 1991; Blankenship 1992) rather than on data revealing sequence similarities. If all RCs

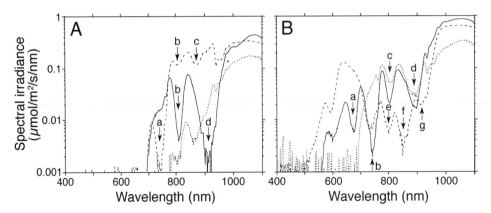

FIGURE 6 Comparative irradiance spectra derived from several different microbial mats with similar irradiance values but very different pigmentation, to illustrate how different pigments affect the light availability within the mat. **A:** Spectrum at depth of 5.0 mm in a thermal mat containing cyanobacteria, *Chloroflexus*, and a filamentous phototroph with Bchl *a* absorbing at 800 and 910 nm (b and d, unbroken line). Spectrum at depth of 2.5 mm in a thermal mat containing cyanobacteria, *Chloroflexus*, and a filamentous phototroph absorbing at 800 and 860 nm (b and c, dotted line). Spectrum at a depth of 3.0 mm in a thermal mat composed entirely of *Chloroflexus*, which contains Bchl *c* absorbing at 740 nm (arrow at a) and Bchl *a* absorbing at 800 and 865 nm (broken line). Note in all cases that all radiation below 700 nm is totally attenuated, leaving only NIR radiation to sustain photosynthesis at these depths. **B:** Spectrum at depth of 5.0 mm in a geyser-spray mat containing cyanobacteria, no *Chloroflexus*, and a filamentous phototroph with Bchl *a* absorbing at 800 and 900 nm (c and d, dotted line). Spectrum at a depth of 0.1 mm in a mat containing purple sulfur bacteria with Bchl *a* absorbing at 790, 860, and 910 nm (e, f, g), *Chloroflexus* with Bchl *c* absorbing at 740 nm (b), and no cyanobacteria (broken line). Spectrum at depth of 1.5 mm in a typical hot-spring mat containing layers of cyanobacteria with Chl *a* absorbing at 670 nm (a), *Chloroflexus* with Bchl *c* absorbing at 740 nm (b), and filamentous bacteria containing Bchl *a* absorbing at 800 and 900 nm (c and d; unbroken line).

evolved from one common ancestor, protein sequences have diverged so greatly over time that finding evidence in sequence homology is difficult.

All extant RCs can be divided into two types depending on the nature of the primary and secondary electron acceptors in photochemistry. The Q-type, or photosystem-II-type, reaction centers found in all purple photosynthetic bacteria, *Chloroflexus aurantiacus,* and in photosystem II of cyanobacteria and chloroplasts, have pheophytin pigments as the primary electron acceptors and quinones as the secondary electron acceptors. The Fe–S RCs found in green sulfur bacteria, heliobacteria, and photosystem I of cyanobacteria and chloroplasts contain chlorophylls, quinones, and bound Fe–S centers as primary and secondary acceptors.

The essential proteins within the Q-type RCs include the L and M subunits in the purple bacteria and *Chloroflexus* and the D1 and D2 subunits in cyanobacterial and chloroplast RCs. The photosystem-I-type RCs are far more complex, and in both cyanobacteria and chloroplasts they contain more subunits and accessory components than the Q-type RCs. The subunits are not yet completely sequenced from the green sulfur bacteria and heliobacteria. Sequence comparisons are therefore difficult to make among all of the RCs, although the question of a common ancestral RC has been probed.

There is little homology between the psaA and psaB polypeptides of the photosystem-I (Fe–S) RCs and the D1 and D2 or L and M polypeptides of photosystem-II RCs. A few conserved features might exist, but if one ignores the functional properties held in common, the necessity to bind similar prosthetic groups, and the essential features of both being integral membrane proteins, one cannot argue very convincingly for a common ancestor (Robert & Moenne-Loccoz 1990; Margulies 1991). If a common ancestor existed, the divergence of the genes encoding the two RCs was a very ancient event.

Sequence homologies exist among the heliobacterial and plant photosystem-I (psaB) polypeptides (Trost *et al.* 1992), and it appears that divergence of these two RCs preceded the gene-duplication event leading to the heterodimeric psaA, psaB RC.

Identification and sequencing of the photosystem-II (Q-type) RC polypeptides has proceeded further, and some very intriguing data have recently become available. Sequences are known for the L and M subunits (the core subunits) from the reaction centers of several purple photosynthetic bacteria and from *C. aurantiacus.* They are also known from the comparable D1 and D2 subunits of cyanobacteria and chloroplasts. The overall sequence identity among all the L, M, D1, and D2 subunits is less than 10% (Michel & Deisenhofer 1988). However, focusing on key regions of the polypeptide reveals that they are probably derived from a common ancestral polypeptide (Beanland 1990; Blankenship 1992). All of the RCs are heterodimers in which the two subunits are clearly related and are encoded by genes that were duplicated and subsequently diverged. The question is how many times independent gene duplications occurred. Parsimony analysis has been applied to these data to construct evolutionary trees for the reaction-center subunits. Beanland (1990) found that his conservative analysis, which withstood rigorous statistical testing, led to the conclusion that the ancestral photosystem-II RC gene diverged deeply to form two lines of descent. One of these had a single gene-duplication event leading to the D1, D2 subunits of the cyanobacteria. The other line also had a single gene-duplication event leading to the L, M subunits of the purple bacteria and *Chloroflexus.* The conclusions have very interesting implications. First, they suggest that the origin of the RC in photosystem II in the cyanobacteria is more ancient than previously thought. Second, they suggest that the RCs in the purple bacteria and *Chloroflexus* are more closely related than is suggested by the 16S rRNA tree. Blankenship (1992) independently produced a very similar evolutionary history for the D1, D2 and L, M RC subunits and has invoked lateral gene transfer of the photosynthetic genes between the purple bacteria and *Chloroflexus* to explain the lack of congruence with the 16S rRNA tree. If this is true, and the transfer of genes was from the purple bacteria to *Chloroflexus,* the argument for the antiquity of photosynthesis in the eubacterial line is substantially weakened. Beanland (1990) has offered a similar interpretation but has also suggested that the phylogenetic relationships expressed in the RNA tree may not have been adequately tested for statistical reliability, calling into question the significance of the distances between branches of the tree. The significance of the relative branching order of the RNA eubacterial tree has also been challenged by Cavalier-Smith (1991b).

The potential antiquity of the cyanobacterial RC and the interesting confusion regarding the antiquity of the RCs of the anoxygenic phototrophs presented above reaffirm the importance of finding more phototrophs to examine for clues into the origin and early evolution of photosynthesis.

Conclusion

The understanding of the evolution of photosynthetic organisms is likely to be improved by more data on the phylogeny of the RC proteins. We are especially in need of data on the phylogeny of the photosystem-I RC proteins. While lateral gene transfer of the photochemical genetic system has been implicated in the evolution of anoxygenic phototrophs with Q-type RCs, all other possible relationships must be carefully evaluated. The argument for the postulated antiquity of photosynthesis in the eubacterial line of descent may be seriously weakened if lateral gene transfer is significant.

The availability of more sequences for Fe–S- (photosystem-I-) type RCs will permit better analysis of the relatedness of these genes to those that encode the Q-type RCs. If indeed it can be shown that these two types of RC had a common ancestral RC, then the interesting question becomes one of whether the gene duplication that led to the two different RCs resulted in the persistence of two functionally similar yet significantly different photosystems in the same cell (the ancestor of modern cyanobacteria), as suggested by Pierson & Olson (1989), or whether the alteration occurrred in different cells. The latter possibility would have resulted in the Q-type RCs being confined to one line of descent and the Fe-S RCs being confined to another and, thus, subsequently would have required another lateral gene transfer to allow for a fusion of the two RCs in an ancestral cyanobacterium (Blankenship 1992). Currently it is impossible to distinguish between these alternative hypotheses.

Perhaps the most intriguing suggestion to emerge from the recent RC data is the possibility that the cyanobacterial RCs could be the most ancient (Beanland 1990). This suggestion has also been made on the basis of the theory of the evolution of chlorophylls based on the Granick hypothesis (see Olson & Pierson 1987a, b; Pierson & Olson 1989), which implies that chlorophyll *a* was the first chlorophyll to be used in photosynthesis. The bacteriochlorophylls would have evolved later, as their biosynthetic pathways proceed from the synthesis of chlorophyll *a*. In an evolutionary scheme proposed by Pierson & Olson (1989) the cyanobacteria were placed on a direct evolutionary line of descent from the ancestral phototroph and may be the least diverged from this ancestor at the level of primary photochemistry. Divergence of all the phototrophs occurred over a very short time span, perhaps because of the intensity of competition for reductants, light, and carbon in the dense microbial-mat environment.

Both the molecular and the geologic records suggest that diversification of phototrophs occurred relatively early, over a short period of time. It may therefore be impossible ever to determine what the very first phototrophs were like. The best approach currently may be to isolate more diverse extant organisms to increase the pool of 16S rRNA data, to increase the pool of RC-sequence data, and to increase our knowledge regarding the use of diverse reductants such as Fe^{2+} and the pathways of CO_2 fixation.

Acknowledgments. – This work was supported by grants from the National Science Foundation (BSR-8521724 and BSR-8818133). I thank Diane Valdez and Mark Larsen for excellent technical assistance.

Note added in proof. – Recently the capacity for photoferrotrophy was demonstrated in anoxygenic bacteria isolated from anoxic sediments, thus supporting the hypothesis presented here that photosynthetic metabolism of iron may have been important in early evolution (Widdel *et al.* 1993).

The origin of eukaryotes and evolution into major kingdoms

Mitchell L. Sogin

Center for Molecular Evolution, Marine Biological Laboratory, Woods Hole, Massachusetts 02543, USA

Molecular phylogenies based upon comparisons of ribosomal RNAs show that geno-typic variation among eukaryotic microorganisms eclipses that seen in the plant, animal and fungal worlds. Five complex eukaryotic evolutionary assemblages make up the "crown" of the eukaryotic subtree. Two novel complex evolutionary assem-blages, the "alveolates" and the "stramenopiles," join the three "higher" kingdoms of plants, animals, and fungi. Alveolates include ciliates, dinoflagellates, and apicom-plexans, while the stramenopiles encompass diatoms, oomycetes, labyrinthulids, brown algae, and chrysophytes. A progression of independent protist branchings, some as ancient as the divergence between the two prokaryotic kingdoms, preceded the nearly coincident separation of the complex eukaryotic assemblages (which occurred approximately one billion years ago). The rRNA phylogenies identify specific protist ancestors for each of the multicellular groups and demonstrate their relative order of branching. A speculative model describing the chimeric origin of the eukaryotic nucleus explains contradictions between the rRNA phylogenetic frameworks and those inferred from genes duplicated before the separation of the eubacterial, archae-bacterial, and eukaryotic lines of descent.

This chapter is about phylogenetic origins. Where did cells with nuclei come from and what are the molecular and ultrastructural innovations that permitted their formation? Who were the microbial ancestors to the multicellular kingdoms of plants, animals, and fungi? What are the taxonomic boundaries separating major groups of eukaryotes, and how are they interrelated? The answers lie within the kingdom Protista whose members dominated the evolutionary history of eukaryotes.

The kingdom Protista – a retrospective view of relationships

The kingdom Protista was originally formalized by Haeckel to represent one of the three primary lines of descent (the others being the Plantae and the Animalia). These evolved from *"Radix communis Organismorum"* – a common group of root organisms (Haeckel 1866). From the beginning an understanding of relationships among protists,

Bengtson, S. (ed.) 1994: *Early Life on Earth. Nobel Symposium No. 84.* Columbia U.P., New York

and between protists and the rest of the biological world, was elusive. At one extreme, Haeckel did not recognize protozoans as separate from monerans (bacteria), fungi, and sponges and treated protists as a sister group to plants and animals (Rothschild 1987). In contrast, mycologists, phycologists, and zoologists split the protists into separate kingdoms, sometimes ancestral to, other times members of, "more highly evolved" kingdoms of plants, animals, or fungi. Today's textbook standard, Whittaker's "Five Kingdom" system (Whittaker 1969; Margulis & Schwartz 1988), fails to place protists in a credible phylogenetic context. The five-kingdom systems (plants, animals, fungi, protists, and bacteria) as well as more complex proposals (Leedale 1974; Ragan & Chapman 1978; Taylor 1978; Cavalier-Smith 1986) treat fungi, plants, and multicellular animals as separate kingdoms more highly evolved than any protozoan, but their relationships to specific protist groups are rarely defined.

Protists are an eclectic assemblage of (predominantly) unicellular eukaryotes but may include macroscopic large brown algae, red algae, and slime molds. They inhabit diverse terrestrial and aquatic environments or parasitize other protists, fungi, plants, or metazoans. Phenotypic variation within the Protista far exceeds that seen in other eukaryotic kingdoms (Margulis *et al.* 1990). Membership and relationships within the Protista are difficult to describe. There are no traits that unify protists to the exclusion of all other eukaryotes, and there is no agreement about the relative importance of different characters for inferring protist relationships. Rather than delineating a cohesive evolutionary assemblage, the concept *Protista* describes levels of organization and is represented by paraphyletic lines of descent.

Until the middle 1960s, 19th century ideas constrained systematics by placing protists into a single, "primitive," relatively unimportant taxonomic unit. Seeds of a new synthesis were sown by the introduction of ultrastructure data, which provided a novel basis for describing eukaryotic microorganisms. The dozen or so traditional taxonomic protist groups soon gave way to schemes that accommodate as many as one hundred protist lineages (Sleigh *et al.* 1984). Despite this major advance over 19th century ideas, ultrastructural studies lack the requisite quantitative component that can define relative branching patterns for these newly recognized assemblages. Protist phylogeny progressed in terms of perceived diversity without a clear understanding of relationship and evolution.

Comparative molecular biology reveals a dimension in phylogeny that is unavailable from comparative studies of phenotypes. Zuckerkandl & Pauling (1965) formalized molecular phylogeny by declaring that macromolecular sequences are "documents of evolutionary history." Since molecular change in genetic information is the source of biological diversity, comparisons of genes or proteins that share a common evolutionary history permit the inference of organismal phylogeny. The "molecular-clock hypothesis" became a metaphor in which the accumulation of genetic change is related to scales of time. But the clock is imprecise. For example, substitution rates in histone or actin coding regions differ by as much as tenfold in separate evolutionary lineages (Sadler & Brunk 1992; Bhattacharya *et al.* 1991). Other coding regions such as ribosomal RNAs (Woese 1987b) and glycerol aldehyde dehydrogenase genes (Michels *et al.* 1991) display a more consistent clocklike behavior. The more proper metric for comparative molecular biology is one of relative order or relative age rather than absolute time. It adopts the quantitative and objective attributes of the "molecular clock" without falling prey to its apparent failings. Phylogenetic frameworks inferred from molecular data are definitions of branching order based on extent of genetic relatedness rather than on reflections of time.

Insights from studies of ribosomal RNA genes

Structural studies of small-subunit ribosomal RNAs (16S-like rRNAs) have transformed the eukaryote–prokaryote dichotomy into a three-kingdom system relating the Eubacteria, the Archaebacteria, and the Eukaryota. These molecular phylogenies redefine boundaries separating major eukaryotic groups and provide evidence of unexpected genetic diversity. It is now clear that the five-kingdom system as popularly conceived fails to reveal the molecular range expressed by the eukaryotes and provides no proper understanding of the interrelationships and origins of the animals, plants, and fungi.

Fig. 1 is a maximum-likelihood, molecular phylogeny (Felsenstein 1981) that is based upon comparisons of all positions that can be unambiguously aligned among full-length 16S-like rRNA sequences from 75 taxa. The horizontal component of segments connecting nodes, or between organisms and nodes, is proportional to the amount of evolutionary change. This analysis concentrates on relationships between the "higher" groups of the five-kingdom scheme, and it corresponds to the great radiation or "crown" of the eukaryotic subtree (see below). It illustrates several important ideas about the definition of major eukaryotic groups and their evolutionary origins.

The crown of the eukaryotic tree

Approximately 1 Ga ago, well after the invention of the nucleus, the Plantae, Animalia, and Fungi diverged nearly simultaneously with two novel evolutionary assemblages, the alveolates and the stramenopiles. Statistical measurements in distance, parsimony, and maximum-likelihood techniques support the phylogenetic boundaries for these groups. The name *alveolates* describes a complex evolutionary assemblage that includes dinoflagellates, apicomplexans, and ciliated protozoans. The stramenopiles (Patterson 1989) include brown algae, labyrinthulids, chrysophytes, xanthophytes, diatoms, and oomycetes. Ultrastructural studies indicate that opalinids, bicosoecids, labyrinthulids, and thraustochytrids are also members of this assemblage, but rRNA sequences are not yet available for these taxa.

The molecular and phenotypic diversity of stramenopiles and alveolates rivals that seen in the multicellular kingdoms. Both assemblages include photosynthetic and nonphotosynthetic groups as well as species that display enormous variation in body plan. Memberships and branching patterns within alveolates and stramenopiles show remarkable congruence with patterns of ultrastructure organization. Detailed branching patterns for ciliates are congruent with evolutionary relationships inferred from comparative analyses of ciliature, ciliary necklace patterns, and rootlet structures. At the species level, there is remarkable agreement between rRNA phylogenies for thirteen species of *Tetrahymena* and relationships inferred from electrophoretic analyses of cytoskeletal proteins (Williams *et al.* 1984; Sogin *et al.* 1986). Many shared ultrastructure features unite the stramenopiles (Andersen 1991; Bhattacharya *et al.* 1992). For example, all members of the group contain tubular, mitochondrial cristae, and many have tripartite mastigonemes not found in any other eukaryotes. This later feature may be the key to evolutionary success of the group. The thrust reversal afforded by this unusual flagellar apparatus dramatically alters feeding mechanisms. It allows capture or ingestion of particles into the flagellated cell.

Despite complications introduced by unequal mutation rates in different evolutionary lineages, actin-coding regions display molecular motifs that distinguish each of the major eukaryotic groups (Bhattacharya *et al.* 1991). Moreover, the correspondence between ultrastructural studies and rRNA phylogenies is impressive. Ciliature and flagella are extensions of the cytoskeleton, but there is no known genetic linkage between coding regions for rRNAs and cytoskeletal proteins. Both serve pivotal roles in complex subcellular structures that are functionally and genetically independent. Agreements between the rRNA-based phylogenies and comparative ultrastructural studies are not surprising; functional constraints may attenuate precipitant changes in molecular components of highly integrated subcellular structures. Collectively these data suggest that the coding regions for certain cytoskeletal proteins might be suitable for molecular-systematics studies in eukaryotes.

The addition of stramenopiles and alveolates to the list of "higher" or complex evolutionary assemblages complicates our understanding of relationships between protists and the multicellular groups. The identification of protist lineages that share the most recent common ancestry with each of the "higher groups" is no longer the only important question. The rRNA-based phylogenies have already demonstrated that (1) higher plants share a recent common ancestry with chlorophytes (exclusive of euglenoids); (2) chytrids, considered by some to be flagellated protists, represent an early-diverging higher fungal lineage; and (3) choanoflagellates lie at the base of the animal radiation. To understand the origins of the complex evolutionary assemblages, we must determine their relative branching orders.

There are numerous theories describing the origins of metazoans. Studies of molecular evolution have only further confused the situation. A frequently cited example is "The Molecular Phylogeny of the Animal Kingdom" by Field *et al.* (1988). This heroic effort (which predates polymerase-chain-reaction techniques) employed sequence methods mediated by reverse transcriptase to characterize portions of 16S-rRNAs from the major animal phyla. This study presented several incongruent phylogenies. Furthermore, different statistical treatments of the Field *et al.* data set produce discordant tree topologies (see Field *et al.* 1988 vs. Lake 1990). A casual interpretation would suggest that animals are parphyletic. In fact, the Field *et al.* study emphasized the rapid divergence and hence unresolved nature of the data sets. If the tree in Fig. 1 accurately recounts evolutionary history, at least two explanations for the confusion surrounding molecular systematic studies of animal origins become obvious. The very early separation of several simple animal lineages occurred suddenly, and the rates of evolution in animal rRNA coding regions accelerated after divergence of the Cnidaria. Accelerated rates of change appear as long segments in molecular phylograms, and they can obscure closely ordered branching patterns (see Felsenstein 1978; Swofford & Olsen

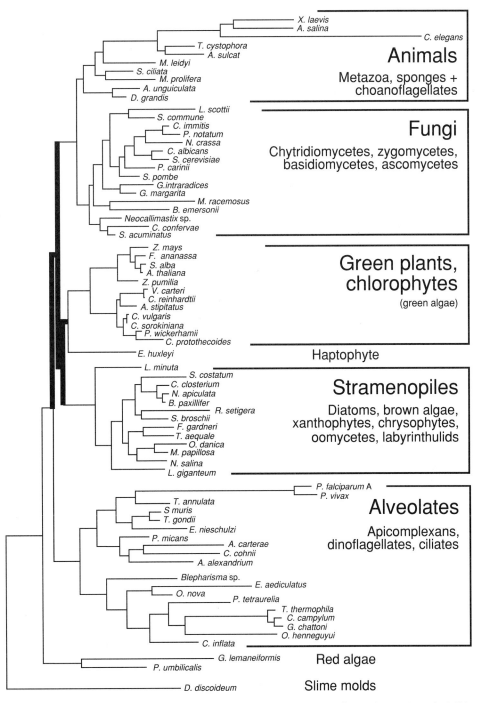

FIGURE 1 Phylogenetic relationships for the crown of the eukaryotic subtree. Approximately 1,600 positions that could be unambiguously aligned were converted to evolutionary distances, and a phylogenetic tree was inferred using maximum-likelihood techniques (Felsenstein 1981). The evolutionary distance between nodes is represented by the horizontal component of their separation. The heavy line indicates lack of statistical support for a preferred branching order.

1990). The reliability of the tight branching pattern is further compromised by use of partial sequences and limitations of "direct" RNA-sequencing techniques; 2–3% of the sequence assignments are ambiguous, and sequence determinations on a single strand are less accurate than redundant analyses of both DNA strands. Since many of the critical nodes in early-diverging animal lineages are separated by fewer than one position change per one hundred positions, incongruous branching patterns in different studies are not surprising. Which of these trees are we to believe? In the final analysis, the reliability of the rDNA frameworks can only be evaluated by the distribution of nonmolecular markers in the molecular trees.

In many regards the branching patterns for the animals represented in Fig. 1 support several consensus ideas about the origins of animals. Choanoflagellates diverged before sponges, sponges before cnidarans, etc. Unlike other proposals based upon partial sequences from the small-subunit and large-subunit rRNAs, the animals are seen to be monophyletic. The unanticipated result in Fig. 1 is the specific relationship between Animalia and Fungi. The five-kingdom schemes removed the Fungi from the Plantae, establishing a separate "higher kingdom." Few proposals suggest a specific relationship between fungi and animals, but Cavalier-Smith (1987c) has called attention to the potential common ancestry between these "higher kingdoms." Both commonly have chitinous exoskeletons, store glycogen but not starch, and lack chloroplasts, and their mitochondrial genetic code employs UGA for tryptophan as opposed to the UGA termination used in plant mitochondria. The Chytridiomycetes, which lie at the base of the fungi, and choanoflagellates, which are at the base of the animals, have flattened, nondiscoidal cristae in their mitochondria and a simple posterior flagellum. The implied relationship between animals and fungi from comparisons of these phenotypic characters agrees with the maximum-likelihood analysis of the rDNA data (Wainright *et al.* 1993).

The relative branching order among the other complex evolutionary assemblages remains unresolved. The heavy lines in the maximum-likelihood tree shown in Fig. 1 indicate lack of confidence in relative branching orders. Other statistical measurements in distance matrix and parsimony analyses are also ambiguous. Less than one nucleotide change per 100 positions separates their basal nodes. This phyletic explosion or crown in the eukaryotic tree represents one of the more significant events in the evolutionary history of eukaryotes. It is important to consider the timing of this event. Estimated times of divergence for multicellular plants and animals from the fossil record can calibrate the mutation rate in rRNA genes. One percent of sequence positions change per 50 Ma (Wilson *et al.* 1987). The identification of bangiophyte red algae in 1.25 Ga old rocks (Butterfield *et al.* 1990) agrees with this estimate. Red algae separate immediately before the crown of the eukaryotic subtree, whose rRNAs are, on average, 80% identical to those of red algae. Protist rRNA genes seem to exhibit clock-like behavior during the last 1.25 Ga of evolution. If these estimated rates of rRNA evolution are correct, the phyletic radiation or eukaryotic crown occurred 1 Ga ago.

The sudden proliferation of morphologically distinct lineages could reflect major environmental changes. For example, a large increase in atmospheric oxygen may have led to the development of new ecological niches. Perhaps a small number of lineages survived a cataclysmic event followed by rapid diversification into unoccupied habitats. Both scenarios predict that there will be radiating patterns in earlier-diverging protist lineages that correspond to the concurrent separation of plants, animals, and fungi. Alternatively, the rate of phenotypic evolution might have accelerated in response to novel mechanisms for managing genetic information. Maybe the invention of homeobox-like mechanisms permitted more complicated patterns of differential gene expression and ultimately tissue differentiation in multicellular organisms. Development of "cis" splicing mechanisms for processing RNA, or invention of vectors for rapidly exchanging genetic information, could have led to shuffling of genetic information and accelerated rates of phenotypic change.

There is renewed optimism that molecular phylogeny will one day provide a clear perspective of the origins of the multicellular kingdoms. It is already possible to define constituency of the major evolutionary assemblages, and the proposed relationship between animals and fungi represents a major advance in phylogenetic understanding. The inclusion of additional taxa in the rDNA data bases, the development of data bases for other gene families representing divergent taxa, and the application of more sophisticated phylogenetic inference techniques to molecular data sets will lead to a more complete understanding of eukaryote evolution. To understand the origins of the eukaryotic line of descent, it is necessary to examine the eukaryotic subtree in the context of "universal" rRNA phylogenies.

The eukaryotic panorama and the universal tree of life

The tree in Fig. 2 summarizes our current perspective of the primary lines of descent. It is an "unrooted" phylogeny inferred by the method of De Soete (1983) and is based upon comparison of 1,020 unambiguously aligned sites (Sogin & Gunderson 1987). This algorithm can accommodate a very large number of taxa (the tree contains 125 species); however, it is particularly sensitive to aberrant rates of change in different lineages. This unrooted tree graphically illustrates relative evolutionary diversity within each of the primary lines of descent.

The rRNA sequences converge upon three major assemblages: the Eubacteria, the Archaebacteria, and the Eukaryota. The 16S-like rRNA sequence diversity within the eukaryotic kingdom eclipses that seen within or between the Archaebacteria and Eubacteria. The diplomonad *Giardia lamblia* represents the earliest-branching eukaryotic lineage in distance matrix analyses. This inference is corroborated by parsimony and the retention of rRNA structural feature's characteristic of archaebacteria and eubacteria; it is the only known nuclear rRNA gene that contains a prokaryote-like Shine-Dalgarno mRNA binding site (Sogin *et al.* 1989). This early divergence is consistent with the lack of mitochondria, the apparent absence of rough endoplasmic reticulum (ER) and Golgi, the lack of sexual life-cycle stages (Feely *et al.* 1984), and the remarkably simple constellation of proteins associated with the *G. lamblia* cytoskeleton (Kabnick & Peattie 1991). Several *G. lamblia* genes, including tubulin (Kirk-Mason *et al.* 1989), a surface protein (Gillen *et al.* 1990), and a heat shock protein (Aggarwal *et al.* 1990) have 5′ regions more characteristic of prokaryotes than eukaryotes. The microsporidian *Vairimorpha necatrix* and the trichomonads *Trichomonas vaginalis* and *Tritrichomonas foetus* represent other early branchings. As in the case of *G. lamblia*, they have adopted parasitic habitats and lack mitochondria. Parasitic lifestyles may have played

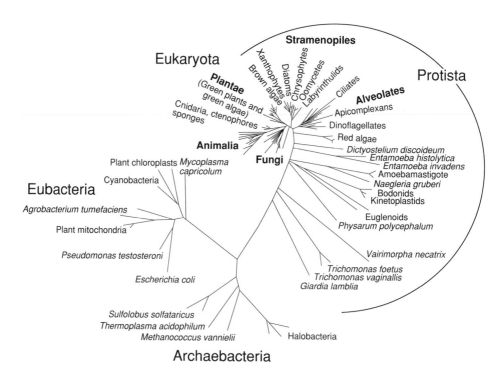

FIGURE 2 Universal phylogenetic framework. Computer-assisted methods were used to align small subunit rRNA sequences from 195 eukaryotic and prokaryotic taxa using a method that considers the phylogenetic conservation of secondary structures. Structural similarities were computed from comparisons of 1,020 sites that could be unambiguously aligned, and these were converted to evolutionary distances. The tree was inferred using the method of De Soete (1983). The line segment lengths represent evolutionary distance between organisms. The large number of taxa included in the analysis precludes complete labeling of terminal nodes.

a role in the survival of these lineages during unfavorable environmental periods, but parasitism alone can not explain their early divergence. Free-living diplomonads related to *G. lamblia* are known, and other parasitic species including *Plasmadium berghei* and *Pneumocystis carinii* display late-branching patterns.

A collage of seemingly unrelated protist lineages follows the earliest branchings. The plasmodial slime mold *Physarum*, the heterolobomseans *Tetramitus*, *Vahlkampfia*, and *Naegleria*, and a lineage leading to euglenoids and kinetoplastids separate nearly simultaneously. Other discrete evolutionary lines include *Entamoeba*, the cellular slime mold *Dictyostelium discoideum*, and red algae. The five complex evolutionary assemblages described above (plants, animals, fungi, stramenopiles, and alveolates), plus several other independent protist lineages, diverged nearly simultaneously at the crown of the eukaryotic subtree. The kingdom of protists is unlike that of plants, animals, and fungi. Instead of being a cohesive phylogenetic group, the Protista describes levels of organization delineated by paraphyletic lines of descent. More significantly, the eukaryotic lineage may be as ancient as its eubacterial and archaebacterial counterparts.

The absence of eukaryotic microbial fossils older than 2 Ga and the evident lack of meaningful biochemical diversity are frequently cited as evidence for a recent origin of eukaryotes. However, such interpretations distort our view of microbial evolution. If a fossil is smaller than 10 μm, it is described as having a prokaryotic origin despite the fact that extant eukaryotes as small as 1 μm in diameter (e.g., *Nanochlorum eukaryotum*) have been described. The lack of paleontological support for extremely ancient eukaryotes may not be surprising. The earliest-diverging lineages in the eukaryotic line of descent are members of the Archezoa (Cavalier-Smith 1987a). These organisms lack skeletons, do not necessarily form spores or cysts, and hence are not well represented in the fossil record. The modest eukaryotic biochemical diversity (relative to prokaryotes) is probably a sampling artifact. Most biochemical studies are confined to the "higher kingdoms," but the majority of eukaryotic diversity by almost any measure is represented by protists that have not been scrutinized at the biochemical level. Molecular studies in protists have often led to observations of phenomena not found in plants, animals, or fungi.

Rooting the universal tree of life

Comparisons to known outgroups normally define branching orders in phylogenetic trees. For example, bacterial outgroups can resolve the relative branching orders for eukaryotic lineages in molecular phylogenies. A similar strategy for determining the root in the universal tree of life is not available, because there are no outgroups for contemporary life forms. An ingenious alternative method for inferring the root of the universal phylogenetic tree takes advantage of ancient, phylogenetically conserved, duplicated gene families (Gogarten *et al.* 1989; Iwabe *et al.* 1989). Since coding regions for elongation factors EF-Tu and EF-G are present in all organisms, these related genes must be products of gene-duplication events that preceded the divergence of the primary lines of descent. Molecular phylogenies can be inferred for the EF-G sequence and the EF-Tu sequences. The EF-G subtree serves as the outgroup for its duplicated gene partner EF-Tu. The EF-G sequences determine relative branching orders in molecular phylogenies based upon comparisons of its duplicated gene partner EF-Tu. This analysis, as well as a parallel study of alpha and beta subunits of ATPase, places the root of the universal tree of life within the eubacterial domain.

Placing the root at this position implies that eukaryotes and archaebacteria shared a common evolutionary history exclusive of eubacteria. This proposal explains similarities between eukaryotic and archaebacterial RNA polymerases (Puhler *et al.* 1989), ribosomal proteins (Hui & Dennis 1985), and the presence of histone-like protein in the thermoacidophilic archaebacterium *Thermoplasma acidophilum* (Searcy 1975). However, placement of the root within the eubacterial line of descent complicates interpretation of the rDNA trees. Fig. 3 summarizes the principal discrepancy between protein-based phylogenies and those inferred from comparisons of ribosomal RNA. If certain protein-coding regions are considered, archaebacteria and eukaryotes appear to be more closely related to each other than either is to the eubacteria. This relationship contradicts the rRNA phylogenies that position archaebacteria as specific relatives to the eubacteria. Accelerated rates of change in all nuclear rRNA coding regions with a concomi-

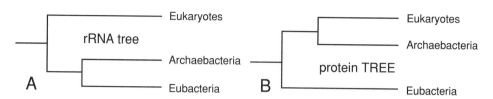

FIGURE 3   Contradictory universal phylogenies based upon (**A**) rRNA and (**B**) proteins. The rRNA topologies indicate that archaebacteria and eubacteria are more closely related to each other than either group is to eukaryotes. This contradicts the topology and proposed root for the universal tree inferred from analysis of conserved protein-coding regions.

tant constancy or even reduction of mutation rates in protein-coding regions might explain the deep eukaryotic branching patterns in the rRNA-based phylogenies. This theory is even more convoluted than it initially appears. It would require acceleration of evolution in rRNA-coding regions within numerous, independent, early-branching eukaryotic lineages. This theory requires a more recent reduction in mutation rates within all independent lineages to account for the clocklike behavior of rDNA genes during the last 1.25 Ga. Invoking changes in mutation rates is not a simple explanation for the incongruent protein and rRNA trees, and it does little to explain the evolutionary processes responsible for the origin of the eukaryotic cell.

The grandest of inventions: The eukaryotic cell

A chimeric model for the formation of the eukaryotic nucleus would explain the contradictory protein and rRNA trees in Fig. 3. The following scenario is presented in the context of the progenote, which Woese (1982, 1987b) originally described as the last common ancestor for archaebacteria, eubacteria, and eukaryotes (however, the model could also be invoked at a later stage of cellular evolution). The progenote was a hypothetical organism or population of organisms, and it was still evolving a link between genotype and phenotype. Its major innovations included the formation of a primitive translation apparatus and informational polymers several hundred nucle-otides in length. RNA-mediated events may have dominated metabolism, and ge-nomes were probably fragmented into many nucleic acid polymers several hundred nucleotides in length. These biopolymers resided within mi-cells that could fuse or separate, allowing for the rapid exchange of macromolecular machinery.

The model presented in Fig. 4 emphasizes that the presence or absence of a nucleus is secondary to existence of the cytoskeleton when differentiating prokaryotes and eukaryotes. It proposes that an anuclear lineage distinct from archaebacteria and eubacteria invented sufficient cytoskeletal complexity to allow the transition from a pro-eukaryotic lineage to cells with nuclei. In this model two primary lines of descent trace their common origins back to the progenote, rather than the three lines originally described by Woese. One lineage developed a sophisticated translation apparatus capable of accurately synthesizing proteins, which became the dominant metabolic machinery in a proto-bacterial lineage. This line of descent may have invented DNA:

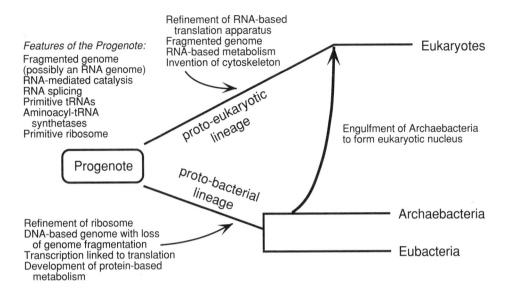

FIGURE 4 Chimeric origin of the nucleus. The proto-eukaryotic lineage and proto-bacterial lineages diverged from the Progenote. The proto-eukaryotic lineage invented the cytoskeleton which permitted engulfment of an archaebacterial genome to form the nucleus.

genome organization switched from extreme fragmentation to formation of a single chromosome, and translation became coupled to transcription before the divergence of the eubacterial and archaebacterial lines of descent.

The second line of descent progressed along a very different evolutionary pathway. It retained the fragmented genome of the progenote but developed sophisticated RNA-splicing mechanisms. It may have relied upon an RNA-dominated metabolism, but its major innovation was the cytoskeleton. Formation of the cytoskeleton confers the ability to engulf other microorganisms, a strategy generally not available to prokaryotes. In this model, the proto-eukaryotic line of descent engulfed an archaebacterial organism whose DNA formed a chimera with the existing proto-eukaryotic genome. This event led to the formation of the earliest nucleated cells. In this model, the nucleus is a chimera; it arose from a symbiosis between a proto-eukaryotic lineage and an ancestral archaebacterial genome. The latter provided most of the structural coding regions, while the pro-eukaryotic genome contributed genes for rRNAs and cytoskeletal proteins. This chimeric origin rationalizes differences between the protein- and rRNA-based phylogenies. Ancestry based upon protein similarities places eukaryotes and archaebacteria as sister lineages, while those based upon comparisons of ribosomal RNA-coding regions would support a closer relationship between archaebacteria and eubacteria.

This hypothesis is fundamentally different from previous models that propose an endosymbiotic origin of the nucleus. It prescribes an anuclear proto-eukaryotic lineage distinct from the archaebacteria and eubacteria, a lineage that gained the nuclear genome through engulfment. In most regards, the host lineage was completely unlike those represented by contemporary organisms, and its origins likely trace back to a Progenote that existed 4–3.5 Ga ago. The model explains sophisticated RNA-splicing mechanisms in eukaryotes that are not present in the prokaryotes, and it is experimentally testable. For example, since invention of cytoskeletal proteins occurred in the proto-eukaryotic lineage, identification of similar proteins in bacterial lineages would be unexpected. To date, there are no convincing descriptions of bacterial protein sequences that are similar to major conserved polypeptides of the eukaryotic cytoskeleton. When bacterial genome sequences become available, it will be possible to test this prediction more thoroughly.

Summary

Phylogenies inferred from comparisons of rRNA sequences depict protists as a progression of independent branchings, some of which appear to be as ancient as the two prokaryotic kingdoms. Other protists are members of previously unrecognized complex evolutionary assemblages that diverged nearly simultaneously with the kingdoms of plants, animals, and fungi. We are just beginning to appreciate the complexity of eukaryotic microbial evolution; many questions remain unanswered. Of the estimated 100 independent protist lineages, only 25 are represented in the 16S-like rRNA data base. Integrating this rapidly expanding molecular data base with studies of ultrastructure evolution will soon provide new insights about the sequence of events that underlay the evolutionary history of the eukaryotic world.

The oldest known records of life: Early Archean stromatolites, microfossils, and organic matter

J. William Schopf

Department of Earth and Space Sciences, IGPP Center for the Study of Evolution and the Origin of Life, and Molecular Biology Institute, University of California, Los Angeles, Los Angeles, California 90024, USA

In comparison with the fossil record known from later geologic time, that of the very early history of life is minuscule. Only two Early Archean (~3.55–~3.0 Ga old) geologic sequences have been investigated paleobiologically in some detail, those of the Swaziland Supergroup of South Africa and the Pilbara Supergroup of Western Australia. Both of these relatively unmetamorphosed sequences contain decipherable evidence of early life: megascopic stromatolites, permineralized microfossils, and biologically produced organic matter. Taken together, these three lines of evidence indicate that microbiological communities, evidently including CO_2-fixing photosynthetic prokaryotes, had become established at least as early as ~3.5 Ga ago. The origin of life on Earth must predate, perhaps substantially, these oldest known records of biologic activity.

Three types of direct evidence have provided insight into the earliest stages of biospheric evolution – the occurrence in the Early Archean (>3.0 Ga old) geologic record of (1) megascopic microbially produced stromatolites; (2) microscopic cellularly preserved microorganisms; and (3) particulate carbonaceous matter (kerogen), identifiable on the basis of its carbon isotopic composition as a product of biological activity. Each of these lines of evidence has recently been reviewed in considerable detail: known Early Archean stromatolites by Walter (1983) and by Hofmann *et al.* (1991); early microfossils by Schopf & Walter (1983) and by Schopf (1992a, 1992b); and the Early Archean carbon isotopic record by Hayes *et al.* (1983) and by Strauss & Moore (1992). As discussed below, these distinct but mutually reinforcing lines of evidence indicate that photoautotroph-containing stromatolitic communities had become established at least as early as ~3.5 Ga ago.

Bengtson, S. (ed.) 1994: *Early Life on Earth. Nobel Symposium No. 84.* Columbia U.P., New York

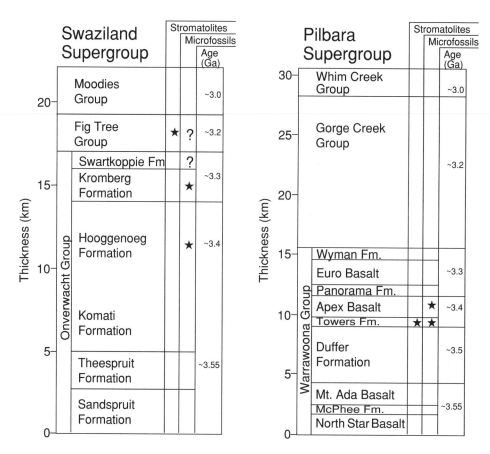

FIGURE 1 Stratigraphic columns for geologic formations of the Swaziland Supergroup, South Africa, and the Pilbara Supergroup, Western Australia.

Stromatolites and microfossils from the Early Archean (~3.55–~3.0 Ga old) Swaziland Supergroup, South Africa

In comparison with the Proterozoic (2.5–~0.55 Ga old) fossil record, that known from the earlier, Archean (>2.5 Ga old) portion of the Precambrian is minuscule. Stromatolites are virtually ubiquitous in Proterozoic carbonate terranes, represented by hundreds of taxonomic occurrences reported from a large number of Proterozoic basins (Walter *et al.* 1992). In contrast, only about two dozen occurrences of Archean stromatolites have as yet been reported (Walter 1983; Hofmann *et al.* 1991). Similarly, although literally hundreds of microfossiliferous formations and nearly 3,000 occurrences of bona fide microfossils have been discovered in Proterozoic-age strata (Mendelson & Schopf 1992b; Schopf 1992c), only about 30 putatively microfossiliferous units (very few of which contain unquestionable microfossils) are known from the Archean (Schopf & Walter 1983).

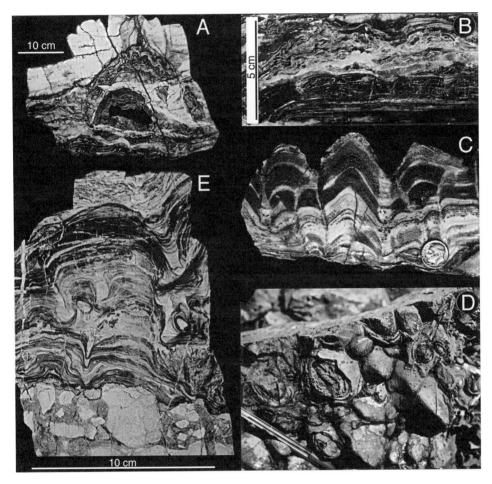

FIGURE 2   Domical (**A**), flat-lying (**B**), and conical (**C–D**) stromatolites from the Pilbara Supergroup (Towers Formation) of Western Australia (Lowe 1980b; Walter *et al.* 1980; Walter 1983), and columnar stromatolites (**E**) from the Swaziland Supergroup (Fig Tree Group) of South Africa (Byerly *et al.* 1986).

Among the notably few Archean sequences to have been investigated paleobiologically in some detail, that of the approximately 23 km thick Early Archean (~3.55–~3.0 Ga old) Swaziland Supergroup (Fig. 1) of the eastern Transvaal, South Africa, has received particular attention. Beginning with the studies of Pflug (1966) and Barghoorn & Schopf (1966), more than 30 publications appeared before 1983 reporting discovery of microfossil-like objects in these strata (Schopf & Walter 1983). Although most of the objects thus reported are now regarded as nonfossils (Schopf & Walter 1983), certain of the simple unicell-like organic spheroids reported from these units are possibly of biologic origin (Muir & Grant 1976; Knoll & Barghoorn 1977; Schopf & Walter 1983).

Recently, biologically more convincing structures have been reported from Swaziland Supergroup sediments. In particular, Byerly *et al.* (1986) have described the wavy laminated stratiform to short-columnar stromatolites shown in Fig. 2E from a thin chert unit of the ~3.2 Ga old Fig Tree Group (in the upper third of the supergroup; Fig. 1)

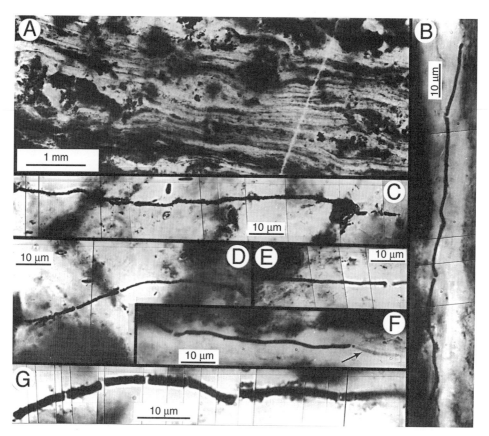

FIGURE 3 Stromatolite-like laminae (**A**) and filamentous microfossils (**B–G**) in cherts of the
Swaziland Supergroup (Kromberg Formation) of South Africa (Walsh & Lowe 1985); arrow in **F**
points to hollow filament; **B–E** and **G** are composite photographs.

southwest of Barberton, South Africa. Although these structures are not known to
contain cellularly preserved remnants of the microorganisms presumably responsible
for their formation, the evidence presented by Byerly *et al.* (1986) seems wholly
consistent with their inferred biogenicity. Moreover, from the stratigraphically underlying Onverwacht Group (Fig. 1) in the same general area, Walsh & Lowe (1985) have
reported the occurrence of threadlike filamentous microfossils in chert units of the
~3.4–~3.3 Ga old Hooggenoeg and Kromberg Formations. As shown in Fig. 3, the
Kromberg filaments occur in association with wavy flat-laminated stromatolite-like
laminae, composed chiefly of amorphous kerogen. Within this stromatolite-like fabric
and commonly oriented parallel or subparallel to the component laminae are thin (1.0–
2.5 µm-diameter), solid or hollow (Fig. 3F, at arrow), kerogenous, and possibly mineral-
encrusted microbe-like filaments (Fig. 3B–G). Not uncommonly, the opaque solid
filaments are broken and disrupted (e.g., Fig. 3C–E); although such breaks are in some
examples rather regularly distributed, suggesting possible cellularity (Fig. 3G), well-
defined cells have not been detected. The biological affinities of these filaments are
uncertain. In general morphology they are similar to many extant prokaryotes. Among

these, they seem more comparable to filamentous bacteria (or to the tubular sheaths of such bacteria) than to most modern cyanobacteria. Together with the carbon isotopic data discussed below, the stromatolites (Fig. 2E) and filamentous microfossils (Fig. 3B–G) of the Swaziland Supergroup sediments provide convincing evidence of Early Archean biologic activity.

Stromatolites and possible microfossils from the Early Archean (~3.55–~3.0 Ga old) Pilbara Supergroup, Western Australia

Paleobiologic studies of the approximately 30 km thick (Fig. 1) ~3.55–~3.0 Ga old Early Archean Pilbara Supergroup of Western Australia were initiated by Dunlop *et al.* (1978) who reported discovery of carbonaceous spheroids in Warrawoona Group sediments (in the lower half of the supergroup; Fig. 1). Although initially regarded as fossil unicells, these microstructures were later reinterpreted by Schopf & Walter (1983) as "most likely to be nonfossils, solid carbonaceous globules of apparently non-biologic origin." Soon after the appearance of the Dunlop *et al.* report, flat-lying, domical, and conical stromatolites were also discovered in Warrawoona strata (Fig. 2A–D; Lowe 1980b; Walter *et al.* 1980; Walter 1983). Although presumably of microbial origin, these structures (like the vast majority of fossil stromatolites) are not known to contain cellularly preserved microfossils, and their biogenicity has therefore been regarded as unproven (Buick *et al.* 1981).

In 1983, what appeared to be a major advance in the search for evidence of Early Archean life was reported by Awramik *et al.* (1983): discovery of cellularly preserved filamentous microbes in cherty strata of the Warrawoona Group. Two categories of filamentous microstructures were described, both from the North Pole Dome area of the Pilbara Block of northwestern Western Australia:

1 *Rosette-like aggregates of fine radiating filaments* were described by Awramik *et al.* (1983) but were interpreted by them only as "possible microfossils" rather than as unquestionable fossil microorganisms. Schopf & Walter (1983) similarly regarded these microstructures as "dubiomicrofossils . . . of possible, but as yet unproven biogenicity," a conclusion also reached by Buick (1984). The biological origin of these microstructures has not been established; they cannot be regarded as firm evidence of Archean life.

2 *Cellular and tubular filaments* were also described by Awramik *et al.* (1983); because of their organic composition and because some of these microstructures are demonstrably cellular, they were regarded both by Awramik *et al.* (1983, 1988) and by Schopf & Walter (1983) as authentic microfossils. However, and although these carbonaceous filaments are evidently bona fide microfossils, the provenance of the microfossiliferous samples is uncertain: despite repeated visits to the outcrop, both by the original collector (Awramik *et al.* 1988) and by others (Buick 1988), the exact source of the fossiliferous cherts (evidently from within the Towers Formation) has not been relocated. Thus, however tantalizing, this discovery has not been recon-

firmed; its relevance to understanding the early history of life has not been established (Schopf 1992a). Fortunately, as summarized below, recent studies have shown that other microfossils, of established Early Archean age, known provenance, and unquestionable biogenicity, are indigenous to Pilbara Supergroup strata.

Authentic Early Archean microfossils from the (~3.4 Ga old) Apex Cherts (Pilbara Supergroup, Warrawoona Group), Western Australia

The recent discovery (Schopf & Packer 1986, 1987; Schopf 1992a, 1993) of cellularly preserved Archean microorganisms in bedded cherts of the ~3.4 Ga old Apex Basalt (in the middle third of the Warrawoona Group; Fig. 1) obviates many of the difficulties noted above. Filamentous kerogenous microfossils (Fig. 4) occur in a bedded 10 m thick chert unit of the Apex Basalt along Chinaman Creek, west of Marble Bar, in a region of northwestern Western Australia well mapped by Hickman & Lipple (1978). Fossiliferous samples have been collected on two occasions from this locality: in June 1982 and in August 1986.

The fossil filaments of the Apex cherts are cellularly preserved (Fig. 4C–G), permineralized within subangular-to-rounded kerogen-rich sedimentary clasts one to a few millimeters in diameter (Fig. 4A). The microfossils occur within these siliceous clasts but not in the surrounding matrix (Fig. 4B). They were preserved initially in an older sedimentary unit that, subsequent to its lithification, was eroded to produce the lithic fragments. These fragments, in turn, were transported and rounded, ultimately being redeposited as primary clastic components of the bedded Apex cherts. Although the Apex microfossils thus predate the sedimentary unit in which they now occur, whether they are greatly older than or are more or less contemporaneous with the bedded Apex chert unit cannot be ascertained from the evidence at hand.

The occurrence of these microfossils in petrographic thin sections (Fig. 4), their presence in the clasts but not in the surrounding chert matrix (Fig. 4B), and their similarity in color and texture to dark-brown-to-black kerogenous particles finely distributed throughout the clasts (Fig. 4A) indicate that they are indigenous to the clasts in which they occur. The microfossils demonstrably are not of secondary (post-depositional) geologic origin, nor are they modern contaminants.

The biogenicity of these filaments is established by their morphological complexity, being composed of well-defined, barrel-shaped, discoidal, or quadrate medial cells (Fig. 4C–G) and exhibiting rounded or conical terminal cells (e.g., Fig. 4D). Some specimens exhibit medial bifurcated cells and paired half-cells (Schopf 1993, Fig. 5H–J) that apparently reflect the original presence of partial septations and, thus, the occurrence of prokaryotic binary cell division. The biological origin of these filaments similarly is indicated by their carbonaceous composition; by the degree of regularity of cell shape and dimensions exhibited by numerous specimens of each of the several taxa recognized in the assemblage (see Schopf 1992a, 1993); and by their obvious morphological similarity to cellular, unbranched, uniseriate prokaryotic trichomes, both extant and fossil.

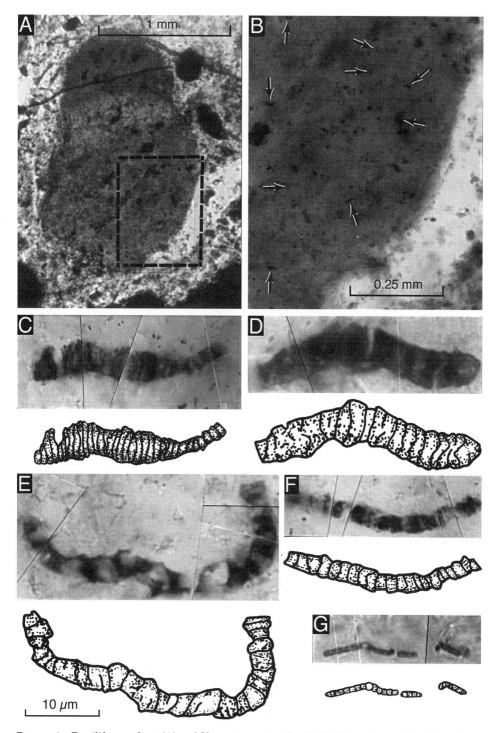

FIGURE 4   Fossiliferous clasts (**A**) and filamentous microfossils (**B–H**) in petrographic thin sections of Apex chert from the Pilbara Supergroup of Western Australia; rectangle in **A** outlines area shown in **B** (in which microfossils are denoted by arrows); **C–G** show composite photomicrographs (scale for all as in **E**). (**A–B, D–G** from Schopf 1993; **C** from Schopf 1992a.)

Thus, the cellular filaments of the Apex cherts meet all of the criteria required of unquestionable Archean microfossils (Schopf & Walter 1983): (1) they occur in sedimentary rocks of established Archean age; (2) they are demonstrably indigenous to these Archean sediments; (3) they occur in lithic components that are assuredly syngenetic with deposition of this sedimentary unit (with the fossils themselves predating deposition of the bedded cherts in which the fossiliferous clasts occur); and (4) they are certainly biogenic. Moreover, to these criteria may now be added a fifth: (5) as demonstrated by replicate sampling of the fossiliferous outcrop, the provenance of these microfossils is firmly established.

The cellular filaments of the Apex cherts provide unequivocal evidence of the existence of biologic systems during Early Archean time. Their prokaryotic nature seems firmly established: they are strikingly similar in morphology to numerous types of filamentous prokaryotes, both modern and fossil, and their cells are evidently divided by a typically prokaryotic pattern of binary cell division. Among known prokaryotes, several of the Apex taxa seem especially comparable to species of modern cyanobacteria (Schopf 1992a, 1993), an interpretation consistent with the occurrence of cyanobacterium-like (chroococcalean-like) colonial unicells in other Pilbara (Towers Formation) sediments (Schopf & Packer 1986, 1987; Schopf 1992a).

Although the precise affinities of Archean microfossils are difficult to establish on the basis of morphology alone (Schopf 1992a), the similarity of the Pilbara fossils to extant photosynthetic prokaryotes and the reported occurrence of stromatolites in both the Pilbara and Swaziland Supergroups suggest strongly that phototactic prokaryotic photoautotrophs – exhibiting bacterial (anoxygenic) and/or cyanobacterial (oxygenic) photosynthesis – are likely to have been represented in the Early Archean biota. Because of kinetic fractionation occurring during the enzyme-mediated CO_2-fixing step of photoautotrophy, the photosynthate produced exhibits a characteristic range of $^{12}C:^{13}C$ ratios, a carbon isotopic signature that is potentially preservable in the carbonaceous (kerogenous) components of relatively unmetamorphosed ancient sediments (Hayes *et al.* 1983; Hayes, this volume). Thus, if photoautotrophs were represented in the Early Archean biota, their presence might be reflected in the isotopic composition of kerogens preserved in the Pilbara and Swaziland sediments. Evidence relating to this possibility is addressed below.

Preservation of the carbon isotopic record

Strauss & Moore (1992) have recently tabulated all carbon isotopic data, both for carbonate carbon and for organic carbon, currently available from Precambrian sediments. Isotopic values ($\delta^{13}C‰$ vs. PDB, the Peedee belemnite standard) for Precambrian carbonate carbon cluster near 0‰, whereas those for organic carbon are typically about –25±5‰ (i.e. the carbonacous matter is enriched in the lighter stable carbon isotope, ^{12}C, by about 25 parts per thousand relative to the PDB standard). (The ranges of isotopic values are indicated by the shaded envelopes in Fig. 7.) Because this isotopic difference between Precambrian carbonate and organic carbon is in the same direction and of the same magnitude as that observed in the modern environment between inorganic oceanic bicarbonate (the carbon source for marine limestones) and the organic products of photoautotrophy, and because in the modern ecosystem this difference is a result of the enzyme-driven isotopic discrimination occurring during photosynthesis, the Precambrian values have been interpreted as strong evidence for the existence of ancient photoautotrophy (Schidlowski *et al.* 1983).

The elemental and isotopic compositions initially encoded in sedimentary organic matter, however, are not immune from alteration by geologic processes. In particular, as preserved organic matter progressively matures – i.e., as it becomes metamorphically altered, principally because of the effects of elevated temperature in its geologic setting – organic-carbon-bound hydrogen is released, and the kerogen residue (the residual particulate carbonaceous matter) gradually approaches the composition of graphite. Thus, it is not surprising that kerogens isolated from Precambrian geologic units of increasingly greater age tend to exhibit progressively decreasing hydrogen-to-carbon (H/C) ratios (Fig. 5); in general, the older the geologic unit, the greater the degree of its metamorphic alteration and resultant loss of kerogen-bound hydrogen.

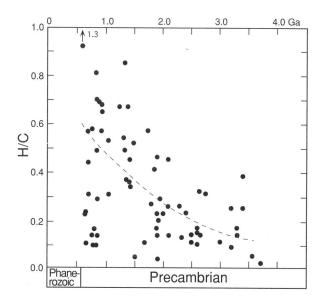

FIGURE 5   H/C ratios of kerogen isolated from Precambrian sediments of various ages (Strauss & Moore 1992).

Isotopic changes also occur as a result of metamorphism. Kerogens preserved in geologic units metamorphically altered to greenschist facies or above (i.e., subjected to temperatures greater than about 275°–300°C) tend to have H/C ratios less than about 0.2 (Hayes *et al.* 1983). At about this stage of alteration, breakage of carbon–carbon bonds in the preserved kerogen begins to occur; because of differences in the geochemical stability of ^{13}C- and ^{12}C-containing covalent bonds, the products released (e.g., CH_4, CO_2) tend to be preferentially enriched in the lighter carbon isotope, ^{12}C. As a result, the residual altered kerogen becomes correspondingly enriched in the heavier carbon isotope, ^{13}C. Fig. 6A (open circles) illustrates the effects of metamorphic alteration on a series of kerogens presumed to have been derived from precursor materials of similar compositions. As shown there, alteration of isotopic compositions is pronounced in those kerogens with H/C ratios less than about 0.2.

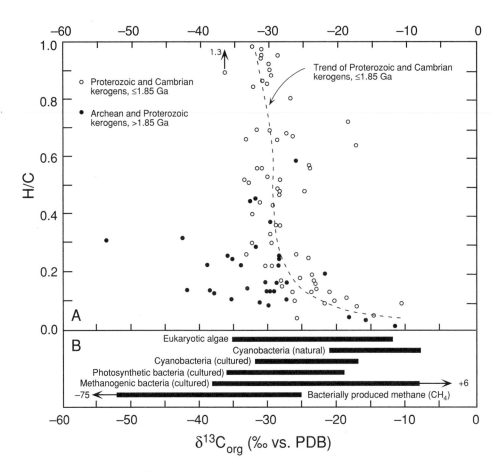

Figure 6   A: H/C and $\delta^{13}C_{organic}$ values for kerogen isolated from Archean, Proterozoic, and Cambrian sediments (McKirdy & Powell 1974; McKirdy & Kantsler 1980; Strauss & Moore 1992). Filled circles: ages >1.85 Ga; open circles: ages ≤1.85 Ga. **B:** $\delta^{13}C_{organic}$ values for extant microorganisms (Schidlowski *et al.* 1983) and bacterially generated methane (Galimov 1980).

Two cardinal points follow from the foregoing observations:

1   Isotopic compositions measured in kerogen residues represent maximum values – kerogen precursor materials may have been initially lighter (more ^{12}C-enriched) but not heavier.

2   The H/C ratios of preserved kerogens serve as an index of their degree of metamorphic alteration – carbon isotopic ratios in kerogens with H/C <0.2 are likely to have been altered from their initial compositions.

The Precambrian carbon isotopic record

Data plotted as open circles in Fig. 6A show the relationship between H/C ratios and $\delta^{13}C_{organic}$ for 65 kerogens isolated from geologic units ranging in age from 1.85 Ga to

Cambrian (McKirdy & Powell 1974; McKirdy & Kantsler 1980; Strauss & Moore 1992). The older of these age limits, 1.85 Ga, was selected as approximately coinciding with the time of development of a stable oxic global environment (Holland 1992, this volume) and the earliest known records of eukaryotic phytoplankton (Schopf 1992d). The younger, Cambrian, age limit was chosen because it is subtantially earlier than the occurrence of the earliest known fossils of vascular ("land") plants (Richardson 1992). Thus, the mix of primary producers giving rise to the precursors of these 65 kerogens (the ultimate biotic source of the kerogens), as well as the aerobic–anaerobic biologic processing of this material before its preservation, were evidently more or less similar.

Filled circles in Fig. 6A show H/C and $\delta^{13}C_{organic}$ data for 32 kerogens isolated from Precambrian geologic units older than 1.85 Ga, strata deposited before the widespread development of oxic conditions and the earliest records of fossil eukaryotes. In comparison with the younger kerogen data shown by the open circles, two principal differences are evident:

1 Kerogens isolated from sediments older than 1.85 Ga tend to have lower H/C ratios (commensurate with a greater degree of metamorphic alteration) than those isolated from younger sediments.

2 Many of these older kerogens are isotopically decidedly lighter (relatively ^{12}C-rich) than those isolated from the younger units.

Although a definitive explanation for the notably ^{12}C-rich nature of certain of the kerogens isolated from sediments >1.85 Ga old remains to be established, three types of possibilities can be suggested.

First, it could be postulated that this pattern of change over time reflects evolutionary changes in carbon-fractionating enzyme systems. If so, however, convincing evidence of such enzymatic evolution has yet to be discovered in the biochemical characteristics of extant photoautotrophs. Moreover, the general uniformity of $\delta^{13}C$ values for both carbonate and organic carbon throughout Precambrian time (Fig. 7), coupled with the well-documented evolutionary conservatism of Precambrian microorganisms (viz., their hypobradytelic rate of morphological, and presumably physiological, evolution; Schopf 1992d), suggest relative constancy, rather than major evolutionary change, in the mechanisms of photosynthetic carbon fixation. Second, as reviewed by Schidlowski, Hayes & Kaplan (1983), unusually light isotopic compositions have been observed in the cellular components of autotrophic microbes (e.g., cyanobacteria, photosynthetic bacteria, and methanogenic bacteria) cultured under high concentrations of CO_2 (viz., >0.5%). Significantly, current models of early atmospheric evolution seem to require similarly high partial pressures of CO_2 (Kasting 1992). Thus, it can be suggested that the ^{12}C-rich kerogens isolated from very ancient sediments may reflect the presence of relatively high concentrations of carbon dioxide in the early environment.

Although plausible, this CO_2-based mechanism is unlikely to explain the presence of the extremely ^{12}C-rich kerogens present in some very ancient sediments. Culture experiments with modern microbes suggest that autotrophic growth at high partial pressures of CO_2 could account for kerogen $\delta^{13}C$ values only as light as about −35 to 40‰ (Fig. 6B). Because substantially lighter kerogens (some having $\delta^{13}C$ values lighter

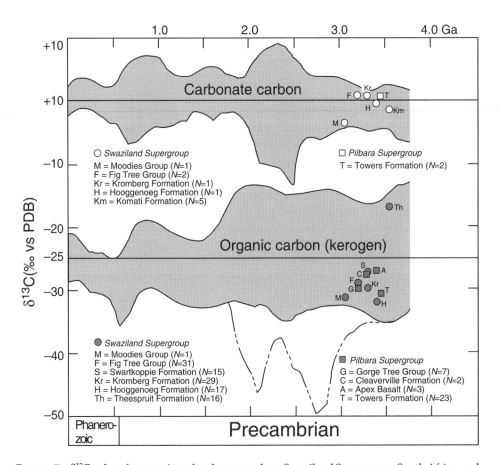

Figure 7　$\delta^{13}C$ values for organic and carbonate carbon, Swaziland Supergroup, South Africa, and Pilbara Supergroup, Western Australia (Strauss & Moore 1992; shaded areas indicate ranges of all measurements of Archean and Proterozoic sediments reported in that study).

than $-50‰$) occur in Early Proterozoic and Archean geologic units (filled circles in Fig. 6A and dashed envelopes in Fig. 7), a third possible explanation needs to be considered.

As shown in Fig. 6B, a promising source for the exceptionally ^{12}C-rich reduced carbon in very ancient sediments (particularly that having $\delta^{13}C$ values lighter than about $-40‰$) is the gaseous CH_4 produced by methanogenic bacteria. Such methane-producing archaebacteria are phylogenetically primitive obligate anaerobes, microbes of a type that may have been globally prevalent before the onset of widespread oxic conditions. However, bacterially generated methane could not have become buried in sediments and ultimately sequestered in the rock record unless it had first been incorporated in the particulate cellular components that make up kerogen-precursor materials. Thus, Hayes (1983) has suggested that the extremely light kerogens of the Early Proterozoic and Archean may be a result of the production of ^{12}C-rich CH_4 by methanogens, the incorporation of this methane into the body mass of CH_4-metaboliz-ing methylotrophic bacteria, and the burial and preservation of the ^{12}C-enriched methylotroph-derived organic matter. This intriguing scenario implies that anaerobic methanogens may have played a major role in the global ecosystem during and before, but not later than, the Early Proterozoic, an interpretation consistent with geologic data (Holland 1992, this volume) that place the onset of widespread oxic conditions at ~1.85 Ga ago.

The carbon isotopic composition of Early Archean (~3.55–~3.0 Ga old) Swaziland and Pilbara Supergroup sediments

As shown by the shaded areas in Fig. 7, the $\delta^{13}C$ values of most Early Archean (>3.0 Ga old) carbonates and kerogens are similar to those known from younger Precambrian units. Values for geologic formations of the ~3.55–~3.0 Ga old Swaziland Supergroup of South Africa and the approximately contemporaneous Western Australian Pilbara Supergroup (Strauss & Moore 1992) are summarized as averages in Fig. 7. Fig. 8 compares the carbon isotopic data from these two Early Archean sequences (Fig. 8A) with isotopic compositions observed in extant microbial photoautotrophs and in the principal modern inorganic-carbon reservoirs (Fig. 8B).

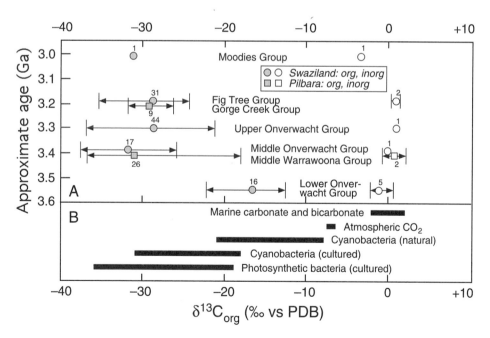

FIGURE 8   Summary of isotopic compositions of organic and inorganic carbon. **A:** Swaziland and Pilbara Supergroup strata. Dots (circles and squares) – average values of samples; numbers above dots – number of samples; horizontal bars – ranges of values. **B:** Extant microbial photoautotrophs and inorganic-carbon reservoirs.

The H/C ratios of kerogens isolated from the Swaziland and Pilbara sediments are relatively low, for each geologic unit having an average value <0.4 (Strauss & Moore 1992). The kerogen preserved in most of these units has been subjected to substantial postdepositional alteration. Of particular note is the average H/C ratio of 0.04 for eight kerogens isolated from the ~3.55 Ga old Theespruit Formation, near the base of the Swaziland Supergroup (Fig. 1), an especially low value consistent with a high degree of metamorphism. Evidently, the $\delta^{13}C_{organic}$ values of the Theespruit kerogens have been reset toward heavier isotopic compositions (Figs. 7 and 8).

With the exception of those of the Theespruit kerogens, $\delta^{13}C_{organic}$ values for Swaziland and Pilbara kerogens are comparable to those observed for extant cyanobacteria and photosynthetic bacteria, particularly to the ^{12}C-rich values characteristic of cultures grown under high partial pressures of CO_2 (compare Fig. 8A and B). Thus, the kerogen isotopic data are consistent with the Early Archean existence of microbial photoautotrophs, thereby supporting inferences based on the independent evidence provided by Early Archean stromatolites and microfossils. Moreover, these isotopic data are also consistent with the occurrence of relatively high concentrations of CO_2 in the Early Archean atmosphere, an interpretation similarly postulated on independent grounds (Kasting 1992). In addition, however, it is notable that many of the Swaziland and Pilbara kerogens have H/C ratios <0.2 (Strauss & Moore 1992). This suggests that the isotopic compositions measured in these samples may have been metamorphically reset from initially lighter values, and that in addition to photoautotroph-derived organic matter the precursor materials of these kerogens may therefore have included a contribution of very ^{12}C-rich carbon, such as that possibly derived from Early Archean methanogens.

Conclusions

Only two relatively well-preserved geologic sequences are now known from the Early Archean, those of the Swaziland Supergroup of South Africa and the Pilbara Supergroup of Western Australia. Both of these ancient sequences, ~3.55–~3.0 Ga in age, contain decipherable evidence of early life: stromatolites, microfossils, and particulate carbonaceous matter identifiable on the basis of its carbon isotopic composition as a product of autotrophy. Taken together, these three mutually consistent paleobiologic lines of evidence seem to establish that prokaryotic, photoautotroph-containing, stromatolitic communities were extant on this planet at least as early at ~3.5 Ga ago. Of particular interest are the carbon isotopic data that suggest that the Early Archean atmosphere may have contained a relatively high partial pressure of CO_2, an interpretation consistent with current models of early environmental evolution. The origin of life on Earth must have occurred earlier (and probably substantially earlier) than the age set by the Early Archean evidence of biologic activity here considered, the oldest such evidence now known in the geologic record.

Acknowledgments. – This article is based on a lecture initially prepared for the Conference on the Frontiers of Life held at Blois, France, in September, 1991. The research reported here was supported by NASA Grant NAGW-2147 and by NSF Grant BSR 86-13583.

Theme 2
The Maturation of Earth and Life

Between the earliest diversification of life in the Archean and the explosive evolution of multicellularity towards the end of the Proterozoic stretch nearly two billion years of Earth's history. How did the Earth and its biosphere evolve during this interval? To what extent did our planet's long Proterozoic maturation condition the world of the succeeding Phanerozoic Eon?

The late Archean growth and stabilization of continents changed the Earth's surface and changed the quality of the sedimentary record that documents biological and environmental history. Veizer considers the environmental consequences of this event, emphasizing that on long time scales, the surface chemistry of the Earth is under tectonic control – a sharp contrast to the widely discussed Gaia model. Hayes, Holland, and Grotzinger further explore various aspects of Earth's surface chemistry and the implications for Proterozoic evolution.

Unlike Archean rocks, those of the Proterozoic provide a wealth of paleontological data. Organic molecules abound, though Ourisson maintains that so far their information value for paleobiology is low. Walter, Runnegar, and Vidal discuss the fossil record of prokaryotes (particularly in the form of stromatolites, fossil microbial mats) and early eukaryotes. The record of the latter gives us important minimum ages of certain key forms, but the understanding of the evolutionary radiation into the major groups of protists must currently be based on living organisms, a topic explored by Taylor.

The key role of endosymbiosis in the origin of the eukaryotic cell, championed in particular by Margulis during the last decades, today seems firmly established, and Margulis & Cohen now stick out their necks to make us consider whether symbiogenesis could not also account for further diversification of eukaryotic taxa. Notwithstanding the importance of eukaryote evolution, Golubic reminds us that prokaryotes such as cyanobacteria continue to play a major role in the biota.

As in studies of the Archean Earth, an intellectually satisfying integration of Proterozoic paleontology and geology with comparative biology is still beyond our grasp, although – again – the path towards integration is becoming discernible.

The Archean–Proterozoic transition and its environmental implications

Ján Veizer

Institut für Geologie, Ruhr Universität, 44780 Bochum, Germany, and Derry/Rust Research Unit, Ottawa–Carleton Geoscience Center, University of Ottawa, Ottawa, Canada K1N 6N5

The geological history of this planet can be viewed as a succession of evolving cycles. The entities of the solid Earth (e.g., tectonic domains) cycle slowly, on time scales of 10^7–10^9 years, and possess therefore a direct residual record for almost 4 Ga. Its deconvolution shows that the bulk of the continental crust has been generated ~2.5±0.7 Ga ago (Archean–Proterozoic transition) from an assembly of volcanic terranes. This, in turn, led to expansion of epeiric seas with their diverse habitats within the photic zone. In contrast to the dominant hierarchy of the solid Earth, the subordinate smaller entities of biosphere, hydrosphere, and atmosphere cycle fast, and their direct record is rapidly lost. Their properties are, however, reflected in a derivative record, such as the chemical and isotopic composition of (bio)chemical sediments. These tracers suggest that the Archean ocean was buffered by large-scale hydrothermal circulation of seawater through oceanic crust. Subsequent to continent formation, the dominant buffering mechanism shifted to rivers. As a result, the redox balance of the exogenic cycle evolved in tripartite manner. In the Archean, the control was endogenic, with the redox balance maintained by C–O–Fe coupling (biosphere–oxygen–magmatic iron). The (early?) Proterozoic represents a transitional stage into a Phanerozoic-like steady state that is decoupled from its endogenic underpinning, with the redox balance at higher oxygen levels maintained by biologically mediated C–O–S coupling.

Life on this Earth and the planet itself coevolved in the course of geological history. To reconstruct their evolution, we have to rely on the record preserved in the rocks. This record is of two types. First, there is a direct record, as manifested in the variety of rocks, basins, mountain ranges, etc. Second, the rocks themselves contain a derivative or proxy record, such as, for example, their isotopic composition, which is inherited from former oceans. In order to decipher past events, we have to read, juxtapose, and interpret both of these records.

Our planet and its subsystems (lithosphere, hydrosphere, atmosphere, and biosphere) are the products of interacting evolutionary and cyclic processes. This interaction can be viewed in a manner similar to that of human life with its directional aging and daily cyclicity, where the aging – or evolution – results from a cumulative effect of the superimposed cycles. The first-order trend, on the time scale of decades, is that of aging. In contrast, on shorter time scales, the directionality of evolution may not even

Bengtson, S. (ed.) 1994: *Early Life on Earth. Nobel Symposium No. 84*. Columbia U.P., New York be discernible, and we may only observe yearly to daily oscillations around a given evolutionary mean. The same is true for the entire planet, albeit at progressively longer time scales (Veizer 1988a).

The rock record and its preservation probabilities

The present-day distribution of rock types in progressively older segments of the Earth's crust, while reflecting the original abundances, is strongly modified by the superimposed recycling. The generation and destruction of packages of rocks – be they volcanic, plutonic, or sedimentary – are in the first instance controlled by the rocks' tectonic setting. The unifying concept that explains this dynamics, the plate tectonics, holds that the Earth's surface is comprised of several plates that are constantly moving at a rate of several centimeters per year, with the result that their boundary interactions generate and destroy ocean basins, volcanic chains, mountain belts, sedimentary basins and platforms, etc. The maturation process that results in the increasing stability of the successor tectonic settings is a sequential phenomenon. The opening and closure of an oceanic basin is a precondition for generation of an island arc, the latter – in turn – melts its deep roots as it thickens to generate granitic plutons of mountain ranges (orogenic belts), erosion of these mountains to an "equilibrium" thickness of ~36 km produces the stable basement on which platformal sediments can accumulate, and so on towards the most stable tectonic unit, the Precambrian shield. At each transition, the bulk of the material involved is simply recycled, and only a subordinate portion is incorporated into the successor tectonic setting. The arrangement can be thus viewed as a nested hierarchy of cycles, more specifically of autocatalytic loops (cf. Odum 1983).

The probability of preservation of rocks within a given tectonic setting (loop) is determined by the rate of recycling. Mathematical systematics of this approach is analogous to that of population dynamics in biology (Veizer & Jansen 1985). The process of generation/destruction (recycling) generates an age structure for the units of a given population, and this age structure, in turn, can be utilized to calculate the rate of recycling as well as the half-life and the life span (or oblivion age) for a given population. The faster the rate of recycling, the smaller the chance of preserving a specific rock association in the ancient rock record. As a rule of thumb, 4–6 half-lives are sufficient to obliterate the record to such a degree that the available data fall within the background noise. This is not to say that, for example, a remnant of an ancient oceanic crust is not recognizable as such even in segments as old as 2.0 Ga (Scott et al. 1991). Yet the quantitative nature and the continuity of the oceanic record is lost by ~160 Ma. Since, in most instances, the inventory of geological entities is no better than ±5–10%, this effectively represents the background noise to quantitative modeling based on rock inventories.

Considering the above uncertainties in the data base for rock inventories of various tectonic settings, and restricting the Archean–Proterozoic transition to a ~2.5±0.8 Ga interval, the theoretical probability of preservation of a reliable quantitative record for anything but continental basement is minimal (Table 1).

As a first-order approximation, the geographical extent of the age provinces within the continental basement is known. However, the uncertainties in higher-order fluctua-

TABLE 1   Summary of half-lives for major global tectonic realms (Veizer 1988b).

Tectonic realm	Theoretical half-life
Active margin basins	27 Ma
Oceanic intraplate basins	51 Ma
Oceanic crust	59 Ma
Passive margin basins	75 Ma
Immature orogenic belts	78 Ma
Mature orogenic belts (roots)	355 Ma
Platforms	361 Ma
Continental basement (depending on the isotopic systematics utilized to define the age provinces)	673–1728 Ma

tions around the general trends are as yet poorly constrained. This is partly due to differing sensitivities of various dating techniques to later resetting (rejuvenation) as well as to cannibalistic incorporation of older segments into subsequent orogenic cycles. It is particularly the old, trailing end of this age pattern – the interval of direct interest to our subject – that is poorly constrained. The present-day surviving age structure, however imperfect, must first be "unrecycled" if we are to arrive at the original extent of continents in the course of geologic history. This can be done by juxtaposition of isotopic dating techniques (DePaolo *et al.* 1991) or purely theoretically (Veizer & Jansen 1979). In both cases, the growth of continental crust approximates a sinusoidal (logistic) pattern, with small continental nuclei in the Archean, bulk of the growth at ~2.5±0.7 Ga ago, and attainment of near present-day extent ~1.8 Ga ago (Fig. 1). Note, however, that production of continental crust from an assembly of volcanic island chains (Lowe, this volume) was not a continuous process but a succession of shorter orogenic episodes. We do not know yet what triggers these episodes. Recently, Peltier & Solheim (1992) demonstrated that the pattern of mantle convection shows features of a nonlinear "chaotic" system that produces sudden drawdowns of the subducting oceanic crust into the mantle. Whatever the details of production of the continental crust, its overall logistic growth pattern is supported also by the temporal

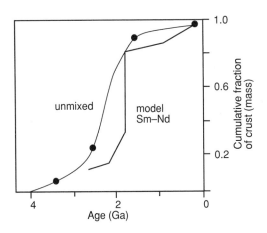

FIGURE 1   Mass–age (and area–age) distribution of continental crust . The "model Sm–Nd" curve represents age distribution in the western United States based on Sm–Nd model ages, but the trends for other cratons are similar. The "unmixed" curve shows this model distribution apportioned to the four most common crystallization ages (orogenies), which are marked by full circles. (Modified after DePaolo *et al.* 1991 and Goodwin 1991.)

distribution of mineral deposits (Veizer *et al.* 1989c), the only other geological entity with appropriate inventories available.

In summary, the Archean–Proterozoic transition was a protracted development of ~1.5 Ga duration that resulted in transformation of the surface of the planet from an assembly of volcanoes projecting from the coeval seas into an Earth that had about one-third of its surface covered by continents. The broad shelves of the latter were inundated by epeiric seas, thus expanding the extent and the diversity of habitats within the photic zone. The above scenario is not an either/or proposition, because both oceanic- and continental-type terranes did coexist on the young Earth and still do today. It is their relative proportions that have changed.

Ocean evolution and proxy record

In the absence of samples of ancient seawater, we have to rely on derivative record inscribed in patchily preserved (bio)chemical sediments, such as carbonate rocks, to elucidate the nature of oceans some 2.5±0.7 Ga ago.

Buffering of seawater composition

The dominant fluxes that control the composition of the present-day ocean are the following: (1) the hydrologic cycle via rivers that reflects the properties of the continental crust, and (2) the hydrothermal flux on submarine oceanic ridges and their flanks that reflects the nature of the oceanic crust and mantle. A volume of water equal to one present-day ocean is reprocessed via cycle 1 every 40,000 years and via cycle 2 every 10 Ma. However, seawater that circulates through the hydrothermal cells has a total salinity about 300 times higher than the river water. Consequently, the two fluxes for many elements (as opposed to water) may be of comparable magnitude. The relative importance of these two fluxes in the geological past could be deciphered if flux-specific tracers were available.

Today, the average strontium isotopic composition of riverine water is 0.711, that of the submarine hydrothermal discharge is 0.703, and of that of ocean water is 0.709 (Veizer 1989). A simple balance calculation shows therefore that, in terms of strontium, the present-day oceanic budget reflects a 75:25 mixing of riverine/hydrothermal inputs. In contrast, the young Earth must have been considerably warmer, because of inherited energy of impacts and particularly because of the heat generated by decay of the short-lived radiogenic isotopes (e.g., Bickle 1978). The dissipation of this internal energy, via ubiquitous volcanism and by hydrothermal circulation on the ocean floor, should have been therefore more important than today. On the other hand, continents (and rivers) were of subordinate importance. If so, the first-order trend in Sr isotopic composition of seawater should indicate a transition from mantle-like values of 0.703 to a more radiogenic river-like composition, a proposition confirmed by experimental data (Fig. 2). The above theoretical and experimental considerations therefore suggest that the most important step in the history of the oceans was a switch in the nature of the dominant buffering mechanism, from submarine warm springs to rivers, coincident with the growth of continents, a sequence of events designated here as the Archean–

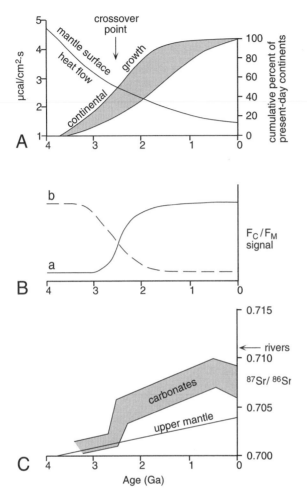

FIGURE 2   **A:** Schematic presentation of mantle surface heat flow and of areal continental growth during geologic history. **B:** The expected signal in seawater composition. This signal should be of type a if the transition results in a gain and of type b if in a loss of an entity. Note that this general shape of the signal will evolve regardless of the details of the two exponential curves of opposing slopes. For radiogenic isotopes, such as Sr, the flat parts of the age curves will be modified by radioactive decay. **C:** Schematic $^{87}Sr/^{86}Sr$ variations in seawater during geologic history. (From Veizer 1984.)

Proterozoic transition. The repercussions of such a development for evolution of the atmosphere and life will be enumerated in the subsequent text.

Ocean temperatures

The proposition by atmospheric modelers that the young Earth could have had a massive CO_2 greenhouse (Kasting 1987), and particularly the oxygen isotope data on ancient cherts and carbonates (e.g., Knauth & Epstein 1976), have been interpreted as being consistent with the existence of warm oceans, with temperatures perhaps in excess of 70°C.

The postulated greenhouse may indeed have been real (see also references for carbon-isotope data in the chapter by Hayes, this volume), and it may have been even essential in order to counteract the theorized lower luminosity of the "faint" young sun that should have resulted in a frozen-over Earth (Walker 1990). The clear geological evidence for ubiquitous aqueous, as opposed to glacial, sediments throughout the

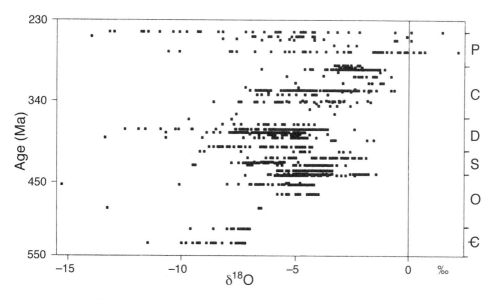

FIGURE 3 The $\delta^{18}O$ secular variations for Paleozoic brachiopods (from Wadleigh & Veizer 1992). An additional set of Ordovician data (Qing & Veizer, unpublished), confirms the general trend towards ^{18}O depletion with age.

entire Precambrian time span clearly contradicts this frozen-Earth scenario. On the other hand, a "massive" greenhouse, providing it was a reality (Walker 1990), does not necessarily mandate a hot ocean.

The depletion in ^{18}O in progressively older (bio)chemical sediments is a definitive experimental observation (Knauth & Epstein 1976; Veizer & Hoefs 1976). However, the maturation of chert via a sequence opal A – opal CT – chert is a result of burial recrystallization of silica at progressively elevated pressures and temperatures (Hesse 1990a, b). The $\delta^{18}O$ of the end product reflects therefore this burial temperature and not the temperature of the ocean water. Similarly, the Precambrian limestones and dolostones were also subjected to diagenetic recrystallization. However, in contrast to chert, this recrystallization happens at minimal burial depths and results in some, but considerably smaller, postdepositional resetting of $\delta^{18}O$. Moreover, a definitive ^{18}O depletion ("hot" oceans) has been documented also for Paleozoic brachiopods (Fig. 3) that have shells from a relatively stable mineral (low-Mg calcite) and that underwent only minor, if any, recrystallization. A coincidence of these ^{18}O-depleted brachiopod samples with several episodes of extensive glaciations argues strongly against seawater temperatures much in excess of the present-day values. Similarly, the postulated hot Precambrian oceans are difficult to reconcile with the existence of the well-documented glaciations some 600–800, 2,300, and perhaps even 2,700 Ma ago (Frakes 1979; Lowe 1992b).

In summary, the existing geological record does not support the postulate of hot Precambrian oceans. A more likely explanation of ^{18}O depletion in Precambrian (bio)chemical sediments is a proposition that the data reflect a combination of postdepositional alteration phenomena with some other factor(s), such as the changing isotopic composition of seawater (Veizer *et al.* 1986).

Geosphere–biosphere coupling

The isotopic signals of past seawater provide us also with proxy evidence for coupling of rock–water–atmosphere–life cycles on geological time scales. The importance of carbon isotope record for understanding the evolution of the biosphere is discussed by Hayes (this volume). As pointed out in that contribution, life in substantial quantities must have been extant since at least 3.5 Ga ago. In the geologic record, this is reflected by the confinement of the $\delta^{13}C$ measurements for ancient carbonates to near present-day values of about $0\pm3\%_o$ PDB. The superimposed second- and higher-order oscillating patterns have as yet to be resolved for the Precambrian, although their reality is clearly indicated even by data from an incomplete and poorly dated rock record. During the Phanerozoic, particularly the last 500 Ma, the $\delta^{13}C$, $^{87}Sr/^{86}Sr$, and $\delta^{34}S$ of marine carbonates and sulfates were tracking each other on time scales of 10^7–10^8 years (Fig. 4). As discussed above, the isotopic composition of marine Sr is controlled solely by inorganic processes. On these time scales, the principal causative factor(s) of all the above isotopic variations must therefore have been of geological nature. Despite the fact that the exact scenario is still enigmatic, tectonic processes likely regulated the river/

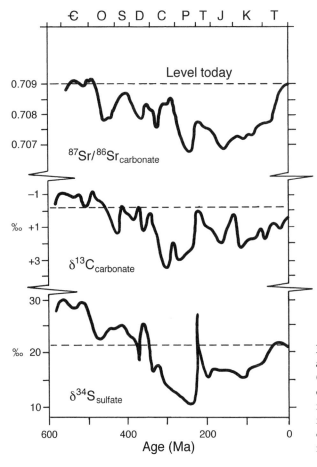

FIGURE 4   Seawater isotopic age curves for the Phanerozoic (latest 550 Ma) as recorded (a) in carbonates and in fossil apatites for Sr isotopes; (b) in carbonates for carbon isotopes; and (c) in evaporite sulfates for sulfur isotopes. (From Veizer 1988a.)

hydrothermal flux ratio ($^{87}Sr/^{86}Sr$) as well as the intensity of organic productivity and burial ($\delta^{13}C$) by, for example, the extent and configuration of shelves, by nutrient supply, or by other controls. In turn, the rate of burial of organic matter regulated the oxidation state of the ocean–atmosphere system. Since the biochemical carbon and sulfur cycles are coupled via an overall redox equation (Garrels & Perry 1974; Holser *et al.* 1988)

$$8SO_4^{2-}+2Fe_2O_3+H^++15CH_2O \rightleftharpoons 4FeS_2+15HCO3^-+8H_2O \qquad (1)$$

the net consequence is a negative correlation between $\delta^{34}S$ and $\delta^{13}C$ (Fig. 4) in Phanerozoic seawater (Veizer *et al.* 1980). This negative correlation also suggests that iron played only a subordinate role in the redox balance of the Phanerozoic exogenic cycle. Note that despite proliferation of the BLAG-type models (e.g., Berner *et al.* 1983), all based on the above negative C/S isotopic correlation, and despite the well-documented observation that the modern diagenetic sulfate reduction by bacteria is coupled to oxidation of organic matter, the geological scenario for the stoichiometric relationship in equation **1** is still an open question. Similarly, a projection of the Phanerozoic oscillatory patterns and periodicities into the Precambrian, particularly if older than 1 Ga (e.g., Worsley & Nance 1989), is not testable with the exisiting data bases.

With these qualifications in mind, it appears nonetheless that the cycles of Sr–C–S–O (and likely other) elements have been coupled and, on time scales of 10^7–10^8 years, controlled by the dynamics of the solid Earth. In other words, the system is hierarchical, and life is constrained by the limits imposed by the dominant hierarchy of the mother Earth (Veizer 1988a), a proposition at odds with the primary role of the biosphere advocated by the Gaia hypothesis (Lovelock 1979). In time scales of $\leq 10^7$ years, lithospheric responses become too sluggish, and the system is dominated by the dynamics of the smaller and faster hierarchies, the hydro-, atmo-, biosphere, all inscribing higher-order wiggles on the observed time series of isotopic signals. As a result, the correlation with Sr is lost, while that of C and S is still maintained to about 10^6 years' resolution. It is only at this, or still higher, hierarchical level that a Gaia-like scenario may apply.

Sulfur and carbon cycles in the Precambrian

The fragmentary Precambrian rock record precludes delineation of the Phanerozoic-like isotopic time series for coeval seawater. The situation is particularly critical for $\delta^{34}S$ in sulfate because of the dearth of sedimentary gypsum and anhydrite in the Precambrian. Yet, even with these limitations, the first-order secular trends for $\delta^{13}C$ and $\delta^{34}S$ differ. Whereas the former appears to have been only slightly, if at all, modified since at least 3.5 Ga ago (Hayes, this volume), the $\delta^{34}S$ data indicate sulfate/sulfide pairs that progressively diverge from basaltic, mantle-like values of 0‰ CDM (Fig. 5). Whatever the details of higher-order oscillations for these isotopic trends, the oxidation/reduction balance of the marine sulfur budget on the young Earth was not mediated by bacterial dissimilatory sulfate reduction. If so, sulfur was not likely to have served as the dominant redox couple for the ancient carbon cycle. Instead, the redox balance in equation **1** was likely maintained by iron. Indeed, ancient carbonates – up to perhaps 1 Ga ago – contain higher Fe^{2+} and Mn^{2+} concentrations than their Phanerozoic coun-

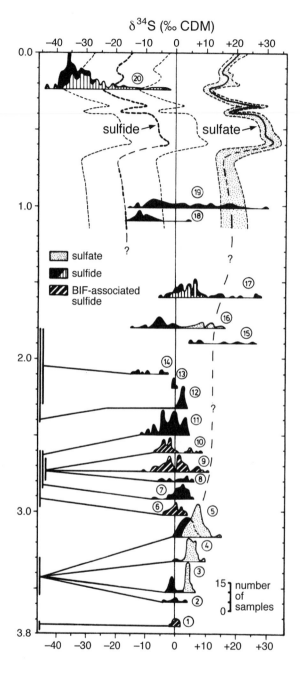

Figure 5 Isotopic composition of sedimentary sulfide and sulfate through time. The isotope age curve of Phanerozoic and late Proterozoic sulfates (younger than about 1.2 Ga) is the better documented part of the record, the stippled area being the estimated uncertainty. The typical feature of the Precambrian record is the predominance of heavy sulfide values, particularly during the Proterozoic. The possible explanations are discussed in Lambert & Donnelly 1991 and Bottomley *et al.* 1992. (From Schidlowski *et al.* 1983.)

terparts (Fig. 6). Some of this Fe^{2+} has been incorporated into carbonate minerals during diagenetic transformation of original sediments into rocks. Nevertheless, the Phanerozoic counterparts have also been subjected to a similar diagenetic stabilization step, yet their Fe^{2+} concentrations are lower. At least some of the "excess" iron in Precambrian carbonates could have been therefore inherited from their depositional environments (oceans), which may have been less oxygenated than their Phanerozoic counterparts.

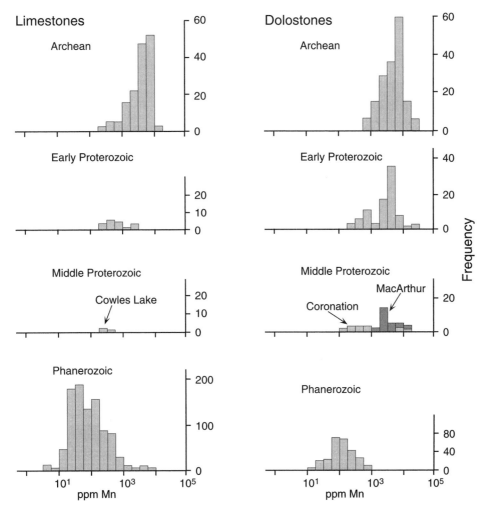

FIGURE 6 Manganese concentrations in Precambrian and Phanerozoic carbonate rocks. Iron concentrations show a similar pattern. (From Veizer *et al.* 1992.)

Biogeochemistry of the Archean–Proterozoic transition

The dominant feature of the Archean–Proterozoic transition appears to have been a change in tectonic style caused by exponentially declining dissipation of the heat from the Earth's mantle. As a result, the original crustal material – morphologically volcanic islands and island chains dotting the seas – has been reprocessed into the near present-day continental mass with its broad shelves and epeiric seas. Concomitant with this lithospheric transition, the dominant buffering of oceans shifted from submarine springs to rivers. This scenario has implications for the redox balance of the atmosphere–ocean system and on the nature of biogeochemical cycles.

Based on the interpretation of the carbon isotope record (Schidlowski *et al.* 1983; Hayes, this volume), the relative sizes of reduced – that is, extant as well as buried "biomass" – and oxidized exogenic reservoirs of carbon have not been markedly different from their present-day steady state since at least 3.5 Ga ago. Such an interpretation does not preclude some second-order fluctuations in the ratio of oxidized/reduced carbon (Fig. 4) or the possibility that both reservoirs may have been growing simultaneously, for example, in response to the growth of continents (Knoll 1984b; Walker 1990) that, in turn, may have resulted in enlargement of the global sedimentary mass (Veizer 1983). The experimental data also do not preclude shifts in relative proportions of extant vs. buried (dead) pools of $C_{organic}$, a development that may have ensued from a progressively more efficient utilization of nutrients (Worsley & Nance 1989).

On the Archean earth, the inventory of $C_{organic}$ has been controlled mostly by cyanobacteria that – in analogy to their modern counterparts – likely possessed photosystem II and thus produced oxygen as a by-product (chapters by Schopf and Towe, this volume). If so, the bulk of photosynthetic oxygen may have been generated already at this stage. Its subsequent history would have been that of recycling by, and repartitioning among, several redox-sensitive biogeochemical cycles. If this were the case, it also follows (Towe 1991, this volume) that "only rapid, oxidative recycling of organic carbon by aerobic heterotrophy could plausibly have kept pace with the oxygen produced annually by net photosynthesis." This probably would have been the case even if the overall inventory of exogenic carbon in the Archean were somewhat less than in the Proterozoic and Phanerozoic. It appears therefore that production/decomposition of organic matter accounted then, as today, for the bulk of biospheric carbon cycling. However, geological evidence – such as the rarity of red beds and ubiquity of detrital uraninite and pyrite (Holland, this volume) – in the terranes older than 1.8 Ga argue for somewhat depressed oxygen levels in the coeval atmosphere/hydrosphere system, with estimates ranging from 0.1% of the present-day concentrations (Towe 1991) to an essentially anoxygenic situation (Kasting 1991). Unfortunately, no available indicator can be calibrated for precise pO_2 monitoring. My personal preference is for concentrations approaching the upper bounds of the above estimates. Whatever the actual pO_2, the progressive rise of oxygen in the course of geologic evolution was probably accomplished via coupling of carbon cycle to cycles of other redox-sensitive elements. As already pointed out, sulfur was the likely redox partner during the Phanerozoic and iron in the Archean, with a crossover at about 1.8 Ga ago.

In an Archean "mantle-buffered" ocean (Veizer *et al.* 1982), seawater composition is controlled mainly by interaction of warm seawater with the basaltic oceanic crust. Today, the principal oxidant in seawater is sulfate, with dissolved oxygen being of subordinate importance, the latter more than compensated for by the reducing power of CH_4 and H_2. This sulfate is quantitatively removed in circulating hydrothermal systems and, if removed permanently, the flux would be sufficient to "scrub" the entire oxidant power in the exogenic sphere in ~60 Ma (Wolery & Sleep 1989). It is not yet clear how much of this sulfate is reduced – oxygen being utilized, for example, for oxidation of Fe^{2+} to Fe^{3+} in basalts – and what proportion is precipitated as sulfate and later redissolved back into seawater. Even if this entire sulfate cycle were a closed loop, hydrothermal fluids discharge sulfide in quantities that are ~10–20% of the sulfate flux. This sulfide is a net reductant regardless of whether it originates from seawater sulfate or from leaching of basalts. If it is accepted that the heat dissipation, hence the rate of seawater circulation via hydrothermal systems, was perhaps six times that of today (Bickle 1978), the drain on the oxidative power of the ocean–atmosphere system must have been considerable. The exponential decay in the rate of hydrothermal circulation, coupled with the growth of continents that enhanced the river flux (Fig. 2), likely were the driving forces responsible for the transition to the more oxygenated steady state attained in the early Proterozoic. The above scenario can also explain the observed S, Sr and C isotopic age patterns. In this understanding, the primary redox balance of carbon cycle has been set at the beginning of, or before, the existing rock record, with coupling to cycles of other elements, such as iron and sulfur, playing only a subordinate role.

The mantle-like S and Sr isotopic values in the oldest rocks (Figs. 2, 5) are consistent with the proposition that these elements were magmatically derived. Subsequently, ~2.7–2.5 Ga ago, the increased overall range in $\delta^{34}S$ values suggests that bacterial dissimilatory sulfate reduction began to play a significant role in the exogenic sulfur cycle, and the shift to radiogenic $^{87}Sr/^{86}Sr$ ratios signifies that rivers started to be felt in the strontium cycle. In the Proterozoic, the biological and atmospheric factors, that is, weathering of continents, became the dominant agents that maintained the redox balance of the exogenic reservoirs. In short, the advancing decoupling of the exogenic system from its endogenic (magmatic) underpinning resulted in progressive buildup of oceanic sulfate concentrations (Walker & Brimblecombe 1985), in enhanced $\delta^{34}S_{sulfate–sulfide}$ fractionation, and in a rise of $^{87}Sr/^{86}Sr$ ratios in seawater (Figs. 2, 5).

Summary

Consideration of seawater isotopic data in the course of geologic history suggests that the exogenic cycle and its redox buffering evolved in a tripartite manner. In the Archean, the control was essentially endogenic, with the redox balance maintained by carbon–oxygen–iron coupling (biosphere–oxygen–magmatic iron). The (early?) Proterozoic represents a transitional stage that eventually evolved into a Phanerozoic-like steady state that, on time scales of $<10^8$ years, is essentially decoupled from the endogenic input and the redox balance of which is maintained by biologically mediated carbon–oxygen–sulfur coupling (Veizer 1988a, b; Walker 1990).

Global methanotrophy at the Archean–Proterozoic transition

John M. Hayes

Biogeochemical Laboratories, Departments of Geological Sciences and of Chemistry, Indiana University, Bloomington, IN 47405-5101, USA

Varying abundances of ^{13}C in sedimentary carbonates and organic materials can be interpreted in terms of (1) mechanisms involved in the production of organic carbon and (2) the division of buried carbon between inorganic and organic forms. Exceptional variations are recorded in the oldest portions of the sedimentary record. Isotopic characteristics of early Archean carbon cycles are consistent with (1) photoautotrophy based on a process of carbon fixation not strongly different from that mediated at present by the rubisco enzyme and (2) recycling of organic material by some process with minimal isotope effects. In marked contrast, those of late Archean and earliest Proterozoic age indicate the coexistence of CO_2 and CH_4 as equal participants in the global carbon cycle. The onset of methane cycling ~2.8 Ga ago indicates the production and consumption of both O_2 and CH_4 in surface environments. The isotopic record suggests that this mode of carbon cycling was of such global significance that paleobiologists working in the Precambrian might refer to an "Age of Methanotrophs" with at least as much justification as those working in the Phanerozoic speak, for example, of the age of reptiles.

An ecosystem functions by processing carbon. Autotrophs capture and store energy by producing organic compounds; heterotrophs remobilize organic carbon and derive energy in the process. As shown in Fig. 1, these biologically mediated flows of carbon occur within Earth's surface environment, through which carbon passes in response to processes that are purely physical. Carbon-containing gases can be injected into the atmosphere and hydrosphere by volcanic or other, less dramatic processes. Organic and inorganic forms of carbon can be eroded from the continents into the oceans. Carbon is withdrawn from the surface inventory by sedimentation and burial of insoluble materials. If the sediments escape deep burial and alteration, this process preserves substances indicative of the status of the carbon cycle at the time of sedimentation.

A "geochemical record" is thus written in the carbon in ancient sediments, and from it we can seek to reconstruct the development of biotic processes within the global carbon cycle. Thinking in biochemical and ecological terms, we might choose any point in time and ask:

- By what mechanisms were autotrophs producing organic carbon?

Bengtson, S. (ed.) 1994: *Early Life on Earth. Nobel Symposium No. 84.* Columbia U.P., New York

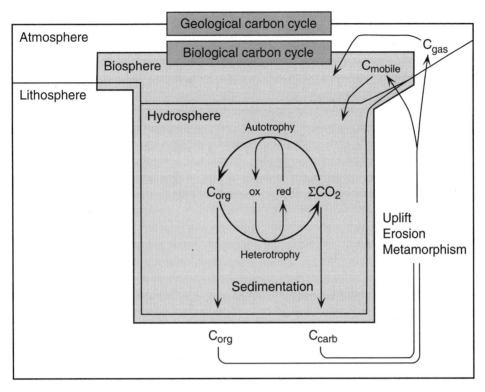

FIGURE 1 The biological (autotrophy ⇌ heterotrophy) and geological (everything else) carbon cycles. It is only for graphic simplicity that this highly schematic view depicts volcanism as an exclusively subaerial process.

- By what mechanisms were heterotrophs remobilizing organic carbon?

- At what rates were these processes occurring?

These are key questions. If we could answer them, we would know much about the development of photosynthesis and of respiratory processes, about the rise of O_2 in Earth's atmosphere, about the structure of ancient ecosystems, and even a bit about the circulation of ancient oceans.

The carbon-geochemical record has multiple aspects: the form (organic vs. inorganic) and abundance of carbon, ratios of ^{13}C to ^{12}C, and structures and abundances of biosynthetically produced carbon skeletons. The last of these is specifically considered by Ourisson (this volume) and will not be discussed here. The first has recently been reviewed by Strauss *et al.* (1992a), who also report numerous new analyses of Proterozoic sedimentary rocks. The results show that amounts of organic carbon comparable to modern levels can be found even in the oldest sediments – there is no evidence that a "preorganic" era has been recorded – and do not suggest any profound changes in the loci of deposition of organic carbon. However, because the sediments preserved represent only a tiny fraction of those initially formed, and because the range of materials preserved is very likely to have been biased during crustal reworking of the debris, little detailed interpretation is possible.

Ratios of ^{13}C to ^{12}C in organic and inorganic materials have varied over geologic time. The mechanisms underlying these changes are largely – though by no means completely – understood, and variations can be interpreted in terms of the histories of processes within the carbon cycle (e.g., Holser *et al.* 1988; Strauss *et al.* 1992b; Des Marais *et al.* 1992). Even if this were not so, the isotopic variations would provide an empirical key to the reconstruction of flows of carbon in ancient environments (Wickman 1956). To explain the significance of isotopic variations more fully, we must begin with a consideration of the carbon cycle and its isotopic aspects.

An isotopic view of the carbon cycle

The "exogenic reaction chamber" in which the biological carbon cycle exists is schematically indicated in Fig. 2. Atoms of carbon cannot be produced or consumed within

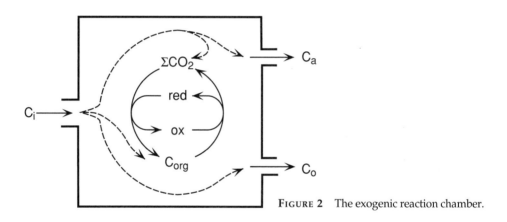

FIGURE 2   The exogenic reaction chamber.

the exogenic cycle, but they can be oxidized and reduced. Amounts of carbon entering and leaving the exogenic reaction chamber must ultimately balance, but electrons can be exchanged with other elements. The mass balance can be specified in both elemental and isotopic terms:

carbon in = carbon out: $\varphi_i = \varphi_a + \varphi_o$ (1)

carbon-13 in = carbon-13 out: $\varphi_i F_i = \varphi_a F_a + \varphi_o F_o$ (2)

where the φ terms represent flows of carbon, moles C per unit of time, and the F terms denote fractional abundances of ^{13}C [i.e. $^{13}C/(^{13}C + ^{12}C)$]. The equations of geochemical arithmetic often involve such combinations of identical terms. Accordingly, the subscripts require attention: i denotes incoming carbon, a represents carbonate (c would be confused with carbon itself), and o represents organic. For isotopic abundances within the range of natural variations, equation 2 can be satisfactorily approximated by

$\varphi_i \delta_i = \varphi_a \delta_a + \varphi_o \delta_o$ (3)

where δ denotes isotopic abundance and is defined by $\delta_x \equiv [(R_x/R_r) - 1]10^3$, where $R \equiv {}^{13}C/{}^{12}C$ and the subscripts x and r denote the carbon form of interest and a fixed reference material, respectively. For carbon, the reference material is calcite (calcium carbonate) from belemnites (fossil squid) of the Pee Dee Formation (Cretaceous, South Carolina, USA, Urey *et al.* 1951; $R_{PDB} = 0.011179$, Bakke *et al.* 1991). Use of this isotopic reference point is specified by noting that δ values have been referred to the "PDB standard." The units for δ are ‰, termed per mill or parts per thousand.

The value of δ_a is controlled by the isotopic composition of carbonate dissolved in seawater. Throughout Earth's history δ_a has varied around the zero point on the PDB scale. The ocean is stirred well enough that the same value of δ_a is found in most open marine environments (Kroopnick 1985). In the absence of secondary alteration, values of δ_a from such environments are thus globally representative. Values of δ_o are more dependent on local conditions, but it is still possible, for most periods of Earth's history, to suggest a globally representative value (Summons & Hayes 1992). In quantitative terms, organic carbon is depleted in ${}^{13}C$ relative to carbonate carbon by 20–60 parts per thousand (i.e. $-20 \geq \delta_o \geq -60‰$). In the modern environment, this isotopic difference arises within the biological carbon cycle and is associated primarily with the initial fixation of CO_2 (e.g., Hayes *et al.* 1989). In general, the magnitude of the isotopic difference, $\Delta_C \equiv \delta_a - \delta_o$, is dependent on (1) the pathways of carbon fixation and mechanisms of biosynthesis employed by autotrophs (Schidlowski *et al.* 1983) and (2) the pathways and mechanisms of remineralization of organic carbon. The magnitude of Δ_C therefore encodes information bearing precisely on the first two questions set out at the beginning of this chapter.

Quite apart from its interpretation in mechanistic terms, the isotopic difference between organic and inorganic carbon is useful when it comes to quantitative consideration of the carbon budget. Combination of equations 1 and 3 yields the expression

$$\delta_i = (1 - f_o)\delta_a + f_o\delta_o \tag{4}$$

where f_o ($\equiv \varphi_o/\varphi_i$) is the fraction of carbon passing through the surface environment and being buried in organic form. This parameter is related to production of organic matter within the ecosystem. If producers were exceptionally successful and *all* carbon were organic, f_o would approach 1.0. The opposing effects of consumer organisms must also be considered, but the significance should be clear: f_o represents some kind of operating point for the carbon cycle. Its value depends not only on productivity but also on the nature of sedimentary environments and communities. If organic matter falling to the seafloor and being incorporated in sediments is aggressively remineralized, f_o will be low. Independent of mechanisms, however, equation 4 can be rearranged so that f_o is the dependent variable:

$$f_o = (\delta_a - \delta_i)/(\delta_a - \delta_o) = (\delta_a - \delta_i)/\Delta_C \tag{5}$$

We may be bathed in ignorance about processes actually occurring at any point in time, but we can determine the operating point of the global carbon cycle from only three δ values. This approach to the determination of f_o, introduced by the Swedish geochemist Wickman (1956), is hugely advantageous in comparison to the alternative of direct assessment, which would involve compilation of the actual quantities of organic and inorganic carbon in sedimentary rocks throughout geologic history and which would be far more laborious and susceptible to systematic errors.

Earliest life

The oldest sedimentary rocks containing organic carbon (as opposed to graphite) are about 3.5 Ga old (Hayes *et al.* 1983). At the time of their formation, liquid water had been present at Earth's surface for at least 300 Ma, and the atmosphere was apparently rich in CO_2 (Walker 1985; Kasting 1987). The differentiation and accumulation of continental crust had just begun. Important nutrients such as phosphorus and oxidizable inorganic species such as ferrous iron and sulfide were being delivered to the surface, and fresh inputs would continue for more than a billion years (Veizer 1988b). Large lithospheric plates had not yet formed. Lifetimes of sediments may have been short, and very little is known about the fates of sedimentary materials.

Questions about O_2 have proven controversial. Although detailed and well supported interpretations of geochemical (Hayes 1983) and micropaleontological (Schopf & Packer 1987) evidence for oxygenic photosynthesis and respiration during the Archean have been presented and noted (e.g., Knoll 1984c; Buick 1992), the levels to which O_2 *accumulated* and the extent to which it was a persistent and ubiquitous component of the atmosphere and hydrosphere have not been pinned down.

The early Archean carbon cycle

Where carbon-bearing sedimentary rocks with ages of 3.45 ± 0.15 Ga are found, primary values of δ_a are near +2.0‰ (Hayes, Kaufman, and Hickman, unpublished results of analyses of massive, bedded carbonates from the Towers Formation, Pilbara Block, Western Australia), and those of δ_o vary widely. The latter variations are inversely correlated with H/C, the atomic ratio of hydrogen to carbon in the organic matter. This trend, a general characteristic of sedimentary organic matter, results from the loss of mobile hydrocarbons depleted in ^{13}C. In principle, the precise relationship between H/C and δ_o should depend on the structure of the organic matter and on the mechanisms of the thermally driven reactions, which may differ between sedimentary units. In fact, however, a review of observed relationships (Hayes *et al.* 1983) shows that the most important isotopic shifts occur when H/C ≤ 1.0 and that all sedimentary organic debris responds similarly to thermal stress in this H/C range, presumably because such low H/C ratios can be provided only by polycondensed aromatic systems. It is, therefore, possible to define a generic relationship between declining H/C and isotopic enrichment (see caption for Fig. 3). When δ values of kerogens are corrected for loss of ^{13}C-depleted hydrocarbons, estimated initial values of δ_o converge to –40‰ for samples ~3.5 Ga in age and to –60‰ for samples ~2.7 Ga in age (Fig. 3). Since variations in δ_a during the same time interval are much smaller (δ_a = +1.5±1.5‰, Veizer *et al.* 1989b), these values of δ_o indicate that the earliest carbon cycle was characterized by $\Delta_C \approx 42‰$, whereas that in the late Archean had $\Delta_C \approx 62‰$.

Interpretation of Δ_C requires a view of processes *inside* the exogenic reaction chamber. A simplified network relating reactants and products is shown in Fig. 4. The arrows represent pathways of carbon flow. Rates, moles/time, and isotopic compositions, ‰, of the carbon flows are denoted by φ and δ terms. The analogy is imperfect, but equilibrium isotope effects – which, for any given temperature, impose a fixed isotopic difference between two chemical species – have been represented by batteries. Kinetic

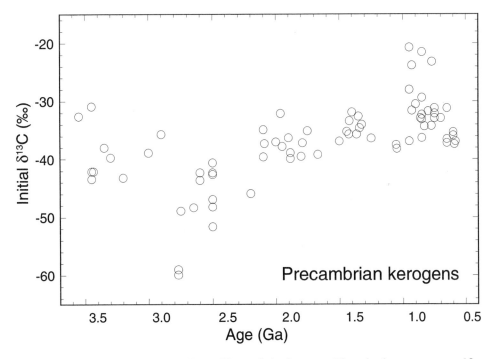

FIGURE 3   Carbon isotopic compositions of Precambrian kerogens. Plotted values are measured δ – postdepositional shift and thus represent isotopic compositions of organic carbon at approximately the time of lithification (i.e. the time of immobilization and removal of carbon from the exogenic reaction chamber). The magnitude of the postdepositional shift from an initial H/C = 1.5 has been estimated as $4.05 - 3.05r + 0.785/r + 0.0165/r^2 - (8.79 \times 10^{-4})/r^3$, where r ≡ the present-day atomic ratio of hydrogen to carbon in the kerogen. This expression is the equation for the broken line in Fig. 5-5 of Hayes et al. (1983). Measured δ values derive from Tables 17.5 and 17.9 of Strauss & Moore (1992) and represent only kerogen preparations containing at least 50% carbon and with r ≥ 0.06. To avoid overrepresentation of intensively studied units, each stratigraphic unit (as tabulated by Strauss & Moore 1992) is represented by a single point, initial δ values having been averaged where multiple analyses are available. Ages are those tabulated by Strauss & Moore (1992), except where more recent information favors adjustment as follows: Khatyspyt Fm., 0.58 Ga; Boonall Dolomite, 0.59 Ga; Biri Fm., 0.6 Ga; Luoquan Fm., 0.6 Ga; Chuanlingguo Fm., 1.79 Ga; Fontano Fm., 1.885 Ga; Rocknest Fm., 1.963 Ga; Union Island Fm., 2.2 Ga; Reivilo Fm., 2.5 Ga; Klipfonteinheuwel Fm., 2.5 Ga; Malmani Dolomite, 2.5 Ga; Gamohaan Fm., 2.5 Ga; Jeerinah Shale, 2.65 Ga; Meetheena Carbonate Mbr., 2.77 Ga; Tumbiana Fm., 2.77 Ga; and Steeprock Fm., 2.9 Ga.

isotope effects – which fractionate isotopes by processing isotopically substituted species at different rates – have been represented by electrical resistors. Magnitudes of the isotope effects, ‰, are represented by ε terms. The significance of the various subscripts is best clarified by the figure itself. For example, P and r, respectively, pertain to production and remineralization (i.e. oxidation) of organic matter.

Carbon enters the system in the form of CO_2 (either directly or by oxidative weathering of recycling organic material) and leaves as sedimentary carbonate or organic material. Isotopic shifts associated with diagenetic stabilization of carbonate minerals have been ignored, but all carbonate equilibria and related isotope effects can be treated adequately in terms of a single, temperature-dependent equilibrium isotope effect (derived from results of Mook *et al.* 1974 and Morse & Mackenzie 1990):

$$\delta_a - \delta_d \approx \varepsilon_a = -14.07 + 7050/T \tag{6}$$

where T is the absolute temperature, °K. Note that, for an equilibrium isotope effect, ε is approximately the isotopic difference between the two species involved in the equilibrium. [The exact relationship is $\alpha_a \equiv R_a/R_d = (1000 + \delta_a)/(1000 + \delta_d)$, where $\alpha_a \equiv 1 + 10^{-3}\varepsilon_a$.]

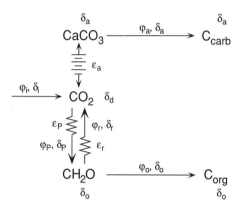

FIGURE 4   Principal processes involved in the isotopic fractionation of carbon in the early Archean carbon cycle. Arrows represent pathways of carbon flow. Fluxes of carbon (moles per unit of time), isotopic compositions (‰), and isotope effects (‰) are indicated by φ, δ, and ε, respectively. Subscripts have been appended to indicate the definitions of terms used in the text. By analogy with electrical circuits, equilibrium isotope effects are symbolized by batteries and kinetic isotope effects by resistors. Contents of ^{13}C (δ) can refer either to a specific reactant or to the carbon flowing through a particular pathway.

Kinetic isotope effects associated with the fixation of CO_2 and synthesis of biomass by autotrophs and with the recycling of organic carbon by heterotrophs are denoted by ε_P and ε_r, respectively. For a kinetic isotope effect, ε is approximately the isotopic difference between the reactant and the instantaneously forming product. Thus $\varepsilon_P \approx \delta_d - \delta_P$ and $\varepsilon_r \approx \delta_o - \delta_r$. The balance of carbon flows to and from CH_2O is given by $\varphi_P\delta_P = \varphi_r\delta_r + \varphi_o\delta_o$. Noting that $\varphi_P = \varphi_r + \varphi_o$ and defining $f_r \equiv \varphi_r/\varphi_P$ then yields

$$\delta_P = f_r\delta_r + (1 - f_r)\delta_o \tag{7}$$

where f_r is the fraction of production that is recycled by heterotrophs. Substituting $\delta_P = \delta_d - \varepsilon_P$, $\delta_d = \delta_a - \varepsilon_a$, and $\delta_r = \delta_o - \varepsilon_r$, rearrangement yields

$$\delta_a - \delta_o = \Delta_C = \varepsilon_a + \varepsilon_P - f_r\varepsilon_r \tag{8}$$

In this expression, Δ_C, the measured isotopic difference, is related to isotope effects associated with specific processes that plausibly occurred in ancient environments. The model is of such simplicity that we may place it in the Archean, but it would serve adequately in describing the many Phanerozoic systems in which recycling of CH_4 has not been a significant source of $C_{organic}$. For example, in sediments of the Greenhorn Formation (Cretaceous, marine, Western Interior Seaway of North America), Hayes *et al.* (1989) were able to evaluate Δ_C, ε_P, and ε_r separately and to show (assuming $f_r \rightarrow 1$) that the observed overall fractionation of 26.0–28.5‰ between carbonate and organic carbon resulted from $19 \leq \varepsilon_P \leq 20‰$, $0.5 \leq \varepsilon_r \leq 2.5‰$, and $\varepsilon_a \approx 10‰$. For the early Archean units with $\Delta_C = 42‰$, we might estimate T $\approx 35 \pm 15$°C (Walker *et al.* 1983) and thus $8 \leq \varepsilon_a \leq 10‰$. It follows that $\Delta_C - \varepsilon_a = \varepsilon_P - f_r\varepsilon_r \approx 33‰$ and thus $\varepsilon_P \geq 33‰$.

What chemical processes should we associate with the early Archean isotopic signal? Values of Δ_C are no more variable than during later intervals of Earth's history and, as in later times, this must reflect consistent *control* of carbon isotopic compositions, presumably through biological catalysis. ε_P can thus be associated with a key isotope effect in the biological system, probably that related to autotrophic production of organic material. Since known chemoautotrophs are ecologically dependent on photo-autotrophs (e.g., for production of O_2), photoautotrophy is more likely. Strikingly, the observed Archean ε_P value is within 10% of the isotope effect characteristic of the modern C-fixing enzyme, rubisco (specifically, the isoenzyme isolated from higher plants [Farquhar *et al.* 1989]). In C_3 plants, that isotope effect is not fully expressed, and overall fractionations are reduced because rates of photosynthesis are partly controlled by transport of CO_2 to the active site (Farquhar *et al.* 1982). In the Archean high-CO_2 environment, however, mass transport was probably not an issue, and full expression of enzymatic isotope effects would be expected. It is therefore possible that the fractionation observed in the early Archean results from operation of a rubisco-like enzyme. The quantitative agreement between the ancient and modern values is, however, not highly significant. Even among modern organisms, several rubisco isoenzymes exist, notably among prokaryotes, and are characterized by different isotope effects (Guy *et al.* 1993). A different comparison is more decisive: the Archean ε_P value is much larger than that associated with operation of the reverse tricarboxylic acid cycle, a C-fixation pathway used by obligately anaerobic green photosynthetic bacteria (Sirevåg *et al.* 1977). In spite of its primitive nature and association with anaerobic photoautotrophy, this pathway therefore appears not to have been dominant in these Archean environments.

Nothing in the isotopic evidence indicates which electron donor (H, S, Fe, O) was involved in fixation of C, but if carbon was being reduced, *something* was being oxidized. And, for that matter, nothing tells us whether or how organic carbon was being handled once it was produced. Was there an aggressive community of hetero-trophs that recycled the organic carbon? If so, by what mechanisms? Before discussing these questions, it will be best to consider events in the late Archean.

The late Archean carbon cycle

Pathways of carbon flow

The model represented by equation 8 would require either a large increase in ε_P or large, negative values for ε_r to produce the 60‰ Δ_C values observed in the late Archean. $\varepsilon_r \ll 0$ is unrealistic. No processes are known in which the ^{13}C content of primary biomass is reduced by 25‰ in a single step (i.e. within the pathways of carbon flow summarized in Fig. 4). $\varepsilon_P \rightarrow 50‰$ cannot be similarly excluded, but there are logical arguments against it. Had it occurred, the isotopic record summarized in Fig. 3 would have to be read as indicating that a system of carbon fixation with an ε_P value that *happened* to be rubisco-like functioned between 3.5 and 2.8 Ga ago, that it was displaced by a mechanism of primary production with a much larger ε_P value 2.8 Ga ago, and that the predecessor of modern rubisco finally became globally dominant sometime after

2.5 and before 2 Ga ago. This hypothesis lacks credibility because (1) there are today no primary autotrophs employing a C-fixation pathway with $\varepsilon \sim 50‰$, even though numerous other minor pathways with distinct ε values have escaped extinction, and (2) it calls for the sequential evolution of three different, globally dominant mechanisms of primary production and requires that the last just happened to have an ε_P value matching that of the first. There is a more plausible alternative, specifically, that the record in Fig. 3 indicates that a single mechanism of primary production has been globally dominant since 3.5 Ga ago and that some process was temporarily *added to* the global carbon cycle about 2.8 Ga ago. This at first caused exceptional depletion of ^{13}C in sedimentary organic carbon (to $-60‰$), but the importance of the additional process waned and, by 2 Ga ago, it no longer exerted an important influence on the ^{13}C content of sedimentary organic matter.

Geochemists familiar with modern systems have no trouble suggesting an additional process that would produce the observed isotopic variation. Schoell & Wellmer (1981) discovered the extreme depletion of ^{13}C in organic carbon from the late Archean of Canada and immediately suggested that it derived from the recycling of methane. To date, no alternatives have been offered. A schematic view of the flows of carbon in such systems is shown in Fig. 5 and will be used here as representative of the key features of the late Archean global carbon cycle. Two pools of organic carbon are shown, one in an aerobic zone, the other in an anaerobic environment. Fermentative processes are

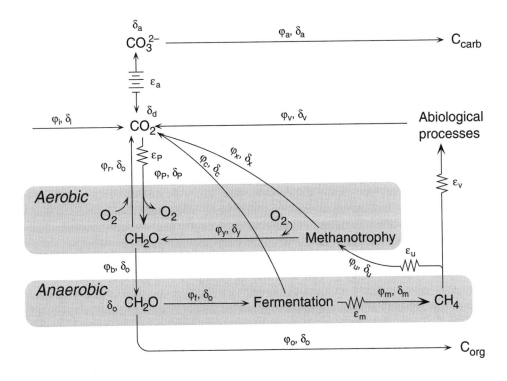

FIGURE 5 Principal processes involved in the isotopic fractionation of carbon in a late Archean carbon cycle involving recycling of methane. See Fig. 4 for explanation of symbols.

important in the latter and, as anyone who has ever poked a stick into late-summer-swamp mud knows, significant quantities of methane can be produced. Due to kinetic isotope effects associated with fermentation, *the CH₄ is strongly depleted in* ^{13}C. It is also a good carbon source for methanotrophic bacteria, which can thrive at CH_4–O_2 interfaces provided that other nutrients (sources of N, P, trace metals, etc.) are available. The methanotrophs convert the methane carbon into biomass, bringing it back into the organic-carbon pool. Even though that pool is symbolized by "CH_2O" in Fig. 5, it represents all organic carbon other than methane, is comprised of biomass and biological debris, and is the precursor of sedimentary $C_{organic}$. Incorporation of methanotrophic biomass and debris decreases the average ^{13}C content. Our task is to consider this process quantitatively in order to gauge the significance of the late-Archean isotopic signal.

Methane cycling and ^{13}C

In Phanerozoic settings, sedimentary organic materials have frequently been well enough preserved that specific molecules representative of methanotrophic organisms can be separated for compound-specific isotopic analysis. When examined, they are found typically to be depleted in ^{13}C by 50‰ relative to coeval primary products (Freeman *et al.* 1990; Collister *et al.* 1992). Therefore, if the isotopic composition of primary materials is given by δ_P and that of methanotrophic products is given by δ_P – 50, the fraction of CH_4-derived carbon in the mixture can be estimated using the expression

$$\delta_o = x(\delta_P - 50) + (1 - x)\delta_P \tag{9}$$

where δ_o is the δ value of the organic mixture deriving from primary and methanotrophic inputs and x is the mole fraction of methane-derived carbon in the mixture. For the late Archean we have $\delta_o = -60‰$. If it is considered that δ_P was approximately $-40‰$ (i.e. the early Archean value), then we find $x = 0.4$, implying that one-third to one-half of the organic carbon derived from recycling of methane.

This value is so high that we must move beyond two-component-mixing calculations and consider a fully balanced accounting of the carbon flows in Fig. 5 (see also Hayes 1983). Fluxes and isotopic compositions are again specified by φ and δ terms, which are defined most succinctly by their placement in the figure. The mass balance for organic carbon in the aerobic zone – the key to determination of δ_o – can then be written as

$$\delta_o = (1 - f_y)\delta_P + f_y\delta_y \tag{10}$$

where $f_y \equiv \varphi_y/(\varphi_r + \varphi_b)$ is the fraction of organic carbon deriving from methanotrophic inputs. For simplicity, the small isotopic fractionation possibly associated with aerobic heterotrophy has been neglected (i.e. φ_r has $\delta = \delta_o$). As before, $\delta_P = \delta_a - \varepsilon_a - \varepsilon_P$. Fermentative communities remobilize carbon by producing both CO_2 and CH_4 without use of external oxidants or reductants. If the input is organic carbon with an average oxidation number of zero, the division of carbon between CO_2 and CH_4 will be roughly 1:1, as required by the balanced equation

$$2CH_2O \rightarrow CH_4 + CO_2 \tag{11}$$

Assigning the subscripts m and c, respectively, to methane and CO_2 in the anaerobic environment and assuming that the initial steps in fermentation are not isotopically selective (i.e. φ_f has $\delta = \delta_o$), we can write the following mass balance

$$\varphi_f\delta_o = \varphi_m\delta_m + \varphi_c\delta_c \tag{12}$$

The isotope effect associated with methanogenesis is given by $\varepsilon_m = \delta_c - \delta_m$, and, because CH_4 and CO_2 are the only mobile products of fermentation and are formed in equal quantities, it follows that $\delta_m = \delta_o - \varepsilon_m/2$.

Determination of δ_o requires consideration of the fates of methane and estimation of δ_y (see Fig. 5 for definitions of subscripts). Measured isotope effects for the methanotrophic processes $CH_4 \rightarrow CO_2$ (Coleman *et al.* 1981) and $CH_4 \rightarrow$ biomass (Summons *et al.* 1994; Hayes & Harder, unpublished) do not differ significantly. The dominant isotope effect associated with methanotrophy (noted in the figure as ε_u) must therefore be associated with the uptake or initial processing of CH_4. Therefore $\delta_y \approx \delta_u \approx \delta_x$. The isotope effect associated with oxidation of CH_4 by HO· is $\varepsilon_v = 5\%o$ (Cantrell *et al.* 1990). Fractionation can, therefore, be associated with the φ_u–φ_v branch point, with

$$\delta_y = \delta_m - (1 - f_u)(\varepsilon_u - \varepsilon_v) \tag{13}$$

where $f_u \equiv \varphi_u/\varphi_m$ is the fraction of methane that is consumed biologically. Substituting in equation 10, we obtain

$$\Delta_C = \varepsilon_a + \varepsilon_P + [f_y/(1 - f_y)][\varepsilon_m/2 + (1 - f_u)(\varepsilon_u - \varepsilon_v)] \tag{14}$$

This relationship is analogous to the much simpler equation 8 in that it relates the observable Δ_C to paleoenvironmental conditions and processes.

Attainable values of Δ_C are summarized graphically in Fig. 6, in which it is emphasized that Δ_C is controlled not only by the isotope effects, for which representative

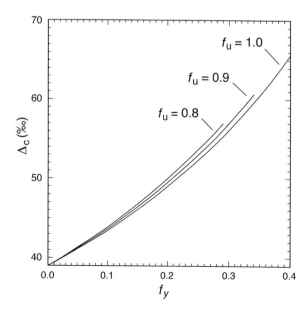

FIGURE 6   Values of Δ_c resulting from operation of the reaction network outlined in Fig. 5. Calculated using equation **14** with $\varepsilon_a = 9\%o$, $\varepsilon_P = 30\%o$, $\varepsilon_m = 80\%o$, $\varepsilon_u = 25\%o$, and $\varepsilon_v = 5\%o$.

values can be specified (see figure caption), but also by f_y and f_u. As indicated, the first of these affects Δ_C strongly. Consideration must therefore be given to the range of ecologically plausible values. First, even if reducing power were perfectly conserved during fermentation, φ_m could not exceed φ_c, and no more than half of the anaerobically recycled organic carbon could appear as CH_4. Second, since methanotrophs obtain energy for biosynthesis by complete oxidation of at least 20% of assimilated methane (Harder & van Dijken 1976), $\varphi_y < 0.8\varphi_u$. Even if all CH_4 is recycled biologically rather than by atmospheric processes, it follows that $\varphi_y < 2(\varphi_P - \varphi_r)/3$ and, incidentally, that methanotrophs could not consume all photosynthetically produced O_2 (that would occur only for $\varphi_y = \varphi_P - \varphi_r$). Consideration of the definition of f_y then yields

$$f_y < 2(\varphi_P - \varphi_r)/3(\varphi_P + \varphi_y) \qquad (15)$$

and, since both φ_y and φ_r must be less than φ_P, the *mathematical* requirement that $f_y < 2/3$. The conditions required for values of f_y approaching 2/3 are, however, unrealistic ($\varphi_r = 0$; no aerobic recycling) and self-defeating ($\varphi_y = 0$; no methanotrophy). On the other hand, $f_y = 1/3$ would be obtained if both φ_r/φ_P and φ_y/φ_P were 1/3. Those requirements are not severe, and it appears that ecologically meaningful maximum values of f_y are near 0.4. If CH_4 is distributed globally (see below), amounts of CH_4 available to methanotrophs will be controlled by competition with atmospheric processes. Limitations on f_y imposed by $f_u < 1.0$ are indicated by the stopping points of the lines in Fig. 6.

Calculation thus demonstrates that, when care is taken to construct a fully balanced carbon budget incorporating realistic values for known isotope effects, the largest Δ_C values observed in the late Archean are near, but within, the limit for an ecosystem in which processes of methane production and consumption proceeded at maximal rates alongside those of CO_2 fixation and O_2 production. Equally, these results indicate that limitation of f_y to submaximal values by restriction of supplies of CH_4 or O_2 would lead to values of Δ_C lower than those observed in the late Archean.

Significance of the isotopic signal

Was recycling of methane local or global? Methanotrophic contributions can be recognized in more recent strata, but the late Archean occurrences are extraordinary. In Phanerozoic systems, computations equivalent to equation **9** nearly always yield $x < 0.1$ (e.g., Freeman *et al.* 1990; Collister *et al.* 1992). Organic materials strongly influenced by seeps of natural gas and with δ values like those observed for 2.75 Ga have been found in Phanerozoic strata (e.g., Kaplan & Nissenbaum 1966), but only in the midst of far more normal kerogens. At the Archean–Proterozoic transition, organic material strongly depleted in ^{13}C is not restricted to a single bed or horizon but occurs uniformly throughout sedimentary successions covering thousands of square kilometers and hundreds of meters of thickness (Schoell & Wellmer 1981; Hayes *et al.* 1983; Strauss 1986; Strauss *et al.* 1992b). It could be suggested that the simultaneous appearance of maximally strong methanotrophic signals in strata from three continents is coincidental (the Canadian occurrences reported by Schoell & Wellmer 1981 and by Strauss 1986 are not represented by points on Fig. 3 because δ values are available only for total organic matter, not kerogen). If so, the isotopic record summarized in Fig. 3

would indicate only that ~2.75 Ga ago local methanotrophic communities happened to be numerous and their products, perhaps for some reason related to developing global tectonic systems, particularly well preserved. But *all* kerogens analyzed thus far from 2.75 Ga reflect maximal methanotrophic contributions. If the signal is local rather than global a second coincidence would therefore be required: evidence of nonmethanotrophic systems, well preserved at earlier and later dates, would have to have been systematically excluded from the record. Chance inevitably affects any record, especially one in which vast intervals of time and space are represented by sparse observations. It is therefore *possible* that evidence of methanotrophy was by chance excluded from the record at times other than 2.7–2.1 Ga and that evidence of normal communities was by chance recorded only outside that interval. But it is more likely that we are confronted by geochemical evidence for an age in which methanotrophy was of global significance.

The case is strengthened by the appearance of the record during the 2.8–2.1-Ga time interval, in which values of δ rise from an initial minimum. Uncertainties about ages of units from which kerogens have been prepared are great enough that it is impossible to be certain about the *rate* of that rise, but the global trend to higher contents of ^{13}C is unmistakable. Each kerogen must reflect local phenomena, but it appears very likely that some variation of global conditions was affecting the cycling of carbon in environments everywhere[1].

A sequence of events

The components of the machine that produced the isotopic signal are oxygenic photosynthesis, methanogenesis, and methanotrophy. The first appearance of the signal could record the addition of one of these to ecosystems in which the other two were already functioning. Early development of fermentative processes has always been regarded as likely (e.g., Bernal 1967; Oparin 1968; Broda 1978; Mason 1991), and methanogens could have played an important role within the carbon cycle from the moment that autotrophy arose. Specifically, in the absence of respiratory production of CO_2, they would provide the only biochemical means of remobilizing carbon that had been fixed by autotrophs. Methane reaching the atmosphere would be oxidized by photochemically driven processes. In the absence of methanotrophy, no ^{13}C-depleted organic material would form and, if the environment were generally anaerobic so that

---

[1] Some views of the the carbon isotopic record (e.g., Schidlowski & Aharon 1992) indicate a second episode of depletion in ^{13}C in $C_{organic}$ 2.1 Ga ago in rocks of the Francevillian Series. No points representative of that unit appear in Fig. 3, because kerogens have never been prepared and, in fact, not even the isotopic analyses of the total organic carbon have been described in detail. The Francevillian Series hosted natural fission reactors (Neuilly *et al.* 1972). The organic matter in the sediments was extensively irradiated, altered, and remobilized (Cortial *et al.* 1990; Nagy *et al.* 1991). Values of δ cover a wide range (from –21.8 to –46.2‰ vs. PDB [Cortial *et al.* 1990; M. Schidlowski, personal communication, 1993]). The significance of this isotopic signal will be uncertain at least until detailed reports are available. Notably, the isotopic variations approximate those observed in other uraniferous strata where radiolytic phenomena have been clearly identified as the cause of the isotopic variations (Leventhal & Threlkeld 1978; Leventhal *et al.* 1987; Landais *et al.* 1990).

methanogens were not required to grow in restricted environments, ^{13}C-enriched carbonates would not form. The occurrence of methanogenesis would go unrecorded. It is therefore permitted by the evidence and, on ecological grounds, quite likely, that methanogens were present long before the Δ_C excursion.

Evidence for photoautotrophy during the early Archean was noted earlier in this paper together with the observation that nothing in the isotopic record indicated what substance was serving as the electron donor. A case can be made (e.g., Schopf *et al.* 1983) that anaerobic photosynthetic processes employing H_2 and reduced sulfur as electron donors are likely to have preceded oxygenic photosynthesis, but the availability of electron donors other than H_2O is likely to have been very limited (Walker & Brimble-combe 1985). More recently, Schopf & Packer (1987) have interpreted the morphology of early Archean microfossils as indicative of oxygenic photoautotrophs (see also Schopf 1993 and this volume), and Buick (1992) has concluded that oxygenic photosynthesis is the only process capable of having produced the organic material found in some Archean strata. The origin of oxygenic photosynthesis is therefore likely to have preceded the late Archean Δ_C excursion by at least 700 Ma. Moreover, it seems very unlikely that autotrophs would have started to produce O_2 without heterotrophs concurrently developing the ability at least to survive in the presence of traces of O_2. The beginnings of respiratory metabolism, if not the development of methanotrophy, are therefore *also* likely to have substantially preceded the Δ_C excursion.

We are left with the possibility that all of the requisite biochemical mechanisms had evolved well before the Δ_C excursion, but that methanotrophy did not become important because either O_2 or CH_4 did not survive long enough for the reactants to come in contact.

Late Archean conditions

It is pertinent to ask what concentrations of CH_4 and O_2 might have been required. Modern methanotrophic bacteria scavenge CH_4 so efficiently that its concentration in water is held to values below 20 nM in aerobic environments where the bacteria are present (Lidstrom & Somers 1984). For seawater at 25 or 40 °C, this concentration would be in equilibrium with an atmospheric pCH_4 of 18 or 22 µatm (Yamamoto *et al.* 1976). Concentrations of O_2 required by the methanotrophs (Harwood & Pirt 1972) are in equilibrium with atmospheric $pO_2 \approx 500$ µatm, or 0.25% of the present level. This value is lower than most estimates of early Proterozoic (~2.2 Ga) O_2 levels (Holland 1984; Kasting 1987) but only establishes a minimum pO_2 for cellular environments and likely provides no information about the late Archean atmosphere. The interesting point is that required concentrations of dissolved CH_4 are so low that atmospheric transport of CH_4 to methane-consuming communities can be considered. In an earlier discussion of the isotopic record, Hayes (1983) pointed out that the extreme depletions of ^{13}C then observed in some late Archean kerogens must indicate methanotrophic consumption of CH_4 and that the gas could have been present during the early and middle Archean but, in the absence of methanotrophy, "geochemically invisible." From a different point of view, Lovelock (1988) has suggested that CH_4 was an important greenhouse gas in the Archean atmosphere, with a fractional abundance of "100 ppm," fivefold higher than the minimum calculated above as required for methanotrophy.

The isotopic signal of methanotrophy appears first and most strongly in stromatolitic units. An association between producers and consumers of O_2 can be envisioned
and is schematically depicted in Fig. 7. Whereas CO_2 and O_2 are globally mobile in the
present system, with CH_4 being the trace gas scavenged at aerobic–anaerobic interfaces,
the roles of O_2 and CH_4 might have been exchanged in late Archean settings. If O_2 were
the trace gas and CH_4 globally distributed, methanotrophs might have grown in close
proximity to oxygenic algae, because it was only there that supplies of O_2 were
adequate. Such associations would be ideally suited for production of organic debris

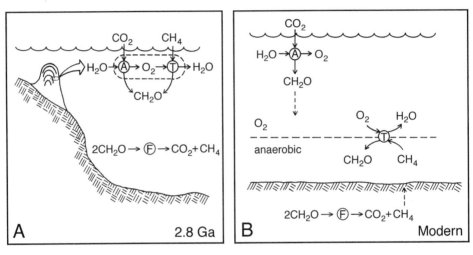

FIGURE 7   Sketches comparing relationships between habitats of oxygenic photoautotrophs (Ⓐ),
methanotrophs (Ⓣ), and fermentative communities producing methane (Ⓕ) during the late Archean
(**A**) and since at least 2.1 Ga ago (**B**).

with maximal methanotrophic contributions. According to this view, the late Archean
Δ_C signal records the time at which sources of O_2, at least those within the microenviron-
ments envisioned in Fig. 7, became strong enough to meet the requirements of
methanotrophic bacteria. As O_2 levels increased, methanogens would be driven into
more restricted environments with poorer supplies of fermentable organic matter, the
strength of the methane source would decrease, methanotrophy would be limited, and
δ values of kerogens would rise. Isotopic evidence for globally significant levels of
methanotrophy is absent by the time atmospheric levels of O_2 reached 2,000 µatm
(Holland & Beukes 1990).

An alternative reading is possible, namely that the reactant missing earlier in the
Archean had been methane rather than O_2. In this case it would be required either that
evolution of methanogens was somehow delayed relative to that of other fermenters or
that some change in the physical environment suddenly increased supplies of methane.
The antiquity of methanogens – prominent members of the Archaea – is more postulated than proven, but the idea that they evolved after oxygenic photoautotrophs is not

attractive. Environmental change (i.e. the second alternative) is, however, inevitable and may have been profound during the reorganization of global tectonics at the Archean–Proterozoic transition. Perhaps changes that would trigger widespread methanotrophy and temporarily suppress normal cycling of carbon will eventually be suggested.

Questions regarding the status of sulfur are relevant and, until now, have been unanswered. In modern environments, sulfate-reducing bacteria are able to out-compete methanogens for available supplies of H_2 and acetate and thus to suppress production of methane (Lovely & Klug 1983). Because some early Archean sulfate evaporites exist (Lambert *et al.* 1978), it is logical to ask whether methanogens could have flourished in the Archean ocean, and some authors have regarded it as significant that the evidence of methanotrophy appears in strata interpreted as lacustrine (Walter 1983; Knoll 1984c; Buick 1992). The interpretation offered here treats recycling of methane as a phenomenon of global rather than local significance and makes that choice because the carbon isotopic record seems to require it. The depletion of ^{13}C in late Archean and early Proterozoic kerogens appears to be global, not a product of rare environments. Indeed, if late Archean continents were small, as many believe (Veizer 1988b; chapters by Lowe and Veizer, this volume), the dominance of the geochemical record by materials derived from very large lakes would be surprising. Is it certain that these units are lacustrine? The existence of some marine dissolved sulfate does not exclude widespread methanogenesis. Significant fractionation of ^{34}S is a feature only of later sulfides, indicating that even if sulfate was a persistent component of Archean seawater, its concentrations were very low (Hayes *et al.* 1992b). Finally, the possibility that sulfate served as an electron *acceptor* in processes of biological methane consump-tion must be recognized (Iversen & Jorgensen 1985). There is, however, no evidence that this process leads to production of biomass depleted in ^{13}C (i.e. it appears that anaerobic oxidation of CH_4 is used as an energy source, but not a carbon source). Accordingly, it cannot be responsible for the observed changes in Δ_C and is not a factor in the present reconstruction of events.

Summary

1 The isotopic characteristics of the early Archean carbon cycle are consistent with the existence of producer organisms employing a process of carbon fixation similar to that catalyzed in modern organisms by rubisco.

2 The existence and characteristics of consumer organisms in the early Archean cannot be discerned from isotopic evidence, but the subsequent development of a global carbon cycle prominently including processes of CH_4 production and consumption strongly suggests the early presence of methanogens.

3 The isotopic characteristics of the late Archean carbon cycle require involvement of methanotrophy. Observed levels of ^{13}C depletion are consistent with derivation of approximately one-third of organic carbon from methane, near the maximum possible for a carbon cycle in which both CO_2 and CH_4 were globally mobile. Consideration of solubilities of gases and metabolic requirements of organisms

indicates that the cycle could be sustained by atmospheric transfer of CH_4 to methane-consuming organisms growing in close association with oxygenic photo-autotrophs.

4 From the onset of methane cycling 2.75 Ga ago, the isotopic difference between sedimentary carbonates and organic materials declined steadily. By 2.1 Ga ago, and possibly much earlier, observed isotopic differences can be accounted for without requiring any recycling of CH_4. This change presumably parallels a decline in atmospheric CH_4 levels brought on by both a weaker methane source (less CH_4 being produced, perhaps because of more aggressive aerobic consumption of organic material) and a more strongly oxidizing atmosphere with significantly lower steady-state concentrations of CH_4.

Acknowledgments. – Supported by the U.S. National Aeronautics and Space Administration, grant NAGW-1940. I appreciate helpful reviews by D.J. Des Marais, discussions with L. Jahnke, M.E. Lidstrom, M. Wahlen, and R.F. Weiss, and the provision of data by M. Schidlowski.

Early Proterozoic atmospheric change

Heinrich D. Holland

Department of Earth and Planetary Sciences, Harvard University, Cambridge, Massachusetts 02138, USA

The oxidation state of paleosols, the nature of uranium ores, the history of red beds, the trace metal content of black shales, the age distribution of banded iron formations, and the evolution of eukaryotes are all consistent with the hypothesis that the O_2 content of the atmosphere increased dramatically between 2.2 and 1.9 Ga ago. The paleosol data suggest that the O_2 content of the atmosphere was ≤1% of the present atmospheric level (PAL) before 2.2 Ga ago and that since 1.9 Ga the O_2 content of the atmosphere has been ≥15% PAL.

The most direct way to define changes in the composition of the atmosphere during the Early Proterozoic would be to find trapped bubbles of air in rocks spanning this period of geologic time. The discovery of trapped air in ice cores has revolutionized the field of atmospheric evolution. Unfortunately, the most ancient ice sampled so far is only about 0.2 Ma old, and it is unlikely that well-preserved air will ever be found in rocks 10,000 times as old as this. At least for the present, direct information regarding the composition of the Early Proterozoic atmosphere is lacking, and inferences must be based on indirect evidence, much of it circumstantial and all of it no better than semiquantitative.

The evidence that we do have is derived from reconstructions of Early Proterozoic surface environments. The intensity and the chemistry of weathering are determined in part by the chemistry of the ambient atmosphere. Paleosol profiles are therefore useful indicators of atmospheric chemistry at the time of their formation, particularly during the period of Earth's history before the evolution of vascular land plants, which exercise a very significant influence on the composition of soil air today. The nature of weathering products preserved in sedimentary rocks is also useful, as is the record of the metals that are released into solution during weathering and that are subsequently trapped in black, organic-rich shales.

Other useful indicators of atmospheric composition include the evidence for the redox state of seawater, the atmospheric requirements of the biosphere, and climate. At present, paleosol profiles seem to be the best indicators of changes in atmospheric composition during the Early Proterozoic. These profiles are therefore discussed first. The other indicators point to the same conclusions and contribute to the establishment of a consistent but by no means complete model of atmospheric evolution during the Early Proterozoic.

Bengtson, S. (ed.) 1994: *Early Life on Earth. Nobel Symposium No. 84.* Columbia U.P., New York

Paleosols as indicators of Proterozoic atmospheric change

Since the evolution of soil profiles is strongly influenced by the composition of the ambient atmosphere, the nature of ancient soil profiles can be used to reconstruct, at least in part, the composition of the ambient atmosphere. To do so, we must first be able to solve three problems: (1) to establish that a particular rock unit is indeed a paleosol, (2) to read through the geologic history of the paleosol and to reconstruct its composition and mineralogy before diagenesis and, if necessary, its metamorphism, and (3) to relate the original composition and mineralogy of the paleosol to the composition of the ambient atmosphere.

Criteria for identifying paleosols developed in place were proposed by Holland & Zbinden (1988), and we have seen no reason to modify these to any great extent. A good deal of progress has been made in reconstructing the diagenetic and metamorphic changes during the conversion of paleosols from soils to rock units. In particular, we have shown that the addition of K^+ and Rb^+ to three of the paleosols that we have studied took place during short events associated with local or regional thermal disturbances (Macfarlane & Holland 1991). The connection between the evolution of soil profiles and the composition of the ambient atmosphere has been explored by Kasting, Holland & Pinto (1985), Pinto & Holland (1988), Holland, Feakes & Zbinden (1989), Feakes, Zbinden & Holland (1989), and Holland & Beukes (1990).

All of the well-studied paleosols that are younger than ca. 1.9 Ga are highly oxidized (for a summary, see Holland 1992). Most, if not all, of the "FeO" present in the parent rocks of these paleosols was oxidized during weathering and was retained within the paleosols as Fe_2O_3 and/or as hydrated Fe_2O_3. In all of the well-studied paleosols that developed on basalts more than 2.2 Ga ago, a large fraction of the "FeO" that was present in the parent rocks was lost from the upper soil horizons during weathering and was partly reprecipitated as one or more hydrated iron silicate minerals in the lower soil horizons. "FeO" loss may not have been as severe in several paleosols older than 2.2 Ga that were developed on more silicic igneous rocks, but none of these paleosols have been studied very carefully.

The difference between paleosols older than 2.2 Ga and those younger than 1.9 Ga is striking and suggests very strongly that weathering during the first few hundred million years of the Proterozoic Era differed considerably from more recent weathering. The difference is almost certainly due to a dramatic increase in the O_2 content of the atmosphere. In the absence of atmospheric O_2, "FeO" released from silicate minerals during weathering by the reaction

$$\text{"FeO"} + H_2O + 2CO_2 \rightarrow Fe^{2+} + 2HCO_3^-$$

moves downward in soils together with the other cations released by reaction with carbonic acid. The pH of soil water rises along this flow path as the concentration of CO_2 decreases and the concentration of HCO_3^- increases. Saturation with respect to one or more ferrous silicate minerals must have occurred to account for the presence of the considerable excess of "FeO" in the chlorite zone of paleosols older than 2.2 Ga. The precipitation of one or more Fe^{2+} silicates rather than siderite, $FeCO_3$, sets an upper limit on the value of pCO_2 in the lower soil zones, and hence, somewhat indirectly, on pCO_2 in the ambient atmosphere. The available, very uncertain thermodynamic data for greenalite, $Fe_3Si_2O_5(OH)_4$, suggest that pCO_2 was $\leq 10^{-2}$ atm in the lower parts of these paleosols. Analyses of groundwaters in sedimentary rocks that contain siderite show that this mineral is indeed stable with respect to iron silicate minerals at CO_2 pressures as low as 10^{-2} atm, but experimental data for the solubility and stability of greenalite and related iron silicates would help considerably to set an upper limit to pCO_2 more than 2.2 Ga ago.

The best estimate for the O_2 content of the atmosphere required to account for the lack of oxidation of Fe^{2+} in paleosols older than 2.2 Ga is $\leq 2 \cdot 10^{-3}$ atm, i.e. 1% PAL (Holland & Beukes 1990). The minimum value of pO_2 that is consistent with the highly oxidized nature of paleosols younger than 1.9 Ga is ~0.03 atm, i.e. 15% PAL. These numbers are somewhat uncertain. They are, however, consistent with estimates based on other lines of evidence for the composition of the Proterozoic atmosphere.

Red beds

Red beds have long been considered evidence for the presence of free O_2 in the atmosphere. In 1968 Cloud (1968b) pointed out that the oldest known thick red beds have an age of about 2.0–1.8 Ga. Since then the age of the oldest red beds has been extended somewhat. The Jatulian red beds in Karelia and particularly in Finland have received a great deal of attention and are now well dated, largely due to the efforts of the Geological Survey of Finland. The Jatulian Group was deposited in a rift setting on 2.44 Ga intrusives (Huhma et al. 1990). They are cut by and intercalated with three groups of diabase dikes and sills of age 2.20, 2.10, and 2.06 Ga (e.g., Perttunen 1985, 1991, and Pekkarinen & Lukkarinen 1991). Red beds have been reported throughout the Jatulian Group. Most seem to be younger than 2.20 Ga, but some are apparently somewhat older. The exact age of the oldest red beds in the sequence is not known, since the time gap between the basement rocks and the basal Jatulian sediments has not been determined. To the extent that red beds are indicators of significant quantities of O_2 in the atmosphere, their distribution in the Jatulian sequence indicates that O_2 was present in the atmosphere somewhat before 2.20 Ga. This is in agreement with an analysis of red beds in the Lower Proterozoic of South Africa (Eriksson & Cheney 1992).

Weathering residues in sedimentary rocks as indicators of Proterozoic atmospheric change

The resistance of some minerals to weathering is strongly affected by the presence of O_2. Uraninite, UO_2, is a case in point. In the presence of O_2, UO_2 oxidizes rather rapidly to UO_3, which dissolves in carbonated water to yield uranyl ions, UO_2^{2+}, and their complexes. The kinetics of UO_2 oxidation and solution have been studied by Grandstaff (1976, 1980). Large ore deposits that contain detrital uraninite, UO_2, occur in the late Archean sediments of the Witwatersrand Basin in South Africa (Robb & Meyer 1990; Robb et al. 1990, 1992) and in early Proterozoic sediments near Blind River, Ontario, in

Canada. No widespread deposits of uraninite have been found in rocks younger than 2.3 Ga. Instead, unconformity-type and roll-front-type uranium deposits are common in younger rocks. In most of these, there is good evidence for solution transport of uranium in the +6 valence state and for the deposition of UO_2 following the reduction of U^{+6} to U^{+4} by a variety of mechanisms. The oldest known ore deposit of this type is the 2.0 Ga Oklo deposit in Gabon.

These observations corroborate the paleosol evidence. The persistence of uraninite during weathering more than 2.3 Ga ago is consistent with a low-O_2 atmosphere during the Archean and the early Proterozoic. The appearance of unconformity- and roll-front-type uranium deposits in more recent sediments demands the presence of oxygenated groundwaters and is therefore consistent with the presence of a significant quantity of oxygen in the atmosphere during the past 2.0 Ga. Unfortunately, the change in the nature of uranium deposits does not add important quantitative constraints regarding the evolution of atmospheric O_2. The survival of uraninite during weathering, transport, and deposition depends on too many imponderables (see for instance Holland 1984, Chapter 7) to set a firm upper limit on atmospheric pO_2 more than 2.3 Ga ago. The presence of enough O_2 in the groundwaters from which the later unconformity- and roll-front-type ores were deposited depends in a complex fashion on the evolution of these solutions and can not be used to set a firm lower limit on atmospheric pO_2 during the past 2.0 Ga.

The rarity of detrital uraninite deposits also limits the usefulness of uranium ores as a tool for dating the increase in atmospheric O_2 during the course of the Proterozoic. Other minerals may prove more useful. Detrital uraninite ores contain a good deal of pyrite, some of which is surely detrital. Detrital pyrite is uncommon in sediments today, because the mineral is easily destroyed during weathering in a highly oxidized atmosphere. Since pyrite is such a common mineral, examining the change in its abundance as a detrital phase in sandstones and conglomerates may turn out to be a useful way to check the timing of the increase in atmospheric pO_2 that is indicated by the paleosol data.

Metals in black shales as indicators of Proterozoic atmospheric change

Most marine black shales contain metals in concentrations that are well above their crustal abundances (see, for instance, Vine & Tourtelot 1970). In many of these shales the pattern of metal enrichment is similar to the pattern of their concentration in seawater (see, for instance, Holland 1984, Chapter 9). The excess of metals in black shales can therefore be used, although only with considerable circumspection, to reconstruct the relative concentration of metals in seawater. The concentration of some metals in seawater is almost certainly related to the oxidation state of the ambient atmosphere. During weathering today, uranium, molybdenum, and a number of other metals are oxidized and owe their relatively high concentrations in seawater in part to the large solubility of salts in which these elements are in their highly oxidized state. In the absence of atmospheric O_2, the concentration of these metals in seawater would almost certainly have been much lower than today, and their concentration in organic-carbon-rich black shales significantly smaller.

Only three Archean, highly carbonaceous, sedimentary units have been studied to date. One is the latest Archean Roy Hill Shale Member of the Jeerinah Formation, which underlies sediments of the Hamersley Group in Western Australia. The age of the Roy Hill Shale is close to 2.7 Ga (Arndt *et al.* 1991). Preliminary chemical analyses show that the concentration of uranium in the Roy Hill Shale is not above its crustal average and that molybdenum is only very slightly enriched in these sediments (Davy & Hickman 1988). On the other hand, metals such as zinc, lead, nickel, and cobalt, which do not change valence during weathering, are significantly enriched in the Roy Hill Shale. These observations are consistent with expectations based on other evidence for a very low O_2 content of the atmosphere during late Archean time. The other carbonaceous sedimentary units are the Mt. McRae Shale and the upper part of the Mt. Sylvia Formation, both members of the 2.5 Ga Hamersley Group of Western Australia. Both have a trace metal chemistry similar to that of the Roy Hill Shale (Davy 1983).

The only early Proterozoic highly carbonaceous sediments for which a considerable number of trace metal analyses are available are the Talvivaara and Outokumpu schists of Finland (Peltola 1960, 1968; Ervamaa & Heino 1983; Loukola-Ruskeeniemi 1990, 1991, 1992). These rock units have an age between 2.10 and ~1.96 Ga. Their trace-metal content is quite different from that of the Roy Hill Shale and is virtually indistinguishable from that of Phanerozoic black shales. They are, however, in part mineralized, and caution must be exercised in interpreting their trace-element composition. If their U, Mo, and V content was not seriously affected by hydrothermal inputs, then the compositional differences between the Roy Hill Shale and the Finnish black schists suggest that pO_2 was much higher 2.1–2.0 Ga ago than 2.7 Ga ago. Much more work must be done on trace-metal concentrations in Precambrian carbonaceous sediments to establish this hypothesis and to link the history of trace metals in carbonaceous sediments to the evolution of atmospheric oxygen. The effort seems well worth making.

Banded iron formations as indicators of Proterozoic atmospheric change

Banded iron formations are the world's major source of iron ore. They are common in Archean and in early Proterozoic rocks (see, for instance, Klein & Beukes 1992). The oldest BIFs, those at Isua in West Greenland, are ~3.8 Ga in age. The youngest of the early Proterozoic BIFs, the Sokoman Iron Formation of the Labrador Trough, has an age just under 1.9 Ga. No BIFs have been found in rocks between 1.9 Ga and ~0.8 Ga. The brief late Proterozoic renaissance of BIFs ended well before the beginning of the Phanerozoic, 0.54 Ga ago.

Iron in BIFs probably owes its origin to the hydrothermal circulation of seawater through oceanic crust, perhaps largely at midoceanic ridges. The metal was then circulated upward into shallow parts of the oceans and was precipitated there by a variety of mechanisms that probably included oxidation by solar UV radiation and, later, by the intrusion of atmospheric O_2 into near-surface seawater (see, e.g., Anbar & Holland 1992 and Beukes & Klein 1992). The efficacy of this mechanism for the formation of BIFs probably depended on the presence of a stratified ocean and certainly on the absence of O_2 in the deeper parts of the oceans. The second of these conditions was endemic before the appearance of significant quantities of O_2 in the atmosphere. It may therefore be more than coincidence that the last of the large early Proterozoic BIFs was formed after the last of the reduced paleosols (the Hekpoort paleosol) and before the first known highly oxidized paleosol (the Flin Flon paleosol). The rise of atmospheric oxygen may have been followed quite closely by the oxygenation of the deep oceans. Thereafter, Fe^{2+} in waters of hydrothermal vents would have been oxidized close to the vents and would have precipitated in the deeper parts of the oceans, much as it does today.

Although this is an attractive hypothesis, it is in conflict with the data discussed above, which suggest that O_2 levels rose more than 2.1–2.0 Ga ago, i.e. 100–200 Ma before the deposition of the Sokoman Iron Formation. The disappearance of BIFs about 1.9 Ga ago may therefore be related more to a change in ocean stratification than to a rise in pO_2. If so, the reappearance of BIFs during the late Proterozoic may be due to a return of severe ocean stratification rather than to a severe drop in atmospheric O_2 levels.

Biological indicators of Proterozoic atmospheric change

Cyanobacteria were probably the dominant O_2-producing organisms during the Archean and the early parts of the Proterozoic (Pierson, this volume). They are remarkable organisms. Their great ability to repair damage due to ultraviolet radiation is probably a memento from an early period in Earth's history, when O_2 levels in the atmosphere were too low to provide a significant ozone shield against solar UV. Their ability to use both photosystems I and II indicates that they lived both in areas where reduced gases were readily available for photosynthesis and in areas where this was not the case. Their ability to fix molecular nitrogen indicates that combined nitrogen was once at a premium in their habitats. All of these accomplishments indicate that cyanobacteria lived at one time in the low- O_2 and/or O_2-free environments suggested by other indicators for the composition of the Archean and early Proterozoic atmosphere.

Autotrophic eukaryotes are quite different. They are obligate aerobes, and they cannot fix molecular nitrogen. Their date of origin is not known with certainty, but fossil evidence suggests that they existed 2.1 Ga ago (Han & Runnegar 1992; Runnegar, this volume). This timing is consistent with the rise of atmospheric O_2 chronicled above. The inability of eukaryotes to fix molecular N_2 suggests that NO_3^- was no longer in as short supply during their evolution as it had been during the evolution of the cyanobacteria. The greater abundance of NO_3^- in the oceans since 2.1 Ga is probably related to the higher O_2 levels in the atmosphere. NO formation in the atmosphere and the input of HNO_3 to the oceans are enhanced by the presence of O_2. At the same time, the use of NO_3^- as an oxidant of organic matter in the oceans is reduced at high O_2 levels in the atmosphere, because the availability of dissolved O_2, which is preferred to NO_3^- as an oxidant, increases in proportion to atmospheric pO_2. At progressively higher O_2 pressures, the production rate of combined nitrogen therefore almost certainly increased and the rate of NO_3^- use decreased. Both trends tended to make NO_3^- more abundant in seawater, and this may explain the success of mid-Proterozoic eukaryotes despite their inability to fix molecular N_2.

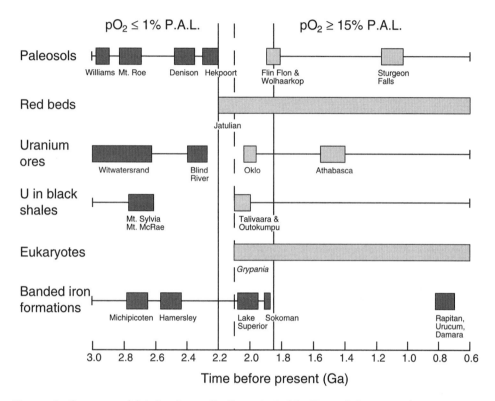

FIGURE 1   Summary of data bearing on the O_2 content of the Precambrian atmosphere.

Climatic indicators of Proterozoic atmospheric change

There is abundant evidence for the presence of liquid water at the surface of the Earth during the past 3.8 Ga, and there is equally strong evidence that during this time period igneous rocks reacted with CO_2 and water during weathering to yield carbonate sediments and detrital residues (Holland 1984, Chapter 5). The hydrologic cycle has clearly been in full swing for at least 3.8 Ga. This implies that Earth's surface temperatures have not been sufficiently low to freeze the oceans, at least not for geologically significant periods of time. The greenhouse effect of the Earth's atmosphere has therefore been sufficiently intense to offset the likely but unproven lower luminosity of the Sun during the early history of the solar system (see, for instance, Kasting 1992).

Several gases have been proposed to account for the rather large greenhouse enhancement that was probably required to prevent freezing of the oceans during the Archean. Of these, CO_2 is the most likely. The gas has been added continuously to the atmosphere as a component of volcanic gases, and the ocean–atmosphere system must have adjusted itself on geologically short time scales so that the CO_2 input by degassing was balanced by the CO_2 output, largely by carbonation reactions during weathering, followed by the deposition of limestones and dolomites (see, for instance, Kump & Volk 1991; H. Holland 1991). Atmospheric CO_2 has probably served as a crude thermostat for the atmosphere. However, precise calculations of the CO_2 pressure in the atmosphere during the past 3.8 Ga are still difficult to make. The absence of siderite and the presence of high-FeO chlorite in the lower, iron-rich zones of late Archean and early Proterozoic paleosols developed on basalts suggest that pCO_2 was $\leq 10^{-2.0}$ atm between 2.75 and 2.2 Ga.

Summary

Much of the data bearing on the evolution of atmospheric O_2 during the Precambrian have been collected in Fig. 1. All lines of evidence are consistent with the view that pO_2 levels were $\leq 1\%$ PAL more than ~2.2 Ga ago, and that pO_2 levels were $\geq 15\%$ PAL less than 1.9 Ga ago. However, only the paleosol data are amenable to semiquantitative interpretation. The other lines of evidence supply information that is largely or entirely qualitative.

The history of atmospheric O_2 between 2.2 and 1.9 Ga is still uncertain. There may have been a very rapid rise of pO_2 close to 2.2 Ga, but it is also possible that pO_2 increased gradually between 2.2 and 1.9 Ga. Recent measurements (Karhu 1993) of the isotopic composition of carbon in early Proterozoic limestones strongly favor a rapid rise of pO_2 between 2.20 and 2.06 Ga.

Trends in Precambrian carbonate sediments and their implication for understanding evolution

John P. Grotzinger

Department of Earth, Atmospheric and Planetary Science, Massachusetts Institute of Technology, Cambridge, Massachusetts 02139, USA

Early Archean carbonate sediments are rare and consist of discontinuous limestone and dolomite units that are often extensively replaced by chert. They accumulated during brief interludes in the otherwise tectonically active history of greenstone belts. Middle Archean carbonates provide the first record of sedimentation on probably small but stable continents. Late Archean carbonates were deposited in cratonic and noncratonic settings and demonstrate the first-order similarity with younger carbonate platforms. Late Archean carbonates contain ubiquitous precipitated carbonates that occur in a variety of facies representing all preserved depositional environments. Early Proterozoic through late Proterozoic carbonates show decreasing evidence for massive precipitation of seafloor cements as tufas, crusts, or botryoidal fans, as well as decrease in taxonomic diversity and density of stromatolites. Evidence of sulfate evaporites is rare or absent in carbonates older than about 1.8 Ga. Archean and early Proterozoic seawater may have been highly oversaturated with respect to calcium carbonate; the ratio of HCO_3^- to Ca^{2+} may have also been increased relative to Phanerozoic seawater. The decline of Proterozoic stromatolites, in part, may be directly related to the proposed decrease in carbonate saturation of seawater. Early marine cementation and precipitation of carbonate cements was an important process in the accumulation of platform carbonates and stromatolites. However, it is not clear whether this precipitation was inorganically or biologically regulated or to what extent this uncertainty applies to micritic sediments that probably were precipitated as whitings. It seems likely that the inception of widespread biocalcification in the early Cambrian did not have an important effect on the saturation state of seawater. Because of the impositions of subsidence and eustasy, biocalcified organisms could not sequester any more carbonate than was possible through inorganic processes.

The analysis of modern carbonate depositional environments has provided important information pertaining to the processes by which ancient Phanerozoic carbonates were formed and distributed. It is widely known that Phanerozoic carbonate production is essentially organic (for review, see Bathurst 1975; Tucker & Wright 1990; Wilson 1975). Calcified organisms contribute to a broad range of loose sediments of varying grain sizes, as well as to the construction of enormous masses of limestone precipitated in situ, such as reefs. Biological evolution has had a profound effect on the

Bengtson, S. (ed.) 1994: *Early Life on Earth. Nobel Symposium No. 84.* Columbia U.P., New York

succession of carbonate-secreting organisms. This has manifested itself in many ways ranging from the mineralogy of skeletons to the impact of reef-building communities.

In this context it is interesting to consider the origin of Precambrian carbonates. What role did organisms play in the construction of Precambrian platforms? Was the appearance of carbonate sediment coincident with the origin of life in the early Archean? Do stromatolites represent the product of microbially influenced precipitation? Are stromatolites carbonate factories in addition to depositories? What changes in carbonate depositional systems demarcate the shift from the microbially dominated Proterozoic environments to metazoan-dominated Paleozoic environments? All of these questions stem from a central problem: did microorganisms have an active role in Precambrian carbonate sedimentation, as metazoans and metaphytes had in the Phanerozoic, or was their involvement only passive?

The goal of this chapter is to provide a framework in which the origin of Precambrian carbonate sediments can be evaluated. A description of the temporal distribution of sediment types is provided first, followed by a discussion of sedimentologic trends and their interpretation. Finally, some speculations are provided on the relative roles of organic versus abiotic processes in the development of Precambrian carbonate facies and their temporal transition into the Phanerozoic.

Archean carbonate sedimentation

Early Archean

Occurrences of well-preserved early Archean carbonates are rare. A few early Archean carbonates are known from the Warrawoona Group of Western Australia and the Swaziland Supergroup of South Africa (Buick & Dunlop 1990; Lowe 1983; Lowe & Knauth 1977; see also Schopf, this volume). These are all extremely thin, consisting of generally discontinuous limestone and dolomite units that are often extensively replaced by chert. None of these occurrences are thought to have accumulated on stable platforms, and they usually represent brief interludes in the otherwise tectonically active history of greenstone belts.

Silicified carbonates from the lower and upper Warrawoona Group (~3.5 Ga) have been described by Lowe (1983) and Buick & Dunlop (1990). They consist primarily of several varieties of chert, chalcedony, and quartz that are interpreted to have replaced primary carbonate mud, "diagenetic" carbonate crystals, rare stromatolites, possible sulfate evaporites, and possible paleospeleothems. The Warrawoona cherts are intimately associated with thick sequences of mafic volcanics, interfingering laterally with volcaniclastic alluvial facies. The units are thought to have formed in shallow, evaporative basins adjacent to eroding volcanic sources, and it seems likely that sedimentation of these orthochemical deposits, now replaced by chert, occurred during brief quiescent periods that punctuated volcanic episodes (e.g., Lowe 1983).

Middle Archean

The middle Archean is significant in that it contains the first record of sedimentation on probably small but stable continents. The Nsuze Group carbonates are possibly the oldest sequence of sediments that can be regarded as evidence for the inception of stable carbonate platform sedimentation.

The Nsuze Group carbonates are up to 30 m thick, forming a thin, laterally discontinuous unit of clastic-textured, stromatolitic, and recrystallized (massive) dolomite. Clastic-textured dolomites display well-developed herringbone cross-bedding, contain large stromatolite-derived intraclasts, and are believed to have formed in tidally influenced environments similar to associated quartz arenites. Ooid and intraclast grainstones occur (Walter 1983). Stromatolites include low-relief structures that generally form undulatory surfaces and are characterized by fine layering, fenestral fabrics, and uncommon fibrous fabrics (Walter 1983). Tidal-flat laminites are also present and contain small ripples, desiccation cracks, and tepee structures. The presence of tepee structures indicates early cementation and lithification of the sediments.

Late Archean

Late Archean carbonates were deposited in cratonic and noncratonic settings. This latter category includes probable atoll-like deposits fringing oceanic seamounts and plateaus developed at sea level, belts of carbonate fringing uplifted and deforming accretionary prisms, and successor basins. Carbonates deposited in these environments are thin (<50 m), and platforms were probably short-lived. In contrast, carbonates deposited in cratonic environments accumulated on stable continental crust and accumulated as a result of rift-related passive thermal subsidence or long-term rises in sea level. In the case of the Campbellrand carbonate belt, it is likely that its initial extent was in excess of 100,000–500,000 km^2 and that it blanketed most of the Kaapvaal craton. The Campbellrand platform is more than 1500 m thick and shows strong shelf-to-basin differentiation of thicknesses and facies (Fig. 1), which is consistent with the geometry of Proterozoic, Phanerozoic, and modern rimmed shelves (Beukes 1987).

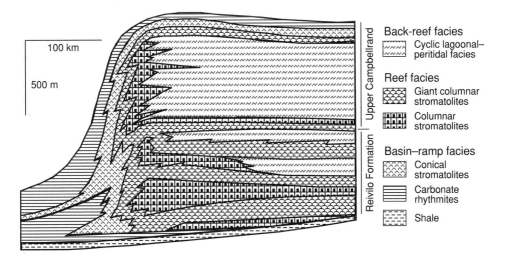

FIGURE 1   Stratigraphic cross-section of the Campbellrand Subgroup. Note strong shelf-to-basin facies gradients. (After Beukes 1987.)

These relationships demonstrate the first-order similarity between Archean and younger carbonate platforms. The evolution of the Campbellrand platform is strikingly similar to that of younger counterparts and indicates that the critical stages of development including growth, diversification, and expansion were not dependent on the presence of carbonate-secreting or other metazoan communities. The same controls that have affected the development of Phanerozoic platforms must have been in place by the late Archean (Grotzinger 1989). These controls include: the mode and tempo of basin subsidence and the rate of siliciclastic influx (controlled by climate, sediment-source composition, etc.); eustatic effects on rim backstepping; incipient to terminal platform drowning; the hierarchical packaging of stratigraphy; and the potential for subaerial exposure and karsting.

However, closer inspection reveals profound differences in the details of individual facies. Late Archean carbonates contain ubiquitous precipitated carbonates that occur in a variety of facies representing all preserved depositional environments. Calcitic marine cements form beds of pure precipitated carbonate, the internal microtexture of many stromatolites, and coatings on depositional surfaces such as breccia clasts and stromatolites. All of the beds and many of the thick coatings of cement have a characteristic texture (Fig. 2A). This texture, which I will here refer to as herringbone cement, is best preserved as calcite and is interpreted as a marine calcite or high Mg-calcite cement. Botryoidal fans (Fig. 2B), of probable aragonite precursor, grew on a variety of depositional surfaces. Fans growing on bedding planes reach up to 50 cm in height, while those growing off the sides of stromatolites commonly form coatings several centimeters thick. Incipient tufa structures, such as colloform stromatolites and associated flat-laminated carbonate, were probably precipitated as well. Identical fabrics are observed in other carbonate sediments of late Archean age, albeit not as well preserved owing to destructive dolomitization or deformation. Large, calcite-crystal fans and herringbone cements occur in the Cheshire Formation, Zimbabwe (Martin *et al.* 1980; Grotzinger *et al.* 1993), at Bulawayo, Zimbabwe (Grotzinger *et al.* 1993), in the late Archean Steeprock Group, Canada (Hofmann 1971), and in the Carrawine Dolomite in Western Australia (Simonson *et al.* 1993).

The precipitation of carbonate cements directly on the seafloor in Late Archean time was a ubiquitous phenomenon independent of tectonic environment. The cement facies are known from at least five different locations on three different continents. Consequently, an analysis of this unique facies has general implications for the state of the Earth during the late Archean.

Precipitation of carbonate cements directly on the seafloor as an actual sedimentary deposit is exceptionally rare during the Phanerozoic. Only one example of *bedded* seafloor cements, precipitated in a normal marine subtidal environment, has been documented (Permian of west Texas [Yurewicz 1977]). In the late Proterozoic, it is rare and only a few examples are known. One is from the Bambui Group of Brazil (Peryt *et al.* 1990), and another is from the Witvlei Group of Namibia (J. Grotzinger, unpublished data). Although not a bedded cement, tussocky microfabric is now recognized as a marine cement distributed as ornamentation on stromatolites (Fairchild *et al.* 1990). In contrast, subtidal bedded marine cements are uncommon but not rare in early Proterozoic sediments. However, bedded marine cements (tufas) are common and perhaps diagnostic of early Proterozoic *tidal flats*. Thus, as will be discussed in more detail below,

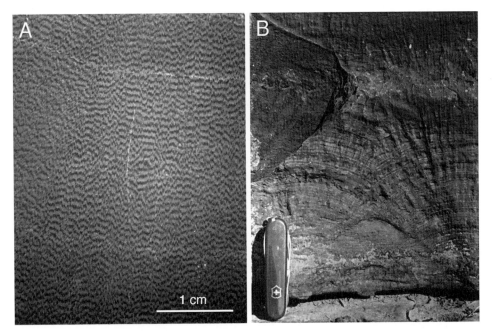

Figure 2   Limestone marine cements of the Campbellrand platform. **A:** Herringbone texture of calcite after probable former (high Mg?) calcite. **B:** Crystal fans of calcite after probable former aragonite.

early Proterozoic carbonates are transitional between the marine cement-dominated late Archean realm and the cement-poor younger record. These tidal-flat tufas declined markedly during the middle Proterozoic and disappeared altogether by the end of that time.

The ubiquitous presence of bedded marine cements in the late Archean carbonate sediments suggests that the seawater was significantly supersaturated with respect to calcium carbonate. The crystal fans are interpreted to represent calcitization of primary botryoidal aragonite and the herringbone cements to represent calcitization of primary high-Mg (or possibly low-Mg) calcite. At this time it is difficult to establish which primary phase was dominant. Nevertheless, it is clear that saturation with respect to calcium carbonate was great enough to overcome all kinetic barriers to precipitation in situ. It also seems clear that this precipitation occurred independently of the presence of stromatolites and, by inference, photosynthetic microbes that could have aided precipitation. However, it seems likely that at least some of the late Archean stromatolites formed as inorganic precipitates; the common botryoidal texture of many inorganic precipitates is a consequence of nucleation kinetics (e.g., caliche, agate, chalcedony, malachite). Layered, domal structures are produced by inorganic processes and mimic features that are often described as stromatolites and interpreted to be the result of benthic microbial activity. The results of recent studies of late Archean carbonates (Grotzinger 1989; Simonson *et al.* 1993; Sumner *et al.* 1991) imply caution in the interpretation of Archean stromatolitic fabrics.

Proterozoic carbonate sedimentation

Early Proterozoic

The growth of large cratonic masses of continental lithosphere, often associated with the Archean–Proterozoic "transition" (e.g., Veizer & Compston 1976), was probably the single greatest event in the evolution of carbonate platforms. It is suggested that the physical, chemical, and/or biological potential for significant carbonate production was present during the Archean but the growth of major platforms was suppressed until substantial continental masses developed. These created spacious, stable (or smoothly subsiding), shallow-water platforms for carbonates to form on. Because this process was diachronous on a global scale, the initial development of carbonate platforms was also diachronous. As with Phanerozoic platforms, the type of basin that Proterozoic carbonates developed in had a strong control on their growth, diversification, zonation, and expansion (Grotzinger 1989).

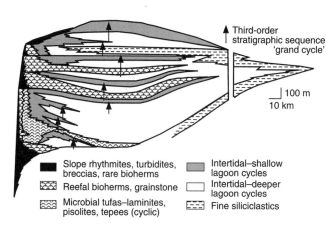

Figure 3 Stratigraphic cross-section of the Rocknest platform, Northwest Territories, Canada. Note strong paleogeographic zonation and shelf-to-basin transition created by narrow reefal rim. (After Grotzinger 1986a, b.)

The Rocknest Formation (<1.97, >1.885 Ga) provides a detailed image of sedimentation on an early Proterozoic rimmed shelf. It is illustrated in Fig. 3. Shelf-edge stromatolite barrier reefs containing minor grainstone generally pass downslope into periplatform carbonate rhythmites, rhythmite breccias, and allodapic megabreccias. Reefal facies pass laterally, through minor grainstone, into shoal-complex facies consisting of peritidal small domal stromatolites, microbial laminites, tufas, and disrupted zones with teepees. Shoal-complex facies then pass into areally extensive but environmentally restricted cyclic, lagoonal–peritidal facies.

The general, first-order facies relationships are similar between early Proterozoic and younger/older platforms. Shelf-to-basin transitions are well represented, and the spectrum of facies that document this transition are surprisingly independent of time. There are several significant differences, however. Relative to late Archean platforms, early Proterozoic platforms do not contain common subtidal, bedded marine cements.

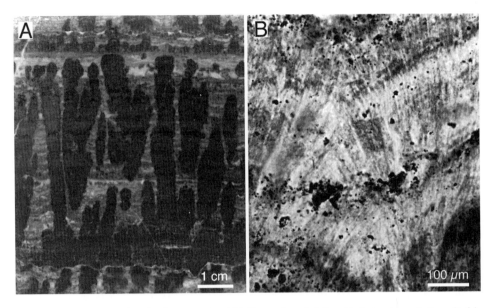

Figure 4 Features of early Proterozoic tidal-flat tufas. A: Microdigitate stromatolites (tufa), Rocknest Formation. B: Thin-section photomicrograph of chert-replaced microdigitate stromatolites of McLeary Formation, Northwest Territories, Canada. Chert replacement predates inversion of primary texture, revealing diagnostic interlocking crystals of former aragonite. Scale bar is 0.1 mm long. (Courtesy of H.J. Hofmann.)

However, they are often well endowed with precipitated tidal-flat tufas (see Grotzinger & Kasting 1993, Fig. 2, for recent tabulation). In turn, this unique facies serves to distinguish early Proterozoic from late Proterozoic and younger carbonates, where occurrences are absent, and from middle Proterozoic carbonates, where occurrences are less common (Grotzinger 1989; Grotzinger & Kasting 1993).

Morphologically, tidal-flat tufas are expressed as millimeter-thick stratiform crusts, colloform crusts, or discrete microdigitate stromatolites (Fig. 4A). Grotzinger & Read (1983) and Grotzinger (1986a, b) determined their likely precipitated origin and classified them as "tufa," similar to modern tufa/travertine. Given their association with stromatolites, it was suggested that they may have been associated with microbial activity, such as photosynthesis, which would have aided in the precipitation process. This interpretation is distinct from previous, wholly biological interpretations of their distinctive radial–fibrous fabric, which was inferred to represent replaced sheaths (e.g., Grey & Thorne 1985). However, the radial–fibrous fabric is only preserved in the least altered cases. Commonly, fabrics consist of pseudospar mosaics, consistent with calcitization or dolomitization of aragonite (Grotzinger & Read 1983). In rare examples the primary fabric is spectacularly well preserved in early chert that predates inversion of the aragonite fabric (Fig. 4B) and provides compelling evidence in favor of inorganically precipitated aragonite (Hofmann & Jackson 1987). In some cases, the tidal-flat tufas are a very important facies, constituting up to 90% of meter-thick beds and forming 20–50% of stratigraphic intervals hundreds of meters thick containing cyclic tidal-flat facies (Rocknest Formation: Grotzinger 1986a, b; Duck Creek Dolomite: P. Hoffman, personal communication, 1990).

The occurrence of precipitated marine cements in early Proterozoic tidal-flat facies is interpreted as a transition between the cement-dominated late Archean carbonate record and the cement-poor late Proterozoic and younger record. Archean carbonates are characterized by both subtidal marine cements *and* tidal-flat tufas; early Proterozoic carbonates contain, at significant levels, only tidal-flat tufas. In late Proterozoic and younger carbonates, marine cementation is almost exclusively manifested in the mode of early lithification of previously deposited sediments. Middle Proterozoic carbonates show transitional textures. Consequently, the shift from Archean to middle Proterozoic carbonates is interpreted as a decrease in the saturation state of the ocean with respect to calcium carbonate. At first, in the late Archean, precipitation of bedded marine cements was possible in unrestricted, subtidal, open marine environments. With time, as more carbonate was deposited on newly formed cratonic platforms, the degree of oversaturation decreased, and "bedded" marine cements could form only on relatively restricted tidal flats as tufas. As saturation further decreased, precipitation of marine cements occurred, except for rare occasions, only in pores between previously deposited sediments. From some point beginning in middle Proterozoic and onward, it is difficult to establish any differences in the history of marine cementation.

Middle Proterozoic

Much of the Earth's inventory of middle Proterozoic carbonate is probably preserved within Siberia and China. Outcrops are remote, and little has been published on the regional facies relationships exhibited by these platforms, many of which rest, undeformed, in the subsurface. However, there a few North American middle Proterozoic carbonates that have been studied in some detail (Grotzinger 1989). One of the best documented is the Dismal Lakes Platform (Donaldson 1976; Kerans & Donaldson 1988; Kerans *et al.* 1981) exposed in northern Canada.

The Dismal Lakes Platform is a classic ramp and is highlighted by well-developed, isolated, stromatolitic buildups (Fig. 5). The Sulky Formation of the Dismal Lakes Group contains a variety of facies developed in upslope as well as downslope positions. A spectacular upslope buildup is formed by giant coniform stromatolites developed as

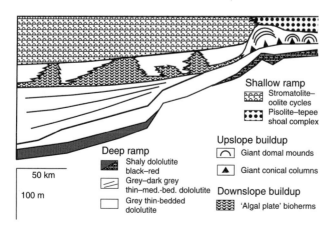

Figure 5 Stratigraphic cross-section of Dismal Lakes platform. Note well-developed upslope and downslope bioherms and overall ramp geometry. (After Kerans & Donaldson 1988.)

a core 25–30 m thick with up to 12 m of synoptic relief on individual columns that have widths of a few meters. Giant conical stromatolites are succeeded by a sequence of giant domal stromatolites with widths of up to 40 m and synoptic relief of 10–15 m. Giant domal stromatolites pass upward into smaller domal stromatolites that form laterally continuous sheets, marking the top of the buildup.

Evaporites – significant first in the middle Proterozoic? – One of the most important constraints on the composition of the Precambrian ocean is provided by the record of evaporites and their relationship to carbonate sediments. The most important point is that *bedded* or *massive* calcium sulfate deposits formed in primary evaporative environments are absent in the Archean and early Proterozoic record (Grotzinger 1989). Significantly, these sediments first appear in the middle Proterozoic. Deposits are reported from the ~1.5 Ga MacArthur Basin (M.J. Jackson *et al.* 1987), the ~1.2 Ga Borden Basin (Jackson & Ianelli 1981), and the ~1.2 Ga Amundsen Basin (Victoria Island, Canada: Young 1981). Before this, only sparse mineral casts are reported from early Proterozoic sediments.

A handful of reputed "evaporites" are known from the Archean. In one instance it has been possible to document the likely former presence of primary gypsum (now preserved as barite [Buick & Dunlop 1990]). Unfortunately, it is likely that there was a considerable influence on the local composition of seawater through influx of continent-derived calcium-rich waters from erosion of highly basaltic surrounding source areas. As pointed out by Hardie (1984), even if a marine depositional setting is established independently, say on sedimentologic evidence from enclosing deposits, the hydrologic restriction required to precipitate evaporites "cannot fail to put a strong non-marine stamp on both the geochemistry and the sedimentology of the deposit." Consequently, evaporite deposits interlayered with marine sediments seldom yield the correct sequence of evaporite minerals predicted by precipitation from normal seawater. Therefore, the "evaporite sequence" reported by Buick & Dunlop (1990) has little bearing on the composition of Archean seawater.

The interpretation of calcitized Archean gypsum has strongly influenced models for the composition of Precambrian seawater. Past recognition of these assumed "gypsum" pseudomorphs in part forms the basis for justification of a uniformitarian model for seawater composition, in which precipitation of calcium carbonate is sequentially followed by calcium sulfate and halite (e.g., Holland 1984; Walker 1983a). However, gypsum precipitation may have been more rare than previously thought and perhaps did not follow as the "normal" precipitate after calcium carbonate (Grotzinger 1989). The appearance of bona fide calcium sulfate evaporites in the middle Proterozoic would therefore represent a major transition in the evolution of Precambrian carbonate platforms and the composition of seawater. The predictions and implications of this transition are discussed below.

Late Proterozoic

Many interesting changes in the character of Precambrian carbonate sediments occur during the late Proterozoic. Of these, one of the most important involves the prominent decline in stromatolites.

Awramik (1971) and Walter & Heys (1985) compiled Proterozoic stromatolite taxa and suggested that the diversity of stromatolites reached a maximum in the middle Riphean, remained constant through the later Riphean, and declined during the Vendian. They attributed this decline to the advent of Ediacaran soft-bodied metazoans that may have been capable of similar grazing and burrowing activities observed by Garrett (1970a) for worms and shelly invertebrates. In addition, it is likely that competitive exclusion by higher algae may also have contributed to decline of Proterozoic stromatolites – evidence for late Proterozoic higher algae is mounting (Butterfield *et al.* 1988; Hofmann 1985a).

A third possible mechanism that may have contributed to the decline of stromatolites is compositional change of seawater, as first suggested by Fischer (1965) and Monty (1973). For lack of evidence, Walter & Heys (1985) had previously dismissed changes in seawater as a cause of stromatolite decline. However, Grotzinger (1989) showed that several secular changes in platform facies could be related to long-term, unidirectional changes in the carbonate chemistry of seawater. As a follow-up study, Grotzinger (1990) constructed a model for how these changes could influence the diversity and abundance of stromatolites through time. The role of in situ precipitation in the development of stromatolitic lamination is interpreted as a time-dependent process. It was most important in the early Proterozoic and possibly negligible in the late Proterozoic. The transition through time is ultimately interpreted to be partially responsible for the decline of Proterozoic stromatolites as a result of their reduced capacity to accrete sediment.

The optimal growth rate of stromatolites is directly related to the rate of sediment production *and* the efficiency of the mechanism by which sediment is accreted as laminae. Accordingly, whatever factors are important in controlling sediment production and sediment accretion will ultimately determine the growth potential of stromatolites. Sediment accretion occurs by precipitation and/or trapping and binding. Precipitation is more efficient than trapping and binding, and early cementation of trapped and bound sediment will prevent subsequent erosion by currents (or burrowing by metazoans in younger stromatolites). Therefore, the efficiency of laminae accretion is interpreted to be directly proportional to the saturation state of seawater with respect to calcium carbonate. As saturation is increased, marine cementation would start sooner, and ultimately cements would be precipitated directly on the sediment surface as laminae.

Consequently, the decline of Proterozoic stromatolites could, in part, be related to a global reduction in carbonate saturation through time. The growth of any stromatolite formed of carbonate sediment would be subject to changes in this condition as discussed above. As saturation decreased, direct precipitation of laminae would have stopped, and eventually whitings would have become more restricted in frequency as well as area. In this manner, the amount of sediment available to become incorporated in stromatolites was decreased, thereby restricting the distribution of stromatolites themselves.

The low density of late Proterozoic stromatolites (Grotzinger 1990) may help to explain the peculiar absence of rimmed shelves at that time. Of the several examples of late Proterozoic carbonate platforms discussed by Grotzinger (1989) and Knoll & Swett (1990) all are ramps, with one exception (see Bertrand-Sarfati & Moussine-Pouchkine

1983). This may be due to the relative rarity of stromatolites in the late Proterozoic. It is well known, based on the study of modern and ancient Phanerozoic platforms, that significant organic buildups are an essential prerequisite to the establishment of rims, which often consist of well-defined reef tracts (see Read 1985). In the absence of organic buildups, which have characteristically high sedimentation rates, the platform will not undergo strong shelf-to-basin differentiation marked by steep gradients and abrupt changes in slope. Instead, transitions between shallow- and deep-water environments are more ramp-like, defined over broad spatial distances and without abrupt changes in slope. The absence of linear stromatolite reefal belts along shelf margins, which are extremely well developed in older parts of the Proterozoic and late Archean, provides a logical and simple explanation for the prevalence of ramps in the late Proterozoic.

Discussion

Biological versus inorganic controls on Precambrian carbonate production

In discussing the constraints on the relative roles of biologic and abiotic processes in the formation of Precambrian carbonates, there are many difficulties in objectively sorting out where the role of biology ends in the accumulation of a carbonate deposit. For example, if microfossils are found in an ancient stromatolite, it is often uncertain whether they had an active role in forming the stromatolite (by binding or inducing precipitation of sediment), or whether they existed passively, as many microbes do today on substrates of all sorts. Similarly, it is hard to imagine an environment where marine cements and organic films are interlayered in which biologic processes had no effect on the kinetics of precipitation. In part, the problem is moot to the extent that it is unreasonable to expect the biological and environmental factors to be mutually exclusive. It is perhaps best viewed as a matter of degrees, and in this context some distinctions can be made, particularly with respect to the establishment of the origin of stromatolitic and other carbonate fabrics.

In the last few years it has become increasingly clear that there are certain fabrics that are entirely based on a precipitation mechanism (e.g., Archean cement beds, including domes). This is based on petrographic and geochemical attributes that are comparable to well-studied ancient Phanerozoic analogs. Prior to this, Precambrian carbonate precipitation had been suggested (e.g., Donaldson 1963; Cloud & Semikhatov 1969; Walter 1972; Serebryakov & Semikhatov 1974) but was unappreciated in terms of its importance because of a lack of definitive criteria. These criteria are now available (e.g., Sandberg 1985), and their use implies a systematic, long-term decrease in the amount of seafloor carbonate precipitation, in the form of both stromatolitic and nonstromatolitic carbonate.

As a consequence, it is necessary to view the record of Precambrian carbonate sedimentation in discrete intervals marked by important yet subtle differences in the style and mode of carbonate production. It is likely that the Archean record of carbonate precipitation was dominated by the inorganic process of seafloor precipitation and the late Proterozoic characterized by microbial trapping and binding of sediment to form micritic stromatolites. Intervening periods were transitional and show features of both. Accordingly, the Archean record is not without examples of micritic stromatolites (late Proterozoic preludes) and the late Proterozoic record free of seafloor cement fans (vestiges of the Archean). Combined with the likely misrepresented history of Precambrian evaporite sedimentation, these changes in the record of carbonate sedimentation provide a warrant for nonuniformitarian models of Earth's evolution and accounts of environmental secular change.

Constraints on ocean composition

Archean and early Proterozoic seawater possibly was highly oversaturated, and any perturbation such as microbially induced uptake of CO_2 might have caused immediate and prolific precipitation of calcium carbonate. The ratio of HCO_3^- to Ca^{2+} may also have been increased, relative to "Phanerozoic" seawater, so that the concentration of bicarbonate rose to twice that of calcium or more. Under such conditions, gypsum could not precipitate before halite. Seawater might have precipitated abundant carbonate in "excess" quantities and induced the "carbonate factory" to exploit new realms such as tidal flats where tufas might form. In the process, most or all available calcium would have been extracted simply by precipitation of carbonate and, during an evaporative situation, little or no calcium sulfate could have precipitated except near sites of continental runoff (e.g., deltas), where influx of additional calcium might be expected. This condition would have been maintained until about 1.8–1.6 Ga, when the HCO_3^-/Ca^{2+} ratio would have changed so that the bicarbonate concentration again was less than twice that of calcium (Grotzinger 1989).

This possibility was explored by Kempe & Degens (1985). They suggest that the Archean ocean was a "soda ocean" in which bicarbonate was the dominant anion, exceeding chloride in concentration. However, they argue that the soda ocean may have persisted until as late as 800 Ma, based on a limited analysis of geological data. This is probably too young an age for the transition, and it is suggested that the crossover point from soda to chlorine ocean occurred much earlier, probably ~1.8–1.4 Ga ago, when the last prominent seafloor precipitates (tidal flat tufas) were replaced by the first prominent sulfate evaporites. The timing of this transition period is very likely related to global deposition of substantial volumes of shallow-water carbonates in response to the stabilization of new cratons formed at the end of the Archean.

This point is of critical importance with respect to interpreting events across the Proterozoic–Phanerozoic boundary. Kaźmierczak et al. (1985) have argued that onset of biocalcification was the direct response to buildup of toxic levels of calcium in the ocean following decay of the soda ocean at ~800 Ma. However, if the analysis presented here is correct, then calcium buildup should have occurred long before the late Proterozoic. Therefore, models of biocalcification based on calcium toxicity should take this timing constraint into account.

Finally, it is possible that the absence of sulfate evaporites may also relate to critically low levels of oceanic sulfate prior to about 1.9 Ga (Grotzinger & Kasting 1993). Given the composition of modern seawater, saturation with respect to gypsum and halite is reached simultaneously for a saturation factor of about 11 (Holland 1984). At this point the product $^mCa^{2+} \cdot {}^mSO_4^{2-} \approx 23$ (mmole/l)^2. This is the approximate minimum value of the product $^mCa^{2+} \cdot {}^mSO_4^{2-}$ to account for the appearance of gypsum before halite. It is possible that this minimum value may not have been attained until after about 1.9 Ga, because of low sulfate concentrations imposed by low atmospheric oxygen. For times prior to about 1.9 Ga, Grotzinger & Kasting (1993) estimate that the concentrations of Ca^{2+} and SO_4^{2-} were on the order of 10 mmole/l and 1 mmole/l, respectively. Consequently, their product would have been less than the minimum value required to precipitate gypsum before halite by about a factor of two. Thus, the possibility that low sulfate concentrations were, at least in part, responsible for the rarity or absence of Archean and early Proterozoic gypsum cannot be discounted.

Effect of biocalcification on calcium carbonate saturation of seawater

As a final issue, it is likely that the calcium carbonate saturation state of seawater did not change substantially following the advent of biocalcification in the terminal Proterozoic and earliest Cambrian. The reason is straightforward.

Inorganic carbon on earth is distributed between the atmosphere, ocean, and crust. The precipitation of calcium carbonate, biologically or inorganically, represents transfer from ocean to crust. Over long periods of time (millions of years), the only way to decrease the oceanic inventory of inorganic carbon is to allow long-term partitioning of carbonate minerals into the crust (Walker 1985). In this manner a new steady state is obtained, in which the oceanic reservoir becomes progressively smaller. The concentration of carbonate in seawater would therefore decrease, other factors being equal.

Prior to the advent of calcareous microplankton in the Jurassic, carbonates were precipitated abundantly only in shallow marine environments. The precipitation of shallow-water carbonates is limited to the space created as a result of sea level rising relative to the land surface (accommodation space). Unlike siliciclastic sediments, carbonate sediments cannot be deposited above sea level, because they are produced in the marine environment (except for volumetrically trivial amounts of lacustrine carbonate). Transgression and onlap commonly result in net carbonate deposition, while regression and offlap result in subaerial exposure and net carbonate dissolution. Consequently, the maximum amount of carbonate that can be extracted from the oceans is directly proportional to the accommodation space over the continents. As shown in this paper and elsewhere (Grotzinger 1989), carbonates have been able to fill the available accommodation space since at least the late Archean. In other words, their growth potential has always been high enough to effectively fill the space created by absolute rises in sea level or accelerations in subsidence. Therefore, the inception of biocalcification may not have had an important effect on the saturation state of seawater; the amount precipitated is still restricted by the available accommodation space. Biocalcification acts only as a catalyst, restrained in its potential to sequester any more carbonate than inorganic means because of the impositions of subsidence and eustasy.

Indeed, it seems that if there was a decrease in the saturation state of Paleozoic seawater, it would be more attributable to long-term flooding of the continents than to the advent of biocalcification. The transgression that started with the breakup of the late Proterozoic supercontinent and culminated in the late Cambrian was responsible for the deposition, and therefore partitioning into the crust, of great volumes of carbonate over all the continents (Bond *et al.* 1989). As a consequence, much inorganic carbon was buried and removed from the oceanic realm.

Conclusions

1 Carbonate platforms with most of the essential features of Phanerozoic platforms were well developed by 2.5 Ga. The evolution of many Proterozoic platforms is strikingly similar to that of younger counterparts and indicates that the critical stages of development, including growth, diversification, and expansion, *were not dependent* on the presence of carbonate-secreting or other metazoan organic communities. Therefore, Phanerozoic platforms can be viewed from a new perspective as Proterozoic *templates*, on which complex organic evolution and diversification took place.

2 Archean carbonates contain large volumes of marine cement precipitated directly on the seafloor in the form of large, botryoidal fans (probably former aragonite) and herringbone-textured sheets (probably former high-Mg calcite). These facies are uncommon in the early Proterozoic and rare in younger times. Late Proterozoic carbonates contain more finer-grained sediments, perhaps reflecting the absence of abundant stromatolites and seafloor precipitates at that time.

3 Evidence of calcium sulfate evaporites is rare or absent in rocks older than middle Proterozoic. True gypsum precipitation was probably more rare than previously thought. The record of evaporites should be reconsidered in the context of the carbonate record, which shows evidence for "surplus" precipitation at times of extremely limited gypsum precipitation.

4 Archean and early Proterozoic seawater may have been highly oversaturated with respect to calcium carbonate; the ratio of HCO_3^- to Ca^{2+} may have also been increased relative to Phanerozoic seawater. The middle Proterozoic probably experienced a substantial decrease in both of these variables. Latest Proterozoic seawater is thought to have had only a marginally higher carbonate saturation than Paleozoic seawater.

5 The decline of Proterozoic stromatolites, in part, may be directly related to the proposed decrease in carbonate saturation of seawater. Precipitation, sediment production, and stromatolite growth rates would have been highest in the early Proterozoic, decreasing progressively through time. This is reflected in the decrease in stromatolite diversity, abundance, and density during the second half of the Proterozoic.

Acknowledgments. – N. Beukes, H. Hofmann, C. Kerans, A. Knoll, B. Simonson, and D. Sumner are gratefully acknowledged for free exchange of ideas, photographs, rocks, and other data. This research was supported by NASA grant NAGW-2795 and NSF grant EAR 90-58199. S. Bengtson and A. Knoll are thanked for reviewing the manuscript.

Biomarkers in the Proterozoic record

Guy Ourisson

Centre de Neurochimie, 5 rue Blaise Pascal, F-67084 Strasbourg, France

Paleontology can be run at the molecular level: the constituents of defunct organisms leave traces, remodeled by microbial activity and chemical reactions in the sediments (molecular taphonomy). Molecular fossils can, by an informed reconstruction à la Cuvier, betray the exact structures of their precursors: mostly membrane constituents. Can one use this type of molecular information to deduce the biochemistry of earlier forms of life? Is it possible to use the results of organic geochemistry to learn something about the Proterozoic world? We shall show that the answer is "Yes and no."

Biomarkers are complex organic substances whose structures imply that they must originate from the maturation in sediments of molecular components of living organisms. A synonym for *biomarker* is *molecular fossil*, and the field is named *organic geochemistry*, or *molecular paleontology*. In what follows, we will mention practically exclusively one large family of biomarkers: terpenoids. In everyday experience, this is a very varied family, comprising substances of similar biosynthesis and of similar structural appearance, but as varied as the scent of the rose, the red of the tomato, the sex hormones of humans, the molting hormones of butterflies, etc. Are they the only biomarkers isolated? Or did we find them only because we knew, from previous work, how to coax them into revealing their structures? The answers are: "No, no!" First, other substances have been isolated and identified, many in Strasbourg – for instance, porphyrins, amino acids, carbohydrates, etc. But the frequent recognition of terpenoids as molecular fossils comes certainly from the fact that they contain the very structural features that would make them resistant to biodegradation: their characteristic accumulation of rings and branched structures is known to lead to such a resistance. This is linked with varied and apparently unrelated facts: that terpenoids have never been found to be reserve constituents of organisms, that people suffering from Refsum's disease are simply not able to get rid of the phytane derivatives derived from green vegetables, that very few microorganisms have been found to attack terpenoids at all, and then very slowly, etc. Furthermore, our own work in organic geochemistry has led us, by an unplanned twist, to recognize that some terpenoids play a universal role in the formation of biomembranes. This role as membrane builders makes them structural elements in a cell, unlike sugars, proteins, nucleic acids, fats, etc., which have a fast turnover. At a molecular level, terpenoids (or rather some of them) are "bones"; this explains why they fare so important in molecular paleontology.

Bengtson, S. (ed.) 1994: *Early Life on Earth. Nobel Symposium No. 84*. Columbia U.P., New York

We shall attempt below to define which could be the most primitive terpenoid-based membranes, and therefore which could be the most primitive molecular fossils; but we first have to give a very superficial survey of the field of molecular paleontology. (Reviews of the work described here, as well as extensive references to the work of others, can be found in the following publications: Ourisson 1987, 1989, 1990; Ourisson & Albrecht 1992; Ourisson & Rohmer 1992; Ourisson *et al.* 1987.)

The biomarkers or molecular fossils

The amount of organic substances accumulated in sediments is incredibly huge. The average content of organic carbon in shales or clays is about 2% – lower in sandstones or limestones, higher of course in oil shales, coals, petroleum mother-rocks, nil in overheated sediments near volcanic intrusions. The grand total amount of fossil *organic* carbon (i.e. excluding carbonates) has been evaluated (with the obvious uncertainties accruing) at about 10^{16} tons of organic C, as compared with 10^{12} tons supposedly present in *all* living organisms put together. This accumulation is entirely due to the death of earlier living organisms and the incompleteness of the recycling into CO_2 of their constituents: a very tiny leak in the carbon cycle, over billions of years.

This fossil organic matter therefore reflects the initial composition of the defunct organisms, but:

- the constituents of defunct organisms support first of all the life of successive populations of macro- and later microorganisms, first aerobic and, at deeper levels of burial, anaerobic;

- and aging in a mineral matrix over millions of years leads to a variety of events: the original constituents are modified by the combined actions of the minerals, of time, and of temperature to give new substances, more or less reminiscent of their precursors.

The study of the molecular structures of organic constituents of sediments, one aspect of organic geochemistry, should therefore give an insight into the nature of constituents of past organisms, provided we could analyze efficiently the hundreds or thousands of these constituents present in any sample (and we can); provided we could, with the minimal amounts usually available, fully define molecular structures (and we can); provided we could understand the nature of the molecular changes brought about by maturation (and we can); and provided we had learned enough about plausible pathways of biosynthesis to derive plausible precursors of the substances thus isolated (and we have).

In more paleontological parlance, we face a problem of molecular taphonomy: we want to isolate defined molecules and reconstruct their "living" precursors from their structures, the known modes of changes they can undergo upon death, and our general knowledge of what Nature can build: to reconstruct a three-legged duck would be a canard.

As in organismic taphonomy, we can deduce nothing from fragments so small or so ambiguous as to yield practically no information. Only *complex* structures (molecular or organismic) have a high enough information content to be characteristic of some living organisms, to be biomarkers or molecular fossils. In information theory, the information content is directly related to complexity, to improbability. For instance, one of the most widespread and abundant organic constituents of sediments is methane, CH_4. This can come either from the action of methanogenic bacteria or from hyper-maturation, the low-temperature–long-duration cracking of any sort of organic matter, and therefore its presence is completely noninformative (with the proviso that the analysis of its isotopic composition can give some indications as to its origin). On the contrary, the frequent isolation, in many sediments, of cholestane (1), a C_{27} molecule of sufficient complexity to have rendered its identification initially difficult, is also improbable enough to have a very high information content: it can only come from precursors of

(1)   (2) Cholesterol

For the benefit of the nonchemist, symbols like these represent molecules: at each end of a segment is implied a carbon atom, saturated with hydrogens within the limit of four bonds in all. The diagrams represent fairly well a planar projection of the skeleton of the molecule. We have not indicated here the stereochemical symbols, necessary for a complete understanding of the geometrical structure, but not in the present context.

similar complexity. In this case cholesterol (2) is the obvious candidate: we know from the accumulated understanding of sedimentary maturation processes that the reductions required to go from **1** to **2** are quite common, and cholesterol itself is a substance known to be an essential component of the membranes of all eukaryotes. Cholestane is therefore a putative molecular fossil of cholesterol – or a putative biomarker, the "putative" being rendered here quite certain, because the structure considered is sufficiently complex.

A number of other molecular fossils have been identified whose origin obviously lies in well-known extant natural substances, substantially modified. For instance, to turn from paleontology for a moment to archaeology and hagiology, the isolation of dehydroabietic acid from balms used in Egyptian mummies or in the coffin of an English saint signals the initial use of pine resin from one of the common *Pinus* species.

CO_2H Dehydroabietic acid CO_2H Abietic acid

Proterozoic biomarkers

There has been much interest in a search for biomarkers in Proterozoic sediments. This has been summarized, e.g., by Summons *et al.* (1988). Of course, the content of recognizable and informative molecular fossils in the very ancient sediments tends to diminish, and the effects of contaminations by permeation over very long times is difficult to evaluate. What emerges from the work of Summons on Australian rocks about 1.7–1 Ga old (and of ourselves on a Francevillian shale from central Africa, about 1.5 Ga old) is that the biomarkers do not differ from those already known from younger sediments. Steranes and 4-Me steranes are abundant in sediments back to about 1 Ga old and decrease in relative importance in older sediments, 1.75 Ga old, while the relative content of hopanes increases. The other biomarkers, n-alkanes, monomethyl branched alkanes, cyclohexyl alkanes, and acyclic isoprenoids, could all be derived from lipids of pro- or eukaryotes similar to the extant taxa. There is no evidence for biomarkers possibly derived from organisms having used extinct metabolic pathways.

Only one group of biomarkers isolated from these Proterozoic sediments cannot be related to lipids of extant organisms: a family of tricyclic polyterpanes. However, as we shall see in the next section, these are also present in much more recent sediments: they are not a sign of ancient organisms but are only a particular case of the "orphan biomarkers," which we shall consider below.

"Orphan biomarkers" and the most important biomarkers, the geohopanoids

To illustrate the difficulties and limitations of the concept of orphan biomarkers, let us consider the case of diacholestane (3). This is a hydrocarbon found in many sediments, alongside with and just as complex as cholestane (1), and therefore just as informative. A plausible precursor could have been 4. However, there is a major difference: no naturally occurring substance based on the skeleton of 3 is (yet) known, neither 4 nor

(3) Diacholestane   (4)

any other one. The isolation of diasteranes like 3 could therefore have called for a search of further cholesterol analogues, and, as long as none would be known, it would be a biomarker (its complexity precluding an abiotic origin), but an orphan one, as it is without known parents. However, in this case, there is a simpler explanation: substances with the skeleton of 3 are easily obtained by heating proximate derivatives of cholesterol with clays; diasteranes similar to 3 are therefore more plausibly products of maturation of cholestane derivatives, of "diagenesis" in the sediment. This fossil, which had been found "with the legs misplaced," had in fact been submitted to taphonomic rearrangement – a less interesting explanation, but a simpler one.

Orphan biomarkers remain therefore in this dismal condition only as long as their parents have not been found, through deliberate search or unintentional discovery. The most interesting case of such temporarily parentless sedimentary derivatives has been that of the geohopanoids (Ourisson & Albrecht 1992).

About 20 years ago, we discovered the universal presence in all sediments studied – whatever their age, their geographic origin, or their nature – of derivatives of one, and only one, family of pentacyclic "extended" triterpenes. By definition, triterpenes are quite common C_{30} substances (e.g., the white of birch bark, or of *Abies alba*), but the geohopanoids range from about C_{24} to C_{35}. While it is conceivable that maturation in a sediment might lead to the *loss* of some carbon atoms, a *gain* was highly improbable: "La vieillesse, quelle déchéance!" has said no lesser expert than Charles de Gaulle. Improbability reached in this case impossibility, as the five supplementary carbon atoms were found always as a straight chain. We promptly became convinced that we were dealing with biomarkers whose parents had to possess the required C_{35} skeleton (5), identified rigorously in every detail. Geohopanoids have been studied in the laboratory of my former student Dr. Pierre Albrecht; more than 200 individual structures have been fully characterized. Geohopanoids, while never *very* abundant in any sediment, are *universally* present in sediments, and therefore their total amount in Nature is huge. We have evaluated it at about 10^{12} tons – they are about as abundant as all "living" carbon put together! Hopane-hunting has become a very successful and standard method in petroleum geochemistry, but this is another story. We left geohopanoids at their early status of orphan biomarkers. In fact, their precursors have *later* been recognized as quite normal, though previously unrecognized, bacterial lipids. They are being studied with

(5) Bacteriohopane

(6) Adenosylhopane

zest and remarkable success by my other former student Professor Michel Rohmer, in Mulhouse, who has described more than three dozen fully identified structures. The most spectacular one is adenosylhopane (6), a hybrid of a triterpene and a nucleoside, but others are hybrids of the same triterpene and peptides, urea derivatives, amino sugars previously known only in antibiotics, etc. The biohopanoids are in fact steroid surrogates, playing in particular in many bacteria the same essential role as cholesterol in eukaryotes: they reinforce membranes. It is probable that the most complex ones, like 6, have also more noble roles; these are still unknown.

We have dubbed the biohopanoids "molecular coelacanths." As is quite clear, however, they are not shedding light on past organisms but have revealed, *through their fossils*, extant essential natural products, an unprecedented process of discovery in chemistry.

In another case, more relevant to the theme of this review, we had also isolated from several sediments hydrocarbons (**7–9**) (Z=H), the first of which, phytane (**7**), could have been derived from chlorophyll. The second one, bisphytane (**8**), was not related to any natural known precursor. It was an orphan only for a short while: **7** and **8** (this time with Z = complex phospholipid-like head group) are now known to be characteristic of the

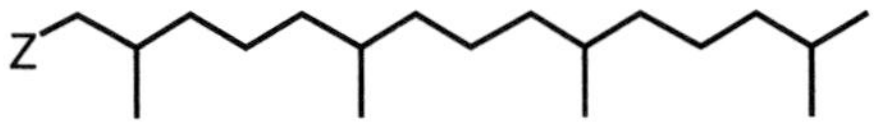

(**7**) Phytane

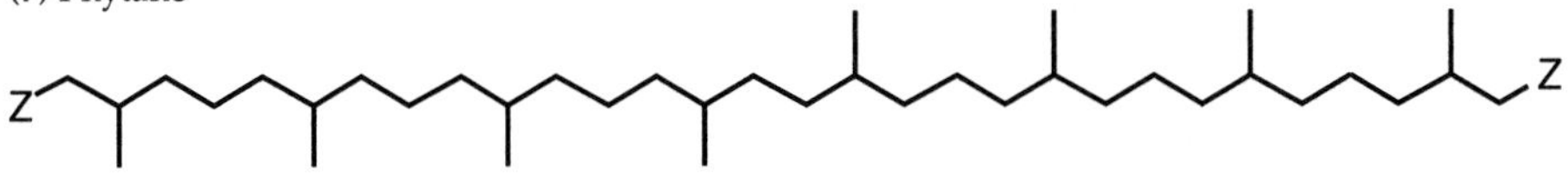

(**8**) Bisphytane (Z=H)

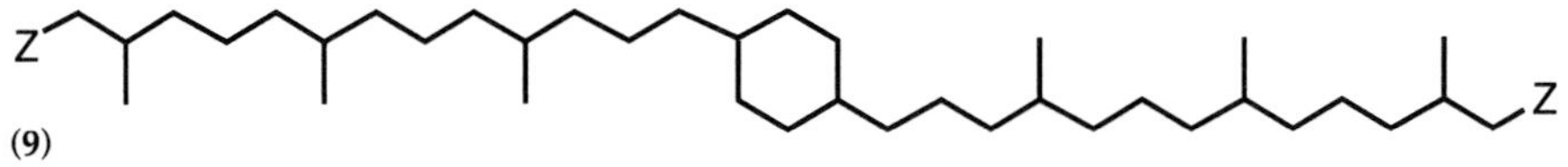

(**9**)

very spectacular membrane lipids of Archaeales, be they thermophilic or methanogenic (in the halophiles, only **7** has been found). As a confirmation of this derivation, intact archaebacterial lipids based on **8** have even been found intact in sediments. The cyclohexane derivative (**9**) (Z=H), however, remains so far an orphan biomarker. Its coexistence with its congeners in many sediments is, however, an unmistakable indication that it *must* originate from still unknown membrane lipids, quite certainly present in some other strain of Archeales.

Another family of orphan biomarkers is that of the tricyclopolyprenanes. These are found in many petroleums and sediments, quite frequently but not universally like the geohopanoids, nor are there as many variants known, by far. Their structures are exemplified by tricyclohexaprenane (**10**). Variants are known with a shorter side-chain,

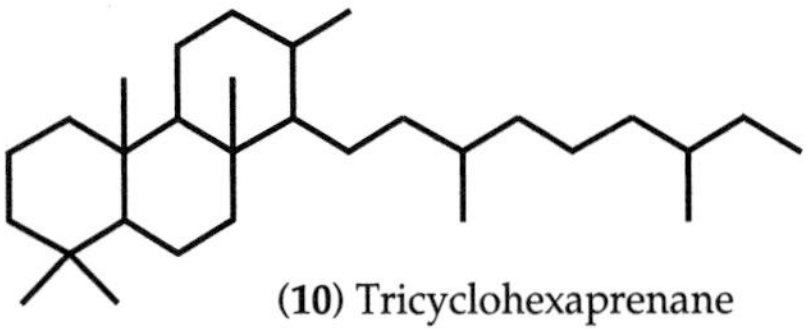

(**10**) Tricyclohexaprenane

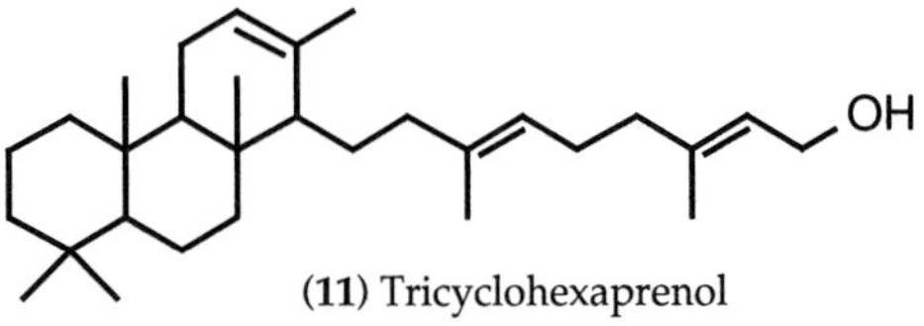

(**11**) Tricyclohexaprenol

and sometimes with a longer one. Tricyclohexaprenane may derive from tricyclohexaprenol (**11**), a C_{30} substance of a type still unknown in any living organism but quite "reasonable," insofar as it could derive from known and widespread precursors, the polyprenols found in every kind of living organism (e.g., **12**, hexaprenol) by well-known and widespread biochemical processes. Tricyclohexaprenol has been synthesized chemically, and we have shown that it plays in synthetic membranes a physical role comparable to that of cholesterol, the biohopanoids, or other cholesterol surrogates.

General biochemical knowledge tells us that a substance such as **12** could be obtained from simple building blocks (the C_2 acetate ion) by completely anaerobic pathways. Furthermore, the reaction required to cyclize **12** into **11** could be a very easy one that could even be reproduced by treating **12** with an appropriate strong acid. It

(12) Hexaprenol

(13) Squalene

(14)

would proceed following the rules of in vitro organic chemistry, in particular the Markovnikof Rule. Note at this stage that, at the place indicated by the arrow in **12**, the arrangement of bonds is very similar to, but different from, that in squalene (**13**). This last C_{30} polyterpene is widespread in nature; it is the biosynthetic precursor of our cholesterol, of the white of birch bark, of the pentacyclic skeleton of the biohopanoids (which require in addition a C_5 linear unit, **14**, and this is again a different story). The cyclization of **13** to the hopane skeleton not only requires help from enzymes like that of **12** to **13**, but this help must furthermore ensure antinatural "anti-Markovnikof" processes at the sites indicated by arrows in **13**. From our hypotheses, this should involve a more evolved enzymatic system than that leading to the tricyclopolyprenols, which could therefore be more primitive membrane constituents than hopanoids.

There is no rational way to search for the microorganisms containing the still unknown tricyclopolyprenols, but they can be partly described: They could be anaerobic, they would have special membranes containing these novel triterpenes, and they could be primitive, even more primitive than the known Archaeales from a certain point of view, as the structure of **11** requires one fewer enzymatic step than those of **7–9**: the last ones contain no double bond, and as their precursors must have contained them, they must have undergone a hydrogenation step, necessitating completely different reactions and therefore completely different enzymes.

Note also that the symmetrical structures of **8** and **9** imply a "tail-to-tail" duplication of a system similar to **12** (but shorter by five carbons), a complicated process from a chemical point of view, and that the structure of squalene (**13**) is also a symmetrical one, but with a "head-to-head" duplication, again a very complicated process to achieve chemically or biochemically. In short, of all the molecular fossils identified so far, the tricyclopolyprenanes (such as **10**) are the ones indicating the nature of the most primitive membrane constituents. But once again, there is no reason to believe that these point to the presence in the past of microorganisms extinct today. The sediments containing tricyclopolyprenanes are not very old ones, certainly not Proterozoic.

All the substances mentioned so far are products of the terpenoid metabolism and take part (presumably in the last case) in membrane formation. As we have already mentioned in passing, still other terpene derivatives play a similar role; they are for instance, many bacterial carotenoids, cycloartenol, etc. We have indeed postulated that this is the universal role of terpenes, which explains why they are always present, in any living organism; from these universal building blocks of membranes would then derive all the other, specific, functions of other terpenoids involved in photosynthesis (in retinal and in chlorophyll), in insect or vertebrate hormones, in plant–insect interactions, etc.

Criteria of biochemical primitivity – criteria for membrane formation

We are now ready to attempt to define which could be the most primitive membranes and therefore which could be the most primitive molecular fossils. But we first have to define criteria of biochemical primitivity and criteria for membrane formation. We have not, in fact, found any useful discussion of the characters indicating primitivity in biochemical processes, and we therefore have to propose a reasonable definition.

We consider that a more primitive biochemical pathway is one involving fewer distinct enzymatic systems to achieve a similar function. Furthermore, we consider that any enzymatic step achieving a particular transformation is more primitive if it simply accelerates a reaction in its "spontaneous" course, the one that would be followed without enzyme, with only in vitro reagents. Both of these assumptions are non-scientific in a Popperian sense, as they cannot be refuted; for this purpose, we should be able to look into the biochemistry of *really* primitive organisms, those of the very early Proterozoic world – or into their fossil signatures, and this is still impossible: the organismic Proterozoic record is not such that we could obtain large amounts of well-identified fossils, grind them, extract them, and isolate from the extract minute amounts of substances not yet known from later forms of life.

In the particular case of the substances considered here, this means, for instance, that any cyclization following the Markovnikof Rule will be more primitive than a similar one leading to similar functions (here, membrane building or strengthening) but running in an anti-Markovnikof way. Let us briefly review the minimal number of distinct chemical steps required to produce the terpenoids mentioned so far. In every case, biosynthesis proceeds from acetate to a C_5 prenyl unit, as the pyrophosphate. This uses a complex series of distinct steps; we shall, however, see at the end of the present essay that the prenyl unit could well have been produced initially much more simply, under abiotic conditions. From the prenyl unit, a polycondensation takes place. This implies a simple reaction:

$$C = C + C^+ \rightarrow C\text{--}C\text{--}C^+ \rightarrow \text{etc.}$$

or, in more precise terms (P = pyrophosphate group):

Acetic acid   Dimethylallyl pyrophosphate   Polyprenyl pyrophosphates

This leads to the various polyprenols existing in organisms: up to more than 20 units linked in a chain, in, for instance, the dolichols.

These polyprenyl chains can be further modified:

- They can undergo reduction to the archaebacterial lipids containing the C_{20} phytanyl chains:

- They have to be duplicated to obtain the C_{40} lipids of the same origin and probably *doubly* duplicated to lead to the cyclohexane lipids, which, we have seen, are still orphan lipids. We have indicated on these formulas "dotted" double bonds, because one does not know whether duplication occurs at the saturated stage, or, much more probably, at the unsaturated one.

- They can also be duplicated not head-to-head, but rather tail-to-tail, to give squalene ($2 \times C_{15}$), the precursor of the triterpenes (and in particular of the most abundant of them, the hopanoids), or to give similarly built $2 \times C_{20}$ precursors of the C_{40} carotenoids.

Squalene or prephytoene

- They can be cyclized, to give the orphan tricyclopolyprenols mentioned above.

It is obvious that all these possibilities thus require more than one type of enzymatic reaction (the last one implies reactions of the general type $C = C + C^+ \rightarrow C–C–C^+$, but in a different mode). Is it possible to use the knowledge accumulated so far, both explicitly and intuitively, to *deduce* what could be really primitive potential membrane constituents, demanding only one type of enzymatic reaction? We believe so and shall deduce it from a very brief summary of the structure of membranes (Fig. 1).

Membranes are essential to the emergence of any kind of three-dimensional life (leaving aside here the two-dimensional hypothetical models, the most elaborate of which are those of Wächtershäuser [1990a, 1992]). An organism *must* be able to distinguish its inside from the outside world, in an environment that is essentially water. This occurs thanks to the self-organization of amphiphilic molecules, provided they are of the right dimensions, carry adequate hydrophilic heads, and can establish enough attractive cooperative van der Waals contacts. In our membranes, this is achieved with straight-chain esters of glycerol, carrying a phosphoric-acid residue itself linked to more polar groups. Such a membrane is adequately stabilized by cholesterol. In bacterial or archaebacterial membranes, the partners vary slightly, but they lead always to double layers approximately 40 Å thick; shorter chains lead to micelles or to solutions, and longer ones to difficulties of self-organization.

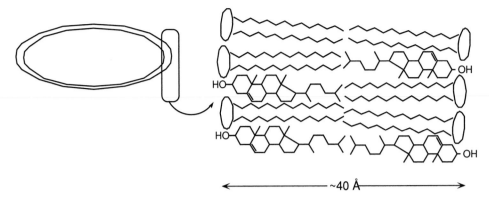

FIGURE 1   Schematic representation of a vesicle, in which the inner water is separated from outside water by a double layer of phospholipids and cholesterol (or a surrogate).

All these amphiphilic molecules contain several molecular modules: the one providing the lipophilic chains; the polar head, which is usually glycerol; and further polar modules attached to it. Nearly all types contain phosphates, which we have seen to intervene also in the biosynthesis of the polyprenic chains.

A model of really primitive membranes

From these considerations, we have come to conceive what could have been the most primitive types of membranes. Their lipophilic part would have to be purely terpenic, not saturated, not duplicated, and their hydrophilic part would have to contain phosphates, and no module implying an independent biosynthetic route. These conditions are most probably met in diesters of phosphoric acid, such as bis(geranylgeranyl) phosphate (**15**). This, and the lower homologues carrying shorter chains (C_{10}: bisgeranyl phosphate, C_{15}: bisfarnesyl phosphate), have now been synthesized. Their phase

(**15**) Bis(geranylgeranyl) phosphate

properties are presently being studied. They form nicely stable vesicles, even with the relatively short geranyl chains (Plobeck *et al.* 1992).

We hope also to be able to investigate whether it is possible to reduce to practice a hypothesis put forward independently, in slightly different forms, by G. Wächtershäuser and us: to obtain the chain elongation from prenyl phosphates on a surface, until they reach the molecular dimensions compelling them to leave the surface, and form vesicles, achieving in this way a reaction selectivity by phase separation. It is even conceivable that the same surface could lead to the required C_5 unit from smaller molecules such as isobutene and formaldehyde, by the so-called Prins reaction:

Conclusion

As we have seen, the Proterozoic record has not yielded any novel biomarker attributable to a form of life no longer existing. Furthermore, if we are right in our conclusions about the most primitive membrane lipids, their molecular fossils would be identical with those or Archaeales and would therefore provide no clue – unless they would have been preserved intact, which would have requiree extraordinarily mild conditions, excluding oxidation, reduction, or hydrolysis. It is therefore quite unlikely that organic geochemistry will directly provide novel information in the search for very primitive organisms.

Acknowledgments. – This essay is based on work initiated under my responsibility but now carried out independently by Pierre Albrecht in Strasbourg and by Michel Rohmer in Mulhouse, as indicated in the text. The study of the molecular aspects of membrane formation is being carried out in cooperation with Yoichi Nakatani in Strasbourg. Alain Milon, now in Toulouse, has taught us to use modern biophysical methods. Marie-Claire Dillenseger has prodded me into leaving the firm ground of established facts and into indulging in Proterozoic dreams.

Stromatolites: The main geological source of information on the evolution of the early benthos

Malcolm R. Walter

School of Earth Sciences, Macquarie University, North Ryde, NSW 2109, Australia

For more than three billion years microbial mats sculpted the seafloor and lakefloors, leaving an abundant fossil record as stromatolites. They built reefs as big as any later built by corals and algae and participated in the construction of continental shelves and ramps with a facies architecture like that of the Phanerozoic. The early Archean mats of hypersaline lagoons were probably built by anaerobic photoautotrophs. By the late Archean cyanobacteria were the likely constructors of stromatolites, and the substantial morphological and microstructural diversity suggests a similar diversity of mat organisms. Paleoproterozoic stromatolites were very diverse, and some were demonstrably built by cyanobacteria. Temporal restriction of numerous stromatolite forms during the Proterozoic seems to reflect both microbial evolution and an evolving environment. There is good evidence for eukaryotic microalgae in latest Neoproterozoic stromatolites, and equivocal evidence back as far as the Paleoproterozoic. Enrichment of cyanobacterial mats by microalgae may explain the high taxonomic diversity of stromatolites in Mesoproterozoic and early Neoproterozoic successions. At some ill-defined time animals invaded the mat ecosystem. Cambrian mats, like their extant analogues, were bored, burrowed, and grazed by metazoans. About 1 Ga ago the diversity of stromatolites in quiet subtidal environments decreased, and 700–600 Ma ago the diversity of all stromatolites declined drastically; this is most likely due to disruption of mats by metazoans, starting in the subtidal realm. Microbial mats are restricted now to a few environments where the pressures of predation and competition are minimal.

For billions of years the shallow seafloor was carpeted with microbial mats. They spread down the continental slope and probably formed huge patches on the deep seafloor. They were widespread in lakes and lagoons. We can speculate that mats also formed extensive subaerial sheets on land, as they still do locally (a Proterozoic terrestrial microbiota has recently been reported by Horodyski [1990]). This was the scene throughout the Proterozoic, a duration of some two billion years.

The Archean was similar, but the distribution of mats was more patchy, with environments suitable for the prolific growth of benthic microorganisms being limited by the effects of a more vigorous tectonic regime. Hyperactive volcanism and mountainbuilding swamped sedimentary depositories with clastic sediment, in which mats rarely persisted (Lowe, this volume).

Bengtson, S. (ed.) 1994: *Early Life on Earth. Nobel Symposium No. 84.* Columbia U.P., New York

The mats trapped, precipitated, and entrained sediment, generating structures preserved as rocks and called stromatolites. In seeking the light, and perhaps for other reasons, the microbes built sedimentary structures with vertical relief, at every scale from the microscopic to the monumental; they made reefs as big as any built later by corals and algae – tens to a hundred meters or more high and up to hundreds of kilometers long. Bizarre forests of giant conical structures like rocket nose cones placed edge to edge were a common feature of the seafloor. Branching forms as diverse as modern corals were abundant.

Microbial mats are complex, layered ecosystems, frequently with several layers of photoautotrophs, each harvesting different wavelengths of light, underlain by layers of aerobic and anaerobic heterotrophs. The chemical interactions between the different physiological types of microbes are intricate. All of this complexity is compressed into less than a centimeter, often only 2–5 mm. The mats act as membranes that mediate the chemical and physical processes at sediment–water and sediment–air interfaces. They are factories for the production and decomposition of organic matter. They are just as productive as higher plant ecosystems, and, in fact, during the Proterozoic the efficiency of their fixation of CO_2 led to the formation of some of the richest petroleum source-rocks known.

And then, for reasons yet to be understood, the visible dominance of microbes ended with the rise to ascendancy of the metaphytes and metazoans, at the dawn of the Phanerozoic. From time to time thereafter, microbes asserted their ability to sculpt the sediment–water interface, as in the great Devonian reefs of Western Australia, but while the role of bacteria as profoundly important geochemical agents continued, their role as mediators of sediment morphology was largely finished.

Ever since the abundance of stromatolites was discovered early this century, their potential as a source of biological information has been recognized. Bringing that potential to fruition has been and continues to be a frustrating experience with limited rewards. It seems clear that an enormous amount of information is encoded in the complex and diverse shapes and internal features of stromatolites but, primarily because extant stromatolites have been very little studied, we can decode only a word here and there.

If we stand back and view the stromatolite record from a distance, some broad patterns are readily perceived. The broadest of these are:

1 Early and middle Archean stromatolites are rare and not very diverse in form; there are none of the complex branching columnar forms so common in the Proterozoic.

2 In the late Archean, about 2.8–2.7 Ga ago, something happened to allow stromatolites to become both more abundant and much more diverse morphologically.

3 Among the columnar forms that rose to prominence in the late Archean and especially in the Paleoproterozoic was an abundance of ministromatolites with a radial fibrous fabric, characteristic of peritidal environments; these declined markedly after the Mesoproterozoic (Grey & Thorne 1985; Grotzinger 1989, 1990).

4 There was an abundance of stromatolites with conical laminae (called *Conophyton* and various other names by biostratigraphers), characteristic of quiet subtidal environments, in the Paleo- and Mesoproterozoic, with a marked decline thereafter (Komar *et al.* 1965; Zhu 1982; Walter & Heys 1985).

5 There was a decline in abundance and diversity of all stromatolites about 700–600 Ma ago (Awramik 1971; Walter & Heys 1985).

6 Thrombolites (unlaminated stromatolites) were rare before the Phanerozoic; these became abundant during the Early Cambrian (Walter & Heys 1985; Kennard & James 1986).

7 Stromatolites are rare after the early Ordovician.

8 The preservation of microscopic microbial fabrics showing traces of cell morphology in carbonate stromatolites is rare in the Archean and Proterozoic but common in the Phanerozoic; this seems to be yet another example of the phenomenon of calcification appearing near the beginning of the Cambrian.

This outline is the coarse framework of stromatolite distribution, familiar to all biostratigraphers who use these fossils. Within this framework many finer patterns have been recognized. It is this that allows a biostratigrapher to walk through a stromatolitic carbonate sequence in the field and make a first rough assessment of the age of the rocks.

Almost all little-metamorphosed limestones, dolomites, and magnesites of Proterozoic age contain stromatolites, and they also occur in phosphorites, iron formations, cherts, and, rarely, in sandstones. Benthic microbial-mat fabrics have also been recognized in shales and siltstones. As a result, a wide range of paleoenvironments is represented in the stromatolite record. Only terrestrial and deep subaqueous (below the photic zone) environments are poorly represented.

Proving that a rock is a fossil

This is the first and most difficult problem when dealing with stromatolites. The term *stromatolite* as used here implies biogenicity, but some authors use a morphological definition that admits abiogenic objects (see discussion in Buick *et al.* 1981). Proterozoic stromatolites are usually presumed to be biogenic, because they are very diverse and abundant, and some are quite clearly biogenic, as demonstrated by their included microfossils, or they have a very high probability of being biogenic because of their complex form. With Archean examples, however, the problem is severe and important. Early Archean stromatolites are rare and of low diversity but are employed as one of our major sources of information about the earliest life on Earth.

The problem of demonstrating biogenicity is an old one and has been discussed many times (e.g., Seward 1931; Young 1929, 1943; Mawson, quoted in Glaessner 1972b; Read 1976; Walter 1976b, 1983; Hofmann 1972; Buick *et al.* 1981; Schopf, this volume). So rather than review the issue I will attempt to demonstrate how much confidence can be attached to the recognition of stromatolites.

One way to approach this problem is to examine some examples that are especially well known. The stromatolites of the Neoproterozoic Bitter Springs Formation of the Amadeus Basin of central Australia were first described by Chewings (1914) and Howchin (1914). More forms were found and described by Mawson & Madigan (1930) and Madigan (1932, 1935); these authors paid particular attention to the origin of stromatolites, an especially contentious matter at the time, and to their biostratigraphic utility. The formation was named and described by Joklik (1955) and Wells *et al.* (1967, 1970). Glaessner *et al.* (1969), Cloud & Semikhatov (1969), and Walter (1972) described nine different forms of distinctive columnar branching stromatolites from the formation, Walter (1972) described several domical forms, and Walter *et al.* (1979) provided some additional taxonomic information. Stewart (1979) interpreted a lagoonal environment for the evaporites of the formation and suggested that the stromatolites formed a barrier to the lagoon. Southgate (1986, 1989) undertook a major sedimentological study of the formation, which led to the recognition of lacustrine, flood-plain, and marine paleoenvironments and showed that many of the stromatolites occur in cyclic repetitions of distinctive lithofacies. Through sequence analysis using both outcrop and seismic data, Lindsay (1987) was able to place the formation in a broader paleoenvironmental context. Meanwhile, microfossils had been discovered in stromatolitic cherts in the formation by Barghoorn & Schopf (1965), and these were subsequently studied in great detail by Schopf (1968), Schopf & Blacic (1971), Oehler (1976, 1977), Oehler *et al.* (1979) and Knoll & Golubic (1979); a comprehensive reference list is given by Mendelson *et al.* (1992). The strontium isotopic composition of the formation has played a key role in interpretations of the evolution of the composition of seawater (Veizer & Compston 1976), as has the presence of thick evaporites (Holland 1984). McKirdy's (1976) pioneering research on the organic geochemistry of some of the stromatolites has been followed by biomarker analyses of parts of the formation (Summons & Walter 1990; Summons & Powell 1991). The formation has been examined as a source rock for petroleum (Summons & Powell 1991) and has been studied by geologists involved in base-metal and petroleum exploration on many occasions over the last thirty years. The evaporites of the formation have been important in the structural development of the Amadeus Basin, and as a result structural geologists have given the unit a lot of attention (e.g., the papers in Korsch & Kennard 1991). The Bitter Springs Formation is a prominent unit in the Amadeus Basin and has been mapped many times over the last 40 years (see references in Korsch & Kennard 1991). All this attention has led to many participants of conference field trips examining the formation and its contained stromatolites.

Sheer weight of numbers means little, but there is no doubt that the stromatolites of the Bitter Springs Formation are among the most intensively studied and examined anywhere. They are accepted as stromatolites, that is, as being biogenic. But can anyone be sure? The answer is no, with only one exception: some of the lacustrine stratiform stromatolites preserved in chert have within them well-preserved microfossils showing the original structure of the microbial mats (e.g., Knoll & Golubic 1979). The same fabric occurs in the carbonate stromatolites within millimeters of the chert lenses, but no microfossils are preserved. All of the other stromatolites in the formation are composed of dolomite or calcite and contain neither microfossils nor even faint remnants of microbial cells. Some of the columnar stromatolites contain chert nodules, but none of my extensive collection of these made some years ago is fossiliferous (Walter, unpublished observations). So all would fail to meet the three most demanding criteria for biogenicity advocated by Buick *et al.* (1981; Table 1 herein), requiring the presence of microfossils or microbial "trace fossils," and would have to be regarded as "possible" stromatolites or, perhaps in the example of the more complex forms, as "probable"

TABLE 1   Criteria for assessing the biogenicity of stromatolite-like objects. (From Buick *et al.* 1981 and Walter 1983; see these references for a detailed discussion.)

Walter	*Buick* ET AL.
1 Orientation in relation to bedding such that syngenicity can be demonstrated	1 Must occur in undoubted sedimentary or metasedimentary rocks
2 Having a macromorphology consistent with a stromatolitic origin	2 Demonstrably synsedimentary
	3 Preponderance of convex-up structures
3 Having a micromorphology consistent with a stromatolitic origin	4 Laminae thicken over crests of flexures
4 Having a chemical composition consistent with a stromatolitic origin	5 If laminated, the laminae should be wavy, wrinkled and/or have several orders of curvature
	6 Microfossils or [microbial] "trace fossils" present
	7 Changes in microfossil assemblages should be accompanied by changes in stromatolite morphology
	8 The microfossils or [microbial] "trace fossils" must be organized in a manner indicating trapping, binding, or precipitation of sediment by living microorganisms

stromatolites. All meet the criteria demanded by Walter (1983; Table 1 herein), and so by those criteria could be called stromatolites. These criteria are more lenient than those of Buick *et al.* (1981), but nonetheless seem to work in excluding all objects except those most sedimentologists and paleontologists would accept as stromatolites. Science by consensus may be dangerous, but in this example it represents the accumulated experience of a century of the studying of stromatolites. I consider that an experienced person can recognize a stromatolite with a high degree of confidence. If all of Walter's (1983) five criteria are met, I judge that level of confidence to be more than 90% (and that seems to be supported by the consensus in interpreting the Bitter Springs stromatolites, which as far as I know is unanimous); meeting all the criteria of Buick *et al.* (1981) would lift confidence to 100%.

The Bitter Springs example is typical of Proterozoic stromatolites; on a per-volume-of-rock basis, much less than 1% of the stromatolites are demonstrably biogenic (judged from measured sections such as Text-fig. 14 of Walter 1972). Yet all are generally accepted as biogenic, because close morphological and microstructural comparisons can be made with microfossiliferous stromatolites and with extant stromatolites, because they occur in appropriate sedimentary rocks, and because in many examples there is no convincing alternative interpretation. That is the current state of stromatolite studies: the word *stromatolite* as used by most specialists carries the connotation of a high (perhaps 90%) level of confidence in biogenicity.

The early Archean benthos

All occurrences of Archean stromatolites known before 1980 were reviewed by Walter (1983). There were just eleven for the whole Archean. Since then, additional information has been provided on some of these (Awramik *et al.* 1988; Buick 1984, 1990; Lowe 1980b, 1983; Hofmann *et al.* 1985; Hofmann & Snyder 1985; Nisbet & Wilks 1988; Beukes & Lowe 1989), and ten new occurrences have been described (Grey 1981; Orpen & Wilson 1981; Byerly *et al.* 1986; Hofmann *et al.* 1991).

The stromatolites of the Warrawoona Group of Western Australia are about 3.5 Ga old and are the oldest convincing examples known (Lowe 1980b, 1983; Walter *et al.* 1980; Buick *et al.* 1981; Groves *et al.* 1981; Walter 1983). Since the published research, additional examples have been discovered (Hickman 1990; Fig. 1 herein). The stromatolites of the Fig Tree Group of South Africa (Byerly *et al.* 1986; Schopf, this volume) may be somewhat younger.

FIGURE 1 Stromatolites from the Towers Formation, Warrawoona Group, Western Australia (locality on the northeastern side of the North Pole Dome). Width of field of view about 30 cm.

The Warrawoona stromatolites (see also Schopf, this volume) are stratiform, pseudocolumnar, and domical (Fig. 1). They occur in cherts that are interpreted as having formed in hypersaline lagoonal environments in a region of active volcanism (Fig. 2). Their biogenicity has been debated at length by all the authors listed above and others who have reviewed the research (e.g., Nisbet 1980). All agree on the presence of at least probable stromatolites, by which they mean that they have the same level of confidence

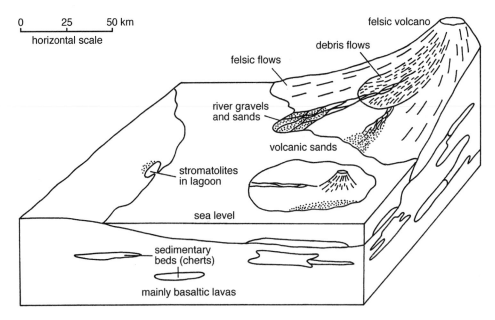

FIGURE 2   Reconstruction of the depositional environment of the Warrawoona stromatolites. (After Groves *et al*. 1981.)

as for Proterozoic stromatolites of similarly simple shapes. Microfossils occur in some of the stratiform examples but not in the pseudocolumnar and domical forms, but faint remnants of microbial filaments have been reported from the latter.

The Onverwacht and Fig Tree stromatolites are best regarded as "possible." The Fig Tree examples (see also Schopf, this volume) look convincing, but they encrust volcanics in an environment where geyserite can be expected, and this can be very difficult to distinguish from stromatolites (Walter 1976a).

The evidence for biogenicity becomes more convincing when the evidence from microfossils and kerogen is added (Schopf 1983b), although this too is disputed (Awramik *et al.* 1988; Buick 1990).

Taken together, the evidence allows a high level of confidence that there are at least some stromatolites known from Early Archean rocks.

The biological implications of this conclusion are as follows (see Walter 1983 for a detailed discussion and Schopf, this volume):

- Life existed 3.55 Ga ago.

- Benthic microbial mats grew in shallow lagoonal environments where the microorganisms had to withstand the rigors of hypersalinity, desiccation, and high light intensity.

- The microbes most probably were filamentous, to judge from the fabric of the stromatolites; there were probably also coccoid forms (and the evidence from microfossils supports these two conclusions [Schopf & Walter 1983; Schopf & Packer 1987; Walsh & Lowe 1985; Schopf 1993, this volume]).

• Carbon isotopic evidence suggests that the microbes were autotrophic (Schidlowski *et al.* 1983); they may have used light as a source of energy, though this is more contentious.

A glimpse of the middle Archean benthos

The stromatolites of the ~3.1 Ga old Insuzi Group of South Africa are the only known sample of the benthos in the interval 3.5–2.8 Ga (Walter 1983; Beukes & Lowe 1989 and earlier references therein). They formed on tidal flats and in adjacent tidal channels (Beukes & Lowe 1989).

Two distinctively biogenic types of fabric occur in these stromatolites. Bulbous stromatolites up to 30 cm wide and high have within them a columnar-layered structure. The laminae are wrinkled, consisting of juxtaposed hemispheroidal lenses of carbonate, some of which have a radial fabric apparently indicating the former presence of radially arranged fine filamentous microorganisms. Several species of extant cyanobacteria form comparable "tussocks."

Other nodular and bulbous stromatolites up to 15 cm wide have within them finely laminated columnar-layered structures in which the columns are 1–5 mm wide. The laminae within the columns are convex to conical. There are also larger isolated columns with conical laminae (Beukes & Lowe 1989). Close comparisons can be made with extant stromatolites built by the oscillatoriacean cyanobacterium *Phormidium tenue*. As a result, it can be suggested that these Archean stromatolites were built by finely filamentous microorganisms that were positively phototactic, i.e. had the ability to move towards the light, and that may have been microaerophilic (see Walter 1983 for a full discussion). Beukes & Lowe (1989) describe several additional forms of stromatolites.

Here for the first time we have structures that because of their sedimentological setting, macro- and microstructural complexity, and comparability to extant stromatolites can be interpreted unequivocally as stromatolites. They provide a record of at least two distinctly different microbial communities.

The late Archean benthos

The rich morphological record of the benthos starts in the late Archean and continues essentially uninterrupted through the Proterozoic and into the Phanerozoic. Biogenicity is no longer a concern (except in some individual examples, as it always will be). The far greater abundance of late Archean rock successions, as compared to those that are older, is reflected in a greater abundance of stromatolites. Counted by group-level rock units, there are 17 occurrences (Grey 1981; Walter 1983; Hofmann *et al.* 1991), and many of them are areally very extensive with abundant and diverse stromatolites.

Of the 17 occurrences, most are likely to be marine, and two or three are probably lacustrine. The marine examples are in "greenstone belts" and are associated with tectonically active settings with abundant volcanism and later intense deformation. They probably formed "thin, short-lived reefs that fringed subsiding volcanos" (Grotzinger 1989). The lacustrine examples are in thick piles of platformal deposits, among the oldest such successions known. The example described by Hofmann *et al.* (1991) may be in the deposits of a volcanic caldera lake.

The stromatolites of the Belingwe Greenstone Belt in Zimbabwe are the best known of the marine examples (Martin *et al.* 1980). They seem to have formed in a lagoonal setting, with evidence of desiccation and hypersalinity. A wide variety of stratiform, domical, and columnar forms is present. Unfortunately the internal fabrics, though apparently well preserved, have not been described or illustrated in detail, so it is impossible to assess the biological significance of these stromatolites. There appear to be several different fabrics representing several kinds of microbial mats, all probably constructed by filamentous organisms, as the fabrics are finely laminated.

Some of the stromatolites of the Steeprock Group of Canada seem to have formed by the trapping of detrital grains; this is indirect evidence of construction by phototactic filamentous organisms (Walter 1983).

The "*Conophyton*-like" stromatolites described by Grey (1981) from Western Australia, and the "*Thyssagetes*-like" forms described by Hofmann *et al.* (1991) from Canada, are especially significant because these distinctive conical forms can be interpreted by analogy with extant forms (Walter *et al.* 1976a). They strongly suggest construction by finely filamentous, positively phototactic microorganisms that may have been micro-aerophilic; the extant forms are constructed by cyanobacteria.

The lacustrine, fluvial, and possibly marine stromatolites of the Fortescue and Ventersdorp Groups of Western Australia and South Africa are superbly preserved (especially in the example of the Fortescue) and have been studied in detail (Buck 1980; Walter 1983; Packer 1990; and earlier references therein).

> *Sediments accumulated within numerous small but deep intermontane graben basins formed by the block faulting of an older andesitic lava terrain. During the early stages of deposition, coarse clastic debris accumulated along the scarped margins of horsts forming talus scree and debris flow alluvial fan deposits, while within basins fine grained terrigenous and chemical sediments, including stromatolites and ooids, accumulated under lacustrine conditions. Later fluvial processes predominated, resulting in the widespread prograding of alluvial fans across basins. The sediments of the succeeding . . . formation accumulated on a regionally extensive alluvial plain dominated by braided rivers. Stromatolites occur intermittently throughout these fluvial sediments, and are interpreted as having developed within pools remaining upon the alluvial plain during intervals between major fluvial discharges.*
>
> Buck (1980)

This interpretation of the Ventersdorp stromatolites is equally applicable to those from the Fortescue (see Packer 1990 for a contrasting interpretation). The stromatolites are stratiform, large domical and small-to-medium-sized columnar forms, including the distinctive *Alcheringa narrina* (Fig. 3). Many contain faint remnants of the constructing filaments. Fabrics recording at least three different microbial communities can be recognized (Walter 1983). One seems to record a *Lyngbya*-like coarsely filamentous oscillatoriacean cyanobacterium that was phototactic or phototropic; another shows evidence of positive phototaxis, perhaps in a finely filamentous oscillatoriacean cyano-bacterium.

FIGURE 3 The stromatolite *Alcheringa narrina* Walter 1972 from the 2.8–2.7 Ga old Fortescue Group, Western Australia (km 211.3 on the Newman – Pt. Hedland railroad). Width of field about 2 cm. Note remnants of biogenic filaments at high angle to growth laminae of the stromatolite.

By 2.8–2.7 Ga ago, then, microbial mats flourished in marginal marine, lacustrine, and fluvial environments. Most were built by filamentous microorganisms with so many similarities to extant cyanobacteria that it is reasonable to conclude that this is what they were.

Reefs and platforms of the Proterozoic

Stromatolites are ubiquitous in Proterozoic sedimentary successions. Limestones and dolomites without them are very rare. Biostratigraphers have documented more than 700 taxa in hundreds of stratigraphic units. This rich record of the benthos lays to rest the conventional view that fossils are rare in the Proterozoic.

Detailed sedimentological studies too numerous to review here have documented the paleoenvironmental distribution and role of Proterozoic stromatolites (see summaries in Walter 1976b; Geldsetzer *et al.* 1988; Valdiya & Tewari 1989; and the review by Grotzinger 1989). A remarkable conclusion, most clearly stated by Grotzinger (1989, this volume), is that the microbes of the Proterozoic built giant barrier reefs, fringing reefs, pinnacle reefs, atolls and so on, and bioherms of many forms, and through this

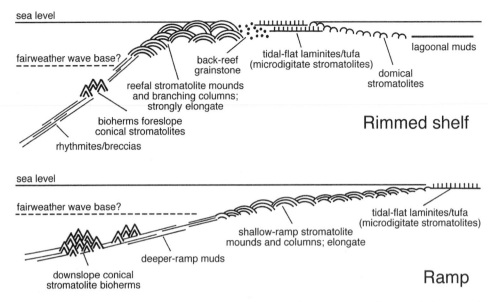

FIGURE 4 **Figure 4**   The role of stromatolites in the construction of Proterozoic carbonate platforms and ramps. (From Grotzinger 1990.)

constructed continental shelves and ramps (Fig. 4) like those later built by corals, calcareous algae, and other skeletal organisms.

Many of the broad patterns of stromatolite distribution referred to in the introduction are focused in the Proterozoic:

- In the Paleoproterozoic and Mesoproterozoic, peritidal environments were characterized by an abundance of ministromatolites ("microdigitate stromatolites") with a radial fibrous fabric; these declined markedly thereafter (Grey & Thorne 1985; Grotzinger 1990).

- In the Paleoproterozoic and Mesoproterozoic, quiet lagoonal and deep subtidal environments were characterised by an abundance of stromatolites with conical laminae (called *Conophyton* and various other names by biostratigraphers); these declined markedly thereafter (Komar *et al.* 1965; Zhu 1982; Walter & Heys 1985).

- All stromatolites declined in abundance and diversity about 600–700 Ma ago (Awramik 1971; Walter & Heys 1985).

- Thrombolites (unlaminated stromatolites) were very rare in the Proterozoic but abundant in the early Paleozoic (Walter & Heys 1985; Kennard & James 1986; Aitken & Narbonne 1989; Pratt & James 1988; Kah & Grotzinger 1992; Kennard, in press).

These broad patterns are complemented by the finer patterns used by biostratigraphers to delineate assemblage zones 50–300 Ma in duration.

The great morphological diversity of stromatolites, comprising the deposits of reefs, lagoons, tidal flats, lakes, and streams, reflects both environmental and biological diversity. A plexus of environmental and biological influences determines the form and fabric of stromatolites. The small-scale features of stromatolites, the "fabric" (lamina shape and microstructure), record the shapes of the constructing microbial colonies and some information on the size, shape, and orientation of the cells in those populations (Hofmann 1975; Monty 1976; Bertrand-Sarfati 1976; Walter 1977; Awramik & Semikhatov 1979; Semikhatov *et al.* 1979).

The most direct approach to extracting biological information from stromatolites is to study any included microfossils. Numerous microfossiliferous stromatolites are now known, but they still represent a tiny proportion of all stromatolites (less than 1% of all stromatolite taxa), and most are simple stratiform and domical stromatolites rather than the morphologically complex columnar forms that are likely to express a rich record of diverse and evolving microbial communities. A second approach is to extrapolate the interpretations from microfossiliferous stromatolites to those that are unfossiliferous but have the same fabric. This is helpful but still excludes most morphologically complex columnar forms. Third, analogies can be made with living stromatolites; this is very informative, but few living stromatolites have been studied in detail, and, furthermore, the diversity of extant stromatolites seems to be very low compared to that known from the Proterozoic. Lastly, inferences can be made from geometric analyses of fabrics; these are limited only by the imagination of the interpreter. This dismal view stands as a challenge: the record exists, but the research has barely begun. Nonetheless, some progress has been made.

The most distinctive stromatolites of "deep" water (but still in the photic zone) are *Conophyton* and other forms with conical laminae (Walter 1977; Kerans & Donaldson 1988; Grotzinger 1989). These range from tiny examples to forms that stood 30 m above the seafloor and edge-to-edge, like a forest of rocket nose cones. They also formed in the shallow water of quiet lagoons (M.J. Jackson *et al.* 1987). As mentioned above, comparison with extant forms strongly suggests construction by finely filamentous, positively phototactic microorganisms that may have been microaerophilic; the extant forms are constructed by cyanobacteria. Also on the deep floors of seas and lakes, and perhaps below the photic zone, were microbial mats that spread across deposits of siliciclastic silt (Hieshima & Pratt 1991); some of these can be compared to the mats currently formed by sulfur-oxidizing bacteria such as *Beggiatoa*.

The view of Gebelein (1976), that all the major groups of stromatolite-building cyanobacteria were already present by the Paleoproterozoic, has been strengthened by later work. The most compelling example is the comparison that can be made between stromatolites built now in the intertidal zone by the coccoid cyanobacterium *Entophysalis* and identical Paleoproterozoic stromatolites (Fig. 5) constructed by *Eoentophysalis* (Golubic & Hofmann 1976). Many other examples are documented by Schopf (1968), Schopf & Blacic (1971), Schopf *et al.* (1977), Oehler (1978), Strother *et al.* (1983), Hofmann & Schopf (1983), Knoll (1985b), Green *et al.* (1989) and Mendelson *et al.* (1992), among numerous other authors. Although microbial preservation in chert is relatively common only in littoral hypersaline sediments (e.g., Knoll 1985a; Southgate 1986), sufficient examples of fossil microbiotas are now known that Gebelein's (1976) conclusion can be accepted in broad terms. This is not to say that there was no evolution in the cyanobacteria during the Proterozoic; the stromatolite record described by biostratigraphers is a strong indication that there was.

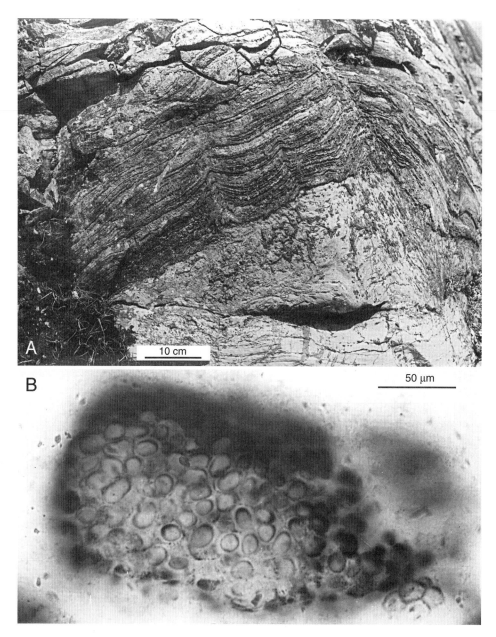

FIGURE 5 Paleoproterozoic stratiform stromatolite (**A**) from the Belcher Islands, Canada, and *Eoentophysalis belcheri*, the cyanobacterium that constructed it (**B**). (Photographs by Hans J. Hofmann.)

Because of the preservational bias mentioned above, much of the probable microbial diversity of the Proterozoic is cryptic. Few stromatolites contain microfossils. Indications of the level of diversity come from studies such as those of Southgate (1989), where the stromatolite taxa of Walter (1972) were put into a detailed paleoenvironmental context. Southgate (1989) documents numerous 1–2 m carbonate cycles that are essentially identical in all respects except that different cycles have taxonomically distinct columnar stromatolites. The environments of deposition of the different cycles must have differed only in very subtle ways. The six different taxa are likely to record a similar number of distinct microbial communities, but none contains microfossils. The hundreds of stromatolite taxa recorded from the Proterozoic (listed or referenced by Bertrand-Sarfati & Walter 1981; Walter & Heys 1985; and Walter *et al.* 1992) can be assumed to reflect a similar diversity of microbial communities.

Several examples of late Neoproterozoic microfossiliferous stromatolites contain abundant probable microalgae as well as cyanobacteria and eubacteria (e.g., Schopf & Sovietov 1976; Schopf *et al.* 1977). These include both coccoid and branched filamentous forms. This is in accord with other evidence from fossils and organic biomarkers (summarized by Summons & Walter 1990) for a substantial contribution by algae (especially chlorophytes) to the biota of these times. Possible calcareous algae occur in stromatolite-like structures in Neoproterozoic patch reefs in Canada (Aitken 1988), and in apparently nonstromatolitic cherts and dolomites in California (Horodyski & Mankiewicz 1990). (See also Riding, this volume.) Evidence for a contribution by microalgae to Mesoproterozoic and early Neoproterozoic mats is particularly contentious, being largely dependent on the interpretation of possible organelles within coccoid cells (e.g., Schopf 1968; Schopf & Blacic 1971; Oehler 1976, 1977; Knoll & Golubic 1979). Fairchild (quoted in Preiss 1987) reports possible algae, including multicellular forms, in early Neoproterozoic stromatolites from Australia. A red alga is known from ~1 Ga(?) old "stratiform laminated" cherts from Canada (Butterfield *et al.* 1990). Some of the more unusual of the microfossils from the richly microfossiliferous stromatolites of the ~2.0 Ga old Gunflint Formation of Canada have been interpreted as algae (see critical discussion in Hofmann & Schopf 1983). Despite the uncertainties, the fossil record of algae goes back to at least 1.7 Ga (Summons & Walter 1990) and perhaps 2.1 Ga (Han & Runnegar 1992; see also chapters in this volume by Knoll, Runnegar, Sun, and Vidal), so we must conclude that microalgae probably inhabited Paleoproterozoic and Mesoproterozoic cyanobacterial mats and may well have had a significant architectural role. Their contribution, along with that of cyanobacteria, may account for the high taxonomic diversity of Mesoproterozoic and early Neoproterozoic stromatolites (Walter & Heys 1985). The diverse cyanobacterial and microalgal communities of the subtidal columnar stromatolites of Shark Bay (Golubic 1976) and the Bahamas (Riding *et al.* 1991) provide extant partial analogues, although these can more informatively be compared with the thrombolites discussed below.

Decline to obscurity

Beginning about 1 Ga ago (Walter & Heys 1985) and culminating before the Vendian about 600 Ma ago (Awramik 1971; Walter & Heys 1985; for a contrasting view see Pratt 1982), stromatolites became progressively less abundant and less diverse (Fig. 6). (A longer-term decline in "density" of stromatolites, a better measure of abundance, has been postulated but not documented by Grotzinger 1990). Whatever the cause or causes of this decline might be, it seems to have been most apparent at first in the quiet subtidal realm that was often inhabited by the conically laminated stromatolite *Cono-*

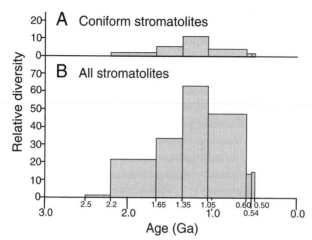

FIGURE 6   A: The relative diversity through time of the subtidal stromatolite *Conophyton* and its relatives. B: The diversity through time of all stromatolites. See Walter & Heys 1985 for data and details of the statistical treatment. The original data have been recalculated to allow for the most recent understanding of the ages of the base of the Cambrian (540 Ma) and the base of the Vendian (600 Ma).

phyton and its relatives and to have spread much later to the peritidal realm. Awramik (1971) and Walter & Heys (1985) considered a number of possible explanations and concluded that grazing and burrowing of the mats by the earliest metazoans may have been the cause. Grotzinger (1990) pointed out that the most recent understanding of metazoan history disputes all records older than the base of the Vendian, about 600 Ma ago, and therefore calls into question any postulated grazing and burrowing before that time. He suggested that the decline is at least partly the consequence of decreasing saturation of the oceans with carbonate and therefore less stabilization of microbial mats by precipitated carbonate. Subsequently, metazoans that are possibly older have been reported by Hofmann *et al.* (1990), and Runnegar (this volume) has interpreted fossil and molecular biological information to suggest that the Metazoa originated more than 1 Ga ago. Valentine (this volume), using other methods, suggests a later origin, some 680–645 Ma ago. A number of authors have suggested that competition with macroalgal benthos led to the decline (e.g., Monty 1973); however, in Shark Bay the stromatolites accrete in harmony with abundant higher algae (Fig. 7). Proterozoic macroalgae are discussed in this volume by Runnegar, Vidal, Hofmann, Sun, and Knoll.

Burrowing of stromatolites is well documented in the early Paleozoic (see references in Walter & Heys 1985), providing direct evidence of the disruption of benthic mats, and it is well known in extant mats. The lack of direct evidence of burrowing in Proterozoic stromatolites parallels the rarity of Proterozoic trace fossils: metazoans were present for at least the last 60 Ma or so (Gehling 1991), and it is not unreasonable to postulate that the meiofauna has a still longer history. The decline has yet to be convincingly explained, but the grazing and burrowing hypothesis is still the simplest and best explanation, in my opinion.

Decreasing carbonate saturation with time seems to be the best explanation of the rarity of microbial tufa (ministromatolites with a radial fibrous fabric) after the Mesoproterozoic (Grey & Thorne 1985; Grotzinger 1990). Yet carbonate precipitation appears to have increased in other mat types during the Cambrian: these are the thrombolites (unlaminated stromatolites with a "clotted" fabric) and the "skeletal"

FIGURE 7   Subtidal stromatolites at Carbla Point, Hamelin Pool, Shark Bay, Western Australia, showing fish and attached macroscopic algae. Exposed portion of vertically oriented spirit level is about 50 cm high.

stromatolites (with microbial fabrics showing calcareous filaments), discussed in this volume by Riding. Until recently, thrombolites were thought to be restricted to the Phanerozoic (Walter & Heys 1985; Kennard & James 1986), but a few Proterozoic examples are now known (Aitken & Narbonne 1989; Kah & Grotzinger 1992). The interpretation that thrombolites formed as a result of the grazing and burrowing of laminated mats (Walter & Heys 1985) has not been generally accepted; detailed petrographic studies (Kennard & James 1986; Pratt & James 1988; Kennard, in press) have tended to indicate that calcification in situ of mats of coccoid cyanobacteria is a more likely explanation, although unequivocal examples of calcified cells have not been found. This "calcification event" is roughly synchronous with that generating skeletons in the Metazoa (Bengtson, this volume). Although calcification of filamentous cyanobacteria and microalgae seems to be most common in the Phanerozoic, Komar (1989) gives many Proterozoic examples of calcified filaments; this same author's examples of calcified coccoid cyanobacteria are less convincing, and many may be simply trapped carbonate peloids.

The subtidal stromatolites of the Paleozoic were parts of complex ecosystems in which metazoans were an integral part. Their reefs and bioherms sheltered trilobites and brachiopods, and were bored and burrowed by many organisms (Playford *et al.* 1976; Kennard & James 1986; Pratt & James 1988; Kennard, in press). Close homologues still live in the subtidal realms of Shark Bay, Western Australia, and in the Bahamas (Golubic 1976; Kennard & James 1986; Riding *et al.* 1991) and have mats not only of

TABLE 2   Census of the protists and animals found in, on, and among the stromatolites of Hamelin Pool, Shark Bay, Western Australia, compiled by members of the former Baas Becking Geobiological Laboratory with the assistance of Warren L. Nicholas and Patrick DeDeckker. The list is certainly not complete, but it serves to illustrate the species-richness of this stromatolite ecosystem, despite the hypersalinity of the waters of Hamelin Pool. Comparable ecosystems must have evolved during the Neoproterozoic and are documented in the early Paleozoic.

	Species	Size range	Food source
Mammals (dugong)	1	~3,000 mm	sea grass
Fish	15	8–1,500 mm	fish, crustaceans, detritus in sand, plankton
Reptiles (sea snake)	1	~1,000 mm	fish
Bivalves	5	2–15 mm	suspended organic matter
Gastropods	5	1–2 mm	cyanobacteria?
Crustaceans	5	0.5–35 mm	plankton and organic detritus
Annelids	2	0.5–5 mm	small animals, suspended organic matter
Coelenterates (medusa)	1	100–300 mm	suspended organic matter
Nematodes	30	<1 mm	bacteria, algae, small animals
Foraminifera	10	<1 mm	bacteria, organic detritus
Sponges	3	5–15 mm	suspended organic matter

Bacteria + Cyanobacteria + Microalgae + Metazoans

Bacteria + Cyanobacteria + Microalgae

Bacteria + Cyanobacteria

Bacteria

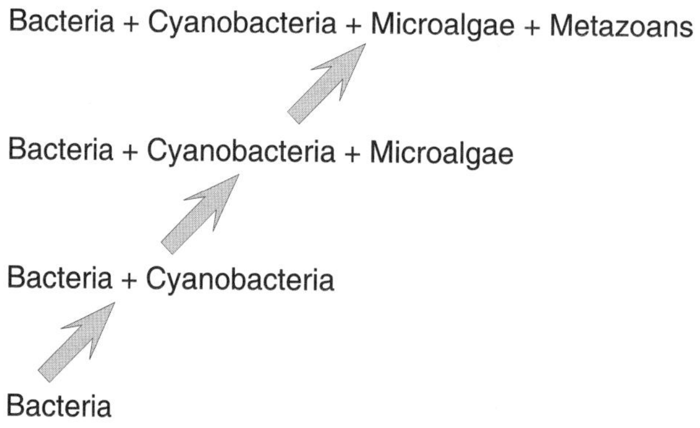

FIGURE 8   The evolutionary succession in the development of stromatolites. The earliest forms may have been constructed by eubacteria. The later addition of cyanobacteria, then microalgae, and finally metazoans to the microbial-mat ecosystem led first to the diversification of stromatolites but ultimately to their decline.

cyanobacteria but also of microalgae of various types. In those places stromatolite reefs and thickets shelter and provide food for sea snakes, fish, gastropods, bivalves, nematodes, and a host of other metazoans and protists (Table 2), and they are the substrate on which many kinds of higher algae ("seaweeds") attach. This ecosystem first became apparent in the Cambrian, but it may well have had a much longer history.

    The evolutionary history of stromatolites is thus to a great extent dependent on the evolution of benthic organisms participating in their construction(Fig. 8), but the details of the interplay between organisms and environment to produce the changes in stromatolite morphology and occurrence that we observe through geological time are still largely conjectural.

Proterozoic eukaryotes: Evidence from biology and geology

Bruce Runnegar

Department of Earth and Space Sciences, Molecular Biology Institute, and Institute of Geophysics and Planetary Physics, University of California, Los Angeles, California 90024-1567, USA

Living eukaryotes are descended from an archaebacterial prokaryote that existed after the origin of methanogenesis (≥2.7 Ga ago) but before the appearance of eukaryotes in the fossil record (~2.1 Ga ago). During this cryptic prehistory, eukaryotes achieved their characteristic architecture (membrane sterols, cytoskeleton, nucleus, Golgi apparatus) and also acquired their endosymbionts (chloroplasts and mitochondria). The oldest fossil thought to be a eukaryote is a probable megascopic alga, *Grypania*, from a 2.1 Ga old banded iron formation in Michigan. Organic-walled microfossils, considered to be the remains of eukaryotic phytoplankters, appeared soon afterwards and then diversified into the late Proterozoic. The close of the Proterozoic is marked by the appearance of the Ediacara fauna and associated trace fossils. There is evidence for the existence of mobile bilateral animals (Bilateria), sessile tube-dwellers (Pogonophora?), foliate "vendobionts," and plausible progenitors of higher metazoans such as arthropods and echinoderms. Trees constructed from ribosomal RNA sequences suggest that most living eukaryotes owe their ancestry to a short-lived radiation that gave rise to the ciliates, red algae, fungi, animals, green algae, and plants. Inadequately dated red algae from the Proterozoic of Canada hint that this radiation may have occurred more than a billion years ago.

Most modern eukaryotes are sizable, complex organisms composed of many different kinds of cells, each containing organelles of endosymbiotic derivation. In addition, eukaroytic cells frequently house prokaryotic or eukaryotic "guests" that form stable or even obligate relationships with their hosts. This remarkable complexity was not achieved instantaneously during eukaryotic evolution, so there is no unique "origin" of the eukaroytic cell. Instead, the characteristics of modern eukaryotes were evolved and assembled over a long period of time, which extended from the late Archean (>2.5 Ga ago) through the Proterozoic to the early Phanerozoic (<550 Ma ago). In this article I review the Proterozoic history of the eukaroytes using evidence from both geology and biology.

Bengtson, S. (ed.) 1994: *Early Life on Earth. Nobel Symposium No. 84.* Columbia U.P., New York

Prehistory of the eukaryotes

Phylogenetic trees constructed from the products of genes that were duplicated before the existence of the last common ancestor of all living organisms have revealed that eukaryotes are more closely related to archaebacteria than they are to eubacteria (Gogarten *et al.* 1989; Iwabe *et al.* 1989). For example, elongation factors are conserved proteins that are involved in protein synthesis in all cells. One kind of elongation factor, EF-1 (EF-Tu or EF1α), assists in the binding of activated tRNAs to ribosomes; another, EF-2 (EF-G or EF-2), is involved in the transfer of amino acids from activated tRNAs to the nascent protein. These two kinds of molecules are homologous (paralogous within organisms), having been produced by a gene duplication in the line leading to the common ancestor of modern life. Phylogenetic trees constructed from unambiguously aligned, conserved parts of EF-1 and EF-2 sequences (about 130 amino acids in length) may therefore be rooted at the gene-duplication event. When this is done, it is seen that the topology of the EF-1 side of the tree is broadly similar to the EF-2 side (Fig. 1A). In particular, the archaebacteria (*Sulfolobus, Thermoplasma, Methanococcus,* and *Halobacterium*) group with the eukaryotes (*Homo* and *Drosophila*) rather than with the eubacteria (*Thermus* and *Escherichia coli*) on both sides of the tree.

Bootstrap resamplings of the data support both the position of the root and the monophyly of the eukaryotes and the eubacteria; the relationships of the principal archaebacterial groups (methanobacteria, halobacteria, and sulfobacteria) to the eukaryotes remain unresolved (Fig. 1A). However, if EF-1 and EF-2 sequences are aligned separately and the conserved regions of each protein are then combined into tandem composites about 600 amino acids in length, it is possible to obtain a more robust solution (Fig. 1B). Furthermore, if only the most distantly related archaebacterial taxa (*Sulfolobus* and *Halobacterium*) are used in the analysis, the resolution of the tree improves (Fig. 1C), revealing some support for Lake's contention that the "eocytes"

FIGURE 1   Phylogenetic trees constructed from conserved parts of the amino acid sequences of elongation-factor proteins EF-1 (EF-Tu or EF1α; aligned by inspection) and EF-2 (EF-G or EF-2; aligned using CLUSTAL V; Higgins *et al.* 1992). After alignment, the amino acid sequences were converted into DNA nucleotides (some ambiguous) by translating the first two positions of each amino acid codon. Maximum parsimony trees were obtained from the translated data using the *branch and bound* option of PAUP 3.0q (D.L. Swofford, Illinois State Natural History Survey); bootstrap values obtained by *heuristic* search (A; 1,000 replicates) or *branch and bound* search (B, C; 1,000 replicatates) are plotted on the trees. Copies of the aligned sequences in the *#nexus* format and the output from PAUP are available by ftp from an anonymous account *igpp.ucla.edu* (use "cd pub/ cseol" to change directories and "get ef.files" to copy files).

   A: Tree based upon highly conserved regions (total length = 128 residues) corresponding to positions 10–111 and 115–140 of the *E. coli* EF-Tu amino acid sequence and to positions 8–33, 36–74, 82–118 and 122–146 of the *E. coli* EF-G sequence. B: Tree based upon conserved regions (total length = 602 residues) corresponding to positions 10–72, 77–111, 115–142, 148–168, 192–251, 256–295, 304–312, 326–339, 351–371, and 374–391 of the *E. coli* EF-Tu amino acid sequence and to positions 9–39, 42–74, 82–162, 278–291, 410–465, 482–490 and 609–676 of the *E. coli* EF-G sequence; each taxon in the tree is represented by a composite, tandem pseudo-sequence. C: Tree based upon the same data set as *B* but with only the most distantly related archaebacteria (*Sulfolobus* and *Halobacterium*) included in the analysis.

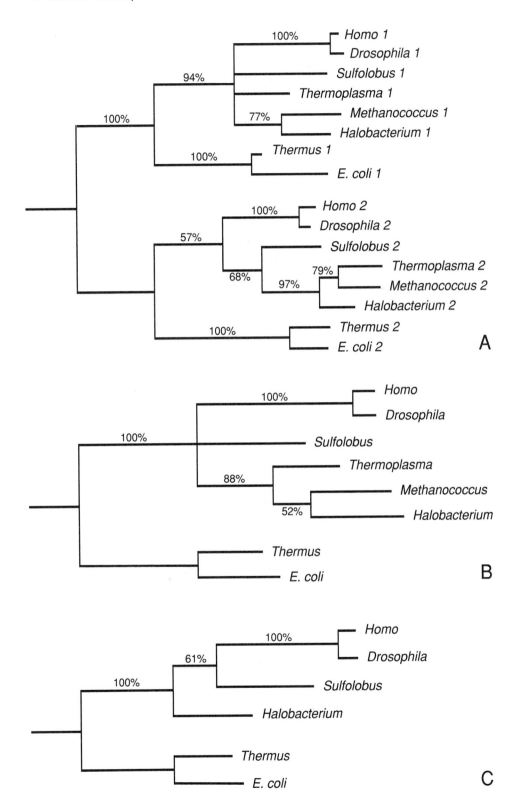

(e.g., *Sulfolobus*) are the closest living prokaryotic relatives of the eukaryotes (Lake 1988). This conclusion is well supported by the presence of a unique eleven-amino-acid insert, which is found in EF-1s from sulfobacteria and eukaryotes but not in EF-2s or any other bacterial EF-1 (Rivera & Lake 1992). The insert is unlikely to have arisen twice by chance and is therefore a synapomorphy of the Sulfobacteria + Eukaryota.

Cavalier-Smith (1987a, b) described a plausible scenario for the origin of the eukaryotic cell from eubacteria that had lost the ability to manufacture murein-based cell walls[1] and therefore needed an internal cytoskeleton plus sterols to stiffen cell membranes. He further suggested that the earliest eukaryotes would have lacked mitochondria and chloroplasts, as these organelles were acquired by endosymbiosis after the divergence of living amitochondriate protists (e.g., *Giardia*) from the line leading to organelle-bearing eukaryotes (Sogin *et al.* 1989; Kabnick & Peattie 1991).

The time of origin of eukaryotes has been estimated using the sizes of organic-walled microfossils (≥1.75, ?2.0 Ga ago [Schopf 1992b]), carbonaceous megafossils (≥1.8 Ga ago [Hofmann & Chen 1981]); biomarker compounds (modified sterols) extracted from unmetamorphosed early Proterozoic rocks (≥1.69 Ga ago [Summons *et al.* 1988; Summons & Walter 1990]); and the molecular clock (1.8±0.4 Ga ago [Doolittle *et al.* 1989]). These estimates are overtaken by the discovery (discussed below) of megascopic fossils resembling *Grypania spiralis* in a 2.1 Ga old banded iron formation in northern Michigan (Han & Runnegar 1992). If the Michigan fossils are correctly interpreted as the remains of eukaryotic algae, the origin of organelle-bearing eukaryotes must have occurred before 2.1 Ga ago. Although this date is ≥300 Ma older than previous estimates, it may also underestimate significantly the times of origin of: (1) the first stem-group eukaryote (latest common ancestor of the monophyletic clade, Eukaryota); (2) the development of eukaryotic organization (membrane sterols, cytoskeleton, nucleus, Golgi apparatus, mitosis, etc.); (3) the first crown-group eukaryote (latest common ancestor of all living eukaryotes); and (4) the endosymbiotic conversion of purple bacteria and cyanobacteria into mitochondria and chloroplasts in an early eukaryote (Cavalier-Smith 1987b). Each of these events took place in the order given above before about 2.1 Ga ago (Fig. 2).

There is limited geological evidence that points to the existence of stem-group eukaryotes in the late Archean. By definition, the first stem-group eukaryote was either the closest coeval relative of the ancestral archaebacterium, if the Archaebacteria is a monophyletic group (Woese *et al.* 1990), or the sister species of the first sulfobacterium, if Lake's "eocytes" are the closest prokaryotic relatives of the eukaryotes. Although the second alternative is probably correct (Fig. 1), the long-standing difficulty of resolving the topology of the node(s) near the base of the archaebacterial branch of the universal tree is an indication that the principal groups of archaebacteria and the stem-group eukaryotes originated at approximately the same time.

Exceptionally light carbon isotope ratios ($\delta^{13}C \leq -40\%_{PDB}$) obtained from organic matter found in late Archean and early Proterozoic rocks (Schidlowski 1988) have been interpreted as evidence for the existence, at those times, of eubacterial methylotrophs that were using methane produced by archaebacterial methanogens (Hayes 1983).

---

[1] Murein is a polydisaccharide, similar to chitin, that is covalently cross-linked by short peptides to form an inextensible net-like bag around the cell.

Thus the original stem-group eukaryote may have lived before 2.7–2.5 Ga ago (Fig. 2). It follows that the divergence between eukaryotes and eubacteria may be much deeper than the ~1.8 Ga date obtained by extrapolating rates of protein evolution (Doolittle *et al.* 1989).

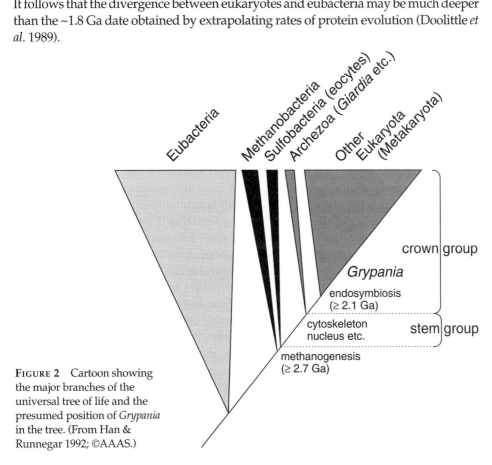

FIGURE 2 Cartoon showing the major branches of the universal tree of life and the presumed position of *Grypania* in the tree. (From Han & Runnegar 1992; ©AAAS.)

The fossil record of Proterozoic eukaryotes

Many different kinds of unicellular and multicellular eukaryotes have been reported from rocks of Proterozoic age. Apart from pseudofossils, misidentified prokaryotes, and unplaced problematica, they include unicellular phytoplankters (acritarchs), microscopic red algae (Butterfield *et al.* 1990), complex protists (Bloeser 1985; Allison & Hilgert 1986[2]), megascopic algae (Butterfield *et al.* 1988; Zhang 1989), calcareous algae (Horodyski & Mankiewicz 1990; Grant *et al.* 1991), *Cloudina* (Conway Morris *et al.* 1990; Grant 1990; Bengtson & Yue 1992), various kinds of organic-walled, annulated tubes (Sokolov 1967; Peat 1984; Sun *et al.* 1986; Chen 1988), trace fossils, and members of the celebrated Ediacara fauna. Only the most informative of these fossils are discussed below.

---

[2] Allison & Hilgert (1986) considered the beds in which these fossils are found to be Early Cambrian in age, but Kaufman *et al.* (1992) have presented evidence that indicates that they are as old as late Riphean (780–620 Ma).

Microscopic algae

Because the size of prokaryotic cells may be limited by the life of their messenger RNAs (Demoulin & Janssen 1981), few living coccoid prokaryotes are larger than about 60 μm in diameter. Consequently, spheroidal organic-walled microfossils (acritarchs) larger than this are considered to be the walls of the vegetative or encystment stages of planktic, eukaryotic algae (Vidal 1984; Schopf 1992b; Mendelson *et al.* 1992).

Early and middle Proterozoic acritarchs are relatively featureless fossils and are therefore difficult to categorize (Peat *et al.* 1978; Horodyski 1980; Zhang 1986). More complex, spinose forms are found in late Proterozoic strata (≤800 Ma old [Butterfield *et al.* 1988; Zang & Walter 1989]), perhaps presaging a major radiation of acanthomorphic and other acritarchs that occurred at the beginning of the Cambrian (Knoll & Swett 1987; Moczydłowska 1991).

Megascopic algae

The recent discovery of fossils resembling *Grypania spiralis* in the 2.1 Ga old Negaunee Iron Formation of Michigan extends the stratigraphic range of *Grypania* by some 700–1,000 Ma from the late middle Proterozoic to the middle early Proterozoic (Han & Runnegar 1992). *Grypania* was a coiled, cylindrical organism that grew to maximum size of about half a meter in length and 2 mm in diameter (Walter *et al.* 1990; Runnegar 1991). It is normally preserved as unbranched, ribbon-like films or impressions on bedding planes, but a unique specimen from India is uncompacted, showing that *Grypania* was originally circular in cross section (Beer 1919; Mathur 1983). Rarely seen terminations are rounded; well-preserved specimens have broad transverse markings. The corkscrew shape of *Grypania* appears to have been due to supercoiling, which was maintained in life and death by helical filaments within the body wall. As a result, *Grypania* is almost always preserved as a compressed coil, sinuous ribbon, or cuspate ribbon. Walter *et al.* (1990) considered the cuspate ribbons to be specimens that had been stretched by currents; it was therefore thought that *Grypania* might have been tethered during life.

Grypania has no certain living relatives but is regarded as a probable eukaryotic alga because of its complexity, structural rigidity, and large size. Spiral-shaped fossil cyanobacteria known as *Obruchevella* and *Spirellus* are widespread in late Proterozoic and early Cambrian strata, but even the exceptionally large Cambrian forms (*Spirellus*) are an order of magnitude smaller than *Grypania*. Similarly, it is unlikely that *Grypania* was the abandoned giant sheath of a sulfide-oxidizing bacterium or an aggregate of cyanobacterial filaments, as has been suggested for late Proterozoic carbonaceous megafossils known as vendotaenids; bacterial sheaths and aggregates of filaments are narrower than Chinese and Indian specimens of *Grypania*, and they lack rounded ends, a coiled morphology, and transverse structures.

The best modern analogue for *Grypania* may be the giant unicellular dasycladacean alga *Acetabularia* (Runnegar 1991). Before the formation of an umbrella-shaped reproductive cap, *Acetabularia* grows as a narrow cylinder about 0.4 mm in diameter and up to 180 mm in length. The single nucleus remains within a holdfast; there is a large, central, sap-filled vacuole that occupies most of the stalk of the alga so that the cytoplasm, which contains numerous chloroplasts and mitochondria, is restricted to the periphery. Although *Acetabularia* is uninucleate, its close relatives have a coenocytic organization, as might have *Grypania*. Few uninucleate organisms have achieved the cytoplasmic volume (~1.5 ml) of Indian specimens of *Grypania*.

Grypania and the Proterozoic acritarchs appear to have been photosynthetic auto-trophs. It is conceivable that they depended upon facultative photosymbionts rather than upon chloroplasts, but there is independent evidence that suggests that the endosymbiotic origin of green choroloplasts occurred before about 2 Ga ago: rocks of that age contain well-preserved cyanobacteria that can be placed in modern families (Hofmann 1976; Walter & Awramik 1979); and trees derived from 16S rRNA sequences suggest that chloroplasts originated during the radiation that produced the modern cyanobacterial groups (Giovannoni *et al.* 1988).

Ediacara fauna and other Vendian metazoans

Long regarded as the earliest evidence for animal life (Cloud & Glaessner 1982; Glaessner 1984), the fossils of the latest Proterozoic (Vendian) Ediacara fauna have come under increasingly close scrutiny as the result of Seilacher's suggestion that the assemblage represents an extinct line of animal life (the Vendozoa) that pioneered a body form and modes of nutrition not seen in the Phanerozoic (Seilacher 1989). More recently, Seilacher and others have suggested that most typical members of the Ediacara fauna were not even animals but instead comprised an extinct group of foliate organisms, the Vendobionta, analogous to algae or fungi, which relied on endogenous photosymbionts or chemosymbionts for food and energy (see Seilacher, this volume).

It would not be surprising if the Ediacaran organisms contained bacterial or eukary-otic endosymbionts, as endosymbiosis at the organelle or unicell level is the norm rather than the exception for many protists and diploblastic metazoans. However, it is difficult to find the evidence to test this interesting hypothesis, because most Ediacaran fossils are merely impressions of soft bodies in sandstones or siltstones (Runnegar 1992).

As there are no clear modern relatives of most of the Ediacaran organisms, it is necessary to resort to logic and experiment to understand them. This has proved to be a formidable task, but some progress is being made with complex taxa such as *Dickinsonia* and *Phyllozoon* (Fig. 3). Although no core member of the Ediacara fauna is unquestionably an early animal, there are more similarities to animals than to algae or fungi (Gehling 1991; Runnegar 1992). There is no evidence at all that the vendozoans are a fourth form of eukaryotic multicellular life.

The best evidence for the existence of late Proterozoic animals comes from trace fossils found with the Ediacara fauna. Looping or spiraling surface trails up to several millimeters in width and strings of fecal pellets point to the presence of soft-bodied bilateria with a well-developed nervous system, anterior–posterior asymmetry, and a one-way gut (Runnegar 1991, 1992). Another kind of trace fossil consists of closely spaced meanders that seem to have been formed by small animals that periodically reversed their direction of motion while grazing the microbial mats, which apparently covered late Proterozoic seafloors in shallow, well-lit environments. Pairs of scratches arranged in fan-shaped sets (Gehling 1991) may be the foraging marks of animals with legs or grasping organs.

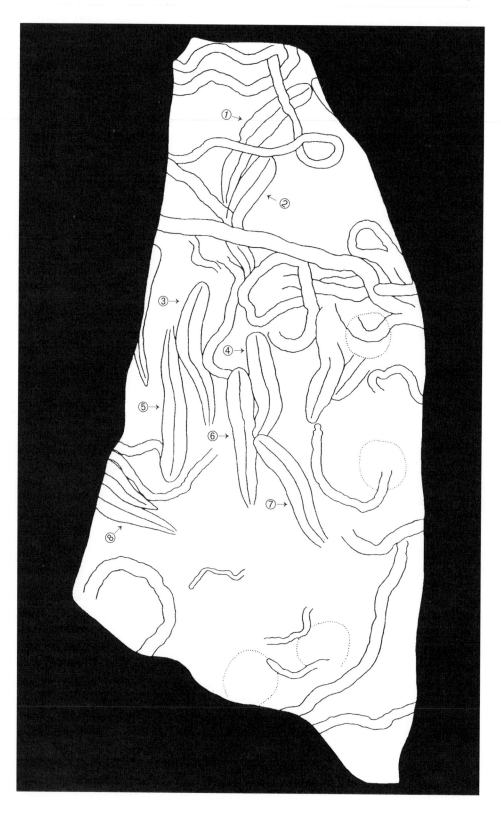

Some bed surfaces are covered with the sandstone casts of unbranched tubes, closed at one end, which may reach 3 cm in width and more than half a meter in length (Fig. 3). Although these fossils are featureless and therefore difficult to interpret, they resemble the tubes of the vestimentiferan pogonophoran *Riftia pachyptila* in size and shape (Jones 1985a). In this context, it is instructive to recall that Sokolov, in consultation with A.V. Ivanov, placed annulated, organic-walled tubes (*Sabellidites, Paleolina, Saarina,* etc.) found in Vendian and Cambrian strata of the Russian Platform into an extinct order of the Pogonophora (Sabelliditida; Sokolov 1967). The cone-in-cone construction of *Saarina* and even of *Cloudina* (Conway Morris *et al.* 1990; Grant 1990; Bengtson & Yue 1992) is seen in many modern pogonophoran tubes (Ivanov 1963).

The triradiate symmetry of several distinctive members of the Ediacara fauna (*Tribrachidium, Skinnera, Rugoconites, Albumares, Anfesta*) is a unifying feature that sets them apart from living metazoans. Furthermore, these were not "foliate pneu constructions" as suggested by Seilacher (1989), nor coelenterate medusae (Fedonkin 1985c), but instead were complex, conical organisms that have been flattened by burial (Gehling 1991). Whether they were related to *Anabarites*, an Early Cambrian trilobed shell (Conway Morris & Chen 1989), or to stem-group echinoderms is unknown.

Although disk-shaped impressions called "medusoids" have normally been interpreted as the remains of cnidarian jellyfish, this now seems unlikely. Most appear to be the casts of sessile, sack-like bodies or the holdfasts of fronds; and radiating structures, previously considered to be "tentacles," may be better thought of as "roots" (Runnegar 1992). Seilacher's (1990, this volume) suggestion that some of these discoidal fossils are the sandy endoskeletons of an extinct group of bottom-dwelling sea anemones (Psammocorallia) is invalidated by the fact that the grain size of the "endoskeletons" is invariably the same as the grain size of the bed that buried the organisms.

That leaves two principal groups of Ediacaran organisms: (1) core members of the Ediacara fauna or Seilacher's Vendobionta – creatures that are sizable, foliate, and composed of "segments" or "modules" arranged in a serial or fractal fashion (*Charnia, Charniodiscus, Dickinsonia, Ernietta, Phyllozoon, Pteridinium, Rangea,* etc.); and (2) smaller, bilaterally symmetrical forms having anterior–posterior asymmetry (*Marywadea, Onega, Parvancorina, Spriggina,* etc.). Whether each of these groups represents a monophyletic clade or an artifical association of unrelated organisms is unclear, but it is likely that the former were sessile, passive collectors of photons, reduced compounds, free amino acids, or particulate food. *Spriggina,* on the other hand, may well have been a mobile scavenger, perhaps an unarmored forerunner of Cambrian sclerite-bearing (cataphract) metazoans or an ancestor of the trilobites.

At present, *Spriggina* is the star of a debate over the nature and affinities of the Ediacara fauna. Its affinities are crucial for an understanding of the time of origin of segmentation in metazoans (Valentine 1989; Jacobs 1990; Saint 1990). It is frustrating

---

FIGURE 3   Sketch of lower surface of sandstone bed covered with impressions of *Phyllozoon hanseni* (better specimens are numbered 1 to 8) and an unnamed tubular body fossil; dashed circles represent poorly preserved specimens of *Dickinsonia costata*. The slab is 1.2 m in length; it was collected by J.G. Gehling and the author from the Ediacara Member, Rawnsley Quartzite, central Flinders Ranges, South Australia. A smaller piece of the same bed is illustrated by Gehling (1991, Pl. 3:2).

that no single piece of evidence is yet able to confirm *Spriggina*'s place in the animal world.

Lastly, it should be noted that *Arkarua adami* (Gehling 1987) has been accepted, with some reservations, by Jefferies (1990) and Smith & Jell (1990) as a crown-group echinoderm. If this were true, it would indicate that the chordate–echinoderm split occurred well before the Cambrian and that early echinoderms were tiny and only slightly mineralized. It also resurrects the possibility that *Tribrachidium* and its allies might have been stem-group echinoderms, for Jefferies (1990) postulated that the five-fold symmetry of crown-group echinoderms was derived from a triadiate organization in the stem lineage.

The Proterozoic radiation of the eukaryotes

Protein- and rRNA-sequence comparisons are being used to tease apart the order in which the numerous groups of living eukaryotes appeared during the Proterozoic (e.g., Sogin *et al.* 1989; Sogin, this volume). Unfortunately, the data are ambiguous because several major clades, including the animals, plants, and fungi, originated at much the same time. It is therefore unknown whether plants (Gouy & Li 1989a), fungi (Bhattacharya *et al.* 1990; Wainwright *et al.* 1993), or plants + fungi (Hendriks *et al.* 1991) constitute the sister group of animals, although the second alternative seems most likely. Nevertheless, the main structure of the radiation of the eukaryotes is becoming clear as old misconceptions are laid to rest: the small-subunit rRNA tree has a series of early branches leading to the diplomonads, microsporidians, euglenoids + kineto-plastids, amoebae, and the slime molds; higher up, a nearly polychotomous set of branches leads to the metazoans, the red algae, the sporozoans, the higher fungi, the ciliates, the green plants, and several minor protistan groups (Hendriks *et al.* 1991; Sogin, this volume).

A minimum date for this radiation of the higher eukaryotes is given by the discovery of a well-preserved, multicellular red alga regrettably from poorly dated, 1.3–0.7 Ga old rocks – in Arctic Canada (Butterfield *et al.* 1990). If the preferred age of 1.2–1.1 Ga is approximately correct, and if multicellularity in rhodophytes is a derived condition, the lineages leading to red algae, fungi, metaphytes, and metazoans might well have diverged from each other more than a billion years ago.

If *Grypania* and the oldest acritarchs were photosynthetic eukaryotes, where do they lie within the rRNA tree? There is no ready answer to this question because of uncertainties that are still under investigation. However, there is mounting evidence (summarized by Palmer 1993) that all modern algal and plant chloroplasts have a monophyletic origin from a free-living cyanobacterium. The great diversity of pigment types seen in modern chloroplasts and their haphazard taxonomic distribution is therefore attributed to subsequent evolution coupled with occasional lateral transfer of the whole organelle (Cavalier-Smith 1992). Furthermore, the origin of green chloro-plasts is inseparable from the radiation that gave rise to the major groups of living cyanobacteria (Giovannoni *et al.*, 1988). As cyanobacteria were well differentiated by about 2 Ga ago (e.g., Hofmann 1976), it follows that chloroplast-containing eukaryotic cells existed at or before that time. It is therefore likely that *Grypania* and the early acritarchs were representatives of photosynthetic eukaryotes that predated the relatively late radiation of the modern algal groups (Sogin, this volume). In contrast to the molecular evidence, the fossil record strongly suggests that a photosynthetic lineage formed the Proterozoic trunk of the eukaryotic tree.

Conclusions

Living eukaryotes are derived from an archaebacterial prokaryote that existed after the origin of methanogenesis (≥ 2.7 Ga ago) but before the appearance of eukaryotes in the fossil record (~2.1 Ga ago). Before their appearance in the fossil record, eukaryotes had acquired their cytological characteristics and their organelles (chloroplasts and mitochondria). The oldest fossil thought to be a eukaryote is a probable megascopic alga (*Grypania*) from 2.1 Ga old strata in Michigan. Organic-walled microfossils, considered to be the remains of eukaryotic phytoplankters, appeared soon afterwards.

The end of the Proterozoic is marked by the appearance of the Ediacara fauna and associated trace fossils. There is evidence for the existence of mobile bilateral animals, sessile tube-dwellers, foliate "vendobionts," and possible ancestors of higher metazoans. Trees constructed from ribosomal RNA sequences suggest that most living eukaryotes owe their ancestry to a short-lived radiation that gave rise to the ciliates, red algae, fungi, animals, green algae, and plants. Inadequately dated red algae from the Proterozoic of Canada hint that this radiation may have occurred more than a billion years ago, long after photosynthetic eukaryotes first appeared in the fossil record.

Acknowledgments. – I thank James A. Lake for assistance with the analysis of the EF sequences. Parts of this research were supported by the National Science Foundation (grant no. EAR 9004601).

Early ecosystems: Limitations imposed by the fossil record

Gonzalo Vidal

*Department of Micropaleontology, Institute of Earth Sciences,
Norbyvägen 22, S-752 36 Uppsala, Sweden*

The end of the Proterozoic Eon, some 550 Ma ago, is marked by the abrupt appearance of mineralized remains belonging to problematic protoctists, calcareous thallophytes, and skeletal marine invertebrates. Mounting evidence indicates that controlled biomineralization may have evolved among marine planktic protoctists as early as ~800 Ma ago. Skeletal constructions produced by marine invertebrates are abundant in the late Neoproterozoic fossil record, some 600 Ma ago. The nature of the pre-Phanerozoic fossil record is biased by numerous factors that contribute to its substantial incompleteness. However, despite major gaps, the increasing level of knowledge attained permits depicting a crude sketch of early life embracing about 3 Ga of Earth's history. For the most part, the record of early life consists of extinct forms of pre-metazoan organization. Evolutionary patterns were largely based on the interaction of prokaryotes and, at a later stage, also eukaryotic protists. Evidence from the fossil record and considerations from molecular biology support the basic proposition that the emergence and evolution of metazoan phyla are intrinsically linked to the origins and patterns of change in protist clades. The fossil record reveals radiation events among photosynthetic planktic protists. An important radiation occurred close in time to the Cambrian radiation event of modern invertebrate phyla. Interpretations of fossil evidence suggest that the thallophytes were extant about 1.7 Ga ago. If correct, these interpretations imply that relatively complex interactions between single-celled prokaryotic and eukaryotic primary producers and consumers may have existed for about 1 Ga prior to the addition of vendobiont and metazoan levels of organization in Neoproterozoic and early Cambrian times.

The contrasting nature of the Phanerozoic and pre-Phanerozoic biotic records was identified at a very early stage in the evolution of the comparatively young disciplines of palaeontology and geology (Darwin 1859). Hence, the beginning of the Phanerozoic Eon is marked by the radiation of coelenterate, acoelomate, procoelomate (Bergström 1989), and coelomate metazoans. In the fossil record, this radiation is signaled by the abrupt appearance of mineralized skeletal remains of problematic protozoans and marine invertebrates representing all modern skeletalized phyla, with the exception of the Bryozoa.

The main purpose of this chapter is to elucidate the nature of the marine ecosystem on which the Phanerozoic radiation was superposed some 550 Ma ago. This involves examining the available fossil record of protist life forms during ~3 Ga of Proterozoic

Bengtson, S. (ed.) 1994: *Early Life on Earth. Nobel Symposium No. 84.* Columbia U.P., New York

and Archean history. Furthermore, questions must be posed concerning the biological and evolutionary significance of these early microbial ecosystems. Answering these questions demands cautious evaluation of the nature and limitations of the paleontological record of early Cambrian, Proterozoic, and Archean life. In so doing, it must be kept in mind that this can be accomplished only through prudent use of actuobiological and modern environmental inferences to interpret early forms of life that interacted under environmental conditions differing substantially from those existing today.

The nature of the paleontological record of early life

Generally, the fossil record of any group of organisms represents a collection of data that are qualitatively and quantitatively strongly limited by a conjunction of conditions, among which habit, habitat, and depositional and burial conditions certainly are but only a few to be considered. In fact, preservational limitations are familiar to paleontologists; perhaps to the level of being quite often forgotten. Keeping this in mind, absence or underrepresentation in the known fossil record could depend on low or non-existing preservational potential. Intrinsically, the completeness of the fossil record is skewed towards the preservation of robust and stable materials and quite often limited to mineralized structures. This limitation is well illustrated by restricted and widely known examples of exceptional occurrences of (again) exceptionally preserved marine invertebrates (e.g., Conway Morris 1989b). Unusual preservation also applies to fossil protists, a feature that is here illustrated with three examples from the fossil record.

The fossil record of plant protists such as dinoflagellates relates to taxa that produced cyst structures made of the extremely stable organic polymer sporopollenin. This selectiveness is important, not only because not all species produce "geologically" preservable cysts, but also because groups of fossil dinoflagellates that in the past have produced preservable cysts do not produce comparable cysts today. One example is the group of ceratioid dinoflagellates, in which there is a 70 Ma gap in the fossil record. It may also seem puzzling that microfossils regarded as bona fide dinocysts occur in late Silurian (~410 Ma) sediments (Evitt 1985), thus implying an approximately 200 Ma absence of dinoflagellates from the fossil record (see also p. 307).

Acritarchs, microfossils constituting possible endocysts and motile life stages of planktic algae, are abundant in marine deposits over the whole span of the Mesoproterozoic, Neoproterozoic, and Paleozoic eons. As even the cysts of some extant dinoflagellate taxa fail to display all the critical morphological criteria of dinoflagellates (Evitt 1985), they would most certainly be regarded as acritarchs if found in the fossil record. In this light, certain early Paleozoic acritarchs with clear operculate excystment openings have been considered as possible (even quite probable) dinoflagellates (Evitt 1985).

The fossil record of prokaryotic microbial organisms could serve as a second illustrative example. Thus, the early fossil record of various groups of cyanobacteria (Fig. 2A–C) appears exceedingly ancient (Fig. 1), although the biogenicity of the oldest putative microfossils (Schopf & Walter 1983) remains questionable. Despite being also bound to uncertainty, stromatolites (Walter 1983) and stable isotopic data (Hayes 1983)

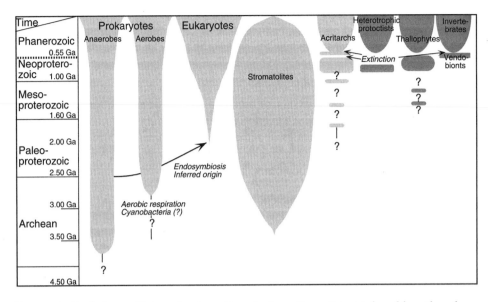

FIGURE 1 Evolutionary history of prokaryotic and eukaryotic protists as inferred from the paleontological record and inferences from nucleotide sequencing (left). The record of stromatolites indirectly implies the Proterozoic expansion of laminated microbial-mat communities (mostly prokaryotic bacteria and cyanobacteria). The fragmentary and discontinuous fossil record of acritarchs (eukaryotic and perhaps also prokaryotic autotrophic aerobes), heterotrophic protoctists, and thallophytes is indicated by dotted areas and question marks. Two inferred terminal Neoproterozoic extinction events are indicated prior to the great thallophyte–metazoan radiation (right). Vendobiont–Metazoan relationships are obscure, and their times of origination (shown as first occurrences in the fossil record) coincide with the patterns of radiation of primary producers at the turn of the Proterozoic (age uncertain). (Time scale modified after Harland *et al.* 1990.)

seem to suggest that photosynthetic cyanobacteria may have formed part of laminated-mat communities in Archean times.

Organic matter present in sedimentary rocks, kerogen, consists of geopolymers, sometimes including microscopically and biochemically recognizable components. Kerogens from all of the Proterozoic yield triterpane and acyclic isoprenoid alkane biomarkers that are consistent with the presence of eubacteria and archaebacteria (Hayes *et al.* 1992a; see also chapters by Hayes and Ourisson, this volume).

Physiological inferences from morphological and actuobiological interpretations are instrumental in establishing the Proterozoic fossil record of cyanobacteria. It appears that cyanobacteria were abundant, and they included many groups that are morphologically, ecologically, and physiologically similar to extant taxa (Golubic & Hofmann 1976). In fact, it seems a paradox that the fossil record of cyanobacteria appears far better illustrated for the Proterozoic Eon than for later periods of Earth's history. Undoubtedly, this feature is an artifact induced by the nature of the fossil record, because of the much wider paleoenvironmental distribution of laminated bacterial biocenoses (as indicated by the distribution of stromatolitic carbonates) during the Proterozoic as compared to later periods (Awramik 1971). The incoming of marine invertebrates considerably reduced the preservation potential of delicate single-celled protists, when grazing (Awramik 1971), bioturbation, and sediment reorganization became common in numerous environments. An additional and probably important factor is that the general absence of other fossils in the Protero-zoic may have directed the attention of paleontologists towards the study of the conspicuous stromatolites.

Early diagenetic silicification in hypersaline environments (Strother *et al.* 1983) constitutes a major portion of the pre-Phanerozoic cyanobacterial record. While being most common in hypersaline environments, feeble cyanobacterial biotas are also well represented in shallow-marine fine-grained detrital deposits (Jankauskas 1989). Any attempt at comparing various microbial assemblages of different Proterozoic ages demands a complete or nearly complete representation of environmental gradients, or at least assuring certainty that the biotic assemblages occupied the same environments (or set of environments). However, in the continuously changing narrow environmen-tal gradients that surround benthic microbial-mat communities, the identification of evolutionary innovation may prove close to impossible. This is probably one reason why even apparently moderately diverse Proterozoic assemblages seem so similar to extant hypersaline mat communities (Strother *et al.* 1983).

Additional examples of unusual preservation under specific taphonomic conditions can be found in the Proterozoic fossil record. Weakly mineralized scale-bearing protists (Allison & Hilgert 1986) and organic-walled thallophyte fossils such as *Tawuia*, (Fig. 2E; an ~800 Ma old problematic thallophyte [Hofmann 1985a]), *Thallophyca*, and *Wengania* (probably rhodophycean thallophytes from the Neoproterozoic ~680 Ma Doushantuo Formation in southern China [Zhang 1989]), and possible cladophoralean green algae (Neoproterozoic Svanbergfjellet Formation in Spitsbergen [Butterfield *et al.* 1988]) are rare, but exceptional preservational circumstances and burial conditions seem to have concurred often enough to allow the preservation of extremely delicate organisms (Butterfield 1990a).

The comparable complexity of the earliest Cambrian life record suggests that certain groups of marine protists and invertebrates may have a much earlier ancestry. For example, the early Cambrian protist record includes representatives of chlorophycean and dasycladacean green algae and agglutinating foraminifers apparently not present even in slightly older Neoproterozoic rocks. Is this absence comparable to the absence of ceratioid dinoflagellates in modern sediments? Despite its obvious incompleteness and allowing for rare occurrences, the record of early life preceding the Cambrian radiation has witnessed a nearly exponential growth during the last three decades. Nevertheless, vast gaps remain unfilled and are likely to remain so because limited preservational potential appears to shape the paleontological record of many impor-tant biotic groups.

Perspectives beyond the fossil record

Sheer logic suggests that the emergence and evolution of metazoan phyla must be viewed as intrinsically linked to the origins and patterns of change in protist clades (Fig. 1). This basic assumption finds much support in 18S rRNA (Woese 1987b), 28S rRNA (Gouy & Li 1989b; Perasso *et al.* 1989) distance-matrix, and cladistic trees. In particular,

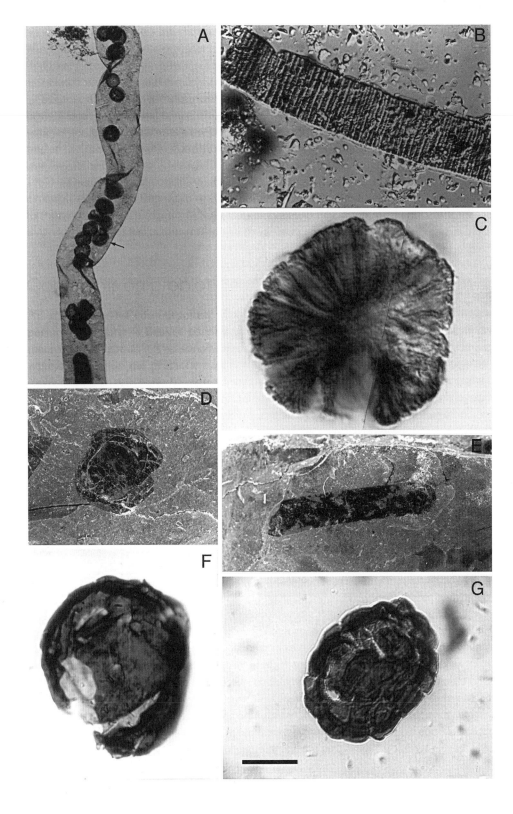

nucleotide sequencing of the large-subunit rRNA molecule seems to suggest that the three main classes of photosynthetic protists radiated during a period of intense diversification close to the metaphyte radiation event (Perasso *et al.* 1989). Moreover, group II introns (common in genes of land plants that emerged ~500–400 Ma ago) in tRNA of members of charophycean green algae were probably acquired by the common ancestor of charophyceans and land plants more than 500 Ma ago (Menhart & Palmer 1990). While limited, the evidence derived from the fossil record seems nevertheless in agreement with molecular data in that the thallophytes, including some problematic forms, may have originated at least 1,700 Ma ago (Walter *et al.* 1990; see also chapters by Runnegar and Hofmann, this volume). Notwithstanding, although the algal affinity of some Proterozoic fossils remains questionable (Vidal 1989; Walter *et al.* 1990), a number of records suggest that metaphytic green and red algae (Hofmann 1985a; Butterfield *et al.* 1988; Zhang 1989; chapters by Hofmann and Sun, this volume) were extant in Neoproterozoic and perhaps in Mesoproterozoic times (Fig. 1).

Neoproterozoic calcified remains of possible rhodophyceans (Grant *et al.* 1991) and problematic cyanobacteria or algae (Horodyski & Mankiewicz 1990) appear to fulfill inferences from the molecular biological record (Perasso *et al.* 1989).

Molecular data seem to have contributed to satisfactorily establishing the branching of the two main divisions of the archaebacteria and eubacteria (Woese *et al.* 1990; Cavalier-Smith 1989a). However, the timing for the major branching of the eukaryota and bacteria is poorly constrained. But once more, sheer logic suggests that bacteria, including at least anaerobic chemoautotrophs, photoautotrophs, chemoheterotrophs, and possibly aerobic photoautotrophs, were extant in Archean times. These levels of biotic organization formed the fundaments of microbial ecosystems preceding the convincingly postulated advent of endosymbiotic plastid and mitochondrial organelles of eukaryotes (Fig. 1; Margulis 1970).

Because of poor fossilization potential and taphonomic factors, the fossil record of bacteria is nearly nonexistent, and we are largely confined to indirect evidence. Hence, the existence of bacteria in the Archean is inferred from the extremely light isotopic

---

FIGURE 2   Proterozoic organic-walled microfossils. **A–B:** Filamentous microfossils interpreted as sheaths of probably oscillatoriacean cyanobacteria (*Palaeolyngbya sphaerocephala* and *Striatella coriacea*; scale bar 33 µm for **A** and 80 µm for **B**). Note remains of trichomal cells (arrow). Specimen at **A** from Neoproterozoic (Upper Riphean) Miroedikha Formation (Turukhansk; Siberia). Specimen at **B** from late Neoproterozoic (Upper Vendian) Redkino Formation (East European Platform; The Ukraine). **C:** Spheroidal microfossil interpreted as mucilagous envelope of a chroococcalean cyanobacterium (*Palaeogomphosphaera* sp.; scale bar 10 µm). From Neoproterozoic (Upper Riphean) Visingsö Group (Sweden). **D:** Compressed spheroidal microfossil (*Chuaria circularis*) interpreted as possible phycoma of prasinophycean green algae (scale bar 1.3 mm). From Neoproterozoic (Upper Riphean) Debengdin Formation (eastern Siberia, Yakutia). **E:** "Sausage"-shaped fossil (*Tawuia*) interpreted as a problematic thallophyte of unknown taxonomic affinity (scale bar 3.3 mm). From Neoproterozoic (Upper Riphean) Debengdin Formation (eastern Siberia, Yakutia). **F:** Distorted spheroidal acritarch interpreted as possible endocyst of eukaryotic phytoplankton (scale bar 10 µm). From the Mesoproterozoic (Lower Riphean) in the Ural Mountains (Barshkiria; Russia). **G:** Slightly flattened spheroidal acritarch; probably a chlorophycean coenobial cell aggregate (scale bar 16 µm). From Neoproterozoic Thule Group (northern Greenland). **A–C, F–G** are interference contrast micrographs. Micrographs at **A** and **B** courtesy of E. Aseeva and T.N. German, respectively.

composition of organic carbon in sedimentary rocks. This probably implies the presence of methanogenic and methylotrophic bacteria and the conceivable coexistence of oxygenic photosynthesis, most likely effected by cyanobacteria (Hayes 1983). Thus the suggested existence of early Archean oxygen levels around 1–2% of PAL (Towe 1990) could have supported aerobic respiration. The much more complex Proterozoic microbial ecosystems were obviously sustained by photosynthetic and chemosynthetic autotrophs and require the recycling of organic carbon and nutrients. This latter feature is particularly significant to the quest of evaluating the evolutionary levels of the ancient microbial ecosystems. Nitrogen compounds are essential to biological processes. There are several reasons why the process of biological nitrogen fixation is exclusive to prokaryotic protists, one being that the fixation of one molecule of nitrogen requires 6–18 molecules of ATP, a process that in the cyanobacteria is undertaken by specialized cells. One may thus argue that compelling evidence suggesting widespread occurrence of cyanobacteria would imply a completely developed system of nutrient recycling leading to propitious conditions for nutrient-dependent eukaryotic protoctists incapable of nitrogen fixation.

Forerunners of animal life

Neontological functional considerations of early Cambrian metazoans and their feeding burrows and trails (ichnofossils; see chapters by Fedonkin and Seilacher, this volume) suggest that infaunal or epifaunal suspension feeders and benthic deposit feeders or grazers were extant.

The sudden rise of diverse biotas with an advanced level of differentiation has come to be known as the Cambrian radiation event (Fig. 1). This radiation of marine invertebrates occurred in the ambit of well-developed ecosystems including primary producing microbes, thalloid algae, and heterotrophs that evolved during the Proterozoic and Archean eons (Fig. 1).

The causes of the early Phanerozoic biotic radiation remain controversial, and the understanding of its total biotic implications may only be obtained from the perspective of the extended evolution of Proterozoic and Archean protist life forms.

A sharp rise in the utilization of organic matter near the water–sediment interface was inferred from the simultaneous appearance of shelly remains (Bengtson 1989) and innovative feeding strategies revealed by ichnofossils (Bergström 1990). From a different perspective, numerous observations indicate that the incoming of the earliest Cambrian faunas is paralleled by major changes among photosynthetic primary producers (acritarchs [Moczydłowska 1991]).

Early metazoan–protist interaction

Viewed either as forerunners of extant metazoan phyla (Fedonkin 1985c), or as an extinct grade unrelated to living invertebrates (Seilacher 1989), the late Vendian Ediacaran fauna (or Vendobionts) must have consisted largely of consumers (see also chapters by Fedonkin and Seilacher, this volume). The same view is taken for the contemporaneous shelly fossil *Cloudina*, which is considered a filter-feeding cnidarian-grade metazoan (Grant 1990). The time of origination of simple medusoid-like fossils (normally regarded as contemporaneous with the radiation of the late Neoproterozoic Ediacaran faunas [Fedonkin 1985c]) was extended by a recent discovery in Neoproterozoic early Vendian interglacial deposits in Canada (Hofmann *et al.* 1990).

The appearance of controlled $CaCO_3$ biomineralization (Grant 1990; Grant *et al.* 1991) in the Neoproterozoic fossil record appears nearly synchronous or slightly younger than the possible advent of filter-feeding, detrital-feeding, cropping, and predatory(?) locomotion (as indicated by the late Vendian trace-fossil record; Bergström 1990; see also Bengtson, this volume).

As we have seen, the fossil record of bacteria and protists must be necessarily incomplete. However, inferences derived from molecular-biological data and chemical fossils seem to match surprisingly well the fossil record of the Neoproterozoic. Hence, the Proterozoic fossil record is dominated by largely benthic, layered-mat communities of cyanobacterial-like microbial forms (see chapters by Golubic and Walter, this volume) and organic-walled phytoplankton (acritarchs; Fig. 2D, F, G; Fig. 3C–H).

Little progress has been achieved in deciphering the taxonomic affinity of acritarchs from the time of their discovery in the 1930s. Nevertheless, the environments that they occupied are well understood, and it is generally agreed that for the most part they represent life stages, either motile or encysted, of prokaryotic and eukaryotic protists.

Late Vendian strata offer some of the most interesting and taxonomically diverse biotic assemblages of large and lavishly ornamented acritarchs (Fig. 4A–B; Awramik *et al.* 1985; Pyatiletov & Rudavskaya 1985; Volkova *et al.* 1980; Yin 1985; Zang & Walter 1989). Late Vendian protists differ radically from their early Cambrian successors (Fig. 4C–I; Moczydłowska 1991), and this contrasting difference may be of some interest in view of the apparent replacement of the Ediacaran vendozoans by "modern" Cambrian faunas (Fig. 1). However, the relative timing of these events is poorly constrained, and all the needed components to obtain acceptable time bracketing are seldom present in one and the same rock sequence.

The sudden disappearance of late Vendian acritarchs has been taken to mark an extinction event (Zang & Walter 1989) perhaps paralleling the decline of the late Vendian Ediacaran fauna (Fig. 1). However, late Vendian microbiotas are generally rare both in carbonate and detrital platforms (Moczydłowska 1991; Vidal & Nystuen 1990b), and more data are needed before long-ranging conclusions can be elaborated. This is more so because large acanthomorphs, once thought to characterize the Neoproterozoic, resurge in the early Cambrian of northern Greenland (Vidal & Peel 1993). Although early Cambrian forms decidedly represent taxa different from the Neoproterozoic ones, this may serve as a sign of caution in interpreting the evolutionary history of protists on the far too fragmentary data available.

Neoproterozoic early Vendian (~650–610 Ma) acritarch assemblages are rare and of low diversity (Vidal & Knoll 1983) and replace the rich and diverse biota characteristic of the late Riphean (~1000–650 Ma) and truncated by the early Vendian Varanger glacial event (Fig. 1; Vidal & Knoll 1983). This feature marks a turning point in Neoproterozoic protist evolution that was interpreted to represent the earliest recorded extinction event (Vidal & Knoll 1983). Notwithstanding, the real nature of this event is hidden behind the preservational bias introduced by the unsuitable conditions of preservation offered by widespread glacial deposits.

The Neoproterozoic Riphean microbiotas comprise a wealth of cyanobacterial and eukaryotic protists (Fig. 2A, D; Fig. 3C–H; Vidal & Knoll 1983; Knoll *et al.* 1991) embracing virtually all represented environments of sediment accumulation in marine basins (Knoll 1985c; Knoll *et al.* 1991; Vidal & Nystuen 1990a).

Carbonaceous fossils interpreted as problematic thallophytes, such as the ~800 Ma old *Tawuia* (Fig. 2E) and possible cladophoralean algae (Butterfield *et al.* 1988), are a notable part of the growing evidence suggesting the perhaps widespread presence of thallophyte-like organisms (Hofmann, this volume). Thus, the association of megascopic spheroidal acritarchs (*Chuaria*) and *Tawuia* (Fig. 2D) has a global distribution in Neoproterozoic rocks. Rocks of this age yield a wide spectrum of sculptured spheroidal acritarchs (Fig. 2F; Jankauskas 1989) accompanied by a variety of more complexly ornamented process-bearing forms with simple conical (Fig. 3C) and complex membrane-bounded processes (Fig. 3F–G) and also yield polygonomorphic acritarchs (Fig. 3E; Butterfield *et al.* 1988; Jankauskas 1989; Knoll 1984a; Knoll *et al.* 1991; Knoll & Calder 1983; Timofeev *et al.* 1976; Vidal 1976, 1981, 1990; Vidal & Ford 1985), demonstrating that complex acritarchs were common in the Neoproterozoic.

Neoproterozoic acritarchs yield important clues to the possible biological affinity of Proterozoic organic-walled phytoplankton at large. Neoproterozoic late Riphean forms possess various structures clearly defining cellular excystment devices, some of which have opercula comparable to that of certain dinocysts (Vidal 1976; Vidal & Ford 1985; Fig. 3C–D). These features alone clearly demonstrate that we are dealing with relatively complex aerobic photosynthetic protists representing a level of biotic organization close to that of present-day chromophytes. While this alone does not constitute compelling proof for affinities with extant eukaryotic flagellate protists, corroboration is added by organic-biomarker data, such as the occurrence of sterane derivatives of eukaryotic membrane lipids (Hayes *et al.* 1992a) and dinosterane (generally associated

---

FIGURE 3 Neoproterozoic microfossils. **A–B:** Vase-shaped microfossils interpreted as possibly mineralized loricas of planktic protoctist micropredators, possibly related to tintinnids. Note the thick (2–3 µm) walls, fractures, and the flaring aboral region (arrows). Ultraviolet-induced fluorescence micrographs (scale bar 49 µm). From Neoproterozoic (Upper Riphean) Visingsö Group (Sweden). **C–D:** Compressed and fractured spheroidal acritarchs interpreted as endocysts of eukaryotic phytoplankton displaying distinctive spiny (specimen at **C**; arrows) and granular (specimen at **D**) operculated excystment openings (scale bar 50 µm for **C** and 10 µm for **D**). From Neoproterozoic (Upper Riphean) Visingsö Group (Sweden; specimen at **C**) and Chuar Group (northern Arizona, USA). **E:** Slightly distorted polygonomorphic acritarch (scanning electron micrograph; scale bar 10 µm). From Neoproterozoic (Upper Riphean) Visingsö Group (Sweden). **F:** Slightly compressed acritarch interpreted as endocyst of eukaryotic phytoplankton displaying operculated excystment opening (arrow) and short, conical, distally flared spines (arrow, lower right corner; scale bar 10 µm). From Neoproterozoic (Upper Riphean) Chuar Group (northern Arizona, USA). **G:** Section of distorted giant acritarch heavily ornamented with club-shaped spines (scale bar 174 µm). From Neoproterozoic Hedmark Group (southern Norway). **H:** Compressed acritarch interpreted as endocyst of eukaryotic phytoplankton (*Valeria lophostriata*) displaying a characteristic sculpture consisting of dense, parallel striations (scale bar 155 µm). From Neoproterozoic (Upper Riphean) Chuar Group (northern Arizona, USA). All except **A**, **D**, **E**, and **G** are interference contrast micrographs; **E** is a scanning electron micrograph; and **G** is a transmitted-light micrograph of a petrographic thin section.

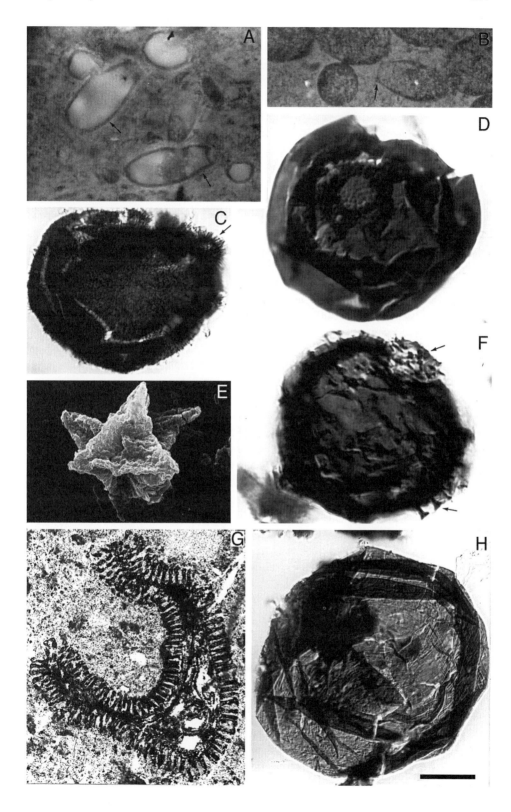

with dinoflagellate cysts; Hayes *et al.* 1992a), that offer a most suggestive combination of evidence.

As we have seen, the preservation potential of motile stages of microscopic algae is variable. The established occurrence of thallophytes in the Proterozoic (see above) suggests that reproductive structures of thallophytes (e.g., cysts, aplanospores, or zygotes) could be possibly represented among some of the relatively rare and larger Neoproterozoic (late Riphean, late Vendian) and early Cambrian acanthomorphic acritarchs (Vidal 1990).

Other significant groups of Neoproterozoic late Riphean microfossils are vase- and gourd-shaped microfossils (Fig. 3A–B; Vidal & Knoll 1983). Worldwide, the fossils are found in uncompressed condition in rocks representing various environments of deposition. Fairchild *et al.* (1978) pointed out that comparable fossils from the Neoproterozoic of Brazil may have possessed firm and presumably mineralized walls comparable to those of tintinnid ciliate protists. Although these fossils generally possess a sturdy organic matrix (Knoll & Vidal 1980), fluorescence microscopy of Swedish material (~800–700 Ma Visingsö Group) reveals smooth and probably lightly mineralized walls of even thickness (Fig. 3A). The original mineralogy cannot be conclusively established, but it seems reasonable to assume that it might have been calcareous. In any event, the new observations reinforce a former interpretation of these microfossils as planktic protist micropredators that produced calcareous loricas and that were possibly related to tintinnids (Knoll & Vidal 1980).

Controlled biomineralization may be a very ancient feature among protoctists. This view is supported by the Mesoproterozoic occurrence of thallophyte-like fossils, perhaps including bangiophyte-like algae (Butterfield *et al.* 1988). It may seem, however, that diagenetic alteration over extended periods of time reduced the possibility of preservation of delicate mineralized protoctists to the level of fortuitous events.

The wealth of cyanobacteria and acritarchs in Neoproterozoic strata has an impoverished counterpart in Mesoproterozoic (1.6–1.0 Ga) deposits. This is perhaps not greatly visible in the record of microbial-mat communities since they display considerable similarity with Neoproterozoic biotas. The form variety of acritarchs, however, is very different. Sequences dated to about 1.2–1.0 Ma are dominated by solitary spheroi-

---

FIGURE 4 Neoproterozoic and early Cambrian organic-walled microfossils. **A–B:** Distorted heavily ornamented giant acritarchs interpreted as endocyst of eukaryotic phytoplankton (scale bars 32 and 36 µm, respectively). From late Neoproterozoic (Yudomian) Talakh Formation at Zapad 742 drillhole (eastern Siberia; Yakutia). **C–F, H–I:** Early Cambrian spinose acritarchs interpreted as endocysts of eukaryotic phytoplankton (scale bar 10 µm for **C, D, F, H**; 14 µm for **E**). Specimens at **C, D, F** from earliest Cambrian Mazowsze Formation (East European Platform; eastern Poland). Specimen at **E** from Lower Cambrian Mickwitzia Sandstone (southern Sweden). Specimen at **H** from Lower Cambrian Buen Formation (north Greenland) **G:** Cellular aggregates interpreted as pleurocapsalean or chroococcalean cyanobacteria or, alternatively, as photosynthetic anoxygenic bacteria (scale bar 10 µm). From Neoproterozoic (Upper Riphean) Brøttum Formation (southern Norway). **I:** Compressed spheroidal tasmanitid acritarch (*Tasmanites volkovae*) interpreted as a possible phycoma stage of prasinophycean green algae (scale bar 44 µm; note characteristic pores). From Lower Cambrian Lingulid Sandstone (southern Sweden). All interference contrast micrographs.

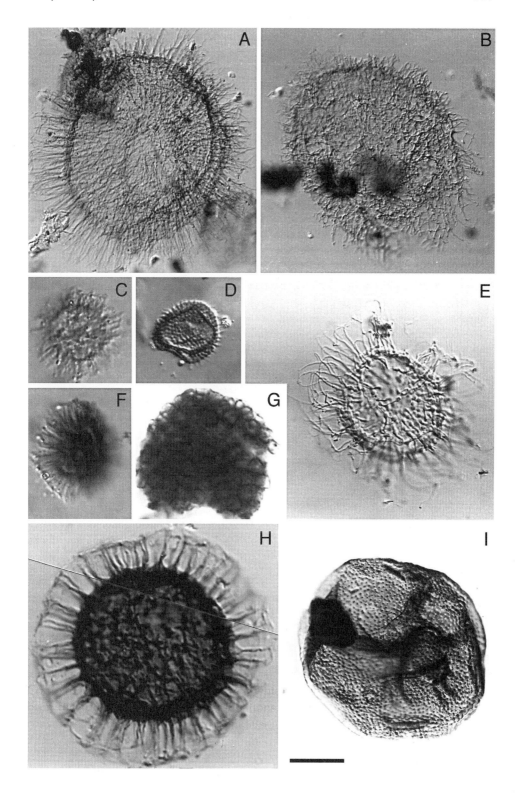

dal and colonial acritarchs that are highly comparable to forms from ~1.4 Ga and 1.7 Ga old rocks in North America (Horodyski 1980) and Australia (Peat *et al.* 1978). Nevertheless, diagnostically sculptured forms occur in about 1.0 Ga-old rocks in central North America (G. Vidal, unpublished data). This substantially lower diversity at the morphological level could obviously hide significant taxonomic diversity behind simple morphology. Furthermore, only forms that produced sporopollenin-like envelopes or cysts resistant to degradation have made their way into the geological record. In trying to evaluate the diversity of Mesoproterozoic acritarchs, one major problem is the enormous discrepancy between the Neoproterozoic and Mesoproterozoic biotic records. All our knowledge of Mesoproterozoic acritarchs can be expressed in a handful of microbiotas. Comparison is therefore very difficult. Nonetheless, there are obvious differences in the composition of the assemblages, suggesting that Mesoproterozoic acritarchs were much less diverse. Despite this, the occurrence of possible Mesoproterozoic thallophytes, although major uncertainty remains about interpretation (Walter *et al.* 1990; Hofmann, this volume; see above), may indicate that considerable diversity may have existed among protists.

The Paleoproterozoic (2.5–1.6 Ga) fossil record contains a variety of bacterial (Strother & Tobin 1987) and cyanobacterial (Awramik & Barghoorn 1977; Lanier 1986; Hofmann 1976) microbial-mat microfossils that at the level of comparable environments of deposition in Mesoproterozoic and Neoproterozoic strata display similar biotic composition. It may then appear that Paleoproterozoic prokaryotic microbiotas were both complex and varied (Hofmann & Schopf 1983).

Acritarch microbiotas from detrital sediments are extremely rare in the Paleoproterozoic and, although morphologically simple (Zhang 1986), they resemble the biotas of 1.6–1.4 Ga old assemblages from the Urals (Jankauskas 1989) and western North America (Horodyski 1980).

Archean microbiotas offer substantial interpretation problems that have to do with taphonomy and preservation (Hofmann & Schopf 1983). Notwithstanding, possible microfossils resembling cyanobacteria were reported from the early (~3.5 Ga [Knoll & Barghoorn 1977; Awramik *et al.* 1983; Schopf & Packer 1987]) and late (3.0–2.5 Ga [Lanier 1986; Klein et. al. 1987; Schopf & Walter 1983]) Archean (Schopf, this volume). Further indirect evidence for Archean prokaryotic diversity comes from reasonably abundant stromatolites (Walter 1983, this volume).

Summary

Past biotic changes are viewed through narrow windows provided by the available fossil record. While the view on complex ecosystems – including the major categories of microbial anaerobe and aerobe chemo- and photosynthesizers, protoctists, seaweeds, and animals – is relatively satisfactory for the early Paleozoic and parts of the Neoproterozoic, the fragmentary nature of the Meso- and Paleoproterozoic and Archean fossil records only allows viewing the contours of slowly evolving ecosystems. Hence, only sporadic glimpses have been revealed of the prokaryotic interactions that constituted the base of biotic change for at least the initial 1 Ga of evolution.

In summarizing this brief synthesis the standing headline is the extremely narrow nature of the pre-Phanerozoic fossil record. In fact, it appears as a sparse sequence of pulses from the remote regions of the Earth's beginnings. In writing the above, my feeling is that this same incomplete record only allows depicting a crude sketch of early life. In fact, the situation for part of the pre-Phanerozoic fossil record is such that one single fossil find could substantially alter our understanding of the level of complexity attained by biotic evolution at a given point (see, e.g., Runnegar, this volume). This is no less disturbing in view of the fact that the approximate length of the Paleoproterozoic is almost twice the length of Phanerozoic time. Obviously, our knowledge of biotic events within the frame of geologic evolution is prone to be biased by numerous phenomena, of which those mentioned in this short essay are but a few. The picture is indeed unclear and dim, but, nonetheless, its resolution increases day by day.

Acknowledgments. – I wish to thank Dr. Małgorzata Moczydłowska for providing micrographs of microfossils in Fig. 4C, D, and F. Drs. E. Aseeva and T.N. German kindly provided micrographs of microfossils in Fig. 2A and B, respectively. My research on Proterozoic – early Palaeozoic biotic change is supported by grants from the Natural Science Research Council (NFR) and the Knut and Alice Wallenberg Foundation.

The role of phenotypic comparisons in the determination of protist phylogeny

F.J.R. "Max" Taylor

Departments of Botany and Oceanography, University of British Columbia, Vancouver, British Columbia, Canada V6T 1Z4

Limitations in the fossil record necessitate the use of "soft-part" comparisons of extant organisms to aid in phylogenetic reconstructions. This paper reviews the main insights provided by ultrastructure and, to a lesser extent, biochemical comparisons. Close links between nonphotosynthetic "protozoans" and photosynthetic "algae" only became recognizable when their traditional separation was disregarded and when ultrastructural characters (mitochondrial type, flagellar features) were used, with subsequent testing by molecular sequencing. Neither ultrastructure nor metabolism were sufficient to root the tree, since the candidates for being most primitive were all distinguished by the lack of a feature, and there was no way of distinguishing primordial lack from subsequent loss. Molecular sequencing has provided positive evidence supporting the primitiveness of some of the amitochondrial protists. Major lineages are becoming clearer through the combined use of phenotypic and genotypic comparisons.

Genotypic change tends to be "clock-like" and measures evolutionary tempo. Phenotypic change is the mode. Woese (1987a)

Phenotypic, together with genotypic, comparisons continue to provide insights into probable phylogenetic relationships between protists and the multicellular lineages that arose from them. The old concepts of protozoa, algae, and fungi, based on demonstrably polyphyletic groupings, are gradually being dropped, at least in phylogenetic studies, even though they are deeply entrenched in texts, journals, and societies. In this contribution, which has a potentially huge scope, focus will be on the key factors that, in my view, have contributed to a revolutionary leap in the understanding of phyletic relations within the "Lower Eukaryotes." This radical change began in the 1970s. I will provide examples of characters that have proved to be most valuable and outline major lineages that are now discernable using a combination of data. Wherever possible vernacular names for groups have been used to make the text more widely accessible.

Bengtson, S. (ed.) 1994: *Early Life on Earth. Nobel Symposium No. 84.* Columbia U.P., New York

The need for "soft-part" data derived from comparisons of living organisms

It is self-evident that, wherever possible, physical or chemical fossils should be used to date the record of past phylogeny. Unfortunately, as detailed elsewhere in these presentations and in numerous other publications, at this time it is not possible to establish the fossil presence of numerous protist lineages before the development of hard parts (Loeblich 1974), and even these may be insufficiently characteristic to determine their affinities. At this time it could be argued that no Precambrian microfossils can be confidently attributed to any extant eukaryotic group.

A hard, critical look at Proterozoic microfossils previously attributed to extant groups of eukaryotes reveals that these attributions were based on very equivocal evidence. For example, the coccoid organism *Eosphaera* was attributed to both red and green algae when there is really no compelling evidence that it was either (Taylor 1981). It would be highly simplistic to accept the yeast-like identity proposed for the 3.4 Ga old *Ramsaysphaera* by Pflug (1978), not only for the lack of hard evidence but also for its discordance with much other data that indicate a much later date for the origin of eukaryotes. In fact, with the rejection of the view that the absence of the "9+2" microtubular arrangement (e.g., flagella, centrioles) is primitive (which would place red algae and fungi at the base of the eukaryotic tree [Taylor 1978]), there is an attractive view that the "true fungi" were represented only by flagellated chytrid-like ancestors until the group became terrestrial more or less following the development of land plants 400 Ma ago (M.R. Berbee, personal communication, 1992).

There is no fossil record at all of groups such as cryptomonads, bodonids, trypanosomes, choanoflagellates, non-testate amoebae, sporozoans, and amitochondrial flagellates, and the record of euglenoids and ciliates is limited to a few loricae (Taylor 1978). As evident elsewhere in this text, great progress is being made in the discovery and interpretation of Precambrian and Paleozoic microfossils, but the record is still largely silent as to the existence of many protist groups before the Mesozoic, even though there is good evidence that they diverged before the late Precambrian origin of the Metazoa. A point that may be obvious but needs stating is that, while the lineage may have diverged early, we often do not know what the early members of it were like, the fungal case above being a good example.

Before the mid-1970s

Views on eukaryotic phylogeny up to this period eventually saw a divergence of "algae," "fungi," and "protozoa" in an unspecified order from a common ancestor, with the Metaphytes arising from green algae and Metazoans from the (most likely ciliate) protozoans. Criteria used came from light microscopy and life-cycle considerations. Attention was focused on macrotaxonomy rather than phylogeny, with free admission that the groupings, particularly "fungi," might be polyphyletic.

H.F.Copeland (1956) and R.H.Whittaker (1969) kept Haeckel's mid-19th-century concept of Protista alive as a midlevel kingdom, but within their schemes one could still find the algal and protozoan assemblages kept distinct from each other within the protists, with the fungi distinct even to the point of their own Kingdom. As late as 1978 Whittaker & Margulis recognized three "branches" that reflected this within the Protista: the Protophyta, Protomycota, and Protozoa. In most schemes all the non-photosynthetic flagellates were grouped as a single group or lineage, the Zoomastigina, even though the eminent electron microscopist Irene Manton liked to assert that "zooflagellates exist only on the blackboards of zoology departments" (K.R. Vickerman, in Corliss 1986).

This comfortably simple view allowed one to make comparisons within only one of these assemblages. Even then, for example, the algae were looked on primarily as the Big Three – reds, greens and browns – with the remaining, largely unicellular, groups regarded in the same way as the zoologists regarded "minor phyla." Protozoa were treated as one group of the Invertebrata, although by 1985 they had graduated to a subkingdom of the Animalia (Lee *et al.* 1985). Emphasis was placed heavily on parasitic groups. Thus there was both an artificially narrowed field of vision and a neglect of groups that were going to turn out to be most evolutionarily informative.

Many trees provided little or no detail of suspected relationships within these groupings, an approach no doubt viewed as properly cautious. Leedale (1974) recognized 18 groupings arising from bacteria (Monera) in his "fan" scheme, but coyly opted out of showing any relationships other than those within the Plantae, Fungi, and Animalia *sensu stricto*.

Of course, this type of treatment was largely the product of the inadequacy of the characters available at the time to allow links to be made, particularly between algal and protozoan subgroups, but it was also undoubtedly a product of the organisms' being studied in different departments, with a jealous guarding of turf. The late microbiologist Roger Stanier once told me that he was eager to teach protistology at his university but would not have been allowed to.

A typical question asked at the time was whether dinoflagellates and euglenoids were plants or animals. This was an astonishingly reverse look down the evolutionary telescope but seemed reasonable to most at the time. Wrong concepts lead to wrong questions!

Several factors came into play in the later 1970s that began to change this, leading to a more phyletically based view of, and interest in, the protists. Corliss (1986) attributed this to a growing awareness of the protist concept, (freeliving) protist diversity, application of better methodologies, and the impact of the Serial Endosymbiosis Theory for the origin of plastids and mitochondria. These and some other factors will be examined below. For a literally exhaustive compilation of references dealing with most of the phyletic questions below, the reader is referred to Corliss (1987).

The rebirth of protistology as a discipline

Several breakthroughs in phyletic linkages arose from disregarding botanical/zoological conventions, for example, those linking euglenoids (phytoflagellates) with kinetoplastids (zooflagellates) and dinoflagellates (phytoflagellates) with ciliates and sporozoans (Taylor 1976). It was the emerging emphasis on similarities, made clear by ultrastructure, rather than differences, that encouraged a holistic view of unicellular eukaryotes and thus permitted these links to be made (Taylor 1976, 1978; Brugerolle & Mignot 1979; Sleigh 1979). This ongoing process is exemplified more recently by Smith & Patterson (1986) and D. Patterson (1989), who continue to link former "zooflagellates," amoeboid forms, and "algae" by ignoring the old barriers.

When I began teaching protistology at the University of British Columbia in 1967 it was viewed as an experiment and ran in parallel with conventional phycological, protozoological, and mycological courses. Early difficulties arose from the lack of both a text that reflected the approach and a "tree" on which to hang the examples used. It was largely this second need that prompted my involvement, along with that of Lynn Margulis, and encouraged by Howard Whisler of the University of Washington, in the formation of the International Society for Evolutionary Protistology in 1975. This appears to have been a major influence in fostering protistology and bringing together authorities on the many disparate groups of protists to deal with phylogeny using cell-biological and molecular tools.

Ultimately this has led to the first compendium of all the protist groups in the work titled the *Handbook of Protoctista* (Margulis *et al.* 1990). From the phylogenetic standpoint expressed here its organization is extremely eccentric (based on the presence or absence of flagella and "complex sexual cycles" à la Margulis & Schwartz 1982), resulting in the wide separation of such obviously linked groups as the Conjugatophyta (a subset of the green algae, recognized at different hierarchical levels by different authors) from the rest of the green algae (instead placed with red algae and two groups of slime molds, while the latter are lumped with the brown algae, diatoms, etc.). Dinoflagellates are far from ciliates and sporozoans. Also, it perpetuates the old polyphyletic zooflagellate grouping (as Zoomastigina). However, it contains a great deal of within-group phyletic information and brings together widely scattered data.

Cladistic ferment

A contributing factor to the resolution of early eukaryote evolution was the religious application of Willi Hennig's methodology to phylogenetics by cladisticians and the controversies that ensued. This revived interest in phylogenetic taxonomy at the macro level, a subject that had been languishing for many years. Lipscomb (1985) was the first to apply this methodology to general protist phylogeny (see Lipscomb 1989 for a more extensive and recent exposition, specifying the 137 different character states, both structural and biochemical, that she used). Once the noise had died down the value of focus on shared derived characters became less controversial and widely used (C. Patterson 1989).

Despite this greater acceptance of cladistic methodology, it is striking that Lipscomb's cladograms are strongly noncongruent in detail with others, particularly the ss rRNA ones (see Sogin, this volume). Why is this? One key factor seems to be her choice of the red algae as the out-group. As explained elsewhere in this chapter, this was one of several possibilities in the 1970s, all based on the lack of a feature. Taylor (1978) chose red algae as the most primitive eukaryote as an explicit working assumption, given the absence of conclusive data. The 5S and ls rRNA parsimony trees tend to place red algae nearer the base of the tree than ss rRNA does. It will be interesting to see what the resolution of these incongruities will be. In this chapter the assumption has been made that some amitochondrial protists are primordially so, in agreement with the ss rRNA trees.

To me, a prime value of cladograms is that the data and the assumptions used to create them are explicit and readily accessible. So, for example, it can be seen that some of the data used by Lipscomb (1989) are wrong. For example, she asserted that the use of an extranuclear spindle in mitosis (her character state 116B), found in only some of her polymastigote flagellates, is shared by oomycetes, euglenoids, dinoflagellates, "chromobionts," and *Chlamydomonas*. In fact this feature is found only in dinoflagellates and trichomonad and hypermastigid polymastigotes.

Williams (1991) has used cladistics to show that the phylogenetic tree used by Cavalier-Smith (1986, 1989b) for the groups within the "kingdom" Chromista is not one of 40 equally parsimonious trees he obtained using the same data. Williams then produced a cladogram using his own larger data set. It is striking that he was unable to see that cryptomonads should not be in the group at all (indicated by his analysis of Cavalier-Smith's data), because the cryptomonad features that are most discordant with chromists are not in his data set! They have flat mitochondrial cristae instead of tubular (see below), produce starch, and have phycobilin pigments (a deceptive chloroplast character for reasons given below), and, although they have a tubular segment to their bipartite flagellar hairs, the hairs are different in detail from the tripartite hairs of the typical "chromists" (the Stramenopiles of D. Patterson [1989]). The mitochondrial cristae suggest a closer relationship of cryptomonads with others possessing flat cristae, such as the green algae.

An interesting exercise using protist ultrastructural characters to compare cladistic and other numerical taxonomic methodologies with an "intuitive tree" was carried out by Smith & Patterson (1986) while they studied heliozoan (polyphyletic) relationships. None of the trees agreed with each other, although some (Wagner parsimony, Dollo parsimony) were closer than others.

Impact of the Serial Endosymbiosis Theory and complications due to intracellular symbioses

In terms of the recognition of major lineages, the Serial Endosymbiosis Theory contributed little directly, since it dealt with the primordial symbiotic origins of plastids and mitochondria. However, the intellectual ferment it created (specifically reviewed by Taylor [1980a]) resulted in the application of powerful tools to evolutionary problems and generated valuable new data. It became legitimate to ask if such events might have occurred more than once. This led to organellar characters, particularly chloroplast features, becoming potentially unreliable indicators of group affinities.

Additionally, the increased awareness of the existence of cells-within-cells resulting from "secondary symbioses" (eukaryotes living within other eukaryotes), one of the extensions to the serial endosymbiosis theory noted by Taylor (1974), allowed characters of the cytobiont (the partner within the host cell) to be distinguished from the host.

The confusion caused by the lack of this recognition (or even of the theory itself) is most clearly seen in the work of Klein & Cronquist (1967).

For example, the presence of fucoxanthin in a few dinoflagellates, a pigment thought to be distinctive of the chrysophyte assemblage, was shown to be due to cytobiotic diatoms (Jeffrey & Vesk 1976). Another example of multiple appearances of an unusual compound in unrelated organisms due to symbiosis is tetrodotoxin; originally reported in puffer fish, but also under other names in salamanders, tree frogs, crabs, and the blue-ringed octopus, it also occurs in bacteria (Do *et al.* 1990 and references therein) and is probably of bacterial origin in all the metazoans above.

A good example of confusion caused by symbiosis has been found in cryptomonads, difficult to place in early trees because of conflicting affinities suggested by their plastids versus the rest of the cell (Taylor 1976). The discovery of a vestigial nucleus, the *nucleomorph,* lying close to the plastid strongly suggested that both came from a eukaryotic (red algal?) cytobiont (Ludwig & Gibbs 1985). This has recently been confirmed by rRNA sequencing (Douglas *et al.* 1991). A similar condition, only this time involving a green endosymbiont, is found in the enigmatic *Chlorarachnion reptans,* which also possesses a nucleomorph (Ludwig & Gibbs 1989). McFadden (1990) has reviewed these types of evolutionary events.

Several authors believe that the presence of more than two membranes around chloroplasts, such as in euglenoids and dinoflagellates (three) or the "chromophytes" (four), is an indication of more ancient secondary symbioses in which the plastid is the only remnant of a eukaryotic, photosynthetic cytobiont (e.g., Whatley *et al.* 1979; Cavalier-Smith 1986).

Such considerations have led to the formation of an entire subdiscipline, endocyto-biology, focused on the evolutionary significance of intracellular symbiosis (for the most recent proceedings see Nardon *et al.* 1990).

The contributions of ultrastructure

Ultrastructure has had a major impact in the rethinking of relationships since the 1970s (Smith & Patterson 1986), revealing that external morphology was often deceptive. Many groups with simple morphology turned out to be polyphyletic (amoebae, sporozoans, fungi, slime molds, heliozoans, etc.).

Because of the potential confusing effect of secondary symbiosis referred to above, when I first approached the topic of flagellate relationships it seemed wisest initially to emphasize "host" features in tracing major lineages, treating plastid and mitochondrial features separately (Taylor 1976). The host features available were primarily nuclear (including mitotic), flagellar, and cortical features.

It was reasonable to expect that nuclei, and particularly mitotic mechanisms, should be highly conservative and reliable indicators of affinity (see reviews by Heath [1980] and Raikov [1982]). This expectation was not fully realized, although it does seem clear that closed mitosis preceded open mitosis and internal spindles preceded external spindles. The strange external spindle of dinoflagellates is also found in the amito-chondrial hypermastigote flagellates. Despite a suggested affinity (Kubai 1975) based on this character alone, nothing else has supported a close relationship between them, and so the feature appears to be the result of convergent evolution. It was found that within groups nuclear features are usually consistent, but in some, notably the green algae, there is considerable variety.

Flagellar features are generally consistent within groups. Details, such as the types of flagellar hairs, have been useful in recognizing intergroup affinities. The best example of this is the presence of tripartite tubular hairs within only the lineage often referred to as chromophytes (Leadbeater 1989), which includes chrysomonads, diatoms, brown algae, and silicoflagellates. These structures also supported views of the affinity of oomycete fungi with the complex (an old idea, based on the admittedly superficial similarity of their reproductive structures with that of the xanthophyte *Vaucheria*). D. Patterson (1989) used the presence of flagella with these hairs to link a number of "zooflagellates" to the chromophytes and has proposed the name *Stramenopiles* (straw hairs) to encompass this larger grouping. Cavalier-Smith (1981, 1989b) used "Chromista" in a roughly equivalent way, although within the assemblage he included cryptomonads that have nontripartite tubular hairs and flat mitochondrial cristae, features that would argue against their inclusion. Within this lineage, a structure known as a transitional helix within the basal region defines several chromophyte groups (Hibberd 1979) and the Oomycetes (a different shape but probably homologous) but can be variably present within the same genus, e.g., *Mallomonas* (Andersen 1991).

A major anomaly in this regard is the conventional view of prymnesiomonads (including coccolithophorids) as being a sister group of the chrysomonads. The flagella of the former group are mostly subequal and smooth, and the relationship was proposed on the basis of other, internal organelles, which may be deceptive (see plastids, below). Furthermore, the transitional region in some of them bears a "stellate structure" like that of green algae, rather than a helix.

The presence of scales on prasinomonad flagella and moss and fern sperm flagella was used as supportive evidence for the origin of the metaphytes from this green flagellate group. Paraxial rods (system-I fibers) within the flagella link euglenoids with kinetoplastids, but a somewhat comparable structure occurs in dinoflagellates (the "striated strand"), pedinellids, and some other flagellates (Cachon *et al*. 1988).

Much of the more recent interest has been on flagellar roots and other cytoskeletal features, although this information has been used largely to study within-group relationships, being particularly well studied in the chlorophyte complex, where it has been described as being "of major importance" in recognizing classes (Mattox & Stewart 1984; for a recent review involving several groups see *Protoplasma 164:1–3*, 1991). Sleigh (1988) has devised a mapping method that facilititates comparisons between groups. From such surveys it is possible to see resemblances between groups considered to be related using other criteria. For example, the euglenoid and kinetoplastid systems are virtually identical. Judging by the overall pattern an early, if not primordial, arrangement seems to be a flagellar pair with four roots, various modifications (mostly reduction) of this being found among the protists.

A good example of cortical organization linking groups hitherto thought unrelated is that of dinoflagellates and ciliates (Corliss 1975; Taylor 1976), where there are many similarities. Each has a single layer of highly ordered cortical alveoli, with rod-like trichocysts interspersed between them. A lesser similarity was observed in the flagellates *Spiromonas perforans* (Brugerolle & Mignot 1979) and in *Perkinsus*, suggesting a link with Apicomplexans (sporozoans). The common lineage of these three has been strongly supported by rRNA sequencing. Cortical, solid, rod-shaped trichocysts (ejectile bodies) are also present in some chloromonads (=raphidophytes), members of the Chromophyte lineage, and this, together with peripheral alveolation, could be used to argue that they are near the divergence with the dinoflagellate–ciliate–sporozoan lineage.

Plastids have not only distinctive pigments but also characteristic ultrastructure in generally good agreement with the former. Furthermore, the membrane structures around the plastid, in addition to the two membranes of the envelope, are distinctive (and, as noted above, perhaps indicative of ancient symbioses), although more at the group recognition level. Thus, the triple-membraned envelopes of dinoflagellates and euglenoids are looked on as independent host responses to ancient symbiotic events. However, the presence of specialized endoplasmic reticulum around the plastids defines a suite of groups in the chromophyte–chromist lineage (see Cavalier-Smith 1986, 1989b; Whatley 1989) and provides one basis for placing the prymnesiomonads within it, together with pigmentation. Green algae, prasinomonads, and metaphytes not only have the same pigments but also share the peculiarity of forming starch as chunks within their plastids, all other starch producers (dinoflagellates, red algae, cryptomonads) producing it outside the plastid.

Several authors have explained plastid diversity in terms of multiple primary symbioses, involving differently pigmented cyanobacteria (chlorophyll a only, $a+b$, $a+c$, etc.), but the question as to how many of these occurred is still open (Taylor 1987). Recently it has been shown that chlorophyll b has evolved independently at least three times in the prokaryotes (Palenik & Haselkorn 1992), requiring relatively little change from chlorophyll a (Taylor 1987). Whatley (1989) has discussed the possibility of multiple plastid origins extensively.

The form of the mitochondrial cristae, the inner projections, has turned out to be surprisingly useful in the recognition of large subsets of the protists and in the linking of non-photosynthetic with photosynthetic groups (Taylor 1976, 1978). It was surprising because the form of these structures was known to be variable, both in higher animals, where different tissues have different cristal forms, and in some amoebae. Under dysaerobic conditions the cristae or the whole mitochondrion may be reduced, as in some trypanosomes or yeasts, respectively. However, they seem to have been remarkably conservative in most protist groups, being fundamentally distinguishable into flat or tubular (platycristate and tubulocristate, respectively), with or without pinched bases (vesicular cristae can be viewed as a modification of the tubular type). Flat, paddle-like cristae are often referred to as discoidal. Major superlineages, supported by other features, can be defined by these characters, as well as a grouping in which mitochondria were primordially absent (the Archezoa of Cavalier-Smith [1981]).

The first use of crista type to recognize group affinity was by Manton (1959) who, despite the horrendous fixation results at the time (she pioneered the ultrastructural study of "algae"), noted that the crista form (flat) of the tiny flagellate then known as *Chromulina pusilla* was inconsistent with the tubular cristae of the other organisms with which it was classified at the time (chrysomonads). A decade later, Taylor *et al.* (1969) were able to distinguish cryptomonad cytobiont cytoplasm by the mitochondrial morphology (platycristate) from host ciliate cytoplasm (elongated tubulocristate). Leadbeater & Manton (1974) used the flat cristae of choanoflagellates to argue that these organisms are not closely related to chrysomonads, as previously thought. The earliest broadscale use of the cristae to recognize large groupings was by Taylor (1976), first on flagellates and then on the protists as a whole (1978). In the latter paper it was also pointed out that the tubulocristate type of ciliates was not consistent with them being the ancestors of the Metazoa (fundamentally platycristate), although the choanoflagellate type is. At the ultrastructural level sponge choanocytes were found to be extremely similar to choanoflagellates, so their affinity is thought to be strong; if other Metazoa do not have a common ancestor with sponges, as so many zoologists assert, then their protistan origin has not been determined yet.

Ultrastructure, particularly mitochondrial crista type, has been instrumental in revealing that some traditional groupings based on external morphology and life-cycles, such as amoebae, slime molds, or heliozoa, are polyphyletic (also Fungi, as traditionally defined). It drew attention to the incongruity of the creature long thought to be the most primitive ciliate, *Stephanopogon*, with other ciliates and, taken with other ultrastructural data, resulted in the removal of this organism and the opalininds to other lineages (summarized by Lipscomb [1989]).

Are there no protistan exceptions to this apparent reliability? A possible case in point is the recent discovery of tubular cristae in *Kathablepharis*, a flagellate thought to be a nonphotosynthetic cryptomonad (Lee & Kugrens 1991). The latter group is usually characterized as having narrow, flat cristae constricted at the bases. In view of other anomalous features, such as flagellar and cortical structure, I agree with Patterson & Zöllfel (1991) in considering *Kathablepharis* a flagellate of "uncertain taxonomic position" whose affinities lie with other groups with tubular cristae. However, this then raises another paradox, for *Kathablepharis* not only possesses coiled, ribbon-like ejectosomes very similar to those of cryptomonads, but even has them in two sizes, arranged as in cryptomonads! Are these bodies an example of incredible convergent evolution at the organellar level, or better indicators of a close relationship between organisms that have them, or similar because of symbiosis? Regarding the last possibility, similar although not identical ribbon structures are known within "kappa" bacteria symbiotic in the ciliate *Paramecium* (Preer 1975) and in prasinomonad flagellates of the genus *Pyramimonas* (Norris & Pearson 1975). The closeness of the latter group to the position of cryptomonads in recent ss rRNA studies (Douglas *et al.* 1991) is suggestive of a common origin for this organelle in these flagellate groups. On the other hand, Schuster (1968) found that antibiotic treatment caused loss of the ejectosomes of the cryptomonad *Cyathomonas*, suggestive of a symbiotic origin, reintroduction of bacteria being necessary to cause their reappearance. Tritiated thymidine labeling produced unconvincing (to me) evidence of DNA within the ejectosomes. This still needs re-examination, despite attention being drawn to it nearly twenty years ago (Taylor 1974).

Tubulocristate groups are shown as having a single common origin in the tree in this paper (Fig. 1). A notable exception is tubulocristate *Acanthamoeba*, which almost invariably appears close to the divergence of green algae and metaphytes in ss rRNA trees (e.g., Gunderson *et al.* 1987.) If the RNA trees are correct, *Acanthamoeba* is an exception to cristal conservatism: if the cristae are reliable indicators of affinity, then there is an artifact that is positioning *Acanthamoeba* in this bizarre location. It has been noted previously that prymnesiomonads are difficult to place because their flagellar features suggest an affinity with green algae, but their plastids and mitochondrial cristal type (tubular) do not.

The limited value of secondary metabolites and metabolic pathways as phylogenetic indicators in eukaryotes

Much early work (exhaustively reviewed by Ragan & Chapman [1978]; see also Rothschild & Heywood 1987) was devoted to the use of metabolic features for recognizing affinities. The classic example of this was the use of photosynthetic pigments to recognize the red, green, and brown algae. Complications arose from the affinities between cyanobacteria and red algae, euglenoids and green algae, crypto-monads and red algae, and dinoflagellates and the chromophyte algae, suggested by their pigments (Klein & Cronquist 1967) but contradicted by other biochemical and structural features. The Serial Endosymbiosis Theory revealed that symbiosis had to be taken account of in these interpretations. Dodge (1979) and others attempted to resolve this by using structure to argue for multiple symbiotic events involving two different primordial cyanobacteria.

The lack of histones in dinoflagellate chromosomes led Dodge (1965) to postulate that these organisms might be the most primitive eukaryotes, "mesokaryotes," since they are also absent in prokaryotes, but current thinking (Dodge 1989), supported by molecular sequencing, indicates that they must have lost their histones.

Characters such as wall composition, which seemed as if they should be good indicators of affinity, turned up in obviously disparate organisms. Cellulose, for example, is present not only in green algae and higher plants but also in dinoflagellates and tunicate animals. Calcium carbonate (dinoflagellates, red and green algae, mol-lusks, etc.) or silica deposition (dinoflagellates, radiolarians, diatoms, chrysomonads) or the presence of chitin or sporopollenin (dinoflagellates, green algae, higher plants) are other examples of occurrences in both related and relatively unrelated groups.

The chromophyte (Green et al. 1989) and chlorophyte complexes (Irvine & John 1984), especially the latter, have turned out to be fairly well circumscribed by pigments and storage products. Chlorophyll c_2 is found also in dinoflagellates (more distantly related to the chromophytes) and cryptomonads (not related). The plastids of prym-nesiomonads may also turn out to be deceptive if their flagellar features are non-convergent with green algae. Rothschild & Heywood (1987) compared the photosyn-thetic enzyme ribulose biophosphate carboxylase from many protist groups but were not able to derive much insight into plastid evolution from it.

Secondary metabolites are, of course, only the phenotypic expression of the final steps of pathways, and it was correctly stressed at an early symposium in this series (Birch 1973) that it should really be the pathways and the improbability of independent development that should be compared. A good example of this is the synthesis of the amino acid lysine, which occurs by one of two pathways, known as the AAA or DAP (also with a shorter variant) pathways. The thousands of species of "true fungi" (not including water molds or slime molds) all use the AAA pathway, unlike most other eukaryotes. Euglenoid flagellates also use this pathway but are thought to be very distant from the fungi. Chitin in their walls also unites the true fungal groups (Zygomycetes, Ascomycetes, Basidiomycetes), which lack any trace of flagella or basal bodies with the posteriorly uniflagellated chytrids. Here, biochemical characters appear to be more reliable than structural ones in indicating affinity.

Bioluminescence is an unusual property that appears to have evolved independently perhaps as many as 30 times (Hastings 1983), with different luciferin molecules usually being found only in single classes or phyla (exceptions being the bacterial, coelenterate cypridina, and dinoflagellate systems). Hastings hypothesized that luciferases evolved late, after the evolution of visual systems.

In the most recent Nobel Symposium to deal with phylogeny, Ragan (1989) has provided a thorough summation of the biochemical pathway data. From this and a comparison with available rRNA trees, he had to explain major discrepancies by the "distressingly frequent invocation of lateral [gene] transfer, multiple losses (e.g., chitin) and homoplasies (sterol biosynthesis)" (Ragan 1989, p. 156).

In general, with the exception of molecular sequencing, biochemical characters seem to have been better for group recognition rather than between-group affinities among the eukaryotes.

Problems with rooting the eukaryotic tree

Since there is little doubt that eukaryotes arose from archaebacteria-like prokaryotes, the rooting of the eukaryote tree should be based on resemblances between them. The development of membrane-bound nuclei and histonally based chromosomes are eukaryotic developments, but there are no known examples of intermediate states that could indicate early divergence on the eukaryotic line. Currently a microtubular mitotic mechanism has not been seen in the amitochondrial giant amoeba *Pelomyxa palustris* (Whatley 1989), although microtubules are plentiful in its cytoplasm. Nuclear divisions are so infrequent that lack of observation so far may not be surprising.

Another eukaryotic trait is the presence of gametic sexuality. Did any living groups diverge before the development of such sexuality? There are groups for which sexuality has not been reported, though it may be present but cryptic (e.g., it may be rare, the gametes may look like normal cells, or it may occur only during the dark cycle). Until quite recently, sexuality was unknown for euglenoids and cryptomonads, although it is now known to exist (references in Kugrens & Lee 1988). Therefore not much use has been made of this feature, although it is possible that it did develop after some or all of the Archezoa.

Unfortunately, therefore, from a structural standpoint the only resemblances between extant eukaryotic groups and prokaryotes were absences: lack of histones (dinoflagellate nuclei), lack of "9+2" microtubular organization (red algae and true fungi), or lack of mitochondria (several amoebae and flagellate groups, plus microsporidia). There was no way in the 1970s to determine by structure or metabolites if these were primordial conditions or subsequent losses.

Early biochemical comparisons, particularly photosynthetic pigments, suggested direct evolution of red algae from cyanobacteria (Klein & Cronquist 1967), but this was typical of the confusion arising from rejection of the Serial Endosymbiosis Theory, which implied that red algal plastids *had been* cyanobacteria, but not the surrounding cytoplasm.

In my opinion the greatest insight that ribosomal RNA sequencing has provided is the identification of *some* of the amitochondrials as the most primitive extant eukaryotes, by the use of positive characters, i.e. shared sequences, while serving to confirm, in most cases, relationships already suggested from ultrastructural studies. Additionally, the depth of divergences can be indicated in a way that is not possible from phenetic comparisons.

The major lineages presently recognizable among the lower eukaryotes

As the result of the application of all available data, several distinct major lineages have become clear, and the positions of at least some groups within these lineages have become noncontroversial. It will be seen that all were evident from ultrastructural features, although their recognition has become more widespread, as they have been found to be "robust" regardless of the type of data. The tree in Fig. 1 represents my current view of relationships among major groups of protists, omitting those whose position is still very equivocal, e.g., foraminiferans and actinopods. The method employed to create it was similar to the Smith–Patterson Intuitive Tree protocol: "a semiparsimonious but subjective evaluation of characters with a variable and unspecified weighting procedure" (Smith & Patterson 1986, p. 329). In the brief verbal outline that follows I will try to avoid the use of formal supraphyla where possible, since I feel that the recognition of major lineages is more important than naming those that are still in a state of flux. In particular, the endings *-phyta* or *-zoa* are inappropriate unless the groups are directly linked to the plant and animal lineages, respectively.

Amitochondrial protists

As noted earlier, amitochondrial protists were viewed as organisms that could have diverged before the acquisition of mitochondria, but there was also the possibility that some, or all, had lost them. It now appears, as the result of molecular sequencing data (Sogin, this volume), that *some* never had mitochondria, namely the amoeboflagellates with single basal bodies associated with their flagella (Archamoebae [Cavalier-Smith 1991a]), diplomonad flagellates (*Giardia*), and, by close structural similarities, retortomonads, microsporidians, and the metamonad flagellates (trichomonads, hypermastigids). The last differ from the others in having golgi bodies and an external rather than internal mitotic spindle and probably evolved later than the others, although the flagellar basal-body arrangement of the trichomonads and retortomonads are rather similar. Some other amitochondrial amoebae, such as *Entamoeba*, apparently lost mitochondria. This is suggested by their position in rRNA trees, indicating the need for

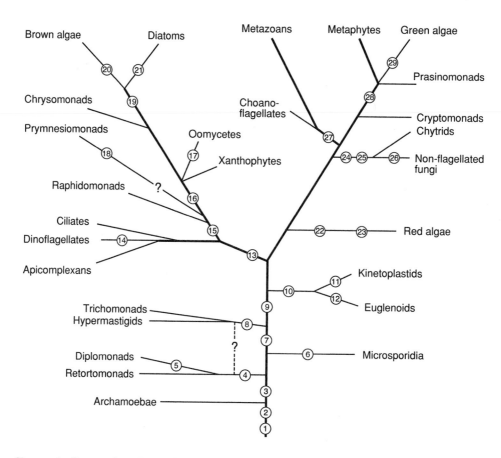

FIGURE 1 Proposed protistan relationships, based primarily on ultrastructure. Numbers indicate examples of key features that define groups at various levels. The placing of several amitochondrial groups at the base of the tree, and cryptomonads, has been based on ss RNA results. **1:** Flagella with single basal bodies. **2:** Closed mitosis with an internal spindle. **3:** Paired flagellar bases. **4:** Quadruple flagella. **5:** Cells "double." **6:** Flagella lost. **7:** Golgi bodies. **8:** External mitotic spindle. **9:** Mitochondria (flat cristae). **10:** Discoidal mitochondrial cristae. **11:** Kinetoplast in mitochondria. **12:** Pellicle (plus green plastid in some). **13:** Mitochondria with tubular cristae. **14:** External mitotic spindle (plus loss of histones in most, acquisition of plastids in many). **15:** Tripartite, tubular flagellar hairs (acquisition of chlorophyll *c* and fucoxanthin-containing plastids). **16:** Transitional helix (also loss of cortical alveolation). **17:** Loss of photosynthesis. **18:** Loss of flagellar hairs. **19:** Loss of transitional helix. **20:** Multicellularity. **21:** Silica frustule. **22:** Loss of flagella, including bases. **23:** Acquisition of phycobilin-containing plastid. **24:** Chitin synthesis. **25:** AAA lysine pathway. **26:** Loss of flagella, including bases. **27:** Collar, vane on flagellum. **28:** Starch in chlorophyll-*b*-containing plastids. **29:** Loss of hairs/scales on flagella.

caution in applying the absence of this organelle as a phylogenetic marker. Many of the amitochondrials are internal parasites, which makes sense in view of these being low-oxygen environments, but at first it seems disturbing that they are living inside animals that evolved very long after them! One can only conclude that they were originally all free-living groups in low-oxygen environments that had to take refuge within hosts once their anaerobic niches shrank.

At present they are viewed as multiple separate early branchings (exceptions being the retortomonad–diplomonad and trichomonad–hypermastigid pairings). None are likely to have left recognizable microfossils.

Euglenoids and kinetoplastids

Ultrastructure alone was insufficient to indicate that euglenoids and kinetoplastids are the most primitive protists containing mitochondria (see Sogin, this volume) although it did lead in the 1970s to the first recognition that they were related to each other, having particularly distinctive paddle-like discoidal mitochondrial cristae (Taylor 1976). Multiple other similarities were soon noted, although they were at odds with the paths of lysine synthesis (see above). The presence of closed mitosis with an internal spindle is considered primitive, also being found in most of the amitochondrials.

The tubular cristate macrolineage

As noted above, protists with tubular mitochondrial cristae show multiple other features (e.g., heterokont flagella and associated structures; see D. Patterson 1989 and Moestrup & Andersen 1991) that indicate that most (all?) are related, as first proposed by Taylor (1976, 1978). Within this assemblage, named the Tubulicristata by Mirabdullaev (1989 and earlier), there appear to be several distinct sublineages.

One main sublineage is defined primarily by bi- or tripartite flagellar hairs (the Stramenopiles of D. Patterson [1989] or Chromista minus cryptomonads of Cavalier-Smith [1989b], including the "chromophyte" algae of Christensen [1966]) and by the presence of chloroplast endoplasmic reticulum (CER) in most of the photosynthetic members. This lineage culminates in diatoms and brown algae and is usually thought to include the coccolithophorids as a subgroup of the prymnesiomonads (= haptophytes). As noted earlier, flagellar features raise doubts as to whether prymnesiomonads belong in this assemblage. Some formerly fungal groups, such as the oomycetes, thraustochytrids, and labyrinthulids, also group with this assemblage, as do various former "protozoa" (D. Patterson 1989). The placing of groups on this lineage may be assisted by assuming a progression from closed, internal spindle mitosis to semiopen with external spindle pole bodies.

Another tubulocristate lineage that is growing in recognition as groups are added to it is the dinoflagellate–ciliate–apicomplexan lineage. Early suggestions that these groups were linked, based on ultrastructure (Corliss 1975; Taylor 1976; Brugerolle & Mignot 1979), have been reinforced by molecular data (e.g., Wolters 1991). Cortical alveolation is a notable feature of this line. Although most members have closed mitosis with an internal spindle, the dinoflagellates *sensu stricto* have acquired an external spindle while gradually losing histones, superficially resembling the apparently unrelated hypermastigids in the former, very unusual feature.

The placing of some amoebae and two tubulocristate groups with a rich fossil history, the actinopods and foraminiferans, within this macrolineage remains enigmatic. On the basis of siliceous internal skeletal elements and capsule-like structures surrounding the nucleus of some dinoflagellates, a possible link with actinopods has been suggested (Taylor 1980b), and the flagellation of the gametes of foraminiferans shows some similarities with "predinoflagellates" such as *Oxyrrhis*.

The upper main platycristate lineage leading to metaphytes and metazoans

The position of red algae has long been problematic, and early trees placed them near the base. Interestingly, 5S and ls rRNA sequencing still suggests this. Here, with the acceptance of the primitiveness of amitochondrial flagellates, red algae must be concluded to have lost flagella and thus are higher up the platycristate branch, near the fungi *sensu stricto*, which also seem to have lost flagella, other than the chytrids. Choanoflagellates also are on this upper platycristate lineage, placed higher than the aforementioned groups because of their probable role in giving rise to the Metazoa. The cryptomonad position is that suggested by ss rRNA (Douglas *et al.* 1991), their cristae and, as noted earlier, their ejectosomes. The divergence of the Metaphyte lineage has long been thought to be from some prasinomonad green flagellates on the basis of biochemical features and ultrastructure. It is striking that three major lineages – plants, animals, and true fungi – arose relatively close together.

Multicharacter congruence rather than single "Golden Keys"

Many authors in search of phylogenetic links have tried to use their favorite character (mitosis, flagellar root features, cytochrome *c*, ss rRNA, etc.) as the one that will reveal all, with varying degrees of success but never complete revelation. For example, a very thorough review of mitotic mechanisms by Heath (1980) produced clusters containing groups that are very discordant with other features included in more recent data. This was one reason why Taylor (1976) examined four cell features, at first independently, then together, in an attempt to obtain a congruent tree, and much more structural detail is now available to test both the phenotypically and genotypically based trees.

It is unlikely, in my opinion, that any character will "work," i.e. be a reliable phylogenetic indicator, over the whole range of organisms, even though it might seem more satisfying theoretically. Congruence between phenotypic and genotypic characters are needed to see the "real" tree: the trees should make biological sense, and perceived anomalies, confirmed by multiple tests, should be explicable (as in the symbiotic cases noted above). It is "robustness" that will provide an ever-solidifying tree of the early lineages of life on Earth.

> *The conflict . . . is not between morphology and molecules, but between methods of analysis.*
>   Colin Patterson (1989)

Acknowledgments. – The author wishes to thank the Nobel Foundation for making his participation possible. Ongoing support from the Natural Sciences and Engineering Research Council of Canada (Grant A 6137) and the National Science Foundation (Grant 91047043, joint with K.R. Roberts) is also acknowledged.

Combinatorial generation of taxonomic diversity: Implication of symbiogenesis for the Proterozoic fossil record

Lynn Margulis and Joel E. Cohen

L.M., Biology Department, University of Massachusetts, Amherst, MA 01003, USA; J.E.C., Rockefeller University, 1230 York Avenue, Box 20, New York, NY 10020-6399, USA

Symbiogenesis is the emergence of new species with identifiably new physiologies and structures as a consequence of stable integration of symbionts. The development of symbiotic associations may lead to evolutionary innovation. Because of the pervasive influence of symbiogenesis in the origin of eukaryotic organisms, the Latin binomials of taxonomy should be recognized as applying to individuals who are greater than single homologous genetic units. All eukaryotes are composite (more than a single organismal type) and should be named and described accordingly. Symbiogenetic recombination of genomes may generate a striking diversity of both higher taxa and individual "species." A small number of symbionts, such as twenty different bacterial strains, individually and in all possible combinations with a host coleopteran, for example, could potentially generate more than a million distinct new species of beetles. A relatively small number of associates potentially can generate as much biological diversity as has been observed. The upper limit for one host and n symbionts is $2^n + n$ combinations. Rejecting the cladistic restriction of taxon origination by dichotomization of lineages, symbiogenesis requires depiction of evolution by anastomosing branches to form net-shaped phylogenies. We infer a relative paucity of symbiogenetically generated diversity in the Archean Eon. A far more significant amount accompanied the origin of undulipodiated and aerobic protoctists in the Proterozoic Eon. Symbiogenesis may be especially significant for the emergence of skeletalized animals in the late Proterozoic and of plants, organisms far more desiccation-resistant than algae, in the Phanerozoic Eon. The polyphyletic acquisition of calcium-precipitating microbial symbionts may underlie the appearance of hard parts at the Proterozoic–Phanerozoic transition.

Symbiosis is the protracted physical association of organisms of different species. Symbiogenesis is the emergence of new species with new structures and physiologies, e.g., mitochondria and oxygen respiration, fish luminous organs, lichens, and oak galls, as a consequence of stable symbiotic associations. The term symbiogenesis was introduced by the Russian biologist Mereschkovsky in 1909 (Khakhina 1979). The importance of symbiogenesis as a mechanism of evolutionary innovation is explored in Margulis & Fester 1991. Symbiogenesis importantly supplements the gradual

Bengtson, S. (ed.) 1994: *Early Life on Earth. Nobel Symposium No. 84.* Columbia U.P., New York

accumulation of base-pair mutations, karyotypic rearrangements, and polyploidization. Yet to be determined is the relative importance of these evolutionary mechanisms, which are not mutually exclusive. Nevertheless, it is recognized that, unlike prokaryotes (bacteria, cyanobacteria, and actinobacteria – the last misnamed "actinomycetes," as if they were fungi), all nucleated organisms (animals, plants, fungi, and protoctists) are evolutionarily derived from early events of symbiogenesis that involved nucleocytoplasm and certain classes of crucial cellular organelles, e.g., oxygen-respiring mitochondria and photosynthetic plastids. The question as to whether other eukaryotic organelles such as peroxisomes (de Duve 1991), hydrogenosomes (Müller 1988; Johnson *et al.* 1990; Lahti & Johnson 1991), and kinetosomes (Margulis & McMenamin 1990) are also symbiotically derived from bacteria is unresolved. We argue here that, because of the pervasive influence of symbiogenesis in the origin of eukaryotic organisms, no individual eukaryotes began with fewer than two types of genomic systems. Hence their Latin binomial names should be reinterpreted as applying to ensembles of genomes, bionts, or symbionts that are integrated to form holobionts.

We also draw attention to the power of symbiogenetic recombination of genomes to generate a diversity of higher taxa and individual "species." We pursue the implications of these ideas for the Archean and Proterozoic fossil record, pointing out that taxonomic practices across the disciplines (bacteriology, mycology, zoology, etc.) are incommensurate.

For quite different perspectives on the forces and patterns of evolution, see, e.g., Nei 1987 (especially Chapter 6 on "Genomic Evolution"), Eldredge 1989, Feldman 1989, and Raup 1991.

Individuals formed by genome integration

Paleontologists face the necessity of devising a useful taxonomy for the geological remains of formerly living communities, such as bioturbated sediments (*Paleodictyon*), stromatolites (such as *Conophyton*), horizontally aligned biogenetic gas holes or burrows (*Skolithos*), and fossil coral reefs (*Axixtes*). They recognize that such structures were most likely generated by communities composed of a great and unknowable diversity of organisms. Each member of the structure may have a distinct genome only remotely related to the others. So the concept of a *form-taxon* is used to describe body and trace fossils, including burrows, tracks, stromatolites, coral reefs, etc., each with a particular set of morphologically distinguishable characteristics. Each form-taxon, with its distinctive characteristics, labels a recognizable, repeatable morphological unit, often called a *morphotype*. The morphotype may even be an entire community. We argue that, notwithstanding the Linnaean claim that Latin binomials refer to individual members of a single species, in many and perhaps the overwhelming majority of cases (e.g., all eukaryotes), species names in contemporary taxonomy also refer to aggregates of individuals with diverse genomes, i.e. communities. For example, the cephalopod mollusk *Euprymna scolopes* forms a light organ with ciliated, microvillous appendages bearing pores that lead to empty spaces. The cilia sweep in bacterial symbionts that will develop into luminous colonies characteristic of this species of squid. When the light organ, which is embedded in the ink sac, has become replete with a dense, single-type luminous bacterial population, the cilia are no longer needed and are absorbed. This sequence of events, repeated each generation, describes the cyclical symbiont integration in the ontogeny of normal *Euprymna* squid (McFall-Ngai & Ruby 1991). The origins of permanently integrated microbial symbionts, like the twenty or so species of *Caedibacter* known in the ciliate genus *Paramecium*, are more difficult to discern. The relation between speciation and genome acquisition in these and other genera is insufficiently investigated.

Combinatorics of symbiotic genomes

The combination of genomes in symbiosis has a power to generate diversity in form-taxa that may not be generally appreciated. With a single host that has no symbionts, only one genomic combination is possible. This is the case with nearly all the bacteria. Exceptions would be *Pelochromatium roseum* or other consortia bacteria (regular associations of a single flagellated heterotroph with clustered anoxygenic photoautotrophs) or encysted *Bdellovibrio* (bdellocyst) that contains, at some points in its development, the genomes of both *Bdellovibrio* and its *Chromatium* host (Tudor & Conti 1977; Tudor & Bende 1986). With a host and one symbiont, three genomic combinations are possible: the host alone, the symbiont alone, and the host and symbiont together. Probable examples are *Giardia* (a diplomonad), *Neocallimastix* (a chytrid), *Retortomonas* (a mastigote), *Vairimorpha* (a microsporidian), calonymphids, and other anaerobic mastigotes that lack mitochondria but display two- or three-componented reproducing karyomastigonts (Kirby 1952; classes Retortomonadida, Diplomonadida, Parabasalia, etc., in Margulis *et al.* 1990). With a host and two symbionts, six combinations are possible: the host alone, each symbiont alone, the host with symbiont 1, the host with symbiont 2, and the host with symbionts 1 and 2 together. In general, the number of genomic combinations that can be generated in this way by a host with n symbionts, assuming that each symbiont in addition can survive by itself, is $2^n + n$. With a host and ten symbionts, the number of potential taxa formed by recombination is 1,034. With a host and 20 symbionts, the number of possible genomic combinations is 1,048,596. With a host and 25 symbionts, the number of possible combinations is 33,554,457. This number approximates the minimal number of species on Earth estimated by some authors (e.g., May 1990; T. Erwin, oral presentation, 1994).

While genomic symbiosis has enormous power to generate diversity, that power may not always be used. For example, the platymonad marine worm *Convoluta* occurs without any photosynthetic symbionts as *Convoluta convoluta*. It is also found in regular and predictable combination with at least two kinds of photosynthetic symbionts, one at a time. With diatoms the yellowish worm is called *Convoluta paradoxa*, and when the symbionts are the green alga *Tetraselmis* (which is the same as *Prasinomonas*), all worms are not only bright green but they are functionally photosynthetic. The green form is called *Convoluta roscoffensis* (Smith & Douglas 1989). Some argue that *C. roscoffensis* should be removed from *Convoluta* to another genus, implying a still more profound effect of the cyclical symbiont integration that is characteristic of these marine worms. It is unlikely that those *Convoluta* occur with more than a single type of photosynthetic symbiont at the same time.

The examples of *Convoluta* and others (Table 1) show that the process of symbiogenesis is currently active at the level of individual species as labeled by conventional Latin binomials. Genomic symbiosis – i.e. acquisition and integration of microbial symbionts – may have played a powerful role in the origin of higher taxa, such as the 33 formally recognized phyla of animals (Margulis & Schwartz 1988). Conventional gradual accumulation of mutations, probably crucial for maintenance of symbionts and emergence of new holobiont properties, may then have differentiated these groups further at the species level (Margulis 1976; Margulis 1993).

TABLE 1   Taxa-specific symbioses: very few examples.

Host	Symbiont 1	Symbiont 2	Symbiosis name	New features, comments
Protists:				
Devescovina[1]	unidentif. fusiform bacterium	peritrichous bacterium	"Rubberneckia"	gliding and swimming motility
Mesodinium	none	none	*Mesodinium album*	heterotrophic mesodinium
Mesodinium	partial cryptomonad	none	*Mesodinium rubrum*	photosynthetic, fast-swimming ciliate
Metopus	methanogen	none	*Metopus contortus*	life in anoxic environments
		?	*Metopus paleoformis*	life in anoxic environments
Paramecium	none	none	*Paramecium aurelia*	ciliate
Paramecium	*Caedibacter*	none	*Paramecium aurelia*	killer-strain ciliate
Paramecium	none	*Chlorella vulgaris*	*Paramecium bursaria*	photosynthetic ciliate
Paramecium	*Caedibacter*	*Chlorella*	no such organism	
Plagiopyla	methanogen	"hydrogenosome"	*Plagiopyla* sp.	life in anoxic environments
Animals:				
Convoluta	none	none	*Convoluta convoluta*	heterotrophic worm
Convoluta	*Tetraselmis*	none	*Convoluta roscoffensis*	photosynthetic worm
Convoluta	none	diatom	*Convoluta paradoxa*	photosynthetic worm
Convoluta	*Tetraselmis*	diatom	no such organism	
Gazza leiognathid fish	vibrio gram-negative bacterium	none	*Gazza minuta* (ponyfish)	gas-bladder light organ, luminous fish
Hydra	none	none	*Hydra* sp.	brown hydra
Hydra	*Chlorella*	none	*Hydra viridis*	photosynthetic hydra
Hydra	*Chlorella*	*Aeromonas*	*Hydra viridis*	photosynthetic hydra
Monastraea[2]	*Symbiodinium*?	none	*M. annularis*, morphotypes I, II & III	carbonate reef formation

[1] Tamm in Margulis 1993
[2] Knowlton *et al.* 1992

The role of an additional symbiont may depend on the number and physiological features of other symbionts, if any, already associated with a given host. For example, when one additional symbiont joins *Convoluta*, the species name changes. By contrast, a domestic cow may have a large number of stably associated rumen ciliates and cellulolytic bacteria and an even larger number of transient rumen ciliates and spore-forming bacteria. When one or another of the transient rumen microorganisms arrives or departs, even in huge numbers, it is more customary to change the description of the "health" of the cow than its species classification.

TABLE 2   Inconsistent names of taxa.

Higher taxa[a]	Name (minimal number of genomes per individual)	Partner (number of genomes per partner)	Basis for name[b]
L	*Heterorhabditis bacteriophora* (3)	*Xenorhabdus 1* (1)	complex
L	*Heterorhabditis luminescens* (3)	*Xenorhabdus 2* (1)	complex
SYM[1]	*Chlorochromatium aggregatum* (2)	*Chlorobium chlorochromatii* (1)	complex
SYM[1]	*Pelochromatium roseum* (2)	brownish chromatium (1)	complex
L	*Paramecium aurelia* (2 + 1 = 3)	*Caedibacter* (1)	L, ST
L	*Paramecium bursaria* (2 + 3 = 5)	*Chlorella* (1)	L, SP
SYM[2]	*Cyanophora paradoxa* (2 + 1 = 3)	cyanobacterium[c] (1)	small
SYM[2]	*Cyanidium caldarium* (2 + 1 = 3)	cyanobacterium[c, d] (1)	small
SYM[3]	*Cladonia cristatella* (2 + 1 = 3)	*Nostoc* (1)	large
SYM[3]	*Cladonia cristatella* (2 + 3 = 5)	*Trebouxia* (3)[e]	large
L	*Glycina max* (3)		
L	*Glycina max* (3 + 1 = 4)	*Rhizobium* (1)	large
SYM[4]	*Microcycas* (3)	*Nostoc* (1)	large
L	*Convoluta convoluta* (2)		
L	*Convoluta paradoxa* (2 + 3 = 5)	*Bacillaria* (3), diatom[e]	large, SP
L	*Convoluta roscoffensis* (2 + 3 = 5)	*Tetraselmis* (3), green alga[e]	large, SP
L	*Homo sapiens* (2)	unknown	large
L	*Homo sapiens* (syphilitic) (2 + 1 = 3)	*Treponema* (1) plus unknown	large

[a] L=large; SYM=smaller symbiont. Higher taxa (families, orders, classes, phyla) based on *large* partner when the presence of the small one is irrelevant to taxonomy and on symbiotic complex (SYM) when the entire higher taxon is defined by traits characteristic of the complex and not of its components. SYM[1] = consortia bacteria; SYM[2] = glaucocystophytes; SYM[3] = lichens; SYM[4] = cycads.

[b] Name of genus based on *large* partner when name is independent of presence or absence of smaller partner; on *small* partner when presence of endosymbiont determines genus name; on *complex* when genus is defined by traits of the partnership. SP = specific name determined by presence of endosymbiont; ST = strain name determined by presence of symbiont.

[c] Sometimes called a cyanelle.

[d] Sometimes called a chloroplast or rhodoplast.

[e] Plastid, mitochondrion, nucleocytoplasm.

In the protoctists – a huge taxon (Kingdom Protista or Protoctista) estimated to encompass 250,000 species – the relative sizes of the symbionts (bionts) that form the "individuals" (holobionts) are far more equal than those of plants, animals, and fungi. Therefore, both the clearly symbiogenetic provenance and the nomenclatorial confusion are far more evident in these eukaryotic microorganisms than in other large taxa (Corliss 1992). Given new results of molecular biology, the taxonomy and practical systematics of the group of former animals (province of zoology), former plants (province of botany), and former fungi (province of mycology) have reached nearly crisis levels (Margulis 1992a). Recognition of the "legitimacy of having distinct high-level ranks for protist species that seem to be widely separated phylogenetically from fellow protists or from eukaryotic assemblages" is fervently pleaded by Corliss (1992).

Not all possible symbiotic combinations are likely to be realized in practice. For example, the presence of one type of photosynthesizer, one hydrogen-sulfide generator, or one dinitrogen fixer probably precludes any selection pressure for a second of the same type. Furthermore, not all of the combinations realized in nature may be distinguishable (Table 2). An open empirical question is to determine the actual relation between the number of possible symbionts associated with a host and the number of "species" conventionally distinguished for the corresponding group of genomic combinations.

The relative poverty of species in the Archean fossil record and their prokaryotic level of organization are well established (Schopf 1983b). This suggests that the major integration of microbial symbionts to form individuals of higher levels of complexity did not occur until the beginning of the Proterozoic Eon associated with the appearance of *Grypania* (Han & Runnegar 1992; Runnegar, this volume) and the later Ediacaran protoctists and animals (McMenamin 1993). The remarkably sudden appearance of large marine animals at the end of the Proterozoic and through the lower Phanerozoic may be related to symbiont acquisition, especially of calcium-precipitating bacteria by soft-bodied animals (Lowenstam & Weiner 1989). This well-known discontinuity in the fossil record may correlate with symbiotic consortia having 7–9 different components and having the capacity to generate hundreds of distinct morphotypes (species). The techniques of molecular biology permit analysis of complex genomes of eukaryotes and recognition of their elemental composition by identification of the original metabolism, morphology, and genomes of microbes that comprise them.

Conclusions

What are the implications of this analysis? First, biologists should recognize explicitly that most of their so-called individuals, including all eukaryotes, are in fact genomic combinations;[1] they should consider the possibility of adopting a consistent large-host nomenclature that appropriately recognizes the integrated genomes. Taxonomic nomenclature should be more consistent across fields; Table 3 illustrates the problem.

Second, the role of symbiogenesis as a driving factor in the diversification of life should be investigated empirically in many more groups than it has been so far. A start

---

[1] E.g., four genomes of algal cells: Nucleocytoplasm, undulipodia, mitochondria, and plastids.

TABLE 3   Phototrophic marine protoctists[a]: Identical organisms[b] (individuals[c]) described by different higher-taxa names.

Taxa assigned	People who use this terminology
phytoplankton, photoplankton, nanoplankton	oceanographers, limnologists
algae, microphytes, phytomonads	phycologists, ecologists, zoologists
aquatic plants	ecologists
green scum	public-at-large
phototrophic protists, photosynthetic eukaryotes	bacteriologists
plants, lower plants, algae, photosynthetic protoctists, chrysophytes, prymnesiophytes, haptomonads, thallophytes	botanists
coccolithophorids, prymnesiophytes, lower plants	paleontologists, geologists
eukaryotic microbes, algae, protists	cell biologists

[a] For detailed classification of these organisms see the *Handbook of Protoctista* (Margulis *et al.* 1990).

[b] Examples: *Chrysochromulina* (dasmotrophic coccolithophorid), *Dunaliella* (motile green alga), *Emiliana* (coccolithophorid), *Mychonastes* (nonmotile encysting green alga of the chlorella type).

[c] If we were to recognize microbiological standards and require growth in pure culture of all the organisms involved, we would not be allowed to name many protoctists, animals, plants, or fungi.

in this direction has been made by McFall-Ngai & Ruby (1991) in their analysis of luminescent squid, by Nealson (1991) in his analysis of "glowworms" (lepidopteran larvae inhabited by nematodes and luminous bacteria), by Schwemmler (1991) in his studies of homopterans such as *Eucelis* with its integrated bacterial symbionts, by Nardon & Grenier (1990) in weevil–bacterial associations, and by Vetter (1991) in his analysis of thiotrophic animals. We predict that between 20 and 22 physiologically distinctive microorganisms (primarily bacteria and fungi) are regularly associated with coleopterans. Genomic combinatorics may explain why, as J.B.S. Haldane observed, God has expressed such an inordinate fondness for His most flamboyant morphotypes: His millions of species of beetles.

Acknowledgments. – L.M. is grateful to the NASA Life Sciences Office (NGR-025-004) and to the Dean of the College of Arts and Sciences, University of Massachusetts, Amherst. J.E.C. acknowledges the support of U.S. National Science Foundation grants BSR87-05047 and BSR92-07293 and the hospitality of Mr. and Mrs. William T. Golden.

This chapter is dedicated to the memory of Heinz A. Lowenstam, pioneer integrator of biological and geological knowledge, a founder of the field of biomineralization.

The continuing importance of cyanobacteria

Stjepko Golubic

Biological Science Center, Boston University, 5 Cummington Street, Boston, Massachusetts 02215, USA

Nucleotide sequence analyses have identified cyanobacteria as a coherent monophyletic group of prokaryotes that introduced oxygenic photosynthesis early in the Earth's history. They also confirmed the hypothesis of the origin of eukaryotic chloroplasts from cyanobacteria. Although their dominant role has been largely replaced by eukaryotes in the course of the Phanerozoic, cyanobacteria still contribute significantly to the primary productivity of modern oceans and continue to build stromatolites.

For many years cyanobacteria have been treated like a primitive and obscure group of algae (blue-green algae), although their kinship with bacteria has long been recognized. In the system first introduced by Ferdinand Cohn in 1857 (see Mollenhauer & Kovacik 1988), the organisms of prokaryotic organization were separated from others on the basis of their reproduction by simple fission. According to that division, a phylum-level distinction separated cyanobacteria as alga-like Schizophyta from the fungus-like Schizomycetes, or common bacteria. This early understanding of the kinship of cyanobacteria with other bacteria, and their taxonomic coherence as a group, was confirmed in modern days by the results of numerous comparative studies, including those of fine structure, biochemical composition, physiological properties, and genetic makeup (Stanier 1977; Castenholz & Waterbury 1989). Electron microscopy revealed their prokaryotic cellular organization. The organization of nucleic acids in cyanobacteria and the mechanisms of DNA replication and protein synthesis follow the patterns common to all bacteria. The cyanobacterial cell wall is typical of gram-negative bacteria. Cyanobacteria are also sensitive to antibiotics that interfere with peptidoglycan synthesis in their walls. These and other strictly prokaryotic properties, including the ability to fix molecular nitrogen and form intracellular gas vesicles, leave no doubt about the bacterial nature of cyanobacteria (Stanier & Cohen-Basire 1977).

The uniqueness and coherence of cyanobacteria as a group and, particularly, their ecological distinctiveness from other bacteria have also been long recognized and reflected by such terms as Myxophyceae Stitzenberger, 1860 (slime-algae), Cyanophyceae Sachs, 1875 (blue-green algae), or Schizophyceae Cohn 1879 (fission-prone algae) (taxonomic references in Gomont 1892). The folded intracellular membranes that form the thylakoids of a cyanobacterial cell are similar to those of other phototrophic

Bengtson, S. (ed.) 1994: *Early Life on Earth. Nobel Symposium No. 84.* Columbia U.P., New York

and chemolithotrophic bacteria but also to the chloroplasts of eukaryotic algae and green plants. The property of oxygenic photosynthesis qualifies cyanobacteria ecologically as *bona fide* primary producers that carry out the same basic ecological functions as algae, mosses, and vascular plants. Accordingly, they often show similar distribution patterns and occupy similar ecological niches.

The structural and functional similarity between the energy-generating systems of cyanobacteria and green plants is not coincidental but rather a consequence of their common origins. The theory of symbiotic origin of the eukaryotic cell was proposed almost a century ago by the Russian biologist K.S. Mereschkowsky (1905). It stated that the chloroplasts and mitochondria of eukaryotic cells have their origins from endosymbiotic cyanobacteria and aerobic bacteria, respectively, whose ancestors were once captured and incorporated by a primitive, anaerobic, heterotrophic host. Scientific documentation of the correctness of these early assumptions had to await modern times. The evidence supporting this hypothesis now includes the presence of a double membrane around chloroplasts and mitochondria, the presence of DNA within them, the size and mode of operation of ribosomes within these organelles (Margulis 1992b), and comparative studies of nucleotide sequences (Woese 1987b; Giovannoni et al. 1988). The last-mentioned results documented unambiguously that chloroplasts in eukaryotes are genetically more similar to free-living cyanobacteria than to the cytoplasm of the eukaryotic cell. Giovannoni et al. (1988) have also shown that modern free-living cyanobacteria constitute a single discrete group, representing one of about ten or more well defined eubacterial phyla (also see Wilmotte & Golubic 1991).

Significant in this context are the discoveries of prochloralean phototrophic oxygenic prokaryotes, characterized by chlorophyll *a* and chlorophyll *b*, a pigment combination found in green algae, mosses, and vascular plants (Lewin 1989); the pigmentation of cyanobacteria, with chlorophyll-a in combination with phycobilins, is more similar to that of red algae and cryptomonads. However, phylogenetic clustering based on nucleotide sequence analyses was not entirely consistent with the phenotypic groupings mentioned above. The proposed origins of prochloralean taxa examined were located on several branches within the cyanobacterial cluster, while the origins of rhodophyte and cryptomonad plastids clustered more distantly. Clarification of the meaning of these data awaits further research.

Taken together, the results support a monophyletic origin of oxygenic photosynthesis derived from cyanobacterial ancestors but suggest that the acquisition of prokaryotic phototrophic endosymbionts in the evolution of eukaryotes may have been a multiple event. They also underline the antiquity as well as the ecological and phylogenetic importance of symbiosis and symbiogenesis in the history of life on Earth. The origin of eukaryotes is thereby placed in the context of coevolution, in which ecological interactions with various degrees of integration, and various degrees of specificity (Rowan 1991), provide the framework for natural selection.

Significance of cyanobacteria to the planet's history

Conclusions about the historic importance of cyanobacteria derived from comparative studies of modern cyanobacteria are consistent with the rich fossil record of the group.

The fossil record confirmed their early presence and diversification in strata more than 3.5 Ga old (Awramik *et al.* 1983; Schopf & Walter 1983; chapters by Schopf and Walter, this volume). It also showed that some ancient forms and/or functions were so successful that they persisted up to the modern times (Golubic & Hofmann 1976; Green *et al.* 1988; Al-Thukair & Golubic 1991). Combined studies of comparative genetics of modern organisms and of the fossil record demonstrate that cyanobacteria as a group can be credited for the introduction of oxygenic photosynthesis early in the planet's history. This metabolic mode became the prevailing and most efficient means of biological energy generation that fueled the primary production of organic matter. This process also introduced significant changes in environmental redox conditions in the Earth's hydrosphere and atmosphere and was responsible for the formation and later maintenance of the highly oxygenated atmosphere of our planet (Holland *et al.* 1986). The environmental changes introduced by oxygenic photosynthesizers were, in turn, a prerequisite for the evolution of the aerobic respiration in heterotrophic bacteria and of the chemolithotrophic metabolic pathways that use oxygen as electron acceptor. Thus, fully functional ecosystems, characterized by efficient nutrient-recycling schemes, must have been established and maintained by prokaryotes alone in Proterozoic marine, freshwater, and terrestrial environments.

Following the evolution of eukaryotes toward the end of the Proterozoic, the eukaryote-dominated ecosystems did not simply replace the preceding prokaryotic ones but rather integrated them into systems of higher structural and functional complexity. Numerous examples of coevolution within these integrated systems have a common underlying theme: one partner evolves responses that accommodate the needs of the other. For example, the jellyfish *Cassiopeia* assumes an inverted position, which exposes its endosymbionts to optimal illumination, and the reef corals provide firm support as well as optimal positioning for their endosymbiotic phototrophic units. When light gets sparse with the ocean's depth, these corals grow flat and leaf-like. In a similar fashion, the stems and leafs of trees can be viewed as devices that provide support and optimal orientation to the incoming light for their endosymbiotic cyanobacteria-turned-chloroplasts. Higher-level integration can lead to vertical differ-entiation of tree canopies that results in complex, stratified communities such as tropical forests. In principle, however, this differentiation process has a lot in common with arrangements in much simpler, stratified, prokaryotic communities such as microbial mats.

The pre-Cambrian sedimentary record has left an impressive array of preserved stromatolitic organosedimentary structures (Awramik 1991). The richness of this record suggests that stromatolites have dominated a wide range of ancient marine (subtidal and intertidal) and freshwater environments throughout most of the Earth's history. Each stromatolite lamina marks the position of an ancient sediment–water interface once occupied by a cyanobacteria-dominated microbial mat (Golubic 1973, 1976), while the vertical profile of these structures, often several meters high ("H"-sensu Hofmann [1969]), indicates the persistence of the microbial communities over time, suggesting indirectly a certain stability of the paleoenvironmental conditions that supported them.

Continued importance of free-living cyanobacteria

Toward the end of the Proterozoic, the diversity of stromatolites had substantially declined (Awramik 1991). The seafloor became inhabited by new functional entities. Cyanobacteria-dominated stromatolitic reefs were gradually replaced by reefs constructed and fortified by crustose red algae, calcareous sponges, and/or coelenterates. Cyanobacteria and stromatolites continued to thrive and dominate in marginal habitats, where harsh environmental conditions excluded competition and/or destruction by eukaryotes (Garrett 1970a). Monty (1973) discussed at length the implications of these changes and hypothesized how cyanobacteria and other stromatolite-building microbes that once occupied optimal ranges in marine and freshwater environments were later displaced by more advanced eukaryotes.

Recent discoveries, however, suggest that the news of the demise of cyanobacteria in environments with optimized conditions has been greatly exaggerated. This point will be illustrated here by two examples of successful cyanobacteria, from the marine plankton and benthos with normal salinity. The contribution of the coccoid cyanobacterial genus *Synechococcus* to the primary production of modern oceans and the role of the filamentous cyanobacterium *Schizothrix* in the construction of modern subtidal stromatolites will be evaluated.

Picoplanctic cyanobacteria

Over the years, the study of natural populations of cyanobacteria has concentrated on freshwater and, to a lesser extent, coastal marine environments. Until recently, the knowledge of the considerable taxonomic and ecological diversity of cyanobacteria of these habitats was contrasted by a perceived subordinate role and paucity of cyanobacteria in the open ocean. Only about half a dozen taxa clustered around the marine cyanobacterial genus *Trichodesmium* have been known from the plankton mostly from tropical oceans. Then, application of new analytical techniques based on fluorescence light microscopy to seawater samples brought about a major discovery (Waterbury *et al.* 1979; Johnson & Sieburth 1979). Fluorescence microscopy revealed the presence of chlorophyll and phycobilins in numerous tiny cells that were otherwise indistinguishable from ordinary bacteria. These small, 0.6–1.8 µm wide, ovoid and rod-shaped cells occur dispersed in the water column, reaching densities of up to 10^6–10^7 cells L^{-1} (up to 10^9 in coastal blooms). In spite of the small size of these organisms, they contribute 10–20% to the primary production of the open ocean. Their light-harvesting apparatus is most efficient, so that their competitiveness and their relative contribution to oceanic primary production increases with depth. They serve as a food source to small protozoan heterotrophs, which exert regular grazing pressure upon them. Biosynthesis and reproduction of both the phototrophic picoplankters and their grazers follows a circadian rhythm (Waterbury *et al.* 1986; Sweeney & Borgese 1989).

Because of the simplicity of their morphology, these tiny cyanobacteria have been provisionally classified within a single genus, *Synechococcus*, although their genetic properties indicate that they represent a diversified group of organisms belonging to at least four separate cyanobacterial genera (Waterbury & Rippka 1989; Wood & Towns-end 1990). *Synechococcus* strains have a global distribution within a wide range of environmental temperatures, being absent only in the frigid polar regions. There is also evidence of functional diversification within the group: the open-ocean strains possess phycoerythrin and are adjusted to low nutrient levels, while the coastal ones have only phycocyanin and respond to elevated nutrient levels (Waterbury *et al.* 1986). Inter-mingled with *Synechococcus* are also picoplanktic prochloralean unicells (Vaulot *et al.* 1990). Picoplanktic cyanobacteria have also been reported from large freshwater bodies such as the Great Lakes of North America (Stockner 1988; Pick 1991; Fahnenstiel & Carrick 1992).

These relatively new findings suggest that a diverse and specialized groups of small coccoid cyanobacteria populated the world's oceans and large freshwater bodies quite early, and that they persisted in these habitats until today. Hypersaline and other extreme environments, on the other hand, were and remain occupied by a different set of specialized cyanobacteria, with their own adaptations and requirements adjusted to the elevated ionic content and osmotic conditions of those environments (Golubic 1980; Montoya & Golubic 1991).

Cyanobacteria in subtidal stromatolites

Early encounters with modern stromatolites in temporary ponds (Black 1933) and in coastal intertidal zones prompted the generalized conclusion that all ancient stromato-lites must have been restricted to intertidal environments (Logan 1961). When sedimen-tological evidence suggested otherwise (Playford & Cockbain 1969; Bertrand-Sarfati & Moussine-Pouchkine 1985), interest turned to modern subtidal stromatolites in the search for appropriate models of the past. The discovery of large, permanently sub-merged stromatolites in tidal channels of the Exuma archipelago, Bahamas (Dravis 1983), offered a model for a presumed wider environmental distribution of Proterozoic stromatolites (Dill *et al.* 1986).

The validity of this model has recently been challenged. Attempting to identify the microbiota responsible for the formation of Bahamian stromatolites, Riding *et al.* (1991) found a complicated picture with a few cyanobacteria and many eukaryotic algae overgrowing the structures. They concluded that the sand-sized sediment grains (pelloids and ooids) found incorporated within these stromatolites (in accordance with sediment sorting in the strong tidal currents surrounding them) were too large for prokaryotes to entrap. They proposed instead that these stromatolites are of eukaryotic and most recent (Cenozoic) origin.

This preliminary picture of modern subtidal stromatolites has since been refined, and the causalities of their genesis could be clarified (K. Browne & S. Golubic, unpublished). The growing fronts of the subtidal (as well as nearby intertidal) Baha-mian stromatolites were found invariably affiliated with a fine filamentous cyano-bacterium of the genus *Schizothrix*, which was recognized as a pioneer of the microbial community. Eukaryotes join the community in a series of successions that parallel a slowdown in accretion rates and a progressive hardening of stromatolitic structures by internal cementation.

Stromatolites in the channels at Lee Stocking Island are located in a rapidly changing environment, in which strong tidal currents generate a "sediment storm" twice a day. Stromatolites are periodically covered and uncovered by moving underwater sand dunes. These dynamics punctuate their development in a cyclical fashion.

Stromatolite accretion is restricted to brief intervals following uncovering of the stromatolite surfaces, while these remain in close proximity to the shifting sands. A fine filamentous cyanobacterium, representing a new species of *Schizothrix*, builds characteristic beige-pink hemispherical protuberances, 1–20 cm in diameter with smooth surface and fine internal lamination. The trapping and binding of sediment grains proceed in a fashion typical of cyanobacterial mats. The surface laminae are characterized by highest densities of *Schizothrix* trichomes, while the interior contains mostly empty sheaths. Vertical sections through the domal protuberances show fine laminations, which illustrate the course of this principal growth phase of the stromatolites. An interstitial cementation process ensues within these structures shortly after their formation, resulting in progressive hardening of the structure.

Correlated with the degree of induration is a decrease in accretion rates, a gradual deterioration of the *Schizothrix* mat, and an increase in density of epiphytic diatoms colonizing mat surfaces. More advanced stages of stromatolite induration are accompanied by settlement of larger eukaryotic algae, specifically *Bathophora* and *Acetabularia*, leading finally to growth of *Sargassum* and crustose red algae. At this point of the development, stromatolites are completely uncovered, with their upper surfaces high above the shifting sands. They are overgrown by a complex benthic community composed of microbes, plants, and animals, which makes these stromatolite surfaces practically indistinguishable from any subtidal hard ground. These assemblages include constructive as well as destructive elements, such as micro- and macroendoliths. The hard-ground community of the subtidal stromatolites is largely destroyed when the stromatolites are buried in the shifting sands, and their reemergence starts a new cycle in their development.

The genesis of these stromatolites is revealed in the initial, accretional stage of the cycle and is linked to cyanobacteria. In this respect, there is no reason to dispute their relevance to Proterozoic stromatolites. Later addition and integration of eukaryotic elements in the formation of these structures can also serve as an instructive metaphor for evolutionary changes at the dawn of the Phanerozoic.

Environmental stress and the future of cyanobacteria

The preceding discussion has argued for a recognition of cyanobacteria as a group of microorganisms ancient in origin and central to life on Earth throughout its history. The impressive diversity of forms and functions that this monophyletic gene pool has generated over time enabled cyanobacteria to occupy a wide range of environments, ranging from the optimal to the extreme. In habitats with optimal conditions cyanobacteria have later been joined by other organisms or entered a symbiotic union with them. Under extreme environmental conditions, however, free cyanobacteria often remained unchallenged as the dominant primary producers.

The increasing deterioration of environmental conditions in recent years, in conjunction with water pollution and eutrophication, has been paralleled by an increased abundance of cyanobacteria. Cyanobacterial blooms accompany eutrophication of lakes and reservoirs, where their odoriferous metabolites (e.g., geosmine) may adversely affect the quality of drinking water (Jüttner *et al.* 1986), and some strains of blooming cyanobacteria (e.g., *Microcystis*) are toxic (Watanabe *et al.* 1986). Elevated phosphate levels resulted in a general increase in primary production until availability of nitrogen compounds became a limiting factor. Such conditions are conducive to the blooming of heterocystous cyanobacteria, which can overcome the limitation by fixing molecular nitrogen. Other cyanobacteria respond favorably to thermal pollution associated with the cooling systems of power plants. Similar responses have been noted in conjunction with increased salt concentrations in waters used for irrigation. In the marine environment, massive blooms of benthic cyanobacteria accompany the onset of anoxic conditions associated with coastal pollution. A marine *Phormidium* has been identified coating and killing corals. There is increasing evidence of the ability of cyanobacteria to suppress their competitors by means of production of allelochemicals and thus to modify species composition and their seasonal succession in aquatic systems (Flores & Wolk 1986; Gross *et al.* 1991)

These and similar observations paint disturbing, nightmarish scenarios of the future of our environment, reminiscent of the fictional image of the "last Adam" with a cockroach crawling over his dead body. Cyanobacteria are just as resilient likely survivors among photosynthetic, but not necessarily palatable, primary producers in ecosystems reduced and impoverished by the extreme conditions of a polluted world. On a more optimistic yet consistent note, I would like to call for a closer attention to cyanobacteria – because changes in the occurrence, density, and composition of their populations may contain important environmental clues that can be used as early warnings, before adverse conditions become irreversible.

Theme 3
Multicellularity and the Phanerozoic Revolution

The most striking event recorded in the fossil record is the explosive diversification of animals near the Proterozoic–Phanerozoic boundary. Fossils tell us quite clearly what has to be explained: macroscopic organisms of uncertain biology were present early in the Proterozoic; "higher" eukaryotes differentiated about 1 billion years ago; metazoans appeared some 600 million years ago; and the main radiation of animals was accomplished with lightning speed near the beginning of the Cambrian Period about 550 million years ago. This record is outlined in the first three chapters of this section (Hofmann; Sun; Fedonkin). Seilacher and Valentine propose intrinsic biological mechanisms for the "Cambrian explosion," and Bengtson and Riding discuss the significance of biomineralization. Knoll considers the late Proterozoic evolutionary events from an environmental context.

The phylogeny of the major groups of animals is a classic can of worms in biology. Conway Morris endeavors to synthesize the new information for extant as well as extinct lineages. Bergström and Christen query some of the mainstream interpretations of available molecular data and propose alternative methods of analyzing them; Rieger likewise suggests alternatives to some of the classical interpretations of metazoan phylogeny. The last word in our book is given to developmental biology and the all-important question as to how new structures and body plans are attained in animal evolution. Raff stresses the flexibility of early and late, in contrast to middle, developmental stages. Miklos & Campbell urge us to look beyond the Modern Synthesis in the search for driving forces of evolution.

Thus we are not short on explanatory hypotheses regarding the Phanerozoic revolution; arguments for causation range from environmental change to ecological interaction and from key biological innovations to a revolution in the genetic control of development. There is no reason to believe that any one of these theories is necessarily wrong and still less to pretend that any is complete. We remain blind men interpreting elephants. But the richness of accumulating data and theories suggest that a deeper understanding is possible – an understanding that will tell us much about how the modern world came to be.

Proterozoic carbonaceous compressions ("metaphytes" and "worms")

Hans J. Hofmann

Department of Geology, University of Montreal, P.O. Box 6128, Station A, Montreal, Quebec, H3C 3J7, Canada

Megascopic carbonaceous films are found in Proterozoic sequences worldwide, ranging from about 2 Ga into the Cambrian. Neoproterozoic occurrences dominate over older ones. The remains are provisionally classified into 13 family-level morphologic categories. Those probably or certainly belonging to the botanical realm are the Chuariaceae, Ellipsophysaceae (fam. nov.), Tawuiaceae, Longfengshaniaceae, Moraniaceae, Beltinaceae (fam. nov.), Vendotaeniaceae, Eoholyniaceae, and Grypaniaceae (fam. nov.). Those with probable metazoan affinities include Sinosabelliditidae (fam. nov.), Protoarenicolidae (fam. nov.), Saarinidae, and Sabelliditidae. The systematic position of most of these taxa with respect to modern groups remains problematic, however, even at the highest hierarchic levels. Some of them may represent nothing more advanced than large colonies of prokaryotic microbes or mat fragments, particularly among the moraniaceans and beltinaceans. The oldest remains thus far reported are spiraliform filaments and round and angulate compressions dating from approximately 2 Ga. More complex filaments with possible anatomical detail occur in approximately 1.4 Ga old rocks, and even more varied complex morphology is exhibited by 800 Ma old taxa. The first annulated organisms, including forms that resemble annelids, may have an age around 700 Ma.

Proterozoic sequences in many localities worldwide have yielded fossils in the form of black carbonaceous films with consistent geometry; some authors regard these as the remains of algae and they, therefore, place them in the Metaphyta. The word *metaphyte* is used to refer to the more highly organized or specialized (multicellular) plants, to distinguish them from the protophytes (primitive unicellular plants). In a comparable way, the word *metazoan* is used to distinguish multicellular animals from protozoans. Such terminology, while suitable for extant organisms, assumes that the metabolism and affinities of the erstwhile life-forms are known (for example, that an organism was an autotrophic photosynthesizer or had a heterotrophic life-style). Unfortunately, paleontologists can only examine that which endures after the organism has died and is left for posterity eons later. What usually remains are the physically and chemically resistant components, their replacements, or their impressions. Mineralized, shelly material is hard to find in Proterozoic rocks (and in fact is used by some as an indicator of the Phanerozoic). Generally, only the pliable organic materials or their

Bengtson, S. (ed.) 1994: *Early Life on Earth. Nobel Symposium No. 84.* Columbia U.P., New York

derivatives are conserved, almost always in compressed form because of sediment compaction, although three-dimensionally preserved occurrences are reported occasionally.

The structures of interest in our present discussion are the black carbonaceous films or related impressions that are visible to the naked eye and are found on bedding planes of Proterozoic siltstones and shales ranging in age from about 2 Ga to the Cambrian. The words *metaphytes* and *worms* in the title are placed in quotation marks because a fair number of the reported structures cannot be confidently assigned to these catego-ries of eukaryotes for lack of conclusive features. In fact, assignments are controversial, some carbonaceous compressions having been considered either metazoans or the remains of colonies of prokaryotic microbes. Nevertheless, most workers prefer to treat the more structured but smooth compressions as algal remains, even though histological attributes are not always in evidence. The presence of true metaphytes in the Proterozoic is corroborated by fossils other than megascopic carbonaceous compressions. Examples are well-preserved bangiophyte colonies (Rhodophyta) in 0.9 Ga old rocks on Somerset Island in northern Canada (Butterfield *et al.* 1990) and relatively complex fossils in the Sinian Doushantuo Formation of southern China, questionably assigned to the Rhodophyta (Zhang 1989). However, such metaphytic fossils are excluded from the present paper, which emphasizes the carbonaceous compressions.

For purposes of discussion, this review is divided into sections on morphology, geographic distribution, and chronologic perspective, followed by a discussion of the possible affinities of the remains.

Morphology and classification

The carbonaceous remains are usually preserved as micrometer-thick, shiny black films visible to the naked eye (>0.2 mm), ranging from less than one to several hundred square millimeters in extent. They are of diverse form, varying from smooth, wrinkled, or twisted slender filaments to irregular angulate and round bodies (Fig. 1). Most are without distinctive ornamentation or internal structure, making them difficult to classify. In many specimens, patterns of folds of compaction origin are prominent, giving some indication of the pliable nature of the original envelopes containing the vital matter but providing few clues to their anatomy. On the other hand, carbonaceous films with regular, closely spaced transverse markings (annulated, segmented?) found associated with the smooth compressions have been interpreted as metazoan remains because of their resemblance to "worms."

For most of the time after they were first reported in the literature in the mid-19th century, little headway was made in classifying the unornamented films because of their lack of distinctive characteristics other than gross shape. The taxonomy of carbonaceous compressions is still in a state of flux. The past decade has seen an explosion of new names at both the genus and the species levels. Names have been introduced, sometimes without knowledge of or reference to previous literature, sometimes without regard to the international rules of nomenclature, and sometimes without appreciation for morphologic variability accounted for by aspects of preser-

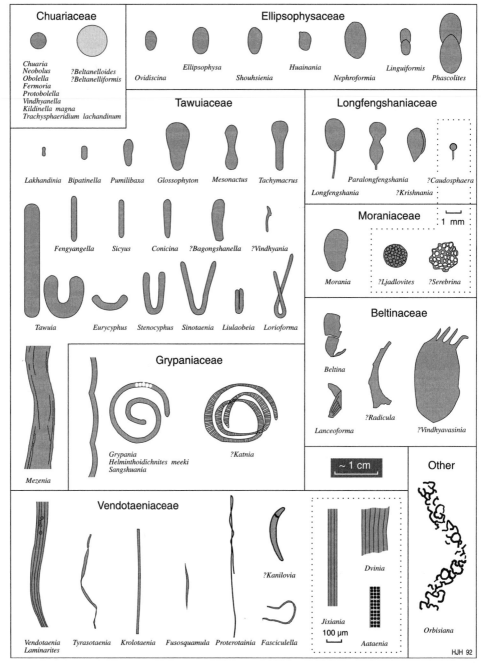

FIGURE 1 Synopsis of genera of black carbonaceous films or their impressions, reported from Proterozoic rocks worldwide. Individual illustrations are simplified representations of fossils reported in the literature. Names shown include synonyms, as well as those that are (or may be) invalid according to the international codes of nomenclature, but they are included here to provide a comprehensive inventory; further study is necessary to resolve the taxonomic status of many. The rugate (segmented?) carbonaceous films and certain microscopic forms are included for comparison. Question marks represent dubious assignment to suprageneric taxa. Dotted lines enclose taxa shown under higher magnification. No illustrations were available for *Qinggounella*.

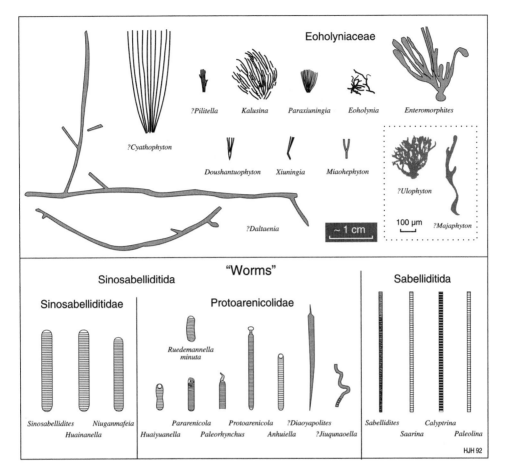

FIGURE 1   (continued)

vation (taphonomy). Altogether about 100 genus-level taxa have been proposed (see Appendix), but this tally includes many synonyms, as well as names that are invalid according to international codes of nomenclature or those whose taxonomic status is at least uncertain (for an assessment of the genera and species up to 1989, see Hofmann 1992a, b).

In an earlier general review (Hofmann 1985b), the forms were tentatively classified in an informal system embracing genus-level taxa, with categories named after the characteristic genus; an emended scheme (Hofmann 1992a) includes additional informal categories. A formal biological classification for carbonaceous remains was used in the monograph by Gnilovskaya *et al.* (1988, pp. 20–21). It follows along the lines used in the informal scheme used by Hofmann (1985b) to group the taxa, but the classification is mainly for Vendian fossils and leaves out many taxa from older rocks.

The matter of synonymy is subjective, a function of the philosophy of how to treat variable morphology in an assemblage (the traditional lumping and splitting of taxa). Much of the observed shape variation in remains of soft organisms can be ascribed to taphonomic factors, particularly if a whole range of intermediate types is available, and therefore fewer taxa need to be recognized. For instance, the sausage-shaped fossils (*Tawuia*) in the Neoproterozoic Little Dal Group of northwest Canada are present in a variety of tomaculate shapes, individual bedding surfaces containing I-, C-, U-, J-, and open S-shaped forms and intermediates (Hofmann & Aitken 1979; Hofmann 1985a). Some workers, studying a similar assemblage in China, have referred such variants to new taxa (Zheng 1980; Duan 1982; Fu 1986). Fu (1986) analyzed collections with abundant curved films in the Neoproterozoic Liulaobei Formation of Anhui Province and concluded that the fossils have parabolic curvature. He introduced two new genera, one for C-shaped and another for U-shaped forms. Such genera can be regarded as synonymous with *Tawuia*. Other authors (e.g., Duan 1982) have viewed C- or U-shaped and rectilinear specimens as separate species (*T. dalensis* and *T. sinensis*, respectively). But even the rectilinear form can be seen as parabolic, with the focus of curvature at infinity. Genera such as *Liulaobeia* are at the other end of this spectrum, with the focus close to the fossil. Parabolic curvature is commonly assumed by slender bodies when stressed. For example, slender flexible objects with ends attached to fixed supports sag to form a parabola; long flexible objects supported near their center of gravity also are pulled into this shape. It would seem feasible to look for a physical explanation for the parabolic shape of soft, flexible bodies such as those responsible for *Tawuia*.

Oval and elliptical specimens have also received different names, as have the filamentous forms, usually based on the presence or absence of a particular attribute, whose presence may even be accidental in some specimens, like tapering and peripheral, concentric, or longitudinal folds. Taphonomic variability may be the best explanation for such taxa, too. Not to be excluded from consideration is the ontogenetic variability, with elliptical or short cylindrical remains representing juveniles and the long ones the mature stages. Such an interpretation for *Tawuia* specimens is supported by the continuity shown by scatter plots of length against width and by the derived growth curves (e.g., Hofmann 1985a, p. 336). These examples serve to highlight some of the difficulties with taxonomy and classification.

Any objective scheme that treats carbonaceous films in Cryptozoic rocks should also accommodate the annulated filaments (Sabelliditida) that developed during the latest Proterozoic but occur mostly in Cambrian deposits. They are usually considered to be Metazoa (Pogonophora) by most authors. The inclusion of the annulated forms in any systematic treatment of carbonaceous films is warranted, because the Neoproterozoic sausage-shaped *Sinosabellidites* and related "worms" occupy an intermediate position, having the size and gross morphology of the unpatterned Tawuiaceae (which most authors regard as vegetal) yet exhibiting the annulation of Sabelliditida (with metazoan affinities).

In order to present a general, all-inclusive, tentative scheme of grouping the various disparate taxa in a relatively objective manner, the following 13 categories of megascopic compressions are here recognized (Fig. 1; see Appendix for more detail):

Chuariaceans   circular discs, usually with concentric or oblique wrinkles, but otherwise unornamented

Ellipsophysaceans   elliptical to ovate forms intermediate between chuariaceans and tawuiaceans

Tawuiaceans	rectilinear to simple curvilinear sausage-shaped forms with broad, rounded termini, usually not twisted, and without transverse elements
Longfengshaniaceans	round to oblong forms with single, narrow, polar stipe
Moraniaceans	irregular round forms, generally without wrinkles; commonly compressed microbial colonies
Beltinaceans	irregular angulate forms; fragmentary
Vendotaeniaceans	individual ribbon-like filaments, generally twisted; some members with associated microscopic discoidal structures
Eoholyniaceans	aggregates of filaments exhibiting branching
Grypaniaceans	slender, curvilinear forms with pronounced coiling tendency, or kinked when drawn out; some specimens with cell-like transverse markings that give uniseriate appearance to filament
Sinosabelliditidans	short to long tomaculate forms with fine, closely spaced transverse elements and broad, round terminus; group includes forms with polar structure such as a "proboscis" or an "aperture"
Sabelliditidans	long, slender tubes or ribbons with regularly spaced, narrow transverse annulation or segmentation
Others	compressions left unassigned to previous categories

Geographic distribution

The distribution of Proterozoic carbonaceous megafossils is global, with most reported localities concentrated in the Northern Hemisphere (Fig. 2). Neoproterozoic occurrences predominate over older ones. The apparent abundance of sites in the northern continents can be explained, at least in part, as being directly related to the number of researchers active in Proterozoic paleontology in a given region: the greater the number of paleontologists, the greater the likelihood that material will be reported in the literature. A second and probably related reason, of course, is the abundance and accessibility of well-preserved Proterozoic sections with paleontologic potential.

The most widespread forms are the chuariaceans and vendotaeniaceans. Tawuiaceans, longfengshaniaceans, and grypaniaceans are much more restricted. Ellipsophysaceans tend to co-occur with chuariaceans, tawuiaceans, and longfengshaniceans, whereas eoholyniaceans associate with vendotaeniaceans, and beltinaceans with moraniaceans. Sinosabelliditidans are very restricted, having been observed only in eastern China, where they are closely associated with tawuiaceans. Sabelliditidans are reported from a few occurrences in the Northern Hemisphere.

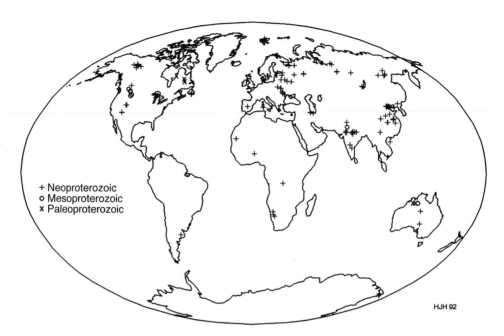

FIGURE 2 Occurrences of Proterozoic carbonaceous megafossils, grouped by eras. More localities have been reported, but geographic coordinates were not available.

Geochronologic perspective

The stratigraphic range of the various categories of carbonaceous films is given in Fig. 3. One would expect that the oldest remains are also the most primitive, that is, the moraniaceans and beltinaceans that most probably represent compressions of colonial microbes, both prokaryotic as well as eukaryotic (typical moraniaceans of Cambrian age are aggregates of filamentous forms). Moraniaceans and beltinaceans do, indeed, occur in Paleoproterozoic shales in North America belonging to geon 19. However, the oldest compressions (replaced) now known are, surprisingly, coiled forms attributed to the grypaniaceans recently recovered (but as yet incompletely studied) from the approximately 2 Ga old Negaunee Formation in Michigan (Han 1991; Han & Runnegar 1992). Much better preserved grypaniaceans are typically Mesoproterozoic and show features interpretable as anatomic detail (Fig. 1). Vendotaeniaceans are known from geon 17 rocks in China (Hofmann & Chen 1981, unpublished new material collected in 1991). Discoidal specimens associated with vendotaeniaceans in this same unit (Tuan-shanzi Formation) were originally attributed to chuariaceans, though their nature has subsequently been questioned inasmuch as they may be moraniaceans, but their chuariacean nature has not yet been disproved. Unquestioned chuariaceans are wide-spread in Neoproterozoic units. Longfengshaniaceans are known from only two, perhaps three, Neoproterozoic occurrences (China, north-western Canada, and, possibly, India); microscopic forms possibly belonging to the longfengshaniaceans (*Caudosphaera*) are present in three regions (Jankauskas 1989, p. 141), including the Late Riphean Lakhanda Group of eastern Siberia. Branching filaments (eoholyniaceans) are

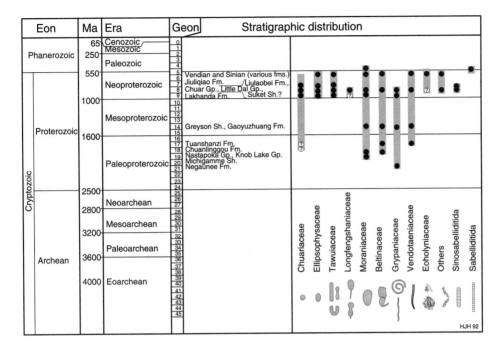

FIGURE 3   Chronologic perspective of major groups of Proterozoic carbonaceous megafossils. Only selected stratigraphic units yielding some of the more important assemblages are mentioned. (The geon scale is from Hofmann 1990.)

also Neoproterozoic and continue into the Phanerozoic. Sinosabelliditidans are even more restricted than other groups, being known only from the Neoproterozoic Qingbaikou sequence in eastern China. Sabelliditidans appear near the very top of the Proterozoic and may in fact be entirely Cambrian in age, depending on correlation of the units with the sequence containing the Precambrian–Cambrian boundary stratotype in Newfoundland.

Discussion

As already mentioned, the higher-level taxonomic positions of the different carbonaceous films are uncertain. Their generally carbonaceous composition has been used to place most of them in the botanical realm. The provisional morphologic classification adopted here does not exclude a possible and even probable polyphyletic nature of members assigned to the same suprageneric taxon. Reviews and discussions of the possible affinities of the compressions are given elsewhere (Hofmann 1985b, 1992a, b; Gnilovskaya *et al.* 1988).

Chuariaceae

The Chuariaceae have a long history of being interpreted in different ways. At one time or another, *Chuaria* has been regarded as a brachiopod, gastropod, hyolithid operculum, trilobite egg, medusoid, chitinous foraminiferan, green alga, megascopic acritarch, and microbial colony (see Spamer 1988 for a comprehensive annotated bibliography). The most recent explanation is that it represents megascopic colonies of cyanobacteria, like modern *Nostoc* balls (Sun 1987); such an explanation has long been applied to the Moraniaceae (Walcott 1919; Tyler *et al.* 1957, p. 1300). Sun's conclusion is based on the presence of a few filaments associated with some specimens of *Chuaria* from China. While such an interpretation is reasonable, it is not clear from the published illustrations whether the filaments are actually inside or outside the envelope. If they are inside and the nostocalean analogy applies, the absence of *Chuaria* in Phanerozoic sequences needs explanation, given the existence of the modern analogue. A further complication is the fact that the size distribution of specimens merges with that of the large sphaeromorph acritarchs, which are devoid of filaments (*Leiosphaeridia (Kildinella), Trachysphaeridium*), as are most other occurrences of *Chuaria* that have been specifically analyzed for cell content by thin sectioning or maceration.

The matter of including the large, concentrically rugate discs referred to *Beltanelloides (Beltanelliformis)* at the other end of the size spectrum within the Chuariaceae (Gnilovskaya *et al.* 1988), or even in the Class Chuariaphyceae or the Metaphyta, is controversial, given the absence of associated carbonaceous matter and the similarity to the gregarious, three-dimensionally preserved *Nemiana*. Specimens of the latter also lack a carbonaceous film, and the genus is commonly interpreted as a primitive sessile metazoan (e.g., Fedonkin 1990a, p. 71). The difference between these two genera appears to be taphonomic, with *Beltanelloides (Beltanelliformis)* having been preserved in pelitic sediment and *Nemiana* in sandy or silty beds.

Tawuiaceae

The Tawuiaceae were originally regarded as probably algal, but possibly metazoan, and placed with the vendotaeniaceans (Hofmann & Aitken 1979). Because *Tawuia* invariably co-occurs with *Chuaria* (see Fig. 3A–C of Sun, p. 361 of this volume), though the converse does not apply, it was thought possible that these two genera (and intermediate forms) are closely related systematically, perhaps representing individuals of the same species at different growth stages, or alternations of generations of eukaryotic algae (Hofmann 1985a). The close affinities were also expressed in the assignment of the two genera to the same family, Chuariaceae, by Duan (1982). More recently, Sun (1987) favored the interpretation that, like *Chuaria*, *Tawuia* represents nostocalean colonies (*Wollea* is a reasonable modern analogue), based on a similar association of filaments. As with *Chuaria*, the evidence is suggestive but not conclusive. Some specimens of *Tawuia* in the type assemblage in north-western Canada have small circular discs (minute *Chuaria*?) associated with them (Hofmann 1985a, Pl. 35:6–7), which could be reproductive structures, making metaphyte affinities more plausible. However, here, too, the position with respect to the envelope is not clear (outside or inside), nor is their primary nature. Some compressed specimens of *Tawuia* have outlines of a dark axis whose function has not been ascertained but which may simply be centripetally condensed content of individual envelopes. Three-dimensionally preserved specimens of both *Chuaria* and *Tawuia* on the same bedding plane in the Little Dal Group, examined in thin section and under the microprobe, have fillings of pure microspar and show no microbial content (Hofmann 1985a, Pl. 37). The affinities remain obscure.

Ellipsophysaceae

The Ellipsophysaceae are an intermediate group, possibly related to the chuariaceans and tawuiaceans on the one hand, or to the longfengshaniaceans on the other. Oval and elliptical forms may simply be deformed chuariaceans or short tawuiaceans. If there is distinguishable rectilinearity to the two lateral margins (either tapering or parallel), the specimens are attributable to the tawuiaceans. The longfengshaniacean connection lies in the size and oval-to-elliptical shapes, which are indistinguishable from those exhibited by the bulb end of individuals of *Longfengshania*; they may represent forms liberated from stalked structures. Specimens of *Phascolites* and *Linguiformis* show evidence of superposition of two individual ellipsoids, possibly accidental, and these genera are therefore grouped here with the ellipsophysaceans rather than treated as a separate group.

Longfengshaniaceae

The Longfengshaniaceae[1] are fossils with a well-differentiated body with large flattened bulb and attached stipe, found in association with chuariaceans and tawuiaceans (Fig. 3 of Sun, p. 361 of this volume). In fact, the broad end resembles individuals placed in those groups, suggesting a possible provenance for some of them. The stipe is viewed as the bulb's anchoring mechanism at the sediment–water interface (e.g., Du & Tian 1985). Zhang (1988) recently proposed possible bryophyte affinities for *Longfengshania*. The arguments are based on the gross morphologic resemblance to the sporophyte of the Devonian *Sporogonites*, but the case has not been convincing, considering the aquatic habitat of the Proterozoic forms and the lack of evidence of attached gametophyte, although a common mass at the center of a bundle of specimens in the Little Dal Group (Hofmann 1985a, Pl. 38:4; Text-fig. 5) might be interpreted in such a way (Zhang 1988). Nor does the lack of differentiated tissue make a better case for an algal interpretation (Phaeophyta?), though this is a possible interpretation. *Longfengshania* somewhat resembles small, mostly microscopic problematic forms described under *Caudosphaera* (German & Timofeev, in Jankauskas 1989, pp. 140–141) and interpreted as possible fungi. These occur in early Neoproterozoic (geon 9) rocks and appear to be somewhat older than the units with *Longfengshania*. Affinities remain obscure.

Moraniaceae

The Moraniaceae are flat, round compressions. Typical *Morania* comprises clusters of tangled filaments – compressed microbial colonies. The group can be considered among the most primitive type of the carbonaceous compressions, possibly representing nothing more advanced than colonies of prokaryotes (Walcott 1919; Tyler *et al.* 1957). Some near-microscopic-sized taxa (*Ljadlovites, Serebrina*) have reticulate patterns suggesting colonial aggregation of microbes.

---

[1] Maithy (1990) proposed *Krishnania* to be a senior synonym of *Longfengshania* (see Appendix). Given the poor preservation of the type material of *Krishnania*, the name Longfengshaniaceae should be conserved at least until the identity of the two genera can be convincingly demonstrated.

Beltinaceae

The Beltinaceae, like the moraniaceans, are most likely fragments of microbial colonies or redistributed fragments of mats, with no eukaryotic affinities implied, though certainly not excluded. They usually are found in the same beds that have the round moraniaceans.

Vendotaeniaceae

The literature on the ribbon-shaped Vendotaeniaceae is extensive, and many detailed descriptions, chemical analyses, and different interpretations are available. A synthesis of the data can be found in Gnilovskaya *et al.* 1988. For a number of reasons, including the megascopic dimensions, composition, and various anatomical details such as longitudinal striations and circular structures interpreted as sporangia, the vendo-taeniaceans are generally placed in the Metaphyta. A divergent opinion was expressed by Vidal (1989), who proposed that vendotaeniaceans are the remains of organotrophic bacteria, illustrating his point with photographs of clumps of the modern *Thioploca* of the Beggiatoaceae, from the deep sea off Peru. His conclusion was that the available data do not allow the identification of metaphytes among the Neoproterozoic Vendo-taeniaceae; he explained the reported anatomical features as differential preservation and fortuitous artifacts. However, while such analogies can cast doubt on the meta-phyte affinities of some members, they do not necessarily disprove them, either; they only provide an alternative working hypothesis. Regularly disposed markings, such as those shown by *Aataenia*, are difficult to explain as bacterial colonies. The vendo-taeniaceans, like the other groups, evidently need further investigation.

Eoholyniaceae

The Eoholyniaceae are filamentous forms exhibiting branching. The presence or absence of branching is not clear in some aggregates, where overlapping unbranched filaments may give the impression of its presence. Large, truly branching thalli, like those of *Enteromorphites*, are very suggestive of phaeophyte algae, but branching in near-microscopic forms is not conclusive of metaphyte affinities, inasmuch as aggre-gates of prokaryotes may also assume such shape. In this respect, forms such as *Majaphyton* need to be considered. These are not like other fossils in that, although they have clear, rounded outlines, they appear to lack a distinct wall and are without folds. They have a spongy or foamy amoeboid appearance that suggests they are compressed, self-coherent aggregates (colonies?) of fine granular elements of uncertain affinities.

Grypaniaceae

The Grypaniaceae are intriguing remains due to their relatively old age, spiraliform habit, and (in some Mesoproterozoic specimens) large, cellular, internal markings (see Fig. 2 of Sun, p. 360 of this volume). The last suggest interpretation as gigantic uniserial filaments with algal affinities, comparable in shape to certain coiled forms found among Proterozoic cryptarchs that are smaller by one or two orders of magnitude (e.g.,

Glomovertella, Leiothrichoides, Tortunema, Volyniella). The inclusion of the "segmented" *Katnia* with the grypaniaceans may at first seem odd, given the reported presence of a proboscis with mouth and jaw-like structures. However, these reported anatomical features (Tandon & Kumar 1977, Fig. 1) have not been convincingly illustrated. This taxon is, therefore, questionably included here because of the resemblance to coiled *Grypania* specimens with cell-like or spiraliform internal features that may be analogous to the "segments" in *Katnia*. *Grypania* has been compared to the modern dasyclad alga *Acetabularia* (Han & Runnegar 1992).

Sinosabelliditida

The Sinosabelliditida include two families, Sinosabelliditidae and Protoarenicolidae (see Fig. 4 of Sun, p. 362 of this volume). Sinosabelliditida are closely tied to tawuiacean occurrences in China; they have annulations like the Sabelliditida, but the gross tomaculate form of Tawuiaceae. Their annulations, whose primary or taphonomic origin is still obscure, make them resemble annelids and leads to their characterization as "worms." However, some authors have also expressed doubts about their metazoan affinities (e.g., Sun *et al.* 1986). Nevertheless, algae with regular distinct annulations are not known among modern taxa, so the systematic position of the Sinosabelliditida remains unclear. One should mention, though, that the transverse features are similar to those in microscopic, transversely rugate filaments assigned to *Plicatidium* (Jankauskas 1989, Pl. 41:3–4), constituting another case where Proterozoic microscopic and megascopic remains have similar form.

Paleontologists are more likely to accept the metazoan affinities of the protoarenicolids and to refer them to a group such as the Annelida. The fossils are more slender than sinosabelliditids and are more commonly curved and twisted; in addition to the general form, size, shape, and annulation (segmentation?), they are provided with a broad, circular aperture or proboscis-like feature at one end interpreted as the anterior, whereas the opposite end is rounded, as both ends are in the sinosabelliditids. Like sinosabelliditids, they, too, occur in beds that have *Tawuia* present. The polar structure is not always evident in illustrations of specimens attributed to the protoarenicolids (e.g., *Ruedemannella*), and one may ask whether such specimens are not better referred to the sinosabelliditids. Possibly also belonging to the protarenicolids, rather than to the grypaniaceans, is *Katnia*.

Sabelliditida

The tubular Sabelliditida are even better candidates for metazoans than are protoarenicolids, and they are classified as pogonophorans by most who have studied these fossils (Glaessner 1979). However, ultrastructural investigation by Urbanek & Mierzejewska (1983) and Ivantsov (1990) to relate the sabelliditids to this group proved inconclusive, leaving their affinities unresolved.

Others

Finally, the category "others" embraces additional forms that are poorly characterized or unassignable to the previous groups.

Summary and conclusions

Proterozoic sequences have yielded diverse forms of carbonaceous compressions that comprise the earliest megascopic fossils other than stromatolites and include probable members of the metaphytes and metazoans. The genera are arranged into 13 distinct morphologic categories at the family level, most of which could be polyphyletic. While many remains are probably nothing more than simple aggregates of prokaryotic microbes, particularly among the moraniaceans and beltinaceans, genera in the other families exhibit more advanced features that indicate eukaryotic organization, and they are therefore of greater biological, biostratigraphic, and evolutionary significance. Among these are the grypaniaceans, longfengshaniaceans, chuariaceans, and tawuiaceans, all of which have shapes that appear to be megascopic replicates of microscopic organisms (grypaniaceans of microscopic filaments, the chuariaceans of sphaeromorph acritarchs, the ellipsophysaceans of *Leiovalia*-like microbes, the tawuiaceans of *Navifusa*-like microbes, the longfengshaniaceans of *Caudosphaera*, and the sinosabelliditids of *Plicatidium*). Together, such similarities in diverse groups are intriguing and worth noting, but the significance of this fact, if any, is not yet understood. Some authors have commented on the exceptionally large cell size attained by Proterozoic organisms, but the reasons for this cell gigantism are speculative, if, indeed, the large size is that of cells and not of colonial aggregates. Abnormally large cell size may have something to do with unusual environmental or histological conditions in the Proterozoic that allowed for efficient metabolism and viability despite large cell size. Further investigation of this problem may be rewarding.

If structural complexity and large cell size are taken to be features of eukaryotes, the grypaniaceans and longfengshaniaceans, if not also some of the others, are eukaryotes; their carbonaceous composition suggests plant affinities. By this reasoning, the recently reported coiled filaments in Michigan (Han 1991; Han & Runnegar 1992) may extend the record of eukaryotes back to 2 Ga or to whatever the true depositional age of the Negaunee Formation is. Vendotaeniaceans are present by the late Paleoproterozoic (geon 17). The stiped longfengshaniaceans of the Neoproterozoic (geon 8) certainly are sufficiently large and complex to warrant consideration as metaphytes, as are the large branching forms of the Eoholyniaceae in the latter part of the Neoproterozoic. The annulated Sinosabelliditidae, whether animal or vegetal, are best viewed as eukaryotes. Protoarenicolidae and Sabelliditida are more readily acceptable as metazoans.

Acknowledgments. – Discussions with Robert J. Horodyski, Malcolm R. Walter, and other members of the Precambrian Paleobiology Research Group – Proterozoic (PPRG–P), headed by J. William Schopf, were of benefit. I thank Du Rulin and Zheng Wenwu for access to their respective collections, discussions, and guidance in the field, which allowed me to get better acquainted with Chinese materials and localities and to collect specimens for comparison with Canadian material. S.M. Mathur kindly provided a photograph of the type specimen of *Vindhyania jonesii*, and Tsu-Ming Han graciously made arrangements for me to examine as yet undescribed 2 Ga old material he discovered in Michigan. Mona Kachaami provided technical assistance. Financial support by the Natural Sciences and Engineering Research Council of Canada (under grant no. A7484) is gratefully acknowledged.

Appendix: Classification scheme for Cryptozoic carbonaceous compressions

The following grouping of form genera incorporates and expands schemes published earlier (e.g., Glaessner 1979; Hofmann 1985b, 1992a; Gnilovskaya *et al.* 1988). The classification should be considered as a tentative working scheme based mainly on the literature; it was possible to examine only a small portion of the type material of the various taxa.

To accommodate genera not included in previous schemes, five new suprageneric taxa are introduced here, based on their respective root type genus – three for metaphytes, and two for "worms." The diagnostic features for the new taxa are as follows: the Family Ellipsophysaceae (fam. nov., based on *Ellipsophysa* Zheng, 1980) is a group of carbonaceous fossils with elliptical and oval shapes, intermediate between circular Chuariaceae and tomaculate Tawuiaceae; the length/ width ratio is typically of the order of 1.3 to 1.7, as compared to typically >2 for Tawuiaceae. The Family Beltinaceae (fam. nov., based on *Beltina* Walcott, 1899) is proposed for angulate carbonaceous films, to separate them from the round Moraniaceae; typical *Morania* specimens from the Cambrian are the remains of megascopic colonies of filamentous microbes, but most Proterozoic material is preserved as smooth, rounded compressions whose microbial nature is only inferred. The Family Grypaniaceae (based on *Grypania* Walter *et al.*, 1976b) is introduced to accommodate slender, megascopic, filamentous forms showing evidence of coiling or corresponding regular kinking when drawn out; forms both with and without regularly spaced, large, cell-like elements are included.

The Family Sinosabelliditidae (fam. nov., based on *Sinosabellidites* Zheng, 1980) embraces carbonaceous films morphologically similar to Tawuiaceae, but bearing regularly and closely spaced transverse markings that have been used to relate the fossils to the "worms" (metazoans). The Family Protoarenicolidae (fam. nov., based on *Protoarenicola* Wang, 1982, emend. Sun *et al.* 1986; Chen 1988) comprises slender carbonaceous films with transverse annulations and a polar structure ("proboscis," "aperture"). Together, these two families are grouped into the new Order Sinosabelliditida (ord. nov., based on the new Family Sinosabelliditidae). The Sinosabelliditida are distinguished from the tubular and much more slender Sabelliditida, usually placed among the Pogonophora, by the more stubby appearance and by the presence of a broad, rounded terminus at one or both ends.

The names for suprageneric taxa in the classification scheme have different suffixes because the botanical and zoological codes of nomenclature have different rules. In the scheme it is assumed that the Sinosabelliditida and Sabelliditida are probable metazoans, and, therefore, the suffixes are those required by the ICZN, whereas the suprageneric names for the presumed metaphyta have endings required by the ICBN. If the Sinosabelliditidae were eventually shown to belong to the botanical realm, the proper corresponding family name would be Sinosabelliditaceae. They can be informally referred to as sinosabelliditans until their true affinities become more evident.

Higher affinities uncertain; probably photosynthesizers

Class Chuariaphyceae Gnilovskaya & Ishchenko, in Gnilovskaya *et al.* 1988
    Order Chuariales Ishchenko, in Gnilovskaya *et al.* 1988
        Family Chuariaceae Wenz, 1938, emend. Duan, 1982 [chuariaceans]
            Chuaria Walcott, 1899
                and synonyms:
                Neobolus Chapman, 1932
                Obolella Chapman, 1933
                Fermoria Chapman, 1935
                Protobolella Chapman, 1935
                Vindhyanella Sahni, 1936
                Kildinella Timofeev, 1969 (*partim*)
                Trachysphaeridium Timofeev, 1969 (*partim*)
            Beltanelloides Sokolov, 1972b, nom. nud.
                and synonym:
                ?*Beltanelliformis* Menner, in Keller *et al.* 1974

Family Ellipsophysaceae Hofmann fam. nov. [ellipsophysaceans]
   Ellipsophysa Zheng, 1980
     and synonyms or possible synonyms:
     Shouhsienia Xing (1979) *ex* Du, 1982
     Ovidiscina Zheng, 1980
     Nephroformia Zheng, 1980
     Huainania Wang *et al.*, 1984, nom. nud.
     Phascolites Duan & Du, in Duan *et al.* 1985
     Linguiformis Chen & Zheng, 1986, nom. nud.
Family Tawuiaceae Ishchenko, in Gnilovskaya *et al.* 1988 [tawuiaceans]
   Tawuia Hofmann, in Hofmann & Aitken, 1979
     and synonyms or possible synonyms:
     Mezenia Sokolov, 1976, nom. nud.
     Pumilibaxa Zheng, 1980
     ?*Vindhyania* Mathur, 1983
     Liulaobeia Wang *et al.*, 1984, nom. nud.
     Glossophyton Duan & Du, in Duan *et al.* 1985
     Eurycyphus Fu, 1986, nom. nud.
     Stenocyphus Fu, 1986, nom. nud.
     Fengyangella Chen & Zheng, 1986, nom. nud.
     Conicina Chen & Zheng, 1986, nom. nud.
     Bipatinella Chen & Zheng, 1986, nom. nud.
     Sicyus Chen & Zheng, 1986, nom. nud.
     Lorioforma Chen & Zheng, 1986, nom. nud.
     ?*Bagongshanella* Chen & Zheng, 1986, nom. nud.
     Sinotaenia [*Sinenia* in caption] Chen & Zheng, 1986, nom. nud.
   Lakhandinia Timofeev & German [Hermann], 1979
   Mesonactus Fu, 1989
   Tachymacrus Fu, 1989
Family Longfengshaniaceae Du & Tian, 1986 (1987) [longfengshaniaceans]
   Longfengshania Du, 1982
   Paralongfengshania Duan & Du, in Duan *et al.* 1985
   ?*Krishnania* Sahni & Srivastava, 1954
   ?*Caudosphaera* German [Hermann] & Timofeev, in Gnilovskaya *et al.* 1988
?Class Chuariaphyceae
  Order Moraniales Gnilovskaya *et al.*, 1988
    Family Moraniaceae Ishchenko, in Gnilovskaya *et al.* 1988, emend. [moraniaceans]
      Morania Walcott, 1919
      ?*Ljadlovites* Ishchenko, 1983 (chuariacean?)
      ?*Serebrina* Ishchenko, in Gnilovskaya *et al.* 1988
    Family Beltinaceae Hofmann fam. nov. [beltinaceans]
      Beltina Walcott, 1899
      Lanceoforma Walter *et al.*, 1976b
      ?*Vindhyavasinia* Tandon & Kumar, 1977
      ?*Radicula* Chen & Zheng, 1986
Class Vendophyceae Gnilovskaya, 1986
  Order Vendotaeniales Gnilovskaya, 1986
    Family Vendotaeniaceae Gnilovskaya, 1986 [vendotaeniaceans]
      Vendotaenia Gnilovskaya, 1971
       and synonym:
       Laminarites Eichwald, 1854
      Tyrasotaenia Gnilovskaya, 1971
      Proterotainia Walter *et al.*, 1976b
      Aataenia Gnilovskaya, 1976
      Fusosquamula Aseeva, 1976
      Dvinia Gnilovskaya, 1979
      Fasciculella Duan, in Duan *et al.* 1985

Jixiania Yan, 1987
Krolotaenia Tewari, 1989
?*Kanilovia* Ishchenko, 1983
Order Eoholyniales Gnilovskaya, 1986
Family Eoholyniaceae Gnilovskaya, 1986 [eoholyniaceans]
Eoholynia Gnilovskaya, 1975
?*Pilitella* Aseeva, 1976
?*Ulophyton* Timofeev & German [Hermann], 1979
?*Majaphyton* Timofeev & German [Hermann], 1979
Enteromophites Zhu & Chen, 1984 (recte *Enteromorphites* Chen & Xiao, 1991)
?*Daltaenia* Hofmann, 1985a
Kalusina Ishchenko, in Gnilovskaya *et al.* 1988
?*Xiuningia* Xing, Bi & Wang, in Xing *et al.* 1989, nom. nud.
?*Paraxiuningia* Xing, Bi & Wang, in Xing *et al.* 1989, nom. nud.
?*Cyathophyton* Xing, Bi & Wang, in Xing *et al.* 1989, nom. nud.
Doushantuophyton Chen, in Chen & Xiao 1991
Miaohephyton Chen, in Chen & Xiao 1991
Class and Order uncertain
Family Grypaniaceae Hofmann fam. nov. [grypaniaceans]
Grypania Walter *et al.*, 1976b, emend. Walter *et al.*, 1990
and synonyms:
Helminthoidichnites Walcott, 1899
Sangshuania Du, Tian & Li, 1986
?*Katnia* Tandon & Kumar, 1977 (may be protoarenicolid)
Affinities uncertain [others]
Orbisiana Sokolov, 1976, nom. nud.
Qinggounella Zhang & Zhai, in Chen & Zheng 1986

Phylum and class uncertain (Metazoa?)

Order Sinosabelliditida Hofmann ord. nov. [sinosabelliditidans]
Family Sinosabelliditidae Hofmann fam. nov. [sinosabelliditids]
Sinosabellidites Zheng, 1980)
and synonym:
Huainanella Wang, 1982
Family Protoarenicolidae Hofmann (fam. nov.) [protoarenicolids]
Protoarenicola Wang, 1982, emend. Chen, 1988
and synonyms:
Anhuiella Yan & Xing, in Xing *et al.* 1985, nom. nud.
Huaiyuanella Xing, Yang & Yin, in Xing 1984
Pararenicola Wang, 1982, emend. Chen 1988
and synonyms:
Paleorhynchus Wang, 1982
Ruedemannella Wang, 1982
?*Niuganmafeia* Chen, in Chen & Xiao 1991
?*Diaoyapolites* Chen, in Chen & Xiao 1991
?*Jiuqunaoella* Chen, in Chen & Xiao 1991

Metazoa

Phylum ?Pogonophora Johansson, 1937
Order Sabelliditida Sokolov, 1965 [sabelliditidans]
Family Saarinidae Sokolov, 1965 [saarinids]
Saarina Sokolov, 1965
Calyptrina Sokolov, 1965
Family Sabelliditidae Sokolov, 1965 [sabelliditids]
Sabellidites Yanishevskij, 1926
Paleolina Sokolov, 1965

Early multicellular fossils

Sun Weiguo

Nanjing Institute of Geology and Palaeontology, Academia Sinica, Nanjing 210008, P.R. China

The sudden appearance of diverse Cambrian faunas was preceded by a long history of multicellular evolution, but in general the fossil record of this process is very sparse. However, coiled carbonaceous megafossils of *Grypania* from the Paleoproterozoic suggest that the initial multicellular evolution dates back to at least 2.1 Ga ago. The early Neoproterozoic *Chuaria–Tawuia* megafossil assemblage, ranging 1,000–700 Ma ago, is composed almost exclusively of multicellular algae, except for a few worm-like forms that are probably of metazoan affinity. The late Neoproterozoic multicellular evolution is best characterized by the introduction of the Ediacara-type metazoans and also simple trace fossils. It seems that the early metazoan evolution may have been constrained by a low level of oxygen in the atmosphere and that, with the increase of oxygen content above a threshold value by the end of Proterozoic, a multitude of evolutionary innovations could have become practically feasible in the earliest Phanerozoic time, as seen in the Early Cambrian Chengjiang biota from Yunnan of China.

The fossil record of life on our planet, spanning nearly 3.5 Ga, can be compared to an iceberg. Demarcated by the abrupt appearance of diverse skeletal fossils near the base of the Cambrian System, about 550 Ma ago, Earth's history consists of two primary divisions: a short, conspicuous, younger (upper) part, termed the Phanerozoic Eon; and a much longer, but essentially invisible, older (lower) part, previously referred to as the "Cryptozoic," now the Archean and Proterozoic Eons.

The Phanerozoic, encompassing less than an eighth of all geologic time, is well documented by a prolific fossil record. Recent studies indicate that virtually all extant metazoan phyla were present already in the earliest Phanerozoic. A spectacular example is the Early Cambrian Chengjiang fauna recently discovered from Yunnan in southern China.

In contrast to the Phanerozoic, multicellular fossils are so sparse in the "Cryptozoic" that, in fact, no evidence for multicellular life was known from the Precambrian in the time of Darwin. The sudden appearance of prolific Cambrian faunas, however, led him (Darwin 1859) to believe that, "if the theory [of evolution] be true," a long history of early multicellular life must have preceded the beginning of the Cambrian Period.

It has taken palaeontologists more than a century to substantiate Darwin's speculation. Owing to the remarkable progress of especially the last three decades, the "Cryptozoic" fossil record is no longer a blank tape. It has started to provide us with growing evidence for early multicellular life (Fig. 1).

Bengtson, S. (ed.) 1994: *Early Life on Earth. Nobel Symposium No. 84.* Columbia U.P., New York

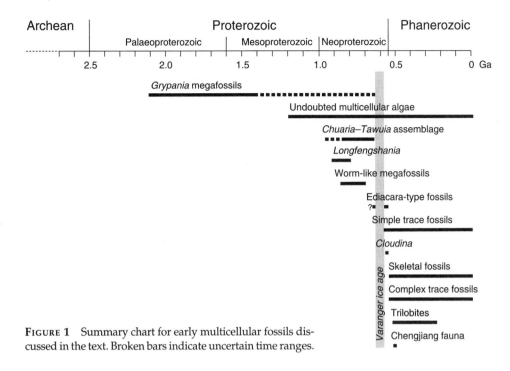

FIGURE 1 Summary chart for early multicellular fossils discussed in the text. Broken bars indicate uncertain time ranges.

Various megafossils, which are megascopic remains of organisms including plants (algae) and animals, have been found from Proterozoic sequences (Hofmann 1987, this volume). They are perhaps the most straightforward evidence of early multicellular life, but many of them are difficult to classify, even at high taxonomic levels, because of poor cellular preservation, lack of distinctive structures, or the absence of analogues in living biotas. Progressively younger biotas show increasing similiarities with living ones, however.

The oldest known multicellular fossils

Coiled, ribbon-like carbonaceous compressions of *Grypania spiralis* (Fig. 2) occur in the 1.4 Ga old Greyson Shale, lower Belt Supergroup, in Montana, USA, and the similarly aged Gaoyuzhuang Formation, upper Changcheng Group, in the Jixian section, northern China (Walter *et al.* 1990). Similar specimens have been reported from the Rohtar Formation of Tikaria, central India (Conway Morris 1989a, Fig. 2B), and the 1,000–800 Ma old Little Dal Group in the Mackenzie Mountains, northwestern Canada (Hofmann 1985a; Walter *et al.* 1990).

The ribbons of *Grypania spiralis* are up to 2 mm wide, 5–15 cm long, and typically preserved in loose coils 0.5–2.5 cm across. No cellular structures are preserved, and distinctive characteristics are apparently absent. Although the possibility of a prokaryotic origin cannot be ruled out completely, *Grypania* seems most likely to have been a multicellular eukaryotic alga on account of its megascopic size, regular shape, and

FIGURE 2 *Grypania spiralis* from the Gaoyuzhuang Formation, upper Changcheng Group, in the Jixian section, Tianjin, northern China. Scale bar represents 5 mm. Specimen deposited at the Nanjing Institute of Geology and Palaeontology, Academia Sinica.

preservation as carbonaceous compressions (Walter *et al.* 1990). The existence of such early multicellular eukaryotes has been supported, for example, by well-preserved algal filaments with distinctive multicellular structures, similar to the living red algae *Bangia*, from silicified carbonate rocks of the approximately 1 Ga old Hunting Formation in Arctic Canada (Butterfield *et al.* 1990).

Recently, the record of *Grypania* was extended to the 2.1 Ga old Negaunee Iron Formation in Michigan (Han & Runnegar 1992), suggesting that the history of multicellular life may be dated back at least to that time.

The second-oldest known type of megafossil assemblage is represented by carbonaceous compressions of the originally spherical *Chuaria* (Fig. 3A) and sausage-shaped *Tawuia* (Fig. 3B), as well as balloon-like *Longfengshania* (Fig. 3D). This assemblage is now known to have a worldwide distribution in late Proterozoic marine sediments, with the most abundant occurrences in the Little Dal Group of northwestern Canada and comparable deposits in northeastern China, representing a time range from 1,000 to 700 Ma (Hofmann 1985a, b; Sun 1987; see also Hofmann, this volume).

Previous interpretations of *Chuaria* range from nonfossils to metazoan fossils, but more recent studies suggest that it could be either a eukaryotic multicellular alga (Hofmann 1985a, b) or a *Nostoc*-like spherical colony of algal filaments (Sun 1987).

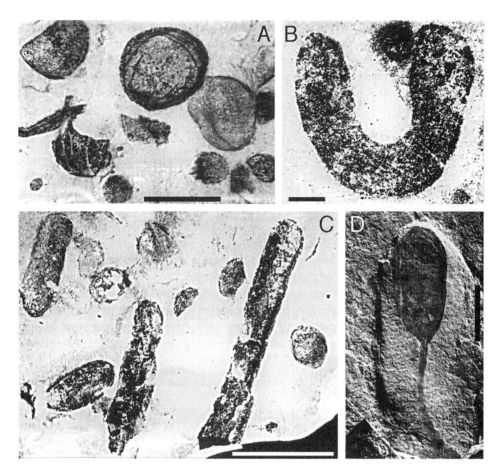

Figure 3   **A,** *Chuaria circularis;* **B,** *Tawuia dalensis,* U-shaped; and **C,** association of *Chuaria* (discoids) and *Tawuia* (ribbons) from the lower Neoproterozoic of the Huainan district, Anhui, China. **D:** *Longfengshania stipata* from the Changlongshan Formation, Qingbaikou Group, northern Hebei Province, China (courtesy of Du Rulin). Scale bars represent 2 mm in **A,** 1 mm in **B,** and 5 mm in **C** and **D.** (**A–C** from Sun 1987; **D** courtesy of Du Rulin, Hebei College of Geology, China.)

Chuaria is more widely distributed and apparently existed longer than *Tawuia,* but where they occur together there is a seemingly gradational variation in size and shape between them (Fig. 3C). Some *Chuaria* and *Tawuia* specimens contain many tiny *Chuaria*-like circular entities within the body. Thus they may be closely related (Hofmann 1985a, b; Sun 1987).

While *Chuaria* and *Tawuia* were probably planktic, *Longfengshania,* with an oval to oblong body (thallus) on a string-like stipe, may have been epibenthic.

Of particular interest are the carbonaceous compressions of megascopic worm-like organisms, which are abundant, but of low diversity, in the *Chuaria–Tawuia* assemblage from the Huainan District of Anhui Province, eastern China (Sun *et al.* 1986). *Sinosabellidites* (Fig. 4A–B), from the 850–800 Ma old Liulaobei Formation, is characterized by

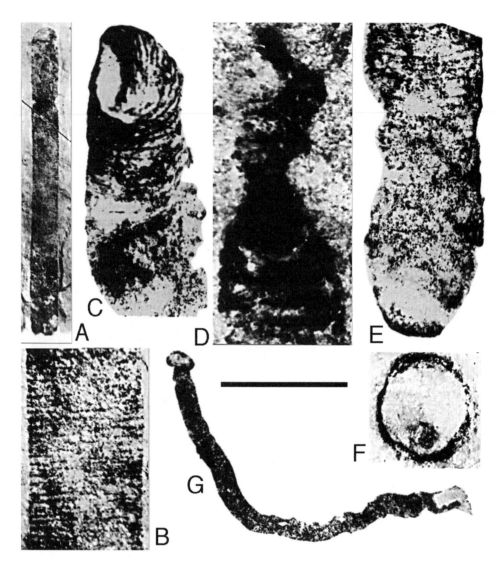

FIGURE 4   Worm-like fossils from the lower Neoproterozoic in Huainan District, Anhui Province, China. **A:** *Sinosabellidites huainanensis* from the Liulaobei Formation, middle part enlarged and shown in **B** for details of annulations. **C–F:** *Pararenicola huaiyunensis* from the Jiuliqiao Formation; **C**, fragment of twisted body; **D**, presumed anterior part; **E**, presumed posterior part; and **F**, naturally preserved cross section, showing a few annulations and a broad cavity with a small circular interior structure of uncertain significance. **G:** *Protoarenicola baiguashania* from the Jiuliqiao Formation, showing slender annulated body with a small ovate bulb (?head or proboscis) at the assumed anterior end. Scale bar 10 mm in **A**, 3 mm in **B**, 1.5 mm in **C–F**, and 5 mm in **G**. (From Sun *et al.* 1986.)

numerous closely spaced fine annulations, but otherwise it cannot be distinguished from the associated algal remains of *Tawuia* (Fig. 3C). *Pararenicola* and *Protoarenicola* (Fig. 4C–G) in the approximately 740 Ma old Jiuliqiao Formation, however, appear remarkably similar to simple worms, showing elongate, cylindrical, flexible bodies with distinct annulations and even a differentiated front end (?head part). Perhaps these worm-like organisms, especially the last two forms, are primitive metazoans (Sun 1986a), but the lack of preserved internal organs makes this interpretation speculative (Cloud 1986; Conway Morris 1989a). For the present, these fossils may be accepted as persuasive but not conclusive evidence of pre-Ediacaran metazoans.

The Ediacaran metazoans

The fossil record of metazoans becomes obvious at long last with the appearance of the Ediacara-type body fossils and simple trace fossils in the terminal Proterozoic Ediacaran interval, ranging about 590–550 Ma ago, between the highest Proterozoic tillites of the Varanger ice age and the Precambrian–Cambrian boundary (Glaessner 1971, 1984; Jenkins 1981; Cloud & Glaessner 1982).

Represented by a prolific assemblage, first found by R.C. Sprigg in the 1940s, from the Pound Quartzite at the Ediacara Hills and in the Flinders Ranges of South Australia, the Ediacara-type metazoans are now known sporadically from more than twenty regions around the world, and occasionally in great abundance. Most remarkable is a Vendian metazoan assemblage from the White Sea Coast region in northern Russia (Fedonkin 1981, this volume; Sokolov & Fedonkin 1984), which is strikingly similar to that from South Australia. Despite their presumed geographic separation, the two assemblages contain many forms in common, even including the most distinctive *Dickinsonia* and *Tribrachidium*. Although certain assemblages differ from the rest, possibly because of relative age or more probably environmental differences, they share a broadly similar level of organization (Glaessner 1984; Conway Morris 1989a; Seilacher 1989, this volume).

Characterizing a significant stage of the early metazoan evolution, almost all Ediacaran metazoans were entirely soft-bodied and moderately large. Their remains are typically preserved as body casts and molds on bedding planes, possibly owing to the absence of scavengers and the lack of deep bioturbation. Coeval trace fossils are chiefly simple surface burrows (see Fedonkin, this volume).

The majority of Ediacara-type body fossils appear to be either ancient cnidarians or of a comparable organizational grade; the rest seem to represent annelid worms, primitive arthropods, and several problematical taxa.

Such assignments, principally established by Glaessner (1984; Conway Morris 1989a, b; Jenkins 1989), are broadly known but have not been universally accepted. Seilacher (1984, 1989, this volume), for example, suggests that the Ediacara-type body fossils cannot be regarded as ancestors of extant metazoans but may be referred to a single extinct group, Vendobionta. In his view, they represent a widespread evolutionary experiment that failed by the end of the Proterozoic, whereas true metazoans may be represented by the coeval trace fossils.

However, the Ediacara-type body fossils as a whole are far more complex than any single body plan or phylum can explain. As the oldest known manifest metazoans, they can be expected to contain many forms that were brand new and differed greatly from any known animal. Some of them could be ancestral to extant groups, and others might represent extinct branches, even at a level as high as phylum.

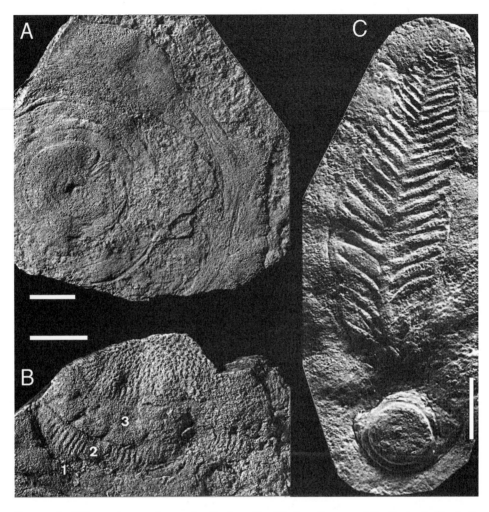

FIGURE 5 Ediacara-type metazoan fossils from the Flinders Ranges of South Australia. **A–B:** *Brachina delicata,* a medusiform fossil of a probable scyphozoan affinity; **A,** assumed exumbrellar side, showing the margin cleft into many small lappets; **B,** supposed subumbrellar side, showing marginal lappets (1), radial canals (2), and ring (coronal) muscle band (3). **C:** *Charniodiscus arboreus,* a possible pennatulacean (sea pen), showing a frond-like main part with a prominent holdfast at the bottom. Scale bars represent 2 cm in **A,** 1 cm in **B,** and 5 cm in **C.** (**A–B** from Sun 1986b; **C** from Jenkins & Gehling 1978.)

Some Ediacaran species are closely comparable with living cnidarian groups, such as bell-shaped cubozoans (e.g., *Kimberella*), annularly chambered chondrophores (e.g., *Ovatoscutum* and *Eoporpita*; see Fig. 1D of Fedonkin, p. 376 in this volume), and holdfast-bearing frond-like pennatulaceans (e.g., *Charniodiscus*, Fig. 5C, and *Glaessnerina*). But some others are less comparable. For example, the discoid *Brachina* (Fig. 5A, B) approaches the morphological complexity of scyphozoans and resembles a coronatan medusa (*Atolla*), from which it differs, however, in its simple radial rather than tetraradial symmetry (Sun 1986b). More striking are medusiform fossils with a well-developed triradial symmetry (e.g., *Skinnera* and *Albumares*; Fig. 4B of Fedonkin, p. 382 in this volume), a feature that finds no parallel in modern cnidarians (Jenkins 1989; Conway Morris 1989a). The enigmatic *Tribrachidium*, characterized by three strong "arms" on a circular disc, also showing a remarkable triradial symmetry (Fig. 4A of Fedonkin, p. 382 in this volume), appears to represent a separate phylum immediate between the lophophorates and echinoderms (Glaessner 1984; Jenkins 1989). Moreover, various multivaned and chambered forms, such as *Pteridinium* and *Phyllozoon*, which were once regarded as possibly of a cnidarian affinity, probably represent an extinct phylum, Petalonamae (Pflug 1972b; Jenkins 1984, 1989).

Also present in the Ediacara-type assemblages are several kinds of segmented creatures apparently belonging to a relatively high structural grade. These are best represented by *Dickinsonia* (Fig. 6), which is a possible annelid worm (to be discussed below), and *Spriggina*, which appears more like a rudimentary arthropod (Conway Morris 1989a) than a polychaete worm (as suggested by Glaessner [1984]). Additionally, recent findings of a trilobite-like animal and some arthropod-made scratch marks have been reported from the Ediacara assemblage in South Australia (Jenkins 1989), confirming the traditional claim that primitive arthropods, probably only a few, existed already during Ediacaran time.

Dickinsonia is remarkable in that it closely resembles the living ectoparasitic annelid *Spinther* in the pattern of segmentation (Glaessner 1984, Fig. 2.4) but differs from the latter in its large size (up to 1 m in length) and very flat foliate shape (probably only a few millimeters thick). Such a peculiar design, with a very high surface-area/volume ratio, could enable *Dickinsonia* to respire in seawater of a relatively low level of oxygen. Taking *Dickinsonia* as an example, Runnegar (1982b, c) has cogently argued for a low oxygen content of the late Proterozoic atmosphere.

The associated or coeval trace fossils seem to have been produced mainly by worm-like, infaunal sediment feeders, few of which have been recognized among the known body fossils. Their systematic feeding patterns, as analyzed by Seilacher (1985), correspond to deep rather than shallow marine environments if compared with those of their Phanerozoic conterparts. As most Ediacara-type creatures actually lived in well-aerated shallow marine environments, this supports the oxygen-deficient nature of the Proterozoic ambient environment.

Although most Ediacaran metazoans were entirely soft-bodied, exceptions did exist. Calcareous tubes of *Cloudina* (see Fig. 2 of Bengtson, p. 417 in this volume), originally described from the Nama Group in Namibia, and various *Cloudina*-like fossils have been found in carbonate deposits of Ediacaran age in a number of regions around the world (Conway Morris *et al.* 1990; Grant 1990). These tubular fossils, typically with cone-in-cone shell structures, may be interpreted as dwelling tubes formed by periodical external secretions of small worm-like suspension feeders, which were either polychaetes (Glaessner 1976, 1984) or at least of a cnidarian grade of organization (Grant 1990). Their global distribution indicates that biomineralization in the late Neoproterozoic was not simply a local phenomenon, yet the principal episode of extensive acquisition of durable skeletons did not start until the beginning of the Phanerozoic.

FIGURE 6 *Dickinsonia costata*, a possible annelid worm, from the Flinders Ranges of South Australia. Silicon cast; original deposited in the Department of Geology and Geophysics, University of Adelaide, South Australia. Scale bar 2 cm.

Early Cambrian multicellular fossils

The onset of the Phanerozoic was signaled by the rather sudden appearance of diverse small skeletal fossils (Qian & Bengtson 1989), along with a striking increase in both abundance and relative complexity of trace fossils (Crimes 1987), in the interval of the Meishucunian Stage across the Precambrian–Cambrian boundary; soon afterwards, in the Qiongzhusian Stage, prevalence was given to the typical Cambrian fossils, such as skeletal remains of trilobites, bradoriids, inarticulate brachiopods, and hyoliths. These data provide strong evidence for major evolutionary radiations, but our knowledge about the magnitude and extent of the "Cambrian explosion" of metazoan diversification would remain very incomplete without the recent discovery of the Chengjiang biota.

One of the most spectacular discoveries in paleontology, the Chengjiang biota is a Burgess-type fossil *lagerstätte* from the Early Cambrian Qiongzhusian (Chiungchussu) Stage in the Kunming region, Yunnan Province, southwestern China (Zhang & Hou 1985; Hou & Sun 1988; Conway Morris 1989b; Chen *et al.* 1991; Hou *et al.* 1991). From mud deposits of what was a seafloor about 530 Ma ago, thousands of specimens have been excavated. More than 70 species of metazoans have been described, including sponges, cnidarians, brachiopods, hyoliths, annelid and priapulid worms, trilobites and a wide variety of other arthropods, as well as several problematic forms of unknown affinities (Fig. 7). There is also an accompanying flora consisting of various megascopic algae (Chen & Erdtman 1991). Not only is the preservation so excellent that even animals without hard parts have been fossilized intact, but this biota lived when skeleton-bearing trilobites (e.g., *Parabadiella* and *Eoredlichia*) and other typical Cambrian fossils first appeared. Because the "ordinary" fossil record is severely biased in favor of organisms with hard parts, the "extraordinarily" preserved Chengjiang soft-bodied biota offers a unique glimpse into the true diversity and complexity reached by the multicellular evolution at the time immediately after the beginning of the Phanerozoic Eon.

Concluding remarks

Despite a sparse fossil record, the history of multicellular life can be traced through time at least back to 2.1 Ga ago, with the appearance of *Grypania*. Metaphytes and metazoans appear to have originated at considerably different times, since unequivocal evidence for pre-Ediacaran metazoans is extremely rare. Perhaps pre-Ediacaran metazoans were too small and/or delicate to survive as body fossils or produce recognizable trace fossils. However, the appearance in the early Neoproterozoic of a few worm-like organisms seems to imply that the initial metazoan evolution may have taken place well before the Ediacaran time. This is also indicated by the levels of evolutionary diversity reached by the Ediacaran metazoans.

The unique composition and preservation of the cnidarian-dominated Ediacara-type assemblages, along with simple trace fossils, reflect special adaptations to the low-oxygen environments of that time. The "Garden of Ediacara" might meet the basic requirements for Ediacaran animals to live, but, perhaps, it could not offer sufficient conditions for more advanced evolutionary experimentations to develop, mainly because of the limited availability of oxygen (Towe 1970). So, not only may the diversity and biomass have been relatively small, but biotic competitions and innovations were probably restricted to a low level compared with Phanerozoic conditions.

It seems that the great success of the Cambrian explosion, as displayed spectacularly by the far-reaching evolutionary radiations of the Chengjiang fauna, was at the cost of displacement or extinction of the Ediacara-type faunas. This fundamental

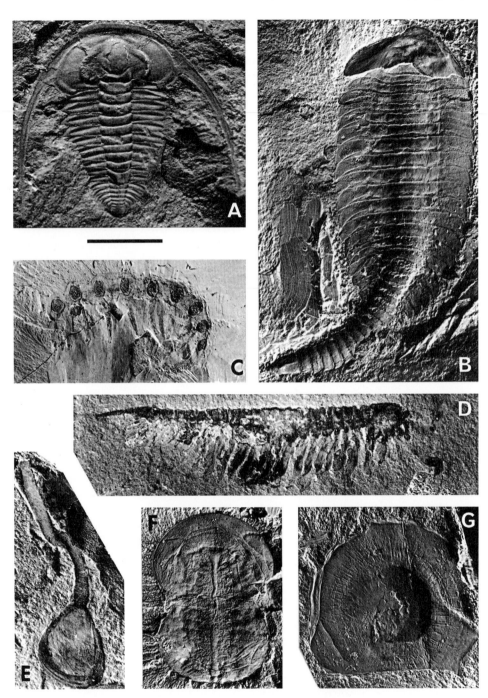

FIGURE 7   Representatives of the Early Cambrian Chengjiang fauna from Yunnan Province, China.
A: *Eoredlichia intermedia*, a trilobite. **B:** *Fuxianhuia protensa*, an arthropod. **C:** *Microdictyon sinicum*, a
lobopod animal. **D:** *Jianfengia multisegmentalis*, an arthropod. **E:** *Lingulella chengjiangensis*, an inarticulate brachiopod, unusually preserved with a long pedicle. **F:** *Naraoia spinosa*, a lightly skeletonized
arthropod. **G:** *Stellostomites eumorphus*, a medusiform fossil of uncertain affinities. Scale bar represents 5 mm in **A**, **E** and **F**; 1 cm in **C**; 4 mm in **D**; and 2 cm in **B** and **G**. (**B** and **D** from Hou 1987a, b.
Specimens in the Nanjing Institute of Geology and Palaeontology, Academia Sinica.)

change appears to have been completed in a rather short interval, and its causes are still obscure.

I would like to present the following scenario for consideration: With the accumulative increase of the atmospheric oxygen content above a threshold value by the end of the Proterozoic, the specialized adaptations of most Ediacaran metazoans may have become redundant, and in the meanwhile, a plethora of previously constrained evolutionary innovations, such as skeletonization, musculization, macrophagous predation, and even the rise of animals with highly diversified body plans could have become practically feasible; furthermore, these evolutionary experimentations must have benefited a great deal from extensive transgressions and the ensuing occurrence of various new ecological niches in the earliest Cambrian time.

Back to the original metaphor of an iceberg, we may find that the Phanerozoic fossil record is so prominent because it is based on a huge but barely visible basement, the history of early life on Earth.

Vendian body fossils and trace fossils

Mikhail A. Fedonkin

Paleontological Institute, Russian Academy of Sciences, Profsoyuznaya ul., 123, 117868 GSP-7 Moscow V-321, Russia

At the beginning of the Vendian Period (610–550 Ma ago), the most intensive glaciation in the Earth's history took place, the Varanger Glaciation. A relatively warm global climate subsequently resulted in a vast transgression of sea over the continents. Extensive shallow-water habitats were occupied by megascopic soft-bodied organisms not known before the Vendian. At first considered the ancestors of the Cambrian invertebrates, these Ediacara organisms were referred to still-existing high-rank taxa. Discoveries of these enigmatic creatures in many places all over the world revealed new peculiarities in their morphology, mode of preservation, and paleoecology. Interpreted as the oldest multicellular animals, the Vendian body fossils demonstrate a cryptic stage in metazoan evolution. Along with a few phylogenetic lineages continuing to the Phanerozoic, there were some problematic forms that can be considered the relicts of the pre-Vendian biota adapted to the primary metazoan biotopes. Diploblastic organisms of the coelenterate grade of organization dominated among the Vendian invertebrates. Less numerous were the bilateral triploblastic animals, including the coelomates. The latter inhabited shallow-water environments and produced diverse and rather complex bioturbations comparable with the ones in Phanerozoic sediments. Active colonization of the sediment by some invertebrates and an acquisition of a mineral skeleton in other groups at the Vendian–Cambrian boundary may have been stimulated by the same external factor(s).

The Late Proterozoic was the time of radical restructuring in the global oceanic ecosystem. The eukaryotic trophic pyramid was built on top of the pre-existing network of biogeochemical interactions between the microbial communities. This additive phase in the evolution of the ocean ecosystem was relatively short. The substitution of individual members in the trophic hierarchy became a major process in the history of the ecosystem, which retained its principal structure.

A crucial role for the evolution of the oceanic ecosystem belonged to the growth of continental plates, which became particularly active between 2.7 and 2.5 Ga (Lowe, this volume). This long-term process resulted in the expansion of shallow-water benthic habitats and increasing upwelling. Both phenomena promoted the recycling of the most important metabolites that enhanced the bioproductivity of the ocean. The most direct consequences of the growth in bioproductivity were the increasing atmospheric oxygen level and the declining carbon dioxide level, as well as the acceleration of biogenic carbonate precipitation and the growth of the carbonate platform, which became the major carbonate reservoir in the Phanerozoic biosphere.

Bengtson, S. (ed.) 1994: *Early Life on Earth. Nobel Symposium No. 84.* Columbia U.P., New York

The withdrawal of a large mass of carbon dioxide from the atmosphere and the subsequent burial of it as biogenic carbonate and organic carbon in the epiplatform basins decreased the greenhouse effect to the level of a very sensitive balance that resulted in a more or less regular alternation of glacial and warm periods since 850 Ma. Glacio-eustatism, as well as active rifting, which was widespread in the interval from 1000 to 600 Ma, changed the physical and chemical parameters in the ocean.

Thus was the background for the following trends and bioevents, as demonstrated by the fossil record (see chapters by Walter, Runnegar, Hofmann, Sun, Vidal, and Knoll, this volume): (1) lateral expansion and growth of diversity in stromatolites with maximum at about 1 Ga; (2) an increase in morphological complexity, systematic diversity, and upper size limit of microfossils during the Late Proterozoic; (3) the origin of multicellularity in eukaryotes, possibly as early as 2.1 Ga ago, and the diversification of megascopic alga-like forms beginning around 1.4 Ga; (4) a gradual decline in abundance and diversity of stromatolites that started after 1 Ga and accelerated after 680 Ma; (5) a mass extinction of microphytoplankton about 610 Ma.

All those events took place against the background of a slow but steady increase in solar luminosity (Kasting 1992). The amount of iron dissolved in the seawater decreased, and the free oxygen in the atmosphere rose. At the same time, the hydrothermal input into the ocean decreased (see Knoll, this volume) and the carbon dioxide content of the atmosphere was lowered a few times, essentially parallel to a decrease in carbonate saturation of the seawater through the Proterozoic (Grotzinger 1990) and increase in the calcium/magnesium ratio of the ocean waters (Rozanov 1986).

Vendian Period in Earth's history

The Varanger glacial episode (610–590 Ma) was the most prominent one in the series of glaciations that took place in the Late Proterozoic (Chumakov 1978). This event, marked by widespread glacial deposits all over the world even at low paleolatitudes, began the 60 Ma long terminal period of Proterozoic history now known as the Vendian (Sokolov & Fedonkin 1984, 1990; Sokolov & Iwanowski 1990).

Type area of the Vendian System is the siliclastic basin of the Russian Platform, where the Laplandian, Redkino, Kotlin, and Rovno stages can be recognized (in ascending order). Except for the essentially glacial Laplandian Stage, the beds contain abundant and diverse fossils: organic-walled microfossils, thalli of megascopic algae, invertebrate body fossils (imprints, casts, and molds of soft-bodied animals and rare remains of mineralized or chitin-like body parts), and trace fossils or bioturbations left by the Vendian animals in the soft-bottom sediment. In addition, Vendian carbonate sediments contain stromatolites, microphytolites, and silicified microfossils commonly known from black cherts.

The Varanger glaciation had a strong effect on the biosphere. The glacio-eustatic drop of ocean level led to the regression of epicontinental seas. The shelf zone became reduced to narrow strips at the edges of the continental slopes.

One could expect that the productivity of the ocean biota decreased during that period, because life occupied mainly the photic zone of the open sea, which is relatively poor in metabolites. The great extinction of microphytoplankton documented by Vidal

& Knoll (1982) at the end of the Riphean seems to be connected with the effect of glaciations on the ocean ecosystem (though Vidal, this volume, p. 305, stresses that preservational bias in fact obscures the nature of this event).

Late Proterozoic glaciations could considerably destabilize benthic shallow-water communities and, in particular, decrease the area of distribution and the diversity of stromatolites. The Proterozoic stromatolite decline (Awramik 1971; Walter & Heys 1985) might be connected also with the appearance of competitive algae, with ecological restrictions by grazing and burrowing metazoans, and with the decrease in the carbonate saturation of seawater through the Proterozoic (Grotzinger 1990).

The post-Varanger glacio-eustatic transgression (the Redkino Stage and its equivalents) saw the rapid radiation of megascopic soft-bodied invertebrates preserved as body and trace fossils. There is no clear sequence in the appearance of the main fossil groups, suggesting that either the post-Varanger radiation was very rapid or invertebrates have a cryptic period of earlier history.

The considerable diversity of the Vendian animal world, phylogenetic interpretation of the molecular-biology data, and the recent discovery of medusa-like fossils in the intertillite beds in northwestern Canada (Hofmann et al. 1990) support a pre-Vendian stage of metazoan evolution. A review of the possible pre-Vendian metazoans is given elsewhere (Fedonkin 1992; see also chapters by Hofmann and Sun, this volume).

The diversities of trace and body fossils decrease remarkably from the Redkino to the Kotlin Stage. At the very end of the Vendian (Rovno Stage), there is a prominent increase in diversity and size of trace fossils, whereas body fossils remain rather scarce. Does this mean that the taphonomic conditions typical for the Phanerozoic (where the fossilization of medusae, for instance, was a quite unusual event) were established well before the Cambrian?

The transition between the Vendian and Cambrian was marked also by a short glacial period at about 550 Ma ago (which was not as great as the previous Varanger glacial period); by a decrease in dolomite precipitation; by the intensive accumulation of sedimentary phosphorite; by the sharp change in the isotopic composition of carbon, sulfur, and oxygen in the seawaters; and by the beginning mass accumulation of natric and potassic salt-bearing deposits (Fedonkin et al. 1987).

Discoveries of the oldest metazoans

Fossil assemblages of Vendian soft-bodied organisms have been discovered on all continents except South America and Antarctica (Glaessner 1984; Fedonkin 1987). The first specimens were collected from Namibia (South-West Africa) in 1908–1914 (Gürich 1929). About thirty years later another locality was discovered at Ediacara in South Australia (Sprigg 1947). Because of the intensive study of the Australian fossil materials (Sprigg 1947, 1949; Glaessner & Wade 1966; Jenkins 1984, 1985; Jenkins & Gehling 1978; Gehling 1987, 1988) the fauna is known worldwide as the Ediacara fauna (Glaessner 1984).

Among the most interesting localities of Vendian metazoans there are England and Wales (Ford 1958, 1963; Cope 1977, 1983), the Avalon Peninsula of southeastern Newfoundland (Anderson & Misra 1968; Anderson & Conway Morris 1982), the

Dniester Valley in the Ukraine (Palij 1969, 1976; Palij, Posti & Fedonkin 1979; Velikanov, Aseeva & Fedonkin 1983; Gureev 1985, 1987, 1988), the Onega Peninsula of the White Sea (Keller *et al.* 1974; Keller & Fedonkin 1976; Fedonkin 1981), the Winter Coast of the White Sea in northern Russia (Fedonkin 1978, 1981, 1985d, 1990a), the Olenek Uplift in northern Yakutia, Siberia (Sokolov & Fedonkin 1984; Fedonkin 1985d, 1990a; Vodanyuk 1989), the Middle Urals (Becker 1977, 1990), northwest Canada (Hofmann 1981; Narbonne & Hofmann 1987; Aitken 1989; Narbonne & Aitken 1990), North Carolina, USA (Gibson *et al.* 1984), southern Nevada, USA (Horodyski 1991), and northeastern Finnmark, Norway (Farmer *et al.* 1991).

Localities with a small number of or even single specimens have been discovered in Iran, China, British Columbia, and Russia (see references in Fedonkin 1987, 1990a; Runnegar & Fedonkin 1992).

Ediacaran fauna: The problem of preservation

The major difference between the Vendian fossil record and the Phanerozoic one is the preservation of soft-bodied invertebrates in well-aerated shallow-water environments. What factors were responsible for such a difference?

M.F. Glaessner supposed that mass preservation of the soft-bodied animals was possible because of absence of the predators and scavengers among the Ediacara metazoans (Glaessner 1984). Though the Vendian coelenterates, which comprise about 70% of the described species, could well be active predators like their recent counterparts, we can confirm that among thousands of body fossils collected from the Vendian there have been no specimens found with traces of damage (bite marks) or regeneration structures.

Among other biotic factors that could promote the preservation of the soft-bodied animals are: (1) high density of bottom populations in zones with active hydrodynamic regimes (rapid burial in the sediment); (2) a large number of attached forms that created preservable sedimentary structures during their life time; (3) a low degree of biological processing of the sediment by deposit feeders and other mobile benthos.

Cyanobacterial films that could grow rapidly over the undisturbed sediment may have played a stabilizing role, preventing the sediment from erosion and thus facilitating the preservation of soft-bodied invertebrates (Gehling 1986; Fedonkin 1987). Moreover, the cyanobacterial films hindered the aeration of the bottom and did not allow decomposition products of the buried animals to leave the sediment. All this could create a reducing microenvironment around the decaying bodies, which resulted in rapid hardening of the sediment. This may be why many Ediacara fossils, both in clastic and in carbonate rocks, are preserved as a "core" or three-dimensional structure.

Sometimes one can see the imprints of microbial films on the bottom of the sandstone beds that cover more clayey lamina deposited in more quiet conditions. Those imprints look like the irregular network of fine wrinkles that could be the counterparts of the miniature crests common for the surface of bacterial mats.

Absence of active filtrators such as sponges, brachiopods, or some mollusks and arthropods in the Vendian biota could also facilitate the preservation of the soft-bodied animals. The role of active filtrators such as copepods in the Recent oceans is well known: they clean the seawater by extracting small food particles. But probably more important is that the nondigested remains are packed into pellets that settle to the bottom much more rapidly than fine particles. This phenomenon makes the seawater clean and the photic zone deep in the Recent ocean.

The situation in the Vendian ocean could have been different. Because of the more or less constant turbidity of the water in the absence of active filtrators, the photic zone was very narrow. The relatively thin upper layer of the water column was warm and enriched by photosynthetic oxygen. Because of the difference in temperature of the upper and lower portions of the water, the vertical circulation (and thus the "ventilation" of the deeper levels of the pelagic zone) was inhibited. Poorly oxygenated or even reducing environments could have existed over those areas of the sea bottom where influence of cold waters from the polar ice or from rivers was minimal. This circumstance restrained the activity of the decomposers and bioturbators in the bottom sediments, thus leaving the buried animal bodies for sufficiently long time to be fossilized.

Morphological diversity and Vendian life-forms

Probably the most surprising phenomenon concerning the Vendian organisms is their large body size. Some forms could be considered giants, especially when compared with the earliest Cambrian small shelly fossils, which rarely exceed a few millimeters. Indeed, the Vendian medusoid *Ediacaria* can reach about one meter in diameter. About the same size are the longest specimens of the leaf-like and ribbed *Dickinsonia*. The feather-like *Charnia* and *Charniodiscus* can have a maximum length of about 1.2 m or more.

The skeletonless nature of the metazoan biota does not seem to be particularly strange if we look at recent multicellular animals. Of 35 metazoan phyla (the number of phyla is debated, but this quantity is more or less widely accepted), about 20 do not have any mineral skeleton, or their tissues are only partially and/or weakly sclerotized. Other calculations show that about 70% of the species and individuals can be considered soft-bodied animals in the Recent marine macrobiota. In some biotopes these organisms are even more numerous, though their preservation potential is extremely low.

The third general peculiarity of the Vendian body fossils is their flat shape. With rare exceptions, the fossils look like discs, leaves, or fronds. This phenomenon could be explained by the flattening of the body during the process of decay under the weight of the sediment above the buried body, as well as by the process of subsequent compaction of the sediment. But the body deformations, including the overfoldings, demonstrate that body thickness is in many cases rather small in comparison with width.

Though animals with cylindrical body were indeed present among the Vendian metazoans (there are rare tubular skeletal fossils and fossil burrows with round cross sections), the majority of the forms are somehow flattened. This can be interpreted in a few ways.

First of all, the flat shape may be the relics of an adaptation to the primary biotope of the first metazoans. The most widespread environment enriched by oxygen and food resources was the surface of the cyanobacterial mats that existed through most of the Archean and Proterozoic during almost 3 Ga. This surface may have been a place of origin for the eukaryotes, because here there existed at least three environments with different geochemistry and microbial population: the marine water above the mat, the surface and the photic zone of the mat (about 1 mm thick) periodically enriched by the photosynthetic oxygen, and the reducing environment below the photic zone of the mat. Each of these environments could have supplied the protoeukaryotes with potential symbiont microorganisms having different biochemical functions.

Another reason for the flat shape of the Vendian organisms could be the presence of endosymbiotic photosynthetic algae. Being widespread among recent invertebrates, especially in nonskeletal forms, this kind of symbiosis could well have existed in the Precambrian (McMenamin 1986).

The flat shape, usually correlated with larger size, may reflect the strategy of defense by those organisms from predators that could swallow the prey as a whole body and did not bite piece by piece.

The reason to have a large body size could be connected with the phenomenon of "reserve biomass" (Ponomarenko 1984). One of the strategies of multicellular animals colonizing new environments is to build up a large individual biomass that promotes survival in conditions to which the colonizer is not quite adapted. The Vendian fauna seems to have played such a pioneer role in the vast new habitats that appeared during the postglacial transgression of the seawaters over the continents.

Actually, a flat shape leads to increase of surface/volume ratio in the organism, which is important for those forms that consume dissolved organic matter (amino acids, glucose, etc.) from the water (Fiala-Medioni 1988) or for the organisms that had permanent ectosymbionts playing an important role in the metabolism of the hosts. High surface/volume ratio is often correlated with the missing or poorly developed circulatory system that seems to be characteristic of the Vendian metazoans.

The most numerous Vendian body fossils are the discoidal imprints and molds that demonstrate concentric or radial structure. In some discoidal fossils one can see combinations of concentric and radial arrangements, while other rare forms demonstrate a combination of concentric and bilateral body organizations. Most disk-shaped fossils are interpreted as medusae or medusoids, as basal (aboral) end of polyps, or as attachments.

The differences in the interpretations of the disk-like fossils are connected with not only the anatomical characters (such as the imprints of the marginal tentacles, gastrovascular channels, and gonads in some medusae) but also with the mode of preservation, the body deformations, and even with the quantity of fossils. For example, the polyps *Nemiana* can be found in thousands, preserved in life position, sitting closely body to body over square meters on the bottom sediment. Free-swimming medusae had incomparably less chance to be preserved, and that is why we do not find them as often as fossil polyps. A special group of sedentary medusae or medusoids (*Cyclomedusa*, *Ediacaria*) were attached to the soft sediment by a short aboral stem, with their oral side turned up (Fig. 1B).

Remarkable representatives of the bottom dwellers were long, frond-like organisms attached to the soft bottom by the stem with a discoidal or bulbous expansion of the lower end. Their elongated upper part was commonly subdivided into a series of

FIGURE 1   A: *Hiemalora stellaris* Fedonkin, Ust-Pinega Formation, Winter Coast of the White Sea, Russia, ×1. **B:** *Cyclomedusa* sp., fragment of aboral side, locality as in **A**, ×0.9. **C:** *Hiemalora* cf. *stellaris* Fedonkin, oral side, Mogilev Formation, Dniester River valley, Ukraine, ×1.3. **D:** *Eoporpita medusa* Wade, oral side, locality as in **A**, ×1.

uniform parts decreasing in size upwards (Fig. 2A, B). This partitioning of the frond-like body resembles the construction of Recent soft corals, in particular living pennatulacean octocorals (sea pens), and the fossils could well belong to the coelenterates.

Along with floating and sedentary medusae and solitary and colonial polyps, there were forms living at the air–water interface like the Recent "by-the-wind-sailors" *Porpita*: e.g., *Eoporpita* (Fig. 1D) and *Protodipleurosoma* (the resistant chitin-like sail of the latter was recently discovered by Fedonkin & Krivosheyev [unpublished]).

Coelenterate-grade Vendian animals seem not only to consist of the familiar life-forms mentioned above; some unusual (from the present-day point of view) sedentary organisms have a multifoliate cabbage-like body (such as *Ernietta*) or an erect body with three ribbed "wings" round the vertical axis (*Pteridinium*, Fig. 2C).

Rare animals can be interpreted as true Bilateria or Triploblastica, e.g., organisms with ectodermal, endodermal, and mesodermal layers of cells at the early individual development. Bilateria have only one plane of symmetry, with pairs of organs situated on both sides of the plane. The existence of such rather complex organisms is demonstrated by trace fossils left by mobile animals in soft sediment (see below).

As to the body fossils, the presence of triploblastic metazoans is more difficult to prove, because even those forms that have been described as annelids or arthropods show some characters that are unusual in later fossil or Recent invertebrates.

The overwhelming majority of Vendian animals that could be considered as true Bilateria had a flattened, segmented body. We cannot prove yet if this segmentation is metamerism such as in Recent annelids or arthropods, i.e. with more or less exact repetition of paired organs in each segment. On the other hand, the Vendian bilaterians include forms with imperfect segmentation or with left and right half-segments shifted one to another (Fig. 5B and D). This body plan is of special importance, demonstrating the possibility of a new, unknown type of body-plan evolution not predicted by the neontologists (Fedonkin 1985c).

Rare animals with a cylindrical body seem to be present among the Vendian metazoans, indicated by trace fossils (*Planolites*, *Medvezhichnus*) as well as by such tubular skeletal fossils as *Cloudina* and *Calyptrina*. The multilayered tubular shell of *Cloudina* (Fig. 2 of Bengtson, this volume) was formed of a rigid $CaCO_3$-impregnated organic-rich material (Germs 1972; Grant 1990). The fossil is interpreted as a filter-feeding metazoan of at least a coelenterate grade of organization. The organic-walled annulated tubes of *Calyptrina*, *Saarina*, (Fig. 3B) and other Sabelliditidae (Sokolov 1965, 1967) were originally interpreted as the tubes of Pogonophora. Later comparative studies of the ultrastructure of the fossil tubes of *Sabellidites* and the tubes of the recent pogonophorans revealed the complex fibrillar and laminar construction of the sabelliditid tubes but did not give a good basis either to prove or to disprove the original hypothesis (Urbanek & Mierzejewska 1983; Ivantsov 1990).

The comb-like 2–3 mm fossils *Redkinia* (Fig. 3C), first discovered in the core of the borehole in the Russian north (Sokolov 1976, 1990) and then found in the Vendian of Ukraine and in other places on the Russian Platform (Velikanov *et al.* 1983), have been interpreted in different ways: (1) as early onychophorans (Sokolov 1976), (2) as scolecodonts resembling jaws of polychaete worms (Sokolov 1990), or (3) as parts of the filter apparatus of an unknown bilateral organism or Triploblastica (Gureev 1987).

FIGURE 3   **A:** *Evmiaksia aksionovi* Fedonkin, Ust-Pinega Formation, Winter Coast of the White Sea, Russia, ×1. **B:** Saarinid tube, borehole Malinovka, depth 360 m, Ust-Pinega Formation, Arkhangelsk Region, Russia, ×4. **C:** *Redkinia spinosa* Sokolov, borehole Nepeitsino, depth 1417–1426 m, Ust-Pinega Formation, Arkhangelsk Region, Russia, ×10. **D:** *Elasenia aseevae* Fedonkin, Mogilev Formation, Dniester River valley, Ukraine, ×3. **E:** *Bonata septata* Fedonkin, locality as in **A**, ×2.

Three approaches to the Vendian body fossils

Since the first years of study of the Vendian body fossils, various authors have differently understood the morphology and the systematics of these oldest megascopic organisms. However, for a long time the opinion dominated among paleontologists that the Precambrian animals belong to existing phyla, classes, and even orders. This

FIGURE 2   **A, B:** *Charnia masoni* Ford, Ust-Pinega Formation, Winter Coast of the White Sea, Russia. **A**, fragment of a large colony, ×0.7; **B**, upper (terminal) part of the same specimen, ×3. **C:** *Pteridinium nenoxa* Keller, fragment with three segmented "wings" preserved, Ust-Pinega Formation, Summer Shore of the White Sea, Russia, ×1.

was partly connected with the strong authority of M.F. Glaessner, the world leader in the study of the Vendian fauna. But I would like to underline the psychological aspect of the approach. Since the time of Darwin, the question of the ancestors of the Cambrian fauna was without answer. This is why the first reaction to the discovery of the Precambrian megascopic organisms should be: "At last! Here they are!".

Another psychological aspect of the approach is that the high-rank system of the major invertebrate groups seemed to be the same from the Cambrian to the recent time. The extremely poor fossil record of the soft-bodied animals made the paleontologists look through the diversity of the recent metazoans for the comparison with the Precambrian ones.

Such a comparison had to be done, but along with attention to the prevailing body plans or the peculiarities of morphology in the norm, one must pay no less attention to the small relict groups, unusual morphological phenomena, or even to the teratology of the invertebrates showing the whole space of morphological possibilities and the life-forms. The norm demonstrates a more narrow spectrum of characters.

According to the classical approach (Glaessner 1984), the Vendian metazoans can be placed into the following taxa: Phylum Coelenterata (classes Hydrozoa, Anthozoa, Scyphozoa, Conulata, medusae of uncertain systematic position, and problematic Petalonamae), Phylum Annelida (class Polychaeta), Phylum Arthropoda (superclass Trilobitomorpha or Chelicerata represented by the uncertain class and superclass Crustacea, class Branchiopoda), Phylum Pogonophora, Phylum Echiurida, as well as some forms of uncertain position even at the level of phylum.

The possibility of a completely different interpretation of the Ediacara fossils developed among paleontologists of the German school. H.D. Pflug in particular invented the high-rank taxon Petalonamae that has become widely accepted by paleontologists (Pflug 1970a, b, 1972a, b, 1973). One can disagree with Pflug's phyloge-netic models, but his morphological interpretation of those enigmatic leaf-like and bag-shaped fossils certainly has some fruitful aspects.

Very interesting and widely discussed now is the approach developed by another German paleontologist, A. Seilacher (1984, 1989, this volume), who offered a new functional, morphological and taxonomic interpretation of the Precambrian megas-copic fossils. Having noted that the Vendian fauna cannot be easily related phylogeneti-cally with later invertebrates, and stressing the fact of unusual preservation of the Ediacara fossils, Seilacher inferred that Precambrian organisms do not have recent analogues and have a unique organization. They are characterized by an extensive body surface that has been developed because of a very complicated relief and by a low body volume by virtue of being relatively flat. The high surface/volume ratio of the body allowed the absorption of oxygen and organic matter dissolved in water by diffusion through the body wall. Thus neither a mouth nor digestive or respiratory organs were necessary.

A third approach was developed by Fedonkin (1983, 1985d, 1987, 1990b). The study of abundant body fossils from the Vendian of Russia and Ukraine, especially from the White Sea fossil localities, exposed the contradictions in the interpretation of the Ediacaran fauna as the representatives of still-existing groups. For practically all species of Ediacaran metazoans one can point out the characters that are in disagreement with the groups of recent invertebrates with which they have been compared.

Based on the comparative body-plan analysis, with special attention to growth patterns, mode of life, and preservation, this approach has led to a different view of the system and the early evolution of metazoans in the Precambrian.

There is no question that the Vendian organisms are different from later (including recent) metazoans. The question is how much they are different. And some authors may ask "Are they metazoans?" (see Seilacher 1989, this volume; Norris 1989; Bergström 1991). What are the major points of doubt in their metazoan nature?

The first point is the preservation of the Vendian soft-bodied organisms in the facies that normally do not demonstrate similar preservation in the Phanerozoic, in particular the moderately shallow-water siliciclastics.

Actually, the situation is overdramatized. Phanerozoic cnidarian resting traces described as sea-anemone burrows or the burrows of solitary coelenterate polyps (*Bergaueria, Astropolithon, Alpertia, Dolopichnus*, etc.) are commonly preserved in abundance in the same facies as their Vendian counterparts that are morphologically similar (*Nemiana, Medusinites, Inaria*). Discoidal attachments of Vendian frond-like *Charnia* and *Charniodiscus*, as well as the sedentary polyps *Cyclomedusa* and *Ediacaria*, seem to belong to the same category of fossils. These organisms deformed the sediment, being in life position, which is why they dominate over the other body fossils.

Free-swimming soft-bodied organisms are preserved relatively rarely in the Vendian rocks. The number of these fossils is a function of different factors, such as the original quantity of the individuals in the particular biotope, the mechanical durability of the body and/or the cover tissues, the resistance of the tissues to the agents of biological (bacterial or fungal) decomposition, peculiarities of the sediment where the body was buried, etc. (Fedonkin 1987). And again, we could not indicate any critical difference between the modes of preservation of Phanerozoic jellyfish and the Vendian medusae.

Though we do not know in the Vendian all the analogues of Phanerozoic Lagerstätten (considered in terms of the paleoenvironments and the sedimentology) we could, nevertheless, say that the spectrum of facies favorable for preservation of soft-bodied organisms was wider in the Vendian. And the relative abundance of the fossils compared with the Phanerozoic can be explained by peculiarities of the biotic and abiotic factors of the marine ecosystem rather than by a special organismic construction of the Vendian creatures.

Another cause of doubt in the metazoan nature of the Vendian body fossils is an apparently weak phylogenetic tie of the Ediacaran fauna with the Cambrian or Paleozoic faunas. This impression is partially linked with the contrast of taphonomic conditions around the Precambrian–Cambrian boundary, as well as with the fact that we are dealing with two different categories of fossils (soft-tissue imprints in the Vendian and skeletal remains in the Phanerozoic). Some groups of the Vendian fauna have become extinct at the end of the period (Fedonkin 1983, 1985d, 1986, 1987; Seilacher 1983a, 1984, 1989).

Nevertheless, there are numerous examples of biotic continuity between the Vendian and the Paleozoic, if we compare the medusiform fossils, sabelliditids, arthropod-like animals, and trace fossils (many of the bioturbation taxa cross the Precambrian–Cambrian boundary), though they indicate a general similarity of body plans, modes of locomotion, or behavioral stereotypes rather than direct morphological continuity.

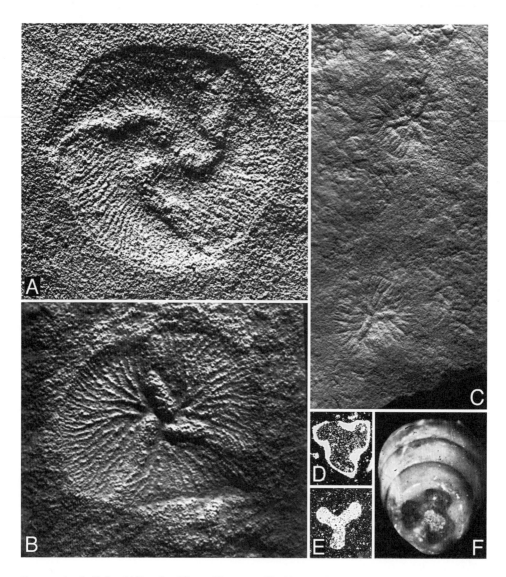

FIGURE 4 **A:** *Tribrachidium heraldicum* Glaessner, Ust-Pinega Formation, Summer Shore of the White Sea, Russia ×6. **B:** *Albumares brunsae* Fedonkin, same locality as in **A**, ×6. **C:** *Anfesta stankovskii* Fedonkin, Ust-Pinega Formation, Winter Coast of the White Sea, ×1.2. **D:** *Anabarites* cf. *tricarinatus* Missarzhevsky, cross section of the shell, Erkeket Formation, Lower Cambrian, Olenyok River, northern Yakutia, Russia, ×30. **E:** *Anabarites ternarius* Missarzhevsky, cross section of the shell, Erkeket Formation, Lower Cambrian, Boyenchime River, northern Yakutia, Russia, ×30. **F:** *Mariochrea sinuosa* Valkov, view from the apical end of the shell, Chabursky Horizon, Lower Cambrian, Salinde River, northern Yakutia, Russia, ×30. (Photographs **D**, **E** and **F** courtesy of A.K. Val'kov.)

With regard to body plan, we might indicate the significant similarity between the medusiform fossils of the Vendian and the Paleozoic. In particular, I would like to stress three major lineages that are typical for the Vendian and less known in the Phanerozoic.

The concentric division of the discoidal fossils that is so typical for the Vendian class Cyclozoa Fedonkin can be observed in the Paleozoic forms interpreted as the floating hydrozoan Porpitidae, Velellidae, and Chondroplidae (G. Stanley 1986), in the *"Brzechowia"* sp. from the Middle Cambrian of Poland (Dzik 1991), in the Lower Cambrian *Rotadiscus* (Sun & Hou 1987a), and others.

The radial body plan with numerous lobes and unstable order of symmetry (Vendian class Inordozoa Fedonkin) is observed in the Lower Cambrian *Medusina* Walcott (1898), *Stellostomites* and *Yunnanomedusa* (Sun & Hou 1987a), in the Late Cambrian *Eomedusa* and *Camptostroma* (Popov 1967) and others.

A threefold radial symmetry, which is rather unique among metazoans as a primary character (though allowed in the coelenterates), can demonstrate the phylogenetic connections between the Vendian class Trilobozoa Fedonkin and the Early Cambrian anabaritids or angustiochreids (Val'kov & Sysoev 1970; Val'kov 1982; Missarzhevskij 1974, 1989; Fig. 4 herein).

Fourfold *Conomedusites*, the only Vendian tentacle-bearing polyp, which has a conical theca with the fine stem-like attachment at the apex, can have some relation to the early Conulata or Scyphozoa (Glaessner 1984).

A few fossils like *Parvancorina, Vendomia, Bomakiella,* and *Mialsemia* (Fig. 5A, C, F) may represent three separate lineages of early arthropods, while the famous *Spriggina*, widely considered as a stem-group arthropod, still requires further detailed study because it may well be neither arthropod nor annelid but something else.

Another group that phylogenetically connects the Vendian and Cambrian biotas are the tubular fossils Sabelliditidae (Sokolov 1965, 1967; Fedonkin 1985b, 1990a).

One can oppose Seilacher's statements concerning the absence of mouth, alimentary channels, or circulatory systems in the Vendian soft-bodied organisms, as well as those concerning the notion of their "air-mattress" construction and the high surface/volume ratio. There is morphological evidence of trophic activity in the Vendian body fossils, e.g., (1) simple round mouth in *Nemiana, Hiemalora, Eoporpita, Cyclomedusa*; (2) feeding grooves falling into three mouth depressions in *Tribrachidium*; (3) tentacles in *Hiemalora, Eoporpita, Ediacaria, Pomoria, Conomedusites*; (4) branching gastrovascular channels in *Albumares*; (5) alternating diverticulae of intestine in *Vendia*; (6) a regular series of pores in the body wall in an enigmatic fossil organism discovered recently (Fedonkin 1992).

The overall morphology of the Vendian body fossils can be interpreted easily in terms of life-forms adapted to passive and active carnivory, benthic and planktic suspension feeding, and epifaunal detrital or bacterial scavenging.

As to the air-mattress or quilted organismic construction, one can find a lot of examples among recent invertebrates, both in colonial forms and in solitary ones. Here we shall find numerous organisms with extremely high surface/volume ratio, such as nudibranchiate mollusks, some primitive annelids, as well as numerous metazoans that take up dissolved organic material through the epidermis and/or have algal endosymbionts.

And even the supposed absence of the mouth or the morphologically developed circulatory system in some Vendian organisms cannot serve as a critical argument against the metazoan nature of the latter. *Riftia* and other pogonophorans are the brightest examples in a set of similar ones.

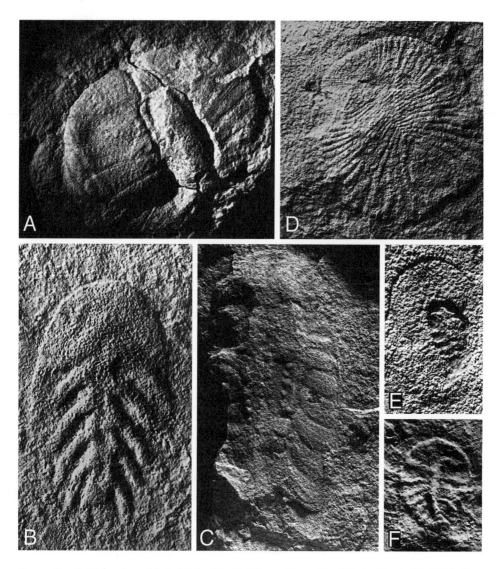

FIGURE 5   A: *Mialsemia semichatovi* Fedonkin, Ust-Pinega Formation, Winter Coast of the White Sea, Russia, ×1. **B:** *Vendia sokolovi* Keller, Ust-Pinega Formation, Winter Coast of the White Sea, ×5. **C:** *Bomakellia kelleri* Fedonkin, Ust-Pinega Formation, Summer Shore of the White Sea, Russia, ×0.7. **D:** *Dickinsonia costata* Sprigg, locality as in **A**, ×1. **E:** *Onega stepanovi* Fedonkin, locality as in **C**, ×5. **F:** *Vendomia menneri* Keller, locality as in **C**, ×5.

This long discussion of the arguments that went into constructing the Vendobionta concept shows that there is no necessity to recognize a new kingdom. Nevertheless, the constructional approach developed by Seilacher can be extremely effective for physiological and ecological interpretation of some Vendian fossils. This is particularly true for forms such as *Phyllozoon* (Jenkins & Gehling 1978). This organism was growing in all directions by addition of uniform body parts and to some extent indeed resembled a quilt (see also Runnegar, this volume).

Trace fossils: Ecological, ethological, and physiological interpretation

Fortunately all paleontologists agree that the trace fossils (tracks, trails, and burrows) are firm evidence of "true" metazoans in the Vendian biocenoses. The annoying side of this evidence is that the producers of the bioturbation remain invisible – there is no correlation between the body fossils and the trace fossils.

The bioturbations were produced in the soft sediments of the sea bottom by animals different from those that are known as body fossils. This situation is not unique for the Vendian, but with rare exceptions it is common for the whole Phanerozoic as well.

Nevertheless, the interpretation of the trace fossils can concern the body morphology and grade of organization of the producer, the mode of locomotion and the feeding habits, some physiological peculiarities and behavioral patterns, and the effect of the animal on the sedimentary fabrics. Paleoichnology gives us extremely valuable information on the evolution of the benthic ecosystems and the history of the colonization of the ocean floor.

Vendian trace-fossil assemblages, or ichnocenoses, include feeding burrows, dwelling burrows, as well as crawling and grazing trails (Fig. 6). Vagile benthic animals mainly occupied a relatively narrow zone at the sediment–water interface. Peculiar characters of the Vendian trace fossils are their small size, shallow penetration into the sediment, and predominance of bioturbations formed by soft-bodied metazoans that used different modes of peristaltic locomotion using hydrostatic skeletons or muscular contraction of the ventral side of the body (pedal wave).

Some of those trace fossils resemble the bioturbations produced by annelid worms or mollusks, but we have to take into account that some Recent coelenterates, for example, sea pens, actinians, and even hydroid polyps, can burrow into soft sediment.

A relatively small number of Precambrian trace fossils can be interpreted as bioturbations left by coelomic-grade animals. Among them we should mention *Neonereites* (series of fecal pellets), *Medvezhichnus* (relatively large burrows with circular cross section and a sinusoidal medial ridge at its sole), as well as some burrows showing relatively complex behavioral patterns (*Palaeopascichnus*, *Nenoxites*, *Yelovichnus*, etc.).

Very few (if any) Vendian trace fossils reflect locomotion by means of parapodia, feet or other appendages. That fact indicates that those Ediacaran animals that are interpreted as arthropods or annelids were swimming rather than crawling creatures.

The majority of the producers of Vendian trace fossils collected small food particles, probably cropping on benthic microorganisms. Some trace fossils can be interpreted as reflecting the function of a collecting apparatus (*Nenoxites*).

A remarkable peculiarity of the Vendian ichnocenoses are the bioturbations reflecting rather complex behavioral patterns of vagile benthic animals that seek to cover the food-enriched area with high efficiency. But the diversity of those patterns remains relatively low during most part of the Vendian. Such traces as *Palaeopascichnus*, *Nenoxites*, *Yelovichnus*, and *Helminthoida* reflect different kinds of sinusoidal (meandering) patterns. Circular trails (*Circulichnus*) or star-like bioturbations are less common.

The substantially two-dimensional (subhorizontal) mode of the Vendian bioturbations left at the sediment–water interface, and their small size, made the preservation

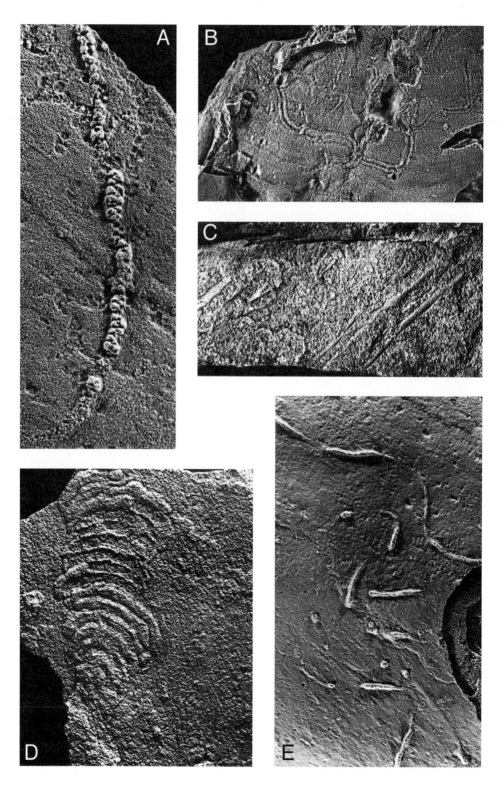

of trace fossils strongly dependent on the properties of the environment. This is why extremely shallow-water deposits demonstrate relatively few trace fossils, especially in the case of homogenous sandy facies. But if sandstone or siltstone is interbedded with clay, then we can meet a rather diverse trace-fossil fauna even in shallow-water facies. In this case we observe bioturbations belonging to different etological groups in one and the same environment. This fact shows a fundamental difference of shallow-water ichnocenoses of the Vendian from the later ones. In the Phanerozoic paleoichnological record there is a so-called bathymetric zonation of trace fossils (Seilacher 1967). It became well established from the Early–Middle Cambrian, but more prominent features can be seen from the Ordovician and later. In the Vendian we observe etologically mixed ichnocenoses in shallow-water environments. This supports an idea that the colonization of the seafloor by metazoans began from the shallow-water biotopes and then spread out to the deeper environments.

The higher number of trace fossils is found in the lower half of the Vendian, which is characterized by abundant imprints of soft-bodied metazoans (Fedonkin 1976, 1981, 1987). About 23 genera and 26 species of trace fossils are described from this level (Redkino Horizon of the Vendian on the Russian Platform). Stratigraphically higher, the diversity of trace fossils has considerably decreased, and metazoan body fossils are extremely rare. This phenomenon can be observed at the Kotlin Horizon of the Vendian on the Russian Platform, in most parts of the St. John and Signal Hill groups (southeastern Newfoundland) and in some other regions where we deal with continuous sequences of rocks crossing the Precambrian–Cambrian boundary. The Kotlin Horizon of the Vendian (V3), for example, contains only seven trace-fossil genera.

Far below the base of the Tommotian stage of the Lower Cambrian, considered to be the Precambrian–Cambrian boundary, one can observe a mass appearance of diverse trace fossils resembling those that are typical for the Paleozoic. This biostratigraphical level – which is identified by the presence of numerous sabelliditids, the first small shelly fossils, peculiar acritarchs, and vendotaenian flora – is known as the Rovno Horizon on the Russian Platform and the Nemakit–Daldyn Horizon in Siberia. At this stratigraphic level we now know 16 genera and 17 species of trace fossils. Many of them are known higher in the stratigraphic column. Ichnocenoses of the uppermost Vendian (Rovno, Nemakit–Daldyn) differ from older trace-fossil assemblages by taxonomic composition, large size, higher diversity of behavioral patterns, and deeper penetration into the sediment. We may state that biological processing of the sediment increased essentially before the Tommotian age.

Although the taxonomic composition of the Vendian Rovno Horizon differs from that of the Redkino or the Kotlin, we should stress some features that are in common, e.g., the mixture of different ecological groups, the dominance of traces left by soft-bodied metazoans, the absence of traces made by parapodia or other appendages, and the occurrence of trace fossils in shallow-water facies.

FIGURE 6   **A:** *Neonereites biserialis* Seilacher, Ust-Pinega Formation, Summer Coast of the White Sea, Russia, ×2. **B:** *Bilinichnus simplex* Fedonkin & Palij, Ust-Pinega Formation, Winter Coast of the White Sea, Russia, ×1. **C:** *Skolithos declinatus* Fedonkin, locality as in **B**, ×1. **D:** *Planolites* cf. *serpens* (Webby), locality as in **A**, ×1. **E:** *Yelovichnus gracilis* Fedonkin, locality as in **B**, ×2.

Conclusions

We are now in the situation Charles Darwin found himself in about 150 years ago. He was puzzled by the absence of the ancestors of the Cambrian invertebrates, considering this fact as a strong argument against his theory of the gradualistic evolution of species. We do not know the ancestors of the Vendian fauna as well, and like the Cambrian biota it appeared suddenly in a "complete state."

The key to this enigma could be a "paradox of transformism," as was formulated by P. Teilhard de Chardin in 1925: the beginning of every phylum or evolutionary stem is not yet a typical component of the latter, and very often it is considered as part of the source phyla or it escapes the attention of the researcher due to the scarce material preserved. This tune is developed in the famous book *The Phenomenon of Man* by the same author:

> *There is nothing more delicate and quick in nature than the beginning. Until the time the zoological group is young its characters stay uncertain. Its structure is frail. Its size is small. It consists of a relatively small number of individuals and they change quickly. Both in time and in space the bud of the living branch has a minimum of differentiation, expansion and resistance. How does time affect the weak zone?*

> *Inevitably destroying what is left . . .*

> *Therefore there is nothing surprising about the fact that retrospectively things seem to us as appearing in complete form.*
>
> Translation from the Russian edition of the book (1987)

Vendian and Early Cambrian faunas can serve as an illustration of this idea.

And though there is little hope to find the ancestors of the Vendian metazoans, we still would like to look into the older times. We shall possibly find in the pre-Vendian deposits quite different organisms that do not look like the Vendian or younger ones. This expectation may now be the most probable, as the early multicellular animals seemed to evolve through series of rapid radiations interrupted by extinctions which eliminated most of the scarce phyletic lineages.

Will the number of characters in the pre-Vendian body fossils be enough to distinguish them as the forerunners of the great world of multicellular animals? Let's wait . . . or go down to the Proterozoic.

Acknowledgments. – I am grateful to the symposium organizers for the opportunity to participate in this fruitful work. My special thanks to Stefan Bengtson and Christina Franzén for editing and improving my manuscript and transferring the text to computer. I further thank three anonymous referees for valuable remarks that helped to improve the chapter.

Early multicellular life: Late Proterozoic fossils and the Cambrian explosion

Adolf Seilacher

Geologisch-Paläontologisches Institut, Sigwartstraße 10, D-72076 Tübingen, Germany; and Department of Geology, Yale University, New Haven, Connecticut 06511, USA

If Ediacara-type body fossils are considered an independent group of probably non-multicellular organisms (Vendobionta), the metazoan record of the Proterozoic is reduced to (1) Psammocorallia, whose actinia-like body was stabilized in sandy bottoms by an internal, organically cemented sand skeleton, and possibly sponges with a similar skeleton; (2) tool marks of various stiff organisms; (3) small trace fossils made by worm-like infaunal deposit feeders. In this view, the last are the ancestral bilaterians that produced a plethora of phyla at the beginning of the Cambrian. In this evolutionary explosion, the broad-scale acquisition of stiffer biomineralized skeletons was not an epiphenomenon but the crucial innovation, because it not only affected the fossilization potential but also modified the soft parts into novel bauplans.

In this symposium I have been assigned the theme "early multicellular life." This would formally relieve me from the task of reiterating my views on Ediacara-type fossils (see contributions by Fedonkin, Runnegar, and Sun in this volume), because I strongly suspect that they were not multicellular. Nevertheless, let me shortly repeat the major arguments:

1 Ediacaran organisms could not function like the jellyfish, sea pens, worms, arthropods, and echinoderms to which they have been traditionally compared because of gross geometric similarities.

2 Instead, they share a very unusual preservation (mere impressions) that has no counterparts in Phanerozoic rocks of equivalent coarsely clastic facies.

3 They also share a foliate and quilted construction, in which the dimension of quiltings is kept fairly constant during growth, either by serial addition of new units or by fractal subdivision of older ones.

Therefore I concluded (Seilacher 1984, 1989, 1992) that Ediacaran fossils represent an extinct category of organisms (Vendobionta, pro Vendozoa, an informal and misleading term) with the following characteristics:

Bengtson, S. (ed.) 1993: *Early Life on Earth. Nobel Symposium No. 84.* Columbia U.P., New York

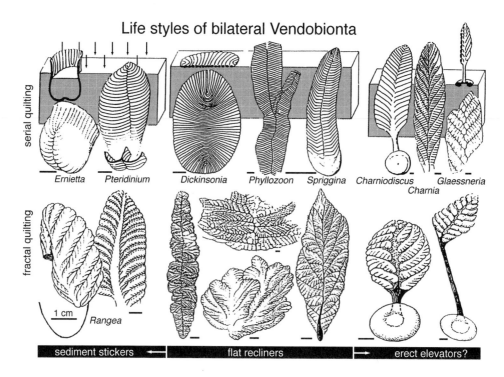

FIGURE 1 Ediacaran Vendobionta are here viewed as nonmetazoan (and probably plasmodial) immobile soft bottom dwellers. Their major subdivision is not by symmetry but by the mode of quilting (serial versus fractal). Otherwise, they diversified into similar ecological guilds (recliners, elevators, sediment stickers) that secondary soft-bottom dwellers derived from sessile metazoans did at later times. Species from Newfoundland (lower right) are as yet unnamed. According to new field evidence from Namibia, the fronds of *Pteridinium* did not emerge above the bottom (as shown in the the figure) but lived completely buried in the sand. (From Seilacher 1992.)

a They were immobile soft-bottom dwellers that radiated in similar directions (flat recliners; erect elevators; sediment stickers), as did later organisms of the same life style (Fig. 1).

b Their integument was flexible enough to deform under tractional or compactional forces but too stiff to develop small-scale wrinkles.

c In the absence of mouth, anus, or other organs, they must have metabolized through the skin either by absorption of dissolved organics or by photo- or chemosymbiosis.

d Since integumental compartmentalization is a general feature of oversized plasmodial organisms (slime molds; larger foraminiferans; *Acetabularia*), I assume that vendobiontic quilting had a similar function; i.e. it not only stiffened the hydrostatic body and maintained its foliate shape, but also allowed large size – without true cell division.

With the Vendobionta thus being removed from metazoan ancestry, where are the roots of the animal phyla that appeared in the Cambrian "explosion"? In my view, true Precambrian metazoans have left their record in three different modes of preservation: (1) storm-smothered sand skeletons, (2) tool marks, and (3) trace fossils. This record must be discussed before we can address the Cambrian explosion.

Sand skeletons: an early substitute for biomineralization

Among the fossils found on bedding planes of Vendian sandstones, some globular forms (largely corresponding to the orders Cyclozoa and Inordozoa of Fedonkin 1985a) differ from associated Vendobionta by the lack of quilting, their high bottom relief, a marginal cleft, and an obvious dorsoventrality: the lower side is hemispherical, sometimes with a central dimple, while the upper is flat, with marked concentric rings.

These are characteristics shared by the problematic fossil *Protolyella* from the Lower Cambrian Mickwitzia Sandstone of Sweden (Jensen 1993; Fig. 2 herein) and from

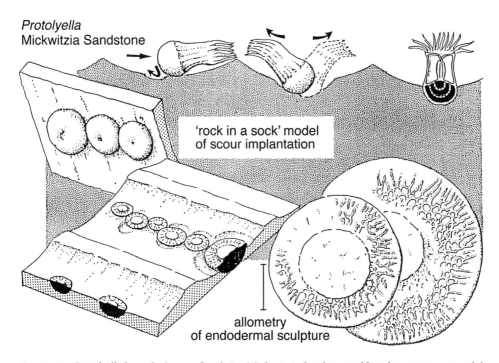

FIGURE 2 *Protolyella* from the Lower Cambrian Mickwitzia Sandstone of Sweden serves as a model for less well-preserved Vendian body fossils. It is here interpreted as an actinia-like anthozoan that stabilized itself on sandy bottoms by an internal sand skeleton. Note the self-organized (allometric) sculpture, reminiscent of stegocephalian skull bones, that increased the digesting endodermal surface. (From Seilacher 1991.) Scale bar 1 cm.

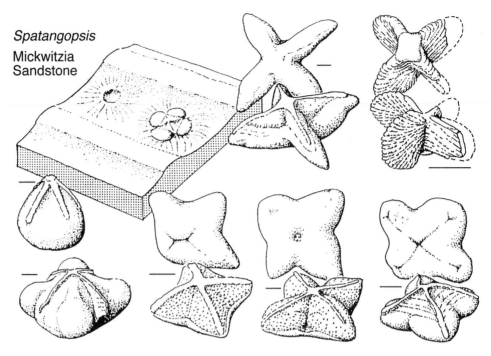

FIGURE 3 *Spatangopsis*, also from the Mickwitzia Sandstone, increased the endodermal surface by radial structures similar to septa of calcareous corals. Because of different fabricational constraints, however, sand septa could be much more robust. Psammocorallia have been exclusively preserved by storm obrution. Still, they are always preserved in life position, which proves the effectiveness of the "rock-in-the-sock" mechanism of stabilization. (From Seilacher 1991). Scale bars 1 cm.

Upper Ordovician sandstones of Jordan (Huckriede 1967; Seilacher 1983b). In the Mickwitzia Sandstone, the round *Protolyella* is also associated with a similarly preserved stellate form (*Spatangopsis*, Fig. 3). A rich material in the Stockholm Museum and in the private collections of Jan Johansson (Örebro) and Holger Buentke (Lugnås), as well as my own work in the field, have yielded the following observations:

- All specimens (including the Jordanian ones) occur on top surfaces of sandy tempestites, where they are concentrated in ripple troughs.

- Not a single specimen has been found overturned. The smooth and convex side always points down, the flat and sculptured side up, although this would not be the mechanically most stable position.

- In a few specimens impressions of radial tentacles are preserved on the bedding surface (Fig. 3).

- Closely crowded specimens touch each other without leaving space for an umbrellar mass around them.

- On the other hand, closely packed bodies never indent each other, suggesting that the sand filling was already cemented by the time of burial.

- After burial, however, infaunal organisms could burrow right across such bodies. So the sand fill had quickly been de-cemented.

- In the Jordanian *Protolyella*, cross sections also reveal concentric laminae that conform to the hemispherical lower surface and strike out on the upper surface as growth rings rather than being horizontally bedded.

These observations confirm that the sand was indeed packed into a soft bag, but not in a gastral cavity and not after death. Rather it has been suggested (Seilacher 1991) that these fossils represent the internal sand skeletons of a kind of sea anemone (Psammocorallia) that was uniquely adapted to living on unstable sandy bottoms. Being situated between the ecto- and endoderm (i.e. in the place of a medusoid mesogloea), this skeleton could simultaneously serve two important functions:

- The upper surface could by its sculpture increase the digesting surface of its endodermal lining. In *Protolyella*, this was done by reticulate ridges of somewhat finer sand. In *Protolyella benderi* concentric growth rings combine with radial ridges of variable size. *Spatangopsis* eventually added sand on the upper as well as on the lower surface and thus grew 4–5 radial folds reminiscent of the septa of later rugose and scleractinian corals.

- In all forms, the bulky sand skeleton stabilized the animal on unstable sand bottoms not only by its weight. As a "rock in the sock" (a term that I owe to Mark McMenamin), it also effected its passive implantation by the scouring action of waves or currents.

This new model has some interesting consequences with regard to the biology and taphonomy of these organisms as well as the phylogeny of coelenterates.

Phylogenetic implications

Because of their simple, gastrula-like construction, coelenterates have long been considered as a very old group that possibly originated apart from other metazoans. Affiliation of most Ediacaran fossils with coelenterates was in line with this view. New molecular data (Diana Bridge, personal communication, 1992) suggest, however, that not jellyfish but anthozoan polyps were the most ancient coelenterates. The sand corals thus suitably fill the gap left by the disclaiming of the Ediacara-type "coelenterates." They also explain why true corals with a calcareous exoskeleton appear only in the Ordovician, long after the Cambrian revolution: they already had a hard skeleton that did not require biomineralization. Of course, Psammocorallia are not considered the direct ancestors of later corals but a branch equivalent to the rugose and scleractinian corals, characterized by their own skeletal construction.

Biologic implications

While the Psammocorallia resembled modern anthozoans in their mode of feeding and reproduction (probably including external budding), their skeleton required the ability to pass sand grains through the endoderm or the ectoderm or both. But it also implied some important ecological constraints: (1) Being internal, it did not allow attachment to hard substrates. (2) It was in the way of active burrowing (which was employed by the maker of the Cambrian trace fossil *Bergaueria*), because the basal part of the body could not be hydrostatically swelled. (3) The internal skeleton and life on soft bottoms probably also blocked the way to coloniality. (4) It did not allow the growth form of accretionary cones.

Taphonomic consequences

The fact that worm burrows could easily penetrate the formerly rigid skeletons suggests that the cement was organic and disintegrated soon after death. Thus, unlike calcareous shells, the remains of Psammocorallia could never be reworked. Neither would they remain visible if embedded in sand of the same grain size. Therefore we find these fossils only in sandy storm layers, where they could either leave their impressions on the sole face or, more commonly, are preserved on the top. In the classification of Fossil-Lagerstätten such occurrences would fall under obrution deposits (Seilacher *et al.* 1985): as the storm waned, sluggish or nonlocomotory animals did survive the deposition of the coarse fraction but became suffocated when – after an interval of nondeposition – the fine mud settled from suspension. In the case of the sand corals it was also essential that the muddy tail of storm sedimentation surrounded the carcasses with material different from their native sand, so that they retained their individuality after the skeletal cement had become disintegrated.

Obrution deposits are unique also in another respect. Since they imply little or no transport, they lack the usual biostratinomic distortion. Rather, they preserve census populations from which standing biomass, age structure, and sometimes spatial distribution can be directly mapped. If we take the bedding surface of Fig. 2 as such a "fossil snapshot," it tells us that Psammocorallia were not evenly distributed on the sandy bottoms but lived in clusters comprising different age groups. Since this makes no sense as a trophic strategy, it is possibly the outcome of an asexual mode of reproduction. If so, one might even approach the heritability of morphological characters within one paleoclone!

Other representatives

We may ask the question whether Psammocorallia were really confined to the sandy habitats in which we now find them, or whether they also invaded silty bottoms. Here, paleontological documentation would be still less likely, because the preservational model just described would not work: silt skeletons in silty tempestites would be very difficult to recognize. Still, there is one fossil that should be mentioned in this context. It is the impression of an actinian-like animal lying parallel to the bedding plane in the famous new Fossil-Lagerstätte of Chengjiang in the Lower Cambrian of South China (Chen & Erdtmann 1991). Like its Middle Cambrian counterpart, the Burgess Shale, this deposit was probably formed in a largely anoxic environment, into which organisms inhabiting adjacent oxic bottoms became imported by storm-induced compensation currents. What makes *Xianguangia* a potential relative of sand corals is the fact that its basal part has a rounded outline and is offset from the tentacled top by a horizontal line at about the level where the internal skeleton would have ended. That it does not preserve its three-dimensional shape should not bother us, since it would have been as fine-grained as the sediment around and thus have suffered much compaction.

Another suspected relative, *Brooksella*, from the Middle Cambrian of Alabama, must be disclaimed. Although it resembles *Spatangopsis* and has been placed in its neighborhood in the coelenterate part of the *Treatise on Invertebrate Paleontology* (Harrington & Moore 1956), it is clearly a trace fossil modified by diagenesis. A large material in the Yale Peabody Museum shows U-shaped tubes radiating from a central shaft, as well as lunate retrusive backfill lamellae around them. The bioturbated mud was later transformed into an ironstone concretion, whose bulging surfaces made the fossil even more jellyfish-like. Thus Jensen's (1991) model is valid, though for *Brooksella* instead of *Spatangopsis*!

Sand sponges?

Since sponges are another ancient group and sometimes do build sand grains into their skeleton even in modern species, the cooccurrence of sponge-like bodies with *Spatangopsis* in the Mickwitzia Sandstone may be of interest. Several specimens in the Stockholm Museum and in the private collection of Jan Johansson (Örebro) have the ear-like shapes characteristic of many sponges. One undescribed species shows irregularly distributed oscula on one surface and smaller ostia on the other side. Another form is a flat funnel incised on one side by a deep notch with a thickened rim. The fact that in one specimen (Fig. 4) drag marks have been carved by this corner suggests that it served

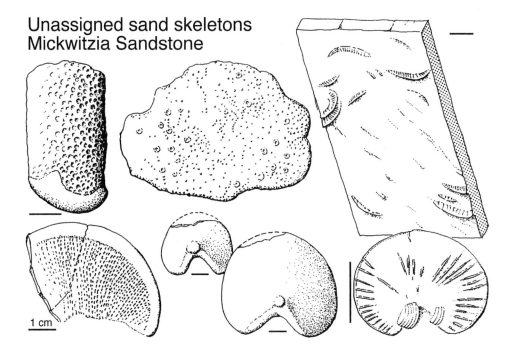

**Unassigned sand skeletons
Mickwitzia Sandstone**

1 cm

FIGURE 4   Possible sand sponges from the Mickwitzia Sandstone. Bounded by the same preservational constraints as the sand corals, they differ by ear-like morphologies. Drag marks in the upper right suggest a stabilizing function and serve as a clue to the origin of the pseudofossil "Eophyton." (From specimens in the collection of Jan Johansson and in Naturhistoriska Riksmuseet, Stockholm.)

as an anchor in the loose sand. Although these examples come from the Lower Cambrian and therefore do not bear directly on the earlier history of the sponges, they might draw attention to similar occurrences in the Precambrian during future expeditions. In any case these forms still coexisted with biomineralized Cambrian sponges that had either nonspicular calcareous (Archaeocyatha) or spicular siliceous skeletons (*Protospongia*).

Distinctive tool marks

Paleontologists often forget that organisms may leave a record not only in the form of body or trace fossils but also when the carcasses merely touch the bottom or drag over it. In order to be preserved, such tool marks should be sand-cast immediately, or, better, be carved through a thin veneer of sand on a mud interface. This restricts them to the bottoms of sandy or silty event-layers, be it turbidites, inundites, or tempestites.

A famous example is "Eophyton." As the name suggests, these bottom markings were originally interpreted as fossil sea weed. It was the merit of Nathorst (1881) to recognize their essentially physical nature. But why is this pseudofossil so common in the Lower Cambrian of the Baltic region that it became the namesake of a stratigraphic unit, the Eophyton Sandstone? The drag marks associated with the body fossil in Fig.

FIGURE 5   The pseudofossil Eophyton in Lower Cambrian sandstones commonly shows pairs or triplets of parallel drag marks. In this slab from the Mickwitzia Sandstone of Lugnås (Riksmuseet, Stockholm) one couplet even made a McDonald's sign as the oscillating storm wave reversed. The common and widespread occurrence of Eophyton suggests that self-stabilized sand skeletons were much more abundant than their exceptional preservation as body fossils would suggest.

5 may serve as a clue: if organisms with heavy sand skeletons were as common on sandy bottoms of that time as the obrutional snapshots suggest, heavier storms must have shuffled them over the eroded mud bottom before they became embedded in sand and disappeared. Since they were built to maintain a certain stable position, they also would glide, rather than roll and leave impact marks (as calcareous shells do in Phanerozoic tempestites). The common occurrence of pairs or triplets of gliding marks and even their reversal with the oscillating current of a wave confirms this view. Thus we have the rare case that a mechanoglyph becomes a stratigraphic index!

While "Eophyton" is again a Lower Cambrian example that still needs to be documented in the Precambrian, this leads us to another kind of problematic fossil that was recently discovered in Late Proterozoic rocks of North Carolina as very delicate impressions on the soles of fine silt laminae. In isolation they look like the biserial trackway of an arthropod or annelid. However, they are too straight and too much confined to a certain length to record locomotion. Therefore the spontaneous association of Tony Rathburn (Duke University), who found the first slab (Fig. 6), was more

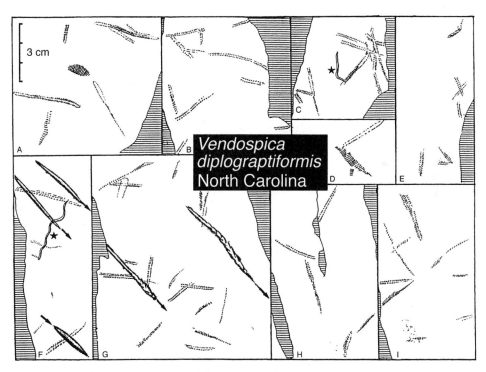

FIGURE 6   Delicate biserial impressions on the bottoms of thin sand laminae in the Upper Protero-zoic of North Carolina are not biserial trackways, as one might gather from individual specimens. These portions from a large slab (found by Tony Rathburn and now housed in the Schiele Museum of Natural History, Gastonia, North Carolina) are drawn in the same relative orientations. They show that the impressions (1) are too straight for trackways (compare true worm burrows with an asterisk!); (2) have uniform lengths not exceeding about 3 cm; (3) sometimes are deformed by dragging or lateral rolling. Therefore they are interpreted as tool marks left by stiff graptolite-like bodies of problematic affiliation. Note that a later turbidity current (arrowed groove casts!) did not reorient the stipes.

adequate: graptolites. This should not be taken literally, of course, but the impressions indeed appear to be produced by stiff stipes with two, or perhaps four, rows of very regularly spaced spines. Whatever they were, we must reckon with some kinds of colonial organisms, perhaps hydroids, that already possessed a probably organic but stiff exoskeleton.

Actinian burrows

In addition to the sand skeletons of Psammocorallia, actinians dwelling on soft bottoms can leave a record also through the active burrowing habits of soft-bodied species. The resulting trace fossils (*Bergaueria*) resemble the bottom side of *Protolyella* (Fig. 2; hemispherical casts with a central depression or knob), but they may also have become duplicated as the animal changed its position. Impressions of this general nature have been variously described from the Precambrian but were interpreted either as medusae or as nonbiogenic structures (sand volcanoes, pot casts). Since physical processes may in fact produce similar features, reported cases must be checked very carefully. For instance, the structures described as abiogenic from levels just above the last tillites in Namibia (Glaessner 1988) are indeed pot casts (F. Pflüger, personal communication, 1993).

Bilaterian burrowers

Having cleared the way by reinterpreting the Vendobionta and discussing the true coelenterates and possibly sponges, we are prepared to approach the prelude to the Cambrian revolution, which now essentially becomes an affair of bilaterian animals. Leaving aside the calcareous tubelets of the problematic genus *Cloudina* (see Bengtson, this volume), Vendian bilaterians have left their record mainly as trace fossils. Of these, an impressive number have been described in recent years from late Proterozoic strata all over the world (summarized in Crimes 1989). Unfortunately, however, Precambrian trace fossils have so far been studied only from a biostratigraphic point of view, so that their careful biomechanical and behavioral analysis is still lacking.

Without going into details, I will try to summarize some salient points:

1 All known Vendian (and perhaps older) trace fossils were made by small worm-like organisms.

2 From preservational details we can conclude that they were made infaunally, though their makers did not penetrate deeply into the sediment.

3 Most forms slavishly followed bedding planes and some of them employed elaborate search programs (meandering, strip-mining, radiating) typical for deposit feeders (see Fedonkin, this volume, Figs. 6E and 1D).

4 As shown by the cooccurrence, in a turbiditic sequence of North Carolina, of the strip-mining trace fossil *Oldhamia* with a few specimens of the vendobiont *Pteridinium* (Gibson 1989; Seilacher & Pflüger 1992), deposit-feeding burrowers had already invaded deep-sea bottoms by Vendian times.

In conclusion, my earlier and often cited statement (Seilacher 1956) that there was a sudden increase in diversity and elaboration of trace fossils at the Precambrian–Cambrian boundary must be modified. Of course, new phyla, such as arthropods (mainly trilobites) and mollusks, did add their ichnological record in the Cambrian. There is also more vertical burrowing (*Skolithos, Diplocraterion, Syringomorpha, Teichichnus*), and burrows do become much larger in the Lower Cambrian than before. Accordingly, the depth and intensity of bioturbational mixing did increase. But even if numbers of ichnogenera were normalized for taxonomic splitting (which tends to be high among Precambrian workers), the break appears much less pronounced now than four decades ago.

This result has consequences for our evolutionary scenario. It suggests that the only truly locomotory organisms of Vendian times were small worm-like bilaterians that lived as shallow infaunal deposit feeders and left the sediment surface to the non-locomotory Vendobionta, Anthozoa, and perhaps sponges. Without a more careful analysis of the described trace fossils it is difficult to link them to particular animal taxa. Ciliary as well as peristaltic mechanisms appear to have been involved in their burrowing, while the sinusoidal movement of nematodes is so far not represented. Nevertheless it is now clear that these "worms" were already anatomically and behaviorally (i.e. neurally) advanced and had conquered marine bottoms at any depth. Yet it is possible that they still lived in a peaceful "Garden of Ediacara" (another metaphor coined by Mark McMenamin [1986]), in which the arms race between animals had not yet begun.

The Cambrian revolution

There has been much speculation about what caused the sudden appearance of bilaterian animal phyla in the Lower Cambrian. Extrinsic factors such as oxygen level, temperature, and seawater chemistry probably played a role. But the major innovation that we can observe is the broad-scale acquisition of hard skeletons made of tanned organic (scleroproteins and polysaccharids) or biomineralized (silica, carbonate, apatite) materials.

Hard skeletons are not simply an addition to pre-existing soft-bodied ancestors. Through their mode of growth (iterative and mutualistic interaction of soft matrices and hard precipitates), stiff skeletons become chief modifiers of organismic bauplans. In other words, there are no such things as soft-bodied protomollusks, protobrachiopods, or protoarthropods, since they could not be morphologically recognized as such. Secondary loss of the skeleton (slugs, octopuses) is a different issue, because by then the bauplan was already genetically and developmentally channeled.

In conclusion, whatever extrinsic factors allowed the simultaneous acquisition of hard skeletons by different groups of worm-like ancestors, this innovation has probably been the proximal cause of the Cambrian taxonomic explosion of bilaterian animals. It allowed the new groups to use hard parts for jaws, legs, or body support (instead of the sediment) and to explore new life styles, starting with the epibenthic mode. The ways in which the new constructional possibility was used and modified during development by the genomic adoption of self-organizing mechanisms (Seilacher 1991) did define what we now recognize as phyla. What followed was primarily a variation of then established themes, driven by niche competition and utilization, arms race, symbiosis, replacement after extinction, and the conquest of new habitats (open water, land, air). In a way the wonderful transformations of animals that we observe during the Phanerozoic are just the epilogue to two earlier events: the emergence (and survival) of multicellularity in the Proterozoic and of hard skeletons and new levels of trophic interaction at the beginning of the Cambrian!

Acknowledgments. – I first thank the organizers of this symposium for providing a platform to discuss these questions and for allowing me to participate. My Precambrian studies were supported by travel funds from the Deutsche Forschungsgemeinschaft (1984, through SFB 53), the National Science Foundation (1990, through Mark Brandon, Yale University) and the National Geographic Society (1991). I also appreciate discussions in the SFB 230 ("Natürliche Konstruktionen," Stuttgart), at the Institute of Biospheric Studies (Leo Buss, Günther Wagner) at Yale University, at UCLA (J. William Schopf and Bruce Runnegar), and at the Phylogenetisches Symposium (1991, Hamburg). Fossils from the Mickwitzia Sandstone were made available by Jan Bergström (Naturhistoriska riksmuseet, Stockholm), Jan Johansson (Örebro), and Holger Buentke (Lugnås). A party including Hans Luginsland and Frieder Pflüger was able to visit outcrops in Sweden through support by SFB 230.

The Cambrian explosion

James W. Valentine

Department of Integrative Biology and Museum of Paleontology, University of California, Berkeley, California 94720, USA

The fossil record of late Precambrian and Early Cambrian time documents the relatively abrupt origination of a wide variety of metazoan body plans. Many of the relatively complex Cambrian clades distinctive at the level of phyla have such important derived features that the last ancestors shared by most any pair of them must have been significantly simpler organisms. Increasing structural complexity implies increasing numbers of cell types. Cell-type numbers required to produce trace and body fossils leading to the Cambrian explosion are inferred from living phyla. Late Precambrian and Phanerozoic rates of evolution of complexity seem comparable and suggest that the Metazoa may have originated about 680 Ma ago. The cascading, hierarchical regulatory systems of metazoan genomes permitted both relatively slow increases in complexity and relatively rapid radiations of body types at given levels of complexity. The Cambrian explosion may be owing to radiations of several stocks beginning at about the 40–50-cell-type level during a few million years.

The explosive appearances of novel clades near the Cambrian–Precambrian boundary are well documented. Whether or not the appearances represent the evolutionary invention of those clades, or are artifacts of an incomplete stratigraphic record, or are owing to an increase in the preservability of invertebrates – to their acquisition of durable skeletons, for example – have been hotly debated questions. Evidence from dating of the Precambrian–Cambrian stratigraphic sequence indicates that the appearances were indeed abrupt, geologically. Evidence from body- and trace-fossil occurrence and diversity and from the appearance of many novel "soft-bodied" taxa in Cambrian *lagerstätten,* suggests that the breadth of the explosion as indicated by shelly forms was not the half of it, figuratively and literally, and that many of the Cambrian clades were new evolutionary inventions (Valentine 1977; Runnegar 1982b; Valentine & Erwin 1987; Valentine *et al.* 1991). It is the purpose of this paper to consider what sorts of evolutionary processes are required to fuel such a profound and unique event.

Several major problems, research on which lies in disparate fields, must be brought much nearer to solution before the Cambrian explosion can be explained in any convincing manner. For example, we need to know the phylogenetic interrelations of the organisms involved: Did the radiation proceed from one or a few ancestral stocks, or was it a parallel event in a great many lineages? We also need to know the body plans of the ancestral stocks from which the radiation(s) proceeded: Were changes in body plans large or small across this interval? With such data, it will be possible to ask

Bengtson, S. (ed.) 1994: *Early Life on Earth. Nobel Symposium No. 84.* Columbia U.P., New York

whether the explosion can have involved the same sorts of genetic changes that are documented among living populations and just what sorts of environmental situations might be necessary to underpin this evolutionary rampage. These are all large questions, and largely unanswered. Still, there is nothing to prevent attempts to fit together some of the hypotheses arising from these various fields; it is possible that the fossil record of the Cambrian events will eventually provide significant clues to these problems.

Metazoan phylogeny and the Cambrian explosion

Speculative phylogenetic trees of phyla have been produced since the late nineteenth century. Most are confined to living phyla, and most attempt to indicate which phyla have given rise to which others. This ancestral–descendant relation is taken literally in many cases; for example, the arthropods are often hypothesized to have descended from the annelids. By this is meant that an actual annelid was the forerunner of arthropods. The sequence of steps along which this change is assumed to have occurred is termed "arthropodization." Consideration of the functional morphologies, fossil records, and molecular-based inferences as to the patterns of relatedness of these phyla render this scenario highly improbable (Manton 1977; Field *et al.* 1988; Lake 1989; Valentine 1989; Turbeville *et al.* 1992; Eernisse *et al.* 1992; Raff, this volume). It seems more likely that the last common ancestor of these two phyla had the body plan of neither of them. It has been hypothesized that the common ancestor, of late Precambrian age, was an elongate, vascularized worm with seriated organ systems (Valentine 1989). The characteristic segmentation of annelids with a hydrostatic coelom, and the cuticular segmentation of arthropods with a hydrostatic hemocoel, probably evolved independently. In arthropods, intramesodermal spaces that qualify as coelomic sometimes occur within larval segments. These spaces eventually either serve as ducts, become confluent with the haeomcoel, or are occluded (Anderson 1973) and may be derived from a common annelid–arthropod ancestor or may be apomorphies.

Mollusca is another group that is commonly postulated to have arisen from a living phylum. At one time Annelida was the preferred ancestor, but more recently the mollusks are believed to have their roots in the phylum Platyhelminthes or at least in organisms of flatworm grade (Vagvolgyi 1967; Stasek 1972; Salvini-Plawen 1985). Although the mollusks are considered coelomic, in the primitive molluscan groups the only intramesodermal spaces that qualify for such status surround organs such as the heart or are ducts. Coelomic spaces in cephalopods, associated with gonads, are relatively capacious, but cephalopods are rather derived organisms with closed circulatory systems and, presumably as a consequence, with relatively small blood volumes (Martin *et al.* 1958). It may well be that the enlargement of coelomic space is a compensation for the reduction in volume of the hemocoel and the loss of associated hydrostatic properties. At any rate, primitive molluscan groups display some organ seriation and may well be vascularized derivatives of flatworms, with some intramesodermal spaces developed to serve their larger, more complex body plan. It seems doubtful that molluscan seriation is reminiscent of the segmentation of either the arthropods or annelids, each of which evolved to support a locomotory system entirely unlike those in mollusks (see Clark 1964; Manton 1977). If a flatworm-like form is a plausible ancestor for the mollusks, it is also a plausible ancestor for the vascularized vermiform clade that is believed to have given rise to both annelids and arthropods. Indeed, the early mollusks were probably a vascularized vermiform clade themselves. The phylogenetic branching patterns among any such plexus of early worms could have had nothing directly to do with the qualities of those organisms as potential ancestors of future body plans.

The best molecular phylogenies of metazoans that are available (Fig. 1), based on 16S and 18S rRNAs (see chapters by Sogin and Raff, this volume), hold no major surprises but do contain a number of small surprises for the comparative anatomist. The evidence

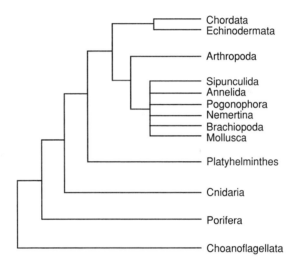

FIGURE 1   Branching pattern of selected metazoan phyla given by 16S-like and 18S rRNA phylogenies. Choanoflagellates, poriferans, and cnidarians were evaluated by maximum-likelihood techniques (Sogin, this volume); the branching from cnidarians to and through the remainder of the metazoan phyla is a consensus of maximum parsimony and distance methods (Raff, this volume; Turbeville *et al.* 1992). Neither the horizontal nor the vertical distances in the diagram are scaled to the molecular data.

suggests that among living phyla, poriferans arose from choanoflagellate ancestors and are a sister group to the other metazoans; that cnidarians branched next and platyhelminths after them; and that the major split between deuterostomes and protostomes then followed. This is a pattern of branching not unlike that suggested by "traditional" phylogenies such as that of Hyman (1940; see reviews of phylogenetic trees in Willmer 1990 and Eernisse *et al.* 1992). Within the protostomes, arthropods appear to be a sister clade to a group of phyla that include annelids, but the branching pattern among this latter group has not yet been resolved (Turbeville *et al.* 1992; Raff, this volume). Among the small surprises are that nemertines and brachiopods fall within this assemblage, although in fact such placements have been postulated by some authorities.

It is highly unlikely that this series of protostome phyla (i.e. body plans) rose directly one from another, so if this pattern is correct, the branching must have occurred within primitive worms. It seems likely that the worms ancestral to these phyla had hemocoels. Some or all of them may even have possessed intramesodermal spaces that would be technically coelomic but that were used to serve organs and were not hydrostatic features. Whatever their anatomical details, it seems that they were a long way from becoming arthropods, annelids, or even mollusks. A possible stem taxon, visualized at a somewhat lower grade, has been designated as the phylum Procoelomata by Bergström (1989). The implication of the molecular phylogeny is that the mollusks are closest in grade to the primitive organisms among which the branchings occurred – they have evolved fewer derived features. This does not imply, however, that mollusks necessarily branched earliest; indeed, the evidence suggests that arthropod ancestors branched earlier. This scenario drives the branching of these phyla back into body plans of organisms that are grades below the level of complexity of most living higher invertebrates.

This situation has interesting consequences for the functional interpretation of the body plans of the phyla. First, the branching that created the clades eventually forming the modern phyla may well have been entirely random with respect to that future. The speciations that produced the branching patterns were probably not at all premonitory of the adaptations associated with the morphological evolution among their descendants that eventually gave rise to new body plans. The branching pattern is certainly of interest, but it may not be the important issue in the origin of most of the phyla. Second, at least some of the anagenetic steps towards living phyla may have involved grades of organisms that are now extinct – essentially acoelomate, hemocoelic, and perhaps seriated worms, for example (Valentine 1991). Third, and of most interest to the subject at hand, it now seems most reasonable to abandon attempts to explain the features of the body plans of higher invertebrates as derived one from another and instead to concentrate on the adaptive steps that lay along evolutionary pathways – sometimes steps of increasing complexity, sometimes of lateral radiations – that led from lower invertebrates to each of the arrays of body types that populated the Phanerozoic oceans. It would appear that, among animals, the Precambrian–Cambrian transition represents the temporal locus of more and of larger adaptive steps than any other comparable period in geologic history.

Body-plan evolution and the Cambrian explosion

The regulation of complex body plans

During ontogeny, the construction of an animal body is orchestrated by genes, including homeobox and zinc-finger genes, that regulate the expression of many of the structural genes (Alberts et al. 1989). The activities of such pattern-formation genes are best known in Drosophila and in Mus, organisms on very distant branches of the phylogenetic tree of metazoans. In Drosophila, the pattern-formation genes operate as a cascade; the expression of genes in an earlier developmental stage is required for the expression of genes in each subsequent stage, with the "earliest" genes in the cascade being maternal, their messages carried by mRNAs in the egg. Additionally, numbers of structural genes are expressed at each stage to produce a sequence of cell phenotypes in appropriate numbers and positions as cleavage and embryogenesis proceed. Hosts of feedback mechanisms, and informational systems that are not directly encoded, also act to regulate the pattern of growth and development. Many of the pattern-formation genes lie next to one another in the genome in the order in which they are expressed; this condition is known best in the gene complexes of the homeotic genes, such as the Antennapedia complex that specifies segment identity in Drosophila.

Some of these genes are so similar in fruit flies and mice that they are assumed to have descended from a common ancestral suite of genes. Even the order of the occurrence of these genes is widely conserved, although the reasons for this are not clear (Alberts *et al.* 1989). Evidently these genes act as transcriptional regulators, much like switches, and when a battery of switches is required for a coordinated sequence of gene activities, a conserved alignment of switch genes is employed. For example, the same sequence of *Antennapedia*-class genes associated with the differentiation of the anterior segments in fruit flies seems to be responsible for the development of subdivisions (rhombomeres) of the hindbrain, among other features, in mice. A duplication of this sequence is also involved in regulating the development of the trunk in mice, although the developmental mechanisms involved seem to be different; evidently the same clusters of transcriptional regulators have been used in several different ways in different developmental contexts within the same organism (Hunt *et al.* 1991a). Indeed, sequences of some of the same genes have been identified in organisms with non-segmented body plans. At least four of the *Antennapedia*-class genes occur in the nematode *Caenorhabditis* in the same order as in *Drosophila* (Kenyon & Wang 1991). The features associated with the expression of these genes are serially aligned in the nematode body, but they are not associated with segmentation, and the genes influence the expression of both repeat and nonrepeat units. *Antennapedia*-type homeobox genes are also reported in cnidarians (Murtha *et al.* 1991).

Thus it is possible that the ancestral homeobox gene sequence evolved simply to specify serial cell diversity (Kenyon & Wang 1991). Schubert *et al.* (1993) have constructed a tree for *Antennapedia*-type genes that suggests that these had diverged from an ancestral gene into three gene clusters before the rise of triploblastic body plans. The origins of the first homeobox-containing gene, or of the first representatives of other transcriptional regulators, are unknown; presumably they lie among unicellular ancestors of metazoans, perhaps for the mediation of aspects of cytoarchitecture. Whenever these genes appeared, they eventually provided the key organizational tools for the construction of bodies involving significant cell differentiation, the basis of multicellular complexity.

Clues to the rise of complexity in body plans

The time of origin of the Metazoa is poorly constrained. The period between the first animals and the Cambrian explosion is thus of uncertain length but witnessed a rise in the complexity of animals from unicells (or protistan colonies) to the immediate ancestors of the Cambrian clades. Absolute dating and correlation of fossiliferous Precambrian rocks are not yet very accurate, but likely metazoan fossils are first found in rocks of about 580 Ma of age, and furrowed trails in the substrate indicate that animals with a creeping locomotion were present at that time. In shallow-water facies of the late Precambrian, as in South Australia, trails appear as early as body fossils (Jenkins 1984), but these horizons have not been dated radiometrically. In Newfoundland, in a deeper-water facies, the earliest metazoan fossils may be on the order of 580 Ma old (Conway Morris 1989c). The fossil record is silent on how much earlier than this the metazoans originated. Radiometric dating now suggests that the main Cambrian explosion may have begun about 530 Ma ago or perhaps somewhat earlier (Compston *et al.* 1992).

Bonner (1965, 1988) has noted that the best single metric of an organism's morphological complexity is the number of cell types that it possesses. The Precambrian history of metazoans involved the establishment and evolution of mechanisms of cell differentiation to produce increasingly complex organisms (with increasing numbers of cell types), culminating in forms well above flatworm grade (Valentine 1991), as suggested earlier for ancestors of the arthropods, annelids, and other phyla. There exists a long tradition of description of cell phenotypes that began with light microscopy and that has been continued and elaborated in ultrastructural studies, which can be applied to the problem of organismal complexity. The cells that are recognized as phenotypically similar do in fact represent some "lumping together" of cells in which different elements of the genome are expressed, if that expression does not produce a phenotypic difference that is deemed important. The ultimate measure of body-plan complexity would presumably be one that reflects the information required to specify the entire body, involving both gene number and the organization of gene expression. Such data are not available, and while the use of somewhat subjective cell types clearly falls short of such a goal, it seems to be the best available method to estimate relative complexities, as Bonner suggested. I am not a competent histologist, but I have attempted to evaluate the histological data of those who are, summarized in Table 1, so as to rate cell types evenly across phyla. The phyla in the table represent a scattering of body plans within a complicated phylogenetic branching pattern, with increasing complexity found along more than one branch. Trends of decreasing complexity must have occurred, especially for body plans of parasitic or miniaturized taxa, but nevertheless the

TABLE 1. Estimates of the approximate cell-type numbers required to construct the simplest individual organism representing various phyla, using a broad definition of type based on cell phenotypes at the ultrastructural level. The references, on the other hand, report cell types in living species but not necessarily the simplest in each phylum. Asterisks indicate references that provide estimates not documented by published histological descriptions; these are not necessarily based on the simplest species with the appropriate body plans.

Phylum	Cell-type number	Reference
Placozoa	4	Ruthmann 1977
Porifera	5	Simpson 1984
Cnidaria	7	D. Chapman 1974
Loricifera	18	Kristensen 1991
Priapulida	19	Storch 1991
Ctenophora	20	Hernandez-Nicaise 1991
Turbellaria	20	Rieger *et al.* 1991b
Nemertinea	35	Turbeville 1991
Annelida	42	Westheide & Hermans 1988
Mollusca	55*	Bonner 1988
Arthropoda	55*	Bonner 1988
Chordata:		
Agnathans (Lampreys)	65	Hardisty & Potter 1971–1982
Average vertebrates	120*	Bonner 1988
Cetaceans	135*	Bonner 1988
Humans	210	Alberts *et al.* 1989

pathway to higher invertebrates certainly involved a succession of forms of generally increasing morphological complexity.

Somewhere along some trend of increasing cell-type diversity, early in metazoan history, the *Antennapedia*-class homeobox gene sequence was assembled. The presence of this regulatory system in distantly related living phyla implies that the basic gene-regulatory mechanisms had become sophisticated enough by the latest Precambrian to produce complex body plans including (or perhaps favoring) body plans with complicated serial systems. The prevalence of serial body plans at the ordinal level during the early stages of the Phanerozoic radiation of metazoans has been documented by Jacobs (1990), who emphasized the role of homeobox selector genes in such architectures. The ability of earliest Cambrian genomes to mediate the evolution of a wide variety of seriated body plans underscores the possibility of independent origins of segmented phyla.

Thus the Cambrian explosion involved the achievement of a certain level of morphological complexity, characterized by perhaps 40–50 or more cell types (Table 1), implying the presence of large numbers of structural genes (to be able to produce all of those cell phenotypes) and of a gene-regulatory apparatus sophisticated enough to mediate their expressions in appropriate amounts, places, and times during ontogeny. It would not be an outlandish speculation to suppose that the organisms responsible for horizontal traces recorded in late Precambrian "Vendian" sediments were at least at the 30-cell-type level. The traces indicate elongate organisms measured in centimeters and capable of displacing and furrowing sediment, which in turn implies a body-wall musculature and some sort of vascularization; such a body, together with a plausible suite of supporting organs, should require a grade of complexity between flatworms and primitive mollusks (of course, the traces could have been formed by more complex organisms as well, or even by some hypothetical diploblastic form, but these possibilities do not seem very plausible). The structural grades represented by the body fossils in late Precambrian sediments have proven difficult to determine. Most of them seem to be at about the cnidarian grade, if they are not actual cnidarians, implying a cell-type number of perhaps 10 or so. Other fossils (such as sprigginids) may be segmented bilaterians, and if so they might belong to a grade with cell-type numbers at least in the middle to high 30s.

Time of origin of metazoans

The origin of new cell types clearly involves more than just adding a number of new structural genes or structural-gene functions to the genome; it also involves adding the information to regulate their expression. Kauffman (1969, 1974) has discussed the origin of new cell types and has suggested that the amount of information required to support multiple cell types may be on the order of the square of the cell-type number. If the information increase required were truly exponential, it should have become increasingly difficult to evolve more complex organisms as complexity increased, and the rate of appearance of organisms of increased complexity would presumably fall off through time. However, in metazoans, much of the regulatory information does not reside in the genome directly but is imparted by induction or other indirect means (Maniatis *et al.* 1987). Furthermore, the system of gene regulation that is emerging from studies of eukaryotes is hierarchical and cascading, and such architectures permit much flexibility in the routing of signals and are capable of expansion without paying the high price required of systems involving brute-force instructions (Alberts *et al.* 1989; for early ideas on such an architecture and how it might relate to the evolution of novel body types see Britten & Davidson 1971; Valentine & Campbell 1975; see also chapters by Raff and by Miklos & Campbell, this volume). Indeed, increasing cell-type diversity might well be in part a function of the number of cell types present. A large number of subroutines of gene expression, which remain silent in most cells only to be switched on when their regulators receive appropriate cues, might well provide a correspondingly large number of opportunities for modification to produce novel cell lines. In this event, the rate of origin of complexity might be expected to increase through time.

Although the rates involved in increasing cell-type number cannot be specified from first principles, rough inferences can be made from the empirical data available from the fossil record. Fig. 2 is an attempt to represent the rate of increase in body-type complexity in metazoans. The more complex organisms are estimated as having about 30 cell types when horizontal trails first appear and 50 cell types at the beginning of the Cambrian, based on the preceding discussion. These estimates yield an average rate of cell-type increase, inferred to occur along some lineage, of about 0.4/Ma during that late Precambrian interval. Advanced mammals have at a minimum some 210 cell types (Alberts *et al.* 1989), when types of nerve cells are lumped. The average rate of increase in cell-type numbers traced along some lineage since the main Cambrian explosion,

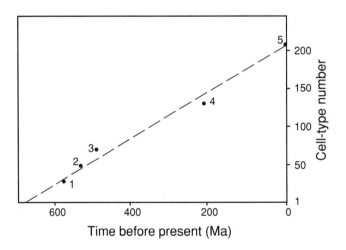

FIGURE 2   Model of cell-type number required by the appearance of selected metazoan taxa during the late Precambrian and Phanerozoic. The cell-type numbers are meant to approximate the minimal numbers that can serve to provide the necessary tissues and organs for the body plans of the various taxa. **1:** 30 cell-type grade inferred from appearance of late Precambrian horizontal trails. **2:** 45–50-cell-type grade inferred for appearance of higher invertebrate clades near the Precambrian–Cambrian boundary. **3:** 60-cell-type grade inferred for appearance of agnathans. **4:** 120–130-cell-type grade inferred for the appearance of mammals. **5:** 210 cell types of humans. Nerve-cell types are lumped. If the rate of increase of cell-type numbers before the first appearance of metazoan fossils was similar to the rates thereafter, then the metazoans originated at about 680 Ma. Sources given in Table 1.

then (more than 160 cell types in 530 Ma), is more than 0.3/Ma. Considering the crudity of these estimates, and considering that the Phanerozoic rate is probably underestimated, little significance can be placed on the differences between these two rates, which would probably not appear as constant anyway if they could be evaluated at narrower time intervals. But as far as can be inferred from the fossil record, the rate of evolution of morphological complexity did not increase or decrease steadily through metaozan history, although it may have varied episodically. Extrapolating from a modern grade of 210 cell types through the 30- and 50-cell-type grades produces an estimate for the origin of metazoans of 680 Ma. I hasten to acknowledge that the number of assumptions underlying these crude estimates is very large indeed. One can imagine sources of error that would lead either to an underestimate, which might be quite large, or to an overestimate, which cannot be very great.

Two other estimates are available that bear on the time of origin of the metazoans and that are based upon extrapolations of rates of evolution. Durham (1971) estimated that the last common ancestor of the deuterostomes lived between 1,700 and 800 Ma ago, based on rates of speciation believed to occur along lineages and on the amount of morphologic change believed to be involved in an average speciation event. Runnegar (1982a) estimated that a radiation of living animal phyla occurred at least 900–1,000 Ma ago, based on a molecular-clock interpretation of rates of change of globin molecules. In both of these cases, the first metazoans would have appeared at some still-earlier date. These methods of estimation thus place the time of origin of metazoans from around one and a half to perhaps three times the age derived from the estimate of the evolution of cell types. However, none of the rates extrapolated in these three estimates need have a clock-like constancy, and indeed it is certain that they are all affected by selection. Although it is quite possible that the evolution of complexity was slowest early in metazoan history, it seems unnecessary to postulate a vast hidden history of animal life. Perhaps the origin of metazoans may be measured only in tens rather than hundreds of millions of years before their appearance in the record.

On the other hand, as the appearance of numerous body plans near the Precambrian–Cambrian transition seems to have been geologically abrupt, is it not reasonable to argue by analogy that earlier radiations might have been equally abrupt – that metazoans arose and diversified only a few million, rather than tens of millions of years, before they enter the record? I believe that the cell-type number argument weighs against this notion, and heavily so. Whatever historical variations there may have been in the rate of origin of novel cell types, it must be a far slower process than cladogenesis. The buildup of complexity that permitted triploblastic body plans of moderate sizes involved a concomitant buildup in the complexity of the regulatory architecture in the genome, permitting a continuing evolution of cell differentiation. Probably, the regulatory mechanisms have their roots in protistans and early metazoans.

The Cambrian explosion is not necessarily associated with a rapid major advance in the complexity of organisms but may result from the more or less gradual achievement of a threshold grade of organization commensurate with the elaboration of the body plans that characterize higher invertebrates, and then radiations at that grade. Diversifications and radiations of metazoans must have been proceeding regularly throughout the history of Precambrian metazoans, and when animals were able to mark the sediments and to leave body fossils, they appear in the fossil record.

By the time of that first metazoan appearance there was probably a wide variety of flatworm-grade organisms in Precambrian seas, together with some more complex forms and, of course, many simpler ones, but even for those late Precambrian times we are privileged to see only those groups that were the more easily fossilizable. At about the Precambrian–Cambrian boundary, however, organisms finally became of such size and complexity that many of the radiating lineages evolved durable skeletons as part of their body plans, creating a *Phanerozoic* fossil record – visible to paleontologists. The early Cambrian radiation, then, is regarded as one that occurred among a number of lineages that simply reached a certain threshold level of complexity within several million years of each other, a level that they rapidly (geologically) exploited in a pattern that probably had many historical precedents; this time, however, the ensuing body plans became richer and more diverse than previously, and many were memorialized in durable remains.

Paleoecology and the Cambrian explosion

The Cambrian explosion certainly involved the expansion of occupied adaptive space, both in terms of the occupation of new habitats and in the creation of new positions within the trophic web and in other webs of ecological interaction. Most of the numerous suggestions as to the causes of the Cambrian explosion invoke environmental triggers (for reviews see Valentine *et al.* 1991; Signor & Lipps 1992; see also chapter by Knoll, this volume). The present suggestion, by contrast, invokes internal evolutionary processes; the environment provides the challenge and the opportunity but need not provoke the explosion through some special change. However, these two sorts of explanations are not mutually exclusive. It is quite possible that the pace of the evolution of organic complexity was stimulated or checked by the evolution of the environment. The rise of atmospheric oxygen concentration is a favorite candidate for controlling the rise of permissible metabolic levels and pathways and therefore of diversification among early animals. Even if some such external environmental trigger is eventually demonstrated, the internal considerations should remain.

Conclusions

Explosions are commonly thought of as involving destruction, and in that sense the metaphor of the Cambrian explosion is most inappropriate, for that event was quite creative. It might even seem that the Cambrian radiation could be considered a major progressive step in evolutionary history. However, if we consider increasing complexity as a criterion of progress and use cell-type number as a metric of complexity, then the Cambrian event may be nothing special insofar as progress is concerned but merely a locus along a geologically more or less continuous rise in complexity that is brought to our attention partly because of its taphonomic consequences. The radiation remains spectacular in terms of its breadth, however, and it is in this sense that the metaphor of an explosion is most apt, for most of the body plans that characterize living phyla, and numbers of extinct ones, probably arose during the Precambrian–Cambrian transition.

The explosion spewed out numerous body plans and types in short order, and whatever is thought of its place in evolutionary progress, it was certainly something special in the history of life. The breadth of the explosion can be attributed partly to its being composed of radiations from a number of stocks, thus multiplying the number of descendants accordingly (Valentine 1973). It then follows that the explosion is rapid partly because many of the stocks involved reached a grade of complexity (and had soft-part anatomies) for which durable skeletons promoted fitness, during a relatively narrow interval of geologic time. And, finally, the metazoan genome with its flexible architecture permitted the geologically rapid exploration of alternate body subplans at similar grades of complexity. Whether or not the explosion was triggered by environmental change remains uncertain.

Acknowledgments. – This paper has been reviewed by Kevin Padian, Department of Integrative Biology, University of California, Berkeley. Discussions with Douglas Erwin, Department of Paleobiology, U.S. National Museum of Natural History, and David Jacobs, American Museum of Natural History, have been most helpful in formulating the ideas herein. The research has been supported by NSF Grant EAR98-15453.

The advent of animal skeletons

Stefan Bengtson

Department of Historical Geology and Paleontology, Institute of Earth Sciences, Norbyvägen 22, S-752 36 Uppsala, Sweden

The advent of animal skeletons coincides with the "Cambrian explosion" of multicellular organisms and so provides the most visible boundary in the fossil record. Whereas the Neoproterozoic has only sporadic records of skeletal organisms, the earliest Cambrian has thousands of taxa, representing most of the known skeletal phyla in addition to a number of groups of uncertain systematic position. The transition from the Proterozoic to the Cambrian is often interpreted as mainly a biomineralization event, but this may be a chimera due to superposition of the skeletal record on the record of general biologic diversity. Biomineralization has a considerably older history and was only one of the prerequisites for skeleton formation. Skeletons evolved using materials that were already largely available. The common skeleton-forming minerals are energetically cheap and feasible for the hardening of organic–inorganic composite skeletal materials. No single environmental event can account for the origin of skeletons, which must be seen as part and parcel of the simultaneous ecologic diversification. The most important factor favoring the appearance of skeletons was selective pressure from predators and parasites; other functions of skeletons are largely secondary.

No event is more conspicuous in the fossil record than the advent of animal skeletons. Sedimentary rocks below the 550 Ma Precambrian–Cambrian boundary present a superficially barren visage, and only recent years' research has revealed that life on Earth has in fact been recorded as fossils in rock almost from its very conception and throughout its long period of maturation (see chapters in this volume by Schopf, Walter, Runnegar, Vidal, Hofmann, Sun, and Fedonkin). Above the same boundary glitters that fossil record of shells, bones, and teeth, which forms the basis for so much paleontological research. The difference could not be more dramatic, and the line between the two worlds is razor-sharp.

Yet paleontology has grown out of its almost exclusive dependence on hard skeletons. The need to deal also with the silent majority of organisms that have no normally fossilizable parts (or with the soft tissues of the skeleton-bearing ones) has directed attention to organic fossils, fossil molecules, trace fossils, and the exceptional preservation of otherwise nonfossilizable soft tissues. The advent of skeletons has thus lost some of its traditional significance. The "Cambrian explosion," often viewed as an explosion of fossils rather than of life, is now recognized as a major biotic revolution in which biomineralized skeletons played only a part. This view was championed by Preston Cloud (1948, 1968a) and has found massive support in more recent studies. Yet there is a tenacious tendency to explain the Cambrian explosion as a biomineralization

Bengtson, S. (ed.) 1994: *Early Life on Earth. Nobel Symposium No. 84.* Columbia U.P., New York

or skeletalization event, either by referring to chemical or physical environmental changes that stimulated biomineralization and/or by invoking skeletons as the key biological innovation that triggered the "explosive" radiation (e.g., chapters by Margulis & Cohen and by Seilacher, this volume).

Skeletons may in fact be overemphasized, because they are so visible in the fossil record; it is quite possible that neither skeletons nor biomineralization played any key role in the grand evolutionary events around the Precambrian–Cambrian boundary. This is not to denigrate the importance of skeletons or biominerals. Hard skeletons are a fundamental character of many protists, plants, and animals, fulfilling a variety of constructional and physiological needs. Biomineralization is an inherent tendency of life (Simkiss 1989). Furthermore, the ability of organisms to make minerals profoundly influences the chemistry of waters and sediments and thus the Earth's face. Research into shells and bones is not only a practical necessity for paleontologists but an endeavor of profound interest for our knowledge of Earth's and life's history. The advent of skeletons, however, may not have been the decisive event in the history of life that it is sometimes construed to be.

The many faces of skeletons

Except for the relative resistance to degradation, hard skeletons have few common denominators. There is a bewildering array of structure, morphology, and composition. Skeletons may serve for support and attachment, as in the spicular skeletons of glass sponges and the calcareous cup of corals, and additionally provide leverage for muscles, as in the jointed ossicles of echinoderms and vertebrates. They may enclose filtration chambers, handle food (teeth, etc.), improve grip on the substrate, serve in sensory organs, and have a score of other functions. They may protect the animal from various perils: predators, parasites, water loss, etc. They may serve as stores for essential elements such as calcium, phosphorous, and silicon.

Whereas all skeletons contain a larger or smaller amount of organic matter, not all contain minerals. Structural organic molecules and complexes – such as collagen and the chitin–protein complex that is the main constituent of arthropod cuticles – can form the framework of entirely organic skeletons that may still be sufficiently resistant to degradation to become readily fossilized. But these molecules often form part of an organic–inorganic composite material (Wainwright et al. 1976; Currey 1990) in which functionally feasible properties are accomplished by impregnation with minerals, sometimes obtained from the animal's surroundings ("agglutinated skeletons") but usually formed within the tissues through biomineralization, a process more or less strongly controlled by the organism and probably involving mediation by non-structural, soluble, acidic macromolecules (Weiner et al. 1983).

Biomineralization, life, and skeletons

To incorporate minerals into a skeleton typically requires several processes, such as concentrating ions into a supersaturated solution, initiating and governing mineral growth, and positioning the mineral bodies in the organic components. The systems that organisms employ to do this are diverse and complex. Our knowledge is still vastly incomplete, but biomineralization is now a prolific field of research: the recent state of the science can be glimpsed in five recent major tomes (Simkiss & Wilbur 1989; Lowenstam & Weiner 1989; Crick 1989; Mann *et al.* 1989; Carter 1990a, b; the last, particularly, contains an extensive bibliography).

Animals use three main types of minerals in skeletons: calcium carbonates ($CaCO_3$, mostly calcite, magnesian calcites, and aragonite), calcium phosphates (mostly apatites, particularly dahllite, $Ca_5(PO_4, CO_3)_3(OH)$), and opal (a hydrated gel of silica, SiO_2). It is becoming increasingly clear that biomineralization is a fundamental part of life processes, a crucial observation with regard to the question of the advent of skeletons. Biomineralization is not equivalent to skeleton formation, however: most biominerals – about 60 are known (Lowenstam & Weiner 1989) – occur only as intra- or extracellular granules or other structures without connection to skeletons.

There is a broad twilight zone between minerals being formed as a secondary result of an organism's metabolic activity – "biologically induced mineralization" of Lowenstam (1981) – and active formation of minerals by an organism for specific purposes – "biologically controlled mineralization" of Mann (1983). When minerals form externally and are not further used by the organism, for example, when sulfate-reducing bacteria cause precipitation of pyrite in the presence of iron in the environment, we are dealing with biologically induced mineralization but not with active biomineralization in the second sense. The distinction becomes considerably less clear, however, when minerals form intracellularly. Whereas bacteria commonly produce sulphur minerals as metabolic waste products, magnetotactic bacteria have been found to produce the ferromagnetic iron-sulfide greigite (Fe_3S_4) as well as iron pyrite (FeS_2) in what appears to be membrane-bound vacuoles (Mann *et al.* 1990). The localized occurrence of chains of greigite crystals in a magnetotactic bacterium indicates controlled biomineralization – a clear parallel to the intracellular magnetite (Fe_3O_4) crystals that function as internal compasses in other magnetotactic bacteria (Blakemore 1975).

The diversity of iron–sulfur biominerals in prokaryotes and the wide distribution of iron oxides in both prokaryotes and eukaryotes (Lowenstam & Weiner 1989) suggest an early origin of iron and sulfur biomineralization. This is in concord with ideas that iron and sulfur were involved in early energy-conversion systems (chapters by Deamer, Stetter, Pierson, Kandler, and Wächtershäuser, this volume; indeed, the last proposes that pyrite formation provided the energy source that made the first life possible). Biologically *controlled* iron biomineralization has a long Proterozoic history, demonstrated by occurrences of 2 Ga old bacterial magnetite (Chang & Kirschvink 1989).

Because most skeletal biominerals are calcium salts, Ca^{2+} plays a major role in most discussions of the origin of mineralized skeletons (e.g., Lowenstam & Margulis 1980; Kaźmierczak *et al.* 1985). Again, formation of calcium carbonates and calcium phosphates is deeply interwoven with basic life processes: the three most common ion pumps – H^+-ATPase, NA^+/K^+-ATPase, and Ca-ATPase – are all involved in fundamental cellular processes, and all can induce calcification (Simkiss & Wilbur 1989, p. 318). The record of biogenic $CaCO_3$ may be traced back to the late Archean, as suggested by calcified cyanobacteria (Klein *et al.* 1987); biologically controlled $CaCO_3$ mineralization probably started in the Neoproterozoic (see below). In view of the fundamental connection of phosphorous with life's origin and early history (chapters by Gedulin & Arrhenius, Baltscheffsky & Baltscheffsky, and Deamer *et al.*, this volume) and the apparent ease by which apatite forms in cells when the Ca^{2+} level goes up (Boyan *et al.* 1984), it would not be unexpected to find early evidence of biologically induced apatite formation. Gedulin & Arrhenius (this volume; see also Arrhenius *et al.* 1993) point out that under normal marine pH and Mg/Ca ratios, likely to have prevailed also in early Archean oceans, all sedimentary apatite should be biogenic; this record goes back to the oldest known sedimentary rocks, those of the 3.8 Ga old Isua Group. We have no undisputable record, however, of biologically *controlled* apatite formation until the Cambrian, though a large number of living prokaryotes and eukaryotes are able to produce crystalline or amorphous phosphatic granules in their tissues (Lowenstam & Weiner 1989; Watabe 1990).

The first skeletal fossils

We thus have evidence in the Archean–Proterozoic for biologically induced mineralization as well as sparse but compelling evidence for biologically controlled mineralization. Bona fide skeletons, however, are more elusive. There are indications of Neoproterozoic calcareous algae (Horodyski & Mankiewicz 1990; Grant *et al.* 1991), and the Neoproterozoic melanocyrillids, vase-shaped protists, may have had a calcified lorica (Vidal, this volume). Siliceous, crysophyte-like microfossils originally reported as being from the Lower Cambrian (Allison & Hilgert 1986) have been redated to the Neoproterozoic (Kaufman *et al.* 1992). Hence, Neoproterozoic eukaryotes seem to have produced both calcareous and siliceous skeletons, but examples are few and do not include metazoans. The proposed impressions of Ediacaran sea-pen spicules (Glaessner & Daily 1959; Jenkins 1985) are too uncertain to be accepted as evidence. Convincing metazoan skeletons do not appear until the very end of the Neoproterozoic, when organic and calcareous tubular fossils herald the Cambrian explosion.

Valentine (this volume) remarks that *explosion* is not a very appropriate term for something so profoundly constructive as the Cambrian radiation. But as an explosion leaves shards, so did the Cambrian simile leave the rocks littered with hard bits and pieces, and, as with shards after an explosion, it takes patient detective work to find out what happened.

What, then, are the shards? No comprehensive treatise has been written, but the early radiation of biomineralizing taxa has been reviewed in some detail recently by Bengtson & Conway Morris (1992), Towe *et al.* (1992a, b), and Bengtson *et al.* (1992). A fuller impression of the nature of the early skeletal biotas can be gathered from recent monographs on the Siberian–Central Asian (Missarzhevsky 1989), Chinese (Qian & Bengtson 1989), and Australian (Bengtson *et al.* 1990) Early Cambrian fossils. Tubicolous organisms have already been mentioned as the first heralds of the skeletal faunas in the latest Proterozoic; fossil tubes of various composition are also a characteristic component of the subsequent earliest Cambrian faunas. Other typical earliest Cambrian fossils are disjunct sclerites of calcium carbonate or calcium phosphate, mollusk-like shells, calcareous and siliceous spicules of sponges and probably soft corals, arthropod carapaces, calcareous cups of sponge-like archaeocyathans, shells of brachiopods or brachiopod-like animals, various tooth-shaped objects, etc.

Thus a rich fauna of fossil metazoans appears with the first Cambrian skeletons. It contains unquestionable members of now-living higher taxa such as arthropods, sponges, mollusks, and brachiopods, as well as numerous more problematic forms that offer few points of comparison with living animals (Bengtson 1977).

Fig. 1 shows the known stratigraphic ranges of groups with mineralized skeletons. Whereas the appearance of various types of skeletal fossils is very dramatic, it has long

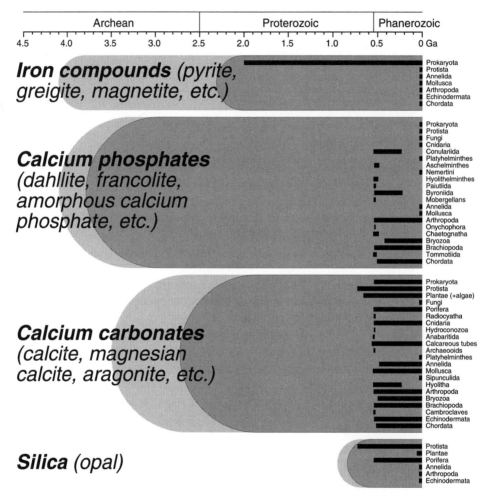

FIGURE 1 History of some major groups of biominerals. Light-shaded fields show hypothetical occurrences of biologically induced mineralization; darker-shaded fields show actual occurrences of minerals where additional evidence (morphology, isotopic composition, etc.) suggests that they are biologically induced; bars show fossil ranges of biologically controlled mineralization broken up into taxonomic groups representing distinct biomineralizing lineages. Some of the groups, such as Protista, represent several lineages that may have evolved controlled biomineralization separately, and some are artificial groups of insufficiently known fossils, for example, "calcareous tubes." The main proliferation of biomineralization coincides with the major radiation event at the Precambrian–Cambrian boundary, about 550 Ma ago. (Data mainly from Lowenstam & Weiner 1989, Towe et al. 1992b, and sources cited in the text.)

been recognized that it is not unique during that time. There is a similar radiation of fossils with nonmineralized or agglutinated skeletons or no skeletons at all; the radiation is as noticeable among trace fossils (Seilacher 1956; Crimes 1987; Bergström 1990), phytoplankton (Moczydłowska 1991), and cyanobacteria (Riding, this volume). Early and Middle Cambrian lagerstätten with soft-bodied preservation suggest that the proportion of mineralized metazoan taxa was about 20%, roughly the same as in today's marine environments (Conway Morris 1986; Chen & Erdtmann 1991).

The diversity of early skeletal fossils is so great (several thousands of taxa) that no short review can do it justice. Instead, I will give a few highlights that in one way or the other have bearing on the ensuing discussion of the causes and consequences of metazoan skeletogenesis.

Highlight 1: Hiding in a tube – the case of CLOUDINA

Cloudina is the earliest known metazoan with a mineralized skeleton, a tubular fossil with a worldwide distribution in beds pre-dating the first Cambrian skeletal assemblages. The most detailed study to date is that by Grant (1990), who concluded that its thin cone-in-cone lamellae were most likely mineralized with a veneer of high-magnesian calcite. Exquisitely preserved material from Shaanxi has revealed new facts about *Cloudina*: a small fraction (2.7%) of the tubes have been penetrated by round holes with diameters strongly correlated to the width of the bored tubes (Fig. 2). This is prime evidence that *Cloudina* was under attack from a shell-boring predator (Bengtson & Yue 1992). Thus, paradoxically, by their occasional *failure* to do the job (though a few holes

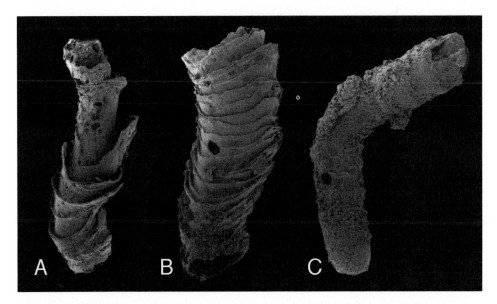

FIGURE 2   Phosphatized specimens of *Cloudina* from the uppermost Proterozoic Dengying Formation, Shaanxi, China. **A:** ×30; **B:** ×30; **C:** ×50. Specimens **B** and **C** show predatorial borings. (From Bengtson & Yue 1992; republished with permission from *Science;* ©AAAS.)

fail to penetrate the inner lamellae), we get an indication that at least one important function of the first mineralized skeletons was to hold off predators and parasites.

Highlight 2: From spicules to shells

One pertinent proposal in mollusk biomineralization studies is that the mollusk shell was initially formed by the coalescence of a spicular coat similar to that found in primitive living mollusks, the Aplacophora and Polyplacophora (Carter & Aller 1975). This hypothesis, though plausible, was long unsubstantiated. When the Cambrian halkieriids were discovered, they were first regarded as conchs of hyoliths, an exclusively Paleozoic group of animals (Poulsen 1967). Later they were recognized as skeletal elements of an animal related to a larger group of Cambrian sclerite-bearing fossils termed *coeloscleritophorans* (Fig. 3A; Bengtson & Missarzhevsky 1981). Bengtson & Conway Morris (1984) reconstructed the halkieriid as a slug-like animal covered with dorsal scales and spines; we concluded that although its affinities most probably lay in

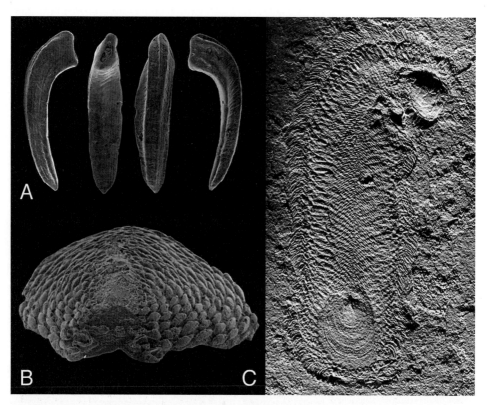

FIGURE 3 Coeloscleritophorans. **A:** Sclerites of *Siphogonuchites* from the Lower Cambrian of Yunnan, China, ×25. **B:** Shell of *Maikhanella* (=*Siphogonuchites*?) from the Lower Cambrian of Yunnan, ×50. **C:** Complete scleritome of *Halkieria* from the Lower Cambrian Buen Formation, northwest Greenland, ×1.8. (**A** and **B** from Qian & Bengtson 1989; republished with permission from the Lethaia Foundation; **C** from Conway Morris & Peel 1990; republished with permission from *Nature*.)

the neighborhood of mollusks, the hollow structure of the spicules precluded a homology between the coeloscleritophoran and the typically solid mollusk spicules.

Then the Rosetta Stone was found: Conway Morris & Peel (1990) unearthed complete specimens of a Cambrian halkieriid from northern Greenland. It was indeed a slug-like animal (Fig. 3C), but in addition to the dorsal scales it carried one anterior and one posterior shell plate that noone had suspected (or at least not dared to suggest) would be there. This led to the proposal (Bengtson 1990a, 1992) that halkieriids were stem-group polyplacophorans that had not yet evolved their six median shell plates and furthermore – following Missarzhevsky's (1989) clue that the scaly shells of *Maikhanella* (Fig. 3B) were in fact built up of aggregates of coeloscleritophoran spicules – that halkieriids and their allies give concrete evidence for the hypothetic evolution of the mollusk exoskeleton: from spicules to shells.

Highlight 3: Encrusting the skin with guano

Tommotiids are prime problematica among the earliest Cambrian shelly faunas: they are represented by phosphatic sclerites that have yet to be found in articulated associations. The best clue to the composite nature of the scleritome is Landing's (1984) discovery that adjacent sclerites occasionally merge during growth.

The tommotiid *Eccentrotheca guano* was a sloppy scleritemaker. The shape of its sclerites (Fig. 4) varies so much that it can hardly be described in a meaningful sense. Had it not been for the fact that these sclerites occur together and share a common structure of irregularly stacked growth increments, they would hardly have been recognized as parts of the same animal. The merging of sclerites (Fig. 4A, C, E) testifies

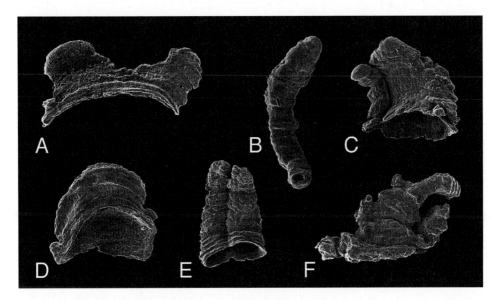

FIGURE 4   Sclerites of *Eccentrotheca guano* from the Lower Cambrian Kulpara Limestone of South Australia, ×50. (From Bengtson *et al.* 1990; republished with permission from the Association of Australasian Palaeontologists.)

to the fact that the scleritome was composite. The skeleton seems best interpreted as phosphate deposited from irregularly distributed growth centers over a wrinkled epithelium (Bengtson *et al.* 1990).

Highlight 4: The silicon alley

Biogenic opal, a noncrystalline, hydrated gel of silicon dioxide (silica), is widespread among protists and sponges. The probable Neoproterozoic record of a diverse assemblage of chrysophyte-like scales was mentioned above. Other siliceous protists were slow in coming, but there is a beginning of a radiolarian record in the Cambrian (Bengtson 1986; Conway Morris & Chen 1990a), which then expanded in the Ordovician (Nazarov & Ormiston 1985). There is, however, a good record of Cambrian siliceous sponges, demosponges as well as hexactinellids (Fig. 5), illustrating that animals having siliceous skeletons radiated simultaneously with those using other minerals.

The shapes we see in the earliest forms of opal-producing organisms are almost indistinguishable from those existing today. Spicules from a modern glass sponge (hexactinellid), for example, can hardly be distinguished from those occurring in the earliest Cambrian (Fig. 5A), though in the late Cambrian, glass sponges may indeed produce variations on this theme that seek their match in later sponges (Fig. 5B–D).

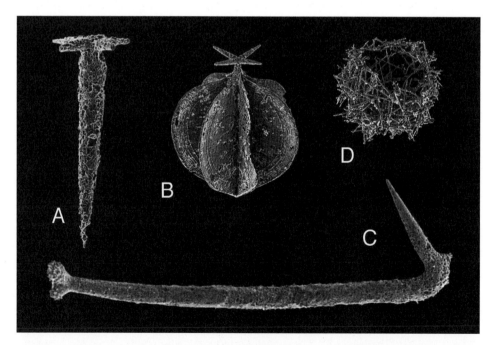

FIGURE 5   Siliceous spicules of Cambrian hexactinellid sponges. **A:** Pentact from the Lower Cambrian Ajax Limestone of South Australia. **B–D:** Pentacts of *Konyrium* (**B**) and *Silicunculus* (**C**) and probable propagule skeleton of *Echidnina* (**D**) from the Upper Cambrian of Queensland. **A, B,** and **D:** ×100; **C:** ×50. (From Bengtson 1986; republished with permission from the Association of Australasian Palaeontologists.)

Why skeletons were made – distinguishing between prerequisites, causes, and consequences

There have been many attempts to give skeletons a star role in the Cambrian explosion by demonstrating that they arose as a consequence of a chemical or physical environmental change or a key biological innovation or that skeletalization itself served to release the evolutionary radiation. These attempts have inherent weaknesses, for because of the tight coupling between biomineralization and fundamental life processes, any proposed trigger of skeleton origination is likely to have had a similar effect on general biotic diversification. Conversely, the appearance of skeletons in the fossil record is imprinted upon the more basic pattern of taxonomic radiation. This means that any testable hypothesis to explain the advent of skeletons must explain patterns other than the mere appearance and rise in diversity.

Most of the hypotheses invoking chemical or physical changes (such as increases in oxygen, phosphate, or temperature) in the environment suffer from these limitations. Consequently, they have also with more or less success been invoked as triggers for the general biotic diversification. The proposal by Degens and his coworkers (Degens 1984; Degens *et al.* 1985; Kaźmierczak *et al.* 1985; Kempe *et al.* 1989), that the onset of skeletal biomineralization at the Precambrian–Cambrian boundary was a detoxification response to an environmental calcium shock, clashes with sedimentological evidence for substantial $CaCO_3$ precipitation already from the Middle Archean (Grotzinger 1989, this volume; Knoll & Swett 1990) and does not account for the simultaneous radiation of animals with noncalcareous skeletons or no skeletons at all. Furthermore, this and other detoxification hypotheses (e.g., Rhodes & Bloxam 1971) provide at best a mechanism to make certain minerals more readily available; they do not explain why it would be selectively advantageous to maintain the excreted minerals in skeletal structures (see also Vermeij 1990). As discussed below, the main physiological cost of producing mineralized skeletons is connected with the organic components, not with the minerals.

Although it is difficult or impossible to prove the causes behind unique historical events such as the advent of skeletons, it is useful to separate possible causes from, one the one hand, prerequisites, and, on the other, subsequent effects. The following discussion is anchored in four general statements that seem justified to make regarding the advent of animal skeletons.

Biomineralization has an ancient history and was only a prerequisite for the advent of skeletons.

This is borne out by the fossil evidence for Archean–Proterozoic biomineralization and by the distribution of biominerals in living organisms, as discussed above. The dramatic rise in skeletal biomineralization at the Precambrian–Cambrian transition directly paralleled the taxonomic radiation, and there is no evidence to suggest that the former was not a subset of the latter. Skeleton formation typically involves a much more sophisticated biomineralization process than mere precipitation. The processes controlling nucleation and growth of biocrystals are diverse, however, and do not bear witness of a single evolutionary breakthrough antecedent to the Cambrian explosion.

Lowenstam & Weiner (1989) suggested that the common occurrence of acidic macro-molecules involved in biomineralization in various animals and protists (radiolarians, foraminiferans, and coccolithophorids) suggests a common origin of biomineralization mechanisms in at least a number of eukaryotes. As the lineages involved probably diverged at least a billion years ago (e.g., Runnegar, this volume), this would in any case not have direct bearing on the timing of the Cambrian evolutionary events.

The scarcity of well-documented cases of biologically controlled mineralization in the Proterozoic thus reflects the fact that there were few organisms able to form recognizable structures incorporating biominerals. Most of the nonskeletal biominerals found in today's life forms would not be possible to recover or recognize from the fossil record. Nevertheless, the evolution from isolated spicules to shells highlighted by the halkieriids suggests that a continued search in Neoproterozoic deposits for mineral grains with "non-inorganic" growth habits would be rewarding; spicular coats are widespread among "soft-bodied" animals (Rieger & Sterrer 1975).

We may also speculate that the apparently lower frequencies of biogenic minerals in Proterozoic sediments compared to those of the Phanerozoic reflect a smaller standing biomass rather than a general inability to biomineralize. In today's oceans there is a massive deposition of biogenic carbonate and silica, largely from protists (coccolitho-phores, foraminifers, diatoms, silicoflagellates, radiolarians), having a significant effect on the chemical composition of seawater (Whitfield & Watson 1983). Neoproterozoic carbonates suggest higher degrees of carbonate supersaturation in the seawater than those of the ensuing Paleozoic (Knoll & Swett 1990). If biomineralization, as opposed to incorporation of minerals into skeletons, did not change character across the Precam-brian–Cambrian boundary, the quantitative rise in the sedimentation of biominerals may be a rough measure of a rise in standing biomass, perhaps connected to a massive release of phosphate into surface waters during that time (Cook & Shergold 1984, 1986; Donnelly *et al.* 1990).

> *Skeletons are constructed using a variety of processes and materials. Minerals are suitable because they give hardness to the composite material, can be produced using preadapted pathways, and are physiologically cheap.*

If, as seems likely, organisms have been physiologically able to make a variety of minerals for billions of years, the advent of skeletons signals the evolution of tissues reinforced with materials that were already available. The common skeletal minerals are, in fact, cheap. The main expense in making a skeleton is for building up the organic framework that controls the shape, structure, and, thus, the properties of the skeleton. Provided that the ions needed to precipitate the minerals are not limited in supply, the cost of that material is comparatively small. Present-day surface waters of the oceans are oversaturated with regard to calcium carbonate, and during much of the Archean and Paleoproterozoic oversaturation appears to have been even more pronounced (Grotzinger 1989, this volume). Modern seawaters are nearly saturated with regard to calcium phosphate; the extensive phosphorite deposits around the Precambrian–Cambrian boundary suggest that neither Ca^{2+} nor PO_4^{3-} were in short supply (Cook & Shergold 1984, 1986). Although silicon is a common element in the Earth's crust, modern seawaters are undersaturated with regard to SiO_2, but this is probably a late effect of the Paleozoic–Mesozoic radiations of radiolarians, diatoms, and silicoflagellates (Broecker 1971) and need not reflect conditions in the Neoproterozoic or Cambrian.

The minerals used in a skeleton are often substantially modified from their inorganic forms, not only in growth directions and size but even in the very texture of the crystal lattice (Berman *et al.* 1993). Direct measurements of the physiological costs related only to biomineralization are therefore difficult, but Palmer (1992) estimated that the cost of producing the calcium carbonate of the shell in certain marine gastropods was only about 5% that of making the corresponding weight of organic matrix. The total energy cost for shells in mollusks may be 10–60% of the cost for somatic growth and 15–150% of that used for producing gametes (Palmer 1992). Biogenic opal is an equally cheap mineral: the cost of producing SiO_2 in plants is less than 4% of that required for the same weight of lignin (Raven 1983). Apatite may form spontaneously in cells if the intracellular Ca^{2+} level is not maintained below environmental levels (Boyan *et al.* 1984). Thus, whereas skeletons as a whole may be physiologically expensive to make, the mineral component as such is not.

Whereas the initial choice of shell mineral usually precludes future evolutionary switches to other minerals (because of the intricate systems developed to modify the growth of the mineral), there is no reliable indication of any regularity in the acquisition of skeletal minerals within or between clades.

An interesting proposal (Brasier 1986) invokes a "biomineral chain reaction" to account for the sequential appearance of different skeletal compositions: organic – aragonite – high-Mg calcite – apatite – low-Mg calcite. But similar to the models proposing an early evolutionary origin of phosphatic shells (Rhodes & Bloxam 1971; Lowenstam & Margulis 1980; Kaźmierczak *et al.* 1985), this one rests on a probable misrepresentation of the pattern of appearance of the skeletal fossils. The first mineralized metazoan skeleton, that of *Cloudina*, was probably high-Mg calcite (Grant 1990), and the first Cambrian assemblages already contained all the varieties in Brasier's chain plus silica, with no clear order of appearance, because of the largely condensed nature of the Precambrian–Cambrian boundary beds in the key regions (Towe *et al.* 1992b).

The claim for a primary role of phosphatic shells is similarly based on a contention that the earliest skeletal fossils were predominantly phosphatic. Though this claim is mistaken (Runnegar 1989), phosphatic skeletons were more widespread among higher taxa in the Early Cambrian than later (Bengtson *et al.* 1992). It is not strange that this has been considered significant, for the Precambrian–Cambrian transition witnessed one of the major global episodes of phosphate deposition (Cook & Shergold 1986).

Phosphate is a limiting resource in most modern environments, and phosphatic skeletons often serve as stores for phosphorous. One proposed trigger for the Cambrian explosion is the sudden release of phosphate from the deep waters when previously stratified oceans became mixed (Cook & Shergold 1984, 1986; Donnelly *et al.* 1990). It seems that such an event could have caused an expansion of biomass that promoted rapid radiation, given that other conditions were right. Did it also encourage phosphate over carbonate biomineralization, as suggested by, e.g., Rhodes & Bloxam (1971) and Cook & Shergold (1984)? This idea could conceivable be tested if we could establish whether the phoshate-secreting lineages arose in areas and times with abundant phosphorous deposition, but at present the various biomineralizing clades appear so widespread from their first appearance that it is not possible. Whereas the "sloppy" phosphatic skeleton seen in *Eccentrotheca guano*, highlighted above, might be taken to indicate that calcium phosphate was a cheap mineral at the time, there being little selective value in economizing, such a conclusion would rest on extremely unsure premises with regard to the total selective value for animals of unknown anatomy and affinities.

The propellant in the evolution of skeletons was organismal interactions, and a primary factor behind the evolution of tubes, shells, sclerites, and spicules was selection pressure from predators.

If no single environmental event accounts for the origin of skeletons, the timing simply reflected the general taxonomic diversification. To ask as a general question why skeletons arose may be no more or no less relevant than to ask why tentacles or eyes arose. They arose in the individual cases because there were developmental conditions that made them possible and because they conferred selective advantages. Are there then no general questions that can be asked?

Metazoan exoskeletons have a number of disadvantages: they are energetically costly, they can be both heavy and bulky, and they often restrict movement, changes in body shape, growth, and the diffusion of gases and nutrients across the body epithelium. Some of these disadvantages can be overcome and even turned to advantages (as in the mechanically efficient exoskeleton of arthropods), but what were the initial advantages to justify the heavy costs?

The case for predators has been argued in particular detail by Vermeij (1987, 1990; see also Stanley 1976 and Simkiss 1989), who used an arms-race metaphor to explain much of skeletal evolution during the Phanerozoic. The need for a soft-bodied organism to protect itself against attacks from predators and parasites may seem obvious but has in fact been a contentious issue:

> *The naive assumption that shells are acquired because they protect soft bodies seems influenced by anthropocentric thinking: man uses shields for protection from aggressors.*
> Glaessner (1984, p. 174)

This almost instinctive dislike of the arms-race analogy is puzzling, for the concept is powerful. On a general plane, it describes the interaction between predators/parasites and prey/hosts in a framework of coevolution. More specifically, it suggests that one of the strategies to avoid predators and parasites is to prevent penetration of the skin or to make the tissues unpalatable. Predation levels were long considered to be very low or absent during the Precambrian–Cambrian transition (e.g., Glaessner 1972a), but predators are now known to have been well established in the early Cambrian (Bengtson 1977; Conway Morris & Jenkins 1985; Conway Morris 1986; Jensen 1990; Babcock & Robison 1989; Qian & Bengtson 1989), where shell-boring also occurs (Conway Morris & Bengtson 1994). The example of the late Precambrian *Cloudina* given above is particularly significant, since it shows that already the first animals with exoskeletons were under attack from predators seeking to penetrate the cover. The calcareous spines, scales, and shells in coeloscleritophorans, as well as the phosphatic cover plates of tommotiids, similarly had a clearly protective function. Also the pure spicular skeletons, such as in the siliceous sponges exemplified above, in addition to their possible support functions, would have made the tissues unpalatable to predators.

Skeletons are, however, only one of several strategies available to animals to avoid predation. Some alternatives are (1) to run away, (2) to hide, (3) to fight back, (4) to become toxic or distasteful, (5) to stay too small or grow too large for a profitable attack.[1] Whereas these strategies would be less obvious in the fossil record, they do provide stimulus for speciation and thus may have helped to fuel the Cambrian explosion together with other ecological interactions leading to diversification. But the availability of the strategies is sufficient explanation as to why only a minority of taxa ever evolved skeletons.

Whereas other selective pressures than predation may have influenced early skeletogenesis in some groups, only attacks from predators or parasites would have such a broad influence as to be able to explain the nearly simultaneous appearance of animal skeletons in a large number of groups. Later modifications of the skeletons (to provide mechanical leverage or attachment, to direct filtration currents, etc.), however successful and significant, must be regarded as "turning necessity into virtue" and not be of the general significance that we have been concerned with here.

Acknowledgments. – I am grateful to Steve Weiner for helpful comments on the manuscript. My work is supported by the Swedish Natural Science Research Council.

---

[1] McMenamin (1992) recently proposed that Proterozoic species responded to predation by themselves turning into predators or parasites and that this provided the mechanism for the Cambrian radiation. I am not aware of any behavior pattern or selective advantage that would produce this effect.

Evolution of algal and cyanobacterial calcification

Robert Riding

Department of Geology, University of Wales, Cardiff CF1 3YE, Wales, UK

Algae appear in the middle Proterozoic, but calcified forms are not definitely known until the late Cambrian and Ordovician, suggesting that groups characterized by calcification evolved relatively late. Interpretation of the apparent scarcity of Proterozoic calcified cyanobacteria is more complicated. Calcified cyanobacteria are known from the early Proterozoic. It is widely believed that cyanobacteria were abundant and diverse during the Proterozoic and that conditions then generally favored calcification. These assumptions may need reassessment, but it is also possible that calcified cyanobacteria have been overlooked in stromatolite microfabrics. Abrupt increase in diversity of calcified cyanobacteria near the Precambrian–Cambrian boundary indicates an evolutionary event paralleling that in invertebrates. Variation in abundance of calcified cyanobacteria through the Proterozoic suggests calcification episodes between 2,500 and 1,890 Ma and between 1,350 and 700 Ma. The Phanerozoic record of marine cyanobacteria is patterned by similar, although shorter, episodes that appear to be temperature-related. Late Proterozoic temperature fall associated with glacial conditions corresponds with decline in calcified cyanobacteria, stromatolites, and acritarchs. During subsequent temperature rise, cyanobacteria, metazoans, and acritarchs diversified in the early Cambrian, and cyanobacterial calcification was restored. It is suggested that although the early Cambrian biotic radiation was determined by preceding evolutionary developments, its timing – and possibly also its speed – was scheduled by temperature rise. It was an event waiting to happen, which was triggered by climatic amelioration.

Shelly calcareous fossils are the most direct index of the change from the Proterozoic to the Palaeozoic. Their abundance in the Cambrian, contrasting with their scarcity in the Precambrian, provides the main paleontological signal for this major biostratigraphical boundary. Yet the biomineralization processes that are responsible for these shells and skeletons are not restricted to animals. Many extant prokaryotes and algae biomineralize (Leadbeater & Riding 1986). So how is it, when cyanobacteria and algae dominate the Precambrian fossil record (Schopf 1983b), that they have not produced abundant shelly fossils well before the Cambrian (Riding 1991a, p. 305)? Does the Precambrian–Cambrian transition constitute an important threshold for biomineralization in plants as well as animals, or do calcified algae and cyanobacteria actually exist in greater abundance in the Precambrian than has been reported?

These questions are significant because it is likely that calcification in cyanobacteria, and probably in some algal groups too, is influenced by extrinsic factors that control

Bengtson, S. (ed.) 1994: *Early Life on Earth. Nobel Symposium No. 84.* Columbia U.P., New York

seawater chemistry (Riding 1982). Consequently, knowledge of the geological record of calcified algae and cyanobacteria may allow inferences to be made concerning changes in environmental factors over long periods of geological time.

Record of algae and cyanobacteria

Precambrian sediments – especially carbonates, associated cherts, and mudrocks – have been intensively searched for fossils since Tyler & Barghoorn (1954) discovered the Gunflint microbiota. Most of those found are microscopic organic-walled cyanobacteria and algae that locally are permineralized in silica (Schopf, this volume). They have been the subject of numerous reviews assessing their biogenicity, syngenicity, affinities, environmental interrelationships, and evolutionary development (see Cloud 1976b; Schopf et al. 1983; Schopf 1992c; Mendelson & Schopf 1992a). Discrete calcified microfossils, as opposed to macroscopic stromatolites, appear to be less common and have certainly been less closely scrutinized.

Algae

Precambrian. – The oldest recorded alga is 2.1 Ga-old *Grypania*, which Han & Runnegar (1992) compare with dasycladalean chlorophytes. Acritarchs 1.8 Ga-old have been reported by Zhang (1986). The earliest red alga known is a well-preserved bangiophyte no older than 1.3 Ga (Butterfield *et al.* 1990). The acritarch *Trachyhystrichosphaera* at 1,000–900 Ma (Timofeev *et al.* 1976) may be the earliest example of a prasinophyte (Knoll *et al.* 1991, p. 539). Carbonaceous filamentous, rounded, and sausage-shaped macroscopic fossils of the middle to late Proterozoic, such as *Vendotaenia* (1.3 Ga), *Chuaria* and *Tawuia* (1,100–700 Ma) are regarded as probable algae (Hofmann 1985b), of which *Vendotaenia* is usually compared with phaeophytes and rhodophytes. These early records of algae – all originally organic-walled forms – date from the middle Proterozoic. They are of low diversity and limited abundance before a significant burst of eukaryote evolution around 1.1–1.0 Ga (Andrew Knoll, personal communication, 1992). The oldest convincing chlorophytes are 800–700 Ma *Cladophora*-like forms (Butterfield *et al.* 1988). Remarkable scales, all now siliceous, resembling those of chrysophytes and prymnesiophytes, were originally reported from the early Cambrian (Allison & Hilgert 1986) but are now believed to be late Riphean (780–620 Ma) in age (Kaufman *et al.* 1992). They may be examples of biomineralization in planktic groups. Thus, a variety of groups, including representatives of rhodophytes, chlorophytes, and prymnesiophytes, which together account for most extant calcified marine algae, are known by the late Proterozoic, and it is likely that further studies will push back the known times of appearance of many of these.

Precambrian reports of possible calcified algae, however, are exceedingly rare and are limited to the very latest Proterozoic (Fig. 1). Horodyski & Mankiewicz (1990) report *Tenuocharta* from the late Proterozoic Pahrump Group (700–600 Ma). They interpret this thin, sheet-like fossil, which in cherts has a cellular structure, as either an alga or a cyanobacterium and consider comparisons with both red algae and colonial chroococcalean cyanobacteria. Horodyski favors an algal affinity for *Tenuocharta* (R.J.

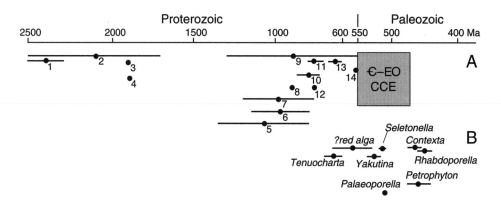

FIGURE 1 Proterozoic – early Palaeozoic record of calcified cyanobacteria (**A**) and calcareous algae (**B**). **1–14** are Proterozoic records of possible calcified cyanobacteria (see Table 1). Paucity of records in the middle and late Proterozoic are the basis for postulated RCCEs (see Fig. 2). *Yakutina* and *Seletonella* may be dasycladaleans, but the earliest definitely known members of this group are weakly calcified *Primicorallina* (middle Ordovician) and heavily calcified *Rhabdoporella* (late Ordovician). Cyclocrinitids, which probably are dasycladaleans, also first appear in the middle Ordovician. *Palaeoporella* (late Cambrian or early Ordovician) is the first record of a halimedacean green alga. *Petrophyton* (middle–late Ordovician) is the first confirmed record of a calcified red alga. Note that horizontal bars indicate uncertainty of age of mainly individual records; they do not represent ranges. (Time-scale modified from Harland *et al.* 1990. Cambrian – early Ordovician CCE from Riding 1992b.)

Horodyski, personal communication, 1992). Grant *et al.* (1991) report a larger, but also foliose, calcified, possible alga from the Nama Group (650–530 Ma) and suggest similarities with late Palaeozoic phylloids and hence with peyssoneliacean and coralline rhodophytes, although no calcified cells are preserved.

Cambrian. – Until quite recently it was accepted that calcified red and green algae were common in the Cambrian (Korde 1973). Reevaluation of the affinities of these fossils (Luchinina 1975), however, indicates that all those of early Cambrian age are more likely to be cyanobacteria (Riding 1991a). The very rare fossils *Yakutina* (Korde 1973), first recorded as *Siberiella* Korde 1957, from the middle Cambrian (Korde 1957, p. 69, Text-fig. 1), and *Seletonella* from the late Cambrian may be the earliest calcareous dasycladaleans (Riding 1991a, p. 327), but the first definite record of a heavily calcified example of this important group is *Rhabdoporella* Stolley (Stolley 1893), probably from the upper Ordovician (see Høeg 1932, p. 75; Roux 1991, Fig. 2; and Fig. 1 herein). The weakly calcified dasycladalean *Primicorallina* Whitfield is from the middle Ordovician, as are the earliest cyclocrinitids, which are also likely to be dasycladaleans (Beadle 1991). The earliest recorded halimedacean green alga is *Palaeoporella* Stolley from the late Cambrian – early Ordovician (Johnson 1954, pp. 51, 66, Pl. 24).

Petrophyton Yabe from the middle or late Ordovician (Høeg 1932, p. 82) is probably the earliest definite record of a Palaeozoic calcified red alga. The possible red algae *Contexta* Gnilovskaya and *Ansoporella* Gnilovskaya are also middle Ordovician (Gnilovskaya 1972, pp. 110–112, 117–119, and table, p. 166). The problematic, possibly algal, receptaculitids *Calathella* Rauff (Nitecki & Debrenne 1979, Pl. 4) and *Calathium* Billings (Toomey 1970, pp. 1323–1324, Fig. 10) from the early Ordovician further emphasize the importance of Ordovician developments in the history of calcareous algae.

Cyanobacteria

Precambrian. – It has been suggested that calcified cyanobacteria are not common in the Proterozoic (Riding 1982, 1991b, p. 75). Even in stromatolites of this age, demonstrably biogenic fabrics appear to be scarce (Hofmann 1969, p. 40; Riding & Voronova 1982; Swett & Knoll 1985, p. 337), although a variety of bushy, peloidal, fibrous, and tussocky structures are known. Certainly, reports of discrete fossils such as *Girvanella*-like filaments (Raaben 1969; Swett & Knoll 1985, Fig. 13; Knoll & Swett 1990, p. 114) are few, although more examples are gradually coming to light. Cement-encrusted stromatolites may be widespread earlier in the Proterozoic. Possible examples include *Pseudogymnosolen–Asperia* microstromatolites (Liang *et al.* 1984; Hofmann & Jackson 1987; see also Donaldson 1963) that Grotzinger (1986a, 1990, p. 93) has termed tidal-flat tufas. But these are not yet known to preserve diagnostic calcified cyanobacterial fossils. Table 1 lists occurrences of possible Proterozoic calcified cyanobacteria known to me, although it should be noted that the precise affinities, mode of calcification, environment of occurrence, and age of many of these are uncertain.

Cambrian. – Reassessment of the affinities of Lower Cambrian calcified microfossils, many of them originally referred to algae (Vologdin 1962; Korde 1973), provides strong evidence that heavily calcified cyanobacteria appear abundantly near the base of the Cambrian (Luchinina 1975; Riding & Voronova 1984) and are remarkably diverse. They include probable rivulariaceans and oscillatoriaceans, together with possible stigonemataleans and coccoid cyanobacteria (Riding 1991a, Table 3) although their affinities, particularly of some of the more common groups such as *Epiphyton* and *Renalcis*, require further resolution.

Discussion

Calcified algae

The available evidence suggests that the geological history of calcified algae is broadly distinct from that of cyanobacteria. Calcified algae seem not to have developed significantly until the late Cambrian and Ordovician. This relatively late development is not expected, but neither is it surprising. Calcified forms constitute a small minority of algae (Pentecost 1991, p. 4). Marine algae that do calcify appear to facilitate carbonate precipitation chemically and/or anatomically (Borowitzka 1989), and major groups, such as coccolithophorids and corallines, have subsequently appeared much later in the Phanerozoic, as have noncalcified groups (Tappan 1980). There is therefore no reason to regard the time of first appearance of calcified algae, after a long phase of noncalcareous algal development and also after the evolution of invertebrate phyla, as being particularly unusual. Nevertheless, current information is no doubt very

TABLE 1   Proterozoic records of possible calcified cyanobacteria (see also Fig. 1).

14   *Obruchevella* (Reitlinger 1959, p. 21; see Riding & Voronova 1984, p. 206). Riding & Voronova (1984) also discuss Nemakit–Daldyn (immediately pre-Tommotian) occurrences of *Renalcis* and *Gemma*.

13   680–600 Ma Vendian "calcareous alga" (Bertrand-Sarfati 1979; Bertrand-Sarfati & Moussine-Pouchkine 1983).

12   >770 Ma Upper Riphean thrombolites containing clotted microfabrics within the mesoclots (Aitken & Narbonne 1989).

11   800–750 Ma late Riphean calcified *Girvanella*-like filaments (Swett & Knoll 1985, Fig. 13; Knoll & Swett 1990, p. 114).

10   870–740, 800–700 Ma *Polybessurus*, regarded by Schopf (1977, Table 2, Figs. 12, 13H) as a potential index fossil for the early late Riphean. Knoll *et al.* (1991) recorded *Polybessus bipartitus*, preserved by dolomitic rinds, from the Draken Conglomerate Formation (800–700 Ma) and placed it in the Pleurocapsales.

9   1,300–550 Ma "algal-patterned fine-structure," probably in the younger part of this range (Cloud *et al.* 1974, Fig. 14).

8   900 Ma *Tarioufetia hemispherica* (Bertrand-Sarfati 1972, p. 56, fig. 17; see also Bertrand-Sarfati & Pentecost 1989) from the Atar Group (for which Fairchild *et al.* [1990] give an age of about 900 Ma).

7   1,200–770 Ma dendriform and lamelliform elements with cellular fabrics from stromatolite reefs (Aitken 1988, pp. 15–16).

6   1,150–800 Ma *Tungussia globulosa* (Bertrand-Sarfati 1970, p. 24, Pl. 12:4; possibly late Riphean, see p. 34; see also Bertrand-Sarfati 1976, pp. 253–255).

5   1,350–800 Ma *Dzhelindia* (middle?–late Riphean; Kolosov 1975, p. 33).

4   1,890 Ma calcareous *Frutexites*, some of which resemble *Renalcis*, from early Proterozoic stromatolites (Hofmann & Grotzinger 1985, Fig. 8).

3   1,900 Ma *Pseudogymnosolen/Asperia* microstromatolites (Hofmann & Jackson 1987; see also Donaldson 1963). Grotzinger (1986a, 1990, p. 93) has termed these tidal-flat tufas.

2   2,500–1,700 Ma "filament molds," cf. *Palaeoleptophycus* (Hofmann & Schopf 1983, Table 14-2, item 13).

1   2,500–2,300 Ma *Siphonophycus transvaalense*, carbonate-encrusted cyanobacterial sheaths (Klein *et al.* 1987, p. 86, Fig. 5, originally incorrectly published as *S. transvaalensis*, corrected by Schopf & Klein [1992]).

incomplete, and further examination of Proterozoic and Cambrian limestones may yet yield earlier firm evidence of calcified algae.

Some algae, notably dasycladalean and halimedacean chlorophytes, do not closely control calcification and appear to be more diverse and abundant during cyanobacterial calcification episodes (Riding 1992a). Although environmental effects on their calcification do not appear to be so profound as in cyanobacteria, it is possible that diversification in calcified green algae was delayed, following their appearance in the Ordovician, until the subsequent cyanobacterial calcification episode in the late Devonian – early Carboniferous.

Cyanobacterial calcification episodes

Cyanobacteria could provide a long-term index of marine calcification (Riding 1982) for three reasons: they are not obligate calcifiers, they appear to respond to environmental conditions favoring carbonate precipitation, and they have a long geological record. Indeed, the Phanerozoic record of marine calcified cyanobacteria shows marked episodicity. They are most common during Cambrian – early Ordovician, late Devonian – early Carboniferous, and middle Triassic – early Cretaceous. These phases when calcified cyanobacteria were common have been termed Cyanobacterial Calcification Episodes (CCEs) and are separated by Reduced Cyanobacterial Calcification Episodes (RCCEs) (Riding 1992b). CCEs are interpreted to reflect enhanced rates of marine carbonate precipitation (Riding 1992b).

Although Precambrian records of possible calcified cyanobacteria are limited and mainly poorly constrained by age, available information (Table 1, Fig. 1) may tentatively be taken to suggest that their Proterozoic history is similarly patterned (Fig. 2):

4   Late Neoproterozoic (700–550 Ma) RCCE.

3   Middle–late Riphean (1,350–700 Ma) CCE.

2   Middle Proterozoic (1,890–1,350 Ma) RCCE.

1   Early Proterozoic (2,500–1,890 Ma) CCE.

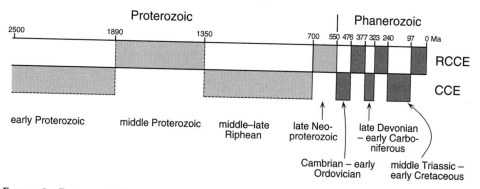

FIGURE 2   Proterozoic–Phanerozoic cyanobacterial calcification episodes (CCEs) and intervening reduced cyanobacterial calcification episodes (RCCEs). The Proterozoic episodes are tentative preliminary assessments based on on the data summarized in Fig. 1 and Table 1. (Time scale modified from Harland *et al.* 1990. Phanerozoic episodes from Riding 1992b.)

The Cambrian – early Ordovician CCE really starts in the very latest Proterozoic (Riding & Voronova 1984; Mankiewicz 1992, Fig. 7.4.8). But if we overlook these latest Vendian occurrences, then the only other reports of possible calcified cyanobacteria in the late Neoproterozoic are those of Bertrand-Sarfati (1979) and Bertrand-Sarfati & Moussine-Pouchkine (1983) from West Africa and that of Cloud *et al.* (1974, Fig. 14) from California, all of which are poorly dated. It should be noted that calcified cyanobacteria continue to occur within the Palaeozoic (middle Ordovician – middle Devonian) RCCE, but at a reduced level of abundance and diversity. Thus, the presence of calcified cyanobacteria in the late Neoproterozoic is not inconsistent with postulation of an RCCE. The critical point is whether they are markedly less abundant than in preceding and subsequent episodes. The three oldest CCEs and RCCEs suggested here between 2,500 and 700 Ma are of substantially greater duration than those recognized in the Phanerozoic. Possibly they each represent an alternating series of CCEs and RCCEs. The CCE in the early Proterozoic, for example, could be broken by an RCCE related to glacial episodes around 2.25 Ga.

In the Phanerozoic there is general correlation between CCEs and abiotic carbonate precipitation in the form of ooids and marine cements (Riding 1992b). Information is less readily available for the Proterozoic. However, early cements (Grotzinger 1986b) occur during the postulated early Proterozoic CCE, "tidal-flat tufas" (Grotzinger 1986a, 1989, this volume) are also common before 1.7 Ga, and pisoids and early cements of 800–700 Ma (Swett & Knoll 1989) occur near the end of the middle–late Riphean CCE. There is qualitative evidence for a decline in stromatolite abundance (Grotzinger 1990, p. 96, Fig. 6) corresponding with the middle Proterozoic RCCE postulated here.

The Precambrian enigma

Tentative recognition of CCEs does not alter the fact that calcified cyanobacteria seem to be scarcer, and also less diverse, than expected in the Proterozoic. If Proterozoic stromatolites were primarily cyanobacterial and marine carbonate precipitation rates high, then calcified cyanobacterial fabrics should be common, particularly within stromatolites. This enigma of Proterozoic scarcity of calcified cyanobacteria (Riding 1989) could be resolved in several different ways.

(1) *Identification.* – A simple solution is that calcified cyanobacteria may be common in the Proterozoic but have gone largely unrecognized. At first this seems unlikely in view of the intensive search that has been carried out for Precambrian microfossils. Nonetheless, it is possible that calcified cyanobacteria have been overlooked in stromatolites and thrombolites, where they may have been regarded as fabrics rather than as discrete fossils (Riding 1992b). Work in progress (Riding, unpublished) suggests that calcified bushy microfabrics closely resembling *Angulocellularia* Vologdin, and similar to those figured by Cloud *et al.* (1974, Fig. 14) are widespread in Proterozoic stromatolites.

(2) *Abundance.* – A more radical alternative is that cyanobacteria in general, not only calcified forms, are actually scarcer in the Proterozoic than has been supposed. The possibility that Precambrian and other ancient stromatolites may have been bacterial has often been mentioned (Walter 1972, 1983, p. 212; Walter *et al.* 1972; Monty 1973; Gebelein 1976, p. 507; Knoll 1979, p. 314; Riding 1992c; Pierson, this volume). Nevertheless, the importance of cyanobacteria for O_2 generation, banded-iron-formation development (Cloud 1976b), and stromatolite formation (Awramik 1971) remains a cornerstone of current Precambrian biosphere and environment models (Schopf *et al.* 1983). The enigma of scarce calcified cyanobacteria underlines the need to review these assumptions (Riding 1989).

(3) *Carbonate-precipitation rates.* – Carbonate-precipitation rates can be expected directly to affect cyanobacterial calcification (Riding 1992b). The possibility that rates were low during the Precambrian, suggested by Riding (1982) to explain the apparent scarcity of calcified cyanobacteria, is currently supported by some views of Precambrian sea-water chemistry (e.g., Garrels 1989, p. 7) and opposed by others (Grotzinger 1989). However, aside from these inferences, there is considerable objective evidence for extensive carbonate precipitation during much of the Proterozoic on scales from individual stromatolites to carbonate platforms (Grotzinger 1989, this volume), which formed despite the virtual absence of skeletal carbonates during this time. Thus, it seems reasonable to expect that conditions favoring carbonate precipitation in this way should also have facilitated cyanobacterial calcification.

(4) *Cyanobacterial diversification.* – If Precambrian cyanobacteria lacked specificity for calcification, this could also account for the paucity of calcified cyanobacteria even during conditions that favored calcification. This in turn implies that ability to calcify was acquired, through evolutionary developments, by the early Cambrian, when calcified cyanobacteria appear in abundance and diversity. It has been suggested that cyanobacteria diversified early (Schopf 1970, p. 339, Fig. 4) and had by the late Proterozoic attained a level of diversity comparable with the Recent (Gebelein 1974; Tappan 1980, p. 80; Knoll 1990, p. 15). Could it be that these estimates are mistaken? In fact, several groups may not be present in the late Proterozoic (Knoll 1985c, p. 411). Calcified cyanobacteria in the Proterozoic (Table 1) are less diverse than in the early Cambrian (Riding 1991a). A straightforward reading of the current record of calcified forms suggests that developments in the early Cambrian mark a significant evolution-ary diversification in cyanobacteria that could include wider acquisition of specificity for calcification.

These changes, coupled with calcification during the Cambrian – early Ordovician CCE, transformed microbial success in carbonate deposition from mainly stromato-lite formation, seen in the middle Proterozoic, into mainly dendrolite and thrombolite formation in the early Palaeozoic (Riding 1991c). Subsequent Phanerozoic CCEs show similar, though possibly less profound, changes in cyanobacteria through time, as they record – through calcification – periodic snapshots of the changing flora (Riding 1992b).

Model

Within the limits of the data it seems that calcified cyanobacteria were generally less abundant, less diverse, and less heavily calcified in the Proterozoic than in the Cam-brian, although it is also likely that they have been overlooked in stromatolite micro-fabrics. Nevertheless, there is some evidence for Proterozoic CCEs and RCCEs. In the Phanerozoic, CCEs appear to correlate with increased global temperature rather than with other possible controlling factors such as pCO_2 and sealevel (Riding 1992b).

It has been suggested that temperature rise following late Proterozoic glaciations stimulated cyanobacterial calcification and diversification close to the Precambrian–Cambrian boundary and also triggered the early Cambrian radiation of marine inver-tebrates (Riding 1992b). The idea is not that temperature rise caused the "Cambrian

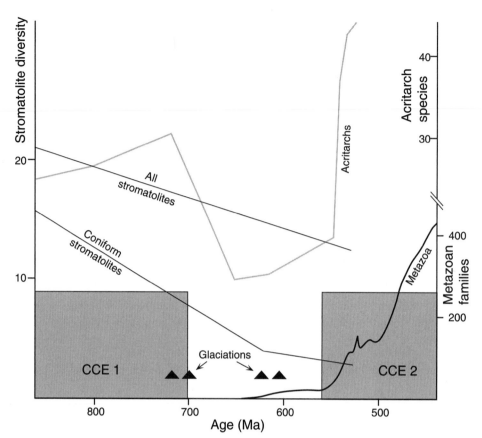

FIGURE 3 Synthesis of events at the Proterozoic–Palaeozoic transition. It is proposed that (a) decline in stromatolite and acritarch diversity and termination of the middle–late Riphean CCE (CCE1) are linked to temperature fall in the latest Proterozoic also reflected by glacial episodes around 700–600 Ma, and (b) commencement of the Cambrian – early Ordovician CCE (CCE2) near the beginning of the Tommotian is linked to temperature rise which also stimulated acritarch, metazoan, and cyanobacterial diversification. CCE1 and CCE2 are separated by an RCCE. During CCE2 calcified cyanobacteria created widespread dendrolite and skeletal thrombolite fabrics in reefs. Biomineralization also took place in a variety of metazoan groups during the early Cambrian. (Sources: acritarchs – Vidal 1984, p. 54; stromatolites – Walter & Heys 1985, Fig. 5B; metazoan diversity – Sepkoski 1979, Fig. 7; glaciations – Harland *et al.* 1990; CCE's – Riding 1992b, this paper.)

explosion" but that it scheduled an event that was waiting to happen by promoting diversification in invertebrates, which, inter alia, involved biomineralization.

In the Proterozoic it is difficult to compare postulated calcification episodes with environmental variables. Estimates of temperature change exist for the Proterozoic–Palaeozoic transition (Karhu & Epstein 1986, Fig. 9) but these only show very broad trends; they do not, for example, resolve fluctuations associated with late Proterozoic glacial episodes. Thus, while it is plausible that CCEs represent responses to environmental variables (Riding 1982; Pentecost & Riding 1986), it remains to be decided whether or not temperature is as likely to have been the controlling factor in the Proterozoic as in the Phanerozoic.

The only way available at present to assess the relevance of a temperature-control hypothesis for Proterozoic CCEs and its implications for developments at the transition to the Phanerozoic is by comparing the history of calcified cyanobacteria and other groups with the record of late Proterozoic glacial episodes (Fig. 3).

This temperature model for the Proterozoic–Palaeozoic transition is developed as follows:

1 Temperature fall in the late Proterozoic resulted in decline in stromatolites, calcified cyanobacteria, and acritarchs.

2 Subsequent temperature rise stimulated Cambrian diversification of invertebrates (and concomitant invertebrate biomineralization) and cyanobacteria, as well as cyanobacterial calcification.

3 Calcified eualgae that did not closely control calcification were also influenced by temperature-related calcification events after their appearance in the Cambrian–Ordovician. Dasycladaleans and halimedaceans, for example, appeared during the Ordovician, during the closing stages of the Cambrian – early Ordovician CCE, and did not then diversify substantially until the following CCE in the late Devonian – early Carboniferous.

Glacial episodes

Glaciations were major global events during the late Proterozoic (Harland 1964; Hambrey & Harland 1985). Two distinct clusters of episodes, near 700 Ma and 600 Ma (Harland *et al.* 1990), correspond closely to the late Neoproterozoic (700–550 Ma) RCCE proposed here (Fig. 3). There is also evidence for further episodes between 950 and 700 Ma (Cloud 1988, Fig. 12.5, pp. 296–303, 308). These glaciations coincide with or follow a marked decline in stromatolite diversity (Awramik 1971) and abundance at about 800–700 Ma (Walter & Heys 1985). Vidal & Knoll (1982, Fig. 59) identified a major (70%) extinction in acritarch taxa at the time of the Varanger glaciation in the middle Vendian. Vidal (1984, p. 57) suggested that acritarch decline resulted from eutrophication connected with glaciation. A possible link between stromatolite decline and glaciations between 850 and 650 Ma has been noted by Fedonkin (1990b, p. 17), but there are also many other possible factors to be considered (Walter & Heys 1985, p. 165): metazoan competition (Garrett 1970b; Awramik 1971), reduced calcification potential, oxygen increase, and competition from algae (Monty 1973). Monty (1973, p. 605; see also Grotzinger 1990, p. 96, and this volume) attributed declining precipitation rates to falling pCO_2 levels. However, variation in pCO_2 does not, in the Phanerozoic, appear to be correlated with RCCEs, whereas temperature decline does (Riding 1992b). Falling temperatures could have reduced abiotic precipitation and simultaneously slowed growth of the organisms forming the stromatolites. Both results would conceivably have reduced diversity and abundance. This change was not unidirectional (Walter & Heys 1985, p. 165) but part of a fluctuation that returned a CCE at the start of the Cambrian. However, the sedimentary manifestations of microbial communities was substantially altered in the process, with typical Proterozoic stromatolites being replaced by early Paleozoic dendrolites and skeletal thrombolites.

Cambrian diversification and biomineralization

Timing. – A variety of environmental factors have been considered as possible controls on the radiation of biomineralizing organisms in the early Cambrian (Glaessner 1984, p. 175; Cloud 1988, p. 327). The likely importance of temperature is not a new suggestion. Harland & Wilson (1956, p. 284), Rudwick (1964, p. 154), and Harland & Rudwick (1964, p. 36) noted that both climatic improvement and sea-level rise could result from increased temperatures following the Vendian glacial episodes and thus provide conditions for evolutionary diversification.

Recognition of Phanerozoic CCEs and their possible link with temperature (Riding 1992b) provides circumstantial evidence bearing on the problem of the Cambrian radiation. Of course, for some workers there is no special problem. If the invertebrate radiation was predetermined by a series of key evolutionary developments that took place during the Proterozoic, then it could follow that the origins of skeletonized taxa "require no special explanation" beyond high rates of speciation (Stanley 1976, p. 56). It has also been suggested that predation explains the origin of skeletons (Vermeij 1990). However, this still leaves the precise timing of the event to be explained.

The commencement and possibly also the speed of the early Cambrian radiation and CCE may both have been scheduled by the prolonged late Proterozoic glaciations (Riding 1992b). The amelioration from the Varangian glaciation (approximately 600 Ma) to the early Cambrian was over a time span comparable in length to the temperature decline leading up to the Quaternary glaciations.

Effects. – It is likely that temperature is a stimulus to biodiversification in general. Temperature has a major influence on organisms, stimulating activity rates (Rose 1967; Precht *et al.* 1973) and presumably increasing growth and development (Valentine 1973, pp. 115–125). Temperature also increases reaction rates and stimulates precipitation and growth of calcium carbonate minerals (Nancollas & Reddy 1971; Wollast 1971, Fig. 125b), which can be expected to promote cyanobacterial carbonate precipitation and also abiotic precipitation.

However, in organisms that more closely control calcification, it is doubtful whether environmental factors will determine biomineralization, although there may well be less profound effects, such as on skeletal composition, for example.

Consequently, the old idea of a direct environmental control on animal biomineralization across the Precambrian–Cambrian transition has been called into question. Hence, Lane's (1917, p. 46) postulation, that Precambrian conditions prevented the formation of skeletons in organisms that existed at that time, has been superseded in the minds of most workers by the suggestion (Cloud 1948, 1968a; Valentine 1973, pp. 415–420) that the dramatic appearance of invertebrate shelly fossils near the Precambrian–Cambrian boundary reflects rapid diversification of new groups rather than acquisition of skeletons by preexisting taxa. Certainly some of the organisms, e.g., brachiopods (Cloud 1948; Valentine 1973, p. 444), have body plans predicated on skeletons. It is now widely believed that skeletonization was merely a by-product of the general diversification (Stanley 1976, p. 73; see also chapters in this volume by Bengtson and, for a contrary opinion, Seilacher).

Nonetheless, the possibility of chemical thresholds affecting invertebrate biomineralization has acquired new life as speculation on the changing character of seawater composition has continued (Garrels 1989, p. 7). Skeletal mineralogy may have responded to changes in sea-water chemistry (Wilkinson 1979; Railsback & Anderson 1987), and calcium stress at the end of the Precambrian has been suggested to have stimulated biocalcification in metazoans (Kaźmierczak et al. 1985; Kempe et al. 1989, p. 39; see also Kretsinger 1983). Although the view has been expressed that it is unlikely that skeleton formation in invertebrates would be strongly determined by environmental factors (Conway Morris 1987, p. 167), the possibility still remains that biomineralization in some animals, such as sponges (Wood 1991b, p. 337), may have been directly promoted by extrinsic factors such as temperature. This interesting possibility remains to be elucidated.

Thus, while it is likely that both diversification and calcification in cyanobacteria were facilitated by temperature, the safest guess at present is that biomineralization in invertebrates, more or less simultaneously in a range of groups, was simply an outcome of general diversification.

The possibility that temperature was the key factor scheduling the Cambrian radiation does not preclude the likelihood that it operated in concert with other positive effects on diversification, including sea-level and pO_2 rise. After all, there is no requisite need to seek a single mechanism to account for the events near the boundary (Glaessner 1984, p. 175). The fundamental cause of the temperature fluctuations presumably relates to changes in the Earth–Sun system (see Williams 1975; Glaessner 1984, pp. 203–206).

Conclusions

1 Calcified cyanobacteria appear early in the Proterozoic, and their subsequent record appears to be patterned by calcification episodes similar to, but longer than, those recognized in the Phanerozoic. However, the abundance of calcified cyanobacteria, even during the most marked CCE postulated for the middle–late Riphean, appears lower than expected. If conventional assumptions are correct, that cyanobacteria were common and carbonate precipitation was favored at this time, then it may be that calcified cyanobacteria have been overlooked in stromatolite microfabrics. However, these assumptions require reassessment. The diversity of Proterozoic calcified cyanobacteria is low, suggesting a significant diversification of calcified forms in the early Cambrian.

2 Convincing reports of Proterozoic calcareous algae are very rare. Dasycladaleans may appear in the middle Cambrian, but their first definite occurrences are middle Ordovician. Halimedaceans appear near the Cambrian–Ordovician boundary. Later Paleozoic diversification of these calcified macroalgae may be related to factors influencing CCEs (Riding 1992a).

3 Lack of information on variations in pCO_2, sealevel, and other environmental factors limits assessment of controls on Proterozoic CCEs. However, comparison with

reported glacial episodes indicates broad correlation of (a) paucity of calcified cyanobacteria (RCCE), (b) stromatolite decline, and (c) acritarch extinctions, with glaciations around 700–600 Ma.

4   It is proposed that rising temperature near the Precambrian–Cambrian boundary was the stimulus both for cyanobacterial evolution, which resulted in acquisition by some cyanobacterial taxa of specificity for calcification, and carbonate-precipitation rates. This combination of events resulted in a marked CCE and produced a wide array of calcified-cyanobacterial deposits, including dendrolites and skeletal thrombolites.

5   Early Cambrian diversification of invertebrates coincides with the start of this major CCE. S. Stanley (1986, p. 294) has commented on the need to account for the "evolutionary delay" of the late Proterozoic. Glacial temperatures could have restrained organic development (Glaessner 1984, p. 203). It is proposed that temperature rise following glaciation stimulated biodiversification in general. Thus temperature determined not the nature of the Cambrian explosion but its timing. For animals this was an event, predicated on prior evolutionary steps, that was waiting to happen (Riding 1992b).

6   Invertebrate diversification incidentally involved biomineralization (Riding 1992b), although the possibility cannot be ruled out that in some groups – such as sponges – calcification was, as in cyanobacteria, stimulated by extrinsic factors related to temperature rise.

Acknowledgments. – Supported by NERC Research Grant GR3/4334, The Royal Society exchange programs with the USSR Academy of Sciences and Academia Sinica, and by NATO Research Grant RG84/0176 to R.R. and Stanley M. Awramik. Martin Brasier, Simon Conway Morris, Klaus Vogel, and two anonymous reviewers critically commented on earlier drafts of parts of the manuscript. Douglas H. Erwin and Robert J. Horodyski reviewed the final manuscript. Andrew Knoll kindly provided advice on Proterozoic algae. Stefan Bengtson guided me in the right direction during preparation of this paper.

Neoproterozoic evolution and environmental change

Andrew H. Knoll

Botanical Museum, Harvard University, Cambridge, Massachusetts 02138, USA

The Neoproterozoic Era (1,000–550 Ma) was an interval of profound biological and environmental change. The era began with a marked morphological diversification that may correspond to the radiation of higher eukaryotes inferred from molecular data. It ended with two distinct metazoan radiations: the Ediacaran diversification of large animals (~580 Ma) and the subsequent Cambrian explosion of eucoelomate invertebrates. These biological events bracket an interval of marked tectonic, biogeochemical, and climatic change. It is difficult to identify physical events that might have triggered the early Neoproterozoic or Cambrian radiations; these may reflect the evolutionary consequences of biological innovations. In contrast, the Ediacaran radiation is closely associated stratigraphically with a complex of interrelated physical events, ultimately driven by tectonics. Beginning about 850 Ma, high fluxes of reduced materials into the oceans may have decreased oxygen levels, increased rates of organic-carbon burial, and drawn down CO_2 levels sufficiently to induce ice ages. A tectonic state shift just prior to the Ediacaran radiation sharply decreased hydrothermal fluxes, permitting resumed high rates of organic-carbon burial to increase O_2 to levels capable of supporting large animals. There are, as yet, no direct indications of latest Proterozoic oxygen concentrations, but accumulating geochemical data force us to consider seriously the hypothesis that during the late Neoproterozoic, biogeochemical cycles linked the physical and biological Earth in changes that ushered in a new world.

Illustrations of geological time are potent visual metaphors for perceptions of Earth's history. Fig. 1A is drawn from a currently popular textbook. The present is at the top and the origin of the Earth at the bottom, with most of the intervening diagram filled by the various periods of the Phanerozoic Eon. The Precambrian, which accounts for nearly 90% of geological time, is represented by a block approximately the same size as the Ordovician. Such a nonlinear depiction of time is common and does not itself cause widespread confusion. But the figure conveys a more subtle and insidious message: everything before the Cambrian is little understood prehistory. Worse, it intimates that the Precambrian can be approached as an entity. Even scientists who recognize distinctions between the Archean and Proterozoic eons risk homogenizing long and complex histories, each nearly four Phanerozoic eons long.

Bengtson, S. (ed.) 1994: *Early Life on Earth. Nobel Symposium No. 84.* Columbia U.P., New York

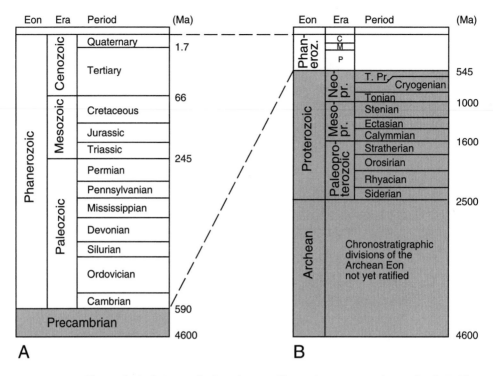

FIGURE 1   **A:** The geological time scale, based on an illustration in a popular textbook. **B:** The geological time scale, redrawn to a linear time scale. The divisions of Proterozoic time used in **B** are those ratified by the International Commission on Stratigraphy (Plumb 1991); with the exception of the initial boundary of the Terminal Proterozoic Period (T Pr), as yet undefined, all boundaries are defined geochronometrically. Comparable divisions for the Archean Eon are under discussion.

Studies of early evolution require a much more refined sense of Precambrian history (Fig. 1B), and for the Neoproterozoic Era (1,000–545 Ma), at least, our understanding of biological and physical events is growing rapidly. It is now clear that the great Cambrian radiation of eucoelomate invertebrates – so clearly marked by its skeleton-ized participants – was preceded by radiations of protists and architecturally simple animals, which occurred near the beginning and the end of the Neoproterozoic, respectively. Independently, geological and geochemical data indicate that the three radiations are nested within a broader pattern of tectonic, biogeochemical, and climatic change. In this paper, I will address a deceptively simple question: Can we relate the great physical and biological events of the Neoproterozoic and Cambrian?

Approaching the question

There are three steps to relating biological and physical events in Earth's history. First, one must document the events under consideration in some detail. Despite a long tradition of relating early animal evolution to an increase in atmospheric oxygen, massive ice ages, the breakup of a supercontinent, sealevel change, and/or changing ocean chemistry, it is only recently that critical data have begun to accumulate on the nature, timing, and magnitude of postulated events. Having established the occurrence and characteristics of evolutionary and environmental changes, it is then necessary to build them into a common stratigraphic framework. We are fortunate that the stratigraphic correlation and radiometric dating of Neoproterozoic successions are improving considerably. The third stage is, of course, the most difficult, and I will not pretend that it has been achieved at present. Once we have demonstrated that events occurred and can be correlated stratigraphically, we must construct hypotheses of cause and effect. In matters of historical reconstruction, certainty is bound to be elusive, but we needn't be satisfied by unconstrained guesswork. Biophysical and biogeochemical models can explore the plausibility of postulated relationships. Well-constructed hypotheses will find tests in geological or biological observations not yet made.

In the limited space available, I can do little more than outline the principal events in question and suggest possible interrelationships. The hope is that these capsule descriptions and bald assertions will stimulate readers to investigate the primary literature or, better, to contribute to research in this fertile field.

Major biological events

The Big Bang of protistan evolution

Molecular phylogenies based on ribosomal RNA sequences suggest that while the Eucarya as a domain may be extremely ancient, most extant higher taxa diversified more recently during a brief evolutionary burst (Sogin *et al.* 1989; Perasso *et al.* 1989). The fossil record exhibits a similar and possibly congruent pattern (Knoll 1992b). Large (40–200 μm) spheroidal microfossils that are plausibly if not unambiguously interpreted as eukaryotic occur in rocks as old as 1.9–1.7 Ga (e.g., Zhang 1986; Hofmann & Chen 1981), but there is little evidence for morphological diversification among protistan microfossils until 1.2–1.0 Ga. At that time, acanthomorphic (spiny) and more modestly sculptured forms began to contribute to a markedly higher diversity of eukaryotic organisms (Jankauskas 1989; Knoll 1992b, 1993; Schopf 1992b). The taxonomic interpretation of many Proterozoic protists is uncertain, although some of the newly appearing forms seem to be prasinophyte green algae (Tappan 1980; Knoll *et al.* 1991). More certain taxonomic interpretation – and, therefore, stronger ties to molecular phylogenies – comes from multicellular green, red, and (possibly) chromophyte algae found in ~750 Ma old shales from Spitsbergen (Butterfield *et al.* 1988; Butterfield 1992), ~950 Ma shales from Siberia (Hermann 1990), and 1,260–950 Ma cherts from arctic Canada (Butterfield *et al.* 1990). The organic geochemical record of this interval is limited, but bitumens analyzed to date suggest a marked Neoproterozoic increase in the abundance and variety of eukaryotic biomarkers (Summons & Walter 1990). There are good reasons to be wary of literal interpretations of the Precambrian fossil record (Knoll 1992b, 1993). Nonetheless, I tentatively accept that the increase in eukaryotic diversity observed near the Mesoproterozoic–Neoproterozoic boundary corresponds to the evolutionary burst inferred from molecular studies. If so, the macroscopic compressions and trace fossils preserved in 1.4–1.2 Ga old rocks from China, Australia, India, and North America (Walter *et al.* 1990; Grey & Williams 1990) must be interpreted as extinct experiments in eukaryotic multicellular or coenocytic organization. In any case, the algae found in 1,200–750 Ma old rocks indicate that the Big Bang of eukaryotic diversification cannot postdate the early Neoproterozoic Era.

The Ediacaran radiation of macroscopic animals

The paleobiology and stratigraphy of the earliest animal fossils have been reviewed repeatedly and well (e.g., Runnegar 1992 and references therein; see also chapters by Fedonkin, Sun, Runnegar, and Seilacher, this volume). I will only emphasize that if the phylogenetic group that includes extant animals diverged from other major taxa during the Big Bang of eukaryotic evolution, then there is necessarily a gap of several hundred million years between the differentiation of the clade and the Ediacaran radiation. Many Ediacaran animals exhibit a broadly cnidarian grade of organization (but see Seilacher, this volume, for a dissenting opinion), although bilaterally symmetric animals are represented by trace and, possibly, body fossils. Diverse Ediacaran faunas are restricted stratigraphically to successions above Varanger glaciogenic rocks and below the base of the Cambrian System; however, small, simple, and in some cases septate disks formed by probable cnidarians occur below Varanger tillites in the Twitya Formation of the Mackenzie Mountains, northwestern Canada (Hofmann *et al.* 1990). While the evolution of multicellularity within the metazoan clade may significantly postdate the differentiation of the clade itself, the presence of simple metazoans in the Twitya Formation indicates that the origin of tissue-grade multicellularity and the beginnings of diversification within the animals both antedate the diverse Ediacaran faunas by 40 million years or more (Narbonne *et al.* 1994). Thus, the main Ediacaran event is properly viewed not as the differentiation of a kingdom but more narrowly as the (polyphyletic) evolution of macroscopic size within the kingdom.

The Cambrian radiation

Even more has been written about the Cambrian radiation (e.g., Valentine *et al.* 1991 and references therein; see also chapters by Valentine, Conway Morris, Seilacher, Miklos & Campbell, and Bengtson in this volume). It appears to record the rapid diversification of eucoelomate invertebrates, including many with mineralized skeletons. Skeletons have understandably been the focus of many discussions of the Cambrian radiation. I will return to this issue below; here it is only necessary to note that while the evolution of skeletons is part and parcel of the diversification, it is not synonymous with it. As the Burgess Shale and other exceptionally preserved faunas make clear, most Cambrian invertebrates did not form mineralized skeletons (Conway Morris 1986). Further, Neoproterozoic fossils document the pre-Cambrian evolution of carbonate and, probably, silica biomineralization in animals (Germs 1972; Grant 1990; Conway Morris *et al.* 1990), multicellular algae (Horodyski & Mankiewicz 1990; Grant *et al.* 1991), and protistan unicells (Allison & Hilgert 1986; redated by Kaufman *et al.* [1992]). The Cambrian explosion does mark the rise of skeletons as quantitatively important elements of the carbon, phosphorus, and silica cycles.

Major physical events

Tectonic activity

The final stages of the Proterozoic Eon were marked by strong tectonic activity. Although this is often described in terms of supercontinental breakup, recent research suggests a more complicated and interesting history. One or a small number of supercontinents may have existed at the beginning of the Neoproterozoic Era, following Grenvillian convergence and collision (Hoffman 1989a). By 850–800 Ma ago, rapidly subsiding extensional basins were initiated on a grand scale. Extension and subsidence were renewed about 650–600 Ma ago, culminating in the opening of Iapetus and other ocean basins near the Proterozoic–Cambrian boundary (Bond *et al.* 1984).

There is evidence, however, for convergence and continental collision during this same period. Pan-African orogenesis began with Cordilleran-style tectonism, with subduction related to ocean expansion in the Avalonian Belt and Arabian Shield (Murphy & Nance 1991). Continental collision that amalgamated half a dozen cratons into the Paleozoic entity of Gondwana took place between 750 and 550 Ma ago, forming the Mozambique Belt of the Pan-African Orogen (Hartnady 1991).

Recent plate reconstructions (Moores 1991; Dalziel 1991; Hoffman 1991) highlight this tectonic dynamism, with Hofmann in particular hypothesizing that western Gondwana (Australia, Antarctica) separated from western North America some 800–750 Ma ago and swept up the cratons of eastern Gondwana (Africa, South America) during the latter part of the era. A single supercontinent may then have existed for a brief period before fragmenting into the Paleozoic continents. Kirschvink (1992), however, suggests that the fusion of Gondwana may not have taken place until after Iapetus had begun to open.

Most paleobiological commentary on these tectonic events has centered on the effects of continental dispersal and transgression on metazoan diversity (e.g., Brasier 1982). Understandably, there has been less speculation on how these events could have influenced the origins of animals or other eukaryotic clades. To quote Valentine *et al.* (1991), "the link between plate tectonics . . . and the origins and radiations of animals remains to be demonstrated."

There is, however, a different way in which tectonic processes could have influenced one or more of the great radiations, and that is through their participation in the biogeochemical cycles that regulate Earth's surface environments (Knoll 1991, 1992a). Evidence for unusually strong hydrothermal influence on Neoproterozoic ocean chemistry comes from the isotopic record of seawater strontium recorded in carbonates. $^{87}Sr/^{86}Sr$ in the oceans is determined by two inputs, that of continental erosion (lithology dependent, but generally high $^{87}Sr/^{86}Sr$) and hydrothermal circulation (low $^{87}Sr/^{86}Sr$). Strontium isotopic ratios recorded in 850–800 Ma old carbonates are extremely low (0.7056), below any value recorded during the Phanerozoic and too low to model effectively without postulating increased hydrothermal input (Veizer *et al.* 1983; Asmerom *et al.* 1991). Large-scale rifting and juvenile crust production correlate with low Neoproterozoic $^{87}Sr/^{86}Sr$. A major shift to unusually *high* $^{87}Sr/^{86}Sr$ (ca. 0.7085) began just after the Varanger ice age; it appears to reflect both reduced hydrothermal flux and increased erosion associated with Pan-African uplift (Asmerom *et al.* 1991).

The Sr isotopic record is complemented by stratigraphically anomalous iron formations (Young 1976) and strata-bound Mn deposits (Lambert & Donnelly 1991). These suggest that strong hydrothermal input was coupled with at least episodic ocean stratification (Knoll 1991; Lambert & Donnelly 1991; Derry *et al.* 1992). Changing fluxes of reduced materials into the oceans (or changing redox potential of volcanic gases; Betts & Holland 1991) could have had a strong impact on environmental oxygen concentrations during this interval.

Climatic change

Another conspicuous feature of 850–590 Ma old successions is the presence of tillites and other glaciogenic lithologies. Stratigraphic uncertainty still clouds discussions of Neoproterozoic climatic change, but most students would agree that there were at least two major ice ages (Sturtian at ~750–725 Ma and Varanger at 610–590 Ma; Harland *et al.* 1990), and possibly as many as four discrete glaciations between 850 and 580 Ma. Despite assertions that ice ages could either facilitate or inhibit metazoan evolution, it is hard to draw convincing connections. Indeed, the end of the Varanger ice age may have preceded the beginning of the Ediacaran radiation by as much as 10–15 Ma (Knoll & Walter 1992).

Biogeochemical changes

One of the more useful and unexpected discoveries of the past decade has been that later (850 Ma and younger) Neoproterozoic carbonates and organic matter commonly have carbon isotopic ratios that are unusually high ($\delta^{13}C_{carb}$ = +5 to +8‰ PDB) relative to those of either younger or older intervals (Knoll *et al.* 1986). A shift of this magnitude and direction occurs in response to a substantial increase in the burial ratio of organic carbon to carbonate carbon, usually (but not necessarily [Derry *et al.* 1992]) attributed to an absolute increase in the rate of organic-carbon burial (Holser 1984). A second feature of the Neoproterozoic isotopic curve is that the intervals of anomalously high $\delta^{13}C$ are punctuated by negative excursions to values closer to those seen today. Significantly, some and perhaps all of the negative excursions correlate stratigraphically with glaciogenic rocks and iron formations (Kaufman *et al.* 1991).

Rates of organic-carbon burial would not in themselves be expected to influence the course of evolution, but they could do so by affecting the concentrations of CO_2 and O_2 in the atmosphere and oceans (Knoll 1991). The secular variation of sulfur isotopes in Neoproterozoic sulfates is more poorly documented, but there does appear to be a major shift in $\delta^{34}S$ to more positive values during the latest Proterozoic (Claypool *et al.* 1980).

Carbonates, phosphorites, and ocean chemistry

A persistent idea is that latest Proterozoic changes in ocean chemistry facilitated the evolution of skeletons (e.g., Riding 1982; Cook & Shergold 1984; Kaźmierczak *et al.*

1985). Three principal lines of evidence have been used to support this suggestion: a decline in the dolomite/calcite ratio across the erathem boundary, the widespread appearance of calcified cyanobacteria at about the same time, and an abundance of phosphorites in latest Proterozoic and Cambrian strata.

In their analysis of more than 1200 carbonate samples, Sochava & Podkovyrov (1992) concluded that Mg/Ca is approximately the same for terminal Proterozoic (Vendian) and Paleozoic carbonates; Mg/Ca is lower than the Paleozoic mean in later Neoproterozoic (850–610 Ma) successions. The clear message is that a marked decrease in the abundance of dolomite across the Proterozoic–Cambrian boundary has yet to be documented. However, this may be misleading. Tucker (1983) and others have pointed out that Proterozoic dolomites are often early diagenetic and fabric-retentive, while younger examples more commonly obliterate sedimentary textures. Therefore, the similarity of dolomite/limestone ratios across the erathem boundary might mask important differences in process or environment. The critical issue, then, is how to account for petrological differences between Proterozoic and Paleozoic dolomites.

Interpretation of petrographic change is also central to the issue of microbial calcification. Calcified microbes are the partial remains of microorganisms, principally the extracellular sheaths of cyanobacteria, that are preserved by carbonate impregnation and/or encrustation. It is true that calcified microbes (and related carbonate build-ups called thrombolites) first become widespread in Cambrian carbonates; however, they are not absent from Proterozoic successions (Knoll *et al.* 1993 and references therein). In fact, the presence of heavily calcified cyanobacteria in Proterozoic carbonates demonstrates that all chemical and biological conditions necessary for microbial calcification existed long before the Proterozoic–Cambrian boundary. Therefore, one must look beyond simple shifts in ocean chemistry to explain the stratigraphic pattern.

Finally, there is the question of phosphorite distribution. Cook & Shergold (1984) drew attention to the immense volume of phosphorites in latest Proterozoic and Cambrian successions, attributing this to the upwelling of previously stratified, anoxic deep waters. They equated the increase in phosphorite deposition to an increase in the fertility of the oceans and hypothesized that higher primary productivity fueled the Cambrian explosion. They also speculated that a temporary increase in marine phosphate availability facilitated the evolution of phosphatic skeletons.

Like dolomite petrology and microbial calcification, phosphorites present a stratigraphic pattern in need of explanation. The concentration of changes near the Proterozoic–Cambrian boundary encourages one to seek a relationship to skeletal evolution. But it does not necessarily follow that chemical shifts induced biological responses. As argued below, the reverse is more likely to have been the case.

Correlation and causation

Fig. 2 shows the stratigraphic relationships among events discussed in preceding paragraphs. Bearing these correlations in mind, there are three possibilities for relating physical and biological change.

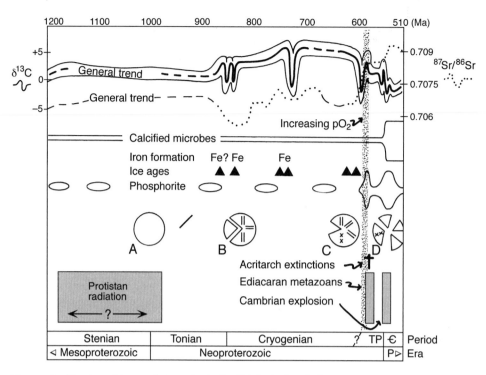

FIGURE 2 Stratigraphic correlation of principal biological and physical events of the late Protero-zoic and Cambrian. Note that isotopic curves before 900 Ma are generalized. The pie diagrams summarize major tectonic events, including **A**: post-Grenville amalgamation of continental blocs, **B**: massive rifting, separation of west Gondwana from North America, and formation of juvenile crust, **C**: the coalescing of eastern and western Gondwana, renewed extension and rapid subsidence; and **D**: continental breakup and dispersal. TP = Terminal Proterozoic Period; C = Cambrian Period; P = Paleozoic Era. The question mark and slanted initial boundary of the Terminal Proterozoic Period indicates that a boundary definition has not yet been accepted by the International Commission on Stratigraphy. (Based on references cited in the text.)

Alternative I: Ships that pass in the night

The great biological and physical events of the Neoproterozoic and Early Cambrian may have nothing to do with one another. This option is neither more nor less parsimonious than those that couple biological and physical events; nor is it easier to support or reject. There is indeed little evidence for concentrated environmental change at the time of the late Mesoproterozoic or early Neoproterozoic diversification of protists. Likewise, although the Cambrian radiation of eucoelomate invertebrates (and a parallel diversification of phytoplankton [Moczydłowska 1991]) is associated in time with continental breakup, there is no compelling theory or data that would link the two in a causal relationship. These radiations may well have been initiated principally by biological innovations. The simultaneity of the invertebrate and phytoplankton radia-tions suggests that ecological interactions also stoked the flames of diversification.

Alternative II: Evolution facilitated by physical events

The one biological event that correlates well with a network of tectonic and bio-geochemical events is the Ediacaran radiation of large metazoans. Many authors have postulated that metazoans could have attained macroscopic size only when atmo-spheric oxygen concentrations crossed some minimum threshold – 3–6% of present-day levels (PAL) if the evolution of circulatory systems preceded attainment of macroscopic size, higher if the reverse (Runnegar 1982b, 1991). The Ediacaran event coincides with major changes in seawater $^{87}Sr/^{86}Sr$ and the carbon isotopic record (Knoll 1991), as well as a sharp drop in phytoplankton diversity (Zang & Walter 1989; Knoll & Butterfield 1989). As discussed below, several independent biogeochemical models suggest that atmospheric oxygen increased substantially at this time. The idea that the Ediacaran radiation was made possible by tectonically driven environmental change is, thus, plausible, but there is at present no direct indication of latest Proterozoic oxygen levels.

Alternative III: Sedimentary and geochemical changes driven by biological evolution

Stratigraphic correlations need not be interpreted as physical stimulus and biological response. Insofar as organisms are important components of biogeochemical systems, evolutionary events can have significant physicochemical consequences. As hinted above, I believe that this alternative provides the best available explanation for litho-logical changes across the Proterozoic–Cambrian boundary.

Before the rise of skeletons to biogeochemical prominence, $CaCO_3$ was routinely deposited as precipitates in coastal water columns or on the seafloor; in restricted platforms or pore fluids such precipitation would have generated residual waters of elevated Mg/Ca and, hence, strong potential for early diagenetic dolomitization (Fairchild et al. 1991).

With the Cambrian radiation, skeletons shifted loci of carbonate deposition away from peritidal environments to more open marine waters well buffered against chemi-cal alteration.

The lithological record documents a consequent change in the conditions of dolo-mite formation. At the same time, the biogeochemical emergence of skeletons reduced the abundance and distribution of crystalline carbonate precipitates, permitting more widespread carbonate nucleation on cyanobacterial sheaths (Knoll et al. 1993). As argued more fully elsewhere (Knoll & Swett 1990; Fairchild et al. 1991; Fairchild 1991; Knoll et al. 1993), Neoproterozoic to Cambrian changes in dolomite formation and microbial calcification are interpreted most parsimoniously as consequences of skeletal evolution.

With less conviction, a similar case can be argued for the phosphorite record. Diversifying animals would have contributed to phosphorite deposition by packaging phosphorus in skeletons, carcasses, and fecal pellets for efficient sedimentation in dysaerobic platform and shelf environments (Knoll 1992a).

There is, however, another aspect of the phosphorite record that requires consider-ation. While uppermost Proterozoic and Cambrian phosphorites are reasonably attributed to continental breakup and ocean turnover, earlier Neoproterozoic breakup and turnover are not associated with phosphogenesis on a comparable scale. Ferric iron, precipitated from seawater as oxyhydroxides, scavenges phosphate (Berner 1973; Froelich et al. 1982). Thus, a decrease in hydrothermal iron flux and increase in oxygen concentrations ~590 Ma ago would have removed a major phosphorus sink. More phosphorus would have been available for recycling. Thus, a latest Proterozoic increase in the fertility of the oceans (relative to earlier Neoproterozoic but not necessarily Mesoproterozoic levels) is plausible (Knoll 1992a), and this *may* have influenced coeval biological events.

Lithological changes in carbonates correlate with the emergence of skeletal invertebrates as major carbonate sinks, *but* they postdate the first appearance of carbonate skeletons. The acme of phosphogenesis also occurred during and after the Cambrian radiation of animals with phosphatic skeletons, although the earlier onset of massive phosphogenesis indicates that, in this case, other factors were important.

I conclude that the secular changes observed in Proterozoic to Cambrian carbonate, chert (Maliva et al. 1987), and (in part) phosphorite deposits are best interpreted as consequences of skeletal evolution. As others have noted, the fact that the Cambrian radiation included organisms with skeletons of silica, phosphorite, calcite, aragonite, organic matter, and agglutinated materials argues against a simple chemical trigger and in favor of biological causes for skeletal evolution. As noted above, skeletons are just one aspect of a broader pattern of diversification (see also Bengtson, this volume).

Suggestions for a complex reality

In closing, I outline a possible course of Neoproterozoic biological and environmental history:

1 Between 1.2 and 1.0 Ga ago, the major clades of higher eukaryotes differentiated in a short-lived burst of evolution. While a physical detonator cannot be ruled out, no presently available evidence supports the idea. What biological innovation(s) might have triggered the radiation is uncertain, but at least some evidence from population genetics supports the proposition that the appearance of sexual population structures could have been key (Schopf et al. 1973; Tibayrenc et al. 1991; Knoll 1992b). It should be made clear that processes of sexual reproduction do exist in lower eukaryotes, but full-blown meiosis is known only from clades that nest in the crown of the eukaryotic tree (Cleveland 1947). Palmer & Logsdon (1991) have also argued that nuclear introns are restricted to crown eukaryotes. In their "introns late" scenario, an increased capacity for exon shuffling could also have contributed to genetic variation.

2 About 850 Ma ago, a major episode of continental extension and juvenile crust formation began, accompanied by high fluxes of reduced materials into the oceans. The carbon cycle was affected, with increases in the proportion of carbon buried as organic matter and, perhaps, absolute rates of organic-carbon burial. It is likely that the flux of reduced materials was sufficient to react the oxygen generated by organic-carbon burial. Indeed, atmospheric oxygen concentrations may have been relatively

low during much of the interval 850–610 Ma, although the strong excursions in the carbon isotopic record may indicate a complex O_2 history (Knoll & Walker 1990). It has been suggested that later Neoproterozoic oceans were stratified for long periods. While this is possible, it is not necessary to explain available data. Brief intervals of stratification in hydrothermally influenced oceans could have reduced atmospheric CO_2 enough to induce ice ages. This may explain the stratigraphic relationships among Neoproterozoic carbon isotopic excursions, tillites, and iron formations (Knoll 1991).

3 At about the time of the Varanger ice age, the system shifted. A strong increase in seawater $^{87}Sr/^{86}Sr$ marks a decrease in the flux of reduced materials and the beginning of a major increase in erosion rates that culminated in the Cambrian (Asmerom *et al.* 1991). Immediately post-Varanger carbon isotope ratios are low, probably reflecting decreased rates of organic carbon burial, the erosion of organic-rich sediments exposed by a drop in sea level, and the mixing of isotopically light deep waters back into the surface ocean (Kaufman *et al.* 1991). However, carbon isotopes record a subsequent spike of organic-carbon burial, which – in the absence of a strong hydrothermal flux – would have resulted in a substantial increase in atmospheric oxygen (Derry *et al.* 1992). The approximately contemporaneous shift in the sulfur isotopic composition of sulfates is less well documented or understood, but its effect would have been to amplify any terminal Proterozoic increase in oxygen. A substantial increase in oxygen may well have facilitated the evolution of macroscopic size in animals.

4 Biological innovations, including more sophisticated genetic control of development as well as the evolution of complex circulatory, respiratory, and neural systems (Valentine *et al.* 1991), ignited further explosive animal evolution near the Proterozoic–Cambrian boundary. Because of their importance in locomotion and protection, mineralized skeletons evolved in numerous clades. The physicochemical consequences of skeletal evolution can be read in the textures and geochemistry of coeval sedimentary rocks.

The relationships outlined here may or may not be correct; they are certainly incomplete. But their specificity accurately reflects the current state of knowledge concerning comparative biology, paleontology, stratigraphy, and the Neoproterozoic geological record. They further make predictions that can be tested in continuing studies, providing a research agenda for aspects of early evolution.

Current research emphases on mass extinction and global change underscore the critical interactions between environment and biota. But extinction is not the only response to environmental change, and global change didn't begin in the Pleistocene. A satisfactory understanding of life's origins and early evolution will become possible only when biological and physical evolution are integrated into a synthetic account of Earth's history.

Acknowledgments. – My research on Neoproterozoic problems is funded by NSF grant BSR 90-17747 and NASA grant NAGW-893. I also thank Gonville and Caius College, Cambridge, for a visiting fellowship that greatly facilitated the preparation of this essay.

Early metazoan evolution: First steps to an integration of molecular and morphological data

Simon Conway Morris

Department of Earth Sciences, University of Cambridge, Downing Street, Cambridge CB2 3EQ, England, UK

Is the study of metazoan phylogeny an intractable problem? To judge by the multiplicity of hypotheses that are available on the basis of morphological data, no obvious solution appears to be in sight. Evidence from molecular biology, however, offers an effectively independent test of relationships. Some results, such as the primitive position of the cnidarians or the protostome–deuterostome division, accord well with current thinking. Other inferences, such as the proposed paraphyly of the arthropods, are contentious because they reverse considered opinion. Molecular biology, however, can tell us nothing about the status of fossil taxa that appear not to fall into accepted schemes of phylogeny, nor is it able to provide information about the morphological and functional transitions that occurred during early metazoan evolution, at a time when the principal body plans were emerging. The fossil record is uniquely valuable in this context. Not only does it show us what intermediate taxa looked like, but it reveals an actual history and shows how the metazoan morphovolume was occupied.

How well do we understand metazoan evolution? The consensus is more or less as follows. Metazoans are late entrants in terms of Earth's history, appearing about 4 Ga after the formation of the planet and first represented by the Ediacaran faunas (~560 Ma ago). Subsequently, near the Precambrian–Cambrian boundary (~550 Ma ago), the so-called Cambrian explosion is initiated, marked most obviously in the fossil record by the widespread appearance of hard parts, diversification of trace fossils, and development of Burgess Shale-type faunas. The principal questions that must be either set against this consensus or at least placed in context are: (1) Do metazoans have a pre-Ediacaran history? (2) Are all Ediacaran taxa metazoans? (3) Can a trigger be identified that initiated the Cambrian explosion? (4) What is the pattern of metazoan evolution in the Cambrian, especially with respect to the so-called problematic taxa?

The origin of the eukaryotic condition in the geological record is contentious. Large cells (acritarch cysts), macroscopic ribbons and sterane biomarkers are consistent with an appearance between about 2.1 and 1.4 Ga (Han & Runnegar 1992). Molecular data are now indicating that although some eukaryotes are very ancient (e.g., Vossbrinck *et al.* 1987), many protistan groups arose as part of a more recent major radiation (e.g., Christen *et al.* 1991a; Christen, this volume). Nearly all protistan groups lack a fossil

Bengtson, S. (ed.) 1994: *Early Life on Earth. Nobel Symposium No. 84.* Columbia U.P., New York

record, and their actual times of diversification depend on molecular methods based on estimating rates of substitution. A time of ~1 Ga for this diversification seems reasonable. Were the first metazoans evolving at this time, multicellular but microscopic, and effectively just one more group of protistans with similar ecologies and environmental preferences?

Direct evidence for such pre-Ediacaran metazoans evolving about 1 Ga ago remains tenuous. Evidence marshaled in favor of this hypothesis includes data from molecular biology (e.g., Runnegar 1985), decline in stromatolite diversity (Walter & Heys 1985), and possible fecal pellets (Glaessner 1984; Robbins et al. 1985; but see Nöthig & Bodungen [1989], who document a process perhaps analogous to pellet production by protistans). How can the hypothesis be tested further? So far as I am aware there are no biomarkers unique to metazoans; this needs to be checked. Perhaps more promising are studies that seek to recognize reliable criteria for the biogenicity of putative trace fossils (Harding & Risk 1986), as well as recognition of bioturbation fabrics (O'Brien 1987; see also Reichelt 1991). This latter approach could be especially valuable if pre-Ediacaran metazoans were microscopic. Two matters require renewed emphasis. First, "search images" based on expectations gleaned from the Phanerozoic record are probably inappropriate. Second, the attempt to find pre-Ediacaran metazoans will never succeed if they never existed.

Space does not permit a detailed review of the status of Ediacaran taxa, especially the debate as to whether they are metazoans (e.g., Glaessner 1984; Gehling 1991; Fedonkin, this volume) or are a separate group, named vendozoans or vendobionts (e.g., Seilacher 1989, 1992, this volume; Bergström 1989, 1991). Several items, nevertheless, call for brief comment. First, the relationships of practically all Ediacaran taxa remain deeply controversial. It is evident that even if traditional assignments to groups such as the scyphozoans (cnidarians) and polychaetes (annelids) are actually correct, then there is evidence that expectations of morphology have influenced some investigators' interpretations. In others words, many Ediacaran taxa show features that appear to have no counterpart in extant taxa. The second point is that some Ediacaran taxa persisted into the Cambrian, where they have been recognized in Burgess Shale-type faunas and possibly elsewhere (Conway Morris 1989b, 1993a). The third point is that our understanding of Ediacaran biotas may be far from complete. Much remains to be learned about the tubicolous cloudinids (Yue et al. 1992; Bengtson & Yue 1992), while new discoveries of possible metazoans in the Doushantuo Formation of China (Chen & Xiao 1991; Xue et al. 1992) are also significant.

Information on Cambrian metazoans and their radiation depends on three principal sources: skeletal remains (e.g., Bengtson & Conway Morris 1992), Burgess Shale-type faunas (Conway Morris 1989b), and trace fossils (Crimes 1987). All are important, but each suffers drawbacks. Fossil skeletons are abundant but are likely to represent only a small fraction of the biota (e.g., Conway Morris 1986) and may be disarticulated (Bengtson 1985). Burgess Shale-type faunas are the richest mine of information, but they are sporadic and may occur in unusual environments, e.g., dysaerobic and deeper-water. Trace fossils are preserved in situ, but they suffer the disadvantage that the precise nature of the trace maker is usually impossible to establish.

The acquisition of hard parts close to the Precambrian–Cambrian boundary still lacks a satisfactory explanation. There is little doubt that shells of sessile organisms and cataphract arrays in others acted to deter predators (e.g., Vermeij 1990). Two observations need to be made. First, the employment of several biominerals, as well as the development of agglutinated skeletons, suggest that biomineralization was polyphyletic, an unsurprising fact given evidence for recurrent evolution of an ability to biomineralize in various other groups during the Phanerozoic (e.g., Wood 1991a; Norton & Behan-Pelletier 1991). However, the number of times that this ability evolved during the Cambrian explosion is not known: it may have been less than sometimes imagined. Until we have a secure phylogeny of early metazoans (see below), it is impossible to ascertain the number of times hard parts evolved. Second, as almost nothing is known about the genetics of biomineralization and relatively little about the organic templates involved, discussion of the history of biomineralization is constrained. (See also Bengtson, this volume.)

Much attention has been devoted to the earliest skeletal record, a significant proportion of it composed of the so-called small skeletal assemblages (small shelly fossils), as well as calcareous sponges (especially archaeocyathids). An era of confusion, due in part to excessive use of form taxa (especially from disarticulated scleritomes), failure to recognize taphonomic variants, and oversplitting of morphological variable populations, is now ending. Several distinct clades are now recognizable, including anabaritids, coeloscleritophorans (including halkieriids), tommotiids, cambroclaves, brachiopods, mollusks, hyoliths, onychophores (as microdictyonids), and sponges. Certainly many taxa remain in limbo as regards wider affinities, but most have been only briefly described and poorly illustrated. Of equal importance is that data on these early skeletal fossils are also being integrated with other evidence from Cambrian paleontology. Properly understood, these assemblages hold the key to a number of questions of early metazoan evolution (see also Conway Morris 1993b).

But how difficult will it be to answer these questions? The orthodox view is that not only did practically all extant body plans evolve during the Cambrian, but they were accompanied by a significant number of extinct body plans. The evolutionary interrelationships between these body plans are mostly enigmatic, but the large number of putative phyla is a clear indication that a wide range of morphologies, at least in comparison with Ediacaran assemblages, evolved over a relatively short period of geological time. This orthodox view, however, is rather static and takes insufficient notice of problems of sampling the fossil record. Despite the range of morphospace occupied during the Cambrian explosion, the number of species was relatively low. This is evident from the many higher taxa represented by few species. But the number of species that survive to be represented in the fossil record is always low in comparison with the original totals. Hence, during times of rapid diversification the incompleteness of the fossil record means that the probability of finding taxa intermediate between supposed phyla is low.

How then can we proceed with the analysis of early metazoan evolution? Phylogenetic analyses, especially those that employ cladistics, will often remain intractable, because so many taxa share few characters. To be sure, progress is being made among some clades, such as the arthropods (Briggs & Fortey 1989; Briggs et al. 1992) and brachiopods (Rowell 1982; Goryansky & Popov 1985), but a cladogram for the early Metazoa remains a daunting task (but see Schram 1991). In part the way forward is simply to collect more data. In particular, new discoveries of small skeletal fossils and Burgess Shale-type faunas (e.g., Conway Morris & Peel 1990; Chen & Erdtmann 1991;

Chen *et al.* 1991; Robison 1991; Hou & Bergström 1991; Hou *et al.* 1991) continue to throw new light on the diversifications. But I believe that at the moment progress is inhibited by overly orthodox interpretations of the fossil record. The impetus for a new approach is now coming from molecular biology, especially sequence data based on ribosomal RNA and other macromolecules (e.g., Christen *et al.* 1991a, b; Ghiselin 1988, 1989; Lake 1990; Field *et al.* 1988; C. Patterson 1989).

The magnitude of the task, however, should not be underestimated. At times of rapid diversification, molecular data may be insufficient to resolve unambiguously the precise order of branching (e.g., Christen *et al.* 1991b; Raff, this volume), and it is arguable whether recovery from the fossil record would ever be complete enough to provide an unambiguous history. Is the cause then a hopeless endeavor? No, and for two reasons. First, any phylogeny is a working hypothesis, and there is no more reason to abandon investigation of early metazoan diversification than any other comparable study. Second, the expected flood of data from other gene sequences and their products (see Miklos & Campbell, this volume) should strongly constrain possible phylogenies. Integration of molecular and morphological data is at present a lively topic (e.g., Patterson 1987). In the context of this paper, the possibility that rapid divergence is paralleled in morphology (the fossil record) and certain molecules (ribosomal RNA) is intriguing on three counts. First, it suggests that during the Cambrian explosion rates of substitution in rRNA were greatly accelerated, although other molecules may show a clock-like behavior during this interval. Second, it may be that rates of evolution in some molecules and morphology need not always be decoupled. Third, this linkage may have a bearing on evidence for developmental flexibility that may have operated in the Cambrian (e.g., Hughes 1991), although no direct involvement of rRNA in developmental genetics appears to have been postulated.

Fortunately, the lessons from many of the former conflicts between paleontologists and molecular biologists, most notably the disagreements about the timing of hominid evolution, have been learned. Nevertheless, while molecular data may offer new frameworks, paleontology remains essential. Not only does it offer an independent test for a molecular phylogeny, but, more importantly, it is only the fossil record that can reveal the actual history of events. Most important are the extinct taxa, intermediate between extant phyla (cf. Bergström 1989), whose functional morphology, anatomy, and ecology must be documented if early metazoan evolution is to be understood fully. Such information may prove crucial, because, as noted above, during such times of rapid diversification establishing the order of branching can prove difficult (e.g., Christen *et al.* 1991b).

The published trees based on molecular data show some points of agreement with established opinion. These include the early origin of cnidarians and ctenophores (e.g., Lake 1990; Christen *et al.* 1991b), possibly consistent with the Ediacaran faunas being dominated by at least the former group. On a coarse level the near-synchronous divergence of many triploblastic groups (e.g., Field *et al.* 1988) also accords with their rapid appearance during the Cambrian, although the molecular trees continue to recognize the classic split between protostomes and deuterostomes (e.g., Lake 1990; but see Bergström 1986 and this volume). But some of the new trees (Lake 1990) also reveal some unexpected features, most notably the appearance of arthropods before mollusks and annelids. If arthropods were among the earliest protostomes, then this may reopen the question of the relationships of the Ediacaran *Spriggina*, and Valentine (1989) has explored some of the other consequences of the first metazoans having a hemocoelic body cavity. For example, he speculates that this anatomy might help to explain the simplicity and two-dimensionality of early burrows.

These molecular trees are not always easy to test against the fossil record, in part because of unresolved problems of times of appearance of the major groups. I suggest, however, that the fossil record can be placed into the context of these new phylogenies, and, more importantly, it can offer unique data to the problem of metazoan diversification.

Two main points require discussion. First, the fossil record is showing unequivocally that some groups that today are relict were of major significance in the Cambrian. Most notable in this context are the onychophores (e.g., Ramsköld & Hou 1991) and priapulids (e.g., Conway Morris & Robison 1986). In neither group is extensive molecular data yet available (but see Ballard *et al.* 1992), presumably because from a neontological perspective their relict nature relegates them to a subsidiary role. From the viewpoint of early metazoan evolution, however, this may be inappropriate if they played a major part in early diversification.

The second significant aspect of the fossil record is the recognition of taxa that establish relationships between supposedly separate major groups. These, of course, are the so-called problematic taxa, and their usefulness will become apparent at a number of taxonomic levels, including phyla and classes. Thus it might be realistic to hope for progress in understanding, for example, annelid–mollusk affinities or relationships within the arthropods. Nevertheless, the obsession with establishing metazoan phylogeny from the perspective of neontological data and a typological emphasis on extant phyla means that a literal interpretation of the known fossil record could confuse the issue. This is because the chances of recovering a complete ancestor–descendant series, colloquially the "missing links," is vanishingly small. Any taxon of interest will show not only synapomorphies that will establish relationships to other major groups but also its own apomorphies acquired during separate descent.

The remainder of this essay is a preliminary attempt to augment the new metazoan phylogenies based on molecular data with the early Phanerozoic fossil record. Runnegar (1992) has offered the outlines of a related exercise, but with fewer data and less explicit nesting. Fig. 1 illustrates the present schema, and because of lack of space here it will be possible to make only brief notes on the disposition of the tree.

1 Sponges are the most primitive metazoans.

2 Coelenterates are the next group to arise, and this position in the tree is consistent with, but does not prove, their dominance in Ediacaran taxa.

3 Extinct groups within the Cnidaria are taken to include the conulariids and their earliest representatives, the carinachitiids, arthrochitids, and hexaconulariids, which are abundant as Lower Cambrian small shelly fossils in south China (Conway Morris & Chen 1992), and more tentatively the anabaritids (Conway Morris & Chen 1989; Bengtson *et al.* 1990).

4 On molecular evidence, platyhelminths also seem to have an ancient origin, but the fossil record is almost silent (Conway Morris 1982, 1985). Their exceptionally low preservation potential as fossils may result from the absence of a cuticle. A recent

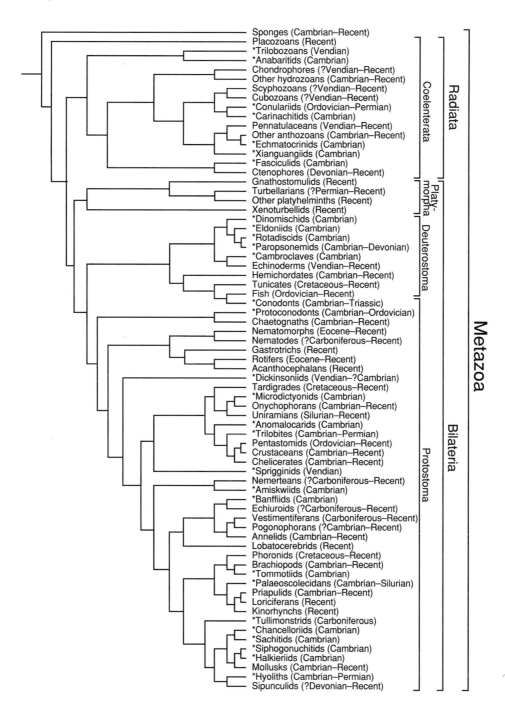

FIGURE 1   A proposal of metazoan phylogeny, based on current schemes of molecular data (rRNA), but with the addition of various extinct major groups principally from the Cambrian (marked with an asterisk). Note that this scheme is tentative and designed to provoke discussion. In addition, a number of taxa – e.g., *Rhombocorniculum* (Cambrian) and *Typhloesus* (Carboniferous) – remain enigmatic and cannot be placed in the existing proposal.

description of a putative turbellarian trace fossil from the Permian of Italy (Alessandrello *et al.* 1988) is a reminder that earlier examples may be recognized.

5   Moving now to the protostome branch, what evidence from the fossil record might support the earliest representatives being hemocoelic prearthropods? The first possibility is the Ediacaran taxon *Spriggina*, which is abundant in the Pound Quartzite of South Australia. Birket-Smith's (1981) reconstruction seems difficult to reconcile with the fossil material.

More promising for further analysis are the anomalocariids, which are segmented animals with prominent lobate extensions arising from the trunk. They are taken to include *Anomalocaris*, a widespread taxon (e.g., Conway Morris & Robison 1988), the closely related *Cassubia* (Dzik & Lendzion 1988), *Opabinia* (Whittington 1975), undescribed anomalocariids and taxa presently recognized as *Hurdia* and *Proboscicaris* from British Columbia (Collins 1992), and related fossils from the Sirius Passet fauna of north Greenland (Budd 1993).

Onychophores, for which there is strong evidence for a substantial Cambrian radiation (e.g., Ramsköld & Hou 1991; Hou *et al.* 1991; Dzik & Krumbiegel 1989), could be closely related to the basal anomalocariid stock (see also Dzik & Krumbiegel 1989). However, Dzik & Krumbiegel's (1989) proposed derivation of these early hemocoelic organisms from priapulids is considered unlikely.

Most significant are the anomalocariids with jointed appendages that, found isolated, were unequivocally assigned to detached portions of arthropods. A shift in developmental processes so that these appendages were duplicated along the body could have given rise to true arthropods. This leaves the problem of the homology of the distinctive plated feeding apparatus, best known in *Anomalocaris*. Is this an apomorphy of derived anomalocariids, or can homologies be found in arthropods? Concomitant with the appearance of serial jointed appendages was sclerotization of other regions of the exterior. The entire fossil record is already providing a series of revealing insights into the evolution of the arthropods, as presently defined (e.g., Briggs & Fortey 1989; Briggs *et al.* 1992; Walossek & Müller 1990). One item of interest is the support from molecular evidence for a relationship between pentastomids and crustaceans (Abele *et al.* 1989), which also finds support in recently described Ordovician fossils (Andres 1989). In this context it may also be worth considering the Chengjiang taxon *Facivermis* (see Chen & Erdtmann 1991), which has an elongate body with five pairs of lobopod-like extensions at the anterior.

6   In the new scheme of metazoan phylogeny, the coelomic condition of annelids appears to be a later development than the anomalocariid–onychophore–arthropod hemocoel radiation. Acquisition of a coelom may have occurred in epifaunal animals, contrary to the classic hypothesis of Clark (1964).

The nature of the first annelids remains speculative. The Cambrian record of polychaetes is reasonable, and taxa such as *Burgessochaeta* may be primitive (Conway Morris 1979). None of the Cambrian fossils appear to approach an oligochaetoid condition, and possible candidates in the form of the palaeoscolecidans (Conway Morris & Robison 1988) are now tentatively assigned to priapulids (see below). Curiously, no annelids appear to have been reported from the Chengjiang Burgess Shale-type fauna.

Pogonophores have been allied with the annelids, either closely (e.g., Southward 1975), or more distantly (e.g., Jones 1985b). This relationship is supported by molecular data (Lake 1990). A recurrent assumption (e.g., Sokolov 1972a) is that sabelliditids, abundant in the Lower Cambrian, represent pogonophoran tubes, but this remains unproven (Urbanek & Mierzejewska 1977).

7 The close relationship of annelids and mollusks is now accepted (e.g., Ghiselin 1988), but the nature of the fossil taxa that linked annelid-grade forms to organisms that evolved to mollusks is speculative. What seems clear, however, is that halkieriids may play a crucial role in understanding early steps in molluscan diversification.

Two issues concerning the relationships of halkieriids require immediate discussion:

(a) Halkieriids have been included in a major clade, the Coeloscleritophora, but is it monophyletic? The case for halkieriids and siphogonuchitids being related seems strong (Peel 1991; Bengtson 1992). Morphological transitions to chancelloriids can be traced via sachitids (Bengtson *et al.* 1990). These suppositions are nevertheless based on sclerite form and structure, and discovery of articulated scleritomes of sipho-gonuchitids and sachitids might reveal some surprises. It is evident from articulated specimens of halkieriids (Conway Morris & Peel 1990) and chancelloriids (Walcott 1920) that if they are related, then the Coeloscleritophora encompasses very diverse anatomies and ecologies.

(b) Is *Wiwaxia*, best known from articulated scleritomes in the Burgess Shale, related to the halkieriids (Bengtson & Conway Morris 1984), or is it a polychaete (Butterfield 1990b)? Reasons to refute the latter suggestion are given by Conway Morris (1992), but the possibility remains that halkieriids are relevant to understanding early annelid evolution, a point not pursued by Butterfield (1990b), who placed *Wiwaxia* in the Chrysopetalidae.

Articulated halkieriids are remarkable because in addition to the dorsal coat of spicule-like sclerites, there is a prominent shell at either end of the body (Conway Morris & Peel 1990). These latter structures clearly grew by lateral accretion, the sclerites seemingly not. In halkieriids there is no obvious indication that the shells are formed by fused or aggregated sclerites, whereas in siphogonuchitids this appears to be the case (Peel 1991; Bengtson 1992 and this volume). However, the secretory tissue of the sclerites and shells in halkieriids was presumably of the same deriva-tion, and transition from sclerite production to shell production may have been straightforward.

Bengtson (1992; see also Conway Morris & Peel 1990) has emphasized how many of the mollusk-like shells, common in early skeletal assemblages (e.g., Qian & Bengtson 1989), might be better assigned to halkieriid-like scleritomes. Testing these hypotheses is often difficult, at least from information available in the published literature. To Bengtson's (1992) list of possibilities, it is worth adding a plausible candidate in the form of the hitherto enigmatic shell *Triplicatella* from the Lower Cambrian of South Australia (Bengtson *et al.* 1990). This taxon is almost invariably associated with the halkieriid *Thambetolepis*, and *Triplicatella* may have formed part of the scleritome. Two features of interest in *Triplicatella* have implications for soft-part anatomy. First, muscle scars on the interior may indicate muscles homologous with the dorsoventral musculature of other mollusks. Second, the marginal folds may reflect sites of inhalant and exhalant currents for a respiratory organ (?ctenidia) located beneath the shell.

8   It is proposed that the priapulid–brachiopod stock arose from a point close to the annelid–mollusk divergence (concerning the position of the brachiopods in the protostomes, see Ghiselin 1989). Molecular evidence from haemerythrin for a relationship between priapulids and inarticulate brachiopods is given by Runnegar & Curry (1992; see also Yano *et al.* 1991). It may be worth considering further homologies between priapulids and brachiopods. Can similarities be established, for instance, between the priapulid trunk and the pedicle of inarticulate brachiopods? Compare, for example, the trunk of *Maotianshania* (Sun & Hou 1987b) to the pedicle of the cooccurring *Lingulepis malongensis* (Jin *et al.* 1991, Pl. 1:3). Both priapulids (Conway Morris 1977; Conway Morris & Robison 1986) and brachiopods (e.g., Koneva 1986; Ushatinskaya 1988, 1990; Roberts & Jell 1990) are a significant component of Cambrian assemblages, and the overall diversity of each clade is underestimated. To the roster of priapulids can probably be added the palaeoscolecidans (Conway Morris & Peel, unpublished data), which include taxa largely or exclusively known from isolated sclerites (e.g., *Milaculum, Utahphospha, Hadimopanella*, see Hinz *et al.* 1990). Ultrastructural comparisons are required, but for the time being the proposal that on the basis of similar ornamentation palaeoscolecidans are related to the early onychophores (Dzik & Krumbiegel 1989; Hou *et al.* 1991) is considered less likely. Not only are Cambrian brachiopods incompletely understood, but the lophophorate radiation may have included the so-called pseudobrachiopods (Bengtson *et al.* 1990). In addition, I would tentatively suggest that the tommotiids also be assigned to this clade, both on the basis of certain morphological (e.g., Bengtson 1977; Laurie 1986) and ultrastructural (Conway Morris & Chen 1990b) similarities.

This re-analysis of the protostome radiation leaves unaddressed the relationships of a number of other phyla, especially the aschelminths. Ghiselin (1989) gives some tentative evidence to ally the nematodes to the arthropod–annelid–mollusk group, but there is an urgent need to confirm other proposed relationships.

The deuterostome branch is more briefly reviewed, with molecular biology supporting the classic dichotomy between chordates and echinoderms. What evidence might the fossil record add? Two preliminary suggestions are made.

First, I speculate that *Eldonia* (presently under detailed re-investigation by D. Friend) may throw significant light on very early deuterostome evolution and that earlier assignments to echinoderms (Durham 1974) may not be entirely incorrect. The possibility that the supposed chondrophore *Rotadiscus*, from the Chengjiang soft-bodied fauna of south China (Sun & Hou 1987a), is related to *Eldonia* may need consideration. More compelling is the likelihood of a close relationship between *Rotadiscus* and *Velumbrella*, the latter from the Middle Cambrian of Poland (see also Conway Morris 1993a). Dzik (1991), who recently reviewed *Velumbrella*, proposed a relationship with *Dinomischus*, which occurs in both the Burgess Shale and Chengjiang faunas. Another avenue that needs investigation is whether *Rotadiscus* and its putative allies are related to the paropsonemids, a distinctive group that survived until the Devonian (see Conway Morris & Robison 1982).

Consequences

The new framework of metazoan phylogeny discussed above has a bearing on various topics presently under discussion:

1 Such analyses move away from metazoan origins depicted as a pectinate arrangement, whereby numerous phyla emerge from a "pool" of common ancestors. For example, Bergström's (1989, 1990) depiction of various phyla emerging from a pool of procoelomates is considered improbable.

2 The discussion above is rather anecdotal and will require more rigorous depiction by cladistic methods. Nevertheless, the problems of such analyses are formidable. In particular, homoplasy combined with insufficient information from the fossil record, rather than providing definitive answers, makes existing proposals useful for discussion.

3 The phylogeny discussed above has no direct bearing on the issue of whether disparity or occupation of morphovolume was greater in the Cambrian than subsequently (e.g., Gould 1991). Curiously, Gould (1991, p.421) writes "I do confess some fears that, *in toto*, the question of morphospace may be logically intractable," and then on the following page reports data by Foote (1990, 1991) that not only shows how morphospace can be measured through time but, more significantly, reaches a conclusion diametrically opposed to that espoused by Gould (1991; see also Briggs *et al.* 1992).

4 The revised view of metazoan phylogeny, combined with information on the fossil record, potentially allows direct inspection of the vexing question of the origin of body plans. The accelerating knowledge of developmental mechanisms operating in various metazoans should allow some insights into the style and degree of developmental programming that accompanied the rise of different morphologies. Such information should assist workers to refrain from vague and tendentious statements such as "congealing of the genome" (Gould 1991) to describe mechanisms that are purported to have operated during metazoan history.

Of particular interest is the recognition of similar genes, especially homeobox domains, concerned with development in groups – e.g., insects, leeches, chordates – whose evolutionary distance is considerable (e.g., Akam 1989; Wedeen & Weisblat 1991; McGinnis & Krumlauf 1992; Weinmaster *et al.* 1991), not to mention even more remote groups such as cnidarians (e.g., Murtha *et al.* 1991; Schierwater *et al.* 1991). It is also clear, however, that although some developmental sequences are strongly conserved, the morphological diversity that arose in the Cambrian can only have resulted from developmental instructions that evolved in specific clades (e.g., P. Holland 1991). By documenting the mosaic of morphotypes that arose in the Cambrian, paleontologists should be able to contribute to the continuing search for an explanation and description of metazoan diversification.

Acknowledgments. – I am grateful to the organizers of this symposium for the invitation to participate and for the warm hospitality of the Nobel Foundation (Alfred Nobel's Björkborn) in Karlskoga. Comments by an anonymous referee are valued. Earth Sciences Publication 2905.

Ideas on early animal evolution

Jan Bergström

Swedish Museum of Natural History, Box 50007,
S-104 05 Stockholm, Sweden

Most attempts to track relationships between metazoan phyla have failed either because of excessive parallelism in somatic evolution or because of the inclusion of superfluous and distorting molecular data. Background noise must be removed from molecular data before any successful phylogenetic analysis can be made on this particular level. The coelomates form a major branch of animals. Paleontological and molecular data indicate that their origination was abrupt, with a bush-like radiation impressed by parallel evolution. This rules out Ediacara organisms from a possible ancestral position and characters like segmentation, tentacles, and oligomery as indicators of relationships between phyla. Poriferans and coelenterates do not share a metazoan ancestor with bilaterian animals.

Analyses of the phylogenetic origins of metazoan phyla, in particular the coelomate ones, have been studded with disappointments. It makes no difference if the starting point has been anatomy and ontogeny or paleontology. The results depend on each author's choice of key characters. This conclusion holds true whatever method is used – cladistic or not – because the basic problem is the multiple parallel origination of even basic designs. To paleontologists, the sudden appearance of the phyla in the Lower Cambrian appears as a major paradox. It is known as the Cambrian explosion or the Cambrian radiation event. The Precambrian Ediacara organisms appear not to be ancestors but to be fundamentally different from bilaterian animals (Seilacher 1989; Campbell 1990; Bergström 1991).

In recent years molecular-sequence techniques have given us new hope that true relationships will ultimately be revealed. However, in many diagrams of molecular evolution, members of individual phyla are planlessly scattered (e.g., C. Patterson 1989, Figs. 2, 4, 5). Phylogenetic trees based on different molecules tend to be strongly dissimilar. The problems are largely caused by the inclusion of distorting data (background noise) in the computer treatment.

The molecular evidence

Provided that the background noise is removed (see below) and a phylogenetic strategy is used, different molecules tend to yield closely comparable phylogenies. When such

Bengtson, S. (ed.) 1994: *Early Life on Earth. Nobel Symposium No. 84.* Columbia U.P., New York

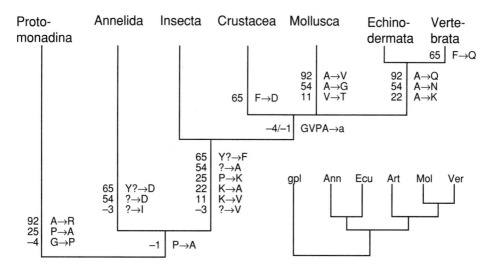

FIGURE 1   Phylogenetic analysis of cytochrome c mutations relevant to radiation of bilaterian phyla (with Protomonadina as outgroup). Note that groups with segmentation form three separate branches; both segmentalization and arthropodization appear to have occurred more than once. Symbols: a, methyl group; A, alanine; D, aspartic acid; F, phenyl alanine; G, glycine; I, isoleucine; K, lysine; M, methionine; N, asparagine; P, proline; Q, glutamine; R, arginine; T, threonine; V, valine; Y, tyrosine. (Redrawn from Bergström 1986).

Lower right: Maximum parsimony analysis of globin sequences. Symbols: gpl, green plants; Ann, Annelida; Ecu, Echiura; Art, Arthropoda; Mol, Mollusca; Ver, Vertebrata. Note similarity with tree based on cytochrome c. (Simplified from Goodman *et al.* 1988.)

phylogenies are compared with those that are well known from other sources, such as the phylogeny of vertebrates, an impressive similarity is seen (Bergström 1986, Fig. 1). This indicates that the molecular data are reliable if we just understand how to interpret them.

Molecule-sequence evidence to separate the coelomate phyla is fairly meager. This indicates a rather short duration of the branching interval. In fact, in some molecules such as 18S rRNA, there seems to be no clear information on the order of branching – all coelomate phyla appear to have radiated from the same point (see conflicting results presented by C. Patterson [1989] and Turbeville *et al.* [1992]; for the latter see also Raff, this volume).

In other molecules, such as cytochrome c, however, there is some information (Fig. 1; Bergström 1986). Long ago, McLaughlin & Dayhoff (1973) used this molecule to construct a phylogenetic tree that is remarkably up-to-date. In this tree, chordates and mollusks form the top, while crustaceans branched off earlier. A single supposedly synapomorphic mutation, a substitution at the N-terminus of a series of amino acids with a hydroxyl group, sets off deuterostomes, mollusks, and crustaceans from insects and annelids further down the tree.

Work on other molecules has supported this interpretation while giving little support for the old idea of a protostomian–deuterostomian dichotomy. Thus, deutero-

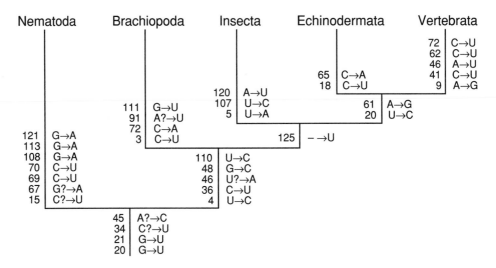

FIGURE 2 Phylogenetic analysis of 5S rRNA mutations relevant to radiation of bilaterian phylum lineages. Note the position of deuterostomes in the top of the protostomian tree. Symbols: A, adenine; C, cytosine; G, guanine; U, uracil. (New, based on Erdmann & Wolters 1986.)

stomes and mollusks form the top of the tree, with arthropods branching off just beneath, in trees based on globins (Goodman *et al.* 1988, Figs. 3, 4), 5S rRNA (Fig. 2 herein), 28S rRNA (Christen *et al.* 1991a, Figs. 3, 4), and small-ribosomal-subunit RNA (srRNA; Hendriks *et al.* 1991). The result is summarized in Fig. 3.

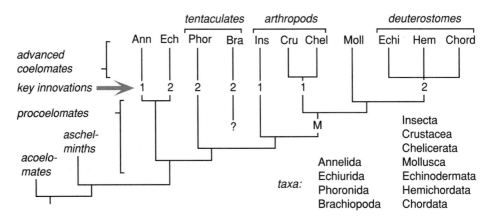

FIGURE 3 Phylogeny of the Bilateria based on molecular evidence, tentative summary. M, methyl group at N terminus in cytochrome c. Figures indicate macroevolutionary events: 1, coordinative introduction of segmentation; 2, paedomorphic introduction of ciliated tentacles. Primitive mollusks retain much of the procoelomate organization. The parallel evolution is the reason why relationships between phyla have been so poorly understood. Several phyla cannot yet be placed.

Removal of noise in sequence comparisons

From a statistical point of view, perhaps about 98% of the molecular differences between extant members of any two coelomate phyla is the result of within-phylum evolution. All this information is irrelevant to the question of the branching order between phyla and forms a distorting background noise. This explains why mathematically objective parsimony analyses often yield such different results and unite animals from different phyla while splitting those from a single phylum (e.g., Field *et al.* 1988; C. Patterson 1989).

A logical way to get rid of this problem is to eliminate the background noise introduced by younger mutations. This can be done if analysis is started within a phylum and if a phylogenetic strategy is used to reconstruct shared ancestral molecular sequences. If the monophyletic origin of a phylum is in question, the analysis can end at a lower level. An alternative and perhaps as effective way is to sort away all those positions in which there is any variation within an accepted phylum. After the sorting decision has been made, the procedure can be fully objective. Very few positions are useful and will remain, while all others are removed from consideration. My experience from a number of trials is most promising: all molecules tested yield virtually the same tree. This result differs markedly from published results based on conventional methods.

Nonsequence evidence

Data from classical fields such as embryology and anatomy have been used extensively for phylogenetic trees. In cases this has been quite successful below the phylum level. However, it is well known that attempts to evaluate the relationships between animal phyla, and particularly between coelomate phyla, have been most discouraging, the results largely depending on the characters chosen.

A basic disagreement concerns the question of whether metazoans form a natural group. Anatomical features, but also the collagen molecule, have been claimed to be synapomorphic features uniting all metazoans (although collagen may have been formed several times during evolution, according to Morris & Cobabe 1991). Conway Morris (1991, p. 19) calls it an "observation" that "metazoans must share a common ancestor," although two papers in the same volume present molecular evidence that metazoans are at least diphyletic (Christen *et al.* 1991a; Bergström 1991). Disagreements on the phylogeny of the coelomates are still more complex.

It has been claimed that it is the lack of a cladistic (phylogenetic) approach that causes the difficulties. However, cladistic approaches using different material in fact give strikingly different results (e.g., Bergström 1991; Schram 1991). As noted above, anatomical data are not useful. The basic problem seems to be that there are very few, if any, distinguishable somatic synapomorphies shared by any two coelomate phyla, deuterostomians excepted. Cladistic analysis on molecular-sequence data works only after removal of background noise.

Schram's (1991) cladistic analysis rightly recognizes the Coelomata but fails to sort this group into phylogenetic units. Instead, his two basal branches recognize two basic modes of life, the sessile and vagile modes. This is ecology and functional morphology, not phylogeny, and the pattern is in complete disagreement with analyses based on molecular sequences.

Is a phylum different from an order?

Did new phyla and classes emerge in the early Phanerozoic, while lower categories continued to emerge until now?

Conway Morris (e.g., 1989b, p. 345) has claimed that groups become more different with time and the disappearance of intermediate forms, until differences reach accepted phylum level; phyla would be phyla only in hindsight. From this point of view, he hypothesizes that there were a great number of phyla in the Cambrian. There is very strong objection to this view (e.g., Briggs 1990; Campbell 1990). The idea was primarily based on the supposedly enormous disparity of the Burgess Shale arthropods, but Briggs *et al.* (1992) have convincingly shown that the disparity is comparable to that of Recent arthropods.

There is absolutely no sign of convergence between phyla as we follow them backwards to the Early Cambrian. They were as widely apart from the beginning as they are today. Hierarchical levels apparently include a biological reality, not only classificatory convention. In fact, the overwhelming taxonomic difficulty is to recognize relationships between phyla, not to distinguish between them. Among classes and lower categories, on the other hand, there may be intermediate forms, and when these become extinct, the gaps will widen between the remaining lineages (Briggs 1990).

The Cambrian radiation event

Animals such as arthropods and brachiopods cannot exist without hard parts. The absence of remains of skeletons and shells in the Precambrian therefore proves that the phyla came into being with the Cambrian, not before, even if the lineages leading to the phyla were separate before the Cambrian. The sudden appearance leaves us with a dilemma. How could phyla have evolved within perhaps 5 million years (Bowring *et al.* 1993), to show no further separation during the subsequent more than 530 million years?

Part of the explanation may be that few if any of the niches occupied by modern coelomates were occupied in the late Precambrian.

I have previously indicated a possible course for a very fast origination of phyla (Bergström 1989, 1991) that was probably important in the filling of these niches and which may explain the short duration of the Cambrian explosion. It includes the multiple origination of new basic life styles tied to parallel inventions of basic new body designs (baupläne; Fig. 3). Once such designs were achieved, they did not reappear (Valentine 1973, 1980).

The primitive existence of pseudosegmentation made repeated acquisition of true segmentation simple. A truly segmented nemertean genus, *Annulonemertes*, among the pseudosegmented nemerteans demonstrates how easily this may come about (Berg 1985). True segmentation was invented independently also in kinorhynchs, tardi-grades, annelids, arthropods, and chordates, and perhaps also in onychophorans. Another main highway for evolution was formation of filter-feeding phyla. This is easily explained by paedomorphosis making use of filter-feeding larvae to create ciliated tentacles. Filter feeding goes together with a U-shaped intestine. The fact that the intestine is turned ventrally in some groups, dorsally in others, reflects the parallel character of the evolution.

Biochemical and paleontological data indicate a paradox: branching of lineages was not fully synchronous, but achievement of phylum characters was a synchronous event. The suggested solution is in full accord with the processes outlined above. The phylum-forming inventions occurred contemporaneously in a number of procoelo-mate branches that had already been phyletically separate for a limited time (Bergström 1989, 1990).

The evolutionary inventions leading to new phyla and classes need not have been principally different from inventions made later on; rather, it was a unique combination of circumstances that resulted in a unique effect.

The reasons for the appearance of so many new lineages in the Cambrian have been extensively discussed in the literature. There is now general agreement that it was a skeletonization event, and there is good evidence to regard it as a radiation event as well. Possible factors triggering the acquisition of skeletons and the radiation include the attainment of a critical level of oxygen partial pressure in the atmosphere (shell secretion is inhibited at low oxygen pressures today), a decrease of carbon dioxide in the atmosphere, a necessity of decreasing the concentration of dissolved carbonate in the body, a need of a protective skeleton, the availability of large shelf areas, and the availability of large amounts of phytoplankton food related to extensive upwelling.

We do not have the final answer, but Vermeij (1990) may be on the right track when suggesting that when extrinsic controlling factors made it possible to secrete minerals, it was predator pressure that forced animals to develop mineralized skeletons. Skel-etons are physiologically expensive, as is collagen. It therefore seems to be a reasonable guess that the partial pressure of oxygen in the atmosphere was an important limiting factor, if not the decisive one. Perhaps this is the reason why cytochrome c seems to be particularly useful in sorting out the radiation: cytochromes take active part in the redox reactions in the cell, and swift changes in the molecule may have been part of the adaptation to the new oxygen-rich environment.

The flexible procoelomate type of animals that must have provided the phylogenetic stock must have dwindled in numbers, ultimately to become extinct. Some of the "small shelly fossil" animals in the Paleozoic may represent procoelomates that had acquired a protective cover.

General conclusions

1 The reliability of a method in phylogenetic analysis is revealed by the distribution of organisms in the tree. If members of monophyletic phyla are mixed, the method does not yield acceptable results.

2 Anatomical data are of little use for understanding relationships between coelomate phyla.

3 The anatomies of coelomate phyla reflect basic adaptations rather than relationships. The origination was a rapid macroevolutionary event that included a great amount of parallel evolution. Lower hierarchical taxa generally originate less abruptly.

4 The origination of coelomate phyla appears to have been the start of the Cambrian radiation event. Three steps can be recognized: (1) a Vendian phyletic radiation of procoelomates; (2) invention of phylum characters in several lineages roughly 530 Ma ago, probably triggered by a worldwide environmental event; (3) a Cambrian radiation within the new coelomate phyla.

5 The coelomate tree has a distinct branching order, with annelids close to the base and arthropods close to the top, the latter consisting of mollusks and deuterostomes. An original deuterostome phylum underwent a second phylum-level radiation leading to the still living hemichordates, echinoderms, and different chordate subphyla.

6 It follows from the preceding paragraph that Protostomia and Deuterostomia do not form two main branches of bilaterians or coelomates, but that protostomian characters are primitive, deuterostomian ones derived.

Acknowledgments. – I am grateful to Stefan Bengtson and to referees for constructive criticism of the manuscript.

Molecular phylogeny and the origin of Metazoa

Richard Christen

Station Zoologique, Observatoire Océanologique, Villefranche sur mer, F-06230 France

Molecular phylogenies have been obtained from the sequences of different molecules and from ribosomal RNA in particular. All analyses suggest that the emergence of Metazoa occurred during a period of intense diversification and in the same region of the protozoan world from which rose vascularized plants and fungi. Moreover, the appearance of an organization in tissues (as opposed to a colony) was followed by a rapid diversification in phyla. Because the internal branches that describe these events are short in regard to the branches that led to living species, their study is relatively difficult. Thorough analysis of protists and reconstruction of ancestral sequences might help to elucidate this problem.

Metazoans have been classified in about 36 phyla. Within each phylum, all organisms share primordial features of body architecture and embryological development and are thought to derive from a single ancient ancestor; even when a large diversification has occurred, the pattern of organization characteristic of each phylum is usually observed and traced back in the fossil record. For similar reasons of shared characters, a common and ancient origin of all metazoan phyla has often been proposed that seemingly dates from before the oldest metazoan fossil ever identified (about 640 million years ago). Although it receives wide support, this scheme is the source of several unresolved questions and controversies:

- Is there really a single common ancestor to all living metazoans?

- Which protozoan lineages are most closely related to the lineage(s) that gave rise to the metazoan organization?

- What was the path (the morphological history) that led from a protozoan organization (single totipotent cells) to a metazoan structure (assembly of specialized tissues)?

In recent years, an approach based on the comparison of homologies observed between sequences of molecules conserved in various species has been presented as an extremely powerful method to determine phylogenies and perhaps to approach the problems mentioned above. Important information has indeed been obtained by these methods, but other data and more sophisticated analyses are now needed to bring further insight into these problems.

Bengtson, S. (ed.) 1994: *Early Life on Earth. Nobel Symposium No. 84.* Columbia U.P., New York

Molecular phylogenies

Differences in the sequences of macromolecules accumulate with time and can be used to reconstruct molecular phylogenies (Felsenstein 1990; Swofford 1990; Swofford & Olsen 1990). Originally obtained from protein sequences (Zuckerkandl & Pauling 1965), molecular phylogenies are now derived from nucleic-acid sequences and rely on an enormous amount of (independent?) characters that can be easily and rapidly obtained. It has also been suggested that the rate of mutation might be constant over time and in various organisms, so that the total number of differences could be used as a molecular clock to date radiations (Kimura 1968; Kimura & Ohta 1974; but see, for example, Dover 1987; Easteal 1987; Sharp & Li 1989).

Ribosomal RNA (rRNA), with a function conserved from bacteria to man, is now mostly used to provide molecular phylogenies. Two large rRNA molecules (the L-rRNA and the S-rRNA) and one small rRNA molecule exist. The smallest rRNA (5S rRNA), a molecule with a length of about 120 nucleotides, has been widely used to deduce phylogenetic relationships as several hundreds of sequences became available. However, this rRNA has a relatively high rate of mutation, and substitutions probably reach a saturation level when sequences of distantly related organisms are compared (but see below). Following the advent of rapid methods to sequence the L-rRNA and the S-rRNA (Qu *et al.* 1983), molecular phylogenies are now preferentially obtained from the comparisons of the sequences of these longer rRNAs (from 1,542 nucleotides in *E. coli* to about 1,900 nucleotides in mammals for the S-rRNA and 2,904 nucleotides in *E. coli* to about 4,500 in mammals for the L-rRNA); the presence of domains of low mutation rates make them particularly suitable for long-range phylogenetic analysis such as the study of metazoan origin. Despite other available sequences or strategies (see Miklos & Campbell, this volume), the analysis presented here will actually review the more recent data based on rRNA sequences.

Metazoan origin

Classical approach

The prevailing view is that metazoans have evolved from simple to more complex body plans, namely: from colony-like organims through diploblastic to triploblastic body plan (see chapters by Raff and Rieger, this volume). Several conflicting hypotheses have been proposed to explain the early diversification of metazoans and the appearance of the various phyla (reviewed by Rieger, this volume). Each hypothesis relies primarily on the choice of particular characters by each author on the belief that they provide stronger phylogenetic information (Grell 1961; Lappan & Morowitz 1974; Valentine 1977; Salvini-Plawen 1978; Whittaker & Margulis 1978; Nielsen 1985; Bergström 1986, this volume).

Extant metazoans share numerous traits (reviewed by Bode & Steele [1989]), and the usual suggestion is a monophyletic origin in the form of a blastula-like organism, followed by the differentiation of phyla of radial symmetry; the animals with bilateral symmetry are thought to be latecomers in the form of flatworms (planarians) originating from one of the coelenterates. Sponges, mesozoans, or placozoans have alternatively been proposed as living representatives of the earliest metazoan organization. On the other hand, Hadzi (1963) proposed that the primitive metazoan was a flatworm that originated from a ciliated protozoan, evolving next into a nematode-like organism that was ancestor to all metazoans including the coelenterates. Poriferans have always been a problem of their own and are often considered the result of an independent radiation. As for the protozoan ancestor to metazoans, a ciliate (the syncytial theory), a flagellate (the colonial theory), or a choanoflagellate (for the similarities in morphologies between choanoflagellates and choanocytes in sponges) have been proposed (but see Rieger, this volume).

These conflicting schemes (among others) demonstrate that it is difficult to find a strongly supported hypothesis for the first morphological forms of the Metazoa as well as for the exact order of radiation of the earliest phyla. Deep phylogenetic relationships are difficult to unravel, because homologies are missing or difficult to analyze, and because our understanding of the mechanisms of morphological evolution is poor (Raff & Kaufman 1983). In addition, a major problem is that fossils that could help to link these phyla are generally missing for events that date from the Precambrian. In this respect, much was expected from molecular data, at least in terms of phylogenetic results, since homologies between sequences are easier to find and to assess.

Molecular analysis

A very large number of 5S rRNA sequences are available (Specht *et al.* 1991), representing all metazoan phyla and many protozoans, and several studies have dealt with the problem of metazoan phylogeny. Hori & Osawa (1987), for example, reached the conclusion that all metazoans including sponges belong to a single cluster and that mesozoans were the earliest radiation, with planarians and nematodes as ancestors to other metazoans; however, the authors noted that the exact order of emergence of the most ancient groups was difficult to assess precisely. By contrast, and using similar data, Hendriks and coworkers reached the conclusion that 5S rRNA sequences are not suited for phylogenetic studies on a time scale of metazoan evolution (Hendriks *et al.* 1987).

Available S-rRNA sequences include few representatives of the diploblastic phyla, with only two cnidarian sequences; mesozoans, poriferans, placozoans, and ctenophores (and nematodes) are absent (Field *et al.* 1988). This set of data was analyzed by several authors (C. Patterson 1989; Lake 1989, 1990; Ghiselin 1989), and the general agreement is now that the Metazoa are monophyletic and that the cnidarian phylum is an early radiation. The question of the identity of the most closely related protozoans was not addressed, since only distant organisms such as yeast, ciliate, and slime mold were used as outgroups.

Finally, a phylogeny was obtained using L-rRNA sequences (Christen *et al.* 1991b). Representatives of several diploblastic phyla had been sequenced (the importance of this point is developed below), but the length of the sequences obtained was relatively short (i.e. about 450 nucleotides), and it is possible that there are too few positions for resolving securely the basis of the metazoan tree. These data, however, clearly showed the classical distinction between the diploblasts and triploblasts and that triploblasts were a monophyletic unit; the surprise was that triploblasts appeared as a deep radiation that occurred before the diversification of diploblasts into the different phyla of sponges, placozoans, cnidarians, and ctenophores. The separation between diploblasts and triploblasts could thus be very ancient. Despite the use of several methods, the authors were unable to resolve the question as to whether the two groups derive from a common ancestor or if they are separated by one or more protozoan lineages (a result that would suggest that metazoans are polyphyletic).

Concerning the relationships to protozoans, the same analysis showed that plants, diploblastic metazoans, and triploblastic metazoans were the results of three closely related radiations that took place at approximately the same time, in a broad cluster that contained all multicellular organisms as well as several protists (Fig. 1): *Chlorogonium elongatum* and *Pyramimonas parkeae* (two chlorophytes), *Cryptomonas ovata* and

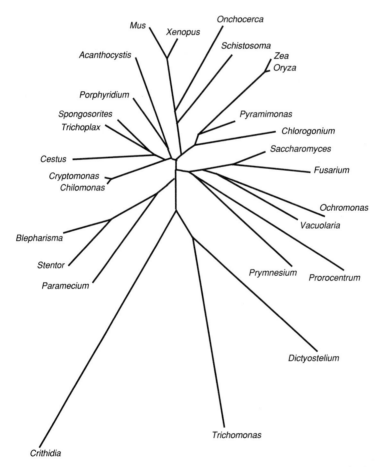

FIGURE 1   Metazoan origin. Molecular phylogeny of the Metazoa and related protists; the figure was obtained with a matrix-distance analysis and partial 28S rRNA sequences. Diploblasts (*Spongosorites*, *Trichoplax*, and *Cestus*) and triploblasts (*Mus*, *Xenopus*, *Onchocerca*, and *Schistosoma*) show deep radiations in a region of the protozoan world also containing the vascularized plants (*Zea*, *Oryza*). Dotted lines represent the region of the tree in which phylogenetic relationships cannot be determined with confidence. (Redrawn from Christen *et al.* 1991b.)

Chilomonas paramecium (two cryptophytes), *Acanthocystis longiseta* (a heliozoan), and *Porphyridium purpureum* (a rhodophyte). In most analyses, the protozoans *Porphyridium* and *Acanthocystis* could not be significantly separated from the diploblast–triploblast unit. All other protozoans analyzed were distantly related.

It is important to stress that the observation of the early origin of the triploblastic organization was possible only because several diploblastic phyla were present in the analysis. If such an early separation between diploblasts and triploblasts is confirmed by later analyses, the monophyletic origin of metazoa will be extremely difficult to demonstrate beyond doubt. Indeed, even if metazoans are demonstrated monophyletic in a particular analysis, it is impossible to rule out that further protozoan data would not show a branching inside the Metazoa. Without the inclusion of closely related protists such as *Acanthocystis* or *Porphyridium*, it would probably have been concluded from the same data that metazoans were monophyletic. It is clear now that further molecular analyses of this problem will have to include representatives of all metazoan phyla as well as many closely related protozoan species.

The origin of multitissular organisms

In addition to reproduction, specialization could be considered a characteristic feature of life: all organisms are permanently adapting to their environment. Bacteria are specialized in terms of metabolism, while eukaryotes present a diversity of forms. Plants and animals can be (presently) viewed as the ultimate form of this process of adaptation by morphological change, since they have "organs," cellular aggregates capable of accomplishing specialized tasks.

An extremely intense diversification is observed in molecular phylogenies as soon as metazoans diverged from the protozoan world. Such a phenomenon, immediately following an evolutionary jump, reveals the important selective advantage conferred by organs. Colonies of unicellular eukaryotes are multicellular organisms that have sometimes elaborate systems of communications, but since the appearance of eukaryotes probably more than two billion years ago, metazoans, plants, and fungi are the only known lineages in which cells have become specialized into permanent organs. Considering the evolutionary advantage conferred by specialized tissues, it is striking that this differentiation has not occurred among the many protozoans that form colonies (and bacteria, for that matter),

It is even more remarkable that the few appearances of tissue specialization have occurred in the same restricted region of the protozoan world. This closeness of origin for metazoans, plants, and fungi could well be one of the most important observations yet obtained from molecular phylogenies. It may suggest that the ability to derive specialized organs was an intrinsic property of a peculiar common ancestral protozoan.

It is possible to propose many explanations as to why tissue differentiation is such a rare event. In particular, one can observe that in tissular organisms, sex and reproduction are tightly linked, whereas the two processes are usually separated in unicellular organisms. There is a definitive short-term selective advantage to being able to reproduce without the need of sex and to having sex independently for the sole purpose of gene shuffling. It has not yet been investigated whether this advantage could be incompatible with tissue specialization and whether tissular organisms in fact derive from some protozan lineage in which the mode of reproduction was restricted to gamete production. It is interesting to note that the appearance of autonomous replication of cancer cells is accompanied by a loss of their differentiation.

However, the many hypotheses proposed for metazoan appearance are more an intellectual game than a real scientific approach based on hard data. It is still impossible to know with certainty the exact order of radiation of the first metazoan phyla, and the relationships between metazoans and their closest protozoan relatives remain fuzzy (but see Sogin, this volume). The major difficulty is that the origin of tissular organisms occurred within a small amount of time a long time ago; one has thus to resolve short internal branches in the presence of very long peripheral branches, and none of the data and methods yet available are satisfactory for this purpose.

Conclusions: Possible progress in molecular analyses

It is usually proposed that a better resolution of molecular trees will be achieved by sequencing longer molecules. On the other hand, it is also agreed that too little attention has been paid to the validity of the data used. The problem of sequences that are too variable and do not contain real phylogenetic information is now well considered by most authors, but even when care has been taken to select regions of appropriate mutation rates, weighting each site individually according to its phylogenetic significance remains a problem. A solution has, for example, been proposed in the PAUP program (Swofford 1990) by the "reweighting" option, but it is probably more a way to increase the consistency of the tree than to determine which characters are appropriate for a phylogenetic analysis.

A different approach has been used by Smith and coworkers (Smith *et al.* 1992) in a study of echinoid phylogeny. In a comparison of molecular and paleontological data, the authors used a parsimony analysis restricted to sites that were invariant in a set of closely related outgroups; the states of the nucleotides in the outgroups were then considered as ancestral (an approach described as "polarized parsimony"). The number of sites that remained after this selection was low, but they gave the correct topology (as judged from the fossil record) when other methods had failed because of large differences in the rates of mutations between the different lineages.

The main idea is that it will always be extremely difficult to resolve the problem of a short internal branch in the presence of long peripheral branches by using sequences from extant organisms (shown as a, b, c, and d in Fig. 2A). In order to increase as much as possible the ratio of internal to peripheral branch lengths, it is thus necessary to obtain sequences of remote ancestors (Fig. 2B). Although DNA has been recovered from some fossils, it is not a general approach, and our attention has been attracted to a different method as described in Fig. 3.

Most analyses have shown that molecular phylogenies from closely related species provide correct internal topologies. In Fig. 3, for example, the respective internal topologies for the a_0, b_0, c_0, and d_0 lines of descent can be easily obtained. The rootings of these topologies can also be obtained by the polarized-parsimony approach described above. The result would produce the sequences for the remote ancestors a_0, b_0, c_0, and d_0, from which the internal nodes would be deduced.

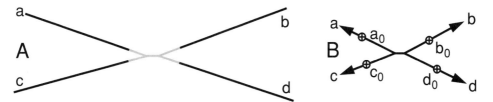

FIGURE 2 Resolving internal nodes from reconstructed ancestral sequences. **A:** The divergence of the four species a, b, c and d occurred in a short interval of time and was followed by a long divergence time. Reconstruction of the internal nodes from the sequences of extant organisms is often very difficult (uncertain relationships are shown in grey lines). **B:** The determination of the position of this internal branch is much easier from sequences of distant ancestors (a_0, b_0, c_0, and d_0) located at short range of the internal nodes.

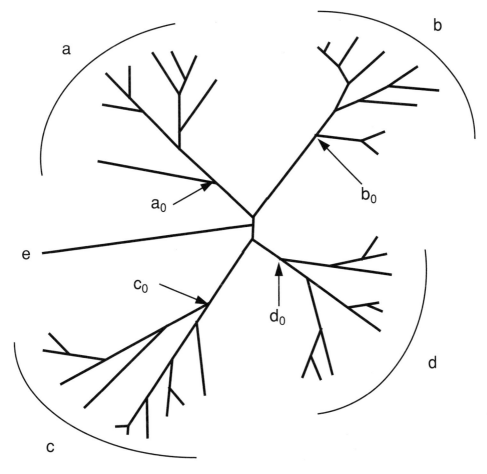

FIGURE 3 Reconstructing molecular phylogenies with ancestral sequences. Groups of organisms such as diploblasts, triploblasts, plants, fungi, and ciliates with numerous radiations are particularly obvious candidates for the approach described in Fig. 2. It will, however, be necessary to combine in the same analysis reconstructed ancestral sequences such as a_0, b_0, c_0, and d_0 with sequences from extant organisms such as e when this organism is the only representative of one deep radiation (possible case of *Acanthocystis* or *Porphyridium* in Fig. 1).

Such analyses of the origin of metazoans will require both elaboration of new computer programs and acquisition of new sequences. As for metazoans, phylogenies have already been worked out to an extraordinarily good resolution from morphological features, and it is relatively easy to choose which animals should now be sequenced. By contrast, considering the large genetic diversity of protozoans and the uncertainty concerning their phylogeny (but see chapters by Sogin and Taylor in this volume), much additional work remains. In particular, we have to find all species that are closely related to metazoans. Sequencing short stretches by the method of polymerase chain reaction (PCR) will allow for a rapid positioning of the numerous species that cannot be cultivated. The scrutiny of fungi and ciliates will also be extremely useful, since they are close to metazoans; their clustering in monophyletic units with numerous branches would probably allow us to reconstruct ancestral sequences more easily.

Acknowledgments. – Particular thanks to Anne Baroin, Roland Perasso, and André Adoutte for sharing their sequence data and experience as well as for stimulating discussions. The idea of reconstructing ancestral sequences to derive internal branches is the common conclusion of numerous discussions with Andrew B. Smith (British Museum, Natural History). This study was supported by grants from the CNRS, the Association Recherche et Partage and the Conseil Général des Alpes Maritimes.

Evolution of the "lower" Metazoa

Reinhard M. Rieger

Institut für Zoologie, Universität Innsbruck, Technikerstraße 25, A-6020 Innsbruck, Austria

Most proposals on the origin and early radiation of the Metazoa that use phenotypic characters suggest that the stem species of the Metazoa, the Eumetazoa, and the Bilateria was a small, ciliated, free-moving, and solitary organism. Relatively few authors have stressed the widespread occurrence of a biphasic life cycle, with a microscopic juvenile or larval stage moving actively by ciliary action and a macroscopic clonal adult feeding passively by taking advantage of water currents. However, recent data, primarily from ultrastructural research, favor the latter model for the origin of most animals. As a consequence of the new phenotypic data on the lower Metazoa and lower Bilateria, the evolution of individuality becomes a central issue in the rise of the Eumetazoa and Bilateria. Two processes are known that can produce solitary individuals from clonal organisms: progenesis and individualization of reproductive modules (zooids). Progenesis has been well defined since Stephen J. Gould's (1977) *Ontogeny and Phylogeny*. Individualization of reproductive modules (zooids) describes the processes by which clonal organisms have lost asexual reproduction by budding or fission. At least three levels (Parazoa, Radiata, Bilateria) of increases in structural design are evident.

The major goal of this chapter is to pinpoint possible events and processes during the early phase of metazoan evolution that have received less attention but lead to alternative concepts for the original life modes of metazoans. Such alternatives can be derived mainly from new data on the ultrastructure of the lower Metazoa – ciliated epithelia, cuticle fine structure, myoepithelial organization of coelomic linings, nephridial structure – and by considering a biphasic life cycle with a microscopic, motile larval stage and a macroscopic, nonmotile adult – as seen in the Porifera and certain coelenterates (e.g., Anthozoa) – as a character state plesiomorphous for the Metazoa.

Considering the present surge of molecular phylogenies (e.g., Erwin 1991; Christen, this volume) and cladistic analysis of metazoan evolution (Ax 1989; Schram 1991; Eernisse *et al.* 1992), this chapter will hopefully add new aspects to the discussion of early animal life.

Traditionally, most phylogenetic discussions emphasize the importance of a microscopic, disk-shaped or vermiform organism, externally similar to *Trichoplax*, the planula-larva, and certain turbellarians, during the initial radiation of the Metazoa. I argue here that such an organism evolved originally as a millimeter-sized dispersive stage (larva) in a biphasic life cycle with a clonal adult stage in the centimeter range. Extant adult forms like the Placozoa or certain Platyhelminthes could be relics from

Bengtson, S. (ed.) 1994: *Early Life on Earth. Nobel Symposium No. 84.* Columbia U.P., New York

early metazoans that lost the original adult phase by progenesis (*sensu* Gould 1977). Such processes are well illustrated in the radiation of the interstitial fauna, particularly of mesopsammic annelids (Westheide 1987).

The origin of metazoan cell colonies

Extant fauna suggests that one of the three known possible processes leading to metazoan cell colonies (see Willmer 1990; Siewing 1985) is much more likely than the others: cell divisions resulting in clones of flagellates (see Salvini-Plawen 1978; Barnes 1985) held together by a common extracellular matrix (ECM). In such "division colonies" (Siewing 1985) the nature of the ECM is crucial for the further evolution of the colony.

An early deviation of this type of organization apparently has occurred within the Parazoa, where it seems that incomplete separation of cells during cytokinesis led to the multinucleated organization of the Symplasma (Hexactinellida; see Bergquist 1985; Mehl & Reiswig 1991; Reiswig & Mehl 1991 and references therein).

Historically, the idea of division colonies as ancestors for the Metazoa goes back at least to Ernst Haeckel. Ultrastructural studies support the idea that division colonies are ancestors to most extant Metazoa (Rieger 1976). The plesiomorphous cell types forming the original metazoan cell colonies were most probably flagellates, perhaps forms similar to extant Choanoflagellida (Barnes 1985). Some cells of such flagellate colonies transformed presumably very early into amoeboid cells and invaded the ECM of the cell colony (Buss 1987).

The other two avenues discussed for the origin of metazoans are:

- The processes of cellularization from a multinucleated cell, as seen in the embryo of *Drosophila* or in the endosperm of angiosperms. This idea was promoted at about the same time that Haeckel formulated his comprehensive Gastrea-theory (1873, 1877).

    Hadzi (1963) and Steinböck (1963, 1966), particularly, have perfected the cellu-larization hypothesis, which gained influence during the 1950s and 1960s. However, ultrastructural studies in the late 1950s (e.g., Pedersen 1961) and in the following decades (Klima 1967a; Kozloff 1972; Tyler & Rieger 1977; Smith 1981; Smith & Tyler 1985) have produced evidence that bears specifically on that question. In spite of the fact that transmission electron microscopy (TEM) established Westblad's "endo-cytium" (central parenchyma) of certain Acoela to be a truly multinucleated tissue (see references in Rieger *et al.* 1991b), Smith & Tyler (1985) demonstrated that such a tissue condition is not plesiomorphous in the Acoelomorpha.

    At the level of the Parazoa, recent electron-microscopy data on hexactinellids established a multinucleated organization with incomplete separation of cells (see above). There is even evidence of such an organization in aggregations of archeo-cytes (Reiswig & Mehl 1991). It is likely that this multinucleated condition is an apomorphous feature within the Parazoa. However, its presence at this primitive level of tissue organization suggests that it originated very early in metazoan evolution. Since it is known from Hexactinellida and Acoela, it surely has developed more than once among the early Metazoa. Its main function in the hexactinellids is most likely that of coordination of responses within cell colonies (Mackie & Singla

1983), while in the Acoela it seems to be involved in the specialized digestive process (for references, see Rieger *et al.* 1991b).

• The aggregation of individual cells as seen in acrasean slime molds. This possibility was advanced by Klima (1967b) and Steinböck's student Reutterer (1969).

   The theory seems less probable for the Metazoa because: first, the ECM is complex and well developed in the majority of groups with the most primitive tissue organization (Parazoa and Coelenterata; Pedersen 1991), and second, the ECM plays a fundamental role during animal embryogenesis (Hay 1981; Caplan 1986; Edelman 1988), a fact that indirectly also supports the view of its use as the major component for cohesion during the origin of metazoan cell colonies.

   As a special case, the idea of cell aggregations formed through symbiosis of different cell lines in metazoan cell colonies still deserves our interest, because the phenomenon of symbiosis with pro- and eukaryotes is widespread even among the extant lower Metazoa (see Ruetzler 1990 for sponges and Margulis & Cohen, this volume, in general).

Division colonies and the extracellular matrix during early radiations of Metazoa

The significant role of the ECM in the evolution of the "lower" Metazoa is not widely discussed in phylogenetic studies but has been pointed out by several investigators (see, e.g., G. Chapman 1974, Wainwright 1988, and Pedersen 1991 for references). It is generally envisioned that macromolecular differences of the ECM of plants, fungi, and animals are likely to be responsible for some of the basic differences in development and evolution of the three multicellular kingdoms. For the Metazoa, developmental studies have established the extracellular matrix as an "instructive component of the organisation and maturation of complex tissues" (Soledad Fernandez *et al.* 1991, p. 46). Since evolution acts primarily on developmental programs of individuals within a population, one can expect the ECM to have played a central role during the early radiation of the Metazoa.

Structure of ECM: Fibrillar proteins and "ground substance"

Fibrous collagen "species" have a high tensile strength and at the same time allow – through complex three-dimensional arrangements – the formation of numerous kinds of flexible, fibrillar networks in the ECM of metazoans (for biomechanical principles, see Wainwright 1988; for biochemical reviews, see Runnegar 1985 and Sarras *et al.* 1991).

   Diversity of collagens is high. Some forms of collagen do not form long fibrils. Collagen IV, for example (see Hudson *et al.* 1989), exhibits a miniature network. Most likely, such networks of collagens (e.g., collagen IV is known from the Porifera [Kurz *et al.* 1991]) as well as collagen fibers (known from larval and adult sponges) were embedded already in the ECM of the metazoan ancestor. But the invention of the biosynthesis of collagen was one of the crucial events in the origin of multicellular animal life (e.g., Towe 1981b). So far collagen appears to be restricted to the Metazoa

(Runnegar 1985). The obligatory link of collagen biosynthesis with molecular oxygen (Towe 1981b) suggests that metazoans could not have evolved in a purely anoxic environment. Interfaces between anoxic and low-oxic conditions – as found today in marine sediments (e.g., Boaden 1975) or in stromatolites (Westphalen 1993; see also references in Walter, this volume) may therefore represent possible sites of the very first steps in the origin of Metazoa.

The original ECM most certainly contained – in addition to fibrillar components such as certain collagens – various classes of macromolecules, e.g., glycoproteins, glycosaminoglycans (GAGs), and proteoglycans. It is, course, difficult to ascertain the different types of biochemical categories of the ECM molecules (glycoproteins, GAGs, proteoglycans) as well as those functional categories of Edelman (CADS, CIDS, CJMS, SAMS) that might have made up the ground substance of the ECM or were present as linking molecules between cells and matrix in Precambrian metazoan ancestors. From the ubiquity of fibronectin (also found in sponges; see Pedersen 1991) it seems likely that this glycoprotein was present from the onset of metazoan evolution, whereas laminin apparently did not occur until the rise of the Eumetazoa (Coelenterata and Bilateria; see Pedersen 1991 for references). The assumption that fibronectin is really an early glycoprotein in metazoan evolution is supported further by embryological data of sea urchins and mammals, where it has been shown to be one of the first glycoproteins to occur in ontogeny as well (Spiegel *et al.* 1983).

While certain glycoproteins are well studied in vertebrates (see Gilbert 1991 and Pedersen 1991 for references), comparatively few studies deal with their invertebrate counterparts (e.g., for fibronectin, see comment by [Schlage 1988, p. 395]). At this time, the field of matrix biology is rapidly advancing (e.g., Edelman 1988; Pedersen 1991) and views about the significance of ECM-molecules will be changing accordingly.

Further differentiation of the ECM and the origin of epithelial tissue

During the early evolution of the Parazoa, cells first "learned" to deposit and to crawl within the extracellular matrix. This is presently still seen in nearly all extant Porifera with their epitheloid (incomplete characteristics of epithelial tissue, see below) pinacoderm and choanoderm on one hand and their complex connective tissue (the mesohyl) on the other hand (Bergquist 1978; Ruetzler 1990).

In the ancestors to the Eumetazoa (Coelenterata and Bilateria, also referred to as Histozoa) two histological advances have been made: (1) The extracellular matrix at the border of the cell colonies differentiated into two regions: the ECM on the apical side of the bordering cell layer, which later formed the cuticle (Rieger 1984), and the ECM at the base of the cell layer, which later became the basal lamina (for a general discussion of the basal matrix of epithelial organization of the Eumetazoa, see Fransen 1982 and Pedersen 1991). (2) Band shaped, cell–cell junctions (apical-junctional complexes) appeared, which began to improve the control of the exchange of interstitial tissue fluid within the cell colony with that of the surrounding environment via the paracellular pathway (Diamond 1977; Bradley & Purcell 1982). In addition, apical-junctional complexes mechanically stabilized the covering cell sheath (Green 1984).

Both processes led to the origin of true epithelial tissues, where molecular distinctions in apical and basolateral cell membrane (Gennis 1989; Rodriguez-Boulan &

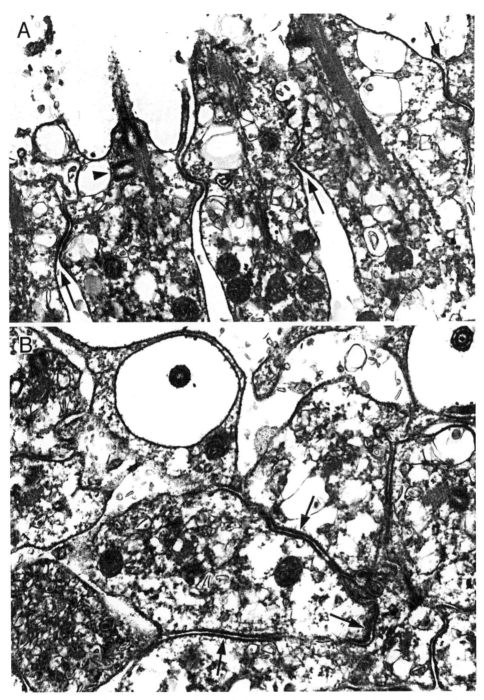

FIGURE 1 Parts of TEM section through a 5 h old larva of the demosponge *Dysidea etheria* from Bermuda. Magnification ×25,000. **A:** Section perpendicular to surface of larva. Observe the apical membrane specialization between flagellated cells (arrows) and the accessory centriol (arrowheads). **B:** Tangential section. Observe the band-shaped apical-junctional specialization (arrows). (Material courtesy Dr. Klaus Ruetzler, Smithsonian Institution, and Wilma Hanton, University of North Carolina, Chapel Hill.)

Powell 1992), together with the restriction of the paracellular pathway to the apical-junctional complex (see Fawcett 1981), enables the organism – varying with the different types of complexes – to control its homeostasis with greater precision.

As a matter of fact, Hertwig & Hertwig recognized as early as 1882 the fundamental histological concept of two structural and functional extremes in the configurations of tissues of Eumetazoa (Rieger 1986a): cells being submerged in the ECM (= connective tissue or apolar tissue), and cells forming sheets of true epithelial layers or polar tissue that separate an apical ECM (= cuticle) from a basal matrix.

Such a theoretical foundation of histology had been largely forgotten until the advent of electron microscopy and molecular biology (see Lemche & Tendall 1977; Hay 1981; Edelman 1988; Rieger 1984, 1986a).

In the placozoan *Trichoplax adhaerens*, apical-junctional complexes have been recently discovered, whereas a basal matrix has not been found (see below). This could be taken as evidence that apical-junctional complexes developed before the differentiation of true cuticles and basal matrices.

However, new findings in adult (references in Pedersen 1991) and larval (see Fig. 1) sponges suggest that apical-junctional complexes might have first appeared in the ancestral metazoan larval form, while a distinct basal matrix may have originated as a mechanical adaptation in the adult sponge-cell colony.

Curious extant "relics" between Parazoa and Eumetazoa – Placozoa and "Mesozoa"

The Placozoa (TRICHOPLAX, TREPTOPLAX)

We lack molecular data on the ECM in this group. Grell & Ruthmann (1991, p. 19) state that the interspace between "dorsal" and "ventral" epithelia contains, besides the so-called fiber cells, "a fluid, probably not very different from sea water."

In addition to the ability for binary fission, *Trichoplax* can form two sorts of asexual swarmers (see Thiemann & Ruthmann 1990, 1991 for references). The various modes of asexual reproduction are generally accepted as another very primitive trait of this extant placozoan, suggesting that asexual strategies at the level of the whole cell colony might have been under heavy selection pressure in Precambrian relatives. The nature of sexual reproduction in *Trichoplax* is not yet fully clarified, however (see references in Haszprunar *et al.* 1991).

Because of the following features, *Trichoplax* appears closer to the Eumetazoa than to the Parazoa:

- Apical-junctional complex (*zonula adherens*-like complex) well developed (Ruthmann *et al.* 1986). However, this chapter reports for the first time in sponges an apical-junction-like specialization in a larva of a Demospongia (Fig. 1).

- Striated ciliary rootlet present (Rieger 1976) – in two recent papers on larvae of calcareous sponges Amano & Hori (1992) and Gallissian & Vacelet (1992) report for the first time the occurrence of striated rootlet fibers in sponges. Thus far, ciliary rootlet fibers of sponges were known as microtubular structures.

- Position of accessory centriol (Rieger 1976; see also Fig. 1 herein). In the larva of calcareous sponges, accessory centriols are not mentioned.

- Feeding by engulfing (see references in Haszprunar *et al.* 1991).

Other characters, however, clearly demonstrate a taxonomic and organizational position of the Placozoa between the Parazoa and the Eumetazoa (Grell & Ruthmann 1991):

- No basal matrix or basal lamina.

- Lack of true neurons or muscle cells.

The "Mesozoa" (Rhombozoa and Orthonectida)

This "phylum" has also been known since the turn of this century (see Haszprunar *et al.* 1991, for references). It comprises about 150 species, all of them endosymbionts or parasites except for the enigmatic genus *Salinella*. The two groups included here may actually be separate phyla with uncertain relationships.

In the Rhombozoa, one of the very few TEM studies did reveal scarce remnants of ECM between outer cell layer and central cells.

Both Orthonectida and Rhombozoa have multiciliated cells with complex rootlet systems. This feature supports the old notion that the "Mesozoa" are simplified relatives of the flatworms. However, the complex rootlet system of the nemerto-dermatid and the acoel turbellarians (Rieger *et al.* 1991b) suggests that transitions from mono- to multiciliated cells might have occurred more than once, very early in the evolution of the Eumetazoa. Such a transition has occurred many times in the evolution of the Bilateria (Rieger 1976; Tyler 1981).

Preliminary freeze-fracture data on the ciliary necklace of rhombozoans suggest differences from all other Metazoa and might reopen the question of their origin from ciliates (Hyman 1940; Bardele 1983).

The peculiar contractile cells described in orthonectids (Kozloff 1971) lack further comparisons with myocytes of Eumetazoa.

The complexity of the life cycle of these organisms is generally attributed to their endosymbiontic or parasitic life style. However, it might as well be a primitive trait that proved well adapted for the evolution of parasitism.

Cnidaria and Ctenophora: "Relics" of the first eumetazoan radiation

We find here the first eumetazoan tissue organization: a digestive-cavity system (gastrovascular system) lined by an epithelium (gastrodermis), acellular or cellular connective tissue (mesogloea), and a covering epithelium at the outer border of the cell colony (epidermis).

The origin of the digestive system is traditionally viewed in light of Haeckel's Gastrea theory or Metschnikoff's Phagocytella theory. Recently, a new model has been developed (Graßhoff 1991 and literature therein) that derives the gastrovascular system from a system of canals reminiscent of that of adult Demospongia.

During this radiation cells obviously perfected their ability to live on the ECM of the central connective tissue, "learning" to use the basal matrix as a guide for their concerted movements.

Perhaps this novelty set the frame for the evolution of muscular foot processes in epithelio-muscle cells. On the other hand, fiber-type muscle cells occur within the mesogloea of Ctenophora. It remains, therefore, an open question whether myoepithelial cells indeed preceded fiber-type muscle cells in evolutionary sequence or whether the two types could have evolved side by side in small larvae (fiber-type) and larger adults (myoepithelial) in a life cycle of eumetazoans. The relationship of the myocyte-like cells in sponges (Bergquist 1978) and eumetazoan myocytes is also part of this puzzle (see Rieger & Lombardi 1987 for literature on the evolution of muscle cells in Bilateria).

With the ability to control the internal ionic environment by means of the covering of epithelial tissues, the stage for the evolution of the nervous system was opened (see variations in the structure of the apical-junctional complex among the Coelenterata in Harrison & Westfall 1991; see also recent reviews concerning the origin of the nervous system, e.g., Arhem 1990, Mackie 1990, and references in Anderson 1991 and Reuter & Gustafsson 1989).

The need for a coordination of the discharge of extrusomes used to capture planktic organisms larger by about one order of magnitude than in the case of the Parazoa (see tabulation of this character in Riedl 1966) may have been one stimulus in the early evolution of the nervous system. In any case, it is odd that cells specialized for prey capture and defense are known only in certain Protista (Hausmann 1978) and among the Coelenterata and some primitive Bilateria (e.g., the rhabdites of rhabditophoran "Turbellaria," saggitocysts, paracnids and rhabdoids of other "lower worms" – see Karling 1966, Sopott-Ehlers 1981, Smith *et al.* 1982, Mamkaev & Kostenko 1991, and Yamasu 1991; or the rhabdoids in the proboscis of nemertines – see Turbeville 1991). This pattern of occurrence suggests a switch in feeding strategies at the rise of the Eumetazoa.

The extant Bilateria: The majority of animal phyla

The origin of the triploblastic organization, with all three germ layers, is still an unresolved question.

Almost all hypotheses advanced to explain the origin of the Bilateria rely on microscopic organisms that use ciliary swimming or gliding to search for food (see Rieger *et al.* 1991a for review of these points). Turbeville & Ruppert (1983) show further that the theories can be divided into two groups: some derive bilaterians from larval forms of prebilaterian stages (e.g., planula, parenchymula, etc.), others derive bilaterians directly through modifications of adult coelenterate stages.

Central for the understanding of these theories is the proposed evolution of a secondary body cavity and of the *entomesoderm* (Rieger 1985). Schizocoely and enterocoely are either viewed as parallel processes (see Anderson 1982), or one of these two mechanisms of the formation of the secondary body cavity is seen as being derived from the other (see, e.g., Salvini-Plawen & Splechtna 1979; Salvini-Plawen 1982).

Equally basic (see, e.g., Valentine 1989; Ruppert 1992) to this question is the assumption about the original *organization of the body cavity* (acoelomate, pseudocoelomate, and coelomate). By far the most widely accepted version is the view that the acoelomate–pseudocoelomate organization preceded the coelomate condition within the Bilateria (see Rieger *et al.* 1991b, b for references). Least common were phylogenetic proposals to regard the pseudocoelomate condition as the original one (but see recent interest in this issue by Nielsen & Nørrevang 1985).

Since Goodrich (1895, 1945), the question of the evolution of the *bilaterian nephridial and osmoregulatory* system figured prominently in theories of the origin of the Bilateria (proto- vs. metanephridia; see recent summary in Bartolomaeus & Ax 1992). In forms below bilaterian organizational level, excretion occurs through general body surfaces. In coelenterates an elaboration of the gastrovascular system might facilitate that function; in the Parazoa, excretory–osmoregulatory processes might still be at the level of individual cells (e.g., contractile vacuoles in pinacoderm cells of freshwater sponges).

Recently Ruppert & Smith (1988; see also Smith & Ruppert 1988) provided evidence that both proto- and metanephridia could have been present in the stem species of the Bilateria. This is an alternative phylogenetic hypothesis to the general paradigm of the plesiomorphy of protonephridia. Bartolomaeus & Ax (1992), on the other hand, claim, from differences in ontogeny and structure of the ciliary funnel of the metanephridial canal system in *Phoronis* and in certain polychaetes, a lack of homology of the meta-nephridial system in these two phyla. In their opinion, this speaks against the proposal mentioned above. The objection is, however, less convincing considering the many known cases of germ-layer substitutions (e.g., Salvini-Plawen & Splechtna 1979).

Other unresolved questions are that of the origin of a center for coordination in the nervous system (the early *brain* [Gustafsson & Reuter 1990]) and that of the origin of the *two-way gut* within the Bilateria (Ax 1989; Willmer 1990; see also Jägersten 1972 and Salvini-Plawen 1978, 1980 for discussions of this important question).

Alternatives to the traditional concepts

Reviewing the older and recent literature on the origin of the Bilateria, Rieger *et al.* (1991a) found an astonishing uniformity of certain assumptions about the bilaterian stem species, namely:

1 microscopic size;

2 direct development;

3 active foraging in the search for food;

4 solitary organism.

Almost all investigators (an exception, e.g., is Jägersten 1972) apparently assume the plesiomorphy of the first three points also for the Metazoa (see Willmer 1990 for review). This is especially true for small body size (see recent review by Conway Morris [1993b, p. 219]).

There is, of course, no doubt that original metazoan cell colonies began as small, multicellular entities with division of labor of cells in a somatic and a reproductive

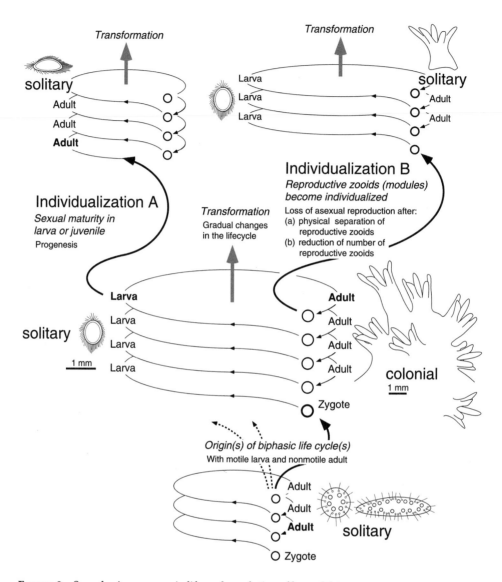

Figure 2   Some basic processes in life-cycle evolution of lower Metazoa.

domain (Fig. 2). But the claim being made in this chapter is that of a much earlier rise of a life cycle with a small (millimeter-range), mobile dispersive stage and with a larger (centimeter-range) reproductive stage. In fact, I propose that such a life cycle was either seen in the common ancestor of all Metazoa or that parallel evolution of that trait occurred early during the first radiation (e.g., in the Parazoa and in the Eumetazoa).

Asexual strategies are rarely discussed in reconstructions of the metazoan stem species (see Rieger 1986b and Smith *et al.* 1986 for the stem species of the Platyhelminthes), although it has been generally assumed that high regenerative power is a plesiomorphous trait within the lower Metazoa (see Steinböck 1966, 1967).

Recently, Jackson *et al.* (1985) and Buss (1987) did provide a new framework for looking at metazoan evolution that explicitly included the consideration of clonal organisms at the origin of the Metazoa.

Independently, ultrastructural research suggests that the traditional assumptions cited above can be questioned (Rieger *et al.* 1991a and references therein), primarily on the basis of the pattern of occurrence of monociliated epithelia in the Eumetazoa (Rieger 1976; Tyler 1981) and the fine structure and differentiation of proto- and metanephridia (Ruppert & Smith 1988, cited in Rieger *et al.* 1991a; Smith & Ruppert 1988).

Considering the size of the oldest known metazoan-like fossils (the Ediacara Fauna; see chapters by Runnegar, Sun, and Fedonkin, and also – for a nonmetazoan interpretation – Seilacher, this volume; see also Conway Morris 1993b) as well as taking into account the body size and life style of the adult Parazoa (histologically very primitive extant Metazoa), it seems odd that the above-listed characteristics should have dominated theories on early metazoan evolution to such an extent. In fact, fossil evidence on the chronology of certain hexactinellid spicules supports the very early occurrence of sizable microscleres, indicative of body sizes in the centimeter range. For the hexactinellid Amphidiscophora, Mostler (1986) has shown that microscleres almost 1 cm in length occurred in the upper Cambrian. In the same paper Mostler (1986, p. 330) suggests that microscleres might have evolved from megascleres, which reiterates the importance of larger (centimeter-range) body size during the early radiation of the sponges.

The data base referred to in the previous paragraph actually suggests rather larger size (centimeter-range), at least in two dimensions, a lack of active locomotion in the search for food, and a high frequency of clonal organization as important characters during early metazoan radiation.

Originally such alternative views were developed for the origin of the Bilateria (Rieger 1986a). For the stem species of this group, the following characteristics have been suggested: a biphasic life cycle with a millimeter-sized, acoelomate or pseudocoelomate larva or juvenile, and a macroscopic (centimeter-range) sedentary or passively floating, filter-feeding and clonal adult. Extant groups similar to this organization are the Bryozoa and the Entoprocta among the nondeuterostomes and the Pterobranchia among the deuterostomes.

In principle, this concept can be expanded to the stem species of the Eumetazoa or of the Metazoa as a whole. Examples for the Eumetazoa would be the anthozoan cnidarians; for the Metazoa, the Porifera.

Another deviation from the traditional view of the bilaterian stem species was the assumption that the acoelomate–pseudocoelomate and the coelomate body-cavity organizations, respectively, did not occur as a phylogenetic sequence (e.g., Bergström 1991). Rather, they developed within the biphasic life cycle as adaptations for a small, actively moving dispersive larva or juvenile, and for a large adult colony with clonal reproduction, respectively.

Clark's (1964) model for the origin of the fluid-filled secondary body cavity (coelom) states that the primary function of this structure was for locomotion and burrowing. If we accept the biphasic life cycle for the bilaterian stem species proposed herein, a different functional origin of the coelom may be suggested: The retractability and eversability of single zooids (polypoids), as seen in bryozoan colonies, could be phylogenetically derived from a similar behavior of zooids in anthozoan colonies retracting and eversing from the gastrovascular cavity. A subsequent division into two functionally different regions within the original gastrovascular cavity (gut and coelom) would be a viable alternative to Clark's model. The origin of the meso- and metacoel would also get a functional explanation other than that for locomotion and burrowing (see Clark 1964, 1981; Mettam 1985). The notion that early coelomates were nonburrowing organisms is consistent with paleontological data on trace fossils (see Valentine 1989, p. 2272; Fedonkin, this volume).

Based on the assumption of a biphasic life cycle for the stem species of the Eumetazoa, one might argue that fiber-type muscle cells (terminology, see Rieger & Lombardi 1987) may have been perfected within the ECM of certain larvae, while the musculature in the adult evolved within the epithelial layers of the gastrodermis and epidermis as epithelio-muscle cells. As in the case of proto- and metanephridia, the same basic genetic information could have been used to evolve fiber-type muscle cells in the larva and myoepithelia in the adult, because of differences in selective forces acting on a millimeter-sized larva (optimization for dispersal) and on a centimeter-sized clonal adult (optimization for asexual and sexual reproductive output).

A consequence of this assumption would be that the fiber-type muscle cell in the Phylum Ctenophora may be derived from larval or juvenile stages of a common metazoan ancestor (via paedomorphosis, *sensu* Gould 1977). Alternatively, muscle cells in the Ctenophora could represent a parallel evolution to the other Eumetazoa, a hypothesis appearing rather unlikely but testable with molecular methods (Rieger 1986a).

Conclusions

The hypothesis of macroscopic adult size (at least in two dimensions), clonal organization, filter feeding on bacteria, and a sedentary or passively drifting existence in the last common ancestor of the Metazoa suggests two evolutionary processes at the organismal level (Fig. 2) and three consecutive levels of adaptive radiation within the "lower" Metazoa (Fig. 3):

Process 1

After the origin ot the biphasic life cycle, selective forces altered the larval and the adult organisms gradually. No formidable change of lifestyle or habitat of the organism would be expected. Following Gould (see Mayr 1988, p. 418), this process could be called transformation. Examples: evolution within certain groups of the Anthozoa and Bryozoa or within the Pterobranchia.

Process 2

As evolution proceeded, clonal organization gradually gave way to solitary individuals. This can be achieved by truncated progenesis (*sensu* Gould 1977) in the juvenile (or larva) – or through the reduction and eventual loss of asexual reproduction. Both are

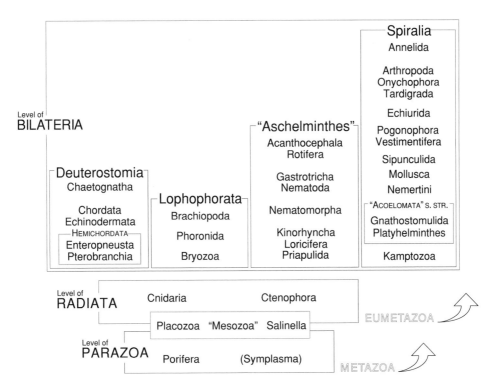

FIGURE 3   Basic levels of organization in extant Metazoa.

processes of individualization (Buss 1987). Examples for the former are: the proposed origin of vertebrates from the tadpole larva of ancestral tunicates (see Willmer 1990 for references). Examples illustrating various aspects of the latter processes may be: certain solitary sea anemones among the Actiniaria or interstitial bryozoans (e.g., Monobryo-zoontidea; see Cook 1988) among their epifaunal relatives.

Phase 1

The origin of metazoan organization (multicellular body with cells embedded within a common ECM containing collagens). Cells "learn" to alter and manipulate their own ECM (especially the glycoproteins, GAGs and proteoglycans) for cell–cell or cell–matrix adhesion, as well as for chemical or physical guidance of their locomotion during development (e.g., Wolpert 1978, Harris *et al.* 1981; Harris *et al.* 1984). The cell colony uses fibrillar collagen as skeletal material, onto which inorganic compounds can be added. Sexual and asexual reproductive-cell lines evolved. Uptake of food was in the form of dissolved organic matter, bacteria, or small protists through passively and/or actively produced water currents (Vogel 1988; Weissenfels 1992). Active movement developed first in small benthic cell colonies (Placozoa?) or in the larval stage of the biphasic life cycle.

Extant relics from that level of organization: Parazoa, Placozoa, and perhaps "Mesozoa."

Phase 2

The origin of the eumetazoan organization. The origin of true epithelial tissue may have two roots: partly within the larva (apical-junctional complex, see above) and partly within the adult of the life cycle (basal matrix known in one group of Porifera; see Pedersen 1991). In any event, the epithelial lining of the central gastrovascular system and the lining of the body wall enables the organisms to control their interior ionic environment much more efficiently and thus sets the stage for the evolution of the nervous system. Epithelial cell-sheaths "learn" to move in concert on the ECM and to use special cell lines for capturing larger prey, predominantly still protists. In that context nerve cells and muscle cells are presumed to have evolved.

Extant relics: Cnidaria and Ctenophora.

Phase 3

The origin of the bilaterian organization. The original gastrovascular system became separated into the actual digestive system and the secondary body cavity (coelom). The latter interacts in various ways early in the evolution of this level with the primary body cavity (see Ruppert 1992) of the larva, thus giving rise to the histological organization of the coelomate, pseudocoelomate, and acoelomate body plans as biomechanical and physiological adaptations within the ancestral, biphasic life cycle. The blood vascular system and the coelomic vascular system become fluid-transport systems separate from the gut (Ruppert & Carle 1983). New organs of osmoregulation and excretion originate as proto- and metanephridia; the former (protonephridia) as adaptations in millimeter-sized larvae or juveniles, the latter (metanephridia) in larger coelomate adults . The larval brain develops into the "early brain" of the adult, as the center for coordination in animals with active locomotion (see Gustafsson & Reuter 1990).

Extant relics, in which clonal organization in the adult may still be plesiomorphous: Kamptozoa, Bryozoa, and Pterobranchia.

As pointed out above, Turbeville & Ruppert (1983) did show that traditional assumptions for the metazoan stem species fall generally in two categories: those that fit with the larval and those that fit with the adult organization in the life cycle of extant "lower" Metazoa. The alternatives advanced here do not represent yet another opposing concept; with Jägersten (1972), one could assume two selective peaks in the life cycle of the metazoan stem species, one for dispersal and for the colonization of new habitats, the other one to increase reproductive output. A synthesis of the opposing concepts could thus be accomplished.

Acknowledgments. – Research supported by FWF grants P7816 Bio and P9138 BIO. For help with the computer illustrations I thank Mag. D. Reiter, University of Innsbruck, and Drs. S. Tyler and M. Shick, University of Maine. For reading all or parts of the manuscipt I am indebted to Dr. G.E. Rieger, Dr. Gerhard Haszprunar, Dr. Thomas Holstein, Professor Helmuth Mostler, and Professor Luitfried v. Salvini-Plawen.

Developmental mechanisms in the evolution of animal form: Origins and evolvability of body plans

Rudolf A. Raff

Institute for Molecular and Cellular Biology and Department of Biology, Indiana University, Bloomington, Indiana 47405, USA

Animals in extant phyla exhibit body plans that have been stable in evolution for a half-billion years. The appearance of these body plans during development provides the commonality in ontogeny among the animals of a phylum. The common stage has been called the phylotypic stage. From the phylotypic stage there then arises the diversity of adults within the phylum. Although the phylotypic stage is evolutionarily conserved, early development can be radically different, even between congeneric species. Thus, ontogenies have an hourglass shape, with evolutionarily diverse early and late stages, but a conserved middle, phylotypic, stage. Surprisingly, the processes that directly produce the phylotypic stage can also differ among related forms. This chapter discusses these observations with respect to what they reveal about the structure of ontogeny and the paradox they pose for explanations of evolutionary stability.

When you're young, all evolution lies before you, every road is open to you, and at the same time you can enjoy the fact of being there on the rock, flat mollusk-pulp, damp and happy. If you compare yourself with the limitations that come afterwards, if you think of how having one form excludes other forms, of the monotonous routine where you finally feel trapped, well, I don't mind saying life was beautiful in those days.

Italo Calvino, Cosmicomics

Estimates of the diversity of living animals run as high as 3–30 million species. Yet this amazing diversity is contained within only about 35 phyla, groupings that reflect the disparity between bauplans (body plans). If we look at the Cambrian world through the window of the Burgess Shale and other well-preserved soft-bodied faunas, it appears that the early radiation of the Metazoa produced a much higher disparity, although much lower diversity, than present today (Whittington 1985; Gould 1989, 1991). The apparent lack of origination of new phyla since the early Paleozoic despite dramatic modifications within bauplans raises three critical questions from the point of view of developmental biology: resolving the origins of the bauplans; understanding

Bengtson, S. (ed.) 1994: *Early Life on Earth. Nobel Symposium No. 84.* Columbia U.P., New York

the developmental bases of modifications of the bauplans during 500 Ma of Phanerozoic history; and determining whether an existing bauplan can be transformed into another. Are bauplans of existing phyla potentially evolvable?

Despite a century of investigation, the relationships between phyla remain unsettled, but they must be resolved to approach our questions. First, the phyla stand distinct morphologically. Some anatomical features such as bilateral symmetry, possession of a coelom, and possession of segmentation have provided important characters for inferring phylogeny. However, not enough synapomorphic anatomical features are available, and our ability to resolve homoplasies is poor. Second, the fossil record presents us with first appearances of phyla, but so far with no intermediates or ancestors. Third, informative data have come from embryological features shared between phyla. However, phylogenies independent of developmental features are needed both for the independent evaluation of the origins and evolution of developmental features and to establish polarities in evolution of these features. Genes are the major remaining source of information.

Most sequence data for phylum-level relationships have come from 18S rDNA, which encodes the major structural RNA of the ribosomal small subunit. The gene is conserved, and about 2,000 bases are available from each species. Our first study utilized partial sequences equivalent to about half of the gene; evaluations of these data showed that coelomates could be distinguished from noncoelomates (Field *et al.* 1988). Major coelomate groups corresponding to arthropods, coelomate protostome phyla, chordates and echinoderms were distinguished (Field *et al.* 1988; Lake 1990). Unfortunately, branch patterns were not robust. With further analysis, a better resolved tree is apparent (Turbeville *et al.* 1992) (Fig. 1). Preliminary inferences drawn from full-length 18S rDNA sequences support this tree (Turbeville *et al.*, unpublished). The tree approximates traditional views (Hyman 1951; Brusca & Brusca 1990). There is poor agreement with partial 28S rDNA sequence trees (Christen *et al.* 1991b; Christen, this volume), but those inconsistencies should be rationalized with more complete 28S rDNA sequences. We also do not yet have a clear picture of the relationships of the "minor" phyla, which make up about half of metazoan diversity. That situation will improve now that polymerase chain reaction (PCR) techniques make the genes of very small animals accessible.

Our 18S rDNA tree indicates three important things. First, metazoan relationships indicated by embryonic and larval features are surprisingly consistent with the gene tree. Second, there was a broad and relatively rapid radiation of coelomate phyla. Third, there was a subsequent, although closely ensuing, radiation of the protostome coelomate phyla. It would be tempting to correlate the coelomate branching pattern with the early Cambrian radiation events (Bengtson 1990b, 1991; Conway Morris & Peel 1990; Ramsköld & Hou 1991; Conway Morris, this volume) and the lowest branch of the tree with the cnidarian-grade Ediacaran radiation (McMenamin & McMenamin 1990; chapters by Runnegar, Sun, and Fedonkin, this volume; for an alternative view, see Seilacher, this volume), but there is not an adequate molecular clock for 18S rDNA evolution.

In order to begin to define the course of regulatory evolution in the metazoan radiation, the results of comparative studies of morphogenetic regulatory genes can be combined with the phylogenetic tree derived from 18S rDNA. Polarity of evolution of

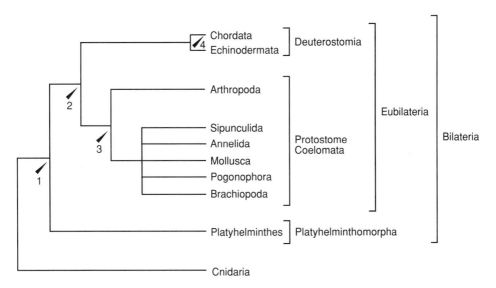

FIGURE 1 Consensus phylogenetic tree for metazoan phyla inferred from 18S rDNA. This tree is derived from analysis of partial 18S rRNA sequences (Field *et al.* 1988; Turbeville *et al.* 1992) and complete 18S rDNA sequences (unpublished data). Both maximum parsimony and distance methods were used in the analyses. The tree shown here is a consensus of the analyses done by both methods. The deepest animal group, the Cnidaria (jellyfish and their relatives), serve as the outgroup for all bilaterally symmetric metazoans. The Platyhelminthes (flatworms) are the sister group to all coelomate metazoans (Eubilateria), which include the major animal phyla characterized by possession of a true coelomic cavity. Protostome and deuterostome superphyla are resolved. Taxa in the protostome coelomates, which are the sister group of the arthropods, are unresolved and shown as a polychotomy. Branch-length differences are not indicated. Arrows indicate some major points of gene-regulatory evolution that can be mapped on the phylogeny. Arrow 1, origin of anterior–posterior axis and central nervous system; arrow 2, shared pattern of expression of homeobox-containing genes in central-nervous-system development between deuterostomes and protostomes; arrow 3, *engrailed* gene shared by coelomate protostomes and arthropods in designation of segmental polarity; arrow 4, additions of anterior head and neural crest with associated genic controls to the chordate bauplan.

morphological features and underlying genetic controls can be assessed. For example, since segmented and nonsegmented phyla are included among the coelomate protostomes, the question arises whether segmentation arose independently in the protostome clade or was derived from a common ancestor shared with the segmented arthropods. Initial results (Patel *et al.* 1989) indicated that the *engrailed* gene involved in arthropod segmentation did not play an equivalent role in annelids. However, studies of the leech shows that *engrailed* does play a role in annelid segmentation similar to that in arthropods (Wedeen & Weisblat 1991). A segmented ancestor utilizing similar genetic mechanisms is implied. Some major morphogenetic regulatory genes are widely shared among coelomates and possibly in all Bilateria. For example, the homeotic genes of arthropods have homologues arranged similarly in mammalian chromosomes that play a role in central-nervous-system determination in both arthropods and vertebrates (Holland 1990).

Definition of phylogenetic relationships among developmental regulatory genes is crucial to determine times and phylogenetic branch points at which regulatory gene systems evolve. The differences in the states of these genes between lineages that share them will allow an assessment of the primitive state of the gene system, as has been inferred for the homeotic gene cluster shared by insects, vertebrates, and nematodes (Holland 1990; Kenyon & Wang 1991). Finally, we can attempt to link regulatory genes to body plans. The presence of shared regulatory genes does not guarantee that they are playing the same role; that requires experimental demonstration. But use of a different genetic program would suggest that two similar body plans are not homologous. That is why the suggestion by Patel *et al.* (1989) that annelids were not using *engrailed* as arthropods do was so striking, and the resolution by Wedeen & Weisblat (1991) so important.

Evolutionary stability of early development

A traditional feature of metazoan phylogeny since Haeckel has been the use of evolutionary conservation of larval forms. For example, mollusks and annelids are distinct in the fossil record from the Early–Middle Cambrian. Both retain similar cleavage, cell determination, and larval forms (Kumé & Dan 1988). There is no doubt that these developmental features are retained for more than 540 Ma from the ancestor of these two phyla. Although most of what we know about evolution of larval features has to be inferred in this way, some surprisingly good fossil evidence also supports long-term conservation. As shown in Fig. 2, fossil larvae from the Upper Cambrian (500 Ma) bear a striking similarity to nauplii of living crustaceans (Müller & Walossek

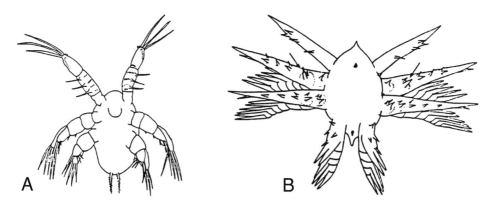

A   B

FIGURE 2 Morphological conservation between Cambrian and Recent nauplii. **A:** First nauplius of the living crustacean *Eurytemora affinis*. **B:** Earliest metanauplius-like larval stage of the Cambrian stem-lineage crustacean *Martinssonia elongata*. The essential features of nauplius structure, an unsegmented body and initially a few pairs of appendages, are conserved. The Cambrian nauplius has four rather than three pairs of appendages and more primitive mouth parts (Walossek & Müller 1990). Additional segments and appendages are added in subsequent growth of both living and Cambrian forms. (Both from Müller & Walossek 1986.)

1986). Well-preserved skeletal rods of sea-urchin plutei from the Upper Jurassic of France provide direct evidence of the pluteus existing 160 Ma ago (Deflandre-Rigaud 1946). Fossil mayfly larvae are known from the Lower Permian (270 Ma) and closely resemble modern mayfly larvae but have a more primitive pattern of attachment of their wings (Kukalová 1968). Finally, fossil pipid-frog tadpoles are known from the Early Cretaceous (120 Ma) (Estes *et al.* 1978). These larvae have persisted with little change from early in the histories of their respective groups.

Living sea urchins represent a relatively well-defined radiation with an excellent fossil record, living representatives of most clades in the radiation, and embryological data from all major clades (Fig. 3). Thus, evolution of sea-urchin pluteus features

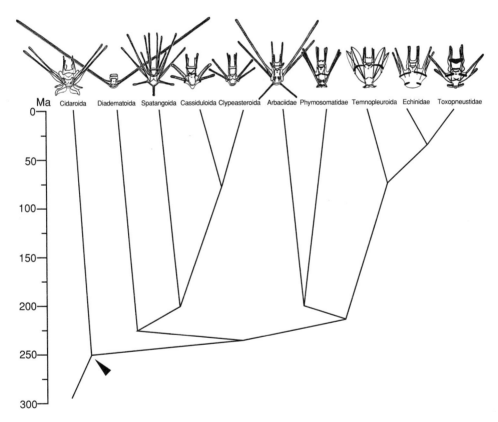

FIGURE 3   The post-Paleozoic radiation of sea-urchin orders and morphological variation among sea-urchin plutei larvae. The crown-group echinoids originated with the survivors of the Permian extinction of nearly all Paleozoic sea-urchin groups. The cidaroids are the sister group of the remaining euechinoids. Branching pattern and times are from cladistic analyses of adult features (Smith 1984, 1988; see also Wray 1992). Mature plutei characteristic of each order are indicated. The range of variation is primarily limited to arm number, details of skeletal structure, and position of ciliary bands. No other style of feeding larva is known among the echinoids. The extensive shared morphological features of these larvae indicates a single origin for them and a highly conserved larval bauplan. Development of the echinus rudiment that results in the juvenile adult is similar in all. (Modified from Wray 1992.)

provides an exceptional opportunity to evaluate the extent of evolution within a larval bauplan (Wray 1992). The echinopluteus larva retains many symplesiomorphies with larvae of other deuterostomes and other echinoderms but also possesses several synapomorphies. Although there are modifications in arm numbers, skeletal morphology, and other features, the basic pluteus architecture is remarkably conserved in all lineages (Fig. 3). The pluteus thus represents a larval bauplan conserved for 250 Ma, which is completely distinct from the adult bauplan generated in the adult rudiment and released at metamorphosis.

It has been suggested that evolutionary conservation of early development is expected on mechanistic grounds, because all subsequent steps of development depend on the correct completion of the initial processes (Arthur 1988). The observation of evolutionary stability in larvae and the theoretical view that development is hierarchical and dependent on early stages suggest strongly that early development must be refractory to substantial evolutionary change. This view, although atttractive, turns out to be wrong, and its wrongness is reflected in our ideas of bauplan stability.

Radical evolutionary changes in early development

Ideas on the evolvability of early development founder upon empirical observation. Despite the prevalence of conserved larval forms, evolutionary changes have occured among closely related species in several groups. In some cases substantial modifications are achieved in a mechanistically conservative manner. An excellent example is provided by *Unio*, a freshwater clam. *Unio* glochidium larvae are modified such that their valves are greatly enlarged and bear powerful teeth. A set of sensory hairs triggers the valves to snap shut upon disturbance, hopefully by a fish, to which the larva must attach to develop until metamorphosis. It was shown in a pioneering study by F.R. Lillie (1895, 1898) that several features of embryonic development are strikingly modified in relative rates of cell division and sizes of embryonic-cell precursors to larval structures (Raff & Kaufman 1983). However, despite the modifications that result in the glochidium larva, the conservative pattern of spiralian cleavage and cell fates is retained.

Members of many other taxa have abandoned larval development for direct development of the adult without going through the ancestral feeding larval stage. The best studied examples are ascidians (Jeffery & Swalla 1992), frogs (Raff & Kaufman 1983; del Pino & Elinson 1983; Elinson 1987; del Pino 1989), and sea urchins (Wray & Raff 1990; Raff 1992). Very significant changes in early development have occurred among closely related species in these groups. Direct-developing ascidians have lost the notochord and somites of the larval tail. Direct-developing frogs have greatly modified cleavage, gastrulation, and relative timing of developmental events. Direct-developing sea urchins (Fig. 4) have remodeled development extensively, with differences in genome size, egg size, sperm-head shape, cleavage patterns, cell-fate maps, blastulation, gastrulation, and heterochronies in aspects of larval development. Larval bauplans that were stable for hundreds of millions of years were readily abandoned in short spans of time (10 Ma or less [Smith *et al.* 1990; McMillan *et al.* 1992]), presumably because ecological demands for a different mode of development provided sufficient selective pressures.

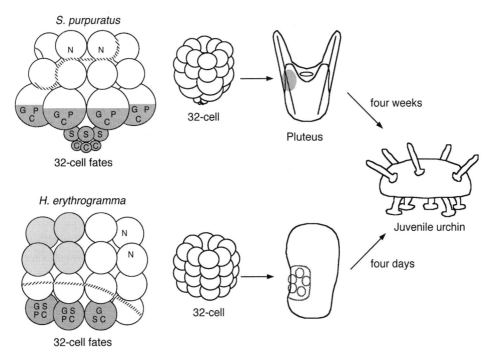

FIGURE 4 Reorganization of early development in a direct-developing sea urchin. The fate maps of the 32-cell stages of the indirect-developing sea urchin *Strongylocentrotus purpuratus* (top left) and the direct-developing *Heliocidaris erythrogramma* (lower left) are shown from their left sides. Animal poles are oriented up. The vegetal fates indicated by dark stipple are C, coelom; G, gut; P, pigment cells; S, skeletogenic mesenchyme cells. In the animal tier, N indicates serotonergic neuron origins, light stippling indicates vestibule, and the circle on the *S. purpuratus* embryo the larval mouth, which is absent in *H. erythrogramma*. The ciliary band, indicated by a heavy stippled line, has shifted from an animal to a vegetal fate in the evolution of the direct developer. Other substantial changes in symmetry, timing, and allocation of cells have also occurred. Details are discussed by Wray & Raff (1990). Major features of the ontogenies for each species are diagrammed to the right of their respective fate maps. Cleavage, larval form, and convergence on the juvenile sea urchin are shown. The pattern exhibited by *S. purpuratus* is primitive, and *H. erythrogramma* features are derived. A detailed discussion is presented in Raff 1992.

Could new phyla have arisen after the Cambrian?

The now-conventional wisdom about the origins of metazoan bauplans is that they had originated by the Cambrian and that no new phyla have originated since, even following mass extinctions such as the terminal Permian event (Gould 1989, 1991; Jacobs 1990). Three kinds of explanations have been advanced: (1) ecologically, new players could not arise because surviving representatives of existing phyla effectively fill all vacant niches (Erwin *et al.* 1987); (2) probabilistically, long jumps to a better adaptive state decrease more rapidly than short jumps on an adaptive landscape (Kauffman 1989); and (3) developmentally, bauplans are too integrated to transform to any other (Jacobs 1990; Gould 1991). This suggests that developmental controls were relatively loose in the Cambrian, allowing for a great deal of experimentation. Then things settled down, and as Gould (1991) puts it, "the key to disparity is not Cambrian lability, but later and active stabilization." Perhaps, but there are problems to consider.

First, half of living phyla (i.e., mesozoans, ctenophores, platyhelminths, rotifers, gnathostomulids, gastrotrichs, acanthocephalans, tardigrades, kinorhynchs, loriciferans, nematodes, nematomorphs, urochordates, phoronids, chaetognaths, echiurids, pogonophorans, sipunculids, entoprocts) either lack a significant or reliable fossil record or are first recorded from times long past the Cambrian (Erwin *et al.* 1987; J. Bergström, personal communication, 1988). These groups have no mineralized hard parts and so have a poor potential for fossilization. They might have arisen in the Cambrian radiation, but no current evidence excludes a later origin. Gene sequences may resolve this issue. Second, we have no direct knowledge of how precisely or loosely organized Cambrian developmental/genetic systems were. Increased constraint with evolution is only an inference from the observed morphological stability of post-Cambrian groups. But evolutionary stability of bauplans does not demonstrate nonevolvability. There are two different logical grounds for not translating an observation of stable body plan into statements about ability to change. Bauplans are distinct, so a transformation to a different one would not be traceable by morphology. Second, although larval body plans are generally conserved, as shown above, they can change radically in a short time. Bauplans may not be so stable in adherence to developmental rules as implicitly assumed.

Bauplans and how they develop

There are three concepts entangled in the word *bauplan* as currently used. The first is the concept of the adult body plan based on the classical idea of homology put forward by Owen and later given an evolutionary context (Desmond 1982). Homologous elements, recognized by spatial and compositional criteria (Riedl 1978), represent similarity as a result of shared ancestry. This concept of a bauplan does not depend on any observable commonality of developmental pattern as long as the same final bauplan results (deBeer 1971; Roth 1988)

Second, many groups develop via a feeding larval stage and then metamorphose to the adult. The larval form can be dramatically and decidedly distinct from the adult form. Since such larval body forms are often highly conserved through long periods of evolutionary time, they can be considered to represent distinct larval bauplans, as real as the adult bauplans. However, they are products of constructional rules distinct from those of the adult bauplan.

A last component of the bauplan concept relates to developmental execution of the adult bauplan. This developmental bauplan concept is the basis of von Baer's Laws. It reflects the processes of ontogeny that produce the "phylotypic" stage that foreshadows the features elaborated in the adult body. The phylotypic stage occurs after organogenesis has begun and is shared by all members of a phylum before the divergence of their later stages. Adult structures derived from further development of the phylotypic stage are indeed homologues in having historical continuity in evolution. However, the phylotypic stage, as I show below, has the troublesome characteristic of being the most evolutionarily conserved stage of development, but of being attainable through nonconserved developmental processes.

We are faced with a paradox. Bauplans are clearly stable over long evolutionary spans. If they were not, we would not recognize persistent higher taxonomic categories such as phyla and classes. Yet basic elements of bauplans are attained by different developmental pathways. They have been extensively modified by additions and, in the evolution of direct-developing species, losses and remodeling of larval bauplans. Adult bauplans have also been remodeled. Since bauplans can be attained by different pathways, why should they be so stable and produce phylotypic stages that apparently channel development through critical morphological elements? A few examples illustrate the point.

Because the vertebrate bauplan is so well known, it is worth seeing the wide variation among different vertebrates in construction of bauplan elements. The vertebrate phylotypic stage is the pharyngula. In the famous diagrams illustrating von Baer's Laws, this is the stage with gill pouches, limb buds, and a tail that all vertebrates from fish to philosopher pass through before they diverge toward their adult forms. A notochord, a dorsal nerve tube, and paired body-muscle bands (somites) are all present in the pharyngula. Although all vertebrates pass through a very similar pharyngula stage, development, including egg size and "yolkiness," cleavage mode, gastrulation, and cellular movements, differs dramatically between classes (Elinson 1987). The vertebrate developmental bauplan arose by accretion to earlier elements – the dorsal neural tube, notochord, and somites – present in the basic chordate bauplan seen in most rudimentary form in ascidians and more fully developed in cephalochordates (*Amphioxus*). As pointed out by Gans (1989), critical vertebrate developmental bauplan apomorphies include the neural-crest cells and neurogenic placodes, such as the optic placode that produces the eye. The neural-crest cells contribute to an amazing array of neuronal, glandular, and pigment cells and to major skeletal and connective tissues of the head.

Structural additions have been accompanied by genetic additions. The branchial region of vertebrates is also present in cephalochordates, where it lies at the anterior end. The anterior parts of the pattern of expression of the homeotic genes are expressed in this region (Hunt *et al.* 1991b), and have apparently been coopted into a new role in patterning the neural crest of vertebrates. One homeobox gene (*Hox 7.1*) expressed in neural crest, limb buds, and visceral arches (Hill *et al.* 1989; Benoit *et al.* 1989) is present as a single gene in ascidians but as three genes in vertebrates (P. Holland 1991). The anterior of the vertebrate head is a new addition (Gans 1989; Hunt *et al.* 1991b). The new anterior head has required evolution of new genetic patterning mechanisms not based on the homeotic genes (Hunt *et al.* 1991a). Both cooption and integration of regulatory genes were required. A different case of gene addition in body-plan elaboration has been the duplications and divergences of muscle-actin genes (Vandekerckhove & Weber 1984). The single muscle-actin gene of ascidians and cephalochordates was duplicated in bony fish to two genes and in amniotes to four each with distinct tissue-specific expression.

Elements of the vertebrate developmental bauplan have also been subjected to extensive modification. Thus, although salamanders and frogs gastrulate similarly, there are major differences. Mesoderm precursors lie beneath the surface in frogs, but in the salamanders the precursors lie on the surface in the early gastrula (Keller 1986; Hanken 1986). The somites, the segmental muscle bundles of the body wall, are present in all chordates, but the formation of somites from mesoderm is different between salamanders and frogs and even among frogs (Radice *et al.* 1989; Malacinski *et al.* 1989). Analogous differences extend to other basic elements, such as the notochord, which arises from a cell sheet in amphibians but from aggregation of mesodermal cells in the chick (Gilbert 1991, pp. 115–246). Other "difficult" modifications to elements of the vertebrate bauplan have occurred in tetrapods, including the formation of the unique turtle trunk, with its superficial ribcage that lies dorsal to the limb girdles (Burke 1989).

Extensive modifications in developmental pathways that build a bauplan are not restricted to vertebrates. Among insects, for example, the common segmental body plan is produced in two different ways (Anderson 1973). The more primitive short-germ-band insects produce abdominal segments progressively from a terminal growth zone, whereas the more advanced long-germ-band insects produce all segments simultaneously. Despite these differences, the short-germ-band insect *Tribolium* carries similar homeotic genes to those of the long-germ-band *Drosophila* and exhibits similar abdominal homeotic gene mutations (Beeman 1987; Stuart *et al.* 1991, 1993).

Stability of phylotypic stages

What should we make of arguments (Gould 1989, 1991; Jacobs 1990) that post-Cambrian arthropods have retained a set of tagmosis patterns that is much more restricted than those of Cambrian arthropods because the surviving classes have stabilized the developmental/genetic mechanisms underlying their segmentation and tagmosis patterns and so lack the developmental flexibility of the Cambrian arthropods? There are three possibilities: (1) The argument is true. (2) The argument is only partially true, because Cambrian developmental/genetic systems were just as tightly constrained as those in living arthropods, but only a few groups survived, resulting in the appearance of a deterministic stability. (3) The argument is false, and living classes could change tagmosis patterns in suitable ecological situations. All three hypotheses are difficult to evaluate, but a consideration of processes in development of the phylotypic stage suggests elements that should allow evolutionary change as well as those that should restrict it.

First, evolutionary changes are facilitated by the duplication and cooption of control genes to carry out new functions that are similar to but distinct from the ancestral gene function (see also Miklos & Campbell, this volume). An example of such a gene family is the steroid-receptor superfamily, which includes many different receptors that play diverse and dramatic roles in morphogenesis and whose misactivation can cause homeotic transformations (Mohanty-Hejmadi *et al.* 1992). Second, regulatory genes are conserved and recombined into novel control pathways containing such elements as receptor-ligand, second-messenger, transcription-factor, and cis-acting domains of responder genes. Finally, many parts of development are modular and can thus be dissociated readily to produce heterochronies and other changes. Conversely, the stabilization of phylotypic stages may well arise because the modularity properties of progressive stages in development are quite different (Raff *et al.* 1991). Early embryos contain few modules and are built from a few informational elements organized along axes established in the egg or by fertilization or other external signals (Fig. 4). Late stages contain many modules, complex within themselves but more or less independent from each other; e.g., developmental programs required to give rise to limb bud, lung rudiment, etc. It is during the mid part of development that modules interact with each other in complex and pervasive ways. Thus, for example, heart mesoderm helps induce the eye (Jacobson 1966). The developmental processes that predominate during these stages will profoundly affect the kinds of evolutionary changes that occur in different

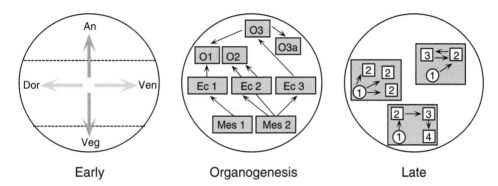

Early   Organogenesis   Late

FIGURE 5   Model of developmental constraints operating in early, mid, and late stages of ontogeny. The model shown here is based on general features of deuterostome development, particularly of echinoderms (Hörstadius 1973; Henry *et al.* 1990; Henry & Raff 1992) and amphibians (Mangold 1961; Ruiz i Altaba & Melton 1990; Cho *et al.* 1991). Events of early development are dominated by axial information systems (animal–vegetal and dorsal–ventral), which establish initial patterning processes and the first localized gene action as cells divide. In these embryos, these axes are global and allow considerable developmental flexibility. The regulative ability of early deuterostome embryos is indicated by the ability of experimentally separated individual cells from 2- or 4-cell embryos to produce normal, albeit small, larvae. Late development also shows considerable developmental flexibility. That flexibility arises because the body is highly modularized by division into separate organ primordia. Genic events (small circles and boxes with arrows indicating process flow) within the primordia (large boxes) are little influenced by genic events in other primordia. Mid development, however, is constrained by a high interconnectivity between elements that will later come to represent separate modules. In this diagram, mesodermal tissues feed information to ectodermal tissues, which in turn interact with sensory placodes. Clearly this represents only a portion of the inductive processes involved in formation of the phylotypic stage.

stages (Fig. 5). Early development has a relatively simple modularity and axial organization. Evolutionary changes in early development are in fact profound (Raff & Kaufman 1983; Elinson 1987). Mid development is more constrained because of complex interactions between what will eventually be discrete and noncontiguous modules. Late development, with its modules within an overall bauplan, will most often undergo localized modifications in anatomy consistent with lower-level taxonomic changes. Modularity of late development will thus also promote mosaic evolution among adult features – a widely recognized phenomenon.

The existence of the phylotypic stage and the apparent complexity of interactions involved in its generation would argue for limits to bauplan evolvability arising from developmental constraints. Yet, if so, one would also predict that the processes producing the constrained phylotypic stage must also be constrained. Instead we find that early development varies widely and that even processes that contribute more proximally to the phylotypic stage vary substantially. In sum, current information on mid development implies a mechanistic component that forces constraint. However, it remains possible that no developmental feature really prevents bauplan transformation but that the ecological constraints arising from competition in Phanerozoic environments have been powerful. We need to be able to run the Cambrian over again, with developmental biologists present this time.

Acknowledgments. – I thank Gregory Wray, J.McClintock Turbeville, and Dieter Walossek for generously providing figures and Jan Bergström for unpublished information on first appearances in the fossil record. J.M. Turbeville and Elizabeth Raff have critically read the manuscript. Research was supported by NIH grant HD 21337 and NSF grant BSR 8818044.

From protein domains to extinct phyla: Reverse-engineering approaches to the evolution of biological complexities

George L. Gabor Miklos and K.S.W. Campbell

G.L.G.M., Centre for Molecular Structure and Function, Australian National University, Canberra, ACT, Australia 2601; K.S.W.C., Department of Geology, Australian National University, Canberra, ACT, Australia 2601

Evolutionary biology is facing two radically different options: to follow a molecular morphoregulatory avenue or to persist with population genetics. The molecular approach reveals that recombinations of protein domains together with massive alterations in the noncoding DNA landscapes that control gene activities have been a major force in early evolution. Informational transmission began in an RNA world, became based on small peptides and DNA, and then proceeded to larger protein domains and finally to multidomain proteins. Shuffling of protein domains, together with the evolution of sophisticated regulatory control regions, can be collectively termed *the recombinational dynamics of modules*. To understand both genomic and organismal evolution, a strong new link needs to be forged between paleontology and molecular biology. Only the fossil record provides the spectrum of designs and the rates at which they evolved; only genetic engineering and protein chemistry will reveal the functional significance of novel designs. Understanding the evolution of metazoan complexities thus requires understanding the evolution and recycling of protein domains as well as the complex multipartite regulatory control regions of genes. There is exciting new knowledge about genes and gene circuits involved in early embryogenesis and neurogenesis from various organisms in different phyla. An understanding of the functional organization of biological systems will only come about by ruthlessly pushing existing genomes to their limits and by constructing novel genomes. In the future we should be able to infer the rough framework of Precambrian genomes and thus glean insights into the genomic engine rooms that helped to power the Cambrian explosion.

We are all specialists and we naturally tend to assume (or even in weak moments to **hope***) that events outside our field will prove irrelevant to our experiments and interests. Such assumptions should be avoided, since they could be dangerous to our scientific health.* Weiner (1987a)

The fossil record often reveals the abrupt appearance of various morphologies, and nowhere is this more apparent than in the emergence of complex multicellular forms from the Precambrian. To the paleontologist has fallen the singular privilege of making sense of this debris of death. The first burst of bizarre designs appeared

---

Bengtson, S. (ed.) 1994: *Early Life on Earth. Nobel Symposium No. 84.* Columbia U.P., New York

abruptly in the Ediacaran, with many organisms having an "air-mattress" or "quilted" construction (Seilacher 1989, this volume). These organisms, however, probably did not provide the ancestors for modern metazoans. This is apparent from both their structural morphologies and their absence from stratigraphic sections between the Ediacaran and Tommotian, where the facies should have been favorable for their preservation. The Ediacaran emergence may have been facilitated by large changes in the Earth's physical environment, including an increase in atmospheric oxygen levels (Knoll 1991, this volume). The second unfolding of different bauplans came at the base of the Cambrian with the small shelly fossils. By the Middle Cambrian, all but one of the 11 skeletonized invertebrate phyla, and 54 of the 56 accepted classes, became visible (Conway Morris 1989b; Jablonski & Bottjer 1990; Bengtson *et al.* 1990).

Two important questions concerning the Ediacaran and the Burgess Shale organisms are as follows:

> *What triggers and constraints underlay and channeled these adaptive radiations?*
> Conway Morris (1989b)

> *Does the rise of so many novel forms indicate the operation of evolutionary mechanisms different to those usually invoked during speciation, or is it due more to opportunistic occupation of an ecological vacuum?*
> Bengtson *et al.* (1990)

The debates in this area have largely been taxonomic ones with two major foci: either the Cambrian organisms represent a morphological continuum and their apparent punctuation is a sampling artefact (Conway Morris 1989b; Briggs & Fortey 1989), or all the phyla received their characteristics instantaneously in a geological sense (Bergström 1986, 1991).

Modern molecular approaches to evolution, on the other hand, have rarely been concerned with the explosion of morphological complexities in the early Paleozoic, because genetic engineers are generally not familiar with these complex morphologies, extinct phyla, and problematica. Instead they commonly ask questions such as: Which phylum is more closely related to which other, and when did eukaryotes and prokaryotes diverge (Raff *et al.* 1989; Doolittle *et al.* 1989)?

The problem with these conventional molecular and paleontological approaches is that even if a phylogeny turns out to be "robust," nothing is revealed of the underlying mechanisms that gave rise to the complex morphologies. Unpalatable as it may seem, paleontological and zoological approaches, and especially that of the "Modern Synthesis," have neither provided explanations of substance for the diversity of forms in the Precambrian and Cambrian nor predicted why certain morphologies arose and others *did not*. Bergström (1991) has correctly concluded, for example, that "classical comparative zoology virtually lacks the power (even) to define inter-phylum relationships."

New recombinant-DNA-based methodologies, however, are now on hand, and their resolving power is awesome. The caveat is that even though Nature has left a DNA encyclopedia for each and every survivor, no encyclopedia comes with instructions on how to proceed from formlessness to form. Furthermore, it is not well known that organisms as diverse as dipterans, cephalopods, and mammals use more than half of their genes for the development and functioning of their nervous systems. With such a huge genetic investment in neural power, animal evolution could more correctly be viewed as the emergence of biological electronics rather than morphologies.

In order to understand the evolution of major biological complexities, it is crucial to know the links between genomes and their phenotypes in a biological-engineering sense. Only then is it possible to ask what new morphological and neuronal systems can conceivably be produced from a given genome. The central issue is to come to grips with evolutionary complexities and then to state at which *level* we seek an understanding of them. The challenge, in fact, is to understand the limits of biological engineering. At the global level, this issue has been formulated by Edelman (1992) as two major questions:

1 *The developmental genetic question: How does the one-dimensional genetic code specify a three-dimensional animal?*

2 *The evolutionary question: How is any proposed answer to the developmental genetic question reconciled with relatively large changes in form occurring in relatively short periods of evolutionary time?*

Genetic engineering: The new biological reductionism

At present biologists are attempting to come to terms with the complexity of biological systems that have unknown design principles. Phase 1 resembles an open-cut-mining operation and includes the genome projects, which are producing the DNA sequences of half a dozen organisms from different phyla. The default output of these different genomic data bases quickly reveals what is common at the DNA-sequence level and at the primary protein-sequence level. Any difficulties are judiciously placed aside to be sifted at later times. Phase 2 involves learning the language of each genomic encyclopedia and deals with the genome–phenotype transition (Brenner 1991). Phase 3 is in its infancy and deals with what can conceivably be built, given the available restrictions of protein structures. Phase 4 involves the highest level of integration. It is the interface between molecular biology and paleontology. It will involve determining if the operational workings of present-day genomes are sufficient to reconstruct the types of organisms that occurred during and after the Cambrian. To obtain an idea of the magnitude of the tasks that lie ahead at these various levels, the informational content of some genomes is shown in Table 1.

The particular organisms in Table 1 are ones whose genomes are being sequenced in the major worldwide genome projects. Clearly, the number of genes in any of them is not unduly large, and it is only a question of time and effort to clone and sequence every

TABLE 1 Approximate gene numbers (transcription units) and genome sizes in different organisms at different levels of morphological and neuronal complexities.

		Genes	Genomes (megabases)
(Eubacteria)	*Escherichia coli*	4,500	4.7
(Fungi)	*Saccharomyces cerevisiae*	7,000	12.5
(Nematoda)	*Caenorhabditis elegans*	15,000	100
(Arthropoda)	*Drosophila melanogaster*	13,000	165
(Chordata)	*Homo sapiens*	50,000	3,300

gene in any genome. For example, the entire chromosome 3 of yeast has already been sequenced. It is 315,000 base pairs long and contains 182 open reading frames (Oliver *et al.* 1992). The entire 2 megabase genome of an archaebacterium, with its 2,000 or so genes, would not be an ambitious undertaking.

More complex questions, however, lie at the level of gene circuitry. Gene number and type is one issue, but how were the genes in hierarchical circuits recruited during evolution to give rise to complex morphologies and nervous systems? As yet we do not know, but we are learning rapidly.

The following examples using antibodies, genes, and transgenic organisms are meant to draw attention to the degree of functional interchangeability between organisms and to highlight the fact that different lines of organisms evolved different levels of complexity and yet used many of the same functional building blocks.

Antibodies. – It comes as a shock to many people to learn that nearly 50% of the monoclonal antibodies made against the brain proteins of *Drosophila melanogaster* cross-react to the brain of human beings (Miller & Benzer 1983). It is also significant that antibodies made to a gap-junction protein expressed in the liver of the rat cross-react with a gap-junction antigen in the cnidarian *Hydra attenuata,* and, in so doing, they eliminate cellular communication (Fraser *et al.* 1987). Since a major form of cellular cross talk occurs through gap junctions, this is an impressive level of functional constraint shared by two phyla.

DNA- and protein-sequence similarities. – There are now literally hundreds of examples of DNA- or protein-sequence similarities between various organisms. Two especially interesting ones are given below.

- A heat-shock protein of the thermophilic archaebacterium *Sulfolobus shibatae* (which is thought to function as a chaperone to assist protein folding) is very similar to polypeptide-1 of yeast and to a gene in the mouse t-complex expressed in developing sperm (Trent *et al.* 1991).

- A protein domain of the peripheral benzodiazepine receptor from rat mitochondria has excellent sequence similarity to a domain of the CrtK protein of *Rhodobacter capsulatus,* a photosynthetic purple bacterium (Baker & Fanestil 1991).

Transgenic organisms. – The ability to make transgenic organisms has meant that DNA from any source can be introduced into a genome and its functional characteristics evaluated. The following examples give some idea of the functional interchangeability between organisms.

- When the *denV* gene of bacteriophage T4 is introduced into mutant *Drosophila melanogaster,* it restores both excision repair and UV resistance. It also restores certain repair capacities to mutant *Escherichia coli,* to yeast, and to hamster and human cell lines (Banga *et al.* 1989).

- When parts of the 26S ribosomal RNA of yeast are replaced by regions from mouse 28S RNA or from an equivalent region of the *E. coli* molecule, ribosomal function in yeast is quite normal, even though it depends on a plethora of interactions with other proteins (Musters *et al.* 1991).

- When eukaryotic pre-mRNAs are cleaved to remove introns, the process occurs in a spliceosome, which is a complex conglomerate of RNAs and proteins. One of its components is a 70K protein. When a chimeric gene consisting of part of a yeast 70K and part of a human 70K gene is constructed, the hybrid protein functions in a yeast cell and is able to rescue a yeast mutant defective in its 70K protein activities (Smith & Barrell 1991).

- Protein traffic within and between cells is a complicated affair. However, an insect luciferase protein, or a chimeric version of it, is correctly imported into peroxisomes of yeast *(S. cerevisiae)*, a plant *(Nicotiana tabacum)*, and peroxisomes of cells of the central nervous system of an amphibian *(Xenopus laevis)* (Gould *et al.* 1990).

These three classes of examples indicate that some RNA and protein domains in different extant phyla must have been present in the Precambrian and have retained their original functional characteristics. How many protein domains then are shared between phyla, when did they arise, and to what uses have they been put in the development of different organisms? One way to answer such questions is to examine present-day prokaryotes and eukaryotes by reverse genetic engineering and to determine how each genome gives rise to its own particular morphologies and how its inventory of protein domains is used in these processes. This will not be an easy task, since some groups of organisms that have interesting developmental characteristics (such as directly developing sea urchins [Wray & Raff 1990]) or intriguing nervous systems (such as minute parasitic Hymenoptera [Miklos 1993]) are not yet amenable to genetic and transgenic analyses.

Molecular developmental biology – dissecting embryogenesis

There has always been much excitement about "higher-order developmental processes" such as movements of sheets of cells at gastrulation and changes in cell shape before invagination and evagination processes, and how such processes are critical to understanding the evolution of shape and form. Indeed they are. However, non-molecular progress in these areas has been very slow, and new reductionist approaches are badly needed (Edelman 1988, 1992). All these important embryological processes initially depend on differential activation and silencing of gene circuits, and some of these are now available for scrutiny, particularly in *D. melanogaster*. In this organism the molecular-genetic approach has resulted in large collections of mutants affecting the processes of early embryogenesis as well as nervous-system development (Perrimon & Mahowald 1987; John & Miklos 1988). Furthermore, the ease of manufacturing transgenic organisms, when combined with sophisticated expression vectors, cloning vehicles such as yeast artificial chromosomes, and "enhancer trap" techniques, has allowed developmental biology to move rapidly from a collection of ad hoc accounts and theories to a firm molecular base, in no small part because of the inputs from *D. melanogaster* (Rubin 1988). This has not been so for organisms that have been traditional objects of embryological studies, such as sea urchins, amphibians, and birds, where mutant and transgenic analyses are in their infancy.

Morphogens and circuits. – For decades, morphogens were nebulous molecules that mysteriously gave rise to gradients that determined polarities from which cells "read" their positional information. The molecular-genetic approach in *D. melanogaster* has provided the first significant glimpses into early embryological circuitry. For example, at least twelve genes are required for a normal dorsal–ventral gradient, with their interactions allowing the gradient of the *dorsal* protein to form so that it has its highest concentration in the ventral region of the early embryo (Stein *et al.* 1991; Jiang *et al.* 1991; Fig. 1 herein).

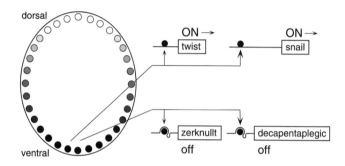

FIGURE 1   The gradient of the *dorsal* protein in the early *D. melanogaster* egg and its influence on gene activities via DNA binding to upstream regulatory regions.

DNA sequence analysis reveals that *dorsal* is a DNA-binding protein and at high concentrations activates the genes *twist* and *snail*, which also encode DNA-binding proteins. The latter two then proceed to activate still other genes responsible for mesoderm differentiation. The concentration of the *dorsal* protein also represses genes such as *zerknullt* and *decapentaplegic,* so that their expression now only occurs in the dorsal part of the embryo and thus allows the differentiation of the dorsal ectodermal derivatives.

The anterior–posterior axis of the early embryo, as another example, is determined partly by the protein product of the *bicoid* gene and the RNA distribution of the *nanos* gene (Driever & Nüsslein-Volhard 1988; Driever *et al.* 1989; Wang & Lehmann 1991). The *bicoid* protein is distributed in a concentration gradient that is highest at the anterior end of the embryo. DNA-sequence analysis reveals that the *bicoid* protein also has a DNA-binding domain (termed the *homeo* domain), and that partly via this domain it activates the *hunchback* gene by binding to several sites upstream of its promoter.

When the *dosage* of the *bicoid* protein is increased, the slope of its gradient is altered, and the cephalic furrow of the embryo can be moved by nearly 20% of the egg length. These data provide the first inklings of how much a critical early embryological process can be modified by genetic engineering.

In summary, many genes are now known to interact in specific circuits and time windows so that their proteins orchestrate early embryogenesis (Table 2). Thus the maternal products lead sequentially to activation of the gap genes, to the pair-rule and segment-polarity genes, to the homeotic genes, to genes involved in gastrulation, and so on. Some of these genes are summarized in Table 2 together with the characteristics of the proteins they encode. Many of them contain domains involved in binding to

DNA. Thus early development in *D. melanogaster* utilizes a large number of executive proteins whose functions are to activate or repress other genes. A future challenge will be to determine how generally the principles of development in this fly apply to organisms in other phyla.

TABLE 2   Genes involved in early embryogenesis in *D. melanogaster*. Proteins with inferred DNA-binding or protein–protein-binding domains are shown in boldface. (Summarized from Perrimon & Mahowald 1987 and John & Miklos 1988.)

Maternal genes: **bicoid**, exuperantia, bicaudal, **caudal**, dorsal, easter, gastrulation defective, nudel, pipe, pelle, snake, spatzle, Toll, tube, windbeutel, cactus, nanos, torso, trunk torso-like, dorsal-like, tudor-like, bicaudal-like, dicephalic, **swallow**

Rescuable maternal effect lethal genes: extra-sex-combs, fs(1)151, almondex, pecanex, fs(1)M53

Late-zygotic lethal genes with maternal-effect lethal phenotypes: pole hole, hopscotch, dishevelled, ultraspiracle, fused,107

Gap genes: **hunchback, Kruppel, knirps, giant, tailless, huckebein**

Pair-rule genes: runt, **hairy, fushi-tarazu, even skipped, paired**, odd paired, **odd skipped**, sloppy paired

Segment polarity genes: gooseberry, wingless, armadillo, **cubitus-interruptus**, fused, hedgehog, naked, patch, engrailed

Homeotic genes: **Ultrabithorax, abdominal A, abdominal B, Antennapedia, Sex combs reduced, Deformed, labial, proboscipedia,** Polycomb

Embryonic lethal genes effecting the dorsal ventral axis: **twist, snail,** Hin-d, zerknullt, decapentaplegic

Neurogenic genes: mastermind, **neuralized,** Delta, Notch, **Enhancer of Split Complex**

Proneural genes: **daughterless, hairy, extramacrochaetae, Achaete-Scute Complex**

Molecular developmental biology – dissecting nervous systems

DNA–protein interactions. – The genetic approach to the development of the embryonic and adult nervous systems of *D. melanogaster* has also been productive. Nervous systems, and hence behavioral repertoires, can also be dismantled using molecular neurogenetics. Brains are not a contorted mass of unanalyzable wiring. They are modular systems that can be systematically taken apart and put together again once the key genes are identified. The *daughterless* gene, for example, is required for the proper formation of neurons of the peripheral nervous system and their associated sensory structures. Deletions of this gene result in the spectacular removal of all the neurons of the embryonic peripheral nervous system yet leave the surrounding epithelium intact (Caudy *et al.* 1988; Vaessin *et al.* 1990; Fig. 2 herein).

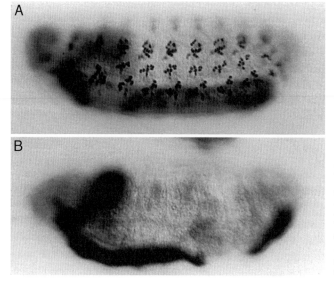

FIGURE 2 **A:** A normal embryo of *D. melanogaster*, which has been stained by a specific antibody to reveal the regular pattern of the developing peripheral nervous system. **B:** A mutant embryo of *D. melanogaster* stained by exactly the same methodology. The embryo is homozygous for a particular allele of the *daughterless* gene. The peripheral nervous system of this embryo is totally absent. (Courtesy of Michael Caudy [see Caudy *et al.* 1988; Vaessin *et al.* 1990] and by permission of the Cold Spring Harbor Laboratory Press.)

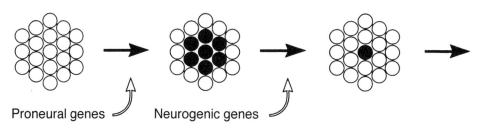

Proneural genes   Neurogenic genes

FIGURE 3 Differentiation of neuroectodermal cell types in an embryonic cell layer and its control by two major gene sets. Under the influence of the proneural genes, some cells (denoted black) enter the neurogenic developmental pathway, leaving the remainder in the epidermogenic one (denoted white). Cell–cell inhibitions mediated via the neurogenic genes suppress the neurogenic capacities of neighboring cells, leading to the differentiated state of a neuronal precursor cell. (Modified from Vaessin *et al.* 1990.)

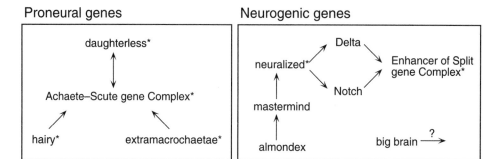

FIGURE 4 Simplified gene-circuit diagrams illustrating the interactions of the proneural and neurogenic genes. The protein products of those genes that encode a DNA-binding or a protein–protein-binding domain are shown by an asterisk. (Modified from Vaessin *et al.* 1990.)

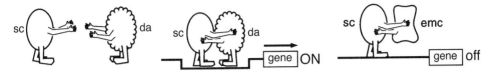

FIGURE 5 The protein–protein and protein–DNA interactions that lead to activation and repression of gene activities that determine external sense-organ formation in the adult of *D. melanogaster*. (Modified from Ellis *et al.* 1990.)

When this phenomenon is examined in terms of early circuitry, two groups of genes are unearthed (Fig. 3). The proneural genes allow cells to choose between the neurogenic and epidermogenic pathways, whereas a second group, termed the neurogenic genes (*mastermind, almondex, neuralized, Notch, Delta, big brain* and the *Enhancer of Split Complex*), are essential for the interaction between neuronal precursor cells and their neighbors (Fig. 4). Thus an undifferentiated sheet of cells achieves a correctly spaced group of neuronal precursor cells for subsequent wiring requirements.

DNA sequence analysis has revealed that *daughterless* and the genes of the *Achaete–Scute Complex* and the *Enhancer of Split Complex* code for proteins containing a *helix–loop–helix* domain important in DNA binding.

Protein–protein interactions. – The modular nature of the nervous system of the adult fly is also revealed by the mutants *lobular–plateless, small optic lobes* and *mini-brain*, which, when mutated, knock out specific neuroanatomical circuits (Fischbach & Heisenberg 1984; Delaney *et al.* 1991). The same principle applies to external sense organs in the adult fly, where the actions of the *extramacrochaetae* protein (*emc*) suppress the development of sensory organs in particular regions on the adult body surface. Normally, the *daughterless* protein (*da*) and proteins of the *Achaete–Scute Complex* bind to each other to form a protein dimer and then bind to DNA to allow transcription of genes involved in sensory organ determination (Fig. 5). The *extramacrochaetae* protein has the ability to bind to other proteins, but does not have DNA-binding ability. Hence when it forms a dimer with a protein of the *Achaete–Scute Complex*, the new dimer is unable to function as a transcriptional regulator and does not allow activation of other genes further downstream in the developmental cascade (Ellis *et al.* 1990). Thus, from an evolutionary perspective, the upstream regulatory region of any gene, with its temporally changing cohort of interacting proteins and their varying concentrations (Mitchell & Tjian 1989), is likely to be as important for the generation of phenotypic complexity as the gene product itself (Dickinson 1988).

These examples highlight the substantial progress made in understanding early embryogenesis and early neurogenesis via the reductionist molecular-genetic approach in a genetically amenable organism. It is necessary to determine, however, the extent to which any particular gene or gene circuit has been conserved or used in different ways in evolution. It is already known that the *helix–loop–helix* domains of the proneural and neurogenic proteins described above have excellent sequence similarities to domains of the human E12 and E47 proteins; the *dorsal* protein has domain similarity to a mammalian DNA-binding protein (*NF-kappaB*); the human homologue of *Notch* has been cloned, and two homologues of the *Achaete–Scute Complex* have been isolated from the rat, where they also function in the nervous system.

Cross-genomic raiding. – One of the biggest biological "industries" at the moment is to determine which genes have readily discernible homologues in different phyla. Nowhere is this better illustrated than in DNA-binding proteins containing a *homeo-domain*. The flow on from the extensive characterization of homeotic genes in *D. melanogaster* has meant that homeodomains are actively being hunted in every phylum. The literature, however, has been colored by the initial observation that the homeotic genes were involved in segmentation processes in arthropods and hence provided a Rosetta stone for the early evolution of segmentation processes. It is a sobering thought that homeodomains have been found in many unsegmented organisms in the phyla Echinodermata, Brachiopoda, Nematoda, and Cnidaria, and in fungi and bacteria. The DNA-binding domain of the *HinR* protein of the bacterium *Salmonella typhimurium*, for example, is structurally related to the eukaryotic homeodomain (Affolter *et al.* 1991). What has undoubtedly occurred is that this same DNA-binding domain has been incorporated into many different proteins during evolution (Holland 1990). While DNA binding is thus assured, the specificity of action of such proteins will depend on what other domains occur in that protein and influence its tertiary structure.

A paleontological perspective

The morphological and geochemical fossil records provide the only definitive evidence of when certain morphological types arose and for how long they were in existence. The DNA record, on the other hand, is always only an inference of what could have been present at a given time. It is possible, however, to make some sensible alignments of these three records without resort to the greasy pole of molecular clocks. It is known from stromatolites as well as from individual fossils that bacterial lineages were certainly in existence at 3.5 Ga and that an overwhelming proportion of these were similar in their morphologies to present-day representatives of the cyanobacteria (Mendelson & Schopf 1992a; Schopf 1993, this volume).

An independent estimate of the times of appearance of prokaryotes and eukaryotes can also be made from compounds collectively termed terpenoid hydrocarbons. The pentacyclic triterpane hydrocarbons are remnants of the bacteriohopane polyols of eubacteria, whereas the acyclic isoprenoids derive from lipids of archaebacteria. Steranes, although not exclusively eukaryotic in origin, are derivatives of eukaryotic sterols (Summons & Walter 1990). When the Proterozoic hydrocarbons are examined, a low but consistent level of steranes is found in a number of formations, the earliest being the Barney Creek Formation of the McArthur Basin in northern Australia (1.69 Ga). Furthermore, the acyclic isoprenoids and bacteriohopane polyols are in abundance. These biomarker signatures thus indicate that the two major bacterial grades and the eukaryotic grade of organization were established by the early Proterozoic (but see Ourisson, this volume).

A molecular–evolutionary perspective

Most proteins have been assembled from smaller domains (Doolittle 1989), and the two most difficult stages in protein evolution are undoubtedly (a) to evolve domains and

(b) to combine them into multidomain complexes (Gilbert 1987). Although there is no formally agreed-upon definition of domains, operationally they can be thought of as units of protein folding or as functional units. Thus, a number of structurally different DNA-binding domains are known. These include homeodomains, *helix–loop–helix* domains, *zinc finger* domains, *leucine zipper* domains, *beta-ribbon* domains, *CTF* domains, an *AP-2* domain, and an *SRF* domain (Mitchell & Tjian 1989; Harrison 1991). There are also transcription-activator domains, the three-dimensional structures of which are not well conserved and which are termed *acid blobs* or *negative noodles*. Many cell-adhesion and matrix-adhesion proteins are constructed of structural domains such as the *immunoglobulin-like, fibronectin type III*, and *EGF-like* domains (Anderson 1990). At the level of tertiary protein structure the *jelly roll* and the *TIM barrel* are two further configurations (Thornton & Gardner 1989).

It is a striking finding that domains from different sources can sometimes be interchanged without loss of function. Thus when alpha helices of a bacterial protein are replaced with alpha helices from other sources, such as from the lysozyme protein of bacteriophage T4, or even replaced by synthetic peptides that have an alpha-helical configuration, protein function is mostly normal (DuBose & Hartl 1989). Furthermore, a "hybrid" protein consisting of the extracellular domain of the human insulin receptor and the cytoplasmic domain of a bacterial aspartate chemoreceptor retains the properties of the individual domains (Ellis *et al.* 1986).

How large, then, is the universe of domains, and what was its rate of evolution, diversification, and recombination? A number of different data bases reveal that a restricted universe of fewer than 10,000 domains was shuffled to produce all extant proteins (reviewed in Miklos & Campbell 1992). While this figure will need to be refined, it is interesting to ask when the shuffling occurred. Was it via a Big Bang or a Continuous Creation? Since the geochemical biomarker data indicate that bacterial lineages and eukaryotic lineages were already independently evolving by 1.7 Ga, comparisons of many proteins between such lineages should reveal which domains were assembled before this time. The *dnaK* protein of *E. coli* has nearly 50% sequence identity to the *Hsp70* protein of *D. melanogaster*, and there is evidence for a homologous gene in an archaebacterium (Bardwell & Craig 1984). Many of the subunits of the proteolipid subunit of ATP synthetase, the c-type cytochromes, and the serine proteases show good conservation between prokaryotes and eukaryotes. Proteins, such as those described above, must all have been assembled before the prokaryotic–eukaryotic split.

Determining how many protein domains are common to the different lineages of organisms is conceptually straightforward, provided a domain can be recognized by either its primary sequence or its three-dimensional configuration. The most reliable signature will of course be its tertiary structure, since this diverges far more slowly than its primary structure. It is known, for example, that up to 75% of the amino acids of a protein can be changed, with only trivial changes in tertiary structure (less than a 2 Å root mean square divergence in backbone positions [Chothia & Lesk 1986]). This means that there will be a number of structural motifs or domains in proteins that can be recognized *only* crystallographically. One such example is that of the soybean trypsin-inhibitor protein and the mammalian interleukin-1 beta protein. They are very different at the level of amino-acid sequences but have the same tertiary structure.

One possibility that should be squarely faced is that few new domains may have evolved since the Archean (Fig. 6). In such a scenario, metazoans would have inherited a fixed number of domains and achieved their present complexity by duplication and shuffling events and by the evolution of more complex gene regulatory sequences. This domain recycling can be readily inferred from the domain duplication and domain arrangements found in cell-adhesion and matrix-adhesion molecules (Anderson 1990; Fig. 7 herein).

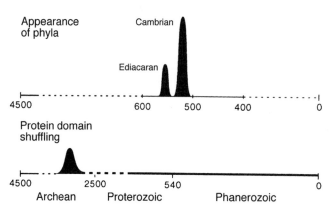

FIGURE 6   Diagrammatic representation of the times of appearance of the major functional designs of organisms and one possible time window for the evolution of protein domains and their major assembling.

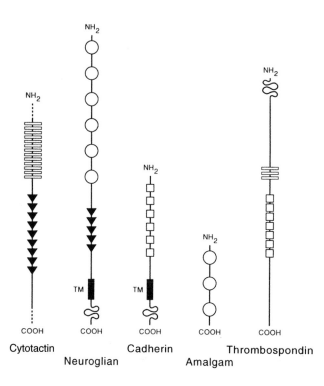

FIGURE 7   Protein domains (or motifs) of a number of cell-adhesion and matrix-adhesion molecules, and the domains of a secreted protein (amalgam). TM denotes the transmembrane spanning region. (Modified from Anderson 1990.)

Furthermore, even without any alteration in the protein itself but only in its time and place of expression, the same protein can be used for an entirely different purpose. Thus an existing enzyme such as lactate dehydrogenase-B has kept its catalytic activities but has been "recruited" into a different developmental pathway. In this case it also acts as the major lens protein in some vertebrate eyes and there fulfills a purely structural role (Piatigorsky & Wistow 1991).

Evolution at the level of protein domains per se may well have been over before the Cambrian explosion, and what may have followed is an increase in the complexity of gene families and the multipartite regulatory regions of individual genes. Alternatively, new protein domains and new combinations may have evolved continuously. Whichever of these two extremes turns out to be the more likely, their resolution will only come from molecular-biological and computer modeling of primary protein sequences.

The fossil record

What helpful restrictions, then, can paleontology place on evolutionary processes, so that different molecular mechanisms can be evaluated in the light of extinct and extant morphologies? Only the fossil record provides rates of morphological change and indicates to us that metazoans did not have a long Precambrian history. The development of complex morphologies and underlying nervous systems was thus very rapid in geological terms. This is beautifully illustrated by, e.g., the Crustacean *Bredocaris* from the Late Cambrian (Müller & Walossek 1988; Fig. 8 herein) and the polymerid trilobite *Olenoides* from the Middle Cambrian Burgess Shale (Whittington 1980).

The nervous systems that controlled organisms such as these, which had well developed eyes and a plethora of sensory hairs on their appendages, must have been substantial. Hence the emergence of sophisticated neural circuitry occuring in conjunction with morphological changes must also have been rapid. Thus paleontology provides discrete time windows within which the evolution of molecular complexity

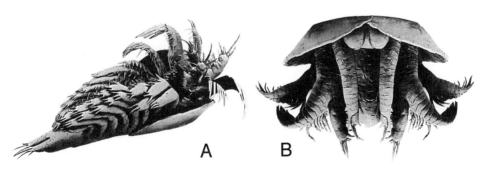

FIGURE 8 Models of the Late Cambrian maxillopod *Bredocaris admirabilis*. Although the adult organism is only 850 μm in length, it reveals exquisite details of morphological differentiation. **A:** Ventrolateral view. **B:** Frontal view. (Courtesy of Müller & Walossek 1988 and by permission of the publishers, Scandinavian University Press, Oslo, Norway.)

has to fit. The Cambrian explosion, occurring in perhaps less than five million years, must have been subtended by gene circuit and neural circuit explosions.

At the level of functional morphology, it is obvious, time and time again, that major innovations occurred early in the history of a phylum. This is clearly seen both in the Echinodermata, where modifications of the body wall for coelomic respiration allow easy tracking of function and morphology, and in the diversification of gnathostomes in the Early Devonian, where a rapid burst of diversification occurred (Campbell 1990).

> *It seems inescapable to us that the major conclusions from the fossil record are the very rapid appearance of functional morphologies inexorably followed by a period that is basically "morphological fiddling at the edges."*

The future

It is evident that it is still too early to provide integrated molecular mechanisms for the major phenotoypic changes that distinguish present-day phyla. This is simply because the molecular cupboard is only just filling up. It should be obvious, however, that it is filling up rather rapidly. Once many of the basic genome homologies, gene and gene-circuit rules, and more tertiary protein structures become known, a more sensible attack can be made on cell lineages and tissue structures.

Finally, it is clear that a number of unhelpful and confusing concepts should be placed aside. First, it is no longer useful in a mechanistic sense to continue with the *current concepts* of macroevolution and microevolution. These should be replaced by a two-phase approach to evolutionary mechanisms: an initial phase in which new functional complexities arise, and a later phase that is, in effect, one of minor modifications. It may well have been more difficult, more time-consuming, and more costly in terms of genetic currency to assemble primitive protein domains than to proceed from fish to human beings. Second, there are no such entities as macromutations and micromutations – they are all mutations. A mutation in a DNA molecule can either give an enormous change in phenotype or none at all, depending on the gene or regulatory sequence involved. Third, "explanations" of the Modern Synthesis are becoming accepted for what they are, namely post hoc statements without any predictive power. The abject failure of this theory to make substantive predictions about the evolution of complexity is highlighted by the following:

> *The mother-lode has been tapped and facts in profusion have been poured into this theory machine. And from the other end has issued – nothing.*
>
> Lewontin (1974)

Closeted as they are in their concept of bean-bag genetics and allelic bookkeeping, advocates of the Modern Synthesis still fail to realize that genes and genomes themselves evolve and, while so doing, change the ground rules for their own subsequent operations. This is beautifully illustrated by the immunoglobulin supergene family (Hood *et al.* 1985), in which the system has evolved exquisitely intricate complexities of gene regulation, complex DNA- as well as RNA-splicing events, protein modifications, and the temporal and spatial deliveries of the protein products. None of this complexity is predicted by population-genetics theory, which is still grappling with the misconception that evolution is about pristine single-copy genes suffering the vicissitudes of mutation and hence giving rise to alternative alleles that inexorably slide into and out of populations. In this mysterious way, complex forms, sculptured and honed by natural selection, gradually emerge from the mist. As we have seen in all the examples in this essay, the reality is very different. If the Modern Synthesis cannot even cope with the origins of genomic novelties (Campbell 1987), it is certainly incapable of predicting the origins of more complex hierarchical systems such as nervous systems (Miklos 1993; Miklos *et al.,* in press).

The contrast between the facts being revealed by molecular embryology and the evolutionary theory expounded by the Modern Synthesis is enormous. The former continues to reveal principles of organismal construction, while the latter persists with the false premise that the manner in which evolutionary mechanisms work at the present time is the same as that in the long-distant past. This is revealed in the erroneous belief (built into endless gradualistic mathematical models) that early evolutionary processes, for example, occurred largely by present-day conventional mutagenic processes to yield single amino acid substitutions and that the only forces operating are mutation, selection, recombination, population size, immigration, and emigration. This view of the triumph of the Modern Synthesis has been forcefully championed by Carson (1987):

> *"**All evolutionary processes** including varying rates and periods of stasis **can be understood** by recourse to **studies of contemporary organisms**. There is no reason to believe that any basic principle can be observed in the fossil record that is not also reflected in population biology"* [our emphases].

This is, in effect, a theory of the origin of species and higher-order categories with *no predictions* about "origins." It is, in essence, a theory that has nothing to say about the emergence of genomic novelties and their effects on developmental processes. It takes no account of the extensively documented turnover processes in present-day genomes (Dover 1987) or in past genomes (Weiner 1987a).

As many have pointed out, genomes and their rules of operation have in fact changed enormously over evolutionary time from that of an original RNA world to one of small peptides, on to multidomain proteins and very sophisticated regulatory hierarchies (see Cold Spring Harbor Symposia on Quantitative Biology 1987; Gilbert 1987; Miklos & Campbell 1992). The one thing that can confidently be said is that the genomes of contemporary organisms are radically different at all levels from those organisms in the not-too-distant past and very different from their single-celled and even-more-primitive ancestors. These molecular data obliterate the major foundation on which the mathematically based Modern Synthesis was founded. The polemics between selectionists and neutralists simply reveal that this synthesis is mired in a mathematical cul-de-sac from which it is incapable of extricating itself. It has congealed into "genetic gridlock" and is making no significant impact on helping us to understand the key processes and rates of early formative events in evolution.

White (1981) has very honestly highlighted these problems.

What, in retrospect should we conclude about the famous "synthesis"? . . . some of its leading figures were so impressed by their achievement that they tended to regard it as a kind of final truth, the details of which were to be expounded but not questioned.

Finally, as one of the field's leading figures has candidly admitted,

By the end of 1932 Haldane, Fisher and Wright had said everything of truly fundamental importance about the theory of genetic change in populations and it is due mainly to man's infinite capacity to make more and more out of less and less, that the rest of us are not currently among the unemployed.   Lewontin (1965)

Astounding as it may seem, it is not well known that for a synthesis that purports to explain evolutionary phenomena, there is nothing, for example, in Fisher's writings that leads to an understanding of speciation or extinction. This reveals the enormous chasm between mathematical theory and the pragmatism that is based upon modern molecular biological data.

We believe that what is now required is a radical new synthesis, but this time a fusion of two entirely different traditions – molecular embryology and paleontology. In this approach, extant systems are dismantled and then reassembled in different ways by transgenic molecular technologies to determine what their functional characteristics may be. Comparisons of present-day genomes from different phyla should allow us to make an assessment of the types of genomes present in the Precambrian and, hence, of what could conceivably be built from them. The organisms of the Early Cambrian, on the other hand, provide us with the limits to which early biological engineering was pushed.

As human beings we have rated biological complexities, entities, and events as *macro* and *micro*, *complex* and *simple*, *important* or *irrelevant*. Our perceptions need bear no relationship whatsoever to the levels of complexity faced by genomes, cells, or organisms. It should be realized that evolution has been a process of the recombinational dynamics of modules, with the environment being permissive and the evolving genomes being proactive.

In other words, at this level the environment should be conceived of as having a passive, receptive role, rather than an active competitive interventionist one.

Campbell (1990)

Evolutionary innovations will only be understood by accessing the genomic engine rooms of different phyla and comparing the outputs to the fossil record (John & Miklos 1988; Miklos & Campbell 1992; Miklos 1993; Miklos *et al.*, in press). They will not be understood by living in a quarantined world where changes in allelic frequencies are the staple commodity and where, as has been pointed out by Franklin (1987), the monuments of mathematical population genetics are polished with finer and finer grades of jewellers rouge.

Acknowledgments. – We wish to thank James Whitehead, Jeffrey Wilson, and Marilyn Miklos for their illustration, photographic, and typing services and Norbert Perrimon, James Posakony, and Michael Caudy for drawing our attention to relevant data sources.

References

Bracketed numbers in italics indicate text pages on which work is referred to.

Abbott, D.H. & Hoffman, S.E. 1984: Archaean plate tectonics revisited. 1. Heat flow, spreading rate, and the age of subducting oceanic lithosphere and their effects on the origin and evolution of continents. *Tectonics 3*, 429–448. [*25, 34*]

Abe, Y. & Matsui, T. 1988: Evolution of an impact-generated H_2O–CO_2 atmosphere and formation of a hot proto-ocean on Earth. *Journal of the Atmospheric Sciences 45*, 3081–3101. [*11*]

Abele, L.G., Kim, W. & Felgenhauer, B.E. 1989: Molecular evidence for inclusion of the phylum Pentastomida in the Crustacea. *Molecular Biology and Evolution 6*, 685–691. [*456*]

Abouchami, W., Boher, M., Michard, A. & Albarede, F. 1990: A major 2.1 Ga event of mafic magmatism in west Africa: An early stage of crustal accretion. *Journal of Geophysical Research 95*, 17605–17629. [*26*]

Achenbach-Richter, L., Gupta, R., Stetter, K.O. & Woese, C.R. 1987a: Were the original Eubacteria thermophiles? *Systematic and Applied Microbiology 9*, 34–39. [*144, 163*]

Achenbach-Richter, L., Stetter, K.O. & Woese, C.R. 1987b: A possible biochemical missing link among archaebacteria. *Nature 327*, 348–349. [*159*]

Affolter, M., Percival-Smith, A., Muller, M., Billeter, M., Qian, Y.Q., Otting, G., Wuthrich, K. & Gehring, W.J. 1991: Similarities between the homeodomain and the Hin recombinase DNA-binding domain. *Cell 64*, 879–880. [*510*]

Aggarwal, A., de la Cruz, V.F. & Nash, T.E. 1990: A heat shock protein in *Giardia lamblia* unrelated to HSFP70. *Nucleic Acids Research 18*, 3409. [*187*]

Aitken, J.D. 1988: Giant "algal" reefs, Middle/Upper Proterozoic Little Dal Group (>770 Ma), Mackenzie Mountains, N. W. T., Canada. In Geldsetzer, H.H.J., James, N.P. & Tebbutt, G.E. (eds.): *Reefs, Canada and Adjacent Areas*, 13–23. Canadian Society of Petroleum Geologists, Memoir 13. [*283, 430*]

Aitken, J.D. 1989: Uppermost Proterozoic fomations in the central Mackenzie Mountains, Northwest Territories. *Geological Survey of Canada, Bulletin 368*. 26 pp. [*373*]

Aitken, J.D. & Narbonne, G.M. 1989: Two occurrences of Precambrian thrombolites from the Mackenzie Mountains, northwestern Canada. *Palaios 4*, 384–388. [*280, 285, 430*]

Akam, M. 1989: *Hox* and HOM: Homologous gene clusters in insects and vertebrates. *Cell 57*, 347–349. [*140, 459*]

Åkermark, B., Eklund-Westlin, U., Bäckström, P. & Löf, R. 1980: Photochemical, metal-promoted reduction of carbon dioxide and formaldehyde in aqueous solution. *Acta Chemica Scandinavica B34*, 27–30. [*13, 21*]

Alberts, B.M. 1986: The function of the hereditary materials: Biological catalyses reflect the cell's evolutionary history. *American Zoologist 26*, 781–796. [*63, 70, 71, 73, 78*]

Alberts, B., Bray, D., Lewis, J., Raff, M., Roberts, K. & Watson, J. 1989: *Molecular Biology of the Cell*. 2d Ed. Garland, New York, N.Y. [*404, 405, 406, 408*]

Alessandrello, A., Pinna, G. & Teruzzi, G. 1988: Land planarian locomotion trail from the Lower Permian of Lombardian pre-Alps. *Atti della Societa italiana di scienze naturali e del Museo civile di storia naturale 129*, 139–145. [*456*]

Allegre, C.J., Staudacher, T. & Sarda, P. 1987: Rare gas systematics: Formation of the atmosphere, evolution and structure of the Earth's mantle. *Earth and Planetary Science Letters 81*, 127–150. [*11*]

Aller, L.H. 1961: *The Abundance of the Elements*. Interscience, London. [*48*]

Allison, C.W. & Hilgert, J.W. 1986: Scale microfossils from the Early Cambrian of northwest Canada. *Journal of Paleontology 60*, 973–1015. [*291, 301, 415, 427, 442*]

Altekar, W. & Rajagopalan, R. 1990: Ribulose bisphosphate carboxylase activity in halophilic Archaebacteria. *Archives of Microbiology 153*, 169–174. [*159*]

Al-Thukair A.A. & Golubic, S. 1991: Five new *Hyella* species from the Arabian Gulf. In Hickel, B., Anagnostidis, K. & Komarek, J. (eds.): *Cyanophyta/Cyanobacteria – Morphology, Taxonomy, Ecology*, 167–197. Algological Studies 64 Schweizerbart'sche, Stuttgart. [*336*]

Altman, S. 1984: Aspects of biochemical catalysis. *Cell 36*, 237–239. [77]

Alvarez, W., Asaro, F. & Montanari, A. 1990: Iridium profile for 10 million years across the Cretaceous–Tertiary boundary at Gubbio (Italy). *Science 250*, 1700–1702. [14, 17]

Amano, S. & Hori, I. 1992: Metamorphosis of calcareous sponges. I. Ultrastructure of free-swimming larvae. *Invertebrate Reproduction and Development 21*, 81–90. [480]

Anbar, A.D. & Holland, H.D. 1992: The photochemistry of manganese and the origin of banded iron formations. *Geochimica et Cosmochimica Acta 56*, 2595–2603. [242]

Anbar, M. 1968: Cavitation during impact of liquid water on water: Geochemical implications. *Science 161*, 1343–1344. [22]

Anders, E. 1989: Pre-biotic organic matter from comets and asteroids. *Nature 342*, 255–257. [37, 108]

Anders, E., Hayatsu, R. & Studier, M.H. 1974: Catalytic reactions in the solar nebula: Implications for interstellar molecules and organic compounds in meteorites. *Origins of Life 5*, 57–67. [37]

Anders, E. & Owen, T. 1977: Mars and Earth: Origin and abundance of volatiles. *Science 198*, 453–465. [11, 46]

Andersen, R.A. 1991: The cytoskeleton of chromophyte algae. *Protoplasma 164*, 143–159. [184, 318]

Anderson, D.T. 1973: *Embryology and Phylogeny in Annelids and Arthropods.* Pergamon, Oxford. [402, 498]

Anderson, D.T. 1982: Origin and relationships among the animal phyla. *Proceedings of the Linnean Society of New South Wales 106*, 151–166. [482]

Anderson, H. 1990: Adhesion molecules and animal development. *Experientia 46*, 2–13. [511, 512]

Anderson, M.M. & Conway Morris, S. 1982: A review, with description of four unusual forms, of the soft-bodied fauna of the Conception and St. John's Groups (Late Precambrian), Avalon Peninsula, Newfoundland. *Third North American Paleontological Convention, Montreal, Proceedings 1*, 1–8. [372]

Anderson, M.M. & Misra, S.B. 1968: Fossils found in the Precambrian Conception Group in Southeastern Newfoundland. *Nature 220*, 680–681. [372]

Anderson, P.A.V. 1991: *Evolution of the First Nervous Systems.* Nato Advanced Studies Institute Series Plenum, New York, N.Y. [482]

Andres, D. 1989: Phosphatisierte Fossilien aus dem unteren Ordoviz von Südschweden. *Berliner Geowissenschaftliche Ahhandlungen. Reihe A. Geologie und Paläontologie 106*, 9–19. [456]

Arculus, R.J. & Delano, J.W. 1980: Implications for the primitive atmosphere of the oxidation state of Earth's upper mantle. *Nature 288*, 72–74. [12]

Arculus, R.J. & Delano, J.W. 1981: Intrinsic oxygen fugacity measurements: Techniques and results for spinels from upper mantle peridotites and megacryst assemblages. *Geochimica et Cosmochimica Acta 45*, 899–913. [12]

Arhem, P. 1990: The evolution of iron channels. In Gustafsson, M.K.S. & Reuter, M. (eds.): *The Early Brain. Proceedings of the Symposium "Invertebrate Neurobiology,"* 95–104. Acta Academiae Åboensis. Ser. B 50 Åbo Academy Press, Åbo. [482]

Armstrong, R.L. 1981: Radiogenic isotopes: The case for crustal recycling on a near-steady-state no-continental-growth Earth. *Philosophical Transactions of the Royal Society of London, A 301*, 443–472. [25]

Arndt, N.T., Nelson, D.R., Compston, W., Trendall, A.F. & Thorne, A.M. 1991: The age of the Fortescue Group, Hamersley Basin, Western Australia, from ion microprobe zircon U–Pb results. *Australian Journal of Earth Sciences 38*, 261–281. [241]

Arrhenius, G. 1952: Sediment cores from the East Pacific. I. Properties of the sediment and their distribution. In Pettersson, H. (ed.): *Reports of the Swedish Deep-Sea Expedition 1947–1948*, Vol. 5, 5–91. [20, 106]

Arrhenius, G. 1984: Minerals with channel structure as substrates for nucleotide and peptide synthesis. In Wickramasinghe, C. (ed.): *Fundamental Studies and the Future of Science*, 301–319. University College Cardiff Press, Cardiff. [94]

Arrhenius, G. 1986: Dysoxic environments as models for primordial mineralisation. In Cairns-Smith, A.G. & Hartman, H. (eds.): *Clay Minerals and the Origin of Life*, 97–104. Cambridge University Press, Cambridge. [18]

Arrhenius, G. 1987: Interaction of hydrous minerals with bioorganic precursor molecules. In *NASA Space Life Sciences Symposium, June 21–26*, 260–261. [94, 95]

Arrhenius, G. 1990: Sources and geochemical evolution of cyanide and formaldehyde. In Bzik, S.E. (ed.): *Fourth Symposium of Chemical Evolution, NASA Ames Research Center*, 31–32. [*92*]

Arrhenius, G., Bachman, J., Gedulin, B., Hui, S. & Paplawsky, W. 1989: Anion selective minerals as concentrators and catalysts for RNA precursor components. *Origins of Life and Evolution of the Biosphere 19*, 235–236. [*95*]

Arrhenius, G., De, B.R. & Alfvén, H. 1974: Origin of the ocean. In Goldberg, E.D. (ed.): *The Sea: Ideas and Observations on Progress in the Study of the Seas*, Vol. 5, 839–861. Wiley, New York, N.Y. [*14, 51*]

Arrhenius, G., Gedulin, B. & Mojzsis, S. 1993: Phosphate in models for chemical evolution. In Ponnamperuma, C. & Chela-Flores, J. (eds.): *Proceedings, Conference on Chemical Evolution and Origin of Life, Trieste, Italy, Oct. 1992*, 1–26. [*92, 106, 415*]

Arthur, M.A. & Natland, J.H. 1979: Carbonaceous sediments in the North Sea and South Atlantic: The role of salinity in stable stratification of Early Cretaceous Basins. In Talwani, M., Hay, W. & Ryan, W.B.F. (eds.): *Deep Drilling Results in the Atlantic Ocean, Continental Margins and Paleoenviro*, 375–401. American Geophysical Union, Maurice Ewing Series 3. [*29*]

Arthur, W. 1988: *A Theory of the Evolution of Development*. Wiley, Chichester. [*494*]

Asada, K., Kanematsu, S., Okaka, S. & Hayakawa, T. 1980: Phylogenetic distribution of three types of superoxide dismutase in organisms and in cell organelles. In Bannister, J.V. & Hill, H.A.O. (eds.): *Chemical and Biochemical Aspects of Superoxide and Superoxide Dismutase*, 136–153. Elsevier, Amsterdam. [*41*]

Aseeva, E.A. 1976: Mikrofitofossilii i vodorosli iz otlozhenij verkhnego dokembriya Volyno-Podolii [Microphytofossils and algae from Upper Precambrian deposits of Volyno-Podolia]. In Shul'ga, P.L. (ed.): *Paleontologiya i Stratigrafiya Verkhnego Dokembriya i Nizhnego Paleozoya Yugo-Zapada Vostochno-Evropejskoj Platformy*, 40–63. Naukova Dumka, Kiev. [*356, 357*]

Asmerom, Y., Jacobsen, S.B., Knoll, A.H., Butterfield, N.J. & Swett, K. 1991: Strontium isotopic variations of Neoproterozoic seawater: Implications for crustal evolution. *Geochimica et Cosmochimica Acta 55*, 2883–2894. [*444, 449*]

Awramik, S.M. 1971: Precambrian columnar stromatolite diversity: Reflection of metazoan appearance. *Science 174*, 825–827. [*254, 272, 280, 283, 284, 300, 301, 372, 432, 435*]

Awramik, S.M. 1984: Ancient stromatolites and microbial mats. In Cohen, Y., Castenholz, R.Z. & Halvorsen, H.O. (eds.): *Microbial Mats: Stromatolites*, 1–22. Alan Liss, New York, N.Y. [*31*]

Awramik, S.M. 1991: Archaean and Proterozoic stromatolites. In Riding, R. (ed.): *Calcareous Algae and Stromatolites*, 289–304. Springer, Berlin. [*336, 337*]

Awramik, S.M. 1992: The oldest records of photosynthesis. *Photosynthesis Research 33*, 75–89. [*163, 166, 167, 168, 172*]

Awramik, S.M. & Barghoorn, E.S. 1977: The Gunflint microbiota. *Precambrian Research 5*, 121–142. [*310*]

Awramik, S.M., Schopf, J.W. & Walter, M.R. 1983: Filamentous fossil bacteria from the Archean of Western Australia. *Precambrian Research 20*, 357–374. [*143, 197, 310, 336*]

Awramik, S.M., Schopf, J.W. & Walter, M.R. 1988: Carbonaceous filaments from North Pole, Western Australia: Are they fossil bacteria in Archaean stromatolites? A discussion. *Precambrian Research 39*, 303–309. [*197, 275, 276*]

Awramik, S.M., Schulte McMenamin, D., Yin, C., Zhou, Z., Ding, Q. & Zhang, S. 1985: Prokaryotic and eukaryotic microfossils from a Proterozoic/Phanerozoic transition in China. *Nature 315*, 655–658. [*305*]

Awramik, S.M. & Semikhatov, M.A. 1979: The relationship between morphology, microstructure, and microbiota in three vertically intergrading stromatolites from the Gunflint Iron Formation. *Canadian Journal of Earth Sciences 16*, 484–495. [*281*]

Ax, P. 1989: Basic phylogenetic systematization of the Metazoa. In Fernholm, B., Bremer, K. & Jörnvall, H. (eds.): *The Hierarchy of Life*, 229–245. Excerpta Medica, Amsterdam. [*475, 483*]

Babcock, L.E. & Robison, R.A. 1989: Preferences of Paleozoic predators. *Nature 337*, 695–696. [*424*]

Bada, J.L. & Miller, S.L. 1968: Ammonium ion concentration in the primitive ocean. *Science 159*, 423–425. [*12*]

Baeza, I., Ibañez, M., Lazcano, A., Santiago, C., Argüello, C., Wong, C. & Oró, J. 1987: Liposomes with polyribonucleotides as models of precellular systems. *Origins of Life and Evolution of the Biosphere* *17*, 321–331. [*56*]

Baker, E.T., Jr., Lavelle, J.W. & Massoth, G.J. 1985: Hydrothermal particle plumes over the southern Juan de Fuca Ridge. *Nature 316*, 342–344. [*20*]

Baker, M.E. & Fanestil, D.D. 1991: Mammalian peripheral-type benzodiazepine receptor is homologous to CrtK protein of *Rhodobacter capsulatus*, a photosynthetic bacterium. *Cell 65*, 721–722. [*504*]

Bakke, E.L., Beaty, D.W. & Hayes, J.M. 1991: Re-evaluation of isotope ratios in the PDB Standard and of algorithms for the calculation of carbon and oxygen isotope abundances. *Geological Society of America, Abstracts with Programs 23*, 150. [*223*]

Ballard, J.W.O., Olsen, G.J., Faith, D.P., Odgers, W.A., Rowell, D.M. & Atkinson, P.W. 1992: Evidence from 12S ribosomal RNA sequences that onychophores are modified arthropods. *Science 258*, 1345–1348. [*454*]

Baltscheffsky, H. 1971: Inorganic pyrophosphate and the origin and evolution of biological energy transformation. In Buvet, R. & Ponnamperuma, C. (eds.): *Chemical Evolution and the Origin of Life*, 466–474. North-Holland, Amsterdam. [*82*]

Baltscheffsky, H., Lundin, M., Luxemburg, C., Nyrén, P. & Baltscheffsky, M. 1986: Inorganic pyrophosphate and the molecular evolution of biological energy coupling. *Chemica Scripta 26B*, 259–262. [*94*]

Baltscheffsky, H. & von Stedingk, L.-V. 1966: Bacterial photophosphorylation in the absence of added nucleotide. A second intermediate stage of energy transfer in light-induced formation of ATP. *Biochemical and Biophysical Research Communications 22*, 722–728. [*84*]

Baltscheffsky, H., von Stedingk, L.-V., Heldt, H.-W. & Klingenberg, M. 1966: Inorganic pyrophosphate: Formation in bacterial photophosphorylation. *Science 153*, 1120–1122. [*84, 94*]

Baltscheffsky, M. 1967: Inorganic pyrophosphate and ATP as energy donors in chromatophores from *Rhodospirillum rubrum*. *Nature 216*, 241–243. [*84*]

Baltscheffsky, M. & Baltscheffsky, H. 1992: Inorganic pyrophosphate and inorganic pyrophosphatases. In Ernster, L. (ed.): *Molecular Mechanisms in Bioenergetics*, 331–348. New Comprehensive Biochemistry Elsevier, Amsterdam. [*85, 87, 89*]

Baltscheffsky, M., Baltscheffsky, H. & von Stedingk, L.-V. 1966: Light-induced energy conversion and the inorganic pyrophosphatase reaction in chromatophores from *Rhodospirillum rubrum*. *Brookhaven Symposia in Biology 19*, 246–257. [*84*]

Baltscheffsky, M., Boork, J., Nyrén, P. & Baltscheffsky, H. 1985: Some basic properties of photosynthetic energy coupling. *Physiologie Végétale 23*, 697–704. [*94*]

Baltscheffsky, M. & Lundin, A. 1979: Flash-induced increase of ATPase activity in *Rhodospirillum rubrum* chromatophores. In Packer, L. & Mukohata, Y. (eds.): *Cation Flux Across Biomembranes*, 209–217. Academic Press, New York, N.Y. [*86*]

Baly, E.C.C. 1924: Photosynthesis. *Industrial and Engineering Chemistry 16*, 1016–1018. [*52*]

Baly, E.C.C., Davies, J.B., Johnson, M.R. & Shanassy, H. 1927: Photosynthesis of naturally occurring compounds: I. Action of ultra-violet light on carbonic acid. *Proceedings of the Royal Society of London, A 116*, 197–202. [*52*]

Banga, S.S., Boyd, J.B., Valerie, K., Harris, P.V., Kurz, E.M. & De Riel, J.K. 1989: *denV* gene of bacteriophage T4 restores DNA excision repair to *mei-9* and *mus201* mutants of *Drosophila melanogaster*. *Proceedings of the National Academy of Sciences, USA 86*, 3227–3231. [*504*]

Bardele, C.F. 1983: Comparative freeze fracture study of the ciliary membrane of Protista in invertebrates and in relation to phylogeny. *Journal of Submicroscopic Cytology 15*, 263–267. [*481*]

Bardwell, J.C.A. & Craig, E.A. 1984: Major heat shock gene of *Drosophila* and the *Escherichia coli* heat-inducible *dnaK* gene are homologous. *Proceedings of the National Academy of Sciences, USA 81*, 848–852. [*511*]

Barghoorn, E.S. & Schopf, J.W. 1965: Microorganisms from the Late Precambrian of Central Australia. *Science 150*, 337–339. [*273*]

Barghoorn, E.S. & Schopf, J.W. 1966: Microorganisms three billion years old from the Precambrian of South Africa. *Science 152*, 758–763. [*195*]

Barley, M.E., Dunlop, J.S.R., Glover, J.E. & Groves, D.I. 1979: Sedimentary evidence for an Archaean shallow-water volcanic-sedimentary facies, eastern Pilbara Block, Western Australia. *Earth and Planetary Science Letters 43*, 74–84. [*10, 27, 28, 29*]

Barnes, R.D. 1985: Current perspectives on the origin and relationships of lower invertebrates. In Conway Morris, S., George, J.D., Gibson, R. & Platt, H.M. (eds.): *The Origins and Relationships of Lower Invertebrates*, 360–368. Clarendon, Oxford. [*476*]

Bar-Nun, A. & Chang, S. 1983: Photochemical reactions of water and carbon monoxide in Earth's primitive atmosphere. *Journal of Geophysical Research 88*, 6662–6672. [*12*]

Bar-Nun, A. & Shaviv, A. 1975: Dynamics of the chemical evolution of Earth's primitive atmosphere. *Icarus 24*, 197–210. [*16*]

Bartolomaeus, T. & Ax, P. 1992: Protonephridia and Metanephridia – their relation within the Bilateria. *Zeitschrift für Zoologische Systematik und Evolutionsforschung 30*, 21–45. [*483*]

Basile, B.P., Middleditch, B.S. & Oró, J. 1978: Polycyclic aromatic hydrocarbons in the Murchison meteorite. *Organic Geochemistry 5*, 211–216. [*119*]

Bathurst, R.G.C. 1975: *Carbonate Sediments and Their Diagenesis*. 2d Ed. Developments in Sedimentology 12 Elsevier, Amsterdam. [*245*]

Bauld, J. D'Amelio, E. & Farmer, J. 1992: Modern microbial mats. In Schopf, J.W. & Klein, C. (eds.): *The Proterozoic Biosphere: A Multidisciplinary Study*, 261–270. Cambridge University Press, Cambridge. [*168*]

Beadle, S.C. 1991: Cyclocrinitids. In Riding, R. (ed.): *Calcareous Algae and Stromatolites*, 114–124. Springer, Berlin. [*428*]

Beanland, T. 1990: Evolutionary relationships between "Q-type" photosynthetic reaction centres: Hypothesis-testing using parsimony. *Journal of Theoretical Biology 145*, 535–545. [*179, 180*]

Beaudry, A.A. & Joyce, G.F. 1992: Directed evolution of an RNA enzyme. *Science 257*, 635–641. [*68*]

Becker, Yu.R. 1977: Pervye paleontologicheskie nakhodki v rife Urala [First paleontological finds in the Riphean of Urals]. *Izvestiya Akademii Nauk SSSR, Seriya Geologicheskaya 3*, 90–100. [*373*]

Becker, Yu.R. 1990: Vendian metazoa from the Urals. In Sokolov, B.S. & Ivanovskij, A.B. (eds.): *The Vendian System. I. Paleontology*, 121–131. Springer, Berlin. [*373*]

Beeman, R.W. 1987: A homoeotic gene cluster in the red flour beetle. *Nature 327*, 247–249. [*498*]

Beer, E.J. 1919: Note on a spiral impression on Lower Vindhyan Limestone. *Geological Survey of India, Records 50*, 139. [*292*]

Beevers, C.A. 1958: The crystal structure of dicalcium phosphate dihydrate, $CaHPO_4 \cdot 2H_2O$. *Acta Crystallographica 11*, 273–277. [*103*]

Bell, M.B., Feldman, P.A., Kwok, S. & Matthews, H.E. 1982: Detection of HC11N in IRC+10 216. *Nature 295*, 389–391. [*49*]

Bengtson, S. 1977: Aspects of problematic fossils in the early Palaeozoic. *Acta Universitatis Upsaliensis 415*. 71 pp. [*416, 424, 458*]

Bengtson, S. 1985: Taxonomy of disarticulated fossils. *Journal of Paleontology 59*, 1350–1358. [*451*]

Bengtson, S. 1986: Siliceous microfossils from the Upper Cambrian of Queensland. *Alcheringa 10*, 195–216. [*420*]

Bengtson, S. 1989: Ecology of the Cambrian explosion. In *Abstract, EUG 5, Strasbourg, 1989*, 199. Terra 1. [*304*]

Bengtson, S. 1990a: *Maikhanella* and *Siphogonuchites*: Problems of skeleton formation in Cambrian metazoans. *Palaeontological Association, Durham 1990, Abstracts*, 3. [*419*]

Bengtson, S. 1990b: The solution to a jigsaw puzzle. *Nature 345*, 765–766. [*490*]

Bengtson, S. 1991: Oddballs from the Cambrian start to get even. *Nature 351*, 184–185. [*490*]

Bengtson, S. 1992: The cap-shaped Cambrian fossil *Maikhanella* and the relationship between coeloscleritophorans and molluscs. *Lethaia 25*, 401–420. [*419, 457*]

Bengtson, S. & Conway Morris, S. 1984: A comparative study of Lower Cambrian *Halkieria* and Middle Cambrian *Wiwaxia*. *Lethaia 17*, 307–329. [*418, 457*]

Bengtson, S. & Conway Morris, S. 1992: Early radiation of biomineralizing phyla. In Lipps, J.H. & Signor, P.W. (eds.): *Origin and Early Evolution of the Metazoa*, 447–481. Plenum, New York, N.Y. [*415, 451*]

Bengtson, S., Conway Morris, S., Cooper, B.J., Jell, P.A. & Runnegar, B.N. 1990: Early Cambrian fossils from South Australia. *Memoirs of the Association of Australasian Palaeontologists 9*. 364 pp. [*415, 419, 420, 454, 457, 458, 502*]

Bengtson, S., Farmer, J.D., Fedonkin, M.A., Lipps, J.H. & Runnegar, B. 1992: The Proterozoic–Early Cambrian evolution of metaphytes and metazoans. In Schopf, J.W. & Klein, C. (eds.): *The Proterozoic Biosphere: A Multidisciplinary Study*, 425–462. Cambridge University Press, Cambridge. [*415, 423*]

Bengtson, S. & Missarzhevsky, V.V. 1981: Coeloscleritophora – a major group of enigmatic Cambrian metazoans. In Taylor, M.E. (ed.): *Short Papers for the Second International Symposium on the Cambrian System 1981*, 19–21. U.S. Geological Survey Open-File Report 81-743. [*418*]

Bengtson, S. & Yue Zhao 1992: Predatorial borings in Late Precambrian mineralized exoskeletons. *Science 257*, 367–369. [*291, 295, 417, 451*]

Benlow, A. & Meadows, A.J. 1977: The formation of the atmospheres of the terrestrial planets by impact. *Astrophysics and Space Science 46*, 293–300. [*14*]

Benoit, R., Sassoon, D., Jacq, B., Gehring, W. & Buckingham, M. 1989: Hox-7, a mouse homeodo box gene with a novel pattern of expression during embryogenesis. *EMBO Journal 8*, 91–100. [*497*]

Berg, G. 1985: *Annulonemertes* gen. nov., a new segmented hoplonemertean. In Conway Morris, S., George, J.D., Gibson, R. & Platt, H.M. (eds.): *The Origins and Relationships of Lower Invertebrates*, 200–209. The Systematics Association Special Volume 28 Clarendon, Oxford. [*465*]

Berg, G.W. 1986: Evidence for carbonate in the mantle. *Nature 365*, 630–633. [*11*]

Berg, W.W., Jr. & Winchester, J.W. 1978: Aerosol chemistry of the marine atmosphere. In Riley, J.P. & Chester, R. (eds.): *Chemical Oceanography*, Vol. 7, 173–231. Academic Press, London. [*21, 22*]

Bergquist, P.R. 1978: *Sponges*. University of California Press, Berkeley, Calif. [*478, 482*]

Bergquist, P.R. 1985: Poriferan relationships. In Conway Morris, S., George, J.D., Gibson, R. & Platt, H.M. (eds.): *The Origins and Relationships of Lower Invertebrates*, 14–28. Clarendon, Oxford. [*476*]

Bergström, J. 1986: Metazoan evolution – a new model. *Zoologica Scripta 15*, 189–200. [*453, 461, 468, 502*]

Bergström, J. 1989: The origin of animal phyla and the new phylum Procoelomata. *Lethaia 22*, 259–269. [*298, 404, 451, 453, 459, 464, 465*]

Bergström, J. 1990: Precambrian trace fossils and the rise of bilaterian animals. *Ichnos 1*, 3–13. [*304, 305, 417, 459, 465*]

Bergström, J. 1991: Metazoan evolution around the Precambrian–Cambrian transition. In Simonetta, A.M. & Conway Morris, S. (eds.): *The Early Evolution of Metazoa and the Significance of Problematic Taxa*, 25–34. Cambridge University Press, Cambridge. [*381, 451, 460, 463, 464, 485, 502*]

Berkner, L.V. & Marshall, L.C. 1965: On the origin and rise of oxygen concentration in the Earth's atmosphere. *Journal of Atmospheric Science 22*, 225–261. [*29*]

Berman, A., Hanson, J., Leiserowitz, L., Koetzle, T.F., Weiner, S. & Addadi, L. 1993: Biological control of crystal texture: A widespread strategy for adapting crystal properties to function. *Science 259*, 776–779. [*423*]

Bernal, J.D. 1949: The physical basis of life. *Proceedings of the Physical Society of London 62(A)*, 537–558. [*95*]

Bernal, J.D. 1954: The origin of life. *New Biologist 16*, 28–40. [*37*]

Bernal, J.D. 1967: *The Origin of Life*. World, New York, N.Y. [*232*]

Berner, R.A. 1973: Phosphate removal from seawater by adsorption on volcanogenic ferric oxides. *Earth and Planetary Science Letters 18*, 77–86. [*106, 448*]

Berner, R.A. & Lasaga, A.C. 1989: Modeling the geochemical carbon cycle. *Scientific American 260*, 74–81. [*51*]

Berner, R.A., Lasaga, A.C. & Garrels, R.M. 1983: The carbonate–silicate geochemical cycle and its effect on atmospheric carbon dioxide over the past 100 million years. *American Journal of Science 283*, 641–683. [*215*]

Berry, W.B.N. & Wilde, P. 1978: Progressive ventilation of the oceans – an explanation for the distribution of the Lower Paleozoic black shales. *American Journal of Science 278*, 257–275. [*29*]

Bertrand-Sarfati, J. 1970: Les édifices stromatolitiques de la série calcaire du Hank (Précambrien supérieur): Déscription, variations latérales, paléoécologie; Sahara occidental, Algérie. *Bulletin de la Société d'Histoire Naturelle de l'Afrique du Nord 61*, 13–38. [*430*]

Bertrand-Sarfati, J. 1972: Paléoécologie de certains stromatolites en récifs des formations du Précambrien supérieur du groupe d'Atar (Mauritanie, Sahara occidental): Création d'espèces nouvelles. *Palaeogeography, Palaeoclimatology, Palaeoecology 11*, 33–63. [*430*]

Bertrand-Sarfati, J. 1976: An attempt to classify Late Precambrian stromatolite microstructures. In Walter, M.R. (ed.): *Stromatolites*, 251–259. Developments in Sedimentology 20 Elsevier, Amsterdam. [*281, 430*]

Bertrand-Sarfati, J. 1979: Une algue inhabituelle verte, rouge ou bleue dans une formation dolomitique présumée d'âge Précambrien supérieur. *Bulletin Centre Recherche Exploration–Production Elf-Acquitaine 3*, 453–461. [*430, 431*]

Bertrand-Sarfati, J. & Moussine-Pouchkine, A. 1983: Platform-to-basin evolution: The carbonates of late Proterozoic (Vendian) Gourma (West Africa). *Journal of Sedimentary Petrology 53*, 275–293. [*254, 430, 431*]

Bertrand-Sarfati, J. & Moussine-Pouchkine, A. 1985: Evolution and environmental conditions of *Conophyton–Jacutophyton* association in the Atar Dolomite (Upper Proterozoic, Mauretania). *Precambrian Research 29*, 207–234. [*338*]

Bertrand-Sarfati, J. & Pentecost, A. 1989: "Tussocky" microstructure in stromatolites: A biological event recorded in the Upper Proterozoic sediments of the West African craton. *Terra Abstracts 1*, 197. [*430*]

Bertrand-Sarfati, J. & Walter, M.R. 1981: Stromatolite biostratigraphy. *Precambrian Research 15*, 353–371. [*283*]

Berzelius, J.J. 1806: *Föreläsningar i Djurkemien, del 1–2*. Stockholm. [*124*]

Betts, J.N. & Holland, H.D. 1991: The oxygen content of ocean bottom waters, the burial efficiency of organic carbon, and the regulation of atmospheric oxygen. *Palaeogeography, Palaeoclimatology, Palaeoecology (Global and Planetary Change Section) 97*, 5–18. [*444*]

Beukes, N.J. 1973: Precambrian iron-formations of southern Africa. *Economic Geology 68*, 960–1004. [*30*]

Beukes, N.J. 1987: Facies relations, depositional environments and diagenesis in a major Early Proterozoic stromatolitic carbonate platform to basinal sequence, Campbellrand Subgroup, Transvaal Supergroup, southern Africa. *Sedimentary Geology 54*, 1–46. [*31, 34, 247*]

Beukes, N.J. & Klein, C. 1990: Geochemistry and sedimentology of a facies transition from micro-banded to granular iron-formation in the early Proterozoic Transvaal Supergroup, South Africa. *Precambrian Research 47*, 99–139. [*40*]

Beukes, N.J. & Klein, C. 1992: Models for iron-formation deposition. In Schopf, J.W. & Klein, C. (eds.): *The Proterozoic Biosphere: A Multidisciplinary Study*, 147–152. Cambridge University Press, Cambridge. [*242*]

Beukes, N.J. & Lowe, D.R. 1989: Environmental control on diverse stromatolite morphologies in the 3000 Myr Pongola Supergroup, South Africa. *Sedimentology 36*, 383–397. [*31, 33, 275, 277*]

Bhattacharjee, S.K. & David, K.A.V. 1977: Unusual resistance to ultraviolet light in dark phase of blue-green bacterium *Anacystis nidulans*. *Nature 265*, 183–184. [*38*]

Bhattacharya, D., Elwood, H.J., Goff, L.J. & Sogin, M.L. 1990: Phylogeny of *Gracilaria lemaneiformis* (Rhodophyta) based on sequence analysis of its small subunit ribosomal RNA coding region. *Journal of Phycology 26*, 181–186. [*296*]

Bhattacharya, D., Medlin, L., Wainright, P.O., Ariztia, E.V., Bibeau, C., Stickel, S.K. & Sogin, M.L. 1992: Algae containing chlorophyll-$a+c$ are paraphyletic: Molecular evolutionary analysis of the Chromophyta. *Evolution 46*, 1801–1817. [*184*]

Bhattacharya, D., Stickel, S.K. & Sogin, M.L. 1991: Molecular phylogenetic analysis of actin genic regions from *Achlya bisexualis* (Oomycota) and *Costariacostata* (Chromophyta). *Journal of Molecular Evolution 33*, 4275–4286. [*182, 184*]

Bickford, M.E. 1988: The accretion of Proterozoic crust in Colorado: Igneous, sedimentary, deformational, and metamorphic history. In Ernst, W.G. (ed.): *Metamorphic and Crustal Evolution of the Western United States*, 411–430. Prentice-Hall, Englewood Cliffs, N.J. [*26*]

Bickle, M.J. 1978: Heat loss from the earth: A constraint on Archean tectonics from the relation between geothermal gradients and the rate of plate production. *Earth and Planetary Science Letters 40*, 301–315. [*211, 219*]

Biebricher, C.K., Diekmann, S. & Luce, R. 1982: Structural analysis of self-replicating RNA synthesized by Qb replicase. *Journal of Molecular Biology 154*, 629–648. [*72*]

Birch, A.J. 1973: Chemistry in botanical classification. In *Proceedings of the 25th Nobel Symposium*, 261–270. [*321*]

Birge, R.R., Cooper, T.M., Lawrence, A.F., Masthay, M.B., Zhang, C.-F. & Zidovetski, R.T. 1989: Spectrophotometric, photocalorimetric, and theoretical investigation of the quantum efficiency of the primary event in bacteriorhodopsin. *Journal of the American Chemical Society 111*, 4063–4074. [*85*]

Birket-Smith, S.J.R. 1981: A reconstruction of the Pre-Cambrian *Spriggina. Zoologische Jahrbücher, Anatomie und Ontogenie der Tiere 105*, 237–258. [*456*]

Black, M. 1933: The algal sediments of Andros Island, Bahamas. *Philosophical Transactions of the Royal Society of London, B 222*, 165–192. [*338*]

Blakemore, R.P. 1975: Magnetotactic bacteria. *Science 190*, 377–379. [*414*]

Blankenship, R. 1992: Origin and early evolution of photosynthesis. *Photosynthesis Research 33*, 91–111. [*177, 179, 180*]

Bloechl, E., Keller, M., Wächtershäuser, G. & Stetter, K.O. 1992: Reactions depending on iron sulfide and linking geochemistry with biochemistry. *Proceedings of the National Academy of Sciences, USA 89*, 8117–8120. [*19*]

Bloeser, B. 1985: *Melanocyrillium*, a new genus of structurally complex late Proterozoic microfossils from the Kwagunt Formation (Chuar Group), Grand Canyon, Arizona. *Journal of Paleontology 59*, 741–765. [*291*]

Boaden, P.J.S. 1975: Anaerobiosis, meiofauna and early metazoan evolution. *Zoologica Scripta 4*, 21–24. [*478*]

Bode, H.R. & Steele, R.E. 1989: Phylogeny and molecular data. *Science 243*, 548–549. [*468*]

Boettcher, A.L., Mysen, B.O. & Modreski, P.J. 1975: Melting in the mantle: Phase relationships in natural and synthetic peridotite–H_2O and peridotite–H_2O–CO_2 systems at high pressures. In Ahrens, L.H., Dawson, J.B., Duncan, A.R. & Erlank, A.J. (eds.): *Physics and Chemistry of the Earth*, 855–867. Pergamon, Oxford. [*11*]

Böggild, O. 1907: Struvit fra Limfjorden. *Dansk Geologisk Forening, Meddelelser 13*, 25–32. [*99*]

Böggild, O. 1909: Struvit von dem Limfjord. *Zeitschrift für Krystallographie und Mineralogie 46*, 608–609. [*99*]

Bonch-Osmolovskaya, E.A., Miroshnichenko, M.L., Kostrikina, N.A., Chernych, N.A. & Zavarzin, G.A. 1990: *Thermoproteus uzoniensis* sp. nov., a new extremely thermophilic archaebacterium from Kamchatka continental hot springs. *Archives of Microbiology 154*, 556–559. [*147*]

Bonch-Osmolovskaya, E.A. & Stetter, K.O. 1991: Interspecies hydrogen transfer in cocultures of thermophilic Archaea. *Systematic and Applied Microbiology 14*, 205–208. [*150*]

Bond, G.C., Kominz, M.A., Steckler, M.S. & Grotzinger, J.P. 1989: Role of thermal subsidence, flexure, and eustasy in the evolution of early Paleozoic passive-margin carbonate platforms. In Crevello, P.D., Wilson, J.L., Sarg, J.F. & Read, J.F. (eds.): *Controls on Carbonate Platform and Basin Development*, 39–61. Society of Economic Paleontologists and Mineralogists, Tulsa, Okl. [*257*]

Bond, G.C., Nickeson, P.A. & Kominz, M.A. 1984: Breakup of a supercontinent between 625 Ma and 555 Ma: New evidence and implications for continental histories. *Earth and Planetary Science Letters 70*, 325–345. [*443*]

Bonner, J.T. 1965: *Size and Cycle*. Princeton University Press, Princeton, N.J. [*406*]

Bonner, J.T. 1988: *The Evolution of Complexity by Means of Natural Selection*. Princeton University Press, Princeton, N.J. [*406*]

Borowitzka, M.A. 1989: Carbonate calcification in algae – initiation and control. In Mann, S., Webb, J. & Williams, R.J.P. (eds.): *Biomineralization*, 63–94. Verlag Chemie, Weinheim. [*429*]

Borowska, Z.K. & Mauzerall, D.C. 1988: Photoreduction of carbon dioxide by aqueous ferrous ion: An alternative to the strongly reducing atmosphere for the chemical origin of life. *Proceedings of the National Academy of Sciences, USA 85*, 6577–6580. [*13, 164, 172, 174*]

Borowska, Z.K. & Mauzerall, D.C. 1991: Corrections and retraction. *Proceedings of the National Academy of Sciences, USA 88*, 4564. [*13*]

Bottomley, D.J., Veizer, J., Nielsen, H. & Moczydłowska, M. 1992: Isotopic composition of disseminated sulfur in Precambrian sedimentary rocks. *Geochimica et Cosmochimica Acta 56*, 3311–3322. [*216*]

Bowring, S.A., Grotzinger, J.P., Isachsen, C.E., Knoll, A.H., Pelechaty, S.M. & Kolosov, P. 1993: Calibrating rates of Early Cambrian evolution. *Science 261*, 1293–1298. [*ix, 464*]

Boyan, B.D., Landis, W.J., Knight, J., Dereszewski, G. & Zeagler, J. 1984: Microbial hydroxyapatite formation as a model of proteolipid-dependent membrane-mediated calcification. *Scanning Electron Microscopy 4*, 1793–1800. [*415, 423*]

Brack, A. & Barbier, B. 1989: Early peptidic enzymes. *Advances in Space Research 9*, 83–87. [*75*]

Bradley, S.E. & Purcell, E.F. 1982: *The Paracellular Pathway*. Yosiah Macy Jr. Foundation, New York, N.Y. [*478*]

Brasier, M.D. 1982: Sealevel changes, facies changes, and the late Precambrian–Early Cambrian evolutionary explosion. *Precambrian Research 17*, 105–123. [*443*]

Brasier, M.D. 1986: Why do lower plants and animals biomineralize? *Paleobiology 12*, 241–250. [*423*]

Braterman, P.S., Cairns-Smith, A.G. & Sloper, R.W. 1983: Photo-oxidation of hydrated Fe^{2+} – significance for banded iron formations. *Nature 303*, 163. [*13, 42, 94*]

Brenner, S. 1991: Summary and concluding remarks. In Osawa, S. & Honjo, T. (eds.): *Evolution of Life: Fossils, Molecules, and Culture*, 453–456. Springer, Tokyo. [*503*]

Bridson, O.K., Fakhrai, H., Lohrmann, R., Orgel, L.E. & van Roode, M. 1981: Template-directed synthesis of oligoguanylic acids: Metal ion catalyses. In Wolman, Y. (ed.): *Origin of Life*, 233–239. Reidel, Dordrecht. [*56*]

Brierley, C.L. & Brierley, J.A. 1973: A chemolithoautotrophic and thermophilic microorganism isolated from an acidic hot spring. *Canadian Journal of Microbiology 19*, 183–188. [*146*]

Briggs, D.E.G. 1990: Early arthropods: Dampening the Cambrian explosion. In Culver, S.J. (ed.): *Arthropod Paleobiology*, 24–43. Short Courses in Paleontology 3 The Paleontological Society, Lawrence, Kans. [*464*]

Briggs, D.E.G. & Fortey, R.A. 1989: The early radiation and relationships of the major arthropod groups. *Science 246*, 241–243. [*452, 456, 502*]

Briggs, D.E.G., Fortey, R.A. & Wills, M.A. 1992: Morphological disparity in the Cambrian. *Science 256*, 1670–1673. [*452, 456, 459, 464*]

Brinkmann, R.T. 1969: Dissociation of water vapor and evolution of oxygen in the terrestrial atmosphere. *Journal of Geophysical Research 74*, 5355–5368. [*46*]

Britten, R.J. & Davidson, E.H. 1971: Repetitive and non-repetitive DNA sequences and a speculation on the origins of evolutionary novelty. *Quarterly Review of Biology 46*, 111–133. [*408*]

Brock, T.D. 1986: Notes on the ecology of thermophilic Archaebacteria. *Systematic and Applied Microbiology 7*, 213–215. [*39*]

Brock, T.D., Brock, K.M., Belly, R.T. & Weiss, R.L. 1972: *Sulfolobus*: A new genus of sulfur-oxidizing bacteria living at low pH and high temperature. *Archives of Microbiology 84*, 54–68. [*146*]

Broda, E. 1978: *The Evolution of the Bioenergetic Processes*. Pergamon, New York, N.Y. [*232*]

Broecker, W.S. 1971: A kinetic model for the chemical composition of sea water. *Quaternary Research 1*, 188–207. [*423*]

Broecker, W.S. & Peng T.H. 1982: *Tracers in the Sea*. Lamont-Doherty Geological Observatory, Palisades, N.Y. [*98*]

Brugerolle, G. & Mignot, J.-P. 1979: Observations sur le cycle l'ultrastructure et la position systématique de *Spiromonas perforans* (*Bodo perforans* Hollande 1938), flagellé parasite de *Chilomonas paramecium*, et relations avec les dinoflagellés et sporozoaires. *Protistologica 15*, 183–196. [*315, 319, 325*]

Brusca, R.C. & Brusca, G.J. 1990: *Invertebrates*. Sinauer, Sunderland, Mass. [*490*]

Buck, S.G. 1980: Stromatolite and ooid deposits within the fluvial and lacustrine sediments of the Precambrian Ventersdorp Supergroup of South Africa. *Precambrian Research 12*, 311–330. [*278*]

Budd, G. 1993: A Cambrian gilled lobopod from Greenland. *Nature 364*, 709–711. [*456*]

Buick, R. 1984: Carbonaceous filaments from North Pole, Western Australia: Are they fossil bacteria in Archaean stromatolites? *Precambrian Research 24*, 157–172. [*197, 275*]

Buick, R. 1988: Carbonaceous filaments from North Pole, Western Australia: Are they fossil bacteria in Archaean stromatolites? A reply. *Precambrian Research 39*, 311–317. [*197*]

Buick, R. 1990: Microfossil recognition in Archaean rocks: An appraisal of spheroids and filaments from 3500 M.Y. old chert-barite unit at North Pole, Western Australia. *Palaios 5*, 441–459. [*163, 275, 276*]

Buick, R. 1992: The antiquity of oxygenic photosynthesis: Evidence from stromatolites in sulphate-deficient Archean lakes. *Science 255*, 74–77. [*168, 172, 224, 233, 235*]

Buick, R. & Dunlop, J.S.R. 1990: Evaporitic sediments of Early Archean age from the Warrawoona Group, North Pole, Western Australia. *Sedimentology 37*, 247–277. [*28, 29, 163, 246, 253*]

Buick, R., Dunlop, J.S.R. & Groves, D.I. 1981: Stromatolite recognition in ancient rocks: An appraisal of irregularly laminated structures in an Early Archaean chert–barite unit from North Pole, Western Australia. *Alcheringa 5*, 161–181. [*197, 272, 273, 274, 275*]

Bujalowski, W. & Porschke, D. 1988: Contributions to selective binding of aromatic amino acid residues to tRNA^{Phe}. *Biophysical Chemistry 30*, 151–157. [*76*]

Burggraf, S., Jannasch, H.W., Nicolaus, B. & Stetter, K.O. 1990: *Archaeoglobus profundus* sp. nov., represents a new species within the sulfate-reducing Archaebacteria. *Systematic and Applied Microbiology 13*, 24–28. [*149*]

Burggraf, S., Olsen, G., Stetter, K.O. & Woese, C.R. 1992: A phylogenetic analysis of *Aquifex pyrophilus*. *Systematic and Applied Microbiology 15*, 352–356. [*39, 144, 149, 151, 155, 157*]

Burggraf, S., Stetter, K.O., Rouviere, P. & Woese, C.R. 1991: *Methanopyrus kandleri*: An archaeal methanogen unrelated to all other known methanogens. *Systematic and Applied Microbiology 14*, 346–351. [*144*]

Burke, A.C. 1989: Development of the turtle carapace: Implications for the evolution of a novel bauplan. *Journal of Morphology 199*, 363–378. [*498*]

Burke, K.C.A. & Dewey, J.F. 1973: An outline of Precambrian plate development. In Tarling, D.H. & Runcorn, S.K. (eds.): *Implications of Continental Drift to the Earth Sciences*, Vol. 2, 1035–1045. Academic Press, London. [*34*]

Burke, K.C.A., Dewey, J.F. & Kidd, W.S.F. 1976: Dominance of horizontal movements, arc and microcontinental collisions during the later permobile regime. In Windley, B.F. (ed.): *The Early History of the Earth*, 113–129. Wiley, London. [*25, 34*]

Buss, L. 1987: *Evolution of Individuality*. Princeton University Press, Princeton, N.J. [*139, 142, 476, 485, 487*]

Butlerow, A. 1861: Formation synthetique d'une substance sucree. *Comptes Rendus de l'Academie des Sciences 53*, 145–147. [*52, 54*]

Butterfield, N.J. 1990a: Organic preservation of non-mineralizing organisms and taphonomy of the Burgess Shale. *Paleobiology 16*, 272–286. [*301*]

Butterfield, N.J. 1990b: A re-assessment of the enigmatic Burgess Shale fossil *Wiwaxia corrugata* (Matthew) and its relationship to the polychaete *Canadia spinosa* Walcott. *Paleobiology 16*, 287–303. [*457*]

Butterfield, N.J. 1992: Studies in Neoproterozoic Paleontology from Svalbard and Arctic Canada. 284 pp. Ph.D. Thesis, Harvard University, Cambridge, Mass., USA. [*441*]

Butterfield, N.J., Knoll, A.H. & Swett, K. 1988: Exceptional preservation of fossils in an upper Proterozoic shale. *Nature 334*, 424–427. [*254, 291, 292, 301, 303, 306, 308, 427, 441*]

Butterfield, N.J., Knoll, A.H. & Swett, K. 1990: A bangiophyte red alga from the Proterozoic of Arctic Canada. *Science 250*, 104–107. [*186, 283, 291, 296, 343, 360, 427, 441*]

Byerly, G.R., Lowe, D.R. & Walsh, M.M. 1986: Stromatolites from the 3,300–3,500 Myr Swaziland Supergroup, Barberton Mountain Land, South Africa. *Nature 319*, 489–491. [*31, 195, 196, 275*]

Cachon, J., Cachon, M., Cosson, M.-P. & Cosson, J. 1988: The paraflagellar rod: A structure in search of a function. *Biology of the Cell 63*, 169–181. [*318*]

Cairns, J., Overbaugh, J. & Miller, S. 1988: The origin of mutants. *Nature 335*, 142–145. [*142*]

Cairns-Smith, A.G. 1965: The origin of life and the nature of the primitive gene. *Journal of Theoretical Biology 10*, 53–88. [*95*]

Calvino, I. 1968: *Cosmicomics*. Harcourt Brace Jovanovich, San Diego, Calif. [*489*]

Cameron, A.G.W. & Benz, W. 1991: The origin of the Moon and the single impact hypothesis.IV. *Icarus 92*, 204–221. [*51*]

Cammack, R., Rao, K.K. & Hall, D.O. 1981: Metalloproteins in the evolution of photosynthesis. *BioSystems 14*, 57–80. [*41*]

Campbell, J.H. 1987: The new gene and its evolution. In Campbell, K.S.W. & Day, M.F. (eds.): *Rates of Evolution*, 283–309. Allen & Unwin, London. [*515*]

Campbell, K.S.W. 1990: Palaeontological contributions to modern evolutionary theory: 1986 Mawson Lecture. *Australian Journal of Earth Sciences 37*, 247–265. [*460, 464, 514, 516*]

Cantrell, C.A., Shetter, R.E., McDaniel, A.H., Calvert, J.G., Davidson, J.A., Lowe, D.C., Tyler, S.C., Cicerone, R.J. & Greenberg, J.P. 1990: Carbon kinetic isotope effect in the oxidation of methane by the hydroxyl radical. *Journal of Geophysical Research 95*, 22455–22462. [*230*]

Canuto, V.M., Levine, J.S., Augustsson, T.R. & Imhoff, C.L. 1982: UV radiation from the young Sun and oxygen and ozone levels in the prebiological palaeoatmosphere. *Nature 296*, 816–820. [*38*]

Caplan, A.I. 1986: The extracellular matrix is instructive. *Bioassays 5*, 129–132. [*477*]

Card, K.D. 1990: A review of the Superior Province of the Canadian Shield, a product of Archean accretion. *Precambrian Research 48*, 99–156. [*26*]

Carson, H.L. 1987: Population genetics, evolutionary rates and neo-Darwinism. In Campbell, K.S.W. & Day, M.F. (eds.): *Rates of Evolution*, 209–217. Allen & Unwin, London. [*515*]

Carter, J.G. (ed.) 1990a: *Skeletal Biomineralization: Patterns, Processes and Evolutionary Trends*, Vol. 1. Van Nostrand Reinhold, New York, N.Y. [*414*]

Carter, J.G. (ed.) 1990b: *Skeletal Biomineralization: Patterns, Processes and Evolutionary Trends*, Vol. 2. Van Nostrand Reinhold, New York, N.Y. [*414*]

Carter, J.G. & Aller, R.C. 1975: Calcification in the bivalve periostracum. *Lethaia 8*, 315–320. [*418*]

Carver, J.H. 1981: Prebiotic atmospheric oxygen levels. *Nature 292*, 136–138. [*46*]

Castenholz, R.W. 1984: Composition of hot spring microbial mats: A summary. In Cohen, Y., Castenholz, R.Z. & Halvorsen, H.O. (eds.): *Microbial Mats: Stromatolites*, 101–119. Alan Liss, New York, N.Y. [*173*]

Castenholz, R.W., Bauld, J. & Jørgensen, B.B. 1990: Anoxygenic microbial mats of hot springs: Thermophilic *Chlorobium* sp. *FEMS Microbiology Ecology 74*, 325–336. [*168*]

Castenholz, R.W., Jørgensen, B.B., D'Amelio, E. & Bauld, J. 1991: Photosynthetic and behavioral versatility of the cyanobacterium *Oscillatoria boryana* in a sulfide-rich microbial mat. *FEMS Microbiology Ecology 86*, 43–58. [*166*]

Castenholz, R.W. & Waterbury, J.B. 1989: Oxygenic photosynthetic bacteria (sect.19), Group I. Cyanobacteria. In Stanley, J.T. (ed.): *Bergey's Manual of Systematic Bacteriology*, Vol. 3, 1710–1799. Williams & Wilkins, Baltimore, Md. [*334*]

Caudy, M., Grell, E.H., Dambly-Chaudière, C., Ghysen, A., Jan, L.Y. & Jan, Y.N. 1988: The maternal sex determination gene *daughterless* has zygotic activity necessary for the formation of peripheral neurons in *Drosophila*. *Genes and Development 2*, 843–852. [*507, 508*]

Cavalier-Smith, T. 1981: Eukaryote Kingdoms: Seven or nine? *BioSystems 14*, 461–481. [*318, 319*]

Cavalier-Smith, T. 1986: The Kingdom Chromista: Origin and systematics. In Round, F.E. & Chapman, D.J. (eds.): *Progress in Phycological Research*, Vol. 4, 309–347. Biopress, Bristol. [*182, 316, 317, 319*]

Cavalier-Smith, T. 1987a: The origin of eukaryote and archaebacterial cells. *Annals of the New York Academy of Sciences 503*, 17–54. [*189, 290*]

Cavalier-Smith, T. 1987b: The simultaneous symbiotic origin of mitochondria, chloroplasts and microbodies. *Annals of the New York Academy of Sciences 503*, 55–71. [*290*]

Cavalier-Smith, T. 1987c: The origin of fungi and pseudofungi. In Rayner, A.D.M., Brasier, C.M. & Moore, D. (eds.): *The Evolutionary Biology of Fungi*, 339–353. British Mycological Society, Symposium 13 Cambridge University Press, Cambridge. [*186*]

Cavalier-Smith, T. 1989a: Archaebacteria and Archezoa. *Nature 339*, 100–101. [*303*]

Cavalier-Smith, T. 1989b: The kingdom Chromista. In Green, J.C., Leadbeater, B.S.C. & Diver, W.L. (eds.): *The Chromophyte Algae: Problems and Perspectives*, 381–407. The Systematics Association Special Volume 38 Clarendon, Oxford. [*316, 318, 319, 325*]

Cavalier-Smith, T. 1991a: Archamoebae: The ancestral eukaryotes. *BioSystems 25*, 25–38. [323]

Cavalier-Smith, T. 1991b: The evolution of cells. In Osawa, S. & Honjo, T. (eds.): *Evolution of Life: Fossils, Molecules, and Culture*, 271–304. Springer, Tokyo. [179]

Cavalier-Smith, T. 1992: The number of symbiotic origins of organelles. *BioSystems 28*, 91–106. [296]

Cech, T.R. & Bass, B.L. 1986: Biological catalysis by RNA. *Annual Review of Biochemistry 55*, 599–629. [57, 62, 71, 76, 132]

Chakoumakos, B.C., Sales, B.C. & Boatner, L.A. 1990: Alpha-decay-induced condensation of phosphate anions in a mineral. *The American Mineralogist 75*, 1447–1450. [98]

Chakrabarti, A. & Deamer, D.W. 1992: Permeability of lipid bilayers to amino acids and phosphate. *Biochimica et Biophysica Acta 1111*, 171–177. [116]

Chameides, W.L. & Davis, D.D. 1983: Aqueous-phase source of formic acid in clouds. *Nature 304*, 427–429. [21]

Chameides, W.L. & Walker, J.C.G. 1981: Rates of fixation by lightning of carbon nitrogen in possible primitive atmospheres. *Origins of Life 11*, 291–302. [12]

Chang, S. 1979: Comets: Cosmic connections with carbonaceous meteorites, interstellar molecules and the origin of life. In Neugebauer, M., Yeomans, D.K. & Brandt, J.C. (eds.): *Space Missions to Comets*, 59–111. NASA Conference Publication 2089 NASA. [14]

Chang, S. 1988: Planetary environments and the conditions of life. *Philosophical Transactions of the Royal Society of London, A 325*, 601–610. [17]

Chang, S., Des Marais, D., Mack, R., Miller, S.L. & Strathearn, G.E. 1983: Prebiotic organic syntheses and the origin of life. In Schopf, J.W. (ed.): *Earth's Earliest Biosphere, Its Origin and Evolution*, 53–92. Princeton University Press, Princeton, N.J. [11, 12, 14, 143]

Chang, S., Williams, J., Ponnamperuma, C. & Rabinowitz, J. 1970: Phosphorylation of uridine with inorganic phosphates. *Space Life Sciences 2*, 144–150. [104]

Chang, S.-B. R. & Kirschvink, J.L. 1989: Magnetofossils, the magnetization of sediments, and the evolution of magnetite biomineralization. *Annual Review of Earth and Planetary Sciences 1989*, 169–195. [414]

Chapman, D.J. & Schopf, J.W. 1983: Biological and biochemical effects on the development of an aerobic environment. In Schopf, J.W. (ed.): *Earth's Earliest Biosphere, Its Origin and Evolution*, 302–320. Princeton University Press, Princeton, N.J. [45]

Chapman, D.M. 1974: Cnidarian histology. In Muscatine, L. & Lenhoff, H.M. (eds.): *Coelenterate Biology, Reviews and New Perspectives*, 2–92. Academic Press, New York, N.Y. [406]

Chapman, F. 1932: In: Palaeontological Department Report. *Records of the Geological Survey of India for 1931 66*, 28–29. [355]

Chapman, F. 1933: In: Palaeontological Department Report. *Records of the Geological Survey of India for 1932 67*, 20–21. [355]

Chapman, F. 1935: Primitive fossils, possible atrematous and neotrematous brachiopods, from the Vindhyans of India. *Records of the Geological Survey of India for 1935–1936 69*, 109–120. [355]

Chapman, G. 1974: The skeletal system. In Muscatine, L. & Lenhoff, H.M. (eds.): *Coelenterate Biology, Reviews and New Perspectives*, 93–128. Academic Press, New York, N.Y. [477]

Chen Junyuan 1988: Precambrian metazoans of the Huai River drainage area (Anhui, E. China): Their taphonomic and ecological evidence. *Senckenbergiana lethaea 69*, 189–215. [291, 355, 357]

Chen Junyuan, Bergström, J., Lindström, M. & Hou Xianguang 1991: Fossilized soft-bodied fauna. *National Geographic Research and Exploration 7*, 8–19. [367, 453]

Chen Junyuan & Erdtmann, B. 1991: Lower Cambrian fossil Lagerstätte from Chengjiang, Yunnan, China: Insights for reconstructing early metazoan life. In Simonetta, A.M. & Conway Morris, S. (eds.): *The Early Evolution of Metazoa and the Significance of Problematic Taxa*, 57–76. Cambridge University Press, Cambridge. [367, 394, 417, 452, 456]

Chen M. & Zheng W. 1986: On the pre-Ediacaran Huainan biota. *Scientia Geologica Sinica 7*, 221–231. (In Chinese, with English abstract.) [356, 357]

Chen Menge & Xiao Zongzheng 1991: Discovery of the macrofossils in the Upper Sinian Doushantuo Formation at Miaohe, eastern Yangtze Gorges. *Scientia Geologica Sinica 4*, 317–324. [357, 451]

Chester, R. 1986: The marine mineral aerosol. In Buat-Menard, P. (ed.): *The Role of Air–Sea Exchange in Geochemical Cycling*, 443–476. Reidel, Dordrecht. [21]

Chewings, C. 1914: Notes on the stratigraphy of Central Australia. *Transactions of the Royal Society of South Australia 38*, 41–52. [*272*]

Cho, K.W.Y., Blumberg, B., Steinbeisser, H. & De Robertis, E.M. 1991: Molecular nature of Spemann's organizer: The role of the *Xenopus* homeobox gene *goosecoid*. *Cell 67*, 1111–1120. [*499*]

Chothia, C. & Lesk, A.M. 1986: The relation between the divergence of sequence and structure in proteins. *EMBO Journal 5*, 823–826. [*511*]

Christen, R., Ratto, A., Baroin, A., Perasso, R., Grell, K.G. & Adoutte, A. 1991a: Origin of metazoans. A phylogeny deduced from sequences of the 28S ribosomal RNA. In Simonetta, A.M. & Conway Morris, S. (eds.): *The Early Evolution of Metazoa and the Significance of Problematic Taxa*, 1–9. Cambridge University Press, Cambridge. [*450, 453, 462, 463*]

Christen, R., Ratto, A., Baroin, A., Perasso, R., Grell, K.G. & Adoutte, A. 1991b: An analysis of the origin of metazoans, using comparisons of partial sequences of the 28S RNA, reveals an early emergence of triploblasts. *EMBO Journal 10*, 499–503. [*453, 469, 470, 490*]

Christensen, T. 1966: Alger. In Bøcher, T.W., Lange, M. & Sorensen, T. (eds.): *Botanik II, Systematisk Botanik 2*. Munksgaard, Copenhagen. [*325*]

Chumakov, N.M. 1978: *Dokembrijskie Tillity i Tilloidy*. Nauka, Moscow. [*371*]

Chyba, C.F. 1987: The cometary contribution to the oceans of the primitive Earth. *Nature 330*, 632–635. [*51*]

Chyba, C.F. & Sagan, C. 1992: Endogenous production, exogenous delivery, and impact-shock synthesis of organic molecules: An inventory for the origins of life. *Nature 355*, 125–131. [*14, 16, 17, 37, 108*]

Chyba, C.F., Thomas, P.J., Brookshaw, L. & Sagan, C. 1990: Cometary delivery of organic molecules to the early Earth. *Science 249*, 366–373. [*14, 51, 108*]

Clark, B.C. 1988: Primeval procreative comet pond. *Origins of Life and Evolution of the Biosphere 18*, 209–238. [*15*]

Clark, R.B. 1964: *Dynamics in Metazoan Evolution*. Clarendon, Oxford. [*403, 456, 485, 486*]

Clark, R.B. 1981: Locomotion and phylogeny of the Metazoa. *Bolletino di Zoologica 48*, 11–28. [*486*]

Claypool, G.E., Holser, W.T., Kaplan. I.R. Sakai, H. & Zak, I. 1980: The age curves of sulfur and oxygen isotopes in marine sulfate and their mutual interpretation. *Chemical Geology 28*, 199–260. [*444*]

Cleveland, L.R. 1947: The origin and evolution of meiosis. *Science 105*, 287–288. [*448*]

Cloud, P.E. 1948: Some problems and patterns of evolution exemplified by fossil invertebrates. *Evolution 2*, 322–350. [*412, 436*]

Cloud, P.E. 1968a: Pre-metazoan evolution and the origins of the Metazoa. In Drake, E.T. (ed.): *Evolution and Environment*, 1–72. Yale University Press, New Haven, Conn. [*412, 436*]

Cloud, P.E. 1968b: Atmospheric and hydrospheric evolution on the primitive earth. *Science 160*, 729–736. [*29, 30, 239*]

Cloud, P. 1973: Paleoecological significance of the banded iron-formation. *Economic Geology 68*, 1135–1143. [*13, 43, 46*]

Cloud, P. 1974: Dating the beginnings of photosynthesis. *American Scientist 62*, 389–390. [*36, 40, 42*]

Cloud, P. 1976a: Major features of crustal evolution. *Geological Society of South Africa, Annexure 79*, 1–33. [*29*]

Cloud, P.E. 1976b: Beginnings of biospheric evolution and their biogeochemical consequences. *Paleobiology 2*, 351–387. [*427, 432*]

Cloud, P. 1983: Early biogeologic history: The emergence of a paradigm. In Schopf, J.W. (ed.): *Earth's Earliest Biosphere, Its Origin and Evolution*, 14–31. Princeton University Press, Princeton, N.J. [*37, 40*]

Cloud, P. 1986: Reflections on the beginnings of metazoan evolution. *Precambrian Research 31*, 405–407. [*363*]

Cloud, P. 1988: *Oasis in Space, Earth History from the Beginning*. Norton, New York, N.Y. [*435, 436*]

Cloud, P. & Glaessner, M. F. 1982: The Ediacarian Period and System: Metazoa inherit the Earth. *Science 217*, 783–792. [*293, 363*]

Cloud, P.E. & Semikhatov, M.A. 1969: Proterozoic stromatolite zonation. *American Journal of Science 267*, 1017–1061. [*255, 273*]

Cloud, P., Wright, L.A., Williams, E.G., Diehl, P. & Walter, M.R. 1974: Giant stromatolites and associated vertical tubes from the Upper Proterozoic Noonday Dolomite, Death Valley region, eastern California. *Geological Society of America, Bulletin 85*, 1869–1882. [*430, 431, 432*]

Cohen, Y. 1984: Oxygenic photosynthesis, anoxygenic photosynthesis, and sulfate reduction in cyanobacterial mats. In Klug, M.J. & Reddy, C.A. (eds.): *Current Perspectives in Microbial Ecology*, 435–441. American Society for Microbiology, Washington, D.C. [*175*]

Cold Spring Harbor Symposia on Quantitative Biology 1987: *Evolution of Catalytic Function*, Vol. 52. Cold Spring Harbor Press, New York. [*515*]

Coleman, D.D., Risatti, J.B. & Schoell, M. 1981: Fractionation of carbon and hydrogen isotopes by methane-oxidizing bacteria. *Geochimica et Cosmochimica Acta 45*, 1033–1037. [*230*]

Collins, D. 1992: Whither *Anomalocaris*? The search in the Burgess Shale continues. *Special Publication of the Paleontological Society (Fifth North American Paleontology Convention Abstracts) 6*, 66. [*456*]

Collister, J.W., Summons, R.E., Lichtfouse, E. & Hayes, J.M. 1992: An isotopic biogeochemical study of the Green River Oil Shale. *Organic Geochemistry 19*, 265–276. [*229, 231*]

Compston, W., Williams, I.S., Kirschvink, J.L., Zhang, Z.-C. & Ma, G.-G. 1992: Zircon U–Pb ages for the Early Cambrian time-scale. *Journal of the Geological Society 149*, 171–184. [*ix, 405*]

Conway Morris, S. 1977: Fossil priapulid worms. *Special Papers in Palaeontology 20*. 95 pp. Palaeontological Association, London. [*458*]

Conway Morris, S. 1979: Middle Cambrian polychaetes from the Burgess Shale of British Columbia. *Philosophical Transactions of the Royal Society of London, B 285*, 227–274. [*456*]

Conway Morris, S. 1982: Parasites and the fossil record. *Parasitology 82*, 489–509. [*454*]

Conway Morris, S. 1985: Non-skeletalized lower invertebrate fossils: A review. In Conway Morris, S., George, J.D., Gibson, R. & Platt, H.M. (eds.): *The Origins and Relationships of Lower Invertebrates*, 343–359. Clarendon, Oxford. [*454*]

Conway Morris, S. 1986: The community structure of the Middle Cambrian Phyllopod Bed (Burgess Shale). *Palaeontology 29*, 423–467. [*417, 424, 442, 451*]

Conway Morris, S. 1987: The search for the Precambrian–Cambrian boundary. *American Scientist 75*, 157–167. [*437*]

Conway Morris, S. 1989a: Early metazoans. *Science Progress, Oxford 73*, 81–99. [*359, 363, 365*]

Conway Morris, S. 1989b: Burgess Shale faunas and the Cambrian explosion. *Science 246*, 339–346. [*299, 363, 367, 451, 464, 502*]

Conway Morris, S. 1989c: South-eastern Newfoundland and adjacent areas (Avalon Zone). In Cowie, J.W. & Brasier, M.D. (eds.): *The Precambrian–Cambrian Boundary*, 7–39. Clarendon, Oxford. [*405*]

Conway Morris, S. 1991: Problematic taxa: A problem for biology or biologists? In Simonetta, A.M. & Conway Morris, S. (eds.): *The Early Evolution of Metazoa and the Significance of Problematic Taxa*, 19–24. Cambridge University Press, Cambridge. [*463*]

Conway Morris, S. 1992: Burgess Shale-type faunas in the context of the "Cambrian explosion": A review. *Journal of the Geological Society, London 149*, 631–636. [*457*]

Conway Morris, S. 1993a: Ediacaran-like fossils in Cambrian Burgess Shale-type faunas of North America. *Palaeontology 36*, 593–635. [*451, 458*]

Conway Morris, S. 1993b: The fossil record and the early evolution of the Metazoa. *Nature 361*, 219–225. [*452, 483, 485*]

Conway Morris, S. & Bengtson, S. 1994: Cambrian predators: Possible evidence from boreholes. *Journal of Paleontology 68*, 1–23. [*424*]

Conway Morris, S. & Chen Menge 1989: Lower Cambrian anabaritids from South China. *Geological Magazine 126*, 615–632. [*295, 454*]

Conway Morris, S. & Chen Menge 1990a: *Blastulospongia polytreta* n.sp., an enigmatic organism from the Lower Cambrian of Hubei, China. *Journal of Paleontology 64*, 26–30. [*420*]

Conway Morris, S. & Chen Menge 1990b: Tommotiids from the Lower Cambrian of South China. *Journal of Paleontology 64*, 169–184. [*458*]

Conway Morris, S. & Chen Menge 1992: Carinachitiids, hexangulaconularids, and *Punctatus*: Problematic metazoans from the early Cambrian of South China. *Journal of Paleontology 66*, 384–405. [*454*]

Conway Morris, S. & Jenkins, R.J.F. 1985: Healed injuries in Early Cambrian trilobites from South Australia. *Alcheringa 9*, 167–177. [*424*]

Conway Morris, S., Mattes, B.W. & Chen Menge 1990: The early skeletal organism *Cloudina*: New occurrences from Oman and possibly China. *American Journal of Science 290-A*, 245–260. [*291, 295, 365, 442*]

Conway Morris, S. & Peel, J.S. 1990: Articulated halkieriids from the Lower Cambrian of North Greenland. *Nature 345*, 802–805. [*418, 419, 452, 457, 490*]

Conway Morris, S. & Robison, R.A. 1982: The enigmatic medusoid *Peytoia* and a comparison of some Cambrian biotas. *Journal of Paleontology 56*, 116–122. [*458*]

Conway Morris, S. & Robison, R.A. 1986: Middle Cambrian priapulids and other soft-bodied fossils from Utah and Spain. *University of Kansas Paleontological Contributions, Paper 117*. 22 pp. [*454, 458*]

Conway Morris, S. & Robison, R.A. 1988: More soft-bodied animals and algae from the Middle Cambrian of Utah and British Columbia. *University of Kansas Paleontological Contributions, Paper 122*. 48 pp. [*456*]

Cook, P.J. & Shergold, J.H. 1984: Phosphorus, phosphorites and skeletal evolution at the Precambrian–Cambrian boundary. *Nature 308*, 231–236. [*422, 423, 444, 445*]

Cook, P.J. & Shergold, J.H. 1986: Proterozoic and Cambrian phosphorites – nature and origin. In Cook, P.J. & Shergold, J.H. (eds.): *Phosphate Deposits of the World. 1. Proterozoic and Cambrian Phosphorites*, 369–386. Cambridge University Press, Cambridge. [*422, 423*]

Cook, P.L. 1988: Bryozoa. In Higgins, R. P. & Thiel, H. (eds.): *Introduction to the Study of Meiofauna*. Smithsonian Institution Press, Washington, D.C. [*487*]

Cope, J.C.W. 1977: An Ediacara fauna from South Wales. *Nature 268*, 624. [*372*]

Cope, J.C.W. 1983: Precambrian fossils of the Carmarthen area, Dyfed. *Nature in Wales June*, 11–16. [*372*]

Copeland, H.F. 1956: *The Classification of Lower Organisms*. Pacific Books, Palo Alto, Calif. [*313*]

Corliss, J.B., Baross, J.A. & Hoffman, S.E. 1981: An hypothesis concerning the relationship between submarine hot springs and the origin of life on Earth. *Oceanologica Acta 26*, 59–69. [*18, 35, 134*]

Corliss, J.O. 1975: Nuclear characteristics and phylogeny in the protistan phylum Ciliophora. *BioSystems 7*, 338–349. [*318, 325*]

Corliss, J.O. 1986: Advances in studies on phylogeny and evolution of protists. *Insect Science and its Application 7*, 305–312. [*314*]

Corliss, J.O. 1987: Protistan phylogeny and eukaryogenesis. *International Review of Cytology 100*, 319–370. [*314*]

Corliss, J.O. 1992: Should there be a separate code of nomenclature for the protists? *BioSystems 27*, 500–560. [*332*]

Cortial, F., Gauthier-Lafaye, F., Lacrampe-Couloume, G., Oberlin, A. & Weber, F. 1990: Characterization of organic matter associated with uranium deposits in the Francevillian Formation of Gabon (lower Proterozoic). *Organic Geochemistry 15*, 73–85. [*232*]

Crick, F.H.C. 1968: The origin of the genetic code. *Journal of Molecular Biology 38*, 367–379. [*63, 76, 138*]

Crick, F. 1981: *Life Itself: Its Origin and Nature*. Simon & Schuster, New York. [*62*]

Crick, R.E. (ed.) 1989: *Origin, Evolution and Modern Aspects of Biomineralization in Plants and Animals*. Plenum, New York, N.Y. [*414*]

Crimes, T.P. 1987: Trace fossils and correlation of late Precambrian and early Cambrian strata. *Geological Magazine 124*, 97–119. [*366, 417, 451*]

Crimes, T.P. 1989: Trace fossils. In Cowie, J.W. & Brasier, M.D. (eds.): *The Precambrian–Cambrian Boundary*, 166–185. Oxford Monographs on Geology and Geophysics 12. [*398*]

Cronin, J.R. 1989: Origin of organic compounds in carbonaceous chondrites. *Advances in Space Research 9*, 54–64. [*50*]

Cronin, J.R., Pizzarello, S. & Cruickshank, D.P. 1988: Organic matter in carbonaceous chondrites, planetary satellites, asteroids and comets. In Kerridge, J.F. & Matthews, M.S. (eds.): *Meteorites and the Early Solar System*, 819–857. University of Arizona Press, Tucson, Ariz. [*109*]

Currey, J.D. 1990: Biomechanics of mineralized skeletons. In Carter, J.G. (ed.): *Skeletal Biomineralization: Patterns, Processes and Evolutionary Trends*, Vol. 1, 11–25. Van Nostrand Reinhold, New York, N.Y. [*413*]

Dalziel, I.W.D. 1991: Pacific margins of Laurentia and East Antarctica–Australia as a conjugate rift pair: Evidence and implications for an Eocambrian supercontinent. *Geology 19*, 598–601. [*443*]

D'Amelio, E.D., Cohen, Y. & Des Marais, D.J. 1987: Association of a new type of gliding, filamentous, purple phototrophic bacterium inside bundles of *Microcoleus chthonoplastes* in hypersaline cyanobacterial mats. *Archives of Microbiology 147*, 213–220. [*173*]

D'Amelio, E.D., Cohen, Y. & Des Marais, D.J. 1989: Comparative functional ultrastructure of two hypersaline submerged cyanobacterial mats: Guerro Negro, Baja California Sur, Mexico, and Solar Lake, Sinai, Egypt. In Cohen, Y. & Rosenberg, E. (eds.): *Microbial Mats: Physiological Ecology of Benthic Microbial Communities*, 97–113. American Society for Microbiology, Washington, D.C. [*168, 173*]

Danchin, R.V. 1967: Chromium and nickel in the Fig Tree Shale from South Africa. *Science 158*, 261–262. [*28*]

Darwin, C. 1859: *The Origin of Species by Means of Natural Selection*. J. Murray, London. [*298, 358*]

Davy, R. 1983: A geochemical study of the Mount McRae shale and the upper part of the Mount Sylvia Formation in Core RD1, Rhodes Ridge, Western Australia. *Geological Survey Record 1983/3*, [*241*]

Davy, R. & Hickman, A.H. 1988: The transition between the Hamersley and Fortescue Groups as evidenced in a drill core. *Geological Survey of Western Australia, Professional Papers, Report 23*, 85–97. [*241*]

Dayhoff, M.O. (ed.) 1972: *Atlas of Protein Sequence and Structure 5*. National Biomedical Research Foundation, Washingon, D.C. [*68*]

Deamer, D.W. 1985: Boundary structures are formed by organic components of the Murchison carbonaceous chondrite. *Nature 317*, 792–794. [*21, 93, 112*]

Deamer, D.W. 1986a: Role of amphiphilic compounds on the evolution of membrane structure on the early Earth. *Origins of Life and Evolution of the Biosphere 17*, 3–25. [*65*]

Deamer, D.W. 1986b: Boundary structures and the non-polar organic components of the Murchison carbonaceous chondrite. *Origins of Life and Evolution of the Biosphere 16*, 363–364. [*93*]

Deamer, D.W. 1992: Polycyclic hydrocarbons: Possible pigment systems in the prebiotic environment. *Advances in Space Research 12*, 183–189. [*121*]

Deamer, D.W. & Barchfeld, G.L. 1982: Encapsulation of macromolecules by lipid vesicles under simulated prebiotic conditions. *Journal of Molecular Evolution 18*, 203–206. [*115*]

Deamer, D.W. & Harang, E. 1990: Light-dependent pH gradients are generated in liposomes containing ferrocyanide. *BioSystems 24*, 1–4. [*117*]

Deamer, D.W. & Oró, J. 1980: Role of lipids in prebiotic structures. *BioSystems 12*, 167–175. [*56, 119, 123*]

Deamer, D.W. & Pashley, R.M. 1989: Amphiphilic components of the Murchison carbonaceous chondrite: Surface properties and membrane formation. *Origins of Life and Evolution of the Biosphere 19*, 21–38. [*56, 112*]

Deamer, D.W., Prince, R. & Crofts, A.R. 1972: Response of fluorescent amines to pH gradients across liposome membranes. *Biochimica et Biophysica Acta 274*, 323–335. [*118, 119*]

deBeer, G. 1971: Homology, an unsolved problem. In Head, J.J. & Lowenstein, O.E. (eds.): *Oxford Biology Readers*, 3–16. Oxford University Press, London. [*496*]

de Duve, C 1987: Selection by differential molecular survival: A possible mechanism of early chemical evolution. *Proceedings of the National Academy of Sciences, USA 84*, 8252–8256. [*82*]

de Duve, C. 1991: *Blueprint For a Cell: The Nature and Origin of Life*. Patterson, Burlington, N.C. [*85, 87, 328*]

Deflandre-Rigaud, M. 1946: Vestiges microscopiques des larves d'Echinoderms de l'Oxfordien de Villers-sur-Mer. *Comptes Rendus de l'Academie des Sciences 222*, 908–910. [*493*]

Degens, E.T. 1978: The protobiosphere. *Chemical Geology 22*, 177–187. [*37*]

Degens, E.T. 1984: Warum verkalken Organismen? *Acta Universitatis Carolinae – Geologica 2*, 109–121. [*421*]

Degens, E.T., Kaźmierczak, J. & Ittekott, V. 1985: Cellular response to Ca2+ stress and its geological implications. *Acta Palaeontologica Polonica 30*, 115–135. [*421*]

Delaney, S.J., Hayward, D.C., Barleben, F., Fischbach, K.-F. & Miklos, G.L.G. 1991: Molecular cloning and analysis of small optic lobes, a structural brain gene of *Drosophila melanogaster*. *Proceedings of the National Academy of Sciences, USA 88*, 7214–7218. [*509*]

del Pino, E.M. 1989: Modifications of oogenesis and development in marsupial frogs. *Development 107*, 169–187. [*494*]

del Pino, E.M. & Elinson, R.P. 1983: A novel developmental pattern for frogs: Gastrulation produces an embryonic disk. *Nature 306*, 589–591. [*494*]

Delsemme, A.H. 1984: The cometary connection with prebiotic chemistry. *Origins of Life 14*, 51–60. [*37*]

Delsemme, A.H. 1992a: Cometary origin of carbon and water on the terrestrial planets. *Advances in Space Science and Technology 12*, 5–12. [*51*]

Delsemme, A.H. 1992b: Cometary Origin of Carbon, Nitrogen and Water on the Earth. *Origins of Life and Evolution of the Biosphere 21*, 279–298. [*51*]

Demoulin, V. & Janssen, M.P. 1981: Relationship between diameter of the filament and cell shape in blue-green algae. *British Phycological Journal 16*, 55–58. [*292*]

DePaolo, D.J. & Linn, A.M. & Schubert, G. 1991: The continental crustal age distribution: Methods of determining mantle separation ages from Sm–Nd isotopic data and application to southwestern United States. *Journal of Geophysical Research 96*, 2071–2088. [*210*]

Deppenmeier, U., Blaut, M., Mahlmann, A. & Gottschalk, G. 1990: Reduced coenzyme F_{420}: Hetero-disulfide oxidoreductase, a proton-translocating redox system in methanogenic bacteria. *Proceedings of the National Academy of Sciences, USA 87*, 9449–9453. [*85*]

Derry, L.A. & Jacobsen, S.B. 1990: The chemical evolution of Precambrian seawater: Evidence from REEs in banded iron formations. *Geochimica et Cosmochimica Acta 54*, 2965–2977. [*13, 40*]

Derry, L.A., Kaufman, A.J. & Jacobsen, S.B. 1992: Sedimentary cycling and environmental change in the Late Proterozoic: Evidence from stable and radiogenic isotopes. *Geochimica et Cosmochimica Acta 56*, 1317–1329. [*444, 449*]

Des Marais, D.J. 1985: Carbon exchange between the mantle and the crust and its effect upon the atmosphere: Today compared to Archean time. In Sundquist, E.T. & Broecker, W.S. (eds.): *Natural Variations in Carbon Dioxide and the Carbon Cycle*, 602–611. American Geophysical Union, Washington, D.C. [*32*]

Des Marais, D.J., Strauss, H., Summons, R.E. & Hayes, J.M. 1992: Carbon isotopic evidence for the stepwise oxidation of the Proterozoic environment. *Nature 359*, 605–609. [*222*]

Desmond, A. 1982: *Archetypes and Ancestors*. University of Chicago Press, Chicago, Ill. [*496*]

De Soete, G.A. 1983: A least squares algorithm for fitting additive trees to proximity data. *Psychometrika 48*, 621–626. [*187, 188*]

de Wit, M.J., Hart, R., Martin, A. & Abbott, P. 1982: Archean abiogenic and probable biogenic structures associated with mineralized hydrothermal vent systems and regional metasomatism, with implications for greenstone belt studies. *Economic Geology 77*, 1783–1801. [*27*]

de Wit, M.J., Roering, C., Hart, R.J., Armstrong, R.A., de Ronde, C.E.J., Green, R.W.E., Tredoux, M., Peberly, E. & Hart, R.A. 1992: Formation of an archean continent. *Nature 357*, 553–562. [*106*]

Diamond, J.M. 1977: The epithelial junction: Bridge, gate and fence. *Physiologist 20*, 10–18. [*478*]

Dickerson, R.E. 1978: Chemical evolution and the origin of life. *Scientific American 239*, 70–109. [*38*]

Dickinson, W.J. 1988: On the architecture of regulatory systems: Evolutionary insights and implications. *BioEssays 8*, 204–208. [*509*]

Dill, R.F., Shinn, E.A., Jones, A.T., Kelly, K. & Steinen, R.P. 1986: Giant subtidal stromatolites forming in normal salinity waters. *Nature 324*, 55–58. [*338*]

Dimroth, E. & Lichtblau, A.P. 1978: Oxygen in the Archean ocean: Comparison of ferric oxide crusts on Archean and Cainozoic pillow basalts. *Neues Jahrbuch für Mineralogie, Abhandlungen 133*, 1–22. [*28, 42*]

Dixon, T.H. & Golombek, M.P. 1988: Late Precambrian crustal accretion rates in northeast Africa and Arabia. *Geology 16*, 991–994. [*26*]

Do, H.K., Kogure, K. & Simidu, U. 1990: Identification of deep-sea-sediment bacteria which produce tetrodotoxin. *Applied and Environmental Microbiology 1990*, 1162–1163. [*317*]

Dodge, J.D. 1965: Chromosome structure in the dinoflagellates and the problem of the mesocaryotic cell. In *2nd International Conference on Protozoology*, 339. Excerpta Medica International Congress Series 91. [*321*]

Dodge, J.D. 1979: The phytoflagellates: Fine structure and phylogeny. In Levandowsky, M. & Hutner, S.H. (eds.): *Biochemistry and Physiology of Protozoa*. 2d Ed., Vol. 1, 7–57. [*321*]

Dodge, J.D. 1989: Phylogenetic relationships of dinoflagellates and their plastids. In Green, J.C., Leadbeater, B.S.C. & Diver, W.L. (eds.): *The Chromophyte Algae: Problems and Perspectives*, 207–227. The Systematics Association Special Volume 38 Clarendon, Oxford. [*321*]

Donaldson, J.A. 1963: Stromatolites in the Denault Formation, Marion Lake, coast of Labrador, Newfoundland. *Geological Survey of Canada, Bulletin 102*. 33 pp. [*255, 429, 430*]

Donaldson, J.A. 1976: Paleoecology of Conophyton and associated stromatolites in the Precambrian Dismal Lakes and Rae Groups, Canada. In Walter, M.R. (ed.): *Stromatolites*, 523–534. Developments in Sedimentology Elsevier, Amsterdam. [*252*]

Donnelly, T.H., Shergold, J.H., Southgate, P.N. & Barnes, C.J. 1990: Events leading to global phosphogenesis around the Proterozoic/Cambrian boundary. In Notholt, A.J.G. & Jarvis, I. (eds.): *Phosphorite Research and Development*, 273–287. Geological Society Special Publications 52 London. [*422, 423*]

Doolittle, R.F. 1989: Similar amino acid sequences revisited. *Trends in Biochemical Sciences 14*, 244–245. [*510*]

Doolittle, R.F. 1990: What we have learned and will learn from sequence databases. In Bell, G.I. & Marr, T.G. (eds.): *Computers and DNA*, 21–31. Addison Wesley, Menlo Park, Calif. [*68*]

Doolittle, R.F., Anderson, K.L. & Feng, D.-F. 1989: Estimating the prokaryote–eukaryote divergence time from protein sequences. In Fernholm, B., Bremer, K. & Jörnvall, H. (eds.): *The Hierarchy of Life*, 73–85. Excerpta Medica, Amsterdam. [*290, 291, 502*]

Dose, K. 1974: Peptides and amino acids in the primordial hydrosphere. In Dose, K., Fox, S.W., Deborin, G.A. & Pavlovskaya, T.E. (eds.): *The Origin of Life and Evolutionary Biochemistry*, 69–77. Plenum, New York, N.Y. [*17*]

Doudna, J.A. & Szostak, J.W. 1989: RNA-catalyzed synthesis of complementary-strand RNA. *Nature 339*, 519–522. [*64, 71*]

Douglas, S.E., Murphy, C.A., Spencer, D.F. & Gray, M.W. 1991: Cryptomonad algae are evolutionary chimeras of two phylogenetically distinct unicellular eukaryotes. *Nature 350*, 148–151. [*317, 320, 326*]

Dover, G.A. 1987: DNA turnover and the molecular clock. *Journal of Molecular Evolution 26*, 47–58. [*468, 515*]

Dravis, J.J. 1983: Hardened subtidal stromatolites, Bahamas. *Science 219*, 385–386. [*338*]

Dressler, G.R. & Gruss, P. 1988: Do multigene families regulate vertebrate development? *Trends in Genetics 4*, 214–219. [*140, 141*]

Drever, J.I. 1974: Geochemical model for the origin of Precambrian banded iron formations. *Geological Society of America, Bulletin 85*, 1099–1106. [*46*]

Driever, W. & Nüsslein-Volhard, C. 1988: A gradient of *bicoid* protein in *Drosophila* embroys. *Cell 54*, 83–93. [*506*]

Driever, W., Thoma, G. & Nüsslein-Volhard, C. 1989: Determination of spatial domains of zygotic gene expression in the *Drosophila* embryo by the affinity of binding sites for the bicoid morphogen. *Nature 340*, 363–367. [*506*]

Drobner, E., Huber, H., Wächtershäuser, G., Rose, D. & Stetter, K.O. 1990: Pyrite formation linked with hydrogen evolution under anaerobic conditions. *Nature 346*, 742–744. [*82, 85, 151, 154*]

Drozd, J.W., Tubb, R.S. & Postgate, J.R. 1972: A chemostat study of the effect of fixed nitrogen sources on nitrogen fixation, membrane, and free amino acids in *Azotobacter chroococcum*. *Journal of General Microbiology 73*, 221–232. [*41*]

Du Rulin 1982: The discovery of the fossils such as *Chuaria* in the Qingbaikou System in Northwestern Hebei and their significance. *Geology Review (Beijing) 28*, 1–6. (In Chinese, with English abstract.) [*356*]

Du Rulin & Tian Lifu 1985: Discovery and preliminary study of mega-alga *Longfengshania* from the Qingbaikou System of the Yanshan Mountain area. *Acta Geologica Sinica 1985*, 183–190. (In Chinese, with English abstract.) [*351*]

Du Rulin & Tian Lifu 1987: The macroalgal fossils of Qingbaikou Period in Yanshan Range. *Chinese Late Precambrian Geology Research Results 12*. 114 pp. Hebei Science and Technology Press. (Date of imprint 1986. In Chinese, with an English abstract.) [*356*]

Du Rulin, Tian Lifu & Li Hanbang 1986: Discovery of megafossils in the Gaoyuzhuang Formation of the Changchengian System, Jixian. *Acta Geologica Sinica 1986*, 115–120. (In Chinese, with English abstract.) [*357*]

Duan, Ch.-H. 1982: Late Precambrian algal megafossils *Chuaria* and *Tawuia* in some areas of China. *Alcheringa 6*, 57–68. [*346, 350, 355, 357*]

Duan C., Xing Y., Du R., Yin Y. & Liu, G. 1985: Macroscopic fossil algae. In Xing, Y. *et al.* (ed.): *Late Precambrian Palaeontology of China.*, 68–77. Ministry of Geology and Mineral Resources, Geological Memoirs, Series 2 2 Geological Publishing House, Beijing. (In Chinese.) [*356*]

DuBose, R.F. & Hartl, D.L. 1989: An experimental approach to testing modular evolution: Directed replacement of µ-helices in a bacterial protein. *Proceedings of the National Academy of Sciences, USA 86*, 9966–9970. [*511*]

Dunbar, C.O. & Rodgers, J. 1957: *Principles of Stratigraphy*. Wiley, New York, N.Y. [*46*]

Dunlop, J.S.R., Muir, M.D., Milne, V.A. & Groves, D.I. 1978: A new microfossil assemblage from the Archaean of Western Australia. *Nature 274*, 676–678. [*197*]

Durham, J.W. 1971: The fossil record and the origin of the Deuterostomia. *North American Paleontological Convention, Chicago, Proceedings H*, 1104–1132. [*409*]

Durham, J.W. 1974: Systematic position of *Eldonia ludwigi*. *Journal of Paleontology 48*, 750–755. [*458*]

Dymek, R.F. & Klein, C. 1988: Chemistry, petrology and origin of banded iron-formation lithologies from the 3800 Ma Isua supracrustal belt, West Greenland. *Precambrian Research 39*, 247–302. [*40, 42*]

Dymond, J. & Roth, S. 1988: Plume dispersed hydrothermal particles: A time-series record of settling flux from the Endeavour Ridge using moored sensors. *Geochimica et Cosmochimica Acta 52*, 2525–2536. [*20*]

Dyson, F. 1985: *Origins of Life*. Cambridge University Press, Cambridge. [*133*]

Dzik, J. 1991: Is fossil evidence consistent with traditional views of early metazoan phylogeny? In Simonetta, A.M. & Conway Morris, S. (eds.): *The Early Evolution of Metazoa and the Significance of Problematic Taxa*, 47–56. Cambridge University Press, Cambridge. [*383, 458*]

Dzik, J. & Krumbiegel, G. 1989: The oldest "onychophoran" *Xenusion*: A link connecting phyla. *Lethaia 22*, 169–181. [*456, 458*]

Dzik, J. & Lendzion, K. 1988: The oldest arthropods of the East European platform. *Lethaia 21*, 29–38. [*456*]

Easteal, S. 1987: The rates of nucleotide substitution in the human and rodent lineages: A reply to Li and Wu. *Molecular Biology and Evolution 4*, 78–88. [*468*]

Edelman, G.M. 1987: *Neural Darwinism*. Basic Books, New York, N.Y. [*139, 142*]

Edelman, G.M. 1988: *Topobiology – An Introduction to Molecular Embryology*. Basic Books, New York, N.Y. [*477, 478, 480, 505*]

Edelman, G.M. 1992: Morphoregulation. *Developmental Dynamics 193*, 2–10. [*503, 505*]

Edwards, P.A. & Jackson, J.B. 1976: The control of the adenosine triphosphatase of *Rhodospirillum rubrum* chromatophores by divalent cations and the membrane high energy state. *European Journal of Biochemistry 62*, 7–14. [*86*]

Eernisse, D.J., Albert, J.S. & Anderson, F.E. 1992: Annelida and Arthropoda are not sister taxa: A phylogenetic analysis of spiralian metazoan morphology. *Systematic Biology 41*, 305–330. [*402, 403, 475*]

Eichwald, E. 1854: *Paleontologiya Rossii, Drevnij Period*. St. Petersburg. [*356*]

Eigen, M. & Schuster, P. 1979: *The Hypercycle*. Springer, Berlin. [*133*]

Elderfield, H. 1988: The oceanic chemistry of the rare-earth elements. *Philosophical Transactions of the Royal Society of London, A 325*, 105–126. [*40*]

Eldredge, N. 1989: *Macroevolutionary Dynamics: Species, Niches and Adaptive Peaks*. McGraw-Hill, New York, N.Y. [*328*]

Elinson, R.P. 1987: Changes in developmental patterns: Embryos of amphibians with large eggs. In Raff, R.A. & Raff, E.C. (eds.): *Development as an Evolutionary Process*, 1–21. Alan Liss, New York, N.Y. [*494, 497, 499*]

Ellis, H.M., Spann, D.R. & Posakony, J.W. 1990: *extramacrochaetae*, a negative regulator of sensory organ development in *Drosophila*, defines a new class of helix–loop–helix proteins. *Cell 61*, 27–38. [*509*]

Ellis, L., Morgan, D.O., Koshland, D.E., Jr., Clauser, E., Moe, G.R., Bollag, G., Roth, R.A., Rutter, W.J. 1986: Linking functional domains of the human insulin receptor with the bacterial aspartate receptor. *Proceedings of the National Academy of Sciences, USA 83*, 8137–8141. [*511*]

Ellsaesser, H.W. 1991: Correspondence to the Editor. *Climatic Change 19*, 431. [*46*]

El-Sayed, S.Z. 1988: Fragile life under the ozone hole. *Natural History 97*, 72–80. [*38*]

Erdmann, V.A. & Wolters, J. 1986: Collection of published 5S, 5.8S and 4.5S ribosomal RNA sequences. *Nucleic Acids Research 14*, r1–r59. (Supplement.) [*462*]

Eriksson, P.G. & Cheney, E.S. 1992: Evidence for the transition to an oxygen-rich atmosphere during the evolution of red beds in the Lower Proterozoic sequences of southern Africa. *Precambrian Research 54*, 257–269. [*239*]

Ernst, W.G. 1983: The early Earth and the Archean rock record. In Schopf, J.W. (ed.): *Earth's Earliest Biosphere, Its Origin and Evolution*, 41–52. Princeton University Press, Princeton, N.J. [*143*]

Ervamaa, P. & Heino, T. 1983: The Ni–Cu–Zn mineralization of Talvivaara, Sotkamo. In *Exogenic Processes and Related Metallogeny in the Svecokarelian Geosynclinal Complex*, 79–83. Geological Survey of Finland, Guide 11. [*241*]

Erwin, D.H. 1991: Metazoan phylogeny and the Cambrian radiation. *TREE 6*, 131–134. [*475*]

Erwin, D.H., Valentine, J.W. & Sepkoski, J.J., Jr. 1987: A comparative study of diversification events: The early Paleozoic versus the Mesozoic. *Evolution 41*, 1177–1186. [*495, 496*]

Escabi-Perez, J.P., Romero, A., Lukak, S. & Fendler, J.H. 1979: Aspects of artificial photosynthesis: Photoionization and electron transfer in dihexadecyl phosphate vesicles. *Journal of the American Chemical Society 101*, 2231–2233. [*120*]

Eschenmoser, A. 1991: Warum Pentose- und nicht Hexose-Nukleinsäuren? *Nachrichten aus Chemie, Technik und Laboratorium 39*, 795. [*93*]

Eschenmoser, A. & Dobler, M. 1992: Warum Pentose- und nicht Hexose-Nukleisäuren? Teil I. Einleitung und Problemstellung, Konformationsanalyse für Oligo-2',3'-dideoxy-glucopyrano-syl-nukleotide ("Homo-DNS") sowie Betrachtungen zur Konformation von A- und B-DNS. *Helvetica Chimica Acta 75*, 218–259. [*93*]

Eschenmoser, A. & Loewenthal, E. 1992: Chemistry of potentially prebiological natural products. *Chemical Society Reviews 21*, 1–16. [*93*]

Estes, R., Spinar, Z.V. & Nevo, E. 1978: Early Cretaceous pipid tadpoles from Israel (Amphibia: Anura). *Herpetologica 34*, 374–393. [*493*]

Evitt, W.R. 1985: *Sporopollenin Dinoflagellate Cysts: Their Morphology and Interpretation*. American Association of Stratigraphic Palynologists, Tulsa, Okla. [*299*]

Fahnenstiel, G.L. & Carrick, H.J. 1992: Phototrophic picoplankton in lakes Huron and Michigan: Abundance, distribution, composition and contribution to biomass and production. *Canadian Journal of Fishery and Aquatic Sciences 49*, 379–388. [*338*]

Fairchild, I.J. 1991: Origins of carbonate in Neoproterozoic stromatolites and the identification of modern analogues. *Precambrian Research 53*, 281–299. [*447*]

Fairchild, I.J., Knoll, A.H. & Swett, K. 1991: Coastal lithofacies and biofacies associated with syndepositional dolomitization and silicification (Draken Formation, Upper Riphean, Svalbard). *Precambrian Research 53*, 165–197. [*447*]

Fairchild, I.J., Marshall, J.D. & Bertrand-Sarfati, J. 1990: Stratigraphic shifts in carbon isotopes from Proterozoic stromatolitic carbonates (Mauritania): Influences of primary mineralogy and diagenesis. *American Journal of Science 290-A*, 46–79. [*248, 430*]

Fairchild, T., Barbour, A.P. & Haralyi, N.L.E. 1978: Microfossils from the "Eopaleozoic" Jacadigo Group at Urucum, Mato Grosso, Southwest Brazil. *Boletim do Instituto de Geologia, Instituto de Geociências, São Paolo 9*, 74–79. [*308*]

Farmer, J.D., Vidal, G., Strauss, H., Moczydłowska, M., Ahlberg, P. & Siedlecka, A. 1991: Ediacaran medusoids from the Innerelv Member (late Proterozoic) Tanafjorden area of N.E. Finnmark. *Geological Society of America, Abstracts with Programs 23*, 97. [*373*]

Farquhar, G.D., Ehleringer, J.R. & Hubick, K.T. 1989: Carbon isotope discrimination and photosynthesis. *Annual Review of Plant Physiology and Plant Molecular Biology 40*, 503–537. [*227*]

Farquhar, G.D., O'Leary, M.H. & Berry, J.A. 1982: On the relationship between carbon isotope discrimination and the intercellular carbon dioxide concentration in leaves. *Australian Journal of Plant Physiology 9*, 121–137. [*227*]

Fawcett, D.W. 1981: *The Cell*. Saunders, Philadelphia, Pa. [*480*]

Feakes, C.R., Zbinden, E.A. & Holland, H.D. 1989: Paleosols at Arisaig, Nova Scotia and the evolution of the atmosphere. *Catena Supplement 16*, 207–232. [*238*]

Fedonkin, M.A. 1976: Sledy mnogokletochnykh iz valdajskoj serii [Traces of multicellular animals in the Valdai Series]. *Izvestiya Akademii Nauk SSSR, Seriya Geologicheskaya 4*, 129–132. [*387*]

Fedonkin, M.A. 1978: Novoe mestonakhozhdenie besskeletnykh Metazoa v vende Zimnego berega [New locality of non-skeletal Metazoa in the Vendian of the Winter Shore]. *Doklady Akademii Nauk SSSR 239*, 1423–1426. [*373*]

Fedonkin, M.A. 1981: Belomorskaya biota venda: Dokembrijskaya besskeletnaya fauna severa Russkoj platformy [White Sea Biota of the Vendian: Precambrian non-skeletal fauna of the northern part of the Russian Platform]. *Trudy Geologicheskogo Instituta Akademii Nauk SSSR 342*. 100 pp. [*363, 373, 387*]

Fedonkin, M.A. 1983: Organicheskij mir venda [Organic world of the Vendian]. *Itogi nauki i tekhniki VINITI Akademii Nauk SSSR, Series Stratigraphy and Paleontology 12*. 128 pp. [*380, 381*]

Fedonkin, M.A. 1985a: Non-skeletal fauna of the Vendian: Promorphological analysis. In Sokolov, B.S. & Ivanovskij, A.B. (eds.): *The Vendian System. Paleontology*, Vol. 1, 7–70. [*391*]

Fedonkin, M.A. 1985b: Sistematicheskoe opisanie vendskikh Metazoa [Systematic description of the Vendian Metazoa]. In Sokolov, B.S. & Ivanovskij, A.B. (eds.): *The Vendian System. Paleontology*, Vol. 1, 70–106. Nauka, Moscow. [*383*]

Fedonkin, M.A. 1985c: Precambrian metazoans: The problems of preservation, systematics and evolution. *Philosophical Transactions of the Royal Society of London, B 311*, 27–45. [*295, 304, 305, 377*]

Fedonkin, M.A. 1985d: Promorfologiya vendskikh Bilateria i problema proizkhozhdeniya metamerii Articulata [Promorphology of the Vendian Bilateria and the problem of the origin of metamerism of Articulata.] In Sokolov, B.S. (ed.): *Problematiki Pozdnego Dokembriya i Paleozoya*, 79–92. Nauka, Moscow. [*373, 380, 381*]

Fedonkin, M.A. 1986: Precambrian problematic animals: Their body plan and phylogeny. In Hoffman, A. & Nitecki, M.H. (eds.): *Problematic Fossil Taxa*, 59–67. Oxford University Press, New York, N.Y. [*381*]

Fedonkin, M.A. 1987: Besskeletnaya fauna venda i ego mesto v ehvolyutsii metazoj [Non-skeletal fauna of the Vendian and its place in the evolution of metazoans]. *Trudy Paleontologicheskogo Instituta Akademii Nauk SSSR 226*. 175 pp. [*372, 373, 380, 381, 387*]

Fedonkin, M.A. 1990a: Systematic description of Vendian Metazoa. In Sokolov, B.S. & Ivanovskij, A.B. (eds.): *The Vendian System: Paleontology*, Vol. 1, 71–120. Springer, Berlin. [*350, 373, 383*]

Fedonkin, M.A. 1990b: Precambrian metazoans. In Briggs, D.E.G. & Crowther, P.R. (eds.): *Palaeobiology, a Synthesis*, 17–24. Blackwell, Oxford. [*380, 435*]

Fedonkin, M.A. 1992: Vendian faunas and the early evolution of Metazoa. In Lipps, J.H. & Signor, P.W. (eds.): *Origin and Early Evolution of the Metazoa*, 87–129. Plenum, New York, N.Y. [*372, 383*]

Fedonkin, M.A., Chumakov, N.M. & Jankauskas, T.V. 1987: Problema global'nykh bioticheskikh i abioticheskikh sobytij v pozdnem dokembrii [The problem of the global biotic and abiotic events in the late Precambrian]. *3rd All-Union Symposium on the Paleontology of the Precambrian and Early Cambrian, Petrozavodsk, Abstracts*, 99–101. [*372*]

Feely D.E., Erlandsen, S.L. & Chase, D.G. 1984: *Giardia and Giardiasis*. Plenum, New York, N.Y. [*187*]

Feely, R.A., Massoth, G.J., Baker, E.T., Cowen, J.P., Lamb, M.F. & Krogslund, K.A. 1990: The effect of hydrothermal processes on midwater phosphorus distributions in the northeast Pacific. *Earth and Planetary Science Letters 96*, 305–318. [*20*]

Fegley, B., Jr., Prinn, R.G., Hartman, H. & Watkins, G.H. 1986: Chemical effects of large impacts on the Earth's primitive atmosphere. *Nature 319*, 305–308. [*12, 16*]

Feldman, M.W. (ed.) 1989: *Mathematical Evolutionary Theory*. Princeton University Press, Princeton, N.J. [*328*]

Felsenstein, J. 1978: Cases in which parsimony or compatibility methods will be positively misleading. *Systematic Zoology 27*, 401–410. [*184*]

Felsenstein, J. 1981: Evolutionary trees from DNA sequences: A maximum likelihood approach. *Journal of Molecular Evolution 17*, 368–376. [*183, 185*]

Felsenstein, J. 1990: *PHYLIP Manual Version 3.3*. University Herbarium University of California Press, Berkeley, Calif. [*468*]

Ferris, J.P. 1987: Prebiotic synthesis: Problems and challenges. *Cold Spring Harbor Symposia on Quantitative Biology 52*, 29–35. [*64, 73*]

Ferris, J.P. & Ertem, G. 1993: Oligomerization reactions of ribonucleotides: The reaction of the 5'-phosphorimidazolide of adenosine with diadenosine pyrophosphate on montmorillonite and other minerals. *Origins of Life and Evolution of the Biosphere 23*, 221–227. [*95*]

Ferris, J.P. & Hagan, W.J., Jr. 1984: HCN and chemical evolution: The possible role of cyano compounds in prebiotic synthesis. *Tetrahedron 40*, 1093–1120. [*92*]

Ferris, J.P., Joshi, P.C., Edelson, E.H. & Lawless, J.G. 1978: HCN: A plausible source of purines, pyrimidines and amino acids on the primitive Earth. *Journal of Molecular Evolution 11*, 293–311. [*54, 55, 108*]

Ferris, J.P., Kamaluddin, M.N. & Ertem, G. 1990: Oligomerization reactions of deoxyribonucleotides on montmorillonite clay: The effect of mononucleotide structure, phosphate activation and montmorillonite composition on phosphodiester bond formation. *Origins of Life and Evolution of the Biosphere 20*, 279–291. [*95*]

Ferris, J.P. & Nicodem, D.E. 1972: Ammonia photolysis and the role of ammonia in chemical evolution. *Nature 238*, 268–269. [*12*]

Ferronskiy, V.I. & Polyakov, V.A. 1982: Origin of the Earth's hydrosphere in the light of isotopic and theoretical studies. *Geochemistry International 19*, 19–27. [*46*]

Fiala, G. & Stetter, K.O. 1986: *Pyrococcus furiosus* sp. nov. represents a novel genus of marine heterotrophic Archaebacteria growing optimally at 100°C. *Archives of Microbiology 145*, 56–61. [*150*]

Fiala, G., Stetter, K.O., Jannasch, H.W., Langworthy, T.A. & Madon, J. 1986: *Staphylothermus marinus* sp. nov. represents a novel genus of extremely thermophilic submarine heterotrophic Archaebacteria growing up to 98°C. *Systematic and Applied Microbiology 8*, 106–113. [*149*]

Fiala-Medioni, A. 1988: Matière organique sissoute et productions secondaire (adultes). *Oceanis 14*, 399–407. [*375*]

Field, K.G., Olsen, G.J., Lane, D.J., Giovannoni, S.J., Ghiselin, M.T., Raff, E.C., Pace, N.R. & Raff, R.A. 1988: Molecular phylogeny of the animal kingdom. *Science 239*, 748–753. [*184, 402, 453, 463, 469, 490, 491*]

Fischbach, K.-F. & Heisenberg, M. 1984: Neurogenetics and behaviour in insects. *Journal of Experimental Biology 112*, 65–93. [*509*]

Fischer, A.G. 1965: Fossils, early life, and atmospheric history. *Proceedings of the National Academy of Sciences, USA 53*, 1205–1215. [*254*]

Fischer, F., Zillig, W., Stetter, K.O. & Schreiber, G. 1983: Chemolithoautotrophic metabolism of anaerobic extremely thermophilic Archaebacteria. *Nature 301*, 511–513. [*147*]

Fisher, J.D. & Volborth, A. 1960: Morinite–apatite–whitlockite. *The American Mineralogist 45*, 645–667. [*102*]

Flegmann, A.W. & Tattersall, R. 1979: Energetics of peptide bond formation at elevated temperatures. *Journal of Molecular Evolution 12*, 349–355. [*20*]

Fleischaker, G.R. 1990: Origins of life: An operational definition. *Origins of Life and Evolution of the Biosphere 20*, 127–137. [*57, 61*]

Flint, E.B. & Suslick, K.S. 1991: The temperature of cavitation. *Science 253*, 1397–1399. [22]

Flores, E. & Wolk, C.P. 1986: Production, by filamentous, nitrogen-fixing cyanobacteria, of a bacteriocin and of other antibiotics that kill related strains. *Archives of Microbiology 145*, 215–219. [340]

Foote, M. 1990: Nearest-neighbor analysis of trilobite morphospace. *Systematic Zoology 39*, 371–382. [459]

Foote, M. 1991: Morphologic patterns of diversification: Examples from trilobites. *Palaeontology 34*, 461–485. [459]

Ford, T.D. 1958: Pre-Cambrian fossils from Charnwood Forest. *Proceedings of the Yorkshire Geological Society 31*, 211–217. [372]

Ford, T.D. 1963: The Pre-Cambrian fossils from Charnwood Forest. *Transactions of the Leicester Literary and Philosophical Society 57*, 57–62. [372]

Forster, T. 1950: *Zeitschrift für Electrochemie 54*, 142. [120]

Fox, S.W. & Dose, K. 1977: *Molecular Evolution and the Origin of Life*. Marcel Dekker, New York, N.Y. [55, 133]

Frakes, L.A. 1979: *Climates Troughout Geologic Time*. Elsevier, Amsterdam. [213]

François, L.M. 1986: Extensive deposition of banded iron formations was possible without photosynthesis. *Nature 320*, 352–354. [42, 43]

Franklin, I.R. 1987: Population biology and evolutionary change. In Campbell, K.S.W. & Day, M.F. (eds.): *Rates of Evolution*, 156–174. Allen & Unwin, London. [516]

Fransen, M.E. 1982: The role of ECM in the development of invertebrates: A phylogeneticist's view. In Hawkes, S. & Wang, J.L. (eds.): *Extracellular Matrix*, 177–181. Academic Press, New York, N.Y. [478]

Fraser, S.E., Green, C.R., Bode, H.R. & Gilula, N.B. 1987: Selective disruption of gap junctional communication interferes with a patterning process in *Hydra*. *Science 237*, 49–55. [504]

Freeman, K.H., Hayes, J.M., Trendel, J.-M. & Albrecht, P. 1990: Evidence from carbon isotope measurements for diverse origins of sedimentary hydrocarbons. *Nature 343*, 254–256. [229, 231]

Friedberg, E.C. 1985: *DNA Repair*. Freeman, San Francisco, Calif. [78]

Froelich, P.N., Bender, M.L., Luedtke, N.A., Heath, G.R. & DeVries, T. 1982: The marine phosphorus cycle. *American Journal of Science 282*, 474–511. [448]

Fu, J. 1986: Cyphomegacritarchs and their mathematical simulation from the Liulaobei Formation, Bagong Mountain, Shouxian County, Anhui Province. *Journal of Northwest University [China] 16*, 76–88. (In Chinese, with English abstract.) [346, 356]

Fu, J. 1989: New materials of Late Precambrian Huainan biota fossil in Shouxian, Anhui. *Acta Palaeontologica Sinica 28*, 72–77. (In Chinese, with English summary.) [356]

Fuchs, J. 1989: Alternative pathways of autotrophic CO_2 fixation. In Schlegel, H.G. & Bowien, B. (eds.): *Autotrophic Bacteria*, 365–382. Science Tech, Madison, Wis. [154, 157, 173, 174]

Fuchs, J., Ecker, A. & Strauss, G. 1992: Bioenergetics and autotrophic carbon metabolism of chemotrophic archaebacteria. In Lunt, G.G., Hough, D.W. & Danson, M.J. (eds.): *The Archaebacteria: Biochemistry and Biotechnology*, 23–39. Portland, Colchester. [154]

Fuhrman, J.A., McCallum, K. & Davis, A.A. 1993: Phylogenetic diversity of subsurface marine microbial communities from the Atlantic and Pacific oceans. *Applied and Environmental Microbiology 59*, 1294–1302. [159]

Gabel, N.W. 1968: Abiotic formation of phosphoric anhydride bonds in dilute aqueous conditions. *Nature 218*, 354. [97]

Gabel, N.W. 1990: A realistic chemical basis for the origin of life. *Evolutionary Theory 9*, 181–209. [97]

Gabel, N.W. & Thomas, V. 1971: Evidence for the occurrence and distribution of inorganic polyphosphates in vertebrate tissues. *Journal of Neurochemistry 18*, 1229–1242. [97]

Galimov, E.M. 1980: C^{12}/C^{13} in kerogen. In Durand, B. (ed.): *Kerogen*, 270–299. Editions Technip, Paris. [202]

Gallissian, M.-F. & Vacelet, J. 1992: Ultrastructure of the oocyte and embryo of the calcified sponge *Petrobiona massilliana* (Porifera Calcarea). *Zoomorphologie 112*, 133–142. [480]

Gans, C. 1989: Stages in the origin of vertebrates: Analysis by means of scenarios. *Biological Reviews 64*, 221–268. [497]

Garcia-Pichel, F. & Castenholz, R.W. 1991: Characterization and biological implications of scyto-nemin, a cyanobacterial sheath pigment. *Journal of Phycology 27*, 395–409. [*166*]

Garcia-Pichel, F., Sherry, N.D. & Castenholz, R.W. 1992: Evidence for a UV sunscreen role of the extracellular pigment scytonemin in the terrestrial cyanobacterium *Cholorogloeopsis* sp. *Photo-chemistry and Photobiology 56*, 17–23. [*166*]

Garrels, R.M. 1987: A model for the deposition of the microbanded Precambrian iron formations. *American Journal of Science 287*, 81–106. [*39, 46*]

Garrels, R.M. 1989: Some factors influencing biomineralization in Earth history. In Crick, R.E. (ed.): *Origin, Evolution, and Modern Aspects of Biomineralization in Plants and Animals*, 1–10. Plenum, New York, N.Y. [*433, 437*]

Garrels, R.M. & Perry, E.A., Jr. 1974: Cycling of carbon, sulfur and oxygen through geologic time. In Goldberg, E.D. (ed.): *The Sea: Ideas and Observations on Progress in the Study of the Seas*, Vol. 5, 303–336. Wiley, New York, N.Y. [*215*]

Garrett, P. 1970a: Phanerozoic stromatolites: Non-competitive ecologic restriction by grazing and burrowing animals. *Science 169*, 171–173. [*254, 337*]

Garrett, P. 1970b: Deposit feeders limit development of stromatolites. *American Association of Petroleum Geologists Bulletin 54*, 848. [*435*]

Garrett, W.D. 1967: The organic chemical composition of the ocean surface. *Deep-Sea Research 14*, 221–227. [*20*]

Garrison, W.M., Morrison, D.C., Hamilton, J.G., Benson, A.A. & Calvin, M. 1951: Reduction of carbon dioxide in aqueous solution by ionizing radiation. *Science 114*, 416–418. [*37*]

Gebelein, C.D. 1974: Biologic control of stromatolite microstructure: Implications for Precambrian time stratigraphy. *American Journal of Science 274*, 575–598. [*433*]

Gebelein, C.D. 1976: The effects of the physical, chemical and biological evolution of the earth. In Walter, M.R. (ed.): *Stromatolites*, 499–515. Developments in Sedimentology 20 Elsevier, Amsterdam. [*281, 432*]

Gehling, J.G. 1986: Algal binding of siliclastic sediments: A mechanism in the preservation of Ediacaran fossils. *12th International Sedimentological Congress. Abstracts*, 117. [*373*]

Gehling, J.G. 1987: Earliest known echinoderm – a new Ediacaran fossil from the Pound Subgroup of South Australia. *Alcheringa 11*, 337–345. [*296, 372*]

Gehling, J.G. 1988: A cnidarian of actinian-grade from the Ediacaran Pound Subgroup, South Australia. *Alcheringa 12*, 299–314. [*372*]

Gehling, J.G. 1991: The case for Ediacaran fossil roots to the metazoan tree. *Memoirs of the Geological Society of India 20*, 181–224. [*284, 293, 295, 451*]

Geldsetzer, H.H.J., James, N.P. & Tebbutt, E. (eds.) 1988: *Reefs, Canada and Adjacent Areas*. Canadian Society of Petroleum Geologists, Memoir 13. [*279*]

Gennis, R.B. 1989: *Biomembranes, Molecular Structure and Function*. Schulz Schirmer Springer Advanced Texts in Chemistry Springer, Heidelberg. [*478*]

German, C.R. & Elderfield, H. 1990: Application of the Ce anomaly as a paleoredox indicator: The ground rules. *Paleoceanography 5*, 823–833. [*40*]

Germs, J.G.B. 1972: New shelly fossils from the Nama Group, South West Africa. *American Journal of Science 272*, 752–761. [*377, 442*]

Getoff, N.G., Scholes, G. & Weiss, J. 1960: Reduction of carbon dioxide in aqueous solution under the influence of radiation. *Tetrahedron Letters 1960*, 17–23. [*13*]

Ghiselin, M.T. 1988: The origin of molluscs in the light of molecular evidence. *Oxford Surveys in Evolutionary Biology 5*, 66–95. [*453, 457*]

Ghiselin, M.T. 1989: Summary of our present knowledge of metazoan phylogeny. In Fernholm, B., Bremer, K. & Jörnvall, H. (eds.): *The Hierarchy of Life*, 261–272. Excerpta Medica, Amsterdam. [*453, 458, 469*]

Gibson, G.G. 1989: Trace fossils from the late Precambrian Carolina Slate Belt, south-central North Carolina. *Journal of Paleontology 63*, 1–10. [*398*]

Gibson, G.G., Teeter, S.A. & Fedonkin, M.A. 1984: Ediacaran fossils from the Carolina Slate Belt. *Geology 12*, 387–390. [*373*]

Gilbert, S.F. 1991: *Developmental Biology*. 3d Ed. Sinauer, Sunderland, Mass. [*478, 498*]

Gilbert, W. 1986: The RNA world. *Nature 319*, 618. [*62, 68, 70, 73, 88, 125, 132*]

Gilbert, W. 1987: The exon theory of genes. *Cold Spring Harbor Symposia on Quantitative Biology 52*, 901–905. [*72, 511, 515*]

Gillen, F.D., Hagblom, P., Harwood, J., Aley, S.B., Reiner, D.S., McCaffery, M., So, M. & Guiney, D. G. 1990: Isolation and expression of the gene for a major surface protein of *Giardialamblia*. *Proceedings of the National Academy of Sciences, USA 87*, 4463–4467. [*187*]

Giovannoni, S.A.J., Revsbech, N.P., Ward, D.M. & Castenholz, R.W. 1987: Obligately phototrophic *Chloroflexus*: Primary production in anaerobic hot spring microbial mats. *Archives of Microbiology 147*, 80–87. [*168*]

Giovannoni, S.A.J., Turner, S., Olsen, G.J., Barns, S., Lane, D.L. & Pace, N.R. 1988: Evolutionary relationships among cyanobacteria and green chloroplasts. *Journal of Bacteriology 170*, 3584–3592. [*293, 296, 335*]

Giovanoli, R. & Arrhenius, G. 1988: Structural chemistry of marine manganese and iron minerals and synthetic model compounds. In Halbach, P., Friedrich, G. & von Stackelberg, E. (eds.): *The Manganese Nodule Belt of the Pacific Ocean*, 20–37. Ferdinand Enke, Stuttgart. [*95*]

Glaessner, M.F. 1971: Geographic distribution and time range of the Precambrian Ediacara fauna. *Geological Society of America, Bulletin 82*, 509–514. [*363*]

Glaessner, M.F. 1972a: Precambrian palaeozoology. In Jones, J.B. & McGowran, B. (eds.): *Stratigraphic Problems of the Later Precambrian and Early Cambrian*, 43–52. University of Adelaide, Center for Precambrian Research, Special Paper 1. [*424*]

Glaessner, M.F. 1972b: Preface. In Walter, M.R. (ed.): *Stromatolites and the Biostratigraphy of the Australian Precambrian and Cambrian*, v–vi. Special Papers in Palaeontology 11 Palaeontological Association, London. [*272*]

Glaessner, M.F. 1976: Early Phanerozoic annelid worms and their geological and biological significance. *Journal of the Geological Society, London 132*, 259–275. [*365*]

Glaessner, M.F. 1979: Precambrian. In Robison, R.A. & Teichert, C. (eds.): *Treatise on Invertebrate Paleontology, Part A.*, 79–118. Geological Society of America, Boulder, Colo. [*353, 355*]

Glaessner, M.F. 1984: *The Dawn of Animal Life: A Biohistorical Study*. Cambridge University Press, Cambridge. [*293, 363, 365, 372, 373, 380, 383, 424, 436, 437, 438, 451*]

Glaessner, M.F. 1988: Pseudofossils explained as vortex structures in sediments. *Lethaia 69*, 275–287. [*398*]

Glaessner, M.F. & Daily, B. 1959: The geology and late Precambrian fauna of the Ediacara fossil reserve. *Records of the South Australian Museum 13*, 369–401. [*415*]

Glaessner, M.F., Preiss, W.V. & Walter, M.R. 1969: Precambrian columnar stromatolites in Australia: Morphological and stratigraphic analysis. *Science 164*, 1056–1058. [*273*]

Glaessner, M.F. & Wade, M. 1966: The Late Precambrian fossils from Ediacara, South Australia. *Palaeontology 9*, 599–628. [*372*]

Gnilovskaya, M.B. 1971: Drevnejshie vodnye rasteniya venda Russkoj platformy [The oldest aquatic plants in the Vendian of the Russian Platform]. *Paleontologicheskij Zhurnal 1971:3*, 101–107. [*356*]

Gnilovskaya, M.B. 1972: *Izvestkovye vodorosli srednego i pozdnego ordovika vostochnogo Kazakhstana [Calcareous algae from the Middle and Late Ordovician of eastern Kazakhstan]*. 196 pp. Nauka, Leningrad. [*428*]

Gnilovskaya, M.B. 1975: Novye dannye o prirode vendotenid [New data on the nature of the vendotaenids]. *Doklady Akademii Nauk SSSR 221*, 258–261. [*357*]

Gnilovskaya, M.B. 1976: Drevnejshie Metaphyta [The oldest Metaphyta]. *International Geological Congress, 25th Session, Reports of Soviet Geologists; Paleontology, Marine Geology*, 10–14. Nauka, Moscow. [*356*]

Gnilovskaya, M.B. 1979: Vendotenidy [vendotaenids]. In Keller, B.M. & Rozanov, A.Yu. (eds.): *Paleontologiya Verkhnedokembrijskikh i Kembrijskikh Otlozhenij Vostochno-Evropejskoj Platformy*, 39–48. Nauka, Moscow. [*356*]

Gnilovskaya, M.B. 1986: Tkanevaya mnogokletochnost' drevnejshikh rastenij [Tissular multicellularity in ancient plants]. In *Aktual'nye Voprosy Sovremennoj Paleoalgologii*, 22–24. Kiev. [*356, 357*]

Gnilovskaya, M.B., Ishchenko, A.A., Kolesnikov, Ch.M., Korenchuk, L.B. & Udal'nov, A.P. 1988: *Vendotenidy Vostochno-Evropejskoj Platformy*. Nauka, Leningrad. [*345, 349, 350, 352, 355, 356, 357*]

Gogarten, J.P., Kibak, H., Dittrich, P., Taiz, L., Bowman, E.J., Bowman, B.J., Manolson, M.F., Poole, R.J., Date, T., Oshima, T., Konishi, J., Denda, K. & Yoshida, M. 1989: Evolution of the vacuolar H+-ATPase: Implications for the origin of eukaryotes. *Proceedings of the National Academy of Sciences, USA 86*, 6661–6665. [*65, 67, 144, 153, 189, 288*]

Goldschmidt, V.M. 1952: Geochemical aspects of the origin of complex organic molecules on the Earth, as precursors to organic life. *New Biologist 12*, 97–105. [*95*]

Golenberg, E.M., Giannasi, D.E., Clegg, M.T., Smiley, C.J., Durbin, M., Henderson, D. & Zurawski, G. 1990: Chloroplast DNA sequence from a Miocene *Magnolia* species. *Nature 344*, 656–658. [*78*]

Golubic, S. 1973: The relationship between blue-green algae and carbonate deposits. In Carr, N. & Whitton B.A. (eds.): *The Biology of Blue-Green Algae*, 434–472. Blackwell, Oxford. [*336*]

Golubic, S. 1976: Organisms that build stromatolites. In Walter, M.R. (ed.): *Stromatolites*, 113–126. Developments in Sedimentology 20 Elsevier, Amsterdam. [*283, 285, 336*]

Golubic, S. 1980: Halophily and halotolerance in cyanophytes. *Origins of Life 10*, 169–183. [*338*]

Golubic, S. & Hofmann, H.J. 1976: Comparison of modern and mid-Precambrian Entophysalidaceae (Cyanophyta) in stromatolitic algal mats: Cell division and degradation. *Journal of Paleontology 50*, 1074–1082. [*281, 300, 336*]

Gomont, M. 1892: Monographie des Oscillariées. *Annales des Sciences Naturelles, Botanique et Biologie Végétale, 7. Ser. 15*, 263–368. [*334*]

Goncharova, N.V. & Goldfeld, M.G. 1990: Magnesium porphyrins as possible photosensitizers of macroergic phosphate bonds formation during prebiotic evolution. *Origins of Life and Evolution of the Biosphere 20*, 309–319. [*85, 86*]

Goodman, M., Pedwaydon, J., Czelusniak, J., Suzuki, T., Gotoh, T., Moens, L., Shishikura, F., Walz, D., & Vinogradov, S. 1988: An evolutionary tree for invertebrate globin sequences. *Journal of Molecular Evolution 27*, 236–249. [*461, 462*]

Goodrich, E.S. 1895: On the Coelom, genital ducts and nephridia. *Quarterly Journal of Microscopical Science 86*, 493–501. [*483*]

Goodrich, E.S. 1945: The study of nephridia and genital ducts since 1895. *Quarterly Journal of Microscopical Science 86*, 113–392. [*483*]

Goodwin, A.M. 1991: *Precambrian Geology: The Dynamic Evolution of the Continental Crust*. Academic Press, London. [*210*]

Gopal, R., Calvo, C., Ito, J. & Sabine, K.W. 1974: Crystal structure of synthetic Mg-whitlockite $Ca_{18}Mg_2H_2(PO_4)_{14}$. *Canadian Journal of Chemistry 52*, 1155–1164. [*102, 106*]

Gorbatschev, R. & Gaal, G. 1987: The Precambrian history of the Baltic Shield. In Kröner, A. (ed.): *Proterozoic Lithospheric Evolution*, 149–159. American Geophysical Union, Series 17. [*26*]

Goryansky, V. Yu. & Popov, L. Yu. 1985: The morphology, systematic position and origin of inarticulate brachiopods with carbonate shells. *Paleontological Journal 1985*, 1–11. [*452*]

Gould, S.J. 1977: *Ontogeny and Phylogeny*. Belknap, Harvard, Mass. [*475, 476, 486*]

Gould, S.J. 1989: *Wonderful Life*. Norton, New York, N.Y. [*141, 489, 495, 498*]

Gould, S.J. 1991: The disparity of the Burgess Shale arthropod fauna and the limits of cladistic analysis: Why we must strive to quantify morphospace. *Paleobiology 17*, 411–423. [*459, 489, 495, 496, 498*]

Gould, S.J., Keller, G.-A., Schneider, M., Howell, S.H., Garrard, L.J., Goodman, J.M., Distel, B. Tabak, H. & Subramani, S. 1990: Peroxisomal protein import is conserved between yeast, plants, insects and mammals. *EMBO Journal 9*, 85–90. [*505*]

Gouy, M. & Li, W.-H. 1989a: Molecular phylogeny of the kingdoms Animalia, Plantae and Fungi. *Molecular Biology and Evolution 6*, 109–122. [*296*]

Gouy, M. & Li, W.-H. 1989b: Phylogenetic analysis based on rRNA sequences supports the archaebacterial rather than the eocyte tree. *Nature 339*, 145–147. [*301*]

Grandstaff, D.E. 1976: A kinetic study of the dissolution of uraninite. *Economic Geology 71*, 1493–1506. [*239*]

Grandstaff, D.E. 1980: Origin of uraniferous conglomerates at Elliot Lake, Canada, and Witwatersrand, South Africa: Implications for oxygen in the Precambrian atmosphere. *Precambrian Research 13*, 1–26. [*29, 46, 239*]

Grandstaff, D.E., Edelman, M.J., Foster, R.W., Zbinden, E. & Kimberley, M.M. 1986: Chemistry and mineralogy of Precambrian paleosols at the base of the Dominion and Pongola Groups (Transvaal, South Africa). *Precambrian Research 32*, 97–131. [*42, 46*]

Grant, S.W.F. 1990: Shell structure and distribution of *Cloudina*, a potential index fossil for the terminal Proterozoic. *American Journal of Science 290-A*, 261–294. [*291, 295, 305, 365, 377, 417, 423, 442*]

Grant, S.W.F., Knoll, A.H. & Germs, G.J.B. 1991: Probable calcified metaphytes in the latest Proterozoic Nama Group, Namibia: Origin, diagenesis, and implications. *Journal of Paleontology 65*, 1–18. [*291, 303, 305, 415, 428, 442*]

Graßhoff, M. 1991: Die Evolution der Cnidaria I. Die Entwicklung der Anthozoen-Konstruktion. *Natur und Museum 121*, 225–282. [*481*]

Gray, M.W. & Schnare, M.N. 1990: Evolution of the modular structure of rRNA. In Hill, W., Dahlberg, A., Garrett, R.A., Moore, P.B., Schelssinger, D. (ed.): *Ribosomes: Structure, Function and Genetics*, 589–597. American Microbiological Society, Washington, D.C. [*77*]

Green, C.R. 1984: Intercellular junctions. In Bereiter-Hahn, J., Matoltsy, A.G. & Richards, K.S. (eds.): *Biology of the Integument. 1. Invertebrates*, 5–16. Springer, Berlin. [*478*]

Green, J.C. 1989: Relationships between the chromophyte algae: The evidence from studies of mitosis. In Green, J.C., Leadbeater, B.S.C. & Diver, W.L. (eds.): *The Chromophyte Algae: Problems and Perspectives*, 189–206. The Systematics Association Special Volume 38 Clarendon, Oxford. [*321*]

Green, J.W., Knoll, A.H. & Swett, K. 1988: Microfossils from oolites and pisolites of the upper Proterozoic Eleonore Bay Group, Central East Greenland. *Journal of Paleontology 62*, 835–52. [*336*]

Green, J.W., Knoll, A.H. & Swett, K. 1989: Microfossils from silicified stromatolitic carbonates of the Upper Proterozoic Limestone–Dolomite "Series," central East Greenland. *Geological Magazine 126*, 567–585. [*281*]

Grell, K.G. 1961: *Trichoplax adhaerens* and the origin of Metazoa. In *Convegno Internazionale: Origine dei Grandi Phyla dei Metazoi*, 107–122. Atti dei Convegni Lincei 49. [*468*]

Grell, K.G. & Ruthmann, A. 1991: Placozoa. In Harrison, F.W. & Westfall, J.A. (eds.): *A Microscopic Anatomy of the Invertebrates*, 5–27. Wiley-Liss, New York, N.Y. [*480, 481*]

Grey, K. 1981: Small conical stromatolites from the Archaean near Kanowna, Western Australia. *Western Australia Geological Survey Annual Report for 1980*, 90–94. [*275, 277, 278*]

Grey, K. & Thorne, A.M. 1985: Biostratigraphic significance of stromatolites in upward shallowing sequences of the early Proterozoic Duck Creek Dolomite, Western Australia. *Precambrian Research 29*, 183–206. [*251, 271, 280, 284*]

Grey, K. & Williams, I.R. 1990: Problematic bedding plane markings from the Middle Proterozoic Manganese Subgroup, Bangemall basin, Western Australia. *Precambrian Research 46*, 307–327. [*442*]

Gross, E.M., Wolk, C.P. & Jüttner, F. 1991: Fischerellin, a new allelochemical from the freshwater cyanobacterium *Fischerella muscicola*. *Journal of Phycology 27*, 686–692. [*340*]

Grotzinger, J.P. 1986a: Cyclicity and paleoenvironmental dynamics, Rocknest platform, northwest Canada. *Geological Society of America, Bulletin 97*, 1208–1231. [*250, 429, 430, 432*]

Grotzinger, J.P. 1986b: Evolution of early Proterozoic passive-margin carbonate platform, Rocknest Formation, Wopmay Orogen, N.W.T., Canada. *Journal of Sedimentary Petrology 56*, 831–847. [*432*]

Grotzinger, J.P. 1989: Facies and evolution of Precambrian carbonate depositional systems: Emergence of the modern platform archetype. In Crevello, P.D., Wilson, J.L., Sarg, J.F. & Read, J.F. (eds.): *Controls on Carbonate Platform and Basin Development*, 79–106. Society of Economic Paleontologists and Mineralogists, Special Publication 44. [*165, 248, 249, 250, 251, 252, 253, 254, 256, 257, 271, 278, 279, 281, 421, 422, 432, 433*]

Grotzinger, J.P. 1990: Geochemical model for Proterozoic stromatolite decline. *American Journal of Science 290-A*, 80–103. [*254, 271, 280, 283, 284, 371, 372, 429, 430, 432, 435*]

Grotzinger, J.P. & Kasting, J.F. 1993: New constraints on Precambrian ocean composition. *Journal of Geology 101*, 235–243. [*251, 256, 257*]

Grotzinger, J.P. & Read, J.F. 1983: Evidence for primary aragonite precipitation, lower Proterozoic (1.9 Ga) dolomite, Wopmay orogen, northwest Canada. *Geology 11*, 710–713. [*251*]

Grotzinger, J.P., Sumner, D.S. & Beukes, N.J. 1993: Archean carbonate sedimentation in an active extensional basin, Belingwe Greenstone Belt, Zimbabwe. *Geological Society of America, Abstracts with Programs 25*, 64. [*248*]

Groves, D.I., Dunlop, J.S.R. & Buick, R. 1981: An early habitat of life. *Scientific American 245*, 64–73. [*10, 28, 29, 275, 276*]

Guerrier-Takada, C., Gardiner, K., Marsh, T., Pace, N. & Altman, S. 1983: The RNA moiety of ribonuclease P is the catalytic subunit of the enzyme. *Cell 35*, 849–857. [*133*]

Gulbrandsen, R.A., Roberson, C.E. & Neil, S.T. 1984: Time and the crystallization of apatite in seawater. *Geochimica et Cosmochimica Acta 48*, 213–218. [*99, 100*]

Gunderson, J.H., Elwood, H., Ingold, A., Kindle, K. & Sogin, M.L. 1987: Phylogenetic relationships between chlorophytes, chrysophytes, and oomycetes. *Proceedings of the National Academy of Sciences, USA 84*, 5823–5827. [*320*]

Gureev, Yu.A. 1985: Vendiata primitivnye dokembrijskie Radialia [Vendiata – primitive Precambrian Radialia]. In *Problematiki Pozdnego Dokembriya i Paleozoya*, 92–103. Nauka, Moscow. [*373*]

Gureev, Yu.A. 1987: Morfologicheskij analiz i sistematika vendiat [Morphological analysis and systematics of the Vendiata]. *Academy of Sciences of the Ukrainian SSR, Institute of Geological Sciences, Preprint 87-15*. 54 pp. [*373, 377*]

Gureev, Yu.A. 1988: Besskeletnaya fauna venda [Non-skeletal fauna of the Vendian]. In *Biostratigrafiya i Paleogeograficheskiye Rekonstruktsii Dokembriya Ukrainy*, 65–80. [*373*]

Gürich, G. 1929: Dis bisland ältesten Spuren von Organismen in Südafrika. *International Geological Congress, Pretoria 15*, 670–680. [*372*]

Gustafsson, M. & Reuter, M. (eds.) 1990: *The Early Brain. Proceedings of the Symposium "Invertebrate Neurobiology."* Acta Academiae Åboensis. Ser. B 50 Åbo Academy Press, Åbo. [*483, 488*]

Guy, R.D., Fogel, M.L. & Berry, J.A. 1993: Photosynthetic fractionation of stable isotopes of oxygen and carbon. *Plant Physiology 101*, 37–47. [*227*]

Hadzi, J. 1963: *The Evolution of Metazoa*. Pergamon, Oxford. [*469, 476*]

Haeckel, E. 1866: *Generelle Morphologie der Organismen*, Vol. 2. Reimer, Berlin. [*181*]

Haeckel, E. 1873: *Natürliche Schöpfungsgeschichte Gemeinverständliche Wissenschaftliche Vorträge über die Entwicklungslehre im Allgemeinen und diejenige von Darwin, Goethe und Lamarck*. Georg Reimer, Berlin. [*476*]

Haeckel, E. 1877: *Biologische Studien. II. Heft: Studien zur Gastreatheorie*. Hermann Dufft, Jena. [*476*]

Haldane, J.B.S. 1954: The origins of life. *New Biologist 16*, 12–27. [*133*]

Hall, B.G. 1990: Spontaneous point mutations that occur more often when they are advantageous than when they are neutral. *Genetics 126*, 5–16. [*142*]

Hambrey, M.J. & Harland, W.B. 1985: The Late Proterozoic glacial era. *Palaeogeography, Palaeoclimatology, Palaeoecology 51*, 255–272. [*435*]

Han, T.-M. 1991: Evidence for megascopic life forms in the Early Proterozoic Negaunee Iron-Formation, Lake Superior region. *Geological Society of America, Abstracts with Programs 23*, 455. [*348, 354*]

Han, T.-M. & Runnegar, B. 1992: Megascopic eukaryotic algae from the 2.1 billion-year-old Negaunee Iron-Formation, Michigan. *Science 257*, 232–235. [*242, 283, 290, 291, 292, 332, 348, 353, 354, 360, 427, 450*]

Handschuh, G.J., Lohrmann, R. & Orgel, L.E. 1973: The effect of Mg^{2+} and Ca^{2+} on urea-catalyzed phosphorylation reactions. *Journal of Molecular Evolution 2*, 251–262. [*97*]

Hanken, J. 1986: Developmental evidence for amphibian origins. *Evolutionary Biology 20*, 389–417. [*498*]

Hanor, J.S. & Baria, L.R. 1977: Controls on the distribution of barite deposits in Arkansas. In Stone, C.G. (ed.): *Symposium on the Geology of the Ouachita Mountains*, Vol. 2, 42–49. Arkansas Geological Commission, Little Rock, Arkansas. [*29*]

Harder, W. & van Dijken, J.P. 1976: Theoretical considerations on the relation between energy production and growth of methane-utilizing bacteria. In Schlegel, H.G., Gottschalk, G. & Pfenning, N. (eds.): *Symposium on Microbial Production and Utilization of Gases (H_2, CH_4, CO)*, 403–418. Goltze, Göttingen. [*231*]

Hardie, L.A. 1984: Evaporites: Marine or non-marine? *American Journal of Science 284*, 193–240. [*253*]

Harding, S.C. & Risk, M.J. 1986: Grain orientation and electron microprobe analyses of selected Phanerozoic trace fossil margins, with a possible Proterozoic example. *Journal of Sedimentary Petrology 56*, 684–696. [*451*]

Hardisty, M.W. & Potter, I.C. (eds.) 1971–1982: *The Biology of Lampreys*. Vols. 1–4B. Academic Press, London. [*406*]

Hargreaves, W.R. & Deamer, D.W. 1978: Liposomes from ionic, single-chain amphiphiles. *Biochemistry 17*, 3759–3768. [*110*]

Hargreaves, W.R., Mulvihill, S. & Deamer, D.W. 1977: Synthesis of phospholipids and membranes in prebiotic conditions. *Nature 266*, 78–80. [*111*]

Harland, W.B. 1964: Critical evidence for a great infra-Cambrian glaciation. *Geologische Rundschau 54*, 45–61. [*435*]

Harland, W.B., Armstrong, R.L., Cox, A.V., Craig, L.E., Smith, A.G. & Smith, D.G. 1990: *A Geologic Time Scale 1989*. Cambridge University Press, Cambridge. [*300, 428, 431, 434, 435, 444*]

Harland, W.B. & Rudwick, M.J.S. 1964: The great infra-Cambrian ice age. *Scientific American 211*, 28–36. [*436*]

Harland, W.B. & Wilson, C.B. 1956: The Hecla Hoek succession in Ny Friesland and Spitsbergen. *Geological Magazine 93*, 265–286. [*436*]

Harrington, H.J. & Moore, R.C. 1956: Protomedusae. In Moore, R.C. (ed.): *Treatise on Invertebrate Paleontology. F. Coelenterata*, F21–F23. [*395*]

Harris, A.K., Stopak, D. & Warner, P. 1984: Generation of spationly periodic patterns by mechanical instability: A mechanical alternative to the Turing model. *Journal of Embryology and Experimental Morphology 80*, 1–20. [*487*]

Harris, A.K., Stopak, D. & Wild, P. 1981: Fibroblast traction as a mechanism for collagen morphogenesis. *Nature 290*, 249–251. [*487*]

Harris, C.M. & Selinger, B.K. 1983: Excited state processes of naphthols in aqueous surfactant solution. *Zeitschrift für Physikalische Chemie 134*, 65–92. [*120*]

Harrison, F.W. & Westfall, J.A. (eds.) 1991: *Microscopic Anatomy of Invertebrates, Vol. 2: Placozoa, Porifera, Cnidaria and Ctenophora*. Wiley-Liss, New York, N.Y. [*482*]

Harrison, S.C. 1991: A structural taxonomy of DNA-binding domains. *Nature 353*, 715–719. [*511*]

Hartman, H. 1984: The evolution of photosynthesis and microbial mats: A speculation on the Banded Iron Formations. In Cohen, Y., Castenholz, R.Z. & Halvorsen, H.O. (eds.): *Microbial Mats: Stromatolites*, 449–453. Alan Liss, New York, N.Y. [*176*]

Hartmann, W.K. & Davis, D.R. 1975: Satellite sized planetesimals and lunar origin. *Icarus 24*, 504–515. [*14*]

Hartnady, C.J.H. 1991: About turn for supercontinents. *Nature 352*, 476–478. [*443*]

Harwood, J.H. & Pirt, S.J. 1972: Quantitative aspects of growth of the methane-oxidizing bacterium *Methylococcus capsulatus* on methane in shake flask and continuous chemostat culture. *Journal of Applied Bacteriology 35*, 597–607. [*233*]

Hastings, J.W. 1983: Biological diversity, chemical mechanisms, and the evolutionary origins of bioluminescent systems. *Journal of Molecular Evolution 19*, 309–321. [*322*]

Haszprunar, G., Rieger, R.M. & Schuchert, P. 1991: Extant "Problematica" within or near the Metazoa. In Simonetta, A.M. & Conway Morris, S. (eds.): *The Early Evolution of Metazoa and the Significance of Problematic Taxa*, 99–107. Cambridge University Press, Cambridge. [*480, 481*]

Hattori, K. & Cameron, E.M. 1986: Archaean magmatic sulphate. *Nature 319*, 45–47. [*155*]

Hausmann, K. 1978: Extrusive organelles in protists. *International Review of Cytology 52*, 197–278. [*482*]

Hawker, J.R. & Oró, J 1981: Cyanamide mediated synthesis of Leu, Ala, Phe peptides under plausible primitive Earth conditions. In Wolman, Y. (ed.): *Origins of Life*, 225–232. Reidel, Dordrecht. [*56*]

Hay, E.D. 1981: *Cell Biology of Extracellular Matrix*. Plenum, New York, N.Y. [*477, 480*]

Hayes, J.M. 1983: Geochemical evidence bearing on the origin of aerobiosis, a speculative hypothesis. In Schopf, J.W. (ed.): *Earth's Earliest Biosphere, Its Origin and Evolution*, 291–301. Princeton University Press, Princeton, N.J. [*42, 45, 204, 224, 229, 233, 290, 299, 304*]

Hayes, J.M., Des Marais, D.J., Lambert, I.B., Strauss, H. & Summons, R. 1992a: Proterozoic biogeochemistry. In Schopf, J.W. & Klein, C. (eds.): *The Proterozoic Biosphere: A Multidisciplinary Study*, 81–134. Cambridge University Press, Cambridge. [*300, 306, 308*]

Hayes, J.M., Kaplan, I.R. & Wedeking, K.M. 1983: Precambrian organic geochemistry, preservation of the record. In Schopf, J.W. (ed.): *Earth's Earliest Biosphere, Its Origin and Evolution*, 93–134. Princeton University Press, Princeton, N.J. [*193, 200, 201, 224, 225, 231*]

Hayes, J.M., Lambert, I.B. & Strauss, H. 1992b: The sulfur-isotopic record. In Schopf, J.W. & Klein, C. (eds.): *The Proterozoic Biosphere: A Multidisciplinary Study*, 129–132. Cambridge University Press, Cambridge. [*235*]

Hayes, J.M., Popp, B.N., Takigiku, R. & Johnson, M.W. 1989: An isotopic study of biogeochemical relationships between carbonates and organic carbon in the Greenhorn Formation. *Geochimica et Cosmochimica Acta 53*, 2961–2972. [*223, 226*]

Heath, I.B. 1980: Variant mitoses in lower eukaryotes: Indicators of the evolution of mitosis? *International Review of Cytology 64*, 1–80. [*317, 326*]

Hedges, J.I. & Hare, P.E. 1987: Amino acid adsorption by clay minerals in distilled water. *Geochimica et Cosmochimica Acta 51*, 255–259. [*17*]

Heinrichs, T.K. & Reimer, T.O. 1977: A sedimentary barite deposit from the Archean Fig Tree Group of the Barberton Mountain Land (South Africa). *Economic Geology 72*, 1426–1441. [*29*]

Helber, J.T., Johnson, T.R., Yarbrough, L.R. & Hirschberg, R. 1988: Effect of nitrogenous compounds on nitrogenase gene expression in anaerobic cultures of *Anabena variabilis*. *Journal of Bacteriology 170*, 558–563. [*41*]

Hendriks, L., De Baere, R., Van de Peer, Y., Neefs, J., Goris, A. & De Wachter, R. 1991: The evolutionary position of the rhodophyte *Porphyra umbilicalis* and the basidiomycete *Leucosporidium scottii* among other eukaryotes as deduced from complete sequences of small ribosomal subunit RNA. *Journal of Molecular Evolution 32*, 167–177. [*296, 462*]

Hendriks, L., Huysmans, E., Vandenberghe, A. & De Wachter, R. 1987: Primary structures of the 5S ribosomal RNAs of 11 arthropods and applicability of 5S RNE to the study of metazoan evolution. *Journal of Molecular Evolution 24*, 103–109. [*469*]

Hennet, J.-C., Holm, N.G. & Engel, M.H. 1992: Abiotic synthesis of amino acids under hydrothermal conditions and the origin of a life: A perpetual phenomenon? *Naturwissenschaften 79*, 361–365. [*19*]

Henry, J.J. & Raff, R.A. 1992: Development and evolution of embryonic axial systems and cell determination in sea urchins. *Seminars in Developmental Biology 3*, 35–42. [*499*]

Henry, J.J., Wray, G.A. & Raff, R.A. 1990: The dorsoventral axis is specified prior to first cleavage in the direct developing sea urchin *Heliocidaris erythrogramma*. *Development 110*, 875–884. [*499*]

Hermann, T.N. 1990: *Organic World Billion Year Ago*. Nauka, Leningrad. [*441*]

Hernandez-Nicaise, M.-L. 1991: Ctenophora. In Harrison, F.W. & Westfall, J.A. (eds.): *Microscopic Anatomy of Invertebrates*, Vol. 2, 359–418. Wiley-Liss, New York, N.Y. [*406*]

Hertwig, O. & Hertwig, R. 1882: Die Coelomtheorie. Versuch einer Erklärung des mittleren Keimblattes. *Jenaische Zeitschrift für Medizin und Naturwissenschaften, 15, neue Folge 8*. 150 pp. [*480*]

Hesse, R. 1990a: Origin of chert: Diagenesis of biogenic siliceous sediments. In McIlreath, I.A. & Morrow, D.W. (eds.): *Diagenesis*, 227–252. Geological Association Canada. [*213*]

Hesse, R. 1990b: Silica diagenesis: Origin of inorganic and replacement cherts. In McIlreath, I.A. & Morrow, D.W. (eds.): *Diagenesis*, 253–275. Geological Association Canada. [*213*]

Hibbard, D.J. 1979: The structure and phylogenetic significance of the flagellar transition region in the chlorophyll *c*-containing algae. *BioSystems 11*, 243–261. [*318*]

Hickman, A.H. 1983: Geology of the Pilbara Block, Western Australia. *Geological Survey of Western Australia, Bulletin 127*. 268 pp. [*33*]

Hickman, A.H. (compiler) 1990: Pilbara and the Hamersley Basin. In Ho, S.E., Glover, J.E., Myers, J.S. & Muhling, J.R. (eds.): *Excursion Guidebook, Third International Archaean Symposium*. University of Western Australia. [*275*]

Hickman, A.H. & Lipple, S.L. 1978: Marble Bar Western Australia, 1:250,000 Geological Series – Explanatory Notes. 24 pp. Geological Survey of Western Australia, Perth. [*198*]

Hieshima, G.B. & Pratt, L.M. 1991: Sulfur/carbon ratios and extractable organic matter of the Middle Proterozoic Nonesuch Formation, North American Midcontinent Rift. *Precambrian Research 54*, 65–79. [*281*]

Higgins, D.G., Bleasby, A.J. & Fuchs, R. 1992: CLUSTAL V: Improved software for multiple sequence alignment. *Cabios 8*, 189–191. [*288*]

Hill, R.E., Jones, P.F., Rees, A.R., Sime, C.M., Justice, M.J., Copeland, N.R., Jenkins, N.A., Graham, E. & Davidson, D.R. 1989: A new family of mouse homeobox-containing gene: Molecular structure, chromosomal location, and developmental expression of Hox-7.1. *Genes and Development 3*, 26–37. [*497*]

Hinkle, K.W., Keady, J.J. & Bernath, P.F. 1988: Detection of C_3 in the interstellar shell of IRC+10 216. *Science 241*, 1319–1322. [*49*]

Hinz, I., Kraft, P., Mergl, M. & Müller, K.J. 1990: The problematic *Hadimopanella, Kaimenella, Milaculum* and *Utahphospha*, identified as sclerites of Palaeoscolecida. *Lethaia 23*, 217–221. [*458*]

Hirschler, A., Lucas, J. & Hubert, J.-C. 1990a: Apatite genesis: A biologically induced or biologically controlled mineral formation process? *Geomicrobiology Journal 7*, 47–57. [*99*]

Hirschler, A., Lucas, J. & Hubert, J.-C. 1990b: Bacterial involvement in apatite genesis. *FEMS Microbiology Ecology 73*, 211–220. [*99*]

Høeg, O. 1932: Ordovician algae from the Trondheim area. *Skrifter utgitt av Det Norske Videnskaps-Akademi i Oslo, I Mat.-Naturv. Klasse 4*, 63–96. [*428*]

Hoffman, P.F. 1989a: Speculations on Laurentia's first gigayear (2.0 to 1.0 Ga). *Geology 17*, 135–138. [*443*]

Hoffman, P.F. 1989b: Precambrian geology and tectonic history of North America. In Bally, A.W. & Palmer, A.R. (eds.): *The Geology of North America – An Overview*, 447–512. The Geology of North America A Geological Society of America, Boulder, Colo. [*25, 26*]

Hoffman, P.F. 1991: Did the breakout of Laurentia turn Gondwanaland inside out? *Science 252*, 1409–1412. [*443*]

Hofmann, H.J. 1969: Attributes of stromatolites. *Geological Survey of Canada, Paper 69-39*. 58 pp. [*336, 429*]

Hofmann, H.J. 1971: Precambrian fossils, pseudofossils, and problematica in Canada. *Geological Survey of Canada, Bulletin 189*, 1–146. [*248*]

Hofmann, H.J. 1972: Precambrian remains in Canada: Fossils, dubiofossils, and pseudofossils. *24th International Geological Congress, Montreal, Section 1*, 20–30. [*272*]

Hofmann, H.J. 1975: Stratiform Precambrian stromatolites, Belcher Islands, Canada: Relations between silicified microfossils and microstructure. *American Journal of Science 275*, 1121–1132. [*281*]

Hofmann, H.J. 1976: Precambrian microflora, Belcher Islands, Canada: Significance and systematics. *Journal of Paleontology 50*, 1040–1073. [*293, 296, 310*]

Hofmann, H.J. 1981: First record of a Late Proterozoic faunal assemblage in the North American Cordillera. *Lethaia 14*, 303–310. [*373*]

Hofmann, H.J. 1985a: The mid-Proterozoic Little Dal macrobiota, Mackenzie Mountains, north-west Canada. *Palaeontology 28*, 331–354. [*254, 301, 303, 346, 350, 351, 357, 359, 360, 361*]

Hofmann, H.J. 1985b: Precambrian carbonaceous megafossils. In Toomey, D.F. & Nitecki, M.H. (eds.): *Paleoalgology, Contemporary Research and Applications*, 20–33. Springer, Berlin. [*345, 349, 355, 360, 361, 427*]

Hofmann, H.J. 1987: Precambrian biostratigraphy. *Geoscience Canada 14*, 135–154. [*359*]

Hofmann, H.J. 1990: Precambrian time units and nomenclature – the geon concept. *Geology 18*, 340–341. [*349*]

Hofmann, H.J. 1992a: Proterozoic carbonaceous films. In Schopf, J.W. & Klein, C. (eds.): *The Proterozoic Biosphere: A Multidisciplinary Study*, 349–358. Cambridge University Press, Cambridge. [*345, 349, 355*]

Hofmann, H.J. 1992b: Proterozoic and selected Cambrian megascopic carbonaceous films. In Schopf, J.W. & Klein, C. (eds.): *The Proterozoic Biosphere: A Multidisciplinary Study*, 957–979. Cambridge University Press, Cambridge. [*345, 349*]

Hofmann, H.J. & Aitken, J. D. 1979: Precambrian biota from the Little Dal Group, Mackenzie Mountains, northwestern Canada. *Canadian Journal of Earth Sciences 16*, 150–166. [*346, 350, 356*]

Hofmann, H.J. & Chen, J. 1981: Carbonaceous megafossils from the Precambrian (1800 Ma) near Jixian, northern China. *Canadian Journal of Earth Sciences 18*, 443–447. [*290, 348, 441*]

Hofmann, H.J. & Grotzinger, J.P. 1985: Shelf-facies microbiotas from the Odjick and Rocknest formations (Epworth Group; 1.89 Ga), northwestern Canada. *Canadian Journal of Earth Sciences 22,* 1781–1792. [*430*]

Hofmann, H.J. & Jackson, G.D. 1987: Proterozoic ministromatolites with radial–fibrous fabric. *Sedimentology 34,* 963–971. [*251, 429, 430*]

Hofmann, H.J., Narbonne, G.M. & Aitken, J.D. 1990: Ediacaran remains from intertillite beds in northwestern Canada. *Geology 18,* 1199–1202. [*284, 305, 372, 442*]

Hofmann, H.J., Sage, R.P. & Berdusco, E.N. 1991: Archean stromatolites in Michipicoten Group siderite ore at Wawa, Ontario. *Economic Geology 86,* 1023–1030. [*193, 194, 275, 277, 278*]

Hofmann, H.J. & Schopf, J.W. 1983: Early Proterozoic microfossils. In Schopf, J.W. (ed.): *Earth's Earliest Biosphere, Its Origin and Evolution,* 321–360. Princeton University Press, Princeton, N.J. [*281, 283, 310, 430*]

Hofmann, H.J. & Snyder, G.L. 1985: Archean stromatolites from the Hartville Uplift, eastern Wyoming. *Geological Society of America, Bulletin 96,* 842–849. [*275*]

Hofmann, H.J., Thurston, P.C. & Wallace, H. 1985: Archean stromatolites from Uchi greenstone belt, northwestern Ontario. In Ayres, L.D., Thurston, P.C., Card, C.D. & Weber, W. (eds.): *Evolution of Archean supracrustal sequences.,* 125–132. Geological Association of Canada, Special Paper 28. [*275*]

Högbom, A.G. 1894: Om sannolikheten för sekulära förändringar i atmosfärens kolsyrehalt. *Svensk Kemisk Tidskrift 6,* 169–177. [*106*]

Holland, H.D. 1973: The oceans: A possible source of iron in iron-formations. *Economic Geology 68,* 1169–1172. [*13, 39, 46*]

Holland, H.D. 1984: *The Chemical Evolution of the Atmosphere and Oceans.* Princeton University Press, Princeton, N.J. [*12, 13, 25, 28, 29, 30, 31, 42, 44, 46, 94, 105, 233, 240, 243, 253, 256, 273*]

Holland, H.D. 1991: The mechanisms that control the carbon dioxide and oxygen content of the atmosphere. In Schneider, S.H. & Boston, P.J. (eds.): *Scientists on Gaia,* 174–179. Massachusetts Institute of Technology Press, Cambridge, Mass. [*244*]

Holland, H.D. 1992: Distribution and paleoenvironmental interpretation of Proterozoic paleosols. In Schopf, J.W. & Klein, C. (eds.): *The Proterozoic Biosphere: A Multidisciplinary Study,* 153–155. Cambridge University Press, Cambridge. [*203, 205, 238*]

Holland, H.D. & Beukes, N.J. 1990: A paleoweathering profile from Griqualand West, South Africa: Evidence for a dramatic rise in atmospheric oxygen between 2.2 and 1.9 b.y.b.p. *American Journal of Science 290-A,* 1–34. [*29, 234, 238, 239*]

Holland, H.D., Feakes, C.R. & Zbinden, E.A. 1989: The Flin Flon paleosol and the composition of the atmosphere 1.8 b.y.b.p. *American Journal of Science 289,* 362–389. [*238*]

Holland, H.D., Lazar, B. & McCaffrey, M. 1986: Evolution of the atmosphere and oceans. *Nature 320,* 27–33. [*336*]

Holland, H.D. & Zbinden, E.A. 1988: Paleosols and the evolution of the atmosphere, Part I. In Lerman, A. & Meybeck, M. (eds.): *Physical and Chemical Weathering in Geochemical Cycles,* 61–82. Kluwer, Dordrecht. [*238*]

Holland, P.W.H. 1990: Homeobox genes and segmentation: Co-option, co-evolution, and convergence. *Seminars in Developmental Biology 1,* 135–145. [*491, 492, 510*]

Holland, P.W.H. 1991: Cloning and evolutionary analysis of *msh*-like homeobox genes from mouse, zebrafish, and ascidian. *Gene 98,* 253–257. [*459, 497*]

Holm, N. 1985: New evidence for a tubular structure of b-iron(III) oxide hydroxide – akaganeite. *Origins of Life and Evolution of the Biosphere 15,* 131–139. [*18*]

Holm, N. 1987: Possible biological origin of banded iron-formations from hydrothermal solutions. *Origins of Life and Evolution of the Biosphere 17,* 229–250. [*94*]

Holm, N.G., Ertem, G. & Ferris, J.P. 1993: The binding and reactions of nucleotides and polynucleotides on iron oxide hydroxide polymorphs. *Origins of Life and Evolution of the Biosphere 23,* 195–215. [*94*]

Holmquist, G. 1989: Evolution of chromosome bands: Molecular ecology of noncoding DNA. *Journal of Molecular Evolution 28,* 469–486. [*142*]

Holo, H. 1989: *Chloroflexus aurantiacus* secretes 3-hydroxypropionate, a possible intermediate in the assimilation of CO_2 and acetate. *Archives of Microbiology 151,* 252–256. [*173*]

Holo, H. & Sirevåg, R. 1986: Autotrophic growth and CO$_2$ fixation of *Chloroflexus aurantiacus*. *Archives of Microbiology 145*, 173–180. [*174*]

Holser, W.T. 1984: Gradual and abrupt shifts in ocean chemistry during Phanerozoic time. In Holland, H.D. & Trendall, A.F. (eds.): *Patterns of Change in Earth Evolution*, 123–144. Springer, Berlin. [*444*]

Holser, W.T., Schidlowski, M., Mackenzie, F.T. & Maynard, J.B. 1988: Geochemical cycles of carbon and sulfur. In Gregor, C.B., Garrels, R.M., Mackenzie, F.T. & Maynard, J.B. (eds.): *Chemical Cycles in the Evolution of the Earth*, 105–173. Wiley, New York, N.Y. [*215, 222*]

Hood, L. & Hunkapiller, T. 1991: Molecular evolution and the immunoglobulin superfamily. In Osawa, S. & Honjo, T. (eds.): *Evolution of Life: Fossils, Molecules, and Culture*, 123–144. Springer, Tokyo. [*141*]

Hood, L., Kronenberg, M. & Hunkapiller, T. 1985: T cell antigen receptors and the immunoglobulin supergene family. *Cell 40*, 225–229. [*514*]

Hori, H. & Osawa, S. 1987: Origin and evolution of organisms as deduced from 5S ribosomal RNA sequences. *Molecular Biology and Evolution 4*, 445–472. [*469*]

Horneck, G., Bücker, H., Reitz, G., Requardt, H., Dose, K., Martens, K.D., Menningmann, H.D. & Weber, P. 1984: Microorganisms in the space environment. *Science 225*, 226–228. [*38*]

Horodyski, R.J. 1980: Middle Proterozoic shale-facies microbiota from the lower Belt Supergroup, Little Belt Mountains, Montana. *Journal of Paleontology 54*, 649–663. [*292, 310*]

Horodyski, R.J. 1990: Isotopic, field and micropaleontologic study of the Proterozoic Beck Spring (Ca.) and Mescal (Az.) formations; evidence for Precambrian terrestrial photosynthetic communities. *Geological Society of America, Abstracts with Programs 22*, 191. [*270*]

Horodyski, R.J. 1991: Late Proterozoic megafossils from southern Nevada. *Geological Society of America, Abstracts with Programs 23*, 163. [*373*]

Horodyski, R.J. & Mankiewicz, C. 1990: Possible late Proterozoic skeletal algae from the Pahrump Group, Kingston Range, southeastern California. *American Journal of Science 290-A*, 149–169. [*283, 291, 303, 415, 427, 442*]

Horowitz, N.H. 1945: On the evolution of biochemical synthesis. *Proceedings of the National Academy of Sciences, USA 31*, 153–157. [*79*]

Hörstadius, S. 1973: *Experimental Embryology of Echinoderms*. Clarendon, Oxford. [*499*]

Hou Xianguang 1987a: Three new large arthropods from Lower Cambrian, Chengjiang, eastern Yunnan. *Acta Palaeontologica Sinica 26*, 272–285. (In Chinese, with English summary.) [*368*]

Hou Xianguang 1987b: Two new arthropods from Lower Cambrian, Chengjiang, eastern Yunnan. *Acta Palaeontologica Sinica 26*, 236–256. (In Chinese, with English summary.) [*368*]

Hou Xianguang & Bergström, J. 1991: The arthropods of the Lower Cambrian Chengjiang fauna, with relationships and evolutionary significance. In Simonetta, A.M. & Conway Morris, S. (eds.): *The Early Evolution of Metazoa and the Significance of Problematic Taxa*, 179–187. Cambridge University Press, Cambridge. [*453*]

Hou Xianguang, Ramsköld, L. & Bergström, J. 1991: Composition and preservation of the Chengjiang fauna – a Lower Cambrian soft-bodied biota. *Zoologica Scripta 20*, 395–411. [*367, 453, 456, 458*]

Hou Xianguang & Sun Weiguo 1988: Discovery of the Chengjiang fauna at Meishucun, Jinning, Yunnan, China. *Acta Palaeontologica Sinica 27*, 1–12. (In Chinese, with English abstract.) [*367*]

Howchin, W. 1914: The occurrence of the genus *Cryptozoon* in the ?Cambrian of Australia. *Transactions of the Royal Society of South Australia 38*, 1–10. [*272*]

Huber, G., Drobner, E., Huber, H. & Stetter, K.O. 1992: Growth by aerobic oxidation of molecular hydrogen in archaea – a metabolic property so far unknown in this domain. *Systematic and Applied Microbiology 15*, 502–504. [*146, 151*]

Huber, G., Spinnler, C., Gambacorta, A. & Stetter, K.O. 1989: *Metallosphaera sedula* gen. and sp. nov., represents a new genus of aerobic, metal-mobilizing, thermoacidophilic Archaebacteria. *Systematic and Applied Microbiology 12*, 38–47. [*146*]

Huber, G. & Stetter, K.O. 1991: *Sulfolobus metallicus*, sp. nov., a novel strictly chemolithoautotrophic thermophilic archaeal species of metal-mobilizers. *Systematic and Applied Microbiology 14*, 372–378. [*146*]

Huber, H., Thomm, M., König, H., Thies, G. & Stetter, K.O. 1982: *Methanococcus thermolithotrophicus*, a novel thermophilic lithotrophic methanogen. *Archives of Microbiology 132*, 47–50. [*149*]

Huber, R., Kristjansson, J.K. & Stetter, K.O. 1987: *Pyrobaculum* gen.nov., a new genus of neutrophilic, rod-shaped Archaebacteria from continental Solfataras growing optimally at 100°C. *Archives of Microbiology 149*, 95–101. [*147*]

Huber, R., Langworthy, T.A., König, H., Thomm, M., Woese, C.R., Sleytr, U.B. & Stetter, K.O. 1986: *Thermotoga maritima* sp. nov. represents a new genus of unique extremely thermophilic Eubacteria growing up to 90°C. *Archives of Microbiology 144*, 324–333. [*150, 157*]

Huber, R., Stoffers, P., Cheminee, J.L., Richnow, H.H. & Stetter, K.O. 1990: Hyperthermophilic Archaebacteria within the crater and open-sea plume of erupting Macdonald Seamount. *Nature 345*, 179–182. [*146*]

Huber, R., Wilharm, T., Huber, D., Trincone, A., Burggraf, S., König, H., Rachel, R., Rockinger, I., Fricke, H. & Stetter, K.O. 1992: *Aquifex pyrophilus* gen.nov. sp. nov., represents a novel group of marine hyperthermophilic hydrogen-oxidizing bacteria. *Systematic and Applied Microbiology 15*, 340–351. [*39, 149, 155, 156, 157*]

Huckriede, R. 1967: *Archeanectritis benderi* n.gen. n.sp. (Hydrozoa), einde Chondrophore von der Wende Ordovizium/Silurium aus Jordanien. *Geologica et Palaeontologica 1*, 101–109. [*392*]

Hudson, B.G., Wieslander, J., Wisdon, B.J., Jr. & Noelken, M.E. 1989: Biology of Disease Goodpasture Syndrome: Molecular architecture and function of basement membrane antigen laboratory. *Investigation 61*, 256–269. [*477*]

Hughes, N.C. 1991: Morphological plasticity and genetic flexibility in a Cambrian trilobite. *Geology 19*, 913–916. [*453*]

Huhma, H., Cliff, R.A., Perttunen, V. & Sakko, M. 1990: Sm–Nd and Pb isotopic study of mafic rocks associated with early Proterozoic continental rifting: The Peräpohja schist belt in Northern Finland. *Contributions to Mineralogy and Petrology 104*, 369–379. [*239*]

Hui, J. & Dennis, P.P. 1985: Characterization of the ribsomal RNA gene cluster in *Halobacterium cutirubrum*. *Journal of Biological Chemistry 260*, 529–533. [*189*]

Hunkapiller, T., Goverman, J., Koop, B.F. & Hood, L. 1989: Implications of the diversity of the immunoglobulin gene superfamily. *Cold Spring Harbor Symposia on Quantitative Biology 54*, 15–29. [*141*]

Hunt, J.M. 1979: *Petroleum Geochemistry and Geology*. Freeman, San Francisco, Calif. [*33*]

Hunt, P., Gulisano, M., Cook, M., Sham, M.-H., Faiella, A., Wilkinson, D., Boncinelli, E. & Krumlauf, R. 1991a: A distinct *Hox* code for the branchial region of the vertebrate head. *Nature 353*, 861–864. [*405*]

Hunt, P., Whiting, J., Muchamore, I., Marshall, H. & Krumlauf, R. 1991b: Homeobox genes as models for patterning the hindbrain and branchial arches. *Development, Supplement 1*, 187–196. [*497*]

Hyman, L.H. 1940: *The Invertebrates: Protozoa through Ctenophora*. McGraw-Hill, New York, N.Y. [*403, 481*]

Hyman, L.H. 1951: *The Invertebrates, Vol. 2: Platyhelminthes and Rhynchocoela. The acoelomate Bilateria*. McGraw-Hill, New York, N.Y. [*490*]

Ingmanson, D.E. & Dowler, M.J. 1977: Chemical evolution and the evolution of the Earth's crust. *Origins of Life 8*, 221–224. [*18*]

Irvine, D.E.G. & John, D.M. (eds.) 1984: *Systematics of the Green Algae*. The Systematics Association Special Volume 27 Academic Press, London. [*321*]

Irvine, W.M. & Knacke, F. 1989: The chemistry of interstellar gas and grains. In Atreya, S.K., Pollack, J.B. & Matthews, M.S. (eds.): *Origin and Evolution of Planetary and Satellite Atmospheres*, 3–34. University of Arizona Press, Tucson, Ariz. [*49*]

Ishchenko, A.A. 1983: K kharakteristike vendskoj vodoroslevoj flory Prednestrov'ya [To the characteristics of the Vendian algal flora of the Dniestr region]. In *Stratigrafiya i Formatsii Dokembriya Ukrainy*, 181–206. Kiev. [*356, 357*]

Ivanov, A.V. 1963: *Pogonophora*. Academic Press, New York, N.Y. [*295*]

Ivantsov, A.Yu. 1990: Novye dannye po ul'trastrukture sabelliditid (Pogonophora?) [New data on the ultrastructure of sabelliditids (Pogonophora?)]. *Paleontologicheskij Zhurnal 1990:4*, 125–128. [*353, 377*]

Iversen, N. & Jorgensen, B.B. 1985: Anaerobic methane oxidation rates at the sulfate-methane transition in marine sediments from the Kattegat and Skagerrak (Denmark). *Limnology and Oceanography 30*, 944–955. [*235*]

Iwabe, N., Kuma, K., Hasegawa, M., Osawa, S. & Miyata, T. 1989: Evolutionary relationship of archaebacteria, eubacteria, and eukaryotes inferred from phylogenetic trees of duplicated genes. *Proceedings of the National Academy of Sciences, USA 86*, 9355–9359. [*67, 77, 144, 145, 153, 189, 288*]

Jablonski, D. & Bottjer, D.J. 1990: The ecology of evolutionary innovation: The fossil record. In Nitecki, M.H. (ed.): *Evolutionary Innovations*, 253–288. University of Chicago Press, Chicago, Ill. [*502*]

Jackson, G.D. & Ianelli, T.R. 1981: Rift-related cyclic sedimentation in the Neohelikian Borden Basin, northern Baffin Island. In Campbell, F.H.A. (ed.): *Proterozoic Basins of Canada*, 269–302. Geological Survey of Canada, Toronto, Ont. [*253*]

Jackson, J.B.C., Buss, L.W. & Cook, R.E. 1985: *Population Biology and Evolution of Clonal Organisms*. Yale University Press, New Haven, Conn. [*485*]

Jackson, M.J., Muir, M.D. & Plumb, K.A. 1987: Geology of the southern McArthur Basin, Northern Territory. *Bureau of Mineral Resources, Geology and Geophysics, Bulletin 220*. 173 pp. [*253, 281*]

Jackson, M.P.A., Eriksson, K.A. & Harris, C.W. 1987: Early Archean foredeep sedimentation related to crustal shortening: A reinterpretation of the Barberton sequence, southern Africa. *Tectonophysics 136*, 197–221. [*28*]

Jacobs, D.K. 1990: Selector genes and the Cambrian radiation of the Bilateria. *Proceedings of the National Academy of Sciences, USA 87*, 4406–4410. [*295, 407, 495, 498*]

Jacobson, A.G. 1966: Inductive processes in embryonic development. *Science 152*, 25–34. [*499*]

Jägersten, G. 1972: *Evolution of the Metazoan Lifecycle. A Comprehensive Theory*. Academic Press, London. [*483, 488*]

Jahnke, R.A., Emerson, S.R., Roe, K.K. & Burnett, W.C. 1983: The present day formation of apatite in Mexican continental margin sediments. *Geochimica et Cosmochimica Acta 47*, 259–266. [*99*]

Jankauskas [Yankauskas], T.V. (ed.) 1989: *Mikrofossilii Dokembriya SSSR*. Nauka, Leningrad. [*301, 306, 310, 348, 351, 353, 441*]

Javoy, M., Pineau, F. & Allegre, C.J. 1982: Carbon geodynamic cycle. *Nature 300*, 171–173. [*11, 32*]

Jefferies, R.P.S. 1990: The solute *Dendrocystoides scoticus* from the Upper Ordovician of Scotland and the ancestry of chordates and echinoderms. *Palaeontology 33*, 631–679. [*296*]

Jeffery, W.R. & Swalla, B.J. 1992: Evolution of alternate modes of development in ascidians. *BioEssays 14*, 219–226. [*494*]

Jeffrey, S.W. & Vesk, M. 1976: Further evidence for a membrane-bound endosymbiont within the dinoflagellate *Peridinium foliaceum*. *Journal of Phycology 12*, 450–455. [*317*]

Jenkins, R.J.F. 1981: The concept of an "Ediacaran Period" and its stratigraphic significance in Australia. *Transactions of the Royal Society of South Australia 105*, 179–194. [*363*]

Jenkins, R.J.F. 1984: Interpreting the oldest fossil cnidarians. *Palaeontographica Americana 54*, 95–104. [*365, 372, 405*]

Jenkins, R.J.F. 1985: The enigmatic Ediacaran (late Precambrian) genus *Rangea* and related forms. *Paleobiology 11*, 336–355. [*372, 415*]

Jenkins, R.J.F. 1989: The "supposed terminal Precambrian extinction event" in relation to the Cnidaria. *Memoirs of the Association of Australasian Palaeontologists 8*, 307–317. [*363, 365*]

Jenkins, R.J.F. & Gehling, J.G. 1978: A review of the frondlike fossils of the Ediacara assemblage. *Records of the South Australian Museum 17*, 347–359. [*364, 372, 384*]

Jensen, R.A. 1976: Enzyme recruitment in evolution of new function. *Annual Review of Microbiology 30*, 409–425. [*80*]

Jensen, S. 1990: Predation by early Cambrian trilobites on infaunal worms – evidence from the Swedish Mickwitzia Sandstone. *Lethaia 23*, 29–42. [*424*]

Jensen, S. 1991: The Lower Cambrian problematicum *Spatangopsis costata* Torell, 1870. *Geologiska Föreningens i Stockholm Förhandlingar 113*, 86–87. [*395*]

Jensen, S. 1993: Trace fossils, body fossils, and Problematica from the Lower Cambrian Mickwitzia Sandstone, South-central Sweden. 171 pp. Ph.D. Thesis, Uppsala University, Sweden. [*391*]

Jiang, J., Kosman, D., Ip, Y.T. & Levine, M. 1991: The *dorsal* morphogen gradient regulates the mesoderm determinant *twist* in early *Drosophila* embryos. *Genes and Development 5*, 1881–1891. [*506*]

Jin Yugan, Wang Huayu & Wang Wei 1991: Palaeoecological aspects of brachiopods from Chiung-chussu Formation of early Cambrian age, eastern Yunnan, China. In Jin Yugun, Wang Jungeng & Xu Shanhong (eds.): *Palaeoecology of China*, Vol. 1, 25–47. Nanjing University Press, Nanjing. [*458*]

Johansson, K.E. 1937: Über *Lamisabella zachsi* und ihre systematische Stellung. *Zoologischer Anzeiger 117*, 23–26. [*357*]

John, B. & Miklos, G.L.G. 1988: *The Eukaryote Genome in Development and Evolution*. Allen & Unwin, London. [*505, 507, 516*]

Johnson, J.H. 1954: An introduction to the study of rock-building algae and algal limestones. *Colorado School of Mines Quarterly 49*. 117 pp. [*428*]

Johnson, P.J., D'Oliveira, C.E., Gorrell, T.E., Müller, M. 1990: Molecular analysis of the hydrogeno-somal ferredoxin of the anaerobic protist *Trichomonas vaginalis*. *Proceedings of the National Academy of Sciences, USA 87*, 6097–6101. [*328*]

Johnson, P.W. & Sieburth, McN. 1979: Chroococcoid cyanobacteria in the sea: A ubiquitous and diverse phototrophic biomass. *Limnology and Oceanography 24*, 928–935. [*337*]

Joklik, G.F. 1955: The geology and mica-fields of the Harts Range, central Australia. *Bureau of Mineral Resources, Geology and Geophysics, Bulletin 26*, [*273*]

Jones, M.L. 1985a: On the Vestimentifera, new phylum: Six new species, and other taxa, from hydrothermal vents and elsewhere. *Biological Society of Washington Bulletin 6*, 117–158. [*295*]

Jones, M.L. 1985b: Vestimentiferan pogonophores: Their biology and affinities. In Conway Morris, S., George, J.D., Gibson, R. & Platt, H.M. (eds.): *The Origins and Relationships of Lower Invertebrates*, 327–342. Clarendon, Oxford. [*457*]

Jones, W.J., Nagle, D.P., Jr. & Whitman, W.B. 1987: Methanogens and the diversity of the Archae-bacteria. *Microbiological Reviews 51*, 135–177. [*39*]

Jørgensen, B.B. & Nelson, D.C. 1988: Bacterial zonation, photosynthesis and spectral light distribu-tion in hot spring microbial mats of Iceland. *Microbial Ecology 16*, 133–147. [*172*]

Joyce, G.F. 1988: Hydrothermal vents too hot? *Nature 334*, 564. [*125*]

Joyce, G.F. 1989: RNA evolution and the origins of life. *Nature 338*, 217–224. [*54, 57, 72, 73, 132, 135*]

Joyce, G.F. 1991: The rise and fall of the RNA world. *The New Biologist 3*, 399–407. [*72, 74, 79*]

Joyce, G.F., Schwartz, A.W., Orgel, L.E. & Miller, S.L. 1987: The case for an ancestral genetic system involving simple analogues of the nucleotides. *Proceedings of the National Academy of Sciences, USA 84*, 4398–4402. [*64, 74*]

Jüttner, F., Höflacher, B. & Wurster, K. 1986: Seasonal analysis of volatile organic biogenic substances (VOBS) in freshwater phytoplankton populations dominated by *Dinobryon*, *Microcystis* and *Aphanizomenon*. *Journal of Phycology 22*, 169–175. [*340*]

Kabnick, K.S. & Peattie, D.A. 1991: *Giardia*: A missing link between prokaryotes and eukaryotes. *American Scientist 79*, 34–43. [*187, 290*]

Kah, L. & Grotzinger, J.P. 1992: Early Proterozoic (1.9 Ga) thrombolites of the Rocknest Formation, Northwest Territories, Canada. *Palaios 7*, 305–315. [*280, 285*]

Kamminga, H. 1992: The structure of explanation in biology and the construction of theories of the origin of life on Earth. *UROBOROS 2*, 47–65. [*61*]

Kandler, O. & König, H. 1985: Cell envelopes of archaebacteria. In Woese, S.R. & Wolfe, R.S. (eds.): *The Bacteria, Vol. 8: Archaebacteria*, 413–457. Academic Press, Orlando, Fla. [*157, 158*]

Kandler, O. & König, H. 1993: Cell envelopes of Archaea: Structure and chemistry. In Kates, M., Kushner, D.J. & Matheson, A.T. (eds.): *The Biochemistry of Archaea (Archaebacteria)*, 223–239. Elsevier, Amsterdam. [*157, 158*]

Kaplan, I.R. & Nissenbaum, A. 1966: Anomalous carbon-isotope ratios in nonvolatile organic material. *Science 153*, 744–745. [*231*]

Karhu, J.A. 1993: Paleoproterozoic evolution of the carbon isotope ratios of sedimentary carbonates in the Fennoscandian shield. *Geological Survey of Finland, Bulletin 371*. 86 pp. [*244*]

Karhu, J. & Epstein, S. 1986: The implication of the oxygen isotope records in coexisting cherts and phosphates. *Geochimica et Cosmochimica Acta 50*, 1745–1756. [*434*]

Karling, T. 1966: On nematocysts and similar structures in turbellarians. *Acta Zoologica Fennica 116*, 1–28. [*482*]

Kasting, J.F. 1987: Theoretical constraints on oxygen and carbon dioxide concentrations in the Precambrian atmosphere. *Precambrian Research 34*, 205–229. [*27, 28, 38, 43, 45, 164, 212, 224, 233*]

Kasting, J.F. 1990: Bolide impacts and the oxidation state of carbon in the Earth's early atmosphere. *Origins of Life and Evolution of the Biosphere 20*, 199–231. [*12, 13, 16*]

Kasting, J.F. 1991: Box models for the evolution of atmospheric oxygen: An update. *Palaeogeography, Palaeoclimatology, Palaeoecology (Global and Planetary Change Section) 97*, 125–131. [*43, 218*]

Kasting, J.F. 1992: Proterozoic climates: The effect of changing atmospheric carbon dioxide concentrations. In Schopf, J.W. & Klein, C. (eds.): *The Proterozoic Biosphere: A Multidisciplinary Study*, 165–168. Cambridge University Press, Cambridge. [*203, 206, 243, 371*]

Kasting, J.F. & Ackerman, T.P. 1986: Climatic consequences of very high carbon dioxide levels in the Earth's early atmosphere. *Science 234*, 1383–1385. [*105*]

Kasting, J.F., Holland, H.D. & Pinto, J.P. 1985: Oxidant abundances in rainwater and the evolution of atmospheric oxygen. *Journal of Geophysical Research 90*, 10497–10510. [*238*]

Kasting, J.F., Liu, S.C. & Donahue, T.M. 1979: Oxygen levels in the prebiological atmosphere. *Journal of Geophysical Research 84*, 3097–3107. [*46*]

Kasting, J.F., Pollack, J.B. & Crisp, D. 1984: Effects of high CO_2 levels on surface temperature and atmospheric oxidation state of the early Earth. *Journal of Atmospheric Chemistry 1*, 403–428. [*11, 13*]

Kasting, J.F. & Walker, J.C.G. 1981: Limits on oxygen concentration in the prebiological atmosphere and the rate of abiotic fixation of nitrogen. *Journal of Geophysical Research 86*, 1147–1158. [*43*]

Kasting, J.F., Zahnle, K.J., Pinto, J.P. & Young A.T. 1989: Sulfur, ultraviolet radiation, and the early evolution of life. *Origins of Life and Evolution of the Biosphere 19*, 95–108. [*12, 21, 164*]

Kasting, J.F., Zahnle, K.J. & Walker, J.C.G. 1983: Photochemistry of methane in the Earth's early atmosphere. *Precambrian Research 20*, 121–148. [*43, 45*]

Katchalski, E. 1951: Poly-α-amino acids. *Advances in Protein Chemistry 6*, 123–125. [*95*]

Katsura, T. & Ito, E. 1990: Melting and subsolidus phase relations in the MgSiO3–MgCO3 system at high pressures: Implications to evolution of the Earth's atmosphere. *Earth and Planetary Science Letters 99*, 110–117. [*11*]

Kauffman, S.A. 1969: Metabolic stability and epigenesis in randomly constructed genetic nets. *Journal of Theoretical Biology 22*, 437–467. [*407*]

Kauffman, S.A. 1974: The large scale structure and dynamics of gene control circuits: An ensemble approach. *Journal of Theoretical Biology 44*, 167–190. [*407*]

Kauffman, S.A. 1989: Cambrian explosion and Permian quiescence: Implications of rugged fitness landscapes. *Evolutionary Ecology 3*, 274–281. [*495*]

Kaufman, A.J., Hayes, J.M., Knoll, A.H. & Germs, G.J.B. 1991: Isotopic compositions of carbonates and organic carbon from Upper Proterozoic successions in Namibia: Stratigraphic variation and the effects of diagenesis and metamorphism. *Precambrian Research 49*, 301–327. [*444, 449*]

Kaufman, A.J., Knoll, A.H. & Awramik, S.M. 1992: Biostratigraphic and chemostratigraphic correlation of Neoproterozoic sedimentary successions: Upper Tindir Group, northwestern Canada, as a test case. *Geology 20*, 181–185. [*291, 415, 427, 442*]

Kaźmierczak, J., Ittekott, V. & Degens, E.T. 1985: Biocalcification through time: Environmental challenge and cellular response. *Paläontologische Zeitschrift 59*, 15–33. [*256, 414, 421, 423, 437, 444*]

Keller, B.M. & Fedonkin, M.A. 1976: Novye nakhodki okamenelostej v valdajskoj serii dokembriya po r. Syuz'me [New organic fossil finds in the Precambrian Valday Series along the Syuz'ma River]. *Izvestiya Akademii Nauk SSSR, Seriya Geologicheskaya N 3*, 38–44. [*373*]

Keller, B.M., Menner, V.V., Stepanov, V.A. & Chumakov, N.M. 1974: Novye nakhodki Metazoa v vendomii Russkoj platformy [New discoveries of Metazoa in the Vendian of the Russian Platform]. *Izvestiya Akademii Nauk SSSR, Seriya Geologicheskaya N 12*, 130–134. [*355, 373*]

Keller, R.E. 1986: The cellular basis of amphibian gastrulation. In Browder, L. (ed.): *Developmental Biology: A Comprehensive Synthesis*, Vol. 2, 241–327. Plenum, New York, N.Y. [*498*]

Kempe, S. & Degens, E.T. 1985: An early soda ocean? *Chemical Geology 53*, 95–108. [*256*]

Kempe, S., Kaźmierczak, J. & Degens, E.T. 1989: The soda ocean concept and its bearing on biotic evolution. In Crick, R.E. (ed.): *Origin, Evolution, and Modern Aspects of Biomineralization in Plants and Animals*, 29–43. Plenum, New York, N.Y. [*421, 437*]

Kennard, J.M., in press: Thrombolites and stromatolites within shale–carbonate cycles, Middle–Late Cambrian Shannon Formation, Amadeus Basin, central Australia. In Monty, C.L.V. (ed.): *Phanerozoic Stromatolites*. [*280, 285*]

Kennard, J.M. & James, N.P. 1986: Thrombolites and stromatolites: Two distinct types of microbial structures. *Palaios 1*, 492–503. [*272, 280, 285*]

Kenyon, C. & Wang, B. 1991: A cluster of *Antennapedia*-class homeobox genes in a nonsegmented animal. *Science 253*, 516–517. [*405, 492*]

Kerans, C. & Donaldson, J.A. 1988: Deepwater conical stromatolite reef, Sulky Formation (Dismal Lakes Group), Middle Proterozoic, N.W.T. In Geldsetzer, H.H.J., James, N.P. & Tebbutt, G.E. (eds.): *Reefs, Canada and Adjacent Areas*, 81–88. Canadian Society of Petroleum Geologists, Memoir 13. [*252, 281*]

Kerans, C., Ross, G.M., Donaldson, J.A. & Geldsetzer, H.J. 1981: Tectonism and deposition of the Helikian Hornby Bay and Dismal Lakes Groups, District of Mackenzie. In Campbell, F.H.A. (ed.): *Proterozoic Basins of Canada*, 157–182. Geological Survey of Canada, Toronto, Ont. [*252*]

Khakhina, L.N. 1979: *Problema Simbiogeneza*. Akademia Nauk, Leningrad. (Translated into English: Margulis, L. & McMenamin, M. [eds.] 1992: *Concepts of Symbiogenesis*. Yale University Press, New Haven, Conn.) [*327*]

Kimura, M. 1968: Evolutionary rate at the molecular level. *Nature 217*, 624–626. [*468*]

Kimura, M. & Ohta, T. 1974: On some principles governing molecular evolution. *Proceedings of the National Academy of Sciences, USA 71*, 2848–2852. [*468*]

Kirby, H. 1952; annotated by Margulis, L. 1994: Harold Kirby's symbionts of termites: Karyomastigont reproduction and calonymphid taxonomy. *Symbiosis 16*, 1–55. [*329*]

Kirby, T.W., Lancaster, J.R., Jr. & Fridovich, I. 1981: Isolation and charaterization of the iron-containing superoxide dismutase of *Methanobacterium bryantii*. *Archives of Biochemistry and Biophysics 210*, 140–148. [*41*]

Kirk-Mason, K.E., Turner, M.J. & Chakraborty, P.R., 1989: Evidence for unusually short tubulin mRNA leaders and characterization of tubulin genes in *Giardia lamblia*. *Molecular and Biochemical Parisitology 36*, 87–100. [*187*]

Kirschvink, J.L. 1992: A paleogeographic model for Vendian and Cambrian time. In Schopf, J.W. & Klein, C. (eds.): *The Proterozoic Biosphere: A Multidisciplinary Study*, 567–582. Cambridge University Press, Cambridge. [*443*]

Kissel, J. & Krueger, F.R. 1987: The organic component in dust from Comet Halley as measured by the PUMA mass spectrometer on board Vega 1. *Nature 326*, 755–760. [*50*]

Klein, C. & Beukes, N.J. 1989: Geochemistry and sedimentology of a facies transition from limestone to iron-formation deposition in the early Proterozoic Transvaal Supergroup, South Africa. *Economic Geology 84*, 1733–1774. [*29, 31, 40, 46*]

Klein, C. & Beukes, N.J. 1992: The distribution, stratigraphy, and sedimentologic setting and geochemistry of Precambrian iron formations. In Schopf, J.W. & Klein, C. (eds.): *The Proterozoic Biosphere: A Multidisciplinary Study*, 139–146. Cambridge University Press, Cambridge. [*241*]

Klein, C., Beukes, N.J. & Schopf, J.W. 1987: Filamentous microfossils in the early Proterozoic Transvaal Supergroup: Their morphology, significance, and paleoenvironmental setting. *Precambrian Research 36*, 81–94. [*310, 414, 430*]

Klein, R.M. & Cronquist, A. 1967: A consideration of the evolutionary and taxonomic significance of some biochemical, micromorphological, and physiological characters in the Thallophytes. *Quarterly Review of Biology 42*, 105–296. [*317, 321, 323*]

Klima, J. 1967a: Zur Feinstruktur des acoelen Süßwasserturbellars *Oligochoerus limnophilus* Ax & Dörjes. *Berichte des Naturwissenschaftlich-Medizinischen Vereines in Innsbruck 55*, 107–124. [*476*]

Klima, J. 1967b: Cytologie: Eine Einführung für die Studierenden der Naturwissenschaften und der Medizin. 342 pp. Gustav Fischer, Stuttgart. [*477*]

Knauth, L.P. & Epstein, S. 1976: Hydrogen and oxygen isotope ratios in nodular and bedded chert. *Geochimica et Cosmochimica Acta 40*, 1095–1108. [*212, 213*]

Knoll, A.H. 1979: Archean photoautotrophy: Some alternatives and limits. *Origins of Life 9*, 313–327. [*34, 432*]

Knoll, A.H. 1984a: Microbiotas of the Precambrian Hunnberg Formation, Nordaustlandet, Svalbard. *Journal of Paleontology 58*, 131–162. [*306*]

Knoll, A.H. 1984b: The Archean/Proterozoic transition: A sedimentary and paleobiological perspective. In Holland, H.D. & Trendall, A.F. (eds.): *Patterns of Change in Earth Evolution*, 221–242. Springer, Berlin. [*218*]

Knoll, A.H. 1984c: *Earth's Earliest Biosphere: Its Origin and Evolution*, a review. *Paleobiology 10*, 286–292. [*224, 235*]

Knoll, A.H. 1985a: Exceptional preservation of photosynthetic organisms in silicified carbonates and silicified peats. *Philosophical Transactions of the Royal Society of London, B 311*, 111–122. [*281*]

Knoll, A.H. 1985b: A paleobiological perspective on sabkhas. In Friedman, G.M. & Krumbein, W.E. (eds.): *Hypersaline Ecosystems*, 407–427. Springer, Berlin. [*281*]

Knoll, A.H. 1985c: The distribution and evolution of microbial life in the Late Proterozoic Era. *Annual Review of Microbiology 39*, 391–417. [*306, 433*]

Knoll, A.H. 1990: Precambrian evolution of prokaryotes and protists. In Briggs, D.E.G. & Crowther, P.R. (eds.): *Palaeobiology, A Synthesis*, 9–16. Blackwell, Oxford. [*433*]

Knoll, A.H. 1991: End of the Proterozoic Eon. *Scientific American 265*, 64–73. [*443, 444, 447, 449, 502*]

Knoll, A.H. 1992a: Biological and biogeochemical preludes to the Ediacaran radiation. In Lipps, J.H. & Signor, P.W. (eds.): *Origin and Early Evolution of the Metazoa*, 53–84. Plenum, New York, N.Y. [*443, 447, 448*]

Knoll, A.H. 1992b: The early evolution of Eukaryotes: A geological perspective. *Science 256*, 622–627. [*441, 448*]

Knoll, A.H. 1993: Precambrian. In Jansonius, J. & McGregor, C.D. (eds.): *Palynology and Stratigraphy*. American Association of Stratigraphic Palynologists, Tulsa, Okla. (In press.) [*441*]

Knoll, A.H. & Barghoorn, E.S. 1977: Archean microfossils showing cell division from the Swaziland System of South Africa. *Science 198*, 396–398. [*195, 310*]

Knoll, A.H. & Bauld, J. 1989: The evolution of ecological tolerance in prokaryotes. *Transactions of the Royal Society of Edinburgh: Earth Sciences 80*, 209–223. [*163, 164, 168*]

Knoll, A.H. & Butterfield, N.J. 1989: New window on Proterozoic life. *Nature 337*, 602–603. [*447*]

Knoll, A.H. & Calder, S. 1983: Microbiotas of the Late Precambrian Ryssö Formation, Nordaustlandet, Svalbard. *Palaeontology 26*, 467–496. [*306*]

Knoll, A.H., Fairchild, I.J. & Swett, K. 1993: Calcified microorganisms in Proterozoic carbonates. *Palaios 8*, 512–525. [*445, 447*]

Knoll, A.H. & Golubic, S. 1979: Anatomy and taphonomy of a Precambrian algal stromatolite. *Precambrian Research 10*, 115–151. [*273, 283*]

Knoll, A.H., Hayes, J.M., Kaufman, A.J., Swett, K. & Lambert, I. 1986: Secular variation in carbon isotope ratios from Upper Proterozoic successions of Svalbard and East Greenland. *Nature 321*, 832–838. [*444*]

Knoll, A.H. & Swett, K. 1987: Micropalaeontology across the Precambrian–Cambrian boundary in Spitzbergen. *Journal of Paleontology 61*, 898–926. [*292*]

Knoll, A.H. & Swett, K. 1990: Carbonate deposition during the late Proterozoic Era: An example from Spitsbergen. *American Journal of Science 290-A*, 104–132. [*254, 421, 422, 429, 430, 447*]

Knoll, A.H., Swett, K. & Mark, J. 1991: Paleobiology of a Neoproterozoic tidal flat/lagoonal complex: The Draken Conglomerate Formation, Spitsbergen. *Journal of Paleontology 65*, 531–570. [*306, 427, 430, 441*]

Knoll, A.H. & Vidal, G. 1980: Late Proterozoic vase-shaped microfossils from the Visingsö Beds, Sweden. *Geologiska Föreningens i Stockholm Förhandlingar 102*, 207–211. [*308*]

Knoll, A.H. & Walker, J.C.G. 1990: Late Proterozoic evolution and environmental change. *Geological Society of America, Abstracts with Programs 22*, 114. [*449*]

Knoll, A.H. & Walter, M.R. 1992: Latest Proterozoic stratigraphy and Earth history. *Nature 356*, 673–678. [*444*]

Knowlton, N., Weil, E., Weigt, L.A. & Guzman, H.M. 1992: Sibling species in *Monastraea annularis*, coral bleaching, and the Coral Climate Record. *Science 255*, 330–333. [*330*]

Koch, A.L. 1985: Primeval cells: Possible energy-generating and cell-division mechanisms. *Journal of Molecular Evolution 21*, 270–277. [*21*]

Koch, A.L. & Schmidt T.M. 1991: The first cellular bioenergetic process: Primitive generation of a proton-motive force. *Journal of Molecular Evolution 33*, 297–304. [*82*]

Kolesnikov, M.P. 1991: Proteinoid microspheres and the process of prebiological photophosphorylation. *Origins of Life and Evolution of the Biosphere 21*, 31–37. [*162*]

Kolosov, P.N. 1975: *Stratigrafiya Verkhnego Dokembriya Yuga Yakutii.* Nauka, Novosibirsk. [*430*]

Komar, V.A. 1989: Classification of the microstructures of the Upper Precambrian stromatolites. In Valdiya, K.S. & Tewari, V.C. (eds.): *Stromatolites and Stromatolitic Deposits*, 229–238. Himalayan Geology 13. [*285*]

Komar, V.A., Raaben, M.E. & Semikhatov, M.A. 1965: Konofitony rifeya SSSR i ikh stratigraficheskoe znachenie [Conophytons in the Riphean of the USSR and their stratigraphic importance]. *Trudy Geologicheskogo Instituta Akademii Nauk SSSR 131.* 72 pp. [*271, 280*]

Koneva, S.P. 1986: A new family of Cambrian inarticulate brachiopods. *Paleontological Journal 1986*, 28–35. [*458*]

Korde, K.B. 1957: Novye predstaviteli sifonnikovykh vodoroslej [New representatives of siphonous algae]. *Osnovy Paleontologii 1*, 67–75. Izdatel'stvo Akademii Nauk SSSR, Moscow. [*428*]

Korde, K.B. 1973: Vodorosli kembriya [Cambrian algae]. *Trudy Paleontologicheskogo Instituta Akademii Nauk SSSR 139.* 349 pp. [*428, 429*]

Kornberg, A. & Baker, T. 1992: *DNA Replication.* 2d Ed. Freeman, San Francisco, Calif. [*66, 78*]

Korsch, R.J. & Kennard, J.M. (eds.) 1991: Geological and geophysical studies in the Amadeus Basin, central Australia. *Bureau of Mineral Resources, Geology and Geophysics, Bulletin 236.* 594 pp. [*273*]

Kozloff, E. 1971: Morphology of the orthonectid *Ciliocincta sabellariae. The Journal of Parasitology 57*, 585–597. [*481*]

Kozloff, E. 1972: Selection of food, feeding, and physical aspects of digestion in the acoel turbellarian *Otocelis luteola. Transactions of the American Microscopical Society 91*, 556–565. [*476*]

Krauskopf, K.B. 1979: *Introduction to Geochemistry.* 2d Ed. McGraw-Hill, New York, N.Y. [*39*]

Kretsinger, R.H. 1983: A comparison of the roles of calcium in biomineralization and in cytosolic signalling. In *Westbroek, P. & de Jong, E.W*, 123–131. Biomineralization and Biological Metal Accumulation Reidel, Dordrecht. [*437*]

Krishnamurthy, K., Epstein, S., Cronin, J., Pizzarello, S. & Yuen, G.U. 1992: Isotopic and molecular analyses of hydrocarbons and monocarboxylic acids of the Murchison meteorite. *Geochimica et Cosmochimica Acta 56*, 4045–4058. [*119*]

Kristensen, R.M. 1991: Loricifera. In Harrison, F.W. & Ruppert, E.E. (eds.): *Microscopic Anatomy of Invertebrates*, Vol. 4, 351–375. Wiley-Liss, New York, N.Y. [*406*]

Kröner, A., Greiling, R., Reischmann, T., Hussein, I.M., Stern, R.J., Durr, S., Kruger, J. & Zimmer, M. 1987: Pan-African crustal evolution in the Nubian segment of northeast Africa. In Kröner, A. (ed.): *Proterozoic Lithospheric Evolution*, 235–258. American Geophysical Union, Geodynamics Series 17. [*26*]

Kröner, A. & Layer, P.W. 1992: Crust formation and plate motion in the Early Archean. *Science 256*, 1405–1411. [*106*]

Kroopnick, P.M. 1985: The distribution of ^{13}C in the world oceans. *Deep-Sea Research 32*, 57–84. [*223*]

Kruger, K., Grabowski, P.J., Zaug, A.J., Sands, J., Gottschling, D.E. & Cech, T.R. 1982: Self-splicing RNA: Autoexcision and autocyclization of the ribosomal RNA intervening sequence of *Tetrahymena. Cell 31*, 147–157. [*133*]

Kubai, D.F. 1975: The evolution of the mitotic spindle. *International Review of Cytology 43*, 167–227. [*317*]

Kugrens, P. & Lee, R.E. 1988: Ultrastructure of fertilization in a cryptomonad. *Journal of Phycology 24*, 385–393. [*322*]

Kukalová, J. 1968: Permian mayfly nymphs. *Psyche 75*, 310–327. [*493*]

Kulaev, I.S. 1979: *The Biochemistry of Inorganic Polyphosphates.* Wiley, New York, N.Y. [*87*]

Kuma, K., Gedulin, B., Paplawsky, B. & Arrhenius, G. 1989: Mixed-valence hydroxides as bio-organic host minerals. *Origins of Life and Evolution of the Biosphere 19*, 573–602. [*94, 95*]

Kumé, M. & Dan, K. 1988: *Invertebrate Embryology.* Garland, New York, N.Y. [*492*]

Kump, L.R. & Volk, T. 1991: Gaia's garden and BLAG's greenhouse: Global biogeochemical climate regulation. In Schneider, S.H. & Boston, P.J. (eds.): *Scientists on Gaia*, 191–199. Massachusetts Institute of Technology Press, Cambridge, Mass. [*244*]

Kuranova, J.P. 1988: Three-dimensional structure of yeast inorganic pyrophosphatase and its active site. *Biokhimiya 53*, 1821–1827. [*89*]

Kurz, E., Holstein, T.W., Perti, B.M., Engel, J. & David, C.N. 1991: Mini-collagens in *Hydra* nematocysts. *Journal of Cell Biology 115*, 1159–1169. [*477*]

Lacey, J.C. & Mullins, D.W. 1983: Experimental studies related to the origin of the genetic code and the process of protein synthesis – a review. *Origins of Life 13*, 3–42. [*76*]

Lagrange-Henri, A.M., Vidal-Majdar, A. & Ferlet, R. 1988: The b Pictoris circumstellar disk, VI: Evidence material falling on to the star. *Astronomy and Astrophysics 190*, 275–282. [*51, 52*]

Lahav, N. & Chang, S. 1976: The possible role of solid surface area in condensation reactions during chemical evolution: Reevaluation. *Journal of Molecular Evolution 8*, 357–380. [*17, 23, 95*]

Lahav, N. & White, D.H. 1980: A possible role of fluctuating clay–water systems in the production of ordered prebiotic oligomers. *Journal of Molecular Evolution 16*, 11–21. [*95*]

Lahti, C. & Johnson, P.J. 1991: *Trichomonas vaginalis* hydrogenosomal proteins are synthesized on free polyribosomes and may undergo processing upon maturation. *Molecular and Biochemical Parasitology 46*, 307–310. [*328*]

Lake, J.A. 1988: Origin of the eukaryotic nucleus determined by rate-invariant analysis of rRNA sequences. *Nature 331*, 184–186. [*39, 290*]

Lake, J.A. 1989: Origin of multicellular animals. In Fernholm, B., Bremer, K. & Jörnvall, H. (eds.): *The Hierarchy of Life*, 273–278. Excerpta Medica, Amsterdam. [*402, 469*]

Lake, J.A. 1990: Origin of the Metazoa. *Proceedings of the National Academy of Sciences, USA 87*, 763–766. [*184, 453, 457, 469, 490*]

Lambert, I.B. & Donnelly, T.H. 1991: Atmospheric oxygen levels in the Precambrian: A review of isotopic and geological evidence. *Palaeogeography, Palaeoclimatology, Palaeoecology (Global and Planetary Change Section) 97*, 83–91. [*216, 444*]

Lambert, I.B., Donnelly, T.H., Dunlop, J.S.R. & Groves, D.I. 1978: Stable isotopic compositions of early Archaean sulphate deposits of probable evaporitic and volcanogenic origins. *Nature 276*, 808–811. [*29, 235*]

Lammers, M. & Follmann, H. 1983: The ribonucleotide reductases – a unique group of metalloenzymes essential for cell proliferation. *Structure and Bonding 54*, 27–91. [*78*]

Lamond, A.I. & Gibson, T.J. 1990: Catalytic RNA and the origin of genetic systems. *Trends in Genetics 6*, 145–149. [*65*]

Landais, P., Dubessy, J., Poty, B. & Robb, L.J. 1990: Three examples illustrating the analysis of organic matter associated with uranium ores. *Organic Geochemistry 16*, 601–608. [*232*]

Landing, E. 1984: Skeleton of lapworthellids and the suprageneric classification of tommotiids (Early and Middle Cambrian phosphatic problematica). *Journal of Paleontology 58*, 1380–1398. [*419*]

Lane, A.C. 1917: Lawson's correlation of the Pre-Cambrian era. *American Journal of Science 43*, 42–48. [*436*]

Lange, M.A. & Ahrens, T.J. 1982: The evolution of an impact-generated atmosphere. *Icarus 51*, 96–120. [*14*]

Langford, F.F. & Morin, J.A. 1976: The development of the Superior province of northwestern Ontario by merging island arcs. *American Journal of Science 276*, 1023–1034. [*26*]

Lanier, W.P. 1986: Approximate growth rates of Early Proterozoic microstromatolites as deduced by biomass productivity. *Palaios 1*, 525–542. [*310*]

Lanier, W.P. & Lowe, D.R. 1982: Sedimentology of the Middle Marker (3.4 Ga), Onverwacht Group, Transvaal, South Africa. *Precambrian Research 18*, 237–260. [*31*]

Lanyi, J.K., Tittor, J., Váró, G., Krippahl, G. & Oesterhelt, D. 1992: Influence of the size and protonation state of acidic residue 85 on the absorption spectrum and photoreaction of the bacteriorhodopsin chromophore. *Biochimica et Biophysica Acta 1099*, 102–110. [*85*]

Lappan, E.A. & Morowitz, H.J. 1974: Characterization of Mesozoa DNA. *Experimental Cell Research 83*, 143–151. [*468*]

Laurie, J.R. 1986: Phosphatic fauna of the early Cambrian Todd River Dolomite, Amadeus Basin, central Australia. *Alcheringa 10*, 431–454. [*458*]

Lawless, J.G. & Yuen, G.U. 1979: Quantitation of monocarboxylic acids in the Murchison carbonaceous meteorite. *Nature 282*, 396–398. [*111*]

Lazcano, A. 1986: Prebiotic evolution and the origin of cells. *Treballs de la Societat Catalana de Biologia 39*, 73–103. [*63, 70, 73, 77*]

Lazcano, A., Fox, G.E. & Oró, J. 1992: Life before DNA: The origin and evolution of early Archean cells. In Mortlock, R.P. (ed.): *The Evolution of Metabolic Function*, 237–295. CRC–Telford, Caldwell, N.J. [*52, 57, 61, 62, 64, 66, 67, 68, 78, 80*]

Lazcano, A., Guerrero, R., Margulis, L. & Oró, J. 1988: The evolutionary transition from RNA to DNA in early cells. *Journal of Molecular Evolution 27*, 283–290. [*78*]

Leadbeater, B.S.C. 1989: Chapter 8. In Green, J.C., Leadbeater, B.S.C. & Diver, W.L. (eds.): *The Chromophyte Algae: Problems and Perspectives*, 145–165. The Systematics Association Special Volume 38 Clarendon, Oxford. [*318*]

Leadbeater, B.S.C. & Manton, I. 1974: Preliminary observations on the chemistry and biology of the lorica of the collared flagellate (*Stephanotheca diplocostata* Ellis). *Journal of the Marine Biological Association of the United Kingdom 54*, 269–276. [*320*]

Leadbeater, B.S.C. & Riding, R. (eds.) 1986: *Biomineralization in Lower Plants and Animals*. The Systematics Association Special Volume 30 Clarendon, Oxford. [*426*]

Lear, J.D., Wasserman, Z.R. & DeGrado, W.F. 1988: Synthetic amphiphilic peptide models for protein ion channels. *Science 240*, 1177–1181. [*66*]

Lee, J.J., Hutner, S.H. & Bovee, E.C. (eds.) 1985: *An Illustrated Guide to the Protozoa*. Society of Protozoologists, Lawrence, Kans. [*314*]

Lee, R.E. & Kugrens, P. 1991: *Katablepharis* [*sic*] *ovalis*, a colorless flagellate with interesting cytological characteristics. *Journal of Phycology 27*, 505–513. [*320*]

Leedale, G.F. 1974: How many are the kingdoms of organisms? *Taxon 23*, 261–270. [*182, 314*]

Le Geros, Z.R. 1981: Apatites in biological systems. *Progress in Crystal Growth and Characterization 4*, 1–45. [*99*]

Leicester, H.M. 1974: *Development of Biochemical Concepts from Ancient to Modern Times*. Harvard Monographs in the History of Science Harvard University Press, Cambridge, Mass. [*52*]

Leiser, M. & Gromet-Elhanan, Z. 1974: Demonstration of acid-base phosphorylation in chromatophores in the presence of a K^+ diffusion potential. *FEBS Letters 43*, 267–270. [*86*]

Lemche, H. & Tendal, O.S. 1977: An interpretation of the sex cells and the early development in sponges, with a note on terms acrocoel and spongocoel. *Zeitschrift für Zoologische Systematik und Evolutionsforschung 15*, 241–252. [*480*]

Lerman, L. 1986: Potential role of bubbles and droplets in primordial and planetary chemistry: Exploration of the liquid–gas interface as a reaction zone for condensation reactions. *Origins of Life and Evolution of the Biosphere 16*, 201–202. [*20, 22*]

Leventhal, J.S., Grauch, R.I., Threlkeld, C.N., Lichte, F.E. & Harper, C.T. 1987: Unusual organic matter associated with uranium from the Claude Deposit, Cluff Lake, Canada. *Economic Geology 82*, 1169–1176. [*232*]

Leventhal, J.S. & Threlkeld, C.N. 1978: Carbon-13/Carbon-12 isotope fractionation of organic matter associated with uranium ores induced by alpha irradiation. *Science 202*, 430–432. [*232*]

Levine, J.S. & Augustsson, T.R. 1985: The photochemistry of biogenic gases in the early and present atmosphere. *Origins of Life and Evolution of the Biosphere 15*, 299–318. [*40, 41*]

Levine, J.S., Boughner, R.E. & Smith, K.A. 1980: Ozone, ultraviolet flux, and the temperature of the paleoatmosphere. *Origins of Life 10*, 199–213. [*38*]

Lewin, R.A. 1989: Oxygenic photosynthetic bacteria (sect.19), Group II: Prochlorales Lewin 1977. In Stanley, J.T. (ed.): *Bergey's Manual of Systematic Bacteriology*, Vol. 3, 1799–1806. Williams & Wilkins, Baltimore, Md. [*335*]

Lewontin, R.C. 1965: The role of linkage in natural selection. In Geerts, S.J. (ed.): *Genetics Today. Proceedings of the XIth International Congress of Genetics*, 517–525. Pergamon, Oxford. [*516*]

Lewontin, R.C. 1974: *The Genetic Basis of Evolutionary Change*. Columbia University Press, New York, N.Y. [*514*]

Liang Yuzuo, Cao Ruiji, Zhang Luyi *et al.* 1984: *Pseudogymnosolenaceae of Late Precambrian in China.* Geological Publishing House, Beijing. (In Chinese, with English abstract.) [*429*]

Lidstrom, M.E. & Somers, L. 1984: Seasonal study of methane oxidation in Lake Washington. *Applied and Environmental Microbiology 47*, 1255–1260. [*233*]

Liebau, F. & Koritnig, S. 1969: Phosphorus. In Wedepohl, K.H. (ed.): *Handbook of Geochemistry.* Springer, Berlin. [*97*]

Lillie, F.R. 1895: The embryology of the Unionidae. *Journal of Morphology 10*, 1–100. [*494*]

Lillie, F.R. 1898: Adaptation in cleavage. In *Biological Lectures of the Marine Biological Laboratory of Woods Hole, Mass.*, 43–67. Ginn & Co., Boston, Mass. [*494*]

Lindsay, J.F. 1987: Upper Proterozoic evaporites in the Amadeus basin, central Australia, and their role in basin tectonics. *Geological Society of America, Bulletin 99*, 852–865. [*273*]

Lipmann, F. 1965: Projecting backward from the present stage of evolution of biosynthesis. In Fox, S.W. (ed.): *The Origins of Prebiological Systems and of Their Molecular Matrices*, 212–226. Academic Press, New York, N.Y. [*84*]

Lipscomb, D.L. 1985: The eukaryote kingdoms. *Cladistics 1*, 127–140. [*315*]

Lipscomb, D.L. 1989: Relationships among the eukaryotes. In Fernholm, B., Bremer, K. & Jörnvall, H. (eds.): *The Hierarchy of Life*, 161–178. Excerpta Medica, Amsterdam. [*315, 316, 320*]

Liss, P.S. 1975: Chemistry of the sea surface microlayer. In Riley, J.P. & Skirrow, G. (eds.): *Chemical Oceanography*, 193–243. Academic Press, London. [*21*]

Loeb, W. 1913: Über das Verhalten des Formamids unter der Wirkung der stillen Entladung: Ein Beitrag zur Frage der Stickstoff-Assimilation. *Berichte der Deutschen Chemischen Gesellschaft 46*, 684–690. [*52*]

Loeblich, A.R., Jr. 1974: Protist phylogeny as indicated by the fossil record. *Taxon 23*, 277–290. [*313*]

Loesche, W.J. 1969: Oxygen sensitivity of various anaerobic bacteria. *Applied Microbiology 18*, 723–727. [*38*]

Logan, B.W. 1961: Cryptozoon and associate stromatolites from the Recent of Shark Bay, Western Australia. *Journal of Geology 69*, 517–533. [*338*]

Loukola-Ruskeeniemi, K. 1990: Kareliden mustaliuskeiden Hili-ja rikkipitoisuudet kerrostumis-sympariston kuvastajina. *Geologi 42*, 95–101. [*241*]

Loukola-Ruskeeniemi, K. 1991: Uranium contents in Finnish Proterozoic black shales. *Geological Survey of Finland, Report M19/3344-91/1/30.* [*241*]

Loukola-Ruskeeniemi, K. 1992: Geochemistry of Proterozoic metamorphosed black shales in eastern Finland, with implications for exploration and environmental studies. (Academic Dissertation, Division of Geology and Mineralogy, Deparment of Geology, University of Helsinki, Geological Survey of Finland, Espoo.) [*241*]

Lovelock, J.E. 1979: *Gaia: A New Look at Life on Earth.* Oxford University Press, New York, N.Y. [*215*]

Lovelock, J.E. 1988: *The Ages of Gaia.* Norton, New York, N.Y. [*43, 45, 233*]

Lovely, D.R. & Klug, M.J. 1983: Sulfate reducers can outcompete methanogens at fresh-water sulfate concentrations. *Applied and Environmental Microbiology 45*, 187–192. [*235*]

Lowe, D.R. 1980a: Archean sedimentation. *Annual Review of Earth and Planetary Sciences 8*, 145–167. [*27, 29, 30*]

Lowe, D.R. 1980b: Stromatolites 3,400-Myr old from the Archaean of Western Australia. *Nature 284*, 441–443. [*195, 197, 275*]

Lowe, D.R. 1982: Comparative sedimentology of the principal volcanic sequences of Archean greenstone belts in South Africa, Western Australia, and Canada: Implications for crustal evolution. *Precambrian Research 17*, 1–29. [*27, 29, 30*]

Lowe, D.R. 1983: Restricted shallow-water sedimentation of 3.4 Byr-old stromatolitic and evaporitic strata of the Strelley Pool Chert, Pilbara Block, Western Australia. *Precambrian Research 19*, 239–283. [*27, 28, 29, 31, 246, 275*]

Lowe, D.R. 1992a: Major events in the geological development of the Precambrian earth. In Schopf, J.W. & Klein, C. (eds.): *The Proterozoic Biosphere: A Multidisciplinary Study*, 67–75. Cambridge University Press, Cambridge. [*25, 26, 33*]

Lowe, D.R. 1992b: Other geological indicators. In Schopf, J.W. & Klein, C. (eds.): *The Proterozoic Biosphere: A Multidisciplinary Study*, 157–158. Cambridge University Press, Cambridge. [*213*]

Lowe, D.R. & Byerly, G.R. 1986: Archaean flow-top alteration zones formed initially in a low-temperature sulphate-rich environment. *Nature 324*, 245–248. [*27, 40*]

Lowe, D.R. & Ernst, W.G. 1992: The Archean geologic record. In Schopf, J.W. & Klein, C. (eds.): *The Proterozoic Biosphere: A Multidisciplinary Study*, 13–20. Cambridge University Press, Cambridge. [*26*]

Lowe, D.R. & Knauth, L.P. 1977: Sedimentology of the Onverwacht Group (3.4 billion years), Transvaal, South Africa, and its bearing on the characteristics and evolution of the early earth. *Journal of Geology 85*, 699–723. [*27, 28, 29, 31, 246*]

Lowenstam, H.A. 1981: Minerals formed by organisms. *Science 211*, 1126–1131. [*414*]

Lowenstam, H.A. & Margulis, L. 1980: Evolutionary prerequisites for early Phanerozoic calcareous skeletons. *BioSystems 12*, 27–41. [*414, 423*]

Lowenstam, H.A. & Weiner, S. 1989: *On Biomineralization*. Oxford University Press, New York, N.Y. [*332, 414, 415, 416, 422*]

Lucas, J. & Prévôt, L. 1981: Synthèse d'apatite à partir de matière organique phosphoreé (ARN) et de calcite par voie bactérienne. *Comptes Rendus de l'Academie des Sciences, Sér. II 292*, 1203–1208. [*99*]

Lucas, J. & Prévôt, L. 1984: Synthèse de l'apatite par voie bactérienne à partir de matière organique phosphatée et de divers carbonates de calcium dans des eaux douce et marine naturelles. *Chemical Geology 42*, 101–118. [*99*]

Lucas, J. & Prévôt, L. 1985: The synthesis of apatite by bacterial activity: Mechanism. *Sciences Géologiques Mémoires 77*, 83–92. [*99*]

Luchinina, V.A. 1975: Paleoal'gologicheskaya kharakteristika rannego kembriya Sibirskoj platformy [Palaeoalgological characteristics of the Early Cambrian of the Siberian Platform]. *Trudy Instituta Geologii i Geofiziki Sibir'skogo Otdeleniya Akademii Nauk SSSR 216*. 98 pp. [*428, 429*]

Ludwig, M. & Gibbs, S.P. 1985: DNA is present in the nucleomorph of cryptomonads: Further evidence that the chloroplast evolved from a eukaryotic endosymbiont. *Protoplasma 127*, 9–20. [*317*]

Ludwig, M. & Gibbs, S.P. 1989: Evidence that the nucleomorphs of *Chlorarachnion reptans* (Chlorarachniophyta) are vestigial nuclei: Morphology, division and DNA–DAPI fluorescence. *Journal of Phycology 25*, 385–394. [*317*]

Lundin, M., Baltscheffsky, H. & Ronne, H. 1991: Yeast *PPA2* gene encodes a mitochondrial inorganic pyrophosphatase that is essential for mitochondrial function. *Journal of Biological Chemistry 266*, 12168–12172. [*89*]

McFadden, G.I. 1990: Evidence that cryptomonad chloroplasts evolved from photosynthetic eukaryotic endosymbionts. *Journal of Cell Science 95*, 303–308. [*317*]

McFall-Ngai, M.J. & Ruby, E.G. 1991: Symbiont recognition and subsequent morphogenesis as early events in an animal–bacterial mutualism. *Science 254*, 1491–1494. [*329, 333*]

Macfarlane, A.W. & Holland, H.D. 1991: The timing of alkali metasomatism in paleosols. *Canadian Mineralogist 29*, 1043–1050. [*238*]

McGinnis, W. & Krumlauf, R. 1992: Homeobox genes and axial patterning. *Cell 68*, 283–302. [*459*]

MacIntyre, F. 1974a: The top millimeter of the ocean. *Scientific American 230*, 62–77. [*21, 22*]

MacIntyre, F. 1974b: Chemical fractionation and sea-surface microlayer processes. In Goldberg, E.D. (ed.): *The Sea: Ideas and Observations on Progress in the Study of the Seas*, Vol. 5., 245–299. Wiley, New York, N.Y. [*21*]

MacIntyre, F. & Winchester, J.W. 1969: Phosphate ion enrichment in drops from breaking bubbles. *Journal of Physical Chemistry 73*, 2163–2169. [*21*]

Mack, E.E. & Pierson, B.K. 1988: Preliminary characterization of a temperate marine member of the Chloroflexaceae. In Olson, J.M., Ormerod, J.G., Amesz, J., Stackebrandt, E. & Trüper, H.G. (eds.): *Green Photosynthetic Bacteria*, 237–241. Plenum, New York, N.Y. [*168*]

Mackie, G.O. 1990: The elementary nervous system revisited. *American Zoologist 30*, 907–920. [*482*]

Mackie, G.O. & Singla, C.L. 1983: Studies on hexactinellid sponges: I. Histology of *Rhabdocalyptus dawsoni*. *Philosophical Transactions of the Royal Society of London, B 301*, 365–400. [*476*]

Mackinnon, I.D.R. & Rietmeijer, F.J.M. 1987: Mineralogy of chondritic interplanetary dust particles. *Reviews of Geophysics 25*, 1527–1553. [*14*]

McKirdy, D.M. 1976: Biochemical markers in stromatolites. In Walter, M.R. (ed.): *Stromatolites*, 163–191. Developments in Sedimentology 20 Elsevier, Amsterdam. [*273*]

McKirdy, D.M. & Kantsler, A.J. 1980: Oil geochemistry and potential source rocks of the Officer Basin, South Australia. *Australian Petroleum Exploration Association Journal 20*, 68–86. [*202, 203*]

McKirdy, D.M. & Powell, T.G. 1974: Metamorphic alteration of carbon isotopic composition in ancient sedimentary organic matter: New evidence from Australia and South Africa. *Geology 2*, 591–595. [*202, 203*]

McLaughlin, P.J. & Dayhoff, M.O. 1973: Eukaryote evolution: A view based on cytochrome *c* sequence data. *Journal of Molecular Evolution 2*, 99–116. [*461*]

McMenamin, M.A.S. 1986: The Garden of Ediacara. *Palaios 1*, 178–182. [*375, 399*]

McMenamin, M.A.S. 1992: The Cambrian transition as a time-transgressive ecotone. *Geological Society of America, Abstracts with Programs 24*, 62. [*425*]

McMenamin, M.A.S. 1993: Osmotrophy in fossil protoctists and early animals. *Invertebrate Reproduction and Development 23*, 165–169. [*332*]

McMenamin, M.A.S. & McMenamin, D.L.S. 1990: *The Emergence of Animals*. Columbia University Press, New York, N.Y. [*490*]

McMillan, W.O., Raff, R.A. & Palumbi, S.R. 1992: Population genetic consequences of reduced dispersal in a direct-developing sea urchin, *Heliocidaris erythrogramma*. *Evolution 46*, 1299–1312. [*494*]

Madigan, C.T. 1932: The geology of the eastern MacDonnell Ranges, Central Australia. *Transactions of the Royal Society of South Australia 56*, 71–117. [*272*]

Madigan, C.T. 1935: The geology of the MacDonnell Ranges and neighbourhood, Central Australia. *Australian and New Zealand Association for the Advancement of Science, Report 21*, 75–86. [*272*]

Maher, K.A. & Stevenson, J.D. 1988: Impact frustration of the origin of life. *Nature 331*, 612–614. [*14*]

Maithy, P.K. 1990: Metaphyte and metazoan remains from the Indian Proterozoic successions. In Jain, K.P. & Tiwari, R.S. (eds.): *Proceedings of Symposium on "Vistas in Indian Palaeobotany,"* 20–38. The Palaeobotanist 38. [*351*]

Malacinski, G.M., Neff, A.W., Radice, G. & Chung, H.-M. 1989: Amphibian somite development: Contrasts of morphogenetic and molecular differentiation patterns between the the laboratory archetype species *Xenopus* (anuran) and axolotl (urodele). *Zoological Science 6*, 1–14. [*498*]

Maliva, R., Knoll, A.H. & Siever, R. 1987: Secular change in chert distribution: A reflection of evolving biological participation in the silica cycle. *Palaios 4*, 519–532. [*448*]

Mamkaev, Yu.V. & Kostenko, A.G. 1991: On the phylogenetic significance of sagittocysts and copulatory organs in acoel turbellarians. In Tyler, S. (ed.): *Turbellarian Biology*, 307–314. Hydrobiologia 227 Kluwer, Dordrecht. [*482*]

Mangold, O. 1961: Grundzüge der Entwicklungsphysiologie der Wirbeltiere mit besonderer Berücksichtigung der Missbildungen auf Grund experimenteller Arbeiten an Urodelen. *Acta Geniticae Medicae et Gemellogiae 10*, 1–49. [*499*]

Maniatis, T., Goodbourn, S. & Fischer, J.A. 1987: Regulation of inducible and tissue-specific gene expression. *Science 236*, 1237–1244. [*407*]

Mankiewicz, C. 1992: Proterozoic and Early Cambrian calcareous algae. In Schopf, J.W. & Klein, C. (eds.): *The Proterozoic Biosphere: A Multidisciplinary Study*, 359–367. Cambridge University Press, Cambridge. [*431*]

Mann, S. 1983: Mineralization in biological systems. *Structure and Bonding 54*, 125–174. [*414*]

Mann, S., Sparks, N.H.C., Frankel, R.B., Bazylinski, D.A. & Jannasch, H.W. 1990: Biomineralization of ferrimagnetic greigite (Fe_3S_4) and iron pyrite (FeS_2) in a magnetotactic bacterium. *Nature 343*, 258–261. [*414*]

Mann, S., Webb, J. & Williams, R.J.P. (eds.) 1989: *Biomineralization. Chemical and Biochemical Perspectives*. Verlag Chemie, Weinheim. [*414*]

Manton, I. 1959: Electron microscopical observations on a very small flagellate: The problem of *Chromulina pusilla* Butcher. *Journal of the Marine Biological Association of the United Kingdom 38*, 319–333. [*319*]

Manton, S.M. 1977: *The Arthropoda*. Clarendon, Oxford. [*402, 403*]

Mar, A. & Oró, J. 1990: Synthesis of the coenzymes ADPG, GDPG, and CDP-ethanolamine under primitive Earth conditions. *Journal of Molecular Evolution 31*, 374–381. [*54*]

Mar, A. & Oró, J. 1991: Synthesis of the coenzymes adenosine diphosphate glucose, guanosine diphosphate glucose, and cytidine diphosphoethanolamine under primitive Earth conditions. *Journal of Molecular Evolution 32*, 201–210. [*54, 72*]

Margulies, M.M. 1991: Sequence similiarity between Photosystems I and II. Identification of a Photosystem I reaction center transmembrane helix that is similar to transmembrane helix IV of the D2 subunit of Photosystem II and the M subunit of the non-sulfur purple and flexible green bacteria. *Photosynthesis Research 29*, 133–147. [*179*]

Margulis, L. 1970: *Origin of Eukaryotic Cells*. Yale University Press, New Haven, Conn. [*303*]

Margulis, L. 1976: A review: Genetic and evolutionary consequences of symbiosis. *Experimental Parasitology 39*, 277–349. [*330*]

Margulis, L. 1992a: Biodiversity: Molecular biological domains, symbiosis and kingdom origins. *BioSystems 27*, 39–51. [*332*]

Margulis, L. 1992b: Symbiosis theory: Cells as microbial communities. In Margulis, L. & Olendzenski, L. (eds.): *Environmental Evolution*, 149–172. M.I.T. Press, Cambridge, Mass. [*335*]

Margulis, L. 1993: *Symbiosis in Cell Evolution: Microbial Communities in the Archean and Proterozoic Eons*. 2d Ed. Freeman, New York, N.Y. [*330*]

Margulis, L., Corliss, J.O., Melkonian, M. & Chapman, D.J. (eds.) 1990: *Handbook of Protoctista: The Structure, Cultivation, Habitats and Life Histories of the Eukaryotic Microorganisms and Their Descendants Exclusive of Animals, Plants and Fungi*. Jones & Bartlett, Boston, Mass. [*182, 315, 329, 333*]

Margulis, L. & Fester, R. (eds.) 1991: *Symbiosis as a Source of Evolutionary Innovation: Speciation and Morphogenesis*. Massachusetts Institute of Technology Press, Cambridge, Mass. [*327*]

Margulis, L. & McMenamin, M. 1990: Kinetosome-centriolar DNA: Significance for endosymbiosis theory. *Treballs de la Societat Catalana de Biologia 41*, 5–16. [*328*]

Margulis, L. & Schwartz, K.V. 1982: *Five Kingdoms: An Illustrated Guide to the Phyla of Life on Earth*. [*315*]

Margulis, L. & Schwartz, K.V. 1988: *Five Kingdoms: An Illustrated Guide to the Phyla of Life on Earth*. 2d Ed. Freeman, New York, N.Y. [*182, 330*]

Margulis, L., Walker, J.C.G. & Rambler, M. 1976: Reassessment of roles of oxygen and ultraviolet light in Precambrian evolution. *Nature 264*, 620–624. [*166*]

Martens, C. & Harriss, R. 1970: Inhibition of apatite precipitation in the marine environment by magnesium ions. *Geochimica et Cosmochimica Acta 34*, 621–625. [*100*]

Martin, A., Nisbet, E.G. & Bickle, M.J. 1980: Archaean stromatolites of the Belingwe Greenstone Belt, Zimbabwe (Rhodesia). *Precambrian Research 13*, 337–362. [*248, 278*]

Martin, A.W., Harrison, F.M., Huston, M.J. & Stewart, D.M. 1958: The blood volumes of some representative molluscs. *Journal of Experimental Biology 35*, 260–279. [*402*]

Martin, D.McB., Stanistreet, I.G. & Camden-Smith, P.M. 1989: The interaction between tectonics and mudflow deposits within the Main Conglomerate Formation in the 2.8–2.7 Ga Witwatersrand Basin. *Precambrian Research 44*, 19–38. [*28*]

Martin, J.H., Knauer, G.A., Karl, D.M. & Broenkow, W.W. 1987: VERTEX: Carbon cycling in the northeast Pacific. *Deep-Sea Research 34*, 267–285. [*43*]

Masinovsky, Z., Lozovaya, G.I., Sivash, A.A. & Drasner, M. 1989: Porphyrin–proteinoid complexes as models of prebiotic photosensitizers. *BioSystems 22*, 305–310. [*162*]

Mason, S.F. 1991: *Chemical Evolution: Origin of the Elements, Molecules, and Living Systems*. Clarendon, Oxford. [*232*]

Mathez, E.A. 1984: Influence of degassing on oxidation states of basaltic magmas. *Nature 310*, 371–375. [*12*]

Mathies, R.A., Lin, S.W., Ames, J.B. & Pollard W.T. 1991: From femtoseconds to biology: Mechanism of bacteriorhodopsin's light-driven proton pump. *Annual Review of Biophysics and Biophysical Chemistry 20*, 491–518. [*85*]

Mathis, P. 1990: Compared structure of plant and bacterial photosynthetic reaction centers: Evolutionary implications. *Biochimica et Biophysica Acta 1018*, 163–167. [*177*]

Mathur, S.M. 1983: A reappraisal of trace fossils described by Vredenburg (1908) and Beer (1919) in rocks of the Vindhyan Supergroup. *Geological Survey of India, Records 113*, 111–113. [*292, 356*]

Matsui, T. & Abe, Y. 1986: Evolution of an impact-induced atmosphere and magma ocean on the accreting Earth. *Nature 319*, 303–308. [*11*]

Matthews, C. & Ludicky, R. 1987: The dark nucleus of Comet Halley: Hydrogen cyanide polymer. In *Proceedings of ESLAB Symposium 20: Exploration of Halley Comet*, 273–277. ESA Publications SP-50. [*50*]

Matthews, P.E. 1967: The pre-Karoo formations of the White Umfolozi Inlier, northern Natal. *Transactions of the Geological Society of South Africa 70*, 39–63. [*31, 33*]

Mattox, K.R. & Stewart, K.D. 1984: Classification of the green algae: A concept based on comparative cytology. In Irvine, D.E.G. & John, D.M. (eds.): *Systematics of the Green Algae*, 29–72. The Systematics Association Special Volume 27 Academic Press, London. [*318*]

Mawson, D. & Madigan, C.T. 1930: Pre-Ordovician rocks of the MacDonnell Ranges, Central Australia. *Transactions of the Royal Society of South Australia 85*, 613–623. [*272*]

May, R.M. 1990: How many species? *Philosophical Transactions of the Royal Society of London, B 330*, 293–304. [*329*]

Mayr, E. 1988: *Eine neue Philosophie der Biologie*. Piper, Munich. [*486*]

Mayr, E. & Ashlock, P.D. 1991: *Principles of Systematic Zoology*. McGraw-Hill, New York, N.Y. [*154, 156, 158*]

Mazghouni, M., Kbir-Ariguib, N., Counioux, J.J. & Sebaoun, A. 1981: Étude des équilibres solide–liquide–vapeur des systèmes binaires K_3PO_4–H_2O et $Mg_3(PO_4)_2$–H_2O. *Thermochimica Acta 47*, 125–139. [*100*]

Mehl, D. & Reiswig, H.M. 1991: The presence of flagellar vanes in choanomeres of Porifera and their possible phylogenetic implications. *Zeitschrift für Zoologische Systematik und Evolutionsforschung 28*, 312–319. [*476*]

Mendelson, C.V., Bauld, J., Horodyski, R.J., Lipps, J.H., Moore, T.B. & Schopf, J.W. 1992: Proterozoic and selected Early Cambrian microfossils: Prokaryotes and protists. In Schopf, J.W. & Klein, C. (eds.): *The Proterozoic Biosphere: A Multidisciplinary Study*, 175–244. Cambridge University Press, Cambridge. [*273, 281, 292*]

Mendelson, C.V. & Schopf, J.W. 1992a: Proterozoic and Early Cambrian acritarchs. In Schopf, J.W. & Klein, C. (eds.): *The Proterozoic Biosphere: A Multidisciplinary Study*, 219–232. Cambridge University Press, Cambridge. [*427, 510*]

Mendelson, C.V. & Schopf, J.W. 1992b: Proterozoic and selected Early Cambrian microfossils and microfossil-like objects. In Schopf, J.W. & Klein, C. (eds.): *The Proterozoic Biosphere: A Multidisciplinary Study*, 865–951. Cambridge University Press, Cambridge. [*194*]

Menhart, J.R. & Palmer, J.D. 1990: The gain of two chloroplast tRNA introns marks the green algal ancestors of land plants. *Nature 345*, 268–270. [*303*]

Mercer-Smith, J. & Mauzerall, D.C. 1984: Photochemistry of porphyrins: A model for the origin of photosynthesis. *Photochemistry and Photobiology 39*, 397–405. [*117*]

Mereschkowsky, K.S. 1905: Über Natur und Ursprung der Chromatophoren im Pflanzenreiche. *Biologisches Zentralblatt 25*, 593–604. [*335*]

Mettam, C. 1985: Functional constraints in the evolution of the Annelida. In Conway Morris, S., George, J.D., Gibson R. & Platt, H.M. (eds.): *The Origins and Relationships of Lower Invertebrates*, 297–310. Clarendon, Oxford. [*486*]

Michel, H. & Deisenhofer, J. 1988: Relevance of the photosynthetic reaction center from purple bacteria to the structure of photosystem II. *Biochemistry 27*, 2–7. [*179*]

Michels, P.A.M., Marchand, M., Kohl, L., Allert, S., Wierenga, R.K. & Opperdoes, R.F. 1991: The cytosolic and glycosomal isoenzymes of glyceraldehyde-3-phosphate dehydrogenase in *Trypanosoma brucei* have a distant evolutionary relationship. *European Journal of Biochemistry 198*, 421–428. [*182*]

Miklos, G.L.G. 1993: Molecules and cognition: The latterday lessons of levels, language and *lac*. *Journal of Neurobiology 24*, 842–890. [*505, 515, 516*]

Miklos, G.L.G. & Campbell, H.D. 1992: The evolution of protein domains and the organizational complexities of metazoans. *Current Opinion in Genetics and Development 6*, 902–906. [*511, 515, 516*]

Miklos, G.L.G., Campbell, K.S.W. & Kankel, D.R. 1994: The rapid emergence of bio-electronic novelty, neuronal architectures and behavioral performance. In *Flexibility and Constraint in Behavioral Systems*. Wiley, Chichester. (In press.) [*515, 516*]

Miller, C.A. & Benzer, S. 1983: Monoclonal antibody cross-reactions between *Drosophila* and human brain. *Proceedings of the National Academy of Sciences, USA 80*, 7641–7645. [*504*]

Miller, S.L. 1953: A production of amino acids under possible primitive Earth conditions. *Science 117*, 528–529. [*52, 124*]

Miller, S.L. 1992: The prebiotic synthesis of organic compounds as a step toward the origin of life. In Schopf, J.W. (ed.): *Major Events in the History of Life*, 1–28. Jones & Bartlett, Boston, Mass. [37]

Miller, S.L. & Bada, J.L. 1988: Submarine hot springs and the origin of life. *Nature 334*, 609–611. [*18, 35*]

Miller, S.L. & Orgel, L.E. 1974: *The Origins of Life on the Earth*. Prentice-Hall, Englewood Cliffs, N.J. [*52, 54, 133*]

Miller, S.L. & Parris, M. 1964: Synthesis of pyrophosphate under primitive earth conditions. *Nature 204*, 1248–1250. [*97*]

Miller, S.L., Urey, H.C. & Oró, J. 1976: Origin of organic compounds on the primitive Earth and in meteorites. *Journal of Molecular Evolution 9*, 59–72. [*108*]

Miller, S.L. & Van Trump, J.E. 1981: The Strecker synthesis in the primitive ocean. In Wolman, Y. (ed.): *Origins of Life*, 135–141. Reidel, Dordrecht. [*14, 105*]

Mills, D.R., Kramer, F.R. & Spiegelman 1973: Complete nucleotide sequence of a replicating RNA molecule. *Science 180*, 916–927. [*72*]

Mimms, L.T., Zampighi, G., Nozaki, Y., Tanford, C. & Reynolds J.A. 1981: Phospholipid vesicle formation and transmembrane protein incorporation using octyl glucoside. *Biochemistry 20*, 833–839. [*116*]

Mirabdullaev, I.M. 1989: Ribosomi, kristi i filogeniya nizshich eukariot (Ribosomes, cristae and phylogeny of lower eukaryotes). *Izvestiya Akademii Nauk SSSR, Seriya Biologicheskaya 1989*, 689–700. [*325*]

Missarzhevskij, V.V. 1974: Novye dannye o drevnejshikh okamenelostyakh rannego kembriya Sibirskoj Platformy [New data on the earliest fossils of Early Cambrian of Siberian Platform]. In *Biostratigrafiya i Paleontologiya Nizhnego Kembriya Evropy i Severnoj Azii*, 179–189. Nauka, Moscow. [*383*]

Missarzhevskij, V.V. 1989: Drevnejshie skeletnye okamenelosti i stratigrafiya pogranichnykh tolshch dokembriya i kembriya. *Trudy Geologicheskogo Instituta Akademii Nauk SSSR 443*, 1–237. [*383, 415, 419*]

Mitchell, P.J. & Tjian, R. 1989: Transcriptional regulation in mammalian cells by sequence-specific DNA binding proteins. *Science 245*, 371–378. [*509, 511*]

Moczydłowska, M. 1991: Acritarch biostratigraphy of the Lower Cambrian and the Precambrian–Cambrian boundary in southeastern Poland. *Fossils and Strata 29*. 127 pp. [*292, 304, 305, 417, 446*]

Moestrup, O. & Anderson, R.A. 1991: Organization of heterotrophic heterokonts. In Patterson, D.J. & Larsen, J. (eds.): *The Biology of Free-living Heterotrophic Flagellates*, 333–360. The Systematics Association Special Volume 45. Clarendon, Oxford. [*325*]

Moffett, J.W. 1990: Microbially mediated cerium oxidation in sea water. *Nature 345*, 421–423. [*40*]

Mohanty-Hejmadi, P., Dutta, S.K. & Mahapatra, P. 1992: Limbs generated at site of tail amputation in marbled balloon frog after vitamin A treatment. *Nature 355*, 352–353. [*498*]

Mollenhauer, D. & Kovacik, L. 1988: Who was who in cyanophyte research – I. In Anagnostidis, K., Golubic, S., Komárek, J. & Lhotsky, O. (eds.): *Cyanophyta (Cyanobacteria): Morphology, Taxonomy, Ecology*, 19–33. Algological Studies 50/53 Schweizerbart'sche, Stuttgart. [*334*]

Montoya, T.H. & Golubic, S. 1991: Morphological variability in natural populations of mat-forming cyanobacteria in the salinas of Huacho, Lima, Peru. In Hickel, B., Anagnostidis, K. & Komarek, J. (eds.): *Cyanophyta (Cyanobacteria): Morphology, Taxonomy, Ecology*, 423–441. Algological Studies 64 Schweizerbart'sche, Stuttgart. [*338*]

Monty, C.L.V. 1973: Precambrian background and Phanerozoic history of stromatolitic communities, an overview. *Annales de la Société Géologique de Belgique 96*, 585–624. [*254, 284, 337, 432, 435*]

Monty, C.L.V. 1976: The origin and development of cryptalgal fabrics. In Walter, M.R. (ed.): *Stromatolites*, 193–249. Developments in Sedimentology 20 Elsevier, Amsterdam. [*281*]

Mook, W.G., Bommerson, J.C. & Staverman, W.H. 1974: Carbon isotope fractionation between dissolved bicarbonate and gaseous carbon dioxide. *Earth and Planetary Science Letters 22*, 169–176. [*226*]

Moore, P.B. 1973: Pegmatite phosphates: Descriptive mineralogy and crystal chemistry. *Mineralogical Record 4*, 103–136. [*97*]

Moore, P.B. 1988: The ribosome returns. *Nature 331*, 223–227. [*76, 77*]

Moores, E.M. 1991: Southwest U.S. – East Antarctic (SWEAT) connection: A hypothesis. *Geology 19*, 325–328. [*443*]

Morowitz, H.J., Heinz, B. & Deamer, D.W. 1988: The chemical logic of a minimum protocell. *Origins of Life and Evolution of the Biosphere 18*, 281–287. [*21, 123*]

Morowitz, H.J., Smith, T. & Deamer, D.W. 1991: Biogenesis as an evolutionary process. *Journal of Molecular Evolution 33*, 207–208. [*123*]

Morris, P. & Cobabe, E. 1991: Cuvier meets Watson and Crick: The utility of molecules as classical homologies. *Biological Journal of the Linnean Society 44*, 307–324. [*463*]

Morse, J.W. & Mackenzie, F.T. 1990: *Geochemistry of Sedimentary Carbonates*. Elsevier, New York, N.Y. [*226*]

Mostler, H. 1986: Beitrag zur stratigraphischen Verbreitung und phylogenetischen Stellung der Amphidiscophora und Hexasterophora (Hexactinellida, Porifera). *Mitteilungen der Österreichischen Geologischen Gesellschaft 78*, 319–359. [*485*]

Moyle, J., Mitchell, R. & Mitchell, P. 1972: Proton-translocating pyrophosphatase of *Rhodospirillum rubrum. FEBS Letters 23*, 233–236. [*86*]

Muir, M.D. & Grant, P.R. 1976: Micropaleontological evidence from the Onverwacht Group, South Africa. In Windley, B.F. (ed.): *The Early History of the Earth*, 595–604. Wiley, London. [*195*]

Mukhin, L.M., Gerasimov, M.V. & Safonova, E.N. 1989: Origin of precursors of organic molecules during evaporation of meteorites and mafic terrestrial rocks. *Nature 340*, 46–48. [*16*]

Müller, D., Pitsch, S., Kittaka, A., Wagner, E., Wintner, C.E. & Eschenmoser, A. 1990: Chemie von Alpha-Aminonitrilen. Aldomerisierung von Glycolaldehyd-phosphat zu racemischen Hexose-2,4,6-triphosphaten und (in Gegenwart von Formaldehyd) racemischen Pentose-2,4-diphosphaten: Rac-Allose-2,4,6-triphosphat und rac-Ribose-2,4-diphosphat sind die Reaktionshauptprodukte. *Helvetica Chimica Acta 73*, 1410–1468. [*54, 64, 73, 74, 92, 97*]

Müller, K.J. & Walossek, D. 1986: Arthropod larvae from the Upper Cambrian of Sweden. *Transactions of the Royal Society of Edinburgh: Earth Sciences 77*, 157–179. [*492*]

Müller, K.J. & Walossek, D. 1988: External morphology and larval development of the Upper Cambrian maxillopod *Bredocaris admirabilis. Fossils and Strata 23*, 1–70. [*513*]

Müller, M. 1988: Energy metabolism of protozoa without mitochondria. *Annual Review of Microbiology 42*, 465–488. [*328*]

Murphy, J.B. & Nance, N.D. 1991: Supercontinent model for the contrasting character of Late Proterozoic orogenic belts. *Geology 19*, 469–472. [*443*]

Murtha, M., Leckman, J. & Ruddle, J. 1991: Detection of homeobox genes in development and evolution. *Proceedings of the National Academy of Sciences, USA 88*, 10711–10715. [*405, 459*]

Musters, W., Gonçalves, P.M., Boon, K., Raue, H.A., van Heerikhuizen, H. & Planta, R.J. 1991: The conserved GTPase center and variable region V9 from *Saccharomyces cerevisiae* 26S rRNA can be replaced by their equivalents from other prokaryotes or eukaryotes without detectable loss of ribosomal function. *Proceedings of the National Academy of Sciences, USA 88*, 1469–1473. [*504*]

Nagy, B., Gauthier-Lafaye, F., Holliger, P., Davis, D.W., Mossman, D.J., Leventhal, J.S., Rigali, M.J. & Parnell, J. 1991: Organic matter and containment of uranium and fissiogenic isotopes at the Oklo natural reactors. *Nature 354*, 472–475. [*232*]

Nancollas, G.H. & Reddy, M.M. 1971: The crystallization of calcium carbonate, II: Calcite growth mechanism. *Journal of Colloid and Interface Science 37*, 824–830. [*436*]

Narbonne, G.M. & Aitken, J.D. 1990: Ediacaran fossils from the Sekwi Brook area, Mackenzie Mountains, northwestern Canada. *Palaeontology 33*, 945–980. [*373*]

Narbonne, G.M. & Hofmann, H.J. 1987: Ediacaran biota of the Wernecke Mountains, Yukon, Canada. *Palaeontology 30*, 647–676. [*373*]

Narbonne, G.M., Kaufman, A.J. & Knoll, A.H. 1994: Integrated chemostratigraphy and biostratigraphy of the upper Windermere Group, Mackenzie Mountains, northwestern Canada. *Geological Society of America, Bulletin.* (In press.) [*442*]

Nardon, P., Gianinazzi-Pearson, V., Grenier, A.M., Margulis, L. & Smith, D.C. (eds.) 1990: *Endocytobiology IV. Proceedings of the 4th International Colloquium on Endocytobiology.* Institut National de la Recherche Agrono. [*317*]

Nardon, P. & Grenier, A.M. 1990: Symbiosis as an important factor for the growth and the evolution of the populations of *Sitophilus oryzae* L. (Coleoptera, Curculionidae). *Endocytobiology IV*, 369–372. [*333*]

Nathorst, A.G. 1881: Om spår av några evertebrerade djur m.m. och deras paleontologiska betydelse. *Kongliga Svenska Vetenskaps-Akademiens Handlingar 18.* 59 pp. [*396*]

Nazarov, B.B. & Ormiston, A.R. 1985: Evolution of Radiolaria in the Paleozoic and its correlation with the development of other marine fossil groups. *Senckenbergiana lethaea 66*, 203–215. [*420*]

Nealson, K.H. 1991: Luminescent bacteria symbiotic with entomopathogenic nematodes. In Margulis, L. & Fester, R. (eds.): *Symbiosis as a Source of Evolutionary Innovation: Speciation and Morphogenesis*, 205–218. Massachusetts Institute of Technology Press, Cambridge, Mass. [*333*]

Nei, M. 1987: *Molecular Evolutionary Genetics.* Columbia University Press, New York, N.Y. [*328*]

Neuilly, M., Bussac, J., Frejacques, C., Nieff, G., Vendryes, G. & Yvon, J. 1972: Sur l'existence dans us passé reculé d'une réaction en chaîne naturelle de fissions dans le gisement d'uranium d'Oklo (Gabon). *Comptes Rendus de l'Academie des Sciences, Sér. D 275*, 1847. [*232*]

Neuman, W.F. & Neuman, M.W. 1964: On the possible role of crystals in the origin of life. *University of Rochester Contribution W-7401-eng-49.*, [*97*]

Neunlist, S. & Rohmer, M. 1985: A novel hopanoid, 30-(5'-adenosyl)hopane, from the purple non-sulphur bacterium *Rhodopseudomonas acidophila*, with possible DNA interactions. *The Biochemical Journal 228*, 769–771. [*72*]

Nicholls, D.G. & Ferguson, S.J. 1992: *Bioenergetics 2.* Academic Press, London. [*85*]

Nielsen, C. 1985: Animal phylogeny in the light of the trochea theory. *Biological Journal of the Linnean Society 25*, 243–299. [*468*]

Nielsen, C. & Nørrevang, A. 1985: The trochea theory: An example of life cycle phylogeny. In Conway Morris, S., George, J.D., Gibson, R. & Platt, H.M. (eds.): *The Origins and Relationships of Lower Invertebrates*, 28–42. Clarendon, Oxford. [*483*]

Nisbet, E.G. 1980: Archaean stromatolites and the search for the earliest life. *Nature 284*, 395–396. [*275*]

Nisbet, E.G. 1985: The geological setting of the earliest life forms. *Journal of Molecular Evolution 21*, 289–298. [*31, 134*]

Nisbet, E.G. & Wilks, M.E. 1988: Archaean stromatolite reef at Steep Rock Lake, Atikokan, northwestern Ontario. In Geldsetzer, H.H.J., James, N.P. & Tebbutt, G.E. (eds.): *Reefs, Canada and Adjacent Areas*, 89–92. Canadian Society of Petroleum Geologists, Memoir 13. [*275*]

Nissenbaum, A., Kenyon, D.H. & Oró, J. 1975: On the possible role of organic melanoidin polymers as matrices for prebiotic activity. *Journal of Molecular Evolution 6*, 253–270. [*17*]

Nitecki, M.H. & Debrenne, F. 1979: The nature of radiocyathids and their relationship to receptaculitids and archaeocyathids. *Géobios 12*, 5–27. [*429*]

Nitschke, W. & Rutherford, A.W. 1991: Photosynthetic reaction centres: Variations on a common structural theme? *Trends in Biochemical Sciences 16*, 241–245. [*177*]

Nocita, B.W. & Lowe, D.R. 1990: Fan-delta sequence in the Archean Fig Tree Group, Barberton Greenstone Belt, South Africa. *Precambrian Research 48*, 375–393. [*28*]

Noller, H.F. 1991: Drugs and the RNA world. *Nature 353*, 302–303. [*57, 63, 76*]

Noller, H.F., Hoffarth, V. & Zimniak, L. 1992: Unusual resistance of peptidyl transferase to protein extraction procedures. *Science 256*, 1416–1419. [*63, 76, 77*]

Noller, H.F., Moazed, D., Stern, S., Powers, T., Allen, P.A., Robertson, J.M., Weiser, B. & Triman, K. 1990: Structure of rRNA and its functional interactions in translation. In Hill, W., Dahlberg, A., Garrett, R.A., Moore, P.B. & Schelssinger, D. (eds.): *Ribosomes: Structure, Function and Genetics*, 73–92. American Microbiological Society, Washington, D.C. [*76*]

Nomura, M. 1987: The role of RNA and protein in ribosome function: A review of early reconstitution studies and prospects for future studies. *Cold Spring Harbor Symposia on Quantitative Biology 52*, 653–663. [*76*]

Nore, B.F., Sakai-Nore, Y., Maeshima, M., Baltscheffsky, M. & Nyrén, P. 1991: Immunological cross-reactivity between proton-pumping inorganic pyrophosphatase of widely phylogenic separated species. *Biochemical and Biophysical Research Communications 181*, 962–967. [*89*]

Norris, R. 1989: Cnidarian taphonomy and affinities of the Ediacara biota. *Lethaia 22*, 381–399. [*381*]

Norris, R.E. & Pearson, B.R. 1975: Fine structure of *Pyramimonas parkeae* sp. nov. (Chlorophyta, Prasinophyceae). *Archiv für Protistenkunde 117*, 192–213. [*320*]

Norton, R.A. & Behan-Pelletier, V.M. 1991: Calcium carbonate and calcium oxalate as cuticular hardening agents in oribatid mites (Acari: Oribatida). *Canadian Journal of Zoology 69*, 1505–1511. [*452*]

Nöthig, E.-M. & Bodungen, B. von 1989: Occurrence and vertical flux of faecal pellets of probably protozoan origin in the southeastern Weddell Sea (Antarctica). *Marine Ecology Progress Series 56*, 281–289. [*451*]

Nyrén, P., Hajnal, K. & Baltscheffsky, M. 1984: Purification of the membrane-bound proton-translocating inorganic pyrophosphatase from *Rhodospirillum rubrum*. *Biochimica et Biophysica Acta 766*, 630–635. [*86*]

Nyrén, P. & Lundin, A. 1985: Enzymatic method for continuous monitoring of inorganic pyrophosphate synthesis. *Analytical Biochemistry 151*, 504–509. [*118*]

Nyrén, P., Nore, B.F. & Baltscheffsky, M. 1986: Studies on photosynthetic inorganic pyrophosphate formation in *Rhodospirillum rubrum* chromatophores. *Biochimica et Biophysica Acta 851*, 276–282. [*86, 87*]

Nyrén, P., Nore, B.F. & Strid, Å. 1991: Proton-pumping N,N'-dicyclohexylcarbodiimide-sensitive inorganic pyrophosphate synthase from *Rhodospirillum rubrum*: Purification, characterization, and reconstitution. *Biochemistry 30*, 2883–2887. [*86*]

O'Brien, N.R. 1987: The effects of bioturbation on the fabric of shale. *Journal of Sedimentary Petrology 57*, 449–455. [*451*]

Oberbeck, V.R. & Aggarwal, H. 1992: Comet impacts and chemical evolution on the bombarded Earth. *Origins of Life and Evolution of the Biosphere 21*, 317–338. [*17*]

Oberbeck, V.R. & Fogelman, G. 1989: Impacts and the origin of life. *Nature 339*, 434. [*14*]

Oberbeck, V.R. & Fogelman, G. 1990: Impact constraints on the environment for chemical evolution and the continuity of life. *Origins of Life and Evolution of the Biosphere 20*, 181–195. [*14*]

Oberbeck, V.R., Marshall, J. & Shen, T. 1991: Prebiotic chemistry in clouds. *Journal of Molecular Evolution 32*, 296–303. [*22*]

Odum, H.T. 1983: *Systems Ecology: An Introduction*. Wiley, New York, N.Y. [*209*]

Oehler, D.Z. 1976: Transmission electron microscopy of organic microfossils from the Late Precambrian Bitter Springs Formation of Australia: Techniques and survey of preserved ultrastructure. *Journal of Paleontology 50*, 90–106. [*273, 283*]

Oehler, D.Z. 1977: Pyrenoid-like structures in Late Precambrian algae from the Bitter Springs Formation of Australia. *Journal of Paleontology 51*, 885–901. [*273, 283*]

Oehler, D.Z. 1978: Microflora of the middle Proterozoic Balbirini Dolomite (McArthur Group) of Australia. *Alcheringa 2*, 269–309. [*281*]

Oehler, D.Z., Oehler, J.H. & Stewart, A.J. 1979: Algal fossils from a late Precambrian, hypersaline lagoon. *Science 205*, 388–390. [*273*]

Oesterhelt, D. 1989: Photosynthetic systems in prokaryotes: The retinal proteins of halobacteria and the reaction centre of purple bacteria. *Biochemistry International 18*, 673–694. [*162*]

Oesterhelt, D., Bräuchle, C. & Hampp, N. 1991: Bacteriorhodopsin: A biological material for information processing. *Quarterly Reviews of Biophysics 24*, 425–478. [*85*]

Oesterhelt, D. & Krippahl, G. 1983: Phototrophic growth of halobacteria and its use for isolation of photosynthetically-deficient mutants. *Annales de Microbiologie (Inst. Pasteur) 134B*, 137–150. [*159*]

Ohmoto, H. & Felder, R.P. 1987: Bacterial activity in the warmer, sulphate-bearing Archaean oceans. *Nature 328*, 244–246. [*28, 29*]

Ohta, T. 1988a: Further simulation studies on evolution by gene duplication. *Evolution 42*, 375–386. [*136, 137*]

Ohta, T. 1988b: Time for acquiring a new gene by duplication. *Proceedings of the National Academy of Sciences, USA 85*, 3509–3512. [*136*]

Oliver, S.G., van der Aart, Q.J.M., Agostoni-Carbone, M.L. *et al.* 1992: The complete DNA sequence of yeast Chromosome III. *Nature 357*, 38–46. [*504*]

Olson, J.M. 1970: Evolution of photosynthesis. *Science 168*, 438–446. [*175*]

Olson, J.M. & Pierson, B.K. 1986: Photosynthesis 3.5 thousand million years ago. *Photosynthesis Research 9*, 251–259. [*164*]

Olson, J.M. & Pierson, B.K. 1987a: Evolution of reaction centers in photosynthetic prokaryotes. *International Review of Cytology 108*, 209–248. [*162, 163, 168, 172, 180*]

Olson, J.M. & Pierson, B.K. 1987b: Origin and evolution of photosynthetic reaction centers. *Origins of Life and Evolution of the Biosphere 17*, 419–430. [*163, 172, 180*]

Oparin, A.I. 1924: *Proizkhozhdenie Zhizni.* Moskovskij Rabochij, Moscow. (English translation published as Appendix in Bernal, J.D. 1967: *The Origin of Life*, 199–234. World, Cleveland, Ohio.) [*52, 68, 124, 133, 143*]

Oparin, A.I. 1968: *Genesis and Evolutionary Development of Life.* Academic Press, New York, N.Y. [*232*]

Orgel, L.E. 1968: Evolution of the genetic apparatus. *Journal of Molecular Biology 38*, 381–393. [*63*]

Orgel, L.E. 1987: Evolution of the genetic apparatus: A review. *Cold Spring Harbor Symposia on Quantitative Biology 52*, 9–16. [*56, 65, 75*]

Orgel, L.E. 1989: The origin of polynucleotide-directed protein synthesis. *Journal of Molecular Evolution 29*, 465–474. [*76*]

Orgel, L.E. & Sulston, J.E. 1971: Polynucleotide replication and the origin of life. In Kimball, A.P. & Oró, J. (eds.): *Prebiotic and Biochemical Evolution*, 89–94. North-Holland, Amsterdam. [*63, 72*]

Oró, J. 1960: Synthesis of adenine from ammonium cyanide. *Biochemical and Biophysical Research Communications 2*, 407–412. [*51, 53, 92*]

Oró, J. 1961: Comets and the formation of biochemical compounds on the primitive Earth. *Nature 190*, 389–390. [*14, 37, 51*]

Oró, J. 1963: Studies in experimental organic cosmochemistry. *Annals of the New York Academy of Sciences 108*, 464–481. [*49, 55*]

Oró, J. 1965: Stages and mechanisms of prebiological organic synthesis. In Fox, S.W. (ed.): *The Origins of Prebiological Systems and of Their Molecular Matrices*, 137–162. Academic Press, New York, N.Y. [*48, 53*]

Oró, J. 1972: Extraterrestrial organic analysis. *Space Life Sciences 3*, 507–550. [*49*]

Oró, J. 1976: Prebiological chemistry and the origin of life: A personal account. In Kornberg, A., Horecker, B., Cornudella, L. & Oró, J. (eds.): *Reflections on Biochemistry in Honour of S. Ochoa*, 423–443. Pergamon, Oxford. [*53, 55*]

Oró, J., Holzer, G. & Lazcano-Araujo, A. 1980: The contribution of cometary volatiles to the primitive Earth. *Life Sciences and Space Research 18*, 67–82. [*37, 51*]

Oró, J., Kimball, A., Fritz, R. & Master, F. 1959: Amino acid synthesis from formaldehyde and hydroxylamine. *Archives of Biochemistry and Biophysics 85*, 115–130. [*53*]

Oró, J. & Lazcano, A. 1990: A holistic precellular organization model. In Ponnamperuma, C. & Eirich, F.R. (eds.): *Prebiological Self Organization of Matter*, 11–34. Deepak, Hampton, Va. [*56, 65*]

Oró, J. & Lazcano-Araujo, A. 1981: The role of HCN and its derivatives in prebiotic evolution. In Vennesland, B., Conn, E.E., Knowles, C.J., Westley, J. & Wissing, F. (eds.): *Cyanide in Biology*, 517–541. Academic Press, London. [*53, 55*]

Oró, J., Miller, S.L. & Lazcano, A. 1990: The origin and early evolution of life on Earth. *Annual Review of Earth and Planetary Sciences 18*, 317–356. [*52, 54, 55, 63, 65, 72, 75, 89*]

Oró, J., Mills, T. & Lazcano, A. 1992a: The cometary contribution to prebiotic chemistry. *Advances in Space Science and Technology 12*, 33–41. [*51*]

Oró, J., Mills, T. & Lazcano, A. 1992b: Comets and the formation of biochemical compounds on the primitive Earth – a review. *Origins of Life and Evolution of the Biosphere 21*, 267–277. [*51*]

Oró, J., Sherwood, E., Eichberg, J. & Epps, D. 1978: Formation of phospholipids under primitive Earth conditions and the role of membranes in prebiological evolution. In Deamer, D.W. (ed.): *Light-Transducing Membranes: Structure, Function and Evolution*, 1–21. Academic Press, New York, N.Y. [*56, 111*]

Orpen, J.L. & Wilson, J.F. 1981: Stromatolites at ~3,500 Myr and a granite–greenstone unconformity in the Zimbabwean Archaean. *Nature 291*, 218–220. [*275*]

Österberg, R., Orgel, L.E. & Lohrmann, R. 1973: Further studies of urea-catalyzed phosphorylation reactions. *Journal of Molecular Evolution 2*, 231–234. [*97*]

Ourisson, G. 1987: A hypothetical phylogeny for membrane reinforcers. *Chimia 6*, 12–14. [*260*]

Ourisson, G. 1989: The evolution of terpenes to sterols. *Pure and Applied Chemistry 1989*, 345–348. [*260*]

Ourisson, G. 1990: The general role of terpenes and their global significance. *Pure and Applied Chemistry 1990*, 1401–1404. [*260*]

Ourisson, G. & Albrecht, P. 1992: The hopanoids, Part I: The geohopanoids, the most abundant natural products on Earth? *Accounts of Chemical Research 25*, 398–402. [*260, 263*]

Ourisson, G. & Rohmer, M. 1992: The hopanoids, Part II: The biohopanoids: a novel family of microbial lipids. *Accounts of Chemical Research 25*, 403–408. [*260*]

Ourisson, G., Rohmer M. & Poralla, K. 1987: Prokaryotic hopanoids and other polyterpenoid sterol surrogates. *Annual Review of Microbiology 41*, 301–333. [*260*]

Owen, T., Cess, R.D. & Ramanathan, V. 1979: Enhanced CO_2 greenhouse to compensate for reduced solar luminosity on early Earth. *Nature 277*, 640–642. [*27, 28*]

Oyaizu, H., Debrunner-Vossbrinck, B., Mandelco, L., Studier, J.A. & Woese, C.R. 1987: The green non-sulfur bacteria: A deep branching in the Eubacterial line of descent. *Systematic and Applied Microbiology 9*, 47–53. [*167, 168*]

Pace, N. 1991: Origin of life – facing up to the physical setting. *Cell 65*, 531–533. [*163*]

Packer, B.M. 1990: Sedimentology, paleontology, and stable-isotope geochemistry of selected formations in the 2.7-billion-year-old Fortescue Group, Western Australia. 187 pp. Ph.D. Thesis, University of California, Los Angeles, Calif. [*278*]

Paerl, H.W. 1984: Cyanobacterial carotenoids: Their roles in maintaining optimal photosynthetic production among aquatic bloom forming genera. *Oecologia 61*, 143–149. [*166*]

Palenik, B. & Haselkorn, R. 1992: Multiple evolutionary origin of prochlorophytes, the chlorophyll *b*-containing prokaryotes. *Nature 355*, 265–267. [*319*]

Palij, V.M. 1969: O novom vide tsiklomeduzy iz venda Podolii [On a new species of cyclomedusae from the Vendian of Podolia]. *Paleontologicheskij Zbornik Lvovskogo Universiteta 6*, 114–117. [*373*]

Palij, V.M. 1976: Ostatki besskeletnoj fauny i sledy zhiznedeyatel'nosti iz otlozhenij verknego dokembriya i nizhnego kembriya [Remains of non-skeletal fauna and traces of life activity from the deposits of the Upper Precambrian and Lower Cambrian of Podolia]. In *Paleontologiya Dokembriya i Nizhnego Paleozoya Yugo-Zapada Vostochno-Evropejskoj Platformy*, 63–77. Naukova Dumka, Kiev. [*373*]

Palij, V.M., Posti, E. & Fedonkin, M.A. 1979: Myagkotelye Metazoa i iskopaemye sledy zhivotnykh venda i rannego kembriya [Soft-bodied Metazoa and fossil traces of animals in the Vendian and Early Cambrian]. In *Paleontologiya Verkhnedokembrijskikh i Kembrijskikh Otlozhenij Vostochno-Evropejskoj Platformy*, 49–82. Nauka, Moscow. [*373*]

Palmer, A.R. 1992: Calcification in marine molluscs: How costly is it? *Proceedings of the National Academy of Sciences, USA 89*, 1379–1382. [*423*]

Palmer, J.A., Phillips, G.N. & McCarthy, T.S. 1987: The nature of the Precambrian atmosphere and its relevance to Archaean gold mineralization. In Ho, S.E. & Groves, D.I. (eds.): *Recent Advances in Understanding Precambrian Gold Deposits*, 327–339. Geology Department, University of Western Australia, Publication 11. [*46*]

Palmer, J.D. 1993: A genetic rainbow of plastids. *Nature 364*, 762–763. [*296*]

Palmer, J.D. & Logsdon, J.M. 1991: The recent origin of introns. *Current Opinion in Genetics and Development 1*, 470–477. [*448*]

Parsot, C. 1987: A common origin for enzymes involved in the terminal step of the threonine and tryptophan biosynthetic pathways. *Proceedings of the National Academy of Sciences, USA 84*, 5207–5210. [*80*]

Patel, N.H., Martin-Blanco, E., Coleman, K., Poole, S.J., Ellis, M.C., Kornberg, T.B. & Goodman, C.S. 1989: Expression of *engrailed* proteins in arthropods, annelids, and chordates. *Cell 58*, 955–968. [*491, 492*]

Patterson, C. (ed.) 1987: *Molecules and Morphology in Evolution: Conflict or Compromise.* Cambridge University Press, Cambridge. [*453*]

Patterson, C. 1989: Phylogenetic relations of major groups: Conclusions and prospects. In Fernholm, B., Bremer, K. & Jörnvall, H. (eds.): *The Hierarchy of Life,* 471–488. Excerpta Medica, Amsterdam. [*315, 326, 453, 460, 461, 463, 469*]

Patterson, D.J. 1989: Stramenopiles: Chromophytes from a protistan perspective. In Green, J.C., Leadbeater, B.S.C. & Diver, W.L. (eds.): *The Chromophyte Algae: Problems and Perspectives,* 357–379. The Systematics Association Special Volume 38. Clarendon, Oxford. [*183, 315, 316, 318, 325*]

Patterson, D.J. & Zölffel, M. 1991: Heterotrophic flagellates of uncertain taxonomic position. In Patterson, D.J. & Larsen, J. (eds.): *The Biology of Free-living Heterotrophic Flagellates,* 425–475. The Systematics Association Special Volume 45. Clarendon, Oxford. [*320*]

Peat, C.J 1984: Precambrian microfossils from the Longmyndian of Shropshire. *Proceedings of the Geologists' Association 95,* 17–22. [*291*]

Peat, C.J., Muir, M.D., Plumb, K.A., McKirdy, D.M. & Norvick, M.S. 1978: Proterozoic microfossils from the Roper Group, Northern Territory, Australia. *Bureau of Mineral Resources, Journal of Australian Geology and Geophysics 3,* 1–17. [*292, 310*]

Pedersen, K.J. 1961: Studies on the nature of planarian connective tissue. *Zeitschrift für Zellforschung und Mikroskopische Anatomie 53,* 596–608. [*476*]

Pedersen, K.J. 1991: Invited review: Structure and composition of basement membranes and other basal matrix systems in selected invertebrates. *Acta Zoologica 72,* 181–201. [*477, 478, 480, 488*]

Peel, J.S. 1991: Functional morphology, evolution and systematics of Early Palaeozoic univalved molluscs. *Grønlands Geologiske Undersøgelse, Bulletin 161.* 116 pp. [*457*]

Pekkarinen, L.J. & Lukkarinen, H. 1991: Paleoproterozoic volcanism in the Kiihtelysvaara-Tohmajärvi district, eastern Finland. *Geological Survey of Finland, Bulletin 357.* 30 pp. [*239*]

Peltier, W.R. & Solheim, L.P. 1992: Mantle phase transitions and layered chaotic convection. *Geophysical Research Letters 19,* 321–324. [*210*]

Peltola, E. 1960: On the black schists in the Outokumpu region in eastern Finland. *Bulletin de la Commission Geologique de Finlande 192.* 107 pp. [*241*]

Peltola, E. 1968: On some geochemical features in the black schists of the Outokumpu area. *Bulletin of the Geological Society of Finland 40,* 39–50. [*241*]

Pentecost, A. 1991: Calcification processes in algae and cyanobacteria. In Riding, R. (ed.): *Calcareous Algae and Stromatolites,* 3–20. Springer, Berlin. [*429*]

Pentecost, A. & Riding, R. 1986: Calcification in cyanobacteria. In Leadbeater, B.S.C. & Riding, R. (eds.): *Biomineralization in Lower Plants and Animals,* 73–90. The Systematics Association Special Volume 30 Clarendon, Oxford. [*434*]

Perasso, R., Baroin, A., Qu, L.H., Bachellerie, J.P. & Adoutte, A. 1989: Origin of the algae. *Nature 339,* 142–144. [*301, 303, 441*]

Perrimon, N. & Mahowald, A.P. 1987: Maternal contributions to early development in *Drosophila.* In Malacinski, G. (ed.): *Primers in Developmental Biology,* 305–328. [*505, 507*]

Perttunen, V. 1985: On the Proterozoic stratigraphy and exogenic evolution of the Peräpohja area, Finland. *Geological Survey of Finland, Bulletin 331,* 133–141. [*239*]

Perttunen, V. 1991: Pre-Quaternary rocks of the Kemi, Karunki, Simo and Runkaus map-sheet areas. *Geological Map of Finland 1:100,000; Sheets 2541, 2542+2524, 2543 and 2544.* 80 pp. Geological Survey of Finland, Espoo. [*239*]

Peryt, T.M., Hoppe, A., Bechstadt, T., Koster, J., Pierre, C. & Richter, D.K. 1990: Late Proterozoic aragonite cement crusts, Bambui Group, Minas Gerais, Brazil. *Sedimentology 37,* 279–286. [*248*]

Pflug, H.D. 1966: Structured organic remains from the Fig Tree Series of the Barberton Mountain Land. *University of the Witwatersrand Economic Geology Research Unit Information Circular 28.* 14 pp. Johannesburg, South Africa. [*195*]

Pflug, H.D. 1970a: Zur Fauna der Nama-Schichten in Südwest-Afrika, I: Pteridinia, Bau und systematische Zugehörigkeit. *Palaeontographica, Abt. A 134,* 226–262. [*380*]

Pflug, H.D. 1970b: Zur Fauna der Nama-Schichten in Südwest-Afrika, II: Rangeidae, Bau und Systematische Zugehörigkeit. *Palaeontographica, Abt. A 135*, 198–231. [*380*]

Pflug, H.D. 1972a: Systematik der jung-präkambrischen Petalonamae Pflug 1970. *Paläontologische Zeitschrift 46*, 56–67. [*380*]

Pflug, H.D. 1972b: Zur Fauna der Nama-Schichten in Südwest-Afrika, III: Erniettomorpha, Bau und Systematik. *Palaeontographica, Abt. A 139*, 134–170. [*365, 380*]

Pflug, H.D. 1973: Zur Fauna der Nama-Schichten in Südwest-Afrika, IV: Mikroskopische Anatomie der Petalo-Organismen. *Palaeontographica, Abt. A 144*, 166–202. [*380*]

Pflug, H.D. 1978: Yeast-like microfossils detected in oldest sediments on Earth. *Naturwissenschaften 65*, 611–615. [*313*]

Piatigorsky, J. & Wistow, G. 1991: The recruitment of crystallins: New functions precede gene duplication. *Science 252*, 1078–1079. [*513*]

Piccirilli, J.A., McConnel, T.S., Zaug, A.J., Noller, H.F. & Cech, T.R. 1992: Aminoacyl esterase acitivity of the *Tetrahymena* ribozyme. *Science 256*, 1420–1424. [*63, 77*]

Pick, F.R. 1991: The abundance and composition of freshwater picocyanobacteria in relation to light penetration. *Limnology and Oceanography 36*, 1457–1462. [*338*]

Pierson, B.K. 1992: Modern mat-building microbial communities: A key to the interpretation of Proterozoic stromatolitic communities: Introduction. In Schopf, J.W. & Klein, C. (eds.): *The Proterozoic Biosphere: A Multidisciplinary Study*, 245–252. Cambridge University Press, Cambridge. [*164, 168*]

Pierson, B.K. & Castenholz, R.W. 1992: The family Chloroflexaceae. In Balows, A., Truper, H.G., Dworkin, M., Harder, W. & Schleifer, K.H. (eds.): *The Prokaryotes*. 2d Ed., 3754–3774. Springer, New York, N.Y. [*171*]

Pierson, B.K., Giovannoni, S.J. & Castenholz, R.W. 1984: Physiological ecology of a gliding bacterium containing bacteriochlorophyll a. *Applied and Environmental Microbiology 47*, 576–584. [*173*]

Pierson, B.K., Mitchell, H.K. & Ruff-Roberts, A.L. 1993: *Chloroflexus aurantiacus* and ultraviolet radiation: Implications for Archean shallow-water stromatolites. *Origins of Life and Evolution of the Biosphere 23*, 243–260. [*167*]

Pierson, B.K. & Olson, J.M. 1989: Evolution of photosynthesis in anoxygenic photosynthetic prokaryotes. In Cohen, Y. & Rosenberg, E. (eds.): *Microbial Mats: Physiological Ecology of Benthic Microbial Communities*, 402–427. American Society for Microbiology, Washington, D.C. [*162, 163, 172, 174, 175, 180*]

Pierson, B.K., Sands, V. & Fredrick, J. 1990: Spectral irradiance and distribution of pigments in a highly layered marine microbial mat. *Applied and Environmental Microbiology 56*, 2327–2340. [*176*]

Pinto, J.P., Gladstone, G.R. & Yung, Y.L. 1980: Photochemical production of formaldehyde in the Earth's primitive atmosphere. *Science 210*, 183–185. [*12, 16, 109*]

Pinto, J.P. & Holland, H.D. 1988: Paleosols and the evolution of the atmosphere, Part II. In Reinhardt, J. & Sigleo, W.R. (eds.): *Paleosols and Weathering Through Geologic Time: Principles and Applications*, 21–34. Geological Society of America, Special Paper 216. [*238*]

Pirie, N.W. 1953: Ideas and assumptions about the origin of life. *Discovery 14*, 238–242. [*63*]

Playford, P.E. & Cockbain, A.E. 1969: Algal stromatolites: Deep-water forms in the Devonian of Western Australia. *Science 165*, 1008–1010. [*338*]

Playford, P.E., Cockbain, A.E., Druce, E.C. & Wray, J.L. 1976: Devonian stromatolites from the Canning Basin, Western Australia. In Walter, M.R. (ed.): *Stromatolites*, 543–563. Developments in Sedimentology 20 Elsevier, Amsterdam. [*285*]

Pley, U., Schipka, J., Gambacorta, A., Jannasch, H.W., Fricke, H., Rachel, R. & Stetter, K.O. 1991: *Pyrodictium abyssi* sp. nov. represents a novel heterotrophic marine archaeal hyperthermophilic growing at 110°C. *Systematic and Applied Microbiology 14*, 245–253. [*149*]

Plobeck, N., Eifler, S., Brisson, A., Nakatani, Y. & Ourisson, G. 1992: Sodium di-polyprenyl phosphates form "primitive membranes." *Tetrahedron Letters 36*, 5249–5252. [*268*]

Plumb, K. 1991: New Precambrian time scale. *Episodes 14*, 139–140. [*440*]

Pohorille, A. & Benjamin, I. 1991: Molecular dynamics of phenol at the liquid–vapor interface of water. *Journal of Chemical Physics 94*, 5599–5605. [*21*]

Ponnamperuma, C. & Chang, S. 1971: The role of phosphates in chemical evolution. In Buvet, R. & Ponnamperuma, C. (eds.): *Chemical Evolution and the Origin of Life*, 216–223. North-Holland, Amsterdam. [87]

Ponomarenko, A.G. 1984: Evolution of ecoysystems: Major events. In *27th International Geological Congress, Sect. C.02 Paleontology: Reports*, 71–74. Nauka, Moscow. [375]

Popov, Yu.N. 1967: Novaya kembrijskaya stsifomeduza [A new Cambrian Scyphomedusa]. *Paleontologicheskij Zhurnal 1967:2*, 122–123. [383]

Postgate, J.R. 1982: *The Fundamentals of Nitrogen Fixation*. Cambridge University Press, Cambridge. [41]

Poulsen, C. 1967: Fossils from the Lower Cambrian of Bornholm. *Matematisk-Fysiske Meddelelser, Kongelige Danske Videnskabernes Selskab, 36*, 1–48. [418]

Pratt, B.R. 1982: Stromatolite decline – a reconsideration. *Geology 10*, 512–515. [283]

Pratt, B.R. & James, N.P. 1988: Early Ordovician thrombolite reefs, St. George Group, Western Newfoundland. In Geldsetzer, H.H.J., James, N.P. & Tebbutt, G.E. (eds.): *Reefs, Canada and Adjacent Areas*, 231–240. Canadian Society of Petroleum Geologists, Memoir 13. [280, 285]

Precht, H., Christophersen, J., Hensel, H. & Larcher, W. 1973: *Temperature and Life*. Springer, Berlin. [436]

Preer, J.R., Jr. 1975: The hereditary symbionts of *Paramecium aurelia*. *Symposia of the Society for Experimental Biology 29*, 125–145. [320]

Preiss, W.V. (compiler) 1987: The Adelaide Geosyncline – late Proterozoic stratigraphy, sedimentation, palaeontology and tectonics. *Bulletin of the Geological Survey of South Australia 53*. 438 pp. [283]

Preuß, A., Schauder, R. & Fuchs, G. 1989: Carbon isotope fractionation by autotrophic bacteria with three different CO_2 fixation pathways. *Zeitschrift für Naturforschung 44c*, 397–402. [174]

Prévôt, L., El Faleh, E.M. & Lucas, J. 1989: Details on synthetic apatites formed through bacterial mediation. Mineralogy and chemistry of the products. *Sciences Géologiques Bulletin 42*, 237–254. [99]

Prévôt, L. & Lucas, J. 1986: Microstructure of apatite-replacing carbonate in synthesized and natural samples. *Journal of Sedimentary Petrology 56*, 153–159. [99]

Prinn, R.G. & Fegley, B., Jr. 1987: Bolide impacts, acid rain, and biospheric traumas at the Cretaceous–Tertia boundry. *Earth and Planetary Science Letters 83*, 1–15. [21]

Puhler, G. Leffers, H., Gropp, F., Palm, P., Klenk, H.P., Lottspeich, F., Garrett, R.A. & Zillig, W. 1989: Archaebacterial DNA-dependent RNA polymerases testify to the evolution of the eukaryotic nuclear genome. *Proceedings of the National Academy of Sciences, USA 86*, 4569–4573. [189]

Pyatiletov, V.G. & Rudavskaya, V.A. 1985: Akritarkhi Yudomskogo kompleksa [Acritarchs from the Yudomian Complex]. In Sokolov, B.S. & Ivanovskij, A.B. (eds.): *Vendskaya Sistema*, Vol. 1, 151–158. Nauka, Moscow. [305]

Qian Yi & Bengtson, S. 1989: Palaeontology and biostratigraphy of the Early Cambrian Meishucunian Stage in Yunnan Province, South China. *Fossils and Strata 24*. 156 pp. [366, 415, 418, 424, 457]

Qu, L.H., Michot, B. & Bachellerie, J-P. 1983: Improved methods for structure probing in large RNAs: A rapid "heterologous" sequencing approach is coupled to the direct mapping of nuclease accessible sites. Application to the 5' terminal domain of eukaryotic 28S rRNA. *Nucleic Acids Research 11*, 5903–5920. [468]

Raaben, M.E. 1969: Stromatolity verkhnego Rifeya (gimnosolenidy) [Upper Riphean stromatolites (gymnosolenids)]. *Trudy Geologicheskogo Instituta Akademii Nauk SSSR 203*. 100 pp. Nauka, Moscow. [429]

Rachel, R., Engel, A.M., Huber, R., Stetter, K.O. & Baumeister, W. 1990: A porin-type protein is the main constituent of the cell envelope of the ancestral eubacterium *Thermotoga maritima*. *FEBS Letters 262*, 64–68. [150]

Radice, G.P., Neff, A.W., Shim, Y.H., Brustis, J.-J. & Malacinski, G.M. 1989: Developmental histories in amphibian myogenesis. *International Journal of Developmental Biology 33*, 325–343. [498]

Raff, R.A. 1992: Direct-developing sea urchins and the evolutionary reorganization of early development. *BioEssays 14*, 211–218. [494, 495]

Raff, R.A., Field, K.G., Olsen, G.J., Giovannoni, S.J., Lane, D.J., Ghiselin, M.T., Pace, N.R. & Raff, E.C. 1989: Metazoan phylogeny based on analysis of 18S ribosomal RNA. In Fernholm, B., Bremer, K. & Jörnvall, H. (eds.): *The Hierarchy of Life*, 247–260. Excerpta Medica, Amsterdam. [*502*]

Raff, R.A. & Kaufman, T.C. 1983: *Embryos, Genes and Evolution*. MacMillan, New York, N.Y. [*139, 469, 494, 499*]

Raff, R.A., Wray, G.A. & Henry, J.J. 1991: Implications of radical evolutionary changes in early development for concepts of developmental constraint. In Warren, L. & Koprowski, H. (eds.): *New Perspectives on Evolution*, 189–207. Wiley-Liss, New York, N.Y. [*498*]

Ragan, M.A. 1989: Biochemical pathways and the phylogeny of the eukaryotes. In Fernholm, B., Bremer, K. & Jörnvall, H. (eds.): *The Hierarchy of Life*, 145–160. Excerpta Medica, Amsterdam. [*322*]

Ragan, M.A. & Chapman, D.J, 1978: *A Biochemical Phylogeny of the Protists*. Academic Press, New York, N.Y. [*182, 321*]

Raikov, I. 1982: *The Protozoan Nucleus*. Springer, New York, N.Y. [*317*]

Railsback, L.B. & Anderson, T.F. 1987: Control of Triassic seawater chemistry and temperature on the evolution of post-Paleozoic aragonite-secreting faunas. *Geology 15*, 1002–1005. [*437*]

Rambler, M. & Margulis, L. 1980: Bacterial resistance to ultraviolet irradiation under anaerobiosis: Implications for pre-Phanerozoic evolution. *Science 210*, 638–640. [*38*]

Ramsköld, L. & Hou Xianguang 1991: New early Cambrian animal and onychophoran affinities of enigmatic metazoans. *Nature 351*, 225–228. [*454, 456, 490*]

Rasool, S.I., Hunten, D.M. & Kaula, W.M. 1977: What the exploration of Mars tells us about Earth. *Physics Today 30*, 23–31. [*37*]

Raup, D.M. 1991: *Extinction: Bad Genes or Bad Luck*. Norton, New York, N.Y. [*328*]

Raven, J.A. 1983: The transport and function of silicon in plants. *Biological Reviews 58*, 179–207. [*423*]

Read, J.F. 1976: Calcretes and their distinction from stromatolites. In Walter, M.R. (ed.): *Stromatolites*, 55–71. Developments in Sedimentology 20 Elsevier, Amsterdam. [*272*]

Read, J.F. 1985: Carbonate platform facies models. *American Association of Petroleum Geologists Bulletin 69*, 1–21. [*255*]

Reanney, D.C. 1982: The evolution of RNA viruses. *Annual Review of Microbiology 36*, 47–73. [*79*]

Reichelt, A.C. 1991: Environmental effects of meiofaunal burrowing. *Symposia of the Zoological Society of London 63*, 33–52. [*451*]

Reimer, T.O. 1990: Archaean baryte deposits of southern Africa. *Journal of the Geological Society of India 35*, 131–150. [*29*]

Reimers, C.E., Kastner, M. & Garrison, R.E. 1991: The role of bacterial mats in phosphate mineralization with particular reference to the Monterey Formation. In Burnett, W.C. & Riggs, S.R. (eds.): *Phosphate Deposits of the World*, Vol. 3, 300–311. Cambridge University Press, New York, N.Y. [*99*]

Reiswig, H.M. & Mehl, D. 1991: Tissue organization of *Farrea occa* (Porifera, Hexactinellida). *Zoomorphologie 110*, 301–311. [*476*]

Reitlinger, E.A. 1959: Atlas mikroskopicheskikh organicheskikh ostatkov i problematiki drevnikh tolshch Sibiri [Atlas of microscopic organic remains and problematica from Siberia's oldest deposits]. *Trudy Instituta Geologii i Geofiziki Sibir'skogo Otdeleniya Akademii Nauk SSSR 269*, 3–22. [*430*]

Retallack, G.J. 1990: Early life on land. In *Soils of the Past*, 351–374. Harper Collins Academic, New York, N.Y. [*164*]

Reuter, M. & Gustafsson, M. 1989: "Neuroendocrine cells" in flatworms – progenitors of metazoan neurons? *Archives of Histology and Cytology 52*, 253–263. [*482*]

Reutterer, A. 1969: Zum Problem der Metazoenabstammung. *Zeitschrift für Zoologische Systematik und Evolutionsforschung 7*, 30–53. [*477*]

Rhodes, F.H.T. & Bloxam, T.W. 1971: Phosphatic organisms in the Paleozoic and their evolutionary significance. In Yochelson, E.L. (ed.): *Phosphate in Fossils*, 1485–1513. North American Paleontological Convention, Chicago, Proceedings K [*421, 423*]

Richardson, J.B. 1992: Origin and evolution of the earliest land plants. In Schopf, J.W. (ed.): *Major Events in the History of Life*, 95–118. Jones & Bartlett, Boston, Mass. [*203*]

Riding, R. 1982: Cyanophyte calcification and changes in ocean chemistry. *Nature 299*, 814–815. [*427, 429, 431, 433, 434, 444*]

Riding, R. 1989: Calcified cyanobacteria in Phanerozoic reefs. *Algae in Reefs Symposium, Granada 1989, Abstracts*, 3–4. [*432*]

Riding, R. 1991a: Cambrian calcareous cyanobacteria and algae. In Riding, R. (ed.): *Calcareous Algae and Stromatolites*, 305–334. Springer, Berlin. [*426, 428, 429, 433*]

Riding, R. 1991b: Calcified cyanobacteria. In Riding, R. (ed.): *Calcareous Algae and Stromatolites*, 55–87. Springer, Berlin. [*429*]

Riding, R. 1991c: Classification of microbial carbonates. In Riding, R. (ed.): *Calcareous Algae and Stromatolites*, 21–51. Springer, Berlin. [*433*]

Riding, R. 1992a: Algal–cyanobacterial calcification episodes and changes in Phanerozoic sea-water chemistry. *9th Meeting of Carbonate Sedimentologists, Liverpool, UK, 21–25 July 1992, Abstracts-Talks*, [*430, 437*]

Riding, R. 1992b: Temporal variation in calcification in marine cyanobacteria. *Journal of the Geological Society, London 149*, 979–989. [*428, 431, 432, 433, 434, 435, 436, 438*]

Riding, R. 1992c: Giant and supergiant stromatolites. *International Symposium on Stromatolites and Plenary Meeting of IGCP 261 Project, Tianjin, China, 14–16 October 1992. Abstracts*, 22. [*432*]

Riding, R., Awramik, S.M., Winsborough, B.M., Griffin, K.M. & Dill, R.F. 1991: Bahamian giant stromatolites: Microbial composition of surface mats. *Geological Magazine 128*, 227–234. [*283, 285, 338*]

Riding, R. & Voronova, L. 1982: Calcified cyanophytes and the Precambrian–Cambrian transition. *Naturwissenschaften 69*, 498–499. [*429*]

Riding, R. & Voronova, L. 1984: Assemblages of calcareous algae near the Precambrian/Cambrian boundary in Siberia and Mongolia. *Geological Magazine 121*, 205–210. [*429, 430, 431*]

Riedl, R. 1966: *Biologie der Meereshöhlen*. Parey, Hamburg. [*482*]

Riedl, R. 1978: *Order in Living Organisms*. Wiley, Chichester. [*496*]

Rieger, R.M. 1976: Monociliated epidermal cells in Gastrotricha: Significance for concepts of early metazoan evolution. *Zeitschrift für Zoologische Systematik und Evolutionsforschung 14*, 198–226. [*476, 480, 481, 485*]

Rieger, R.M. 1984: Evolution of the cuticle in the lower Metazoa. In Bereiter-Hahn, J., Matoltsy, A.G. & Richards, K.S. (eds.): *Biology of the Integument, Vol. 1: Invertebrates*, 389–399. Springer, Berlin. [*478, 480*]

Rieger, R.M. 1985: The phylogenetic status of the acoelomate organization within the Bilateria. In Conway Morris, S., George, J.D., Gibson, R. & Platt, H.M. (eds.): *The Origins and Relationships of Lower Invertebrates*, 101–123. Clarendon, Oxford. [*482*]

Rieger, R.M. 1986a: Über den Ursprung der Bilateria: Die Bedeutung der Ultrastrukturforschung für ein neues Verstehen der Metazoenevolution. *Verhandlungen der Deutschen Geologischen Gesellschaft 79*, 31–50. [*480, 485, 486*]

Rieger, R.M. 1986b: Asexual reproduction and the turbellarian archetype. *Hydrobiologia 132*, 35–45. [*484*]

Rieger, R.M., Haszprunar, G. & Schuchert, P. 1991a: On the origin of the Bilateria: Traditional views and recent alternative concepts. In Simonetta, A.M. & Conway Morris, S. (eds.): *The early Evolution of Metazoa and the Significance of Problematic Taxa*, 107–113. Cambridge University Press, Cambridge. [*482, 483, 485*]

Rieger, R.M. & Lombardi, J. 1987: Ultrastructure of coelomic lining in echinoderm podia: Significance for concepts in the evolution of muscle and peritoneal cells. *Zoomorphologie 107*, 191–208. [*482, 486*]

Rieger, R.M. & Sterrer, W. 1975: New spicular skeletons in Turbellaria, and the occurrence of spicules in marine meiofauna. *Zeitschrift für Zoologische Systematik und Evolutionsforschung 13*, 207–278. [*422*]

Rieger, R.M., Tyler, S., Smith, J.P.S., III & Rieger, G.E. 1991b: Platyhelminthes: Turbellaria. In Harrison, F.W. & Bogitsch, B.J. (eds.): *Microanatomy of the Invertebrates, Vol. 3: Platyhelminthes and Nemertea*, 7–140. Wiley-Liss, New York, N.Y. [*406, 476, 477, 481, 483*]

Rivadeneyra, M.A., Pérez-García, I. & Ramos-Cormenzana, A. 1993: Ammonium and bacterial struvite production. *Geomicrobiology Journal 10*, 125–137. [*99*]

Rivera, M.C. & Lake, J.A. 1992: Evidence that eukaryotes and eocyte prokaryotes are immediate relatives. *Science 257*, 74–76. [*290*]

Robb, L.J., Davis, D.W. & Kamo, S.L. 1990: U–Pb ages on single detrital zircon grains from the Witwatersrand Basin, South Africa: Constraints on the age of sedimentation and on the evolution of granites adjacent to the basin. *Journal of Geology 98*, 311–328. [*239*]

Robb, L.J., Davis, D.W., Kamo, S.L. & Meyer, F.M. 1992: Ages of altered granites adjoining the Witwatersrand Basin with implications for the origin of gold and uranium. *Nature 357*, 667–680. [*239*]

Robb, L.J. & Meyer, F.M. 1990: The nature of the Witwatersrand hinterland: Conjectures on the source area problem. *Economic Geology 85*, 511–536. [*239*]

Robbins, E.I., Porter, K.G. & Haberyan, K.A. 1985: Pellet microfossils: Possible evidence for metazoan life in early Proterozoic time. *Proceedings of the National Academy of Sciences, USA 82*, 5809–5813. [*451*]

Robert, B. & Moenne-Loccoz, P. 1990: Is there a proteic substructure common to all photosynthetic reaction centers? *Current Research in Photosynthesis 1*, 65–68. [*179*]

Roberts, J. & Jell, P.A. 1990: Early Middle Cambrian (Ordian) brachiopods of the Coonigan Formation, western New South Wales. *Alcheringa 14*, 257–309. [*458*]

Robison, R.A. 1991: Middle Cambrian biotic diversity: Examples from four Utah Lagerstätten. In Simonetta, A.M. & Conway Morris, S. (eds.): *The Early Evolution of Metazoa and the Significance of Problematic Taxa*, 77–98. Cambridge University Press, Cambridge. [*453*]

Rodriguez, L. & Orgel, L.E. 1991: Template-directed extension of a guanosine 5'-phosphate covalently attached to an oligodeoxycytidylate. *Journal of Molecular Evolution 33*, 477–482. [*74*]

Rodriguez-Boulan E. & Powell S. K. 1992: Polarity of epithelial and neuronal cells. *Annual Review of Cell Biology 8*, 395–427. [*479*]

Rolfe, R.D., Hentges, D.J., Campbell, B.J. & Barrett, J.T. 1978: Factors related to the oxygen tolerance of anaerobic bacteria. *Applied and Environmental Microbiology 36*, 306–313. [*38*]

Rose, A.H. 1967: *Thermobiology*. Academic Press, London. [*436*]

Roth, V.L. 1988: The biological basis of homology. In Humphries, C.J. (ed.): *Ontogeny and Systematics*, 1–26. Columbia University Press, New York, N.Y. [*496*]

Rothschild, L.J. 1991: A model for diurnal patterns of carbon fixation in a Precambrian microbial mat based on a modern analog. *BioSystems 25*, 13–23. [*174*]

Rothschild, L.J. & Heywood, P. 1987: Protistan phylogeny and chloroplast evolution: Conflicts and congruence. *Progress in Protistology 2*, 1–68. [*182, 321*]

Rothschild, L.J. & Mancinelli, R.L. 1990: Model of carbon fixation in microbial mats from 3,500 Myr ago to the present. *Nature 345*, 710–712. [*43, 174*]

Rouse, R.C., Peacor, D.R. & Freed, R.L. 1988: Pyrophosphate groups in the structure of canaphite, $CaNa_2P_2O_7 \cdot 4H_2O$: The first occurrence of a condensed phosphate as a mineral. *The American Mineralogist 73*, 168–171. [*87, 97*]

Roux, A. 1991: Ordovician algae and global tectonics. In Riding, R. (ed.): *Calcareous Algae and Stromatolites*, 335–348. Springer, Berlin. [*428*]

Rowan, R. 1991: Molecular systematics of symbiotic algae. *Journal of Phycology 27*, 661–666. [*335*]

Rowell, A.J. 1982: The Cambrian brachiopod radiation – monophyletic or polyphyletic origins? *Lethaia 15*, 299–307. [*452*]

Rozanov, A.Yu. 1986: *Chto Proizoshlo 600 Millionov Let Nazad?*. Nauka, Moscow. [*371*]

Rubey, W.W. 1951: Geologic history of seawater: An attempt to state the problem. *Geological Society of America, Bulletin 62*, 1111–1148. [*27*]

Rubey, W.W. 1955: Development of the hydrosphere and atmosphere, with special reference to probable composition of the early atmosphere. In Poldervaart, A. (ed.): *Crust of the Earth*, 631–650. Geological Society of America, New York, N.Y. [*27*]

Rubin, G.M. 1988: *Drosophila melanogaster* as an experimental organism. *Science 240*, 1453–1459. [*505*]

Rudwick, M.J.S. 1964: The infra-Cambrian glaciation and the origin of the Cambrian fauna. In Nairn, A.E.M. (ed.): *Problems in Palaeoclimatology*, 150–155. Wiley, London. [*436*]

Ruetzler, K. (ed.) 1990: *New Perspectives in Sponge Biology*. Smithsonian Institution Press, Washington, D.C. [*477, 478*]

Ruiz i Altaba, A. & Melton, D.A. 1990: Axial patterning and the establishment of polarity in the frog embryo. *Trends in Genetics 6*, 57–64. [*499*]

Runnegar, B. 1982a: A molecular-clock date for the origin of the animal phyla. *Lethaia 15*, 199–205. [*409*]

Runnegar, B. 1982b: The Cambrian explosion: Animals or fossils? *Journal of the Geological Society of Australia 29*, 395–411. [*365, 401, 447*]

Runnegar, B. 1982c: Oxygen requirements, biology and phylogenetic significance of the Late Precambrian worm *Dickinsonia*, and the evolution of the burrowing habit. *Alcheringa 6*, 223–239. [*365*]

Runnegar, B. 1985: Collagen gene construction and evolution. *Journal of Molecular Evolution 22*, 141–149. [*451, 477, 478*]

Runnegar, B. 1989: The evolution of mineral skeletons. In Crick, R.E. (ed.): *Origin, Evolution, and Modern Aspects of Biomineralization in Plants and Animals*, 75–94. Plenum, New York, N.Y. [*423*]

Runnegar, B. 1991: Precambrian oxygen levels estimated from the biochemistry and physiology of early eukaryotes. *Palaeogeography, Palaeoclimatology, Palaeoecology (Global and Planetary Change Section) 97*, 97–111. [*38, 41, 292, 293, 447*]

Runnegar, B. 1992: Evolution of the earliest animals. In Schopf, J.W. (ed.): *Major Events in the History of Life*, 65–93. Jones & Bartlett, Boston, Mass. [*293, 295, 442, 454*]

Runnegar, B. & Curry, G.B. 1992: Amino acid sequences of hemerythrins from *Lingula* and a priapulid worm and the evolution of oxygen transport in the Metazoa. *29th International Geological Congress, Kyoto, Abstracts*, 346. [*458*]

Runnegar, B. & Fedonkin, M.A. 1992: Proterozoic metazoan body fossils. In Schopf, J.W. & Klein, C. (eds.): *The Proterozoic Biosphere: A Multidisciplinary Study*, 369–390. Cambridge University Press, Cambridge. [*373*]

Ruppert, E.E. 1992: Introduction to the Aschelminth phyla. A consideration of mesoderm, body cavities and cuticle. In Harrison, F.W. & Ruppert, E.E. (eds.): *Microscopic Anatomy of Invertebrates*, Vol. 4, 1–17. Wiley-Liss, New York, N.Y. [*483, 488*]

Ruppert, E.E. & Carle, K.J. 1983: Morphology of metazoan circulatory systems. *Zoomorphologie 103*, 193–208. [*488*]

Ruppert, E.E. & Smith, P.R. 1988: The functional organization of filtration nephridia. *Biological Reviews 63*, 231–258. [*483, 485*]

Ruthmann, A. 1977: Cell differentiation, DNA content and chromosomes of *Trichoplax adhaerens* F.E. Schulze. *Cytobiologie 15*, 58–64. [*406*]

Ruthmann, A., Behrendt, G., Wahl, R. 1986: The ventral epithelium of *Trichoplax adhaerens* (Placozoa): Cytoskeletal structures, cell contacts and endocytosis. *Zoomorphologie 106*, 115–122. [*480*]

Sackett, W. 1978: Suspended matter in sea-water. In Riley, J.P. & Chester, R. (eds.): *Chemical Oceanography*, Vol. 7, 127–171. Academic Press, London. [*23*]

Sadler, L.A. & Brunk, C.F. 1992: Phylogenetic relationships and unusual diversity in histone H4 proteins within the *Tetrahymena pyriformis* complex. *Molecular Biology and Evolution 9*, 70–85. [*182*]

do myedit 1973: Ultraviolet selection pressure on the earliest organisms. *Journal of Theoretical Biology 39*, 195–200. [*38*]

Sagan, C. & Mullen, G. 1972: Earth and Mars: Evolution of atmospheres and surface temperatures. *Science 177*, 52–56. [*24*]

Sahni, M.R. 1936: *Fermoria minima*: A revised classification of the organic remains from the Vindhyan of India. *Records of the Geological Survey of India 69*, 458–468. [*355*]

Sahni, M.R. & Srivastava, R.N. 1954: New organic remains from the Vindhyan System and the probable systematic position of *Fermoria*, Chapman. *Current Science 23*, 39–41. [*356*]

Saint, R. 1990: Homoeobox genes in morphogenesis and tissue pattern formation. *Today's Life Science 2*, 14–21. [*295*]

Sales, B.C., Chakoumakos, B.C., Boatner, L.A. & Ramey, J.O. 1992: Structural evolution of the amorphous solids produced by heating crystalline $MgHPO_4 \cdot 3H_2O$. *Journal of Materials Research 7*, 1–5. [*98, 104, 106*]

Sales, B.C., Chakoumakos, B.C., Boatner, L.A. & Ramey, J.O. 1993: Structural properties of the amorphous phases produced by heating crystalline $MgHPO_4 \cdot 3H_2O$. *Journal of Non-Crystalline Solids 159*, 121–139. [*104*]

Sales, B.C., Ramsey, R.S., Bates, J.B. & Boatner, L.A. 1986: Investigation of the structural properties of lead–iron phosphate glasses using liquid chromatography and raman scattering spectroscopy. *Journal of Non-Crystalline Solids 87*, 137–158. [*98*]

Salvini-Plawen, L. 1978: On the origin and evolution of lower metazoa. *Zeitschrift für Zoologische Systematik und Evolutionsforschung 16*, 40–88. [*468, 476, 483*]

Salvini-Plawen, L. von 1980: Phylogenetischer Status und Bedeutung der mesenchymaten Bilateria. *Zoologische Jahrbücher. Abteilung für Anatomie und Ontogenie der Tiere 103*, 354–373. [*483*]

Salvini-Plawen, L. von 1982: A paedomorphic origin of the oligomerous animals? *Zoologica Scripta 11*, 77–81. [*482*]

Salvini-Plawen, L. von 1985: Early evolution and the primitive groups. In Trueman, E.R. & Clarke, M.R. (eds.): *Mollusca, Vol. 10: Evolution*, 59–150. Academic Press, London. [*402*]

Salvini-Plawen, L. von & Splechtna, H. 1979: Zur Homologie der Keimblätter. *Zeitschrift für Zoologische Systematik und Evolutionsforschung 17*, 10–30. [*482, 483*]

Sampson, J.R., Sullivan, F.X., Behlen, L.S., DiRenzo, L.S. & Uhlenbeck, O.C. 1987: Characterization of two RNA-catalyzed RNA cleavage reactions. *Cold Spring Harbor Symposia on Quantitative Biology 52*, 267–275. [*72*]

Sanchez, R.A., Ferris, J.P. & Orgel, L.E. 1967: Studies in prebiotic synthesis, II: Synthesis of purine precursors and amino acids from aqueous hydrogen cyanide. *Journal of Molecular Biology 30*, 223–253. [*92*]

Sandberg, P.A. 1985: Nonskeletal aragonite and pCO_2 in the Phanerozoic and Proterozoic. In Sundquist, E.T. & Broecker, W.S. (eds.): *The Carbon Cycle and Atmospheric CO_2: Natural Variations Archean to Present*, 585–594. American Geophysical Union, Washington, D.C. [*255*]

Sarafian, V., Kim, Y., Poole, R.J. & Rea, P.A. 1992: Molecular cloning and sequence of cDNA encoding the pyrophosphate-energized vavuolar membrane proton pump (H^+-PPase) of *Arabidopsis thaliana*. *Proceedings of the National Academy of Sciences, USA 89*, 1775–1779. [*89*]

Sarras, M.P., Jr., Madden, M.E., Zhang, X., Gundwar, S., Huff, J.K. & Hudson, B.G. 1991: Extracellular matrix (mesogloea) of *Hydra vulgaris*, I: Isolation and characterization. *Developmental Biology 148*, 481–494. [*477*]

Schau, M. & Henderson, J.B. 1983: Archean chemical weathering at three localities on the Canadian Shield. *Precambrian Research 20*, 189–224. [*42*]

Scherer, S., Chen, T.W. & Böger, P. 1988: A new UV-A/B protecting pigment in the terrestrial cyanobacterium *Nostoc commune*. *Plant Physiology 88*, 1055–1057. [*166*]

Schidlowski, M. 1976: Archaean atmosphere and evolution of the terrestrial oxygen budget. In Windley, B.F. (ed.): *The Early History of the Earth*, 525–535. Wiley, London. [*29*]

Schidlowski, M. 1988: A 3,800-million year old isotopic record of life from carbon in sedimentary rocks. *Nature 333*, 313–318. [*31, 38, 42, 43, 45, 174, 290*]

Schidlowski, M. & Aharon, P. 1992: Carbon cycle and carbon isotope record: Geochemical impact of life over 3.8 Ga of earth history. In Schidlowski, M., Golubic, S., Kimberley, M.M., McKirdy, D.M. & Trudinger, P.A. (eds.): *Early Organic Evolution*, 147–175. Springer, Berlin. [*232*]

Schidlowski, M., Hayes, J.M. & Kaplan, I.R. 1983: Isotopic inferences of ancient biochemistries: Carbon, sulfur, hydrogen, and nitrogen. In Schopf, J.W. (ed.): *Earth's Earliest Biosphere, Its Origin and Evolution*, 149–186. Princeton University Press, Princeton, N.J. [*201, 202, 203, 216, 217, 223, 277*]

Schierwater, B., Murtha, M., Dick, M., Ruddle, F.H. & Buss, L.W. 1991: Homeoboxes in cnidarians. *Journal of Experimental Zoology 260*, 415–416. [*459*]

Schlage, W.K. 1988: Isolation and characterization of a fibronectin from marine coelenterates. *European Journal of Cell Biology 47*, 395–403. [*478*]

Schlesinger, G. & Miller, S.L. 1983: Prebiotic synthesis in atmospheres containing CH_4, CO, and CO_2. II: Hydrogen cyanide, formaldehyde and ammonia. *Journal of Molecular Evolution 19*, 383–390. [*92*]

Schneider, K.C. & Benner, S.A. 1990: Oligonucleotides containing flexible nucleoside analogs. *Journal of the American Chemical Society 112*, 453–455. [*64*]

Schoell, M. & Wellmer, F.W. 1981: Anomalous ^{13}C depletion in early Precambrian graphites from Superior Provence, Canada. *Nature 290*, 696–699. [*228, 231*]

Schopf, J.W. 1968: Microflora of the Bitter Springs Formation, Late Precambrian, Central Australia. *Journal of Paleontology 42*, 651–688. [*273, 281, 283*]

Schopf, J.W. 1970: Precambrian micro-organisms and evolutionary events prior to the origin of vascular plants. *Biological Reviews 45*, 319–352. [*433*]

Schopf, J.W. 1975: Precambrian paleobiology: Problems and perspectives. *Annual Review of Earth and Planetary Sciences 3*, 213–249. [*29*]

Schopf, J.W. 1977: Biostratigraphic usefulness of stromatolitic Precambrian microbiotas: A preliminary analysis. *Precambrian Research 5*, 143–173. [*430*]

Schopf, J.W. 1983a: How to read this book. In Schopf, J.W. (ed.): *Earth's Earliest Biosphere, Its Origin and Evolution*, 3–13. Princeton University Press, Princeton, N.J. [*36, 46*]

Schopf, J.W. (ed.) 1983b: *Earth's Earliest Biosphere*. Princeton University Press, Princeton, N.J. [*viii, 31, 276, 332, 426*]

Schopf, J.W. 1992a: Paleobiology of the Archean. In Schopf, J.W. & Klein, C. (eds.): *The Proterozoic Biosphere: A Multidisciplinary Study*, 25–39. Cambridge University Press, Cambridge. [*193, 198, 199, 200*]

Schopf, J.W. 1992b: The oldest fossils and what they mean. In Schopf, J.W. (ed.): *Major Events in the History of Life*, 29–63. Jones & Bartlett, Boston, Mass. [*193, 290, 292, 441*]

Schopf, J.W. 1992c: Proterozoic prokaryotes: Affinities, geologic distribution, and evolutionary trends. In Schopf, J.W. & Klein, C. (eds.): *The Proterozoic Biosphere: A Multidisciplinary Study*, 195–218. Cambridge University Press, Cambridge. [*194, 427*]

Schopf, J.W. 1992d: Evolution of the Proterozoic biosphere: Benchmarks, tempo, and mode. In Schopf, J.W. & Klein, C. (eds.): *The Proterozoic Biosphere: A Multidisciplinary Study*, 583–600. Cambridge University Press, Cambridge. [*203*]

Schopf, J.W. 1993: Microfossils of the Early Archean Apex Chert: New evidence of the antiquity of life. *Science 260*, 640–646. [*143, 163, 168, 198, 199, 200, 233, 276, 510*]

Schopf, J.W. & Blacic, J.M. 1971: New microorganisms from the Bitter Springs Formation (Late Precambrian) of the north-central Amadeus Basin, Australia. *Journal of Paleontology 45*, 925–959. [*273, 281, 283*]

Schopf, J.W., Dolnik, T.A., Krylov, I.N., Mendelson, C.V., Nazarov, B.B., Nyberg, A.V., Sovietov, Yu.K. & Yakshin, M.S. 1977: Six new stromatolitic microbiotas from the Proterozoic of the Soviet Union. *Precambrian Research 4*, 269–284. [*281, 283*]

Schopf, J.W., Haugh, B.N., Molnar, R.E. & Satterthwait, D.F. 1973: On the development of metaphytes and metazoans. *Journal of Paleontology 47*, 1–9. [*448*]

Schopf, J.W., Hayes, J.M. & Walter, M.R. 1983: Evolution of Earth's earliest ecosystems: Recent progress and unsolved problems. In Schopf, J.W. (ed.): *Earth's Earliest Biosphere, Its Origin and Evolution*, 361–384. Princeton University Press, Princeton, N.J. [*143, 151, 233, 427, 432*]

Schopf, J.W. & Klein, C. (eds) 1992: *The Proterozoic Biosphere*. Cambridge University Press, Cambridge. [*viii, 430*]

Schopf, J.W. & Packer, B.M. 1986: Newly discovered Early Archean (3.4–3.5 Ga old) microorganisms from the Warrawoona Group of Western Australia. *5th Meeting International Society for the Study of the Origins of Life and the 8th International Conference on the Origin of Life, Abstracts*, 163–164. [*198, 200*]

Schopf, J.W. & Packer, B.M. 1987: Early Archean (3.3-billion to 3.5-billion-year-old) microfossils from Warrawoona Group, Australia. *Science 237*, 70–73. [*31, 38, 42, 143, 163, 166, 167, 168, 172, 198, 200, 224, 233, 276, 310*]

Schopf, J.W. & Sovietov, Yu.K. 1976: Microfossils in *Conophyton* from the Soviet Union and their bearing on Precambrian biostratigraphy. *Science 193*, 143–146. [*283*]

Schopf, J.W. & Walter, M.R. 1983: Archean microfossils: New evidence of ancient microbes. In Schopf, J.W. (ed.): *Earth's Earliest Biosphere, Its Origin and Evolution*, 214–239. Princeton University Press, Princeton, N.J. [*163, 194, 195, 197, 200, 276, 299, 310, 336*]

Schram, F.R. 1991: Cladistic analysis of metazoan phyla and the placement of fossil problematica. In Simonetta, A.M. & Conway Morris, S. (eds.): *The Early Evolution of Metazoa and the Significance of Problematic Taxa*, 35–46. Cambridge University Press, Cambridge. [*452, 463, 475*]

Schrauzer, G.N., Guth, T.D. & Palmer, M.R. 1979: Nitrogen reducing solar cells. In Hautala, R.R., King, R.B. & Kutal, C. (eds.): *Solar Energy: Chemical Conversion and Storage*, 261–269. Humana, Clifton, N.J. [*12*]

Schrauzer, G.N., Strampach, N., Hui, L.N., Palmer, M.R. & Salehi, J. 1983: Nitrogen photoreduction on desert sands under sterile conditions. *Proceedings of the National Academy of Sciences, USA 80*, 3873–3876. [*12*]

Schubert, F.R., Nieselt-Struwe, K. & Gruss, P. 1993: The Antennapedia-type homeobox genes have evolved from three precursors separated early in metazoan evolution. *Proceedings of the National Academy of Sciences, USA 90*, 143–147. [*405*]

Schuster, F.L. 1968: The gullet and trichocysts of *Cyathomonas truncata. Experimental Cell Research 49*, 277–284. [*320*]

Schwartz, A., Joosten, H. & Voet, A.B. 1992: Prebiotic adenine synthesis via HCN oligomerization in ice. *BioSystems 15*, 191–193. [*92*]

Schwartz, A.W. & Orgel, L.E. 1985: Template-directed synthesis of novel, nucleic acid-like structures. *Science 228*, 585–587. [*56, 74*]

Schwartz, A.W., Visscher, J., Bakker, C.G. & Niessen, J. 1987: Nucleic acid-like structures, II: Polynucleotide analogues as possible primitive precursors of nucleic acids. *Origins of Life and Evolution of the Biosphere 17*, 351–357. [*93*]

Schwemmler, W. 1991: Symbiogenesis in insects as a model for morphogenesis, cell differentation, and speciation. In Margulis, L. & Fester, R. (eds.): *Symbiosis as a Source of Evolutionary Innovation: Speciation and Morphogenesis*, 178–204. Massachusetts Institute of Technology Press, Cambridge, Mass. [*333*]

Scott, D.J., St. Onge, M.R., Lucas, S.B. & Holmstaedt, H. 1991: Geology and chemistry of the Early Proterozoic Partunik ophiolite, Cape Smith Belt, northern Quebec, Canada. In Peters, T., Nicholas, A. & Coleman, R.G. (eds.): *Ophiolite Genesis and Evolution of the Oceanic Lithosphere*, 817–849. Kluwer, Dordrecht. [*209*]

Searcy, D.G. 1975: Histone-like protein in the prokaryote *Thermoplasma acidophilum. Biochimica et Biophysica Acta 395*, 535–547. [*189*]

Searcy, D.G. & Whatley, F.R. 1984: *Thermoplasma acidophilum:* Glucose degradative pathways and respiratory activities. *Systematic and Applied Microbiology 5*, 30–40. [*159*]

Segerer, A.H., Neuner, A., Kristjansson, J.K. & Stetter, K.O. 1986: *Acidianus infernus* gen.nov., sp. nov., and *Acidianus brierleyi* comb.nov.: Facultatively aerobic, extremely acidophilic thermophilic sulfur-metabolizing archaebacteria. *International Journal of Systematic Bacteriology 36*, 559–564. [*146*]

Segerer, A.H., Trincone, A., Gahrtz, M. & Stetter, K.O. 1991: *Stygiolobus azoricus* gen. nov., sp. nov. represents a novel genus of anaerobic, extremely thermoacidophilic Archaebacteria of the Order Sulfolobales. *International Journal of Systematic Bacteriology 41*, 495–501. [*146*]

Seilacher, A. 1956: Der Beginn des Kambriums als biologische Wende. *Neues Jahrbuch für Geologie und Paläontologie, Abhandlungen 103*, 155–180. [*399, 417*]

Seilacher, A. 1967: Bathymetry of trace fossils. *Marine Geology 5*, 413–428. [*387*]

Seilacher, A. 1983a: Precambrian metazoan extinctions. *Geological Society of America, Abstracts with Programs 15*, 683. [*381*]

Seilacher, A. 1983b: Paleozoic sandstones in southern Jordan: Trace fossils, depositional environments and biogeography. In Abed, A.M. & Khaled, H.M. (eds.): *Geology of Jordan.* Proceedings of the First Jordanian Geological Conference. Jordan Geologist Association. [*392*]

Seilacher, A. 1984: Late Precambrian and Early Cambrian Metazoa: Preservational or real extinction? In Holland, H.D. & Trendall, A.F. (eds.): *Patterns of Change in Earth Evolution (Dahlem Konferenzen 1984)*, 159–168. Springer, Berlin. [*363, 380, 381, 389*]

Seilacher, A. 1985: Discussion of Precambrian Metazoa. *Philosophical Transactions of the Royal Society of London, B 311*, 47–48. [*365*]

Seilacher, A. 1989: Vendozoa: Organismic construction in the Proterozoic biosphere. *Lethaia 22*, 229–239. [*293, 295, 304, 363, 380, 381, 389, 451, 460, 502*]

Seilacher, A. 1990: Lost constructions: Vendozoa and Psammocorallia. *Geological Society of America, Abstracts with Programs 22*, 128–129. [*295*]

Seilacher, A. 1991: Self-organizing mechanisms in morphogenesis and evolution. In Schmidt-Kittler, N. & Vogel, K. (eds.): *Constructional Morphology and Evolution*, 251–271. Springer, Berlin. [*391, 392, 393, 400*]

Seilacher, A. 1992: Vendobionta and Psammocorallia: Lost constructions of Precambrian evolution. *Journal of the Geological Society, London 149*, 607–613. [*389, 390, 451*]

Seilacher, A. & Pflüger, F. 1992: Trace fossils from the Late Proterozoic of North Carolina: Early conquest of deep-sea bottoms. *5th North American Paleontological Convention, Chicago, Abstracts*, 265. [*398*]

Seilacher, A., Reif, W.-E. & Westphal, F. 1985: Sedimentological, ecological and temporal patterns of fossil Lagerstätten. In Whittington, H. & Conway Morris, S. (eds.): *Extraordinary Fossil Biotas: Their Ecological and Evolutionary Significance*, 5–24. Philosophical Transactions of the Royal Society of London, B 311. [*394*]

Semikhatov, M.A., Gebelein, C.D., Cloud, P., Awramik, S.M. & Benmore, W.C. 1979: Stromatolite morphogenesis – progress and problems. *Canadian Journal of Earth Sciences 16*, 992–1015. [*281*]

Sepkoski, J.J., Jr. 1979: A kinetic model for Phanerozoic taxonomic diversity, II: Early Phanerozoic families and multiple equilibria. *Paleobiology 5*, 222–251. [*434*]

Serebryakov, S.N. & Semikhatov, M.A. 1974: Riphean and Recent stromatolites: A comparison. *American Journal of Science 274*, 556–574. [*255*]

Seward, A.C. 1931: *Plant Life Through the Ages*. Cambridge University Press, Cambridge. [*272*]

Shapiro, R. 1986: *Origins – A Skeptic's Guide to the Creation of Life on Earth*. Summit, New York, N.Y. [*64, 73*]

Sharp, P.M. & Li, W.H. 1989: On the rate of DNA sequence evolution in *Drosophila*. *Journal of Molecular Evolution 28*, 398–402. [*468*]

Shegelski, R.J. 1980: Archean cratonization, emergence and red bed development, Lake Sheban-dowan area, Canada. *Precambrian Research 12*, 331–347. [*42, 46*]

Shen, C., Lazcano, A. & Oró, J. 1990c: The enhancement activities of histidyl–histidine in some prebiotic reactions. *Journal of Molecular Evolution 31*, 445–452. [*56, 65, 75*]

Shen, C., Miller, S.L. & Oró, J. 1990a: Prebiotic synthesis of histidine. *Journal of Molecular Evolution 31*, 167–174. [*54*]

Shen, C., Mills, T. & Oró, J. 1990b: Prebiotic synthesis of histidyl–histidine. *Journal of Molecular Evolution 31*, 175–179. [*56*]

Sherwood, E., Joshi, A. & Oró, J. 1977: Cyanamide mediated synthesis under plausible primitive Earth conditions, II: The polymerization of deoxythymidine 5'-triphosphate. *Journal of Molecular Evolution 10*, 192–209. [*56*]

Shock, E.L. 1990a: Do amino acids equilibrate in hydrothermal fluids? *Geochimica et Cosmochimica Acta 54*, 1185–1189. [*18, 19*]

Shock, E.L. 1990b: Geochemical constraints on the origin of organic compounds in hydrothermal systems. *Origins of Life and Evolution of the Biosphere 20*, 331–367. [*18, 19*]

Shock, E.L. 1992: Chemical environments of submarine hydrothermal systems. *Origins of Life and Evolution of the Biosphere 22*, 67–108. [*18, 19*]

Shvedova, T.A., Korneeva, G.A., Otroschenko, V.A. & Venkstern, T.V. 1987: Catalytic activity of the nucleic acid component of the 1,4-alpha-glucan branching enzyme from rabbit muscles. *Nucleic Acids Research 15*, 1745–1752. [*71*]

Siewing, R. (ed.) 1985: *Lehrbuch der Zoologie, Bd. 2, Systematik*. Gustav Fischer, Stuttgart. [*476*]

Signor, P.W. & Lipps, J.H. 1992: Origin and early radiation of the Metazoa. In Lipps, J.H. & Signor, P.W. (eds.): *Origin and Early Evolution of the Metazoa*, 3–23. Plenum, New York, N.Y. [*410*]

Sillén, L.G. 1965: Oxidation state of Earth's ocean and atmosphere, I: A model calculation on earlier states. The myth of the "prebiotic soup." *Arkiv för Kemi 24*, 431–456. [*94*]

Simkiss, K. 1989: Biomineralization in the context of geological time. *Transactions of the Royal Society of Edinburgh: Earth Sciences 80*, 193–199. [*413, 424*]

Simkiss, K. & Wilbur, K.M. 1989: *Biomineralization. Cell Biology and Mineral Deposition*. Academic Press, San Diego, Calif. [*414*]

Simonson, B.M., Schubel, K.A. & Hassler, S.W. 1993: Carbonate sedimentology of the early Precambrian Hamersley Group of Western Australia. *Precambrian Research 60*, 287–335. [*248, 249*]

Simpson, T.L. 1984: *The Cell Biology of Sponges*. Springer, New York, N.Y. [*406*]

Singer, S.L. & Nicolson, G.L. 1972: The fluid mosaic model of the structure of cell membranes. *Science 175*, 720–731. [*110*]

Sirevåg, R., Buchanan, B.B., Berry J.A. & Troughton, J.H. 1977: Mechanisms of CO_2 fixation in bacterial photosynthesis studied by the carbon isotope technique. *Archives of Microbiology 112*, 35–38. [227]

Siskin, M. & Katritzky, A.R. 1991: Reactivity of organic compounds in hot water: Geochemical and technological implications. *Science 254*, 231–237. [20]

Siu, P.M.L. & Wood, H.G. 1962: Phosophoenol pyruvic carboxytransphosphorylase, a CO_2 fixation enzyme from propionic acic bacteria. *Journal of Biological Chemistry 237*, 3044–3051. [84]

Sivash, A.A., Masinovsky, Z. & Lozovaya, G.I. 1991: Surfactant micelles containing protoporphyrin IX as models of primitive photocatalytic systems: A spectral study. *BioSystems 25*, 131–140. [162]

Sleep, N.H., Zahnle, K.J., Kasting, J.F. & Morowitz, H.J. 1989: Annihilation of ecosystems by large asteroid impacts on the early Earth. *Nature 342*, 139–142. [14, 34]

Sleigh, M.A. 1979: Radiation of the Eukaryote Protista. In House, M.R. (ed.): *The Origin of Major Invertebrate Groups*, 23–54. The Systematics Association Special Volume 12 Academic Press, London. [315]

Sleigh, M.A. 1988: Flagellar root maps allow speculative comparisons of root patterns and of their ontogeny. *BioSystems 21*, 277–282. [318]

Sleigh, M.A., Dodge, J.D. & Patterson, D.J. 1984: "Kingdom Protista." In Barnes, R.S.K. (ed.): *A Synoptic Classification of Living Organisms*, 25–88. Blackwell, Oxford. [182]

Sloper, R.W., Braterman, P.S., Cairns-Smith, A.G., Truscott, T.G. & Craw, M. 1983: Direct observation of hydrated electrons in the U.V. photo-oxidation of $(Fe(H_2O)_6)^{2+}$. *Journal of the Chemical Society, Chemical Communications 270*, 488–489. [94]

Smith, A.B. 1984: *Echinoid Paleobiology*. Allen & Unwin, London. [493]

Smith, A.B. 1988: Phylogenetic relationship, divergence times, and rates of molecular evolution for camarodont sea urchins. *Molecular Biology and Evolution 5*, 345–365. [493]

Smith, A.B. & Jell, P.A. 1990: Cambrian edrioasteroids from Australia and the origin of starfishes. *Memoirs of the Queensland Museum 28*, 715–778. [296]

Smith, A.B., Lafay, B. & Christen, R. 1992: Comparative variation of morphological and molecular evolution through geologic time: 28S ribosomal RNA versus morphology in echinoids. *Philosophical Transactions of the Royal Society of London, B 338*, 365–382. [472]

Smith, B.A. & Terrile, R.J. 1984: A circumstellar disk around β Pictoris. *Science 226*, 1421–1424. [51]

Smith, D.C. & Douglas, A. 1989: *The Biology of Symbiosis*. Arnold, London. [329]

Smith, J.P.S. 1981: Fine structural observations on the central parenchyma in *Convoluta* sp. *Hydrobiologia 84*, 259–265. [476]

Smith, J.P.S. & Tyler, S. 1985: The acoel turbellarians: Kingpins of metazoan evolution or a specialized offshoot? In Conway Morris, S., George, J.D., Gibson, R. & Platt, H.M. (eds.): *The Origins and Relationships of Lower Invertebrates*, 123–143. Clarendon, Oxford. [476]

Smith, J.P.S., Tyler, S. & Rieger, R.M. 1986: Is the Turbellaria polyphyletic? *Hydrobiologia 132*, 13–21. [484]

Smith, J.P.S., Tyler, S., Thomas, M.B. & Rieger, R.M. 1982: The nature of turbellarian rhabdites: Phylogenetic implications. *Transactions of the American Microscopical Society 101*, 209–228. [482]

Smith, M.J., Boom, J.D.G. & Raff, R.A. 1990: Single copy DNA distances between two congeneric sea urchin species exhibiting radically different modes of evelopment. *Molecular Biology and Evolution 7*, 315–326. [494]

Smith, P.R. & Ruppert, E.E. 1988: Nephridia. In Westheide, W. & Hermans, C.O. (eds.): *The Ultrastructure of the Polychaeta*, 231–262. Microfauna Marina 4. [483, 485]

Smith, R.McK. & Patterson, D.J. 1986: Analyses of heliozoan interrelationships: An example of the potentials and limitations of ultrastructural approaches to the study of protistan phylogeny. *Proceedings of the Royal Society of London, B 227*, 325–366. [315, 316, 317, 323]

Smith, V. & Barrell, B.G. 1991: Cloning of a yeast U1 snRNP 70K protein homologue: Functional conservation of an RNA-binding domain between humans and yeast. *EMBO Journal 10*, 2627–2634. [505]

Snetsinger, K.G., Ferry, G.V., Russell, P.B., Pueschel, R.F., Oberbeck, V.R., Hayes, D.M. & Fong, W. 1987: Effects of El Chichon on stratospheric aerosols late 1982 to early 1984. *Journal of Geophysical Research 92*, 14761–14771. [21]

Sochava, A.V. & Podkovyrov, V.N. 1992: The evolution of carbonate rocks composition during Meso- and Neoproterozoic. *29th International Geological Congress, Kyoto, Abstracts 1*, 241. [*445*]

Sogin, M.L. 1991: Early evolution and the origin of eukaryotes. *Current Opinion in Genetics and Development 1*, 457–463. [*160*]

Sogin, M.L. & Gunderson, J.H. 1987: Structural diversity of eukaryotic small subunit ribosomal RNAs: Evolutionary implications. *Endocytobiology, III: Annals of the New York Academy of Sciences 503*, 125–140. [*187*]

Sogin, M.L., Gunderson, J.H., Elwood, H.J., Alonso, R.A. & Peattie, D.A. 1989: Phylogenetic meaning of the kingdom concept: An unusual ribosomal RNA from *Giardia lamblia. Science 243*, 75–77. [*187, 290, 296, 441*]

Sogin, M.L., Ingold, A., Karlok, M., Nielsen, H. & Engberg, J. 1986: Phylogenetic evidence for the acquisition of ribosomal RNA introns subsequent to the divergence of some of the major *Tetrahymena* groups. *EMBO Journal 5*, 3625–3630. [*184*]

Sokolov, B.S. 1965: Drevnejshie otlozheniya rannego kembriya i sabelliditidy [The oldest deposits of the Early Cambrian and the sabelliditids]. *Vsesoyuznyj Simpozium po Paleontologii Dokembriya i Rannego Kembriya, 25–30 oktyabrya 1965*, 78–91. [*357, 377, 383*]

Sokolov, B.S. 1967: Drevneyshie pogonofory. *Doklady Akademii Nauk SSSR 177*, 201–204. [*291, 295, 377, 383*]

Sokolov, B.S. 1972a: Vendian and Early Cambrian Sabellitida (Pogonophora) of the USSR. *1968 Proceedings of the International Paleontological Union, 23rd International Geological Congress*, 79–86. [*457*]

Sokolov, B.S. 1972b: Vendskij ehtap v istorii zemli [The Vendian period in Earth history]. *Paleontologiya – Doklady Sovetskikh Geologov 7*, 114–124. Nauka, Moscow. [*355*]

Sokolov, B.S. 1976: Organicheskij mir Zemli na puti k fanerozojskoj differentsiatsii [Organic world of the Earth on the way to Phanerozoic differentiation]. *Vestnik Akademii Nauk SSSR 1*, 126–143. [*377, 356, 357*]

Sokolov, B.S. 1990: Vendian Polychaeta. In Sokolov, B.S. & Ivanovskij, A.B. (eds.): *The Vendian System, Vol. 1: Paleontology*, 244–246. Springer, Berlin. [*377*]

Sokolov, B.S. & Fedonkin, M. A. 1984: The Vendian as the terminal system of the Precambrian. *Episodes 7*, 12–19. [*363, 371, 373*]

Sokolov, B.S. & Fedonkin, M.A. (eds.) 1990: *The Vendian System, Vol. 2: Regional Geology*. Springer, Berlin. [*371*]

Sokolov, B.S. & Iwanowski, A.B. (eds.) 1990: *The Vendian System, Vol. 1: Paleontology*. Springer, Berlin. [*371*]

Soledad Fernandez, M., Dennis, J.E., Drushel, R.F., Carrino, D.A., Kimata, K., Yamagata, M. & Caplan, A.I. 1991: The dynamics of compartmentalization of embryonic muscle by extracellular matrix molecules. *Developmental Biology 147*, 46–61. [*477*]

Sopott-Ehlers, B. 1981: Ultrastructural observations on paracnids, I: *Coelogynopora axi* Sopott (Turbellaria Proseriata). *Hydrobiologia 84*, 253–257. [*482*]

Souillard, N., Magot, M., Possot, O. & Sibold, L. 1988: Nucleotide sequence of regions homologous to *nif*H (nitrogenase Fe protein) from the nitrogen-fixing archaebacteria *Methanococcus thermolithotrophicus* and *Methanobacterium ivanovii*: Evolutionary implications. *Journal of Molecular Evolution 27*, 65–76. [*41*]

Southgate, P.N. 1986: Depositional environment and mechanism of preservation of microfossils, upper Proterozoic Bitter Springs Formation, Australia. *Geology 14*, 683–686. [*273, 281*]

Southgate, P.N. 1989: Relationship between cyclicity and stromatolite form in the Late Proterozoic Bitter Springs Formation, Australia. *Sedimentology 36*, 323–339. [*273, 282*]

Southward, E.C. 1975: Fine structure and phylogeny of the Pogonophora. *Symposia of the Zoological Society of London 36*, 235–251. [*457*]

Spamer, E.E. 1988: Geology of the Grand Canyon, 3. Part III. An annotated bibliography of the world literature on the Grand Ganyon type fossil *Chuaria circularis* Walcott, 1899, an index fossil for the Late Proterozoic. *Geological Society of America, Microfilm Publication 17*, [*350*]

Specht, T., Wolters, J. & Erdmann, V.A. 1991: Compilation of 5S rRNA and 5S rRNA gene sequences. *Nucleic Acids Research 19*, 2189–2191. [*469*]

Spiegel, E., Burger, M.M. & Spiegel, M. 1983: Fibronectin and laminin in the extracellular matrix and basement membrane of sea urchin embryos. *Experimental Cell Research 144*, 47–55. [*478*]

Spiegelman, S. 1971: An approach to the experimental analysis of precellular evolution. *Quarterly Reviews of Biophysics 4*, 215–253. [*72*]

Spirin, A.S. 1986: *Ribosome Structure and Protein Biosynthesis*. Benjamin Cummings, Menlo Park, Calif. [*77*]

Sprigg, R.C. 1947: Early Cambrian (?)jellyfishes from the Flinders Ranges, South Australia. *Transactions of the Royal Society of South Australia 71*, 212–234. [*372*]

Sprigg, R.C. 1949: Early Cambrian "jellyfishes" of Ediacara, south Australia, and Mount John, Kimberley District, western Australia. *Transactions of the Royal Society of South Australia 73*, 72–99. [*372*]

Stahl, F.W. 1988: A unicorn in the garden? *Nature 335*, 112–113. [*142*]

Stanier R.Y. 1977: The position of cyanobacteria in the world of phototrophs. *Carlsberg Research Communications 42*, 77–98. [*334*]

Stanier, R.Y. & Cohen-Basire, G. 1977: Phototrophic prokaryotes: The cyanobacteria. *Annual Review of Microbiology 1977*, 225–274. [*334*]

Stanley, G.D. 1986: Chrondrophorine hydrozoans as problematic fossils. In Hoffman, A. & Nitecki, M.H. (eds.): *Problematic Fossil Taxa*, 68–86. Oxford University Press, New York, N.Y. [*383*]

Stanley, S.M. 1976: Fossil data and the Precambrian–Cambrian evolutionary transition. *American Journal of Science 276*, 56–76. [*424, 436*]

Stanley, S.M. 1986: *Earth and Life through Time*. Freeman, New York, N.Y. [*438*]

Stasek, C.R. 1972: The molluscan framework. In Florkin, M. & Sheer, B.J. (eds.): *Chemical Zoology*, Vol. 7. Academic Press, New York, N.Y. [*402*]

Stein, D., Roth, S., Vogelsang, E. & Nüsslein-Volhard, C. 1991: The polarity of the dorsoventral axis in the *Drosophila* embryo is defined by an extracellular signal. *Cell 65*, 725–735. [*506*]

Steinböck, O. 1963: Origin and affinities of the lower Metazoa. In Dougherty, E.C. (ed.): *The Lower Metazoa*, 45–54. University of California Press, Berkeley, Calif. [*476*]

Steinböck, O. 1966: Die Hofsteniiden (Turbellaria Acoela), grundsätzliches zur Evolution der Turbellarien. *Zeitschrift für Zoologische Systematik und Evolutionsforschung 4*, 58–195. [*476, 484*]

Steinböck, O. 1967: Regenerationsversuche mit *Hofstenia giselae* Steinb. (Turbellaria Acoela). *Roux Archiv für Entwicklungsmechanik 158*, 394–458. [*484*]

Stetter, K.O. 1986: Diversity of extremely thermophilic archaebacteria. In Brock, T.D. (ed.): *Thermophiles: General, Molecular and Applied Microbiology*, 39–74. Wiley, New York, N.Y. [*144, 149*]

Stetter, K.O. 1988: *Archaeoglobus fulgidus* gen. nov, sp. nov.: A new taxon of extremely thermophilic Archaebacteria. *Systematic and Applied Microbiology 10*, 172–173. [*149*]

Stetter, K.O. 1992: Life at the upper temperature border. In Trân Thanh Vân, J. & K., Mounolou, J.C., Schneider, J. & McKay, C. (eds.): *Frontiers of Life*, 195–211. Proceedings of the 3d Rencontres de Blois, France [*144, 153*]

Stetter, K.O., Fiala, G., Huber, R. & Segerer, A. 1990: Hyperthermophilic microorganisms. *FEMS Microbiology Reviews 75*, 117–124. [*144, 147, 152, 163*]

Stetter, K.O., König, H., & Stackebrandt, E. 1983: *Pyrodictium* gen.nov., a new genus of submarine disc-shaped sulphur reducing Archaebacteria growing optimally at 105°C. *Systematic and Applied Microbiology 4*, 535–551. [*144, 149*]

Stetter, K.O., Lauerer, G., Thomm, M. & Neuner, A. 1987: Isolation of extremely thermophilic sulfate reducers: Evidence for a novel branch of Archaebacteria. *Science 236*, 822–824. [*39, 149, 155, 159*]

Stetter, K.O., Segerer, A., Zillig, W., Huber, G., Fiala, G., Huber, R. & König, H. 1986: Extremely thermophilic sulfur-mobilizing archaebacteria. *Systematic and Applied Microbiology 7*, 393–397. [*146*]

Stetter, K.O., Thomm, M., Winter, J., Wildgruber, G., Huber, H., Zillig, W., Janecovic, D., König, H., Palm, P. & Wunderl, S. 1981: *Methanothermus fervidus*, sp. nov., a novel extremely thermophilic methanogen from an Icelandic hot spring. *Zentralblatt für Bakteriologie und Hygiene. I. Abteilung Originale. C 2*, 166–178. [*148*]

Stevenson, D.J. 1983: The nature of the Earth prior to the oldest known rock record: The Hadean Earth. In Schopf, J.W. (ed.): *Earth's Earliest Biosphere, Its Origin and Evolution*, 14–29. Princeton University Press, Princeton, N.J. [*11*]

Stewart, A.J. 1979: A barred basin marine evaporite in the Upper Proterozoic of the Amadeus Basin, central Australia. *Sedimentology 26*, 33–62. [*273*]

Stillwell, W. 1980: Facilitated diffusion as a method for selective accumulation of materials from the primordial oceans by a lipid-vesicle protocell. *Origins of Life 10*, 277–292. [*21*]

Stockner, J.G. 1988: Phototrophic picoplankton: An overview from marine and freshwater ecosystems. *Limnology and Oceanography 33*, 765–775. [*338*]

Stolley, E. 1893: Über silurische Siphoneen. *Neues Jahrbuch für Mineralogie, Geologie und Paläontologie 1893*, 135–146. [*428*]

Stolz, J.F. 1983: Fine structure of the stratified microbial community at Laguna Figueroa, Baja Califonia, Mexico, I: Methods of an in situ study of the laminated sediments. *Precambrian Research 20*, 479–492. [*168, 173*]

Storch, V. 1991: Priapulida. In Harrison, F.W. & Ruppert, E.E. (eds.): *Microscopic Anatomy of Invertebrates*, Vol. 4, 333–350. Wiley-Liss, New York, N.Y. [*406*]

Strauss, G., Eisenreich, W., Bacher, A. & Fuchs, G. 1992: ^{13}C-NMR study of autotrophic CO_2 fixation pathways in the sulfur-reducing Archaebacterium *Thermoproteus neutrophilus* and in the phototrophic Eubacterium *Chloroflexus aurantiacus*. *European Journal of Biochemistry 205*, 853–866. [*154, 173*]

Strauss, H. 1986: Carbon and sulfur isotopes in Precambrian sediments from the Canadian Shield. *Geochimica et Cosmochimica Acta 50*, 2653–2662. [*231*]

Strauss, H., Des Marais, D.J., Hayes, J.M. & Summons, R.E. 1992a: Concentrations of organic carbon and maturities and elemental compositions of kerogens. In Schopf, J.W. & Klein, C. (eds.): *The Proterozoic Biosphere: A Multidisciplinary Study*, 95–99. Cambridge University Press, Cambridge. [*221*]

Strauss, H., Des Marais, D.J., Hayes, J.M. & Summons, R.E. 1992b: The carbon-isotopic record. In Schopf, J.W. & Klein, C. (eds.): *The Proterozoic Biosphere: A Multidisciplinary Study*, 117–127. Cambridge University Press, Cambridge. [*222, 231*]

Strauss, H. & Moore, T.B. 1992: Abundances and isotopic compositions of carbon and sulfur species in whole rock and kerogen samples. In Schopf, J.W. & Klein, C. (eds.): *The Proterozoic Biosphere: A Multidisciplinary Study*, 709–798. Cambridge University Press, Cambridge. [*193, 200, 201, 202, 203, 204, 205, 206, 225*]

Stribling, R. & Miller, S.L. 1987: Energy yields for hydrogen cyanide and formaldehyde syntheses: The HCN and amino acid concentrations in the primitive ocean. *Origins of Life and Evolution of the Biosphere 17*, 261–273. [*12, 14, 17, 110*]

Strid, Å., Nyrén, P., Boork, J. & Baltscheffsky, M. 1986: Kinetics of the membrane-bound inorganic pyrophosphatase from *Rhodospirillum rubrum* chromatophores: Effect of transmembrane electric potential on the rate constants. *FEBS Letters 196*, 337–340. [*86*]

Strother, P.K., Knoll, A.H. & Barghoorn, E.S. 1983: Micro-organisms from the Late Precambrian Narssârssuk Formation, north-western Greenland. *Palaeontology 26*, 1–32. [*281, 301*]

Strother, P.K. & Tobin, M. 1987: Observation on the genus *Huroniospora*: Implications for the paleoecology of the Gunflint chert. *Precambrian Research 36*, 323–333. [*310*]

Stuart, J.J., Brown, S.J., Beeman, R.W. & Denell, R.E. 1991: A deficiency of the homoeotic complex of the beetle *Tribolium*. *Nature 350*, 72–74. [*498*]

Stuart, J.J., Brown, S.J., Beeman, R.W. & Denell, R.E. 1993: The *Tribolium* homoeotic gene Abdominal is homologous to abdominal-A of the *Drosophila* Bithorax Complex. *Development 117*, 233–243. [*498*]

Studier, M.H., Hayatsu, R. & Anders, E. 1972: Origin of organic matter in early solar system, V: Further studies of meteoritic hydrocarbons and a discussion of their origin. *Geochimica et Cosmochimica Acta 36*, 189–215. [*111*]

Summers, D. & Chang, S. 1993: Prebiotic ammonia from reduction of nitrite by iron (II) on the early Earth. *Nature 365*, 630–633. [*14*]

Summons, R.E. & Hayes, J.M. 1992: Principles of molecular and isotopic biogeochemistry. In Schopf, J.W. & Klein, C. (eds.): *The Proterozoic Biosphere: A Multidisciplinary Study*, 83–93. Cambridge University Press, Cambridge. [*223*]

Summons, R.E., Jahnke, L.L. & Roksandic, Z. 1994: Carbon isotope fractionation in lipids from methanotrophic bacteria: Relevance for interpretation of the geochemical record of biomarkers. *Geochimica et Cosmochimica Acta 58*, (In press.) [*230*]

Summons, R.E. & Powell, T.G. 1991: Petroleum source rocks of the Amadeus Basin. In Korsch, R.J. & Kennard, J.M. (eds.): *Geological and Geophysical Studies in the Amadeus Basin, Central Australia*, 511–524. Bureau of Mineral Resources, Geology and Geophysics, Bulletin 236. [*273*]

Summons, R.E., Powell, T.G. & Boreham, C.J. 1988: Petroleum geology and geochemistry of the Middle Proterozoic McArthur Basin, northern Australia: III. Composition of extractable hydrocarbons. *Geochimica et Cosmochimica Acta 52*, 1747–1763. [*262, 290*]

Summons, R.E. & Walter, M.R. 1990: Molecular fossils and microfossils of prokaryotes and protists from Proterozoic sediments. *American Journal of Science 290-A*, 212–244. [*273, 283, 290, 441, 510*]

Sumner, D.Y., Beukes, N.J. & Grotzinger, J.P. 1991: Massive marine cementation of the Archean Campbellrand–Malmani carbonate platform. *Geological Association of Canada, Program with Abstracts 16*, 120. [*249*]

Sun Weiguo 1986a: Are there pre-Ediacarian metazoans? *Precambrian Research 31*, 409–410. [*363*]

Sun Weiguo 1986b: Late Precambrian scyphozoan medusa *Mawsonites randellensis* sp. nov. and its significance in the Ediacara metazoan assemblage, South Australia. *Alcheringa 10*, 169–181. [*364, 365*]

Sun Weiguo 1987: Palaeontology and biostratigraphy of Late Precambrian macroscopic colonial algae: *Chuaria* Walcott and *Tawuia* Hofmann. *Palaeontographica, Abt. B. 203*, 109–134. [*350, 360, 361*]

Sun Weiguo & Hou Xianguang 1987a: Early Cambrian medusae from Chengjiang, Yunnan, China. *Acta Palaeontologica Sinica 26*, 257–271. (In Chinese, with English summary.) [*383, 458*]

Sun Weiguo & Hou Xianguang 1987b: Early Cambrian worms from Chengjiang, Yunnan, China: *Maotianshania* gen. nov. *Acta Palaeontologica Sinica 26*, 299–305. (In Chinese, with English summary.) [*458*]

Sun Weiguo, Wang Guixiang & Zhou Benhe 1986: Macroscopic worm-like body fossils from the upper Precambrian (900–700 Ma), Huainan district, Anhui, China and their stratigraphic and evolutionary significance. *Precambrian Research 31*, 377–403. [*291, 353, 355, 361, 362*]

Suslick, K.S. 1989: The chemical effects of ultrasound. *Scientific American 260*, 80–86. [*22*]

Suzuki, S., Hori, Y. & Koga, O. 1979: Decomposition of ozone on natural sand. *Bulletin of the Chemical Society of Japan 52*, 3103–3104. [*37*]

Sweeney, B.M. & Borgese, M.B. 1989: A circadian rhythm in cell division in a prokaryote, the cyanobacterium *Synechococcus* WH7803. *Journal of Phycology 25*, 183–186. [*337*]

Swett, K. & Knoll, A.H. 1985: Stromatolitic bioherms and microphytolites from the late Proterozoic Draken Conglomerate Formation, Spitsbergen. *Precambrian Research 28*, 327–347. [*429, 430*]

Swett, K. & Knoll, A.H. 1989: Marine pisolites from Upper Proterozoic carbonates of east Greenland and Spitsbergen. *Sedimentology 36*, 75–93. [*432*]

Swofford, D.L. 1990: PAUP: Phylogenetic analysis using parsimony, version 3.0. Computer program distributed by the Illinois Natural History Survey, Champaign, Ill. [*468, 472*]

Swofford, D.L. & Olsen, G.J. 1990: Phylogeny reconstruction. In Hillis, D.M. & Moritz, C. (eds.): *Molecular Systematics*, 411–501. Sinauer, Sunderland, Mass. [*184, 468*]

Sylvester-Bradley, P.C. 1971: Environmental parameters for the origin of life. *Proceedings of the Geologists' Association 82*, 87–136. [*37*]

Talley, F.P., Stewart, P.R., Sutter, V.L. & Rosenblatt, J.E. 1975: Oxygen tolerance of fresh clinical anaerobic bacteria. *Journal of Clinical Microbiology 1*, 161–164. [*38*]

Tandon, K.K. & Kumar, S. 1977: Discovery of annelid and arthropod remains from lower Vindhyan rocks (Precambrian) of central India. *Geophytology 7*, 126–129. [*353, 356, 357*]

Tanford, C. 1978: The hydrophobic effect and the organization of living matter. *Science 200*, 1012–1018. [*56*]

Tappan, H. 1968: Primary production, isotopes, extinctions and the atmosphere. *Palaeogeography, Palaeoclimatology, Palaeoecology 4*, 187–210. [*43*]

Tappan, H. 1980: *The Paleobiology of Plant Protists*. Freeman, San Francisco, Calif. [*429, 433, 441*]

Taylor, F.J.R. 1974: Implications and extensions of the Serial Endosymbiosis Theory of the origin of eukaryotes. *Taxon 23*, 229–258. [*316, 320*]

Taylor, F.J.R. 1976: Flagellate phylogeny: A study in conflicts. *Journal of Protozoology 23*, 28–40. [*314, 315, 317, 318, 319, 320, 325, 326*]

Taylor, F.J.R. 1978: Problems in the development of an explicit hypothetical phylogeny of the lower eukaryotes. *BioSystems 10*, 67–89. [*182, 313, 315, 319, 320, 325*]

Taylor, F.J.R. 1980a: The stimulation of cell research by endosymbiotic hypotheses for the origin of eukaryotes. In Schwemmler, W. & Schenk, H.E.A. (eds.): *Endocytobiology*, 917–942. De Gruyter, Berlin. [*316*]

Taylor, F.J.R. 1980b: On dinoflagellate evolution. *BioSystems 13*, 65–108. [*325*]

Taylor, F.J.R. 1981: Review of Tappan, H.: *The Paleobiology of Plant Protists*. *Science 211*, 1413–1414. [*313*]

Taylor, F.J.R., Blackbourn, D.J. & Blackbourn, J. 1969: Ultrastructure of the chloroplasts and associated structures within the marine ciliate *Mesodimium rubrum* (Lochmann). *Nature 224*, 819–821. [*319*]

Taylor, S.R. & McLennan, S.M. 1985: *The Continental Crust: Its Composition and Evolution*. Blackwell, Oxford. [*25*]

Teilhard de Chardin, P. 1925: Le paradoxe transformiste. *Revue des questions scientifiques 7*, 53–80. [*388*]

Teilhard de Chardin, P. 1955: *Le Phénomène Humain*. (English translation 1959; Russian translation 1987.) [*388*]

Tewari, V.C. 1989: Upper Proterozoic – Lower Cambrian stromatolites and Indian stratigraphy. *Himalayan Geology 13*, 143–180. [*357*]

Theng, B.K.G. 1974: *The Chemistry of Clay – Organic Reactions*. Wiley, New York, N.Y. [*95*]

Thiemann, M. & Ruthmann, A. 1990: Spherical forms of *Trichoplax adhaerens* (Placozoa). *Zoomorphology 110*, 37–45. [*480*]

Thiemann, M. & Ruthmann, A. 1991: Alternative modes of asexual reproduction in *Trichoplax adhaerens* (Placozoa). *Zoomorphologie 110*, 165–174. [*480*]

Thomsen, L. 1980: ^{129}Xe on the outgassing of the atmosphere. *Journal of Geophysical Research 85*, 4374–4378. [*11*]

Thomson, J., Carpenter, M.S.N., Colley, S., Wilson, T.R.S., Elderfield, H. & Kennedy, H. 1984: Metal accumulation rates in northwest Atlantic pelagic sediments. *Geochimica et Cosmochimica Acta 48*, 1935–1948. [*40*]

Thornton, J.M. & Gardner, S.P. 1989: Protein motifs and data-base searching. *Trends in Biochemical Sciences 14*, 300–304. [*511*]

Tibayrenc, M., Kjellberg, F., Arnaud, J., Oury, B., Breniere, S.F., Darde, M.-L. & Ayala, F.J. 1991: Are eukaryotic microorganisms clonal or sexual? A population genetics vantage. *Proceedings of the National Academy of Sciences, USA 88*, 395–411. [*448*]

Timofeev, B.V. 1969: *Sferomorfidy Proterozoya*. Nauka, Leningrad. [*355*]

Timofeev, B.V. & German [Hermann], T.N. 1979: Dokembrijskaya mikrobiota Lakhandinskoj svity [Precambrian microbiota of the Lakhanda Formation]. In Sokolov, B.S. (ed.): *Paleontologiya Dokembriya i Rannego Kembriya*, 137–147. Nauka, Leningrad. [*356, 357*]

Timofeev, B.V., German [Hermann], T.N. & Mikhailova, N.S. 1976: Mikrofitofossilii dokembriya, kembriya i ordovika [Microphytofossils from the Precambrian, Cambrian and Ordovician]. *Trudy Instituta Geologii i Geochronologii Dokembriya, Leningrad*. 106 pp. [*306, 427*]

Tingle, T.N., Green, H.W. & Finnerty, A.A. 1988: Experiments and observations bearing on the solubility and diffusivity of carbon in olivine. *Journal of Geophysical Research 93*, 15289–15304. [*11*]

Tomizawa, J. 1993: Evolution of functional structures of RNA. In Gestland, R. & Atkins, J. (eds.): *The RNA World*, 419–445. Cold Spring Harbor Laboratory Press, Plainview, N.Y. [*140*]

Toomey, D.F. 1970: An unhurried look at a Lower Ordovician mound horizon, southern Franklin Mountains, West Texas. *Journal of Sedimentary Petrology 40*, 1318–1334. [*429*]

Towe, K.M. 1970: Oxygen–collagen priority and the early metazoan fossil record. *Proceedings of the National Academy of Sciences, USA 65*, 781–788. [*367*]

Towe, K.M. 1978: Early Precambrian oxygen: A case against photosynthesis. *Nature 274*, 657–661. [*29*]

Towe, K.M. 1981a: Environmental conditions surrounding the origin and early evolution of life. *Precambrian Research 16*, 1–10. [*37*]

Towe, K.M. 1981b: Biochemical keys to the emergence of complex life. In Billingham, J. (ed.): *Life in the Universe*, 297–307. Massachusetts Institute of Technology Press, Cambridge, Mass. [*477, 478*]

Towe, K.M. 1985: Habitability of the early Earth: Clues from the physiology of nitrogen fixation and photosynthesis. *Origins of Life and Evolution of the Biosphere 15*, 235–250. [*41*]

Towe, K.M. 1988: Early biochemical innovations, oxygen, and earth history. In Broadhead, T.W. (ed.): *Molecular Evolution and the Fossil Record*, 114–129. Paleontological Society, Short Course 1. [*38, 39, 41*]

Towe, K.M. 1990: Aerobic respiration in the Archean? *Nature 348*, 54–56. [*44, 172, 304*]

Towe, K.M. 1991: Aerobic carbon cycling and cerium oxidation: Significance for Archean oxygen levels and banded iron-formation deposition. *Palaeogeography, Palaeoclimatology, Palaeoecology (Global and Planetary Change Section) 97*, 113–123. [*40, 44, 172, 218*]

Towe, K.M., Bengtson, S., Fedonkin, M.A., Hofmann, H.J., Mankiewicz, C. & Runnegar, B. 1992a: Proterozoic and earliest Cambrian carbonaceous remains, trace and body fossils. In Schopf, J.W. & Klein, C. (eds.): *The Proterozoic Biosphere: A Multidisciplinary Study*, 343–424. Cambridge University Press, Cambridge. [*415*]

Towe, K.M., Bengtson, S., Fedonkin, M.A., Hofmann, H.J., Mankiewicz, C. & Runnegar, B. 1992b: Described taxa of Proterozoic and selected earliest Cambrian carbonaceous remains, trace and body fossils. In Schopf, J.W. & Klein, C. (eds.): *The Proterozoic Biosphere: A Multidisciplinary Study*, 953–1054. Cambridge University Press, Cambridge. [*415, 416, 423*]

Trent, J.D., Nimmesgern, E., Wall, J.S., Hartl., F.-U. & Horwich, A.L. 1991: A molecular chaperone from a thermophilic archaebacterium is related to the eukaryotic protein t-complex polypeptide-1. *Nature 354*, 490–493. [*504*]

Trost, J., Brune, D. & Blankenship, R. 1992: Protein sequences and redox titrations indicate that the electron acceptors in reaction centers from heliobacteria are similar to Photosystem I. *Photosynthesis Research 32*, 11–22. [*179*]

Tseng, R.-S., Viechnicki, J.T., Slop, R.A. & Brown, J.W. 1992: Sea-to-air transfer of surface-active organic compounds by bursting bubbles. *Journal of Geophysical Research 97*, 5201–5206. [*21*]

Tucker, M.E. 1983: Diagenesis, geochemistry and origin of a Precambrian dolomite: The Beck Spring Dolomite of eastern California. *Journal of Sedimentary Petrology 53*, 1097–1119. [*445*]

Tucker, M.E. & Wright, V.P. 1990: *Carbonate Sedimentology*. Blackwell, Oxford. [*245*]

Tudor, J.J. & Bende, S.M. 1986: The outer cyst wall of *Bdellovibrio* bdellocycts is made de novo and not from preformed units from the prey wall. *Current Microbiology 13*, 185–189. [*329*]

Tudor, J.J. & Conti, S.F. 1977: Characterization of bdellocysts of *Bdellovibrio* sp. *Journal of Bacteriology 131*, 314–322. [*329*]

Turbeville, J.M. 1991: Nemertinea. In Harrison, F.W. & Bogitsch, B.J. (eds.): *Microscopic Anatomy of Invertebrates, Vol. 3: Platyhelminthes and Nemertea*, 285–328. Wiley-Liss, New York, N.Y. [*406, 482*]

Turbeville, J.M., Field, K.G. & Raff, R.A. 1992: Phylogenetic position of Phylum Nemertini, inferred from 18S rRNA sequences: Molecular data as a test of morphological character homology. *Molecular Biology and Evolution 9*, 235–249. [*402, 403, 461, 490, 491*]

Turbeville, J.M. & Ruppert, E.E. 1983: Epidermal muscle and peristalitic burrowing in the Carinoma tremaphoros (Nemertini): Correlates of effective burrowing without segmentation. *Zoomorphologie 103*, 103–120. [*482, 488*]

Tyler, S. 1981: Development of cilia in embryos of the turbellarian *Macrostomum*. *Hydrobiologia 84*, 231–239. [*481, 485*]

Tyler, S.A. & Barghoorn, E.S. 1954: Occurrence of structurally preserved plants in pre-Cambrian rocks of the Canadian Shield. *Science 119*, 606–608. [*427*]

Tyler, S.A., Barghoorn, E.S. & Barret, L.P. 1957: Anthracitic coal from Precambrian Upper Huronian black shale of the Iron River district, northern Michigan. *Geological Society of America, Bulletin 68*, 1293–1304. [*350, 351*]

Tyler, S. & Rieger, R.M. 1977: Ultrastructural evidence for the systematic position of the Nemertodermatida (Turbellaria). *Acta Zoologica Fennica 54*, 193–207. [*476*]

Uhlenbeck, O.C. 1987: A small catalytic oligoribonucleotide. *Nature 328*, 596–600. [*64, 73*]

Urbanek, A. & Mierzejewska, G. 1977: The fine structure of zooidal tubes in Sabelliditida and Pogonophora with reference to their affinity. *Acta Palaeontologica Polonica 22*, 223–240. [*457*]

Urbanek, A. & Mierzejewska, G. 1983: The fine structure of zooidal tubes in Sabelliditida and Pogonophora. In Urbanek, A. & Rozanov, A.Yu. (eds.): *Upper Precambrian and Cambrian Palaeontology of the East European Platform*, 100–111. Wydawnictwa Geologiczne, Warsaw. [*353, 377*]

Urey, H.C. 1952: On the early chemical history of the Earth and the origin of life. *Proceedings of the National Academy of Sciences, USA 38*, 351–363. [*52*]

Urey, H.C. 1957: Origin of tektites. *Nature 179*, 556–557. [*51*]

Urey, H.C., Lowenstam, H.A., Epstein, S. & McKinney, C.R. 1951: Measurement of paleotemperatures and temperatures of the Upper Cretaceous of England, Denmark, and Southeastern United States. *Geological Society of America, Bulletin 62*, 399–416. [*223*]

Ushatinskaya, G.T. 1988: Obolellidy (brakhiopody) s zamkovym sochleneniem stvorok iz nizhnego kembriya Zabajkal'ya [Obolellids (brachiopods) with articulate shell condition from the Lower Cambrian of Transbaikalia]. *Paleontologicheskij Zhurnal 1988:1*, 32–38. [*458*]

Ushatinskaya, G.T. 1990: Cambrian inarticulate brachiopods with phosphatic-calcium shell. *Byulleten' Moskovskogo obshchestva ispytatelei prirody 65*, 47–59. [*458*]

Vaessin, H.V., Caudy, M., Bier, E., Jan, L.Y. & Jan, Y.N. 1990: The role of helix–loop–helix proteins in *Drosophila* neurogenesis. In *The Brain*, 239–245. Cold Spring Harbor Symposia on Quantitative Biology 55. [*507, 508*]

Vagvolgyi, J. 1967: On the origin of the molluscs, the coelom, and coelomic segmentation. *Systematic Zoology 16*, 153–168. [*402*]

Valdiya, K.S. & Tewari, V.C. 1989: Stromatolites and stromatolitic deposits. *Himalayan Geology 13*. 289 pp. [*279*]

Valentine, J.W. 1973: *Evolutionary Paleoecology of the Marine Biosphere*. Prentice-Hall, Englewood Cliffs, N.J. [*411, 436, 464*]

Valentine, J.W. 1977: General patterns of metazoan evolution. In Hallam, A. (ed.): *Patterns of Evolution as Illustrated by the Fossil Record*, 27–57. Elsevier, Amsterdam. [*140, 401, 468*]

Valentine, J.W. 1980: Determinants of diversity in higher taxonomic categories. *Paleobiology 6*, 444–450. [*464*]

Valentine, J.W. 1989: Bilaterians of the Precambrian–Cambrian transition and the annelid–arthropod relationship. *Proceedings of the National Academy of Sciences, USA 86*, 2272–2275. [*295, 402, 453, 483, 486*]

Valentine, J.W. 1991: Major factors in the rapidity and extent of the metazoan radiation during the Proterozoic–Phanerozoic transition. In Simonetta, A.M. & Conway Morris, S. (eds.): *The Early Evolution of Metazoa and the Significance of Problematic Taxa*, 11–13. Cambridge University Press, Cambridge. [*404, 406*]

Valentine, J.W., Awramik, S.M., Signor, P.W. & Sadler, P.M. 1991: The biological explosion at the Precambrian–Cambrian boundary. *Evolutionary Biology 25*, 279–356. [*401, 410, 442, 443, 449*]

Valentine, J.W. & Campbell, C.A. 1975: Genetic regulation and the fossil record. *American Scientist 63*, 673–680. [*408*]

Valentine, J.W. & Erwin, D.H. 1987: Interpreting great developmental experiments: The fossil record. In Raff, R.A. & Raff, E.C. (eds.): *Development as an Evolutionary Process*, 71–107. Alan Liss, New York, N.Y. [*401*]

Val'kov, A.K. 1982: *Biostratigrafiya Nizhnego Kembriya Vostoka Sibirskoj Platformy*. Nauka, Moscow. [*383*]

Val'kov, A.K. & Sysoev, V.A. 1970: Angustiokreidy kembriya Sibiri [Cambrian angustiochreids of Siberia]. In *Stratigrafiya i Paleontologiya Proterozoya i Kembriya Vostoka Sibirskoj Platformy*, 94–100. Yakutsk. [*383*]

Van Cappellen, P. & Berner, R.A. 1991: Fluorapatite crystal growth from modified seawater solutions. *Geochimica et Cosmochimica Acta 55*, 1219–1233. [*100*]

Vandekerckhove, J. & Weber, K. 1984: Chordate muscle actins differ distinctly from invertebrate muscle actins. The evolution of the different vertebrate muscle actins. *Journal of Molecular Biology 179*, 391–413. [*497*]

Vaulot, D., Partensky, F., Neveux, J., Fauzi, C., Mantoura, C. & Llevellyn, C.A. 1990: Winter presence of prochlorophytes in surface waters of the northwestern Mediterranean Sea. *Limnology and Oceanography 35*, 1156–1164. [*338*]

Veizer, J. 1978: Secular variations in the composition of sedimentary carbonate rocks: II. Fe, Mn, Ca, Mg, Si and minor constitutents. *Precambrian Research 6*, 381–413. [*42*]

Veizer, J. 1983: Geologic evolution of the Archean–Early Proterozoic Earth. In Schopf, J.W. (ed.): *Earth's Earliest Biosphere, Its Origin and Evolution*, 240–259. Princeton University Press, Princeton, N.J. [*13, 218*]

Veizer, J. 1984: The evolving earth: Water tales. *Precambrian Research 25*, 5–12. [*212*]

Veizer, J. 1988a: The earth and its life: Systems perspective. *Origins of Life and Evolution of the Biosphere 18*, 13–39. [*209, 214, 215, 219*]

Veizer, J. 1988b: The evolving exogenic cycle. In Gregor, C.B., Garrels, R.M., Mackenzie, F.T. & Maynard, J.B. (eds.): *Chemical Cycles in the Evolution of the Earth*, 175–220. Wiley, New York, N.Y. [*25, 210, 219, 224, 235*]

Veizer, J. 1989: Strontium isotopes in seawater through time. *Annual Review of Earth and Planetary Sciences 17*, 141–167. [*211*]

Veizer, J. & Compston, W. 1976: $^{87}Sr/^{86}Sr$ in Precambrian carbonates as an index of crustal evolution. *Geochimica et Cosmochimica Acta 40*, 905–914. [*250, 273*]

Veizer, J., Compston, W., Clauer, N. & Schidlowski, M. 1983: $^{87}Sr/^{86}Sr$ in late Proterozoic carbonates: Evidence for a mantle event at 900 Ma ago. *Geochimica et Cosmochimica Acta 47*, 295–302. [*443*]

Veizer, J., Compston, W., Hoefs, J. & Nielsen, H. 1982: Mantle buffering of the early oceans. *Naturwissenschaften 69*, 173–180. [*25, 218*]

Veizer, J., Fritz, P. & Jones, B. 1986: Geochemistry of brachiopods: Oxygen and carbon isotopic record of Paleozoic oceans. *Geochimica et Cosmochimica Acta 50*, 1679–1696. [*213*]

Veizer, J. & Hoefs, J. 1976: The nature of O^{18}/O^{16} and C^{13}/C^{12} secular trends in sedimentary carbonate rocks. *Geochimica et Cosmochimica Acta 40*, 1387–1395. [*213*]

Veizer, J., Hoefs, J., Lowe, D.R. & Thurston, P.C. 1989b: Geochemistry of Precambrian carbonates: 2. Archean greenstone belts and Archean sea water. *Geochimica et Cosmochimica Acta 53*, 859–871. [*25, 27, 224*]

Veizer, J., Hoefs, J., Ridler, R.H., Jensen, L.S. & Lowe, D.R. 1989a: Geochemistry of Precambrian carbonates: 1. Archean hydrothermal systems. *Geochimica et Cosmochimica Acta 53*, 845–857. [*27*]

Veizer, J., Holser, W.T. & Wilgus, C.K. 1980: Correlation of $^{13}C/^{12}C$ and $^{34}S/^{32}S$ secular variations. *Geochimica et Cosmochimica Acta 44*, 579–587. [*215*]

Veizer, J. & Jansen, S.L. 1979: Basement and sedimentary recycling and continental evolution. *Journal of Geology 87*, 341–370. [*25, 26, 45, 210*]

Veizer, J. & Jansen, S.L. 1985: Basement and sedimentary recycling, 2: Time dimension to global tectonics. *Journal of Geology 93*, 625–643. [*25, 26, 43, 209*]

Veizer, J., Laznicka, P. & Jansen, S.L. 1989c: Mineralization through geologic time: Recycling perspective. *American Journal of Science 289*, 484–524. [*211*]

Veizer, J., Plumb, K.A., Clayton, R.N., Hinton, R.W. & Grotzinger, J. 1992: Geochemistry of Precambrian carbonates: 5. Late Paleoproterozoic (1.8±0.2) sea water. *Geochimica et Cosmochimica Acta 56*, 2487–2501. [*217*]

Velikanov, V.A., Aseeva, E.A. & Fedonkin, M.A. 1983: *Vend Ukrainy*. Naukova Dumka, Kiev. [*373, 377*]

Vermeij, G.J. 1987: *Evolution and Escalation. An Ecological History of Life*, 1–527. Princeton University Press, Princeton, N.J. [*424*]

Vermeij, G.J. 1990: The origin of skeletons. *Palaios 4*, 585–589. [*421, 424, 436, 452, 465*]

Vetter, R. 1991: Symbiosis and the evolution of novel trophic strategies: Thiotrophic organisms at hydrothermal vents. In Margulis, L. & Fester, R. (eds.): *Symbiosis as a Source of Evolutionary Innovation: Speciation and Morphogenesis*, 219–245. Massachusetts Institute of Technology Press, Cambridge, Mass. [*333*]

Vidal, G. 1976: Late Precambrian microfossils from the Visingsö Beds in southern Sweden. *Fossils and Strata 9*. 57 pp. [*306*]

Vidal, G. 1981: Micropalaeontology and biostratigraphy of the Upper Proterozoic and Lower Cambrian sequences in East Finnmark, northern Norway. *Norges Geologiske Undersøkelse 362.* 53 pp. [306]

Vidal, G. 1984: The oldest eucaryotic cells. *Scientific American 250,* 48–57. [292, 434, 435]

Vidal, G. 1989: Are late Proterozoic carbonaceous megafossils metaphytic algae or bacteria? *Lethaia 22,* 375–379. [303, 352]

Vidal, G. 1990: Giant acanthomorph acritarchs from the Upper Proterozoic in southern Norway. *Palaeontology 33,* 287–298. [306, 308]

Vidal, G. & Ford, T.D. 1985: Microbiotas from the Late Proterozoic Chuar Group (Northern Arizona) and Uinta Mountain Group (Utah) and their chronostratigraphic implications. *Precambrian Research 28,* 349–489. [306]

Vidal, G. & Knoll, A.H. 1982: Radiations and extinctions of plancton in the late Proterozoic and Early Cambrian. *Nature 297,* 57–60. [372, 435]

Vidal, G. & Knoll, A.H. 1983: Proterozoic plankton. *Geological Society of America, Memoir 161,* 265–277. [305, 306, 308]

Vidal, G. & Nystuen, J.P. 1990a: Micropalaeontology, depositional environment and biostratigraphy of the Upper Proterozoic Hedmark Group, southern Norway. *American Journal of Science 290-A,* 170–211. [306]

Vidal, G. & Nystuen, J.P. 1990b: Lower Cambrian acritarchs and Proterozoic–Cambrian boundary in southern Norway. *Norsk Geologisk Tidsskrift 70,* 191–222. [305]

Vidal, G. & Peel, J.S. 1993: Acritarchs from the Lower Cambrian Buen Formation in North Greenland. *Grønlands Geologiske Undersøgelse, Bulletin 35.* 35 pp. [305]

Vidal-Majdar, A., Hobbs, L.M., Ferlet, R., Gry, C. & Albert, C.E. 1986: The circumstellar gas cloud around Beta Pictoris. II. *Astronomy and Astrophysics 167,* 325–332. [51]

Vine, J.D. & Tourtelot, E.B. 1970: Geochemistry of black shale deposits: A summary report. *Economic Geology 65,* 253–272. [240]

Virgo, D., Luth, R.W., Moats, M.A. & Ulmer, G.C. 1988: Constraints on the oxidation state of the mantle: An electrochemical and ^{57}Fe Mossbauer study of mantle-derived ilmenites. *Geochimica et Cosmochimica Acta 52,* 1781–1794. [12]

Vodanyuk, S.A. 1989: Ostatki besskeletnykh Metazoa iz khatyspytskoj svity Olenyokskogo podnyatiya [Remains of non-skeletal metazoans from the Khatyspyt Formation of the Olenyok Uplift]. In *Pozdnij Dokembrij i Rannij Paleozoj Sibiri,* 61–74. Trudy Instituta Geologii i Geofiziki Sibir'skogo Otdeleniya Akademii Nauk SSSR [373]

Vogel, S. 1988: *Life's Devices – The Physical World of Animals and Plants.* Princeton University Press, Princeton, N.J. [487]

Völkl, P., Huber, R., Drobner, E., Rachel, R., Burggraf, S., Trincone, A.R. & Stetter, K.O. 1993: *Pyrobaculum aerophilum* sp. nov., a novel nitrate-reducing hyperthermophilic archaeum. *Applied and Environmental Microbiology 59,* 2918–2926. [147]

Volkova, N.A., Kirjanov, V.V., Pyatiletov, V.G., Rudavskaya, V.A., Treshchetenkova, A.A., Faizulina, Z.Kh. & Jankauskas, T.V. 1980: The Upper Precambrian microfossils of the Siberian Platform. *Izvestiya Akademii Nauk SSSR, Seriya Geologicheskaya 1,* 23–29. [305]

Vologdin, A.G. 1962: *Drevnejshie Vodorosli SSSR.* Nauka, Moscow. [429]

von Ahsen, U., Davis, J. & Schroder, R. 1991: Antibiotic inhibition of group I ribozyme function. *Nature 353,* 368–370. [76]

Vossbrinck, C.R., Maddox, J.V., Friedman, S., Debrunner-Vossbrinck, B.A. & Woese, C.R. 1987: Ribosomal RNA sequence suggests microsporidia are extremely ancient eukaryotes. *Nature 326,* 411–413. [450]

Wächtershäuser, G. 1988a: Pyrite formation, the first energy source for life: A hypothesis. *Systematic and Applied Microbiology 10,* 207–210. [18, 82, 85, 126, 151, 154, 155, 159, 163]

Wächtershäuser, G. 1988b: Before enzymes and templates: Theory of surface metabolism. *Microbiological Reviews 52,* 452–484. [18, 76, 85, 126, 127, 131, 143, 150, 154, 155, 159, 163]

Wächtershäuser, G. 1990a: Evolution of the first metabolic cycles. *Proceedings of the National Academy of Sciences, USA 87,* 200–204. [18, 127, 154, 174, 267]

Wächtershäuser, G. 1990b: The case for the chemoautotrophic origin of life in an iron-sulfur world. *Origins of Life and Evolution of the Biosphere 20*, 173–176. [*126, 163, 174*]

Wächtershäuser, G. 1991: Biomolecules: The origin of their activity. *Medical Hypotheses 36*, 307–311. [*126*]

Wächtershäuser, G. 1992: Groundworks for an evolutionary biochemistry – The iron–sulfur world. *Progress in Biophysics and Molecular Biology 58*, 85–201. [*128, 267*]

Wadleigh, M.A. & Veizer, J. 1992: $^{18}O/^{16}O$ and $^{13}C/^{12}C$ in Lower Paleozoic brachiopods: Isotopic evolution of sea water. *Geochimica et Cosmochimica Acta 56*, 431–443. [*213*]

Wainright, P.O., Hinkle, G., Sogin, M.L. & Stickel, S.K. 1993: The monophyletic origins of the Metazoa; an unexpected evolutionary link with Fungi. *Science 260*, 340–243. [*186, 296*]

Wainwright, S.A. 1988: *Axis and Circumferences*. Harvard University Press, Cambridge, Mass. [*477*]

Wainwright, S.A., Biggs, W.D., Currey, J.D. & Gosline, J.M. 1976: *Mechanical Design in Organisms*. Edward Arnold, London & Halstead, N.Y. [*413*]

Walcott, C.D. 1898: Fossil medusae. *U.S. Geological Survey Monographs 30*. 201 pp. [*383*]

Walcott, C.D. 1899: Pre-Cambrian fossiliferous formations. *Geological Society of America, Bulletin 10*, 199–244. [*355, 356, 357*]

Walcott, C.D. 1919: Cambrian geology and paleontology IV, No. 5: Middle Cambrian algae. *Smithsonian Miscellaneous Collections 67*, 217–260. [*350, 351, 356*]

Walcott, C.D. 1920: Cambrian geology and paleontology IV, No. 6: Middle Cambrian sponges. *Smithsonian Miscellaneous Collections 67*, 261–364. [*457*]

Walker, J.C.G. 1976: Implications for atmospheric evolution of the inhomogeneous accretion model of the origin of the Earth. In Windley, B.F. (ed.): *The Early History of the Earth*, 537–546. Wiley, London. [*46*]

Walker, J.C.G. 1977: *Evolution of the Atmosphere*. MacMillan, London. [*16*]

Walker, J.C.G. 1978: Oxygen and hydrogen in the primitive atmosphere. *Pure and Applied Geophysics 116*, 222–231. [*46*]

Walker, J.C.G. 1982: Climatic factors on the Archean earth. *Palaeogeography, Palaeoclimatology, Palaeoecology 40*, 1–11. [*28*]

Walker, J.C.G. 1983a: Possible limits on the composition of the Archean ocean. *Nature 302*, 518–520. [*253*]

Walker, J.C.G. 1983b: Carbon geodynamic cycle. *Nature 303*, 730–731. [*11*]

Walker, J.C.G. 1985: Carbon dioxide on the early Earth. *Origins of Life and Evolution of the Biosphere 16*, 117–127. [*13, 224, 257*]

Walker, J.C.G. 1987: Was the Archaean biosphere upside down? *Nature 329*, 710–712. [*30*]

Walker, J.C.G. 1990: Precambrian evolution of the climate system. *Palaeogeography, Palaeoclimatology, Palaeoecology 82*, 261–289. [*27, 28, 29, 33, 212, 213, 218, 219*]

Walker, J.C.G. & Brimblecombe, P. 1985: Iron and sulfur in the pre-biologic ocean. *Precambrian Research 28*, 205–222. [*12, 13, 42, 219, 233*]

Walker, J.C.G., Klein, C., Schidlowski, M., Schopf, J.W., Stevenson, D.J. & Walter, M.R. 1983: Environmental evolution of the Archean – Early Proterozoic Earth. In Schopf, J.W. (ed.): *Earth's Earliest Biosphere, Its Origin and Evolution*, 260–290. Princeton University Press, Princeton, N.J. [*29, 143, 226*]

Wall, J.D., Rapp-Giles, B.J., Brown, M.F. & White, J.A. 1990: Response of *Desulfovibrio desulfuricans* colonies to oxygen stress. *Canadian Journal of Microbiology 36*, 400–408. [*38*]

Wallace, G.T., Jr. & Duce, R.A. 1978: Open-ocean transport of particulate trace metals by bubbles. *Deep-Sea Research 25*, 827–835. [*23*]

Walossek, D. & Müller, K.J. 1990: Upper Cambrian stem-lineage crustaceans and their bearing upon the monophyletic origin of Crustacea and the position of *Agnostus*. *Lethaia 23*, 409–427. [*456, 492*]

Walsh, M.M. 1989: Carbonaceous cherts of the Swaziland Supergroup, Barberton Mountain Land, South Africa. 199 pp. Ph.D. thesis, Louisiana State University, Baton Rouge, Louis. [*31, 32*]

Walsh, M.M. & Lowe, D.R. 1985: Filamentous microfossils from the 3,500-Myr-old Onverwacht Group, Barberton Mountain Land, South Africa. *Nature 314*, 530–532. [*31, 163, 196, 276*]

Walter, M.R. 1972: Stromatolites and the biostratigraphy of the Australian Precambrian and Cambrian. *Special Papers in Palaeontology 11*. 190 pp. Palaeontological Association, London. [*255, 273, 274, 279, 282, 432*]

Walter, M.R. 1976a: Geyserites of Yellowstone National Park: An example of abiogenic "stromatolites." In Walter, M.R. (ed.): *Stromatolites*, 87–112. Developments in Sedimentology 20 Elsevier, Amsterdam. [*276*]

Walter, M.R. (ed.) 1976b: *Stromatolites*. Developments in Sedimentology 20 Elsevier, Amsterdam. [*272, 279*]

Walter, M.R. 1977: Interpreting stromatolites. *American Scientist 65*, 563–571. [*281*]

Walter, M.R. 1983: Archean stromatolites: Evidence of the Earth's earliest benthos. In Schopf, J.W (ed.): *Earth's Earliest Biosphere, Its Origin and Evolution*, 187–213. Princeton University Press, Princeton, N.J. [*32, 163, 164, 168, 193, 194, 195, 197, 235, 247, 272, 274, 275, 276, 277, 278, 299, 310, 432*]

Walter, M.R. & Awramik, S.M. 1979: *Frutexites* from stromatolites of the Gunflint Iron Formation of Canada, and its biological significance. *Precambrian Research 9*, 23–33. [*293*]

Walter, M.R., Bauld, J. & Brock, T.D. 1972: Siliceous algal and bacterial stromatolites in hot spring and geyser effluents of Yellowstone National Park. *Science 178*, 402–405. [*432*]

Walter, M.R., Bauld, J. & Brock, T.D. 1976a: Microbiology and morphogenesis of columnar stromatolites (*Conophyton, Vacerrilla*) from hot springs in Yellowstone National Park. In Walter, M.R. (ed.): *Stromatolites*, 273–310. Developments in Sedimentology 20 Elsevier, Amsterdam. [*278*]

Walter, M.R., Buick, R. & Dunlop, J.R.S. 1980: Stromatolites 3,400–3,500 Myr old from the North Pole area, Western Australia. *Nature 284*, 443–445. [*40, 195, 197, 275*]

Walter, M.R., Du Rulin & Horodyski, R.J. 1990: Coiled carbonaceous megafossils from the middle Proterozoic of Jixian (Tianjin) and Montana. *American Journal of Science 290-A*, 133–148. [*292, 303, 310, 357, 359, 360, 442*]

Walter, M.R., Grotzinger, J.P. & Schopf, J.W. 1992: Proterozoic stromatolites. In Schopf, J.W. & Klein, C. (eds.): *The Proterozoic Biosphere: A Multidisciplinary Study*, 253–260. Cambridge University Press, Cambridge. [*163, 194, 283*]

Walter, M.R. & Heys, G.R. 1985: Links between the rise of the Metazoa and the decline of stromatolites. *Precambrian Research 29*, 149–174. [*254, 271, 272, 280, 283, 284, 285, 372, 434, 435, 451*]

Walter, M.R., Krylov, I.N. & Preiss, W.V. 1979: Stromatolites from Adelaidean (Late Proterozoic) sequences in central and South Australia. *Alcheringa 3*, 287–305. [*273*]

Walter, M.R., Oehler, J.H. & Oehler, D.Z. 1976b: Megascopic algae 1300 million years old from the Belt Supergroup, Montana: A reinterpretation of Walcott's *Helminthoidichnites*. *Journal of Paleontology 50*, 872–881. [*355, 356, 357*]

Wang, C. & Lehmann, R. 1991: Nanos is the localized posterior determinant in *Drosophila*. *Cell 66*, 637–647. [*506*]

Wang, F., Chen, Q. & Zhao, X. 1984: New information on Sinian acritarch from SW China and its significance. *Kexue Tongbao 29*, 656–659. [*356*]

Wang, G.X. 1982: Late Precambrian Annelida and Pogonophora from the Huainan of Anhui Province. *Tianjin Institute of Geology and Mineral Resources, Bulletin 1982*, 9–22. (In Chinese, with English abstract.) [*355, 357*]

Ward, D.M., Bauld, J., Castenholz, R.W. & Pierson, B.K. 1992: Modern phototrophic microbial mats: Anoxygenic, intermittently oxygenic/anoxygenic, thermal, eukaryotic and terrestrial. In Schopf, J.W. & Klein, C. (eds.): *The Proterozoic Biosphere: A Multidisciplinary Study*, 309–324. Cambridge University Press, Cambridge. [*168*]

Ward, D.M., Roland, W. & Bateson, M. 1990a: 16S rRNA sequences reveal numerous uncultured microorganisms in a natural community. *Nature 344*, 63–65. [*172*]

Ward, D.M., Roland, W. & Bateson, M.M. 1990b: 16S rRNA sequences reveal uncultured inhabitants of a well studied thermal community. *FEMS Microbiology Reviews 75*, 105–116. [*172*]

Ward, D.M., Weller, R., Shiea, J., Castenholz, R.W. & Cohen, Y. 1989: Hot spring microbial mats: Anoxygenic and oxygenic mats of possible evolutionary significance. In Cohen, Y. & Rosenberg, E. (eds.): *Microbial Mats: Physiological Ecology of Benthic Microbial Communities*, 3–13. American Society for Microbiology, Washington, D.C. [*168*]

Watabe, N. 1990: Calcium phosphate structures in invertebrates and protozoans. In Carter, J.G. (ed.): *Skeletal Biomineralization: Patterns, Processes and Evolutionary Trends*, Vol. 1, 35–44. Van Nostrand Reinhold, New York, N.Y. [*415*]

Watanabe, M.F., Oishi, S., Watanabe, Y. & Watanabe, M. 1986: Strong probability of lethal toxicity in the blue-green alga *Microcystis viridis* Lemmermann. *Journal of Phycology 22*, 552–556. [*340*]

Waterbury, J.B. & Rippka, R. 1989: Subsection I. Order Chroococcales Wettstein 1924, emend. Rippka *et al.*, 1979. In Stanley, J.T. (ed.): *Bergey's Manual of Systematic Bacteriology*, Vol. 3, 1728–1746. Williams & Wilkins, Baltimore, Mld. [*338*]

Waterbury, J.B., Watson, S.W., Guillard, R.R.L & Brand, L.E. 1979: Widespread occurrence of a unicellular, marine planktonic cyanobacterium. *Nature 277*, 293–294. [*337*]

Waterbury, J.B., Watson, S.W., Valois, W. & Franks D.G. 1986: Biological and ecological characterization of the marine unicellular cyanobacterium *Synechococcus*. In Platt, T & Li W.K.W. (eds.): *Photosynthetic Picoplankton*, 71–120. Canadian Bulletin of Fishery and Aquatic Sciences 214. [*337, 338*]

Watson, J.D., Hopkins, N.H., Roberts, J.W., Steitz, J.A. & Weiner, A.M. 1987: *Molecular Biology of the Gene*. 4th Ed. The Benjamin, Menlo Park, Calif. [*137*]

Wedeen, C.J. & Weisblat, D.A. 1991: Segmental expression of an *engrailed*-class gene during early development and neurogenesis in an annelid. *Development 113*, 805–814. [*459, 491, 492*]

Weiner, A.M. 1987a: Summary. *Cold Spring Harbor Symposia on Quantitative Biology 52*, 933–941. [*501, 515*]

Weiner, A.M. 1987b: The origins of life. In Watson, J.D., Hopkins, N.H., Roberts, J.W., Steitz, J.A. & Weiner, A. (eds.): *Molecular Biology of the Gene*, Vol. 2, 1098–1163. Benjamin Cummings, Menlo Park, Calif. [*79, 132*]

Weiner, A.M., Deininger, P.L. & Efstratiadis, A. 1986: Nonviral retroposons: Genes, pseudogenes, and transposable elements generated by the reverse flow of genetic information. *Annual Review of Biochemistry 55*, 631–661. [*139*]

Weiner, A.M. & Maizels, N. 1987: tRNA-like structures tag the 3' ends of genomic RNA molecules for replication: Implications for the origin of protein synthesis. *Proceedings of the National Academy of Sciences, USA 84*, 7383–7387. [*76*]

Weiner, A.M. & Maizels, N. 1991: The genomic tag model for the origin of protein synthesis: further evidence from the molecular fossil record. In Osawa, S. & Honjo, T. (eds.): *Evolution of Life: Fossils, Molecules, and Culture*, 51–66. Springer, Tokyo. [*135, 137*]

Weiner, S., Traub, W. & Lowenstam, H.A. 1983: Organic matrix in calcified exoskeletons. In Westbroek, P. & de Jong, E.W. (eds.): *Biomineralization and Biological Metal Accumulation*, 205–224. Reidel, Dordrecht. [*413*]

Weinmaster, G., Roberts, V.J. & Lemke, G. 1991: A homolog of *Drosophila Notch* expressed during mammalian development. *Development 113*, 199–205. [*459*]

Weissenfels, N. 1992: The filtration apparatus for food collecting in fresh water sponges (Porifera, Spongillidae). *Zoomorphologie 112*, 51–56. [*487*]

Welhan, J.A. 1988: Origins of methane in hydrothermal systems. *Chemical Geology 71*, 183–198. [*16*]

Wells, A.T., Forman, D.J., Ranford, L.C. & Cook, P.J. 1970: Geology of the Amadeus Basin, central Australia. *Bureau of Mineral Resources, Geology and Geophysics, Bulletin 100*. 222 pp. [*273*]

Wells, A.T., Ranford, L.C., Stewart, A.J., Cook, P.J. & Shaw, R.D. 1967: The geology of the northeastern part of the Amadeus Basin, Northern Territory. *Bureau of Mineral Resources, Geology and Geophysics, Bulletin 113*. 97 pp. [*273*]

Wenz, W. 1938: Gastropoda. Allgemeiner Teil und Prosobranchia. In Schindewolf, O.H. (ed.): *Handbuch der Paläozoologie*, Vol. 6, 1–240. Bonträger, Berlin. [*355*]

Westheide, W. 1987: Progenesis as a principle in meiofauna evolution. *Journal of Natural History 21*, 843–854. [*476*]

Westheide, W. & Hermans, C.O. (eds.) 1988: *The Ultrastructure of Polychaeta*. Microfauna Marina 4. [*406*]

Westheimer, F.H. 1986: Polyribonucleic acids as enzymes. *Nature 319*, 534–536. [*77*]

Westheimer, F.H. 1987: Why nature chose phosphates. *Science 235*, 1173–1178. [*93, 119*]

Westphalen, D. 1993: Stromatolitoid microbial nodules from Bermuda – a special microhabitat for meiofauna. *Marine Biology*. [478]

Wetherill, G.W. 1980: Formation of the terrestrial planets. *Annual Reviews of Astrophysics 18*, 77–113. [14]

Wetherill, G.W. 1990: Formation of the Earth. *Annual Review of Earth and Planetary Sciences 18*, 205–256. [50]

Whatley, J.M. 1989: Chromophyte chloroplasts – a polyphyletic orgin? In Green, J.C., Leadbeater, B.S.C. & Diver, W.L. (eds.): *The Chromophyte Algae: Problems and Perspectives*, 125–144. The Systematics Association Special Volume 38 Clarendon, Oxford. [319, 322]

Whatley, J.M. & Whatley, F.R. 1979: From extracellular to intracellular: The establishment of mitochondria and chloroplasts. *Proceedings of the Royal Society of London, B 204*, 165–187. [317]

Wheelis, M.L., Kandler, O. & Woese, C.R. 1992: On the nature of global classification. *Proceedings of the National Academy of Sciences, USA 89*, 2930–2934. [152]

White, H.B. 1976: Coenzymes as fossils of an earlier metabolic state. *Journal of Molecular Evolution 7*, 101–104. [131]

White, H.B. 1982: Evolution of coenzymes and the origin of pyridine nucleotides. In Everse, J., Anderson, B. & You, K.S. (eds.): *The Pyridine Nucleotide Coenzymes*, 2–17. Academic Press, New York, N.Y. [63, 72, 73]

White, M.J.D. 1981: Tales of long ago – The birth of evolutionary theory as a scientific discipline. *Paleobiology 7*, 287–291. [516]

Whitesides, G.M., Mathias, J.P. & Seto, C.T. 1991: Molecular self-assembly and nanochemistry: A chemical strategy for the synthesis of nanostructures. *Science 254*, 1312–1319. [21]

Whitfield, M. & Watson, A.J. 1983: The influence of biomineralisation on the composition of seawater. In Westbroek, P. & de Jong, E.W. (eds.): *Biomineralization and Biological Metal Accumulation*, 57–72. Reidel, Dordrecht. [422]

Whittaker, R.H. 1969: New concepts of kingdoms of organisms. *Science 163*, 150–160. [182, 313]

Whittaker, R.H. & Margulis, L. 1978: Protist classification and the kingdoms of organisms. *BioSystems 10*, 3–18. [314, 468]

Whittington, H.B. 1975: The enigmatic animal *Opabinia regalis*, Middle Cambrian, Burgess Shale, British Columbia. *Philosophical Transactions of the Royal Society of London, B 271*, 1–43. [456]

Whittington, H.B. 1980: Exoskeleton, small and moult stage, appendage morphology, and habits of Middle Cambrian trilobite *Olenoides serratus*. *Palaeontology 23*, 171–204. [513]

Whittington, H.B. 1985: *The Burgess Shale*. Yale University Press, New Haven, Conn. [489]

Wickman, F.E. 1956: The cycle of carbon and the stable carbon isotopes. *Geochimica et Cosmochimica Acta 9*, 136–153. [222, 223]

Widdel, F., Schnell, S., Heising, S., Ehrenreich, A., Assmus, B. & Schink, B. 1993: Ferrous iron oxidation by anoxygenic phototrophic bacteria. *Nature 362*, 834–836. [180]

Wiebols, J.H. 1955: A suggested glacial origin for the Witwatersrand conglomerates. *Transactions of the Geological Society of South Africa 58*, 367–382. [28]

Wilde, P. & Berry, W.B.N. 1982: Progressive ventilation of the oceans – Potential for return to anoxic conditions in the post-Paleozoic. In Schlanger, S.O. & Cita, M.B. (eds.): *Nature and Origin of Cretaceous Carbon-rich Facies*, 209–224. Academic Press, London. [29]

Wilkinson, B.H. 1979: Biomineralization, paleoceanography, and the evolution of calcareous marine organisms. *Geology 7*, 524–527. [437]

Williams, D.M. 1991: Phylogenetic relationships among the Chromista: A review and preliminary analysis. *Cladistics 7*, 141–156. [316]

Williams, G.E. 1975: Late Precambrian glacial climate and the Earth's obliquity. *Geological Magazine 112*, 441–465. [437]

Williams, N.E., Buhse, H.E. & Smith, M.G. 1984: Protein similarities in the genus *Tetrahymena* and a description of *Tetrahymena leucophrys* sp. *Journal of Protozoology 31*, 313–321. [184]

Williams, R.J.P. 1988: Proton circuits in biological energy interconversions. *Annual Review of Biophysics and Biophysical Chemistry 17*, 71–97. [85]

Willmer, P. 1990: *Invertebrate Relationships. Patterns in Animal Evolution*. Cambridge University Press, Cambridge. [403, 476, 483, 487]

Wilmotte, A. & Golubic, S. 1991: Morphological and genetic criteria in taxonomy of Cyanophyta/ Cyanobacteria. In Hickel, B., Anagnostidis, K. & Komarek, J. (eds.): *Cyanophyta/Cyanobacteria – Morphology, Taxonomy, Ecology*, 1–24. Algological Studies 64 Schweizerbart'sche, Stuttgart. [*335*]

Wilson, A.C., Ochman, H. & Prager, E.M. 1987: Molecular time scale for evolution. *Trends in Genetics* 3, 341–347. [*186*]

Wilson, J.L. 1975: *Carbonate Facies in Geologic History*. Springer, Berlin. [*245*]

Wilson, M.A. & Pohorille, A. 1991: Interaction of monovalent ions with the water liquid–vapor interface: A molecular dynamics study. *Journal of Chemical Physics* 95, 6005–6013. [*21*]

Windberger, E., Huber, R., Trincone, A., Fricke, H. & Stetter, K.O. 1989: *Thermotoga thermarum* sp. nov. and *Thermotoga neapolitana* occurring in African continental Solfataric springs. *Archives of Microbiology* 151, 506–512. [*148*]

Wise, W.S. 1977: Mineralogy of the Champion Mine, White Mountains, California. *Mineralogical Record* 8, 478–486. [*97*]

Woese, C.R. 1967: *The Genetic Code: The Molecular Basis for Genetic Expression*. Harper & Row, New York, N.Y. [*63, 76, 77*]

Woese, C.R. 1979: A proposal concerning the origin of life on the planet Earth. *Journal of Molecular Evolution* 13, 95–101. [*143*]

Woese, C.R. 1980: Just so stories and Rube Goldberg machines: Speculations on the origin of the protein synthetic machine. In Chambliss, G., Crave, G.R., Davies, J., Davis, K., Kahan, L. & Nomura, M. (eds.): *Ribosomes: Structure, Function and Genetics*, 357–373. University Park Press, Baltimore, Md. [*76, 77*]

Woese, C.R. 1982: Archaebacteria and cellular origins: An overview. *Zentralblatt für Bakteriologie und Hygiene. I. Abteilung Originale. C* 3, 1–17. [*156, 190*]

Woese, C.R. 1987a: Macroevolution in the microscopic world. In Patterson, C. (ed.): *Molecules and Morphology in Evolution: Conflict or Compromise.*, 177–202. Cambridge University Press, London. [*312*]

Woese, C.R. 1987b: Bacterial evolution. *Microbiological Reviews* 51, 221–271. [*39, 66, 144, 152, 156, 162, 163, 182, 190, 301, 335*]

Woese, C.R., Achenbach, L., Rouviere, P. & Mandelco, L. 1991: Archaeal phylogeny: Reexamination of the phylogenetic position of *Archaeoglobus fulgidus* in light of certain composition-induced artifacts. *Systematic and Applied Microbiology* 14, 364–371. [*149*]

Woese, C.R. & Fox, G.E. 1977: Phylogenetic structure of the prokaryotic domain: The primary kingdoms. *Proceedings of the National Academy of Sciences, USA* 74, 5088–5090. [*144*]

Woese, C.R., Gutell, R., Gupta, R. & Moller, H.F. 1983: Detailed analysis of the higher-order structure of 16S-like ribosomal ribonucleic acids. *Microbiological Reviews* 47, 621–669. [*144*]

Woese, C.R., Kandler, O. & Wheelis, M.L. 1990: Towards a natural system of organisms: Proposal for the domains Archaea, Bacteria, and Eucarya. *Proceedings of the National Academy of Sciences, USA* 87, 4576–4579. [*62, 67, 144, 145, 152, 153, 159, 290, 303*]

Woese, C.R., Sogin, M., Stahl, D.A., Lewis, B.J. & Bonen, L. 1976: A comparison of the 16S ribosomal RNAs from mesophilic and thermophilic bacilli. *Journal of Molecular Evolution* 7, 197–213. [*144*]

Wolery, T.J. & Sleep, N.H. 1976: Hydrothermal circulation and geochemical flux at mid-ocean ridges. *Journal of Geology* 84, 249–275. [*13*]

Wolery, T.J. & Sleep, N.H. 1989: Interactions of geochemical cycles with the mantle. In Gregor, C.B., Garrels, R.M., Mackenzie, F.T. & Maynard, J.B. (eds.): *Chemical Cycles in the Evolution of the Earth*, 77–103. Wiley, New York, N.Y. [*218*]

Wollast, R. 1971: Kinetic aspects of the nucleation and growth of calcite from aqueous solutions. In Bricker, O.P. (ed.): *Carbonate Cements*, 264–273. Johns Hopkins University Press, Baltimore, Md. [*436*]

Wolpert, L. 1978: Pattern formation in biological development. *Scientific American* 239, 154–164. [*487*]

Wolters, J. 1991: The troublesome parasites – molecular and morphological evidence that Apicomplexa belong to the dinoflagellate–ciliate clade. *BioSystems* 25, 75–83. [*325*]

Wong, J.T.-F. 1981: Coevolution of genetic code and amino acid biosynthesis. *Trends in Biochemical Sciences* 6, 33–36. [*137*]

Wong, J.T.-F. 1991: Origin of genetically encoded protein synthesis: A model based on selection for RNA peptidation. *Origins of Life and Evolution of the Biosphere 21*, 165–176. [*76*]

Wood, A.M. & Townsend, D. 1990: DNA polymorphism within the WH7803 serogroup of marine *Synechococcus* spp. (Cyanobacteria). *Journal of Phycology 26*, 576–585. [*338*]

Wood, B.J. & Virgo, D. 1989: Upper mantle oxidation state: Ferric iron contents of lherzolite spinels by ^{57}Fe Mossbauer spectroscopy and resultant oxygen fugacities. *Geochimica et Cosmochimica Acta 53*, 1277–1291. [*12*]

Wood, H.G. 1985: Inorganic pyrophosphate and polyphosphates as sources of energy. *Current Topics in Cellular Regulation 26*, 355–369. [*84*]

Wood, R.A. 1991a: Problematic reef-building sponges. In Simonetta, A.M. & Conway Morris, S. (eds.): *The Early Evolution of Metazoa and the Significance of Problematic Taxa*, 113–124. Cambridge University Press, Cambridge. [*452*]

Wood, R.A. 1991b: Non-spicular biomineralization in calcified demosponges. In Reitner, J. & Keupp, H. (eds.): *Fossil and Recent Sponges*, 322–340. Springer, Berlin. [*437*]

Worrell, G.F. 1985: Sedimentology and mineralogy of silicified evaporites in the basal Kromberg Formation, South Africa. 152 pp. Master's thesis, Louisiana State University, Baton Rouge, Louis. [*28, 29*]

Worsley, T.R. & Nance, R.D. 1989: Carbon redox and climatic control through earth history: A speculative reconstruction. *Palaeogeography, Palaeoclimatology, Palaeoecology 75*, 259–282. [*215, 218*]

Wray, G.A. 1992: The evolution of larval morphology during the post-Paleozoic radiation of echinoids. *Paleobiology 18*, 258–287. [*493, 494*]

Wray, G.A. & Raff, R.A. 1990: Novel origins of lineage founder cells in the direct-developing sea urchin *Heliocidaris erythrogramma*. *Developmental Biology 141*, 41–54. [*494, 495, 505*]

Wu, J. 1981: Evidence of sea spray produced by bursting bubbles. *Science 212*, 324–326. [*21*]

Xing Yusheng 1979: The Sinian of China. *Institute of Geology, Academy of Geological Sciences, Collected Papers, International Exchange of Geological Sciences, Vol. 2: Stratigraphy, Paleontology*, 1–12. Geological Publishing House, Beijing. [*356*]

Xing Yusheng 1984: Description of a new worm family – Huaiyuanellidae Xing from the Upper Sinian of North Anhui, China. *Bulletin of the Institute of Geology, Chinese Academy of Geological Sciences 9*, 151–154. (In Chinese.) [*357*]

Xing Y.-S., Duan C.-H., Liang Y.-Z., Cao R.-G. *et al.* 1985: Late Precambrian palaeontology of China. *Geological Memoirs of the Ministry of Geology and Mineral Resources, Peoples Republic of China, Ser 2 2*. 243 pp. (In Chinese.) [*357*]

Xing Y.-S., Gao Z.-J., Liu G.-Zh., Qiao X.-F., Wang Z.-Q., Zhu H., Chen Y.-Y. & Quan Q.-Q. 1989: The Upper Precambrian of China. *Stratigraphy of China 3*. 314 pp. Geological Publishing House, Beijing. [*357*]

Xue Yaosong, Tang Tianfu & Yu Congliu 1992: Discovery of oldest skeletal fossils from Upper Sinian Doushantuo Formation in Wengian, Guizhou, and its significance. *Acta Palaeontologica Sinica 31*, 530–539. (In Chinese, with English summary.) [*451*]

Yamagata, Y., Watanabe, H., Saitoh, M. & Namba, T. 1991: Volcanic production of polyphosphates and its relevance to chemical evolution. *Nature 352*, 516–519. [*21, 87, 98*]

Yamamoto, S., Alcauskas, J.B. & Crozier, T.E. 1976: Solubility of methane in distilled water and seawater. *Journal of Chemical and Engineering Data 21*, 78–80. [*233*]

Yamasu, T. 1991: Fine structure and function of ocelli and sagittocysts of acoel flatworms. In Tyler, S. (ed.): *Turbellarian Biology*, 273–282. Kluwer, Dordrecht. [*482*]

Yan Y.-Zh. 1987: The tubular algal fossils from the Xiamaling Formation in Jixian County. *Tianjin Institute of Geology and Mineral Resources, Bulletin 16*, 165–169. [*357*]

Yanagawa, H. & Kobayashi, K. 1992: An experimental approach to chemical evolution in submarine hydrothermal systems. *Origins of Life and Evolution of the Biosphere 22*, 147–160. [*19, 20*]

Yanagawa, H., Ogawa, Y., Kojima, K. & Ito, M. 1988: Construction of protocellular structures under simulated primitive Earth conditions. *Origins of Life and Evolution of the Biosphere 18*, 179–207. [*19*]

Yanishevskij, M.E. 1926: Ob ostatkakh trubchatykh chervej iz kembrijskoj Sinej gliny [On remains of tube-dwelling worms from the Cambrian Blue Clay]. *Ezhegodnik Vsesoyuznogo Paleontologicheskogo Obshchestva 4*, 99–111. [*357*]

Yano, H., Satake, K., Ueno, Y., Kondo, K. & Tsugita, A. 1991: Amino acid sequence of the hemerythrin subunit from *Lingula unguis*. *Journal of Biochemistry 110*, 376–380. [*458*]

Yarus, M. 1988a: Specificity of arginine binding by the *Tetrahymena* intron. *Biochemistry 28*, 980–988. [*63*]

Yarus, M. 1988b: A specific amino acid binding site composed of RNA. *Science 240*, 1751–1758. [*76*]

Yin Leiming 1985: Microfossils of the Doushantuo Formation in the Yangtze Gorge district, Western Hubei. *Palaeontologia Cathayana 2*, 229–249. [*305*]

Young, G.M. 1976: Iron-formation and glaciogenic rocks of the Rapitan Group, Northwest Territories. *Precambrian Research 3*, 137–158. [*444*]

Young, G.M. 1981: The Amundsen embayment, Northwest Territories: Relevance to the upper Proterozoic evolution of North America. In Campbell, F.H.A. (ed.): *Proterozoic Basins of Canada*, 203–218. Geological Survey of Canada, Toronto, Ont. [*253*]

Young, R.B. 1929: Pressure phenomena in the dolomitic limestones in the Campbell Rand Series. *Transactions of the Geological Society of South Africa 31*, 157–165. [*272*]

Young, R.B. 1943: The domical–columnar structure and other minor deformations in the Dolomite Series. *Transactions of the Geological Society of South Africa 46*, 91–106. [*272*]

Yue Zhao, Bengtson, S. & Grant, S.W.F. 1992: Biology and functional morphology of *Cloudina*, the earliest known metazoan with a mineralized skeleton. *Special Publication of the Paleontological Society (Fifth North American Paleontology Convention Abstracts) 6*, 325. [*451*]

Yurewicz, D.A. 1977: Origin of the massive Capitan Limestone (Permian), Guadalupe Mountains, New Mexico and West Texas. In *1977 Field Conference Guidebook: SEPM Permian Basin Section*, 45–92. Society of Economic Paleontologists and Mineralogists, Tulsa, Okl. [*248*]

Zahnle, K.J. 1986: Photochemistry of methane and the formation of hydrocyanic acid (HCN) in the Earth's early atmosphere. *Journal of Geophysical Research 91*, 2819–2834. [*12*]

Zahnle, K.J. 1990: Atmospheric chemistry by large impacts. In Sharpton, V. & Ward, P. (eds.): *Global Catastrophes in Earth History*, 271–288. Geological Society of America, Special Paper 247. [*12*]

Zahnle, K.J., Kasting, J.F. & Pollack, J.B. 1988: Evolution of a steam atmosphere during Earth's accretion. *Icarus 74*, 62–97. [*11*]

Zang Wenlong & Walter, M.R. 1989: Latest Proterozoic plankton from the Amadeus Basin in Central Australia. *Nature 337*, 642–645. [*292, 305, 447*]

Zhang Wentang & Hou Xianguang 1985: Preliminary notes on the occurrence of the unusual trilobite *Naraoia* in Asia. *Acta Palaeontologica Sinica 24*, 591–595. (In Chinese, with English abstract.) [*367*]

Zhang Yun 1989: Multicellular thallophytes with differentiated tissues from Late Proterozoic phosphate rocks of South China. *Lethaia 22*, 113–132. [*291, 301, 303, 343*]

Zhang Zhonying 1986: Clastic facies microfossils from the Chuanlinggou Formation (1800 Ma) near Jixian, North China. *Journal of Micropalaeontology 5*, 9–16. [*292, 310, 427, 441*]

Zhang Zhonying 1988: *Longfengshania* Du emend.: An earliest record of bryophyte-like fossils. *Acta Palaeontologica Sinica 27*, 416–426. (In Chinese, with English summary.) [*351*]

Zhao, M. & Bada, J.L. 1989: Extraterrestrial amino acids in Cretaceous/Tertiary boundary sediments at Stevns Klint, Denmark. *Nature 339*, 463–465. [*17*]

Zheng, W. 1980: A new occurrence of fossil group of *Chuaria* from the Sinian System in North Anhui and its geologic meaning. *Chinese Academy of Geological Sciences, Bulletin, Series 6 1*, 49–69. (In Chinese, with English abstract.) [*346, 355, 356, 357*]

Zhu, S. 1982: An outline of studies on the Precambrian stromatolites of China. *Precambrian Research 18*, 367–396. [*271, 280*]

Zhu W. & Chen M. 1984: On the discovery of macrofossil algae from the Late Sinian in the eastern Yangtze Gorges, South China. *Acta Botanica Sinica 26*, 558–560. [*357*]

Zillig, W., Gierl, A., Schreiber, G., Wunderl, S., Janecovic, D., Stetter, K.O. & Klenk, H.P. 1983a: The archaebacterium *Thermofilum pendens* represents a novel genus of the thermophilic, anaerobic, sulfur respiring Thermoproteales. *Systematic and Applied Microbiology 4*, 79–87. [*147*]

Zillig, W., Holz, I., Janecovic, D., Schäfer, W. & Reiter, W.D. 1983b: The archaebacterium *Thermococcus celer* represents a novel genus within the thermophilic branch of the Archaebacteria. *Systematic and Applied Microbiology 4*, 88–94. [*150*]

Zillig, W., Palm, P. & Klenk, H.P. 1992: The nature of the common ancestor of the three domains of life and the origin of the Eucarya. In Trân Thanh Vân, J. & K., Mounolou, J.C., Schneider, J. & McKay, C. (eds.): *Frontiers of Life*, 181–193. Proceedings of the 3d Rencontres de Blois, France [160]

Zillig, W., Stetter, K.O., Prangishvilli, D., Schäfer, H., Wunderl, S., Janecovic, D., Holz, I. & Palm, P. 1982: Desulfurococcaceae, the second family of the extremely thermophilic anaerobic sulfur respiring Thermoproteales. *Zentralblatt für Bakteriologie und Hygiene. I. Abteilung Originale. C 3*, 304–317. [148]

Zillig, W., Stetter, K.O., Schäfer, W., Janecovic, D., Wunderl, S., Holz, I. & Palm, P. 1981: Thermoproteales: A novel type of extremely thermoacidophilic anaerobic Archaebacteria isolated from Icelandic Solfataras. *Zentralblatt für Bakteriologie und Hygiene. I. Abteilung Originale. C 2*, 205–227. [147]

Zuckerkandl, E. & Pauling, L. 1965: Molecules as documents of evolutionary history. *Journal of Theoretical Biology 8*, 357–366. [66, 144, 152, 182, 468]

Index